Anion Receptor Chemistry

MONOGRAPHS IN SUPRAMOLECULAR CHEMISTRY

Series Editor: J Fraser Stoddard, FRS
University of California at Los Angeles, USA

This series has been designed to reveal the challenges, rewards, fascination and excitement in this new branch of molecular science to a wide audience and to popularize it among the scientific community at large.

Other titles in this series:

Calixarenes
By C. David Gutsche, *Washington University, St Louis, USA*

Cyclophanes
By François Diederich, *University of California at Los Angeles, USA*

Crown Ethers and Cryptands
By George W. Gokel, *University of Miami, USA*

Container Molecules and Their Guests
By Donald J. Cram and Jane M. Cram, *University of California at Los Angeles, USA*

Membranes and Molecular Assemblies: The Synkinetic Approach
By Jürgen-Hinrich Fuhrhop and Jürgen Köning, *Freie Universität Berlin, Germany*

Calixarenes Revisited
By C. David Gutsche, *Texas Christian University, Fort Worth, USA*

Self-assembly in Supramolecular Systems
By Len Lindoy, *The University of Sydney, Australia,* and Ian Atkinson, *James Cook University, Townsville, Australia*

How to obtain future titles on publication

A subscription is available for this series. This will bring delivery of each new volume immediately on publication. For further information visit www.rsc.org/Publishing/Books/MOSC/ or write to:

Sales and Customer Care, Royal Society of Chemistry
Thomas Graham House, Science Park, Milton Road, Cambridge, CB4 0WF, UK
Telephone: +44(0) 1223 432360, Fax: +44(0) 1223 426017, E-mail: sales@rsc.org

Anion Receptor Chemistry

Jonathan L. Sessler
University of Texas, Austin, Texas, USA

Philip A. Gale
University of Southampton, Southampton, UK

Won-Seob Cho
University of Texas, Austin, Texas, USA

RSC Publishing

ISBN-10: 0-85404-974-6
ISBN-13: 978-0-85404-974-5

A catalogue record for this book is available from the British Library

Published by The Royal Society of Chemistry,
Thomas Graham House, Science Park, Milton Road,
Cambridge CB4 0WF, UK

Registered Charity Number 207890

For further information see our web site at www.rsc.org

Typeset by Macmillan India Ltd, Bangalore, India
Printed by Henry Lings, Dorchester, Dorset, UK

Dedication

This book is dedicated to Carol, Nittaya, and Hye-Young

Preface

Supramolecular chemistry, as the readers of these monographs likely appreciate with enthusiasm, involves the fascinating and almost infinite world of molecular interactions that extend beyond the limits of simple covalent bonds. Understanding these interactions and exploiting them in new ways continues to define the essence of the field. Not surprisingly, the challenges and opportunities associated with these goals are continuing to attract increasing numbers of ever more talented researchers into the area, while fueling its nearly exponential growth as a venue for intellectual discovery and useful, real-world-relevant contribution. This is particularly true in the area of anion recognition chemistry. Here, growth is being driven by an increasing appreciation for the importance of noncovalent anion–molecule interactions in biology as well as by a tangible sense that anion recognition chemistry has a beneficial role to play in the areas of physiology, medicine, synthetic chemistry, materials development, analyte sensing, and waste remediation. Progress is also being abetted by the fact that basic principles that allow for controlled anion recognition are becoming increasingly well codified. As a result, promises of practical application are starting to drive the field even though it remains animated in large measure by its inherent aesthetic appeal. The goal of this monograph is to provide a concise overview of supramolecular anion chemistry with a focus on the chemistry of synthetic anion receptor design. The hope is to enunciate the guiding principles of anion binding while highlighting the prospects for ultimate practical utility. Towards this end, this book is divided into eight main chapters. These chapters are designed to introduce the most important research themes currently animating the field of anion recognition chemistry and to provide an entry point into the relevant literature. These eight chapters are prefaced by an introductory chapter that is intended to highlight in an anecdotal fashion, the importance of anions in everyday life and the critical role anion recognition chemistry plays in the biological world. As part of this introduction, the anion recognition field is traced from an historical perspective. The hope here is to provide the reader with a quick overview of where the field has been, where it stands now, and where it might be going. This same overview is also intended to introduce the readers to many of the more common anion recognition motifs that have emerged as "key players" in both the synthetic and biological anion recognition worlds.

The authors are grateful to the various granting agencies that made work on this book and related experimental projects possible. The National Institutes of Health (grant GM 58907 to J.L.S.) and the Department of Energy (grant DE-FG02-04ER63741 to J.L.S.) are explicitly thanked. P.A.G. would like to thank the Royal Society for a University Research Fellowship, EPSRC for funding and the Royal Society of Chemistry for an International Journals Grant.

Jonathan L. Sessler
Philip A. Gale
Won-Seob Cho

Table of Contents

Glossary

A^-	anion (generic representation)
AA	amino acid
Aba	3-amino-benzoic acid
ADP	adenosine diphosphate
AMP	adenosine monophosphate
ANS	8-anilinonaphthalene-1-sulfonate
ATP	adenosine triphosphate
AZT	azidothymidine
AZTTP	3′-azido-2′-deoxythymidine 5′-triphosphate
BETMP	bis(ethylthio)methylenepropanedioate
bipy	2,2′-bipyridine
2,3-BPG	2,3-bisphosphoglycerate
BPN	2,7-bis(l*H*-pyrrol-2-yl)ethynyl-l,8-naphthyridine
bptz24	3,6-bis(2-pyridyl)-1,2,4,5-tetrazine
Bu	*n*-butyl
CD	circular dichroism or cyclodextrin (depending on context)
CFTR	cystic fibrosis transmembrane conductance regulator
CHEMFET	chemically modified field effect transistor
CMP	cytosine monophosphate
CPK	Corey-Pauling-Koltun
CTP	cytidine triphosphate
CV	cyclic voltammetry or cyclic voltammetric
DABCO	1,4-diaza[2.2.2]-bicycloocatane
dach	trans(±) –1,2-diaminocyclohexane
ddTTP	2′,3′-dideoxythymidine 5′-triphosphate
DMSO	dimethyl sulfoxide
DNA	deoxyribonucleic Acid
DNNO	2,4-dinitronaphthalen-1-olate
DMF	dimethylformamide
DOP	dioctyl phthalate
dppa	*N*,*N*′-bis(diphenylphosphino)amine
dppm	bis(diphenylphosphino)methane
DPQ	dipyrrolyquinoxaline
DQF-COSY	double-quantum filtered correlated spectroscopy
EDTA	ethylenediaminetetraacetic acid
ESI	electrospray ionization

FAB	fast-atom bombardment
G	guest
GABA	γ-amino butyric acid
GMP	guanosine monophosphate
H	host
HEPES	*N*-2-hydroxyethylpiperazine-*N′*-2-ethanesulfonic acid
HPLC	high performance liquid chromatography
HpztBu	3{5}-tert-butylpyrazole
IP_3	inositol-1,4,5-trisphosphate
ISE	ion-selective electrodes
ITC	isothermal titration calorimetry
L	ligand
MeCN	acetonitrile
MES20	(10 mM 2-(*N*-morpholino)ethanesulfonic acid, 200 mM NaCl, 1 mM EDTA, pH 6.2)
MS	mass spectrometry
NAD	nicotinamide adenine dinucleotide
NADP	nicotinamide adenine dinucleotide phosphate
NBS	*N*-bromosuccinimide
NCP	*N*-confused porphyrins
N.D.	no binding detected
NMR	nuclear magnetic resonance
NOESY	nuclear Overhauser effect spectroscopy
OG	(disodisodium-7-hydroxy-8-phenylzaonaphthalene-l,3-disulfonate)
Orange G	(disodisodium-7-hydroxy-8-phenylzaonaphthalene-1,3-disulfonate)
OGA	octyl β-D-galactpyranoside
OGU	octyl β-D-glucopyranoside
ox	oxidation
PBP	phosphate-binding protein
PCET	proton coupled electron transfer
PCP	*Pneumocystis carinii* pneumonia
Ph	phenyl
PhCN	benzonitrile
PDT	photodynamic therapy
POPC	1-palmitoyl-2-oleoyl-phosphatidylcholine
PPNC1	bis(triphenylphosphoranylidene)ammonium chloride
PVC	poly(vinyl chloride)
py	pyridine
red	reduction
RNA	ribonucleic acid
ROESY	rotating-frame NOE spectroscopy
SBP	sulfate-binding protein
SLM	supported liquid membrane
StCIC	chloride channel CIC protein from *Salmonella serovar typhimurium*
TDDMAC1	tridodecylmethylammonium chloride
TBA	tetrabutylammonium

TCQ	tetrachloroquinone
TEA	tetraethylammonium
TfO^-	triflate (trifluoromethanesulfonate)
TFQ	tetrafluoroquinone
THF	tetrahydrofuran
Tips	triisopropylsilyl
TMA	tetramethylammonium
TNB	trinitrobenzene
TNS	6-*p*-toluidinonaphthalene-2-sulfonate
TOCSY	total correlated spectroscopy
Tren	*N*1, *N*1-bis(2-aminoethyl)ethane-1,2-diamine
TREN	tris(2-aminoethyl)amine
Tris	tris(hydroxymethyl)aminomethane
TsO^-	tosylate (4-methylbenzenesulfonate)
UMP	uridine monophosphate
UV/vis	ultraviolet/visible
YFP	yellow fluorescent protein

CHAPTER 1

Introduction

1.1 Importance of Anions in the Modern World

Although often overlooked in terms of their importance, anions are ubiquitous in the natural world. Chloride anions are present in large quantities in the oceans; nitrate and sulfate are present in acid rain; and carbonates are key constituents of biomineralized materials. Anthropogenic anions, including pertechnetate, a radioactive product of nuclear fuel reprocessing, and phosphate and nitrates from agriculture and other human activities, constitute major pollution hazards.

Anions are also critical to the maintenance of life as we know it. Indeed, without exaggeration, the recognition, transport, or transformation of anions is involved at some level in almost every conceivable biochemical operation. It is essential in the formation of the majority of enzyme–substrate and enzyme–cofactor complexes as well as in the interaction between proteins and RNA or DNA. ATP, phosphocreatine, and other high-energy anionic phosphate derivatives are at the centre of power processes as diverse and important as biosynthesis, molecular transport, and muscle contraction. They also serve as the energy currency for a host of enzymatic transformations. Anion channels and carriers are involved in the transport of small anions such as chloride, phosphate, and sulfate and thus serve to regulate the flux of key metabolites into and out of cells while maintaining osmotic balance.

On a less salubrious level, mis-regulation of various anion transport mechanisms can have serious consequences. For instance, a malfunctioning of the CFTR chloride transport channel is implicated in cystic fibrosis, one of the most commonly inherited diseases among Caucasians. Likewise, the so-called ATP binding cassette transport systems, multispecific organic anion transporters, can confer resistance to a variety of modern pharmaceutical agents and are responsible in part for one of the most pressing problems in medicine, namely multidrug resistance. On very different level, an inability to process or catabolize effectively xenobiotic anions, including such chemically simple species as cyanide, oxalate, arsenate, or nitrite, can produce symptoms of chronic or acute toxicity. Poor processing of naturally occurring phosphate and sulfate is also a serious problem for patients with renal failure. Likewise, an inability to remove excess superoxide and peroxynitrite is considered responsible for many of the symptoms associated with reperfusion injury following heart attacks and strokes. On the other hand, anionic species either administered directly as in the case of fluoride used to prevent dental caries, or produced *in vivo* from a wide range of prodrugs, running the gamut in complexity from aspirin to AZT, are key features of modern medicine. This dichotomy underscores the complexity and importance of anion recognition in biology; it also highlights the need for, and potential utility of, synthetic anion receptor chemistry.

Later in this chapter we provide a brief selection of biologically relevant paradigms. Needless to say, in a work of this size, an exhaustive treatment is not possible. However, it is hoped that the examples chosen as highlights will help illustrate how nature uses many of the tools, such as hydrogen bonding, electrostatics, size/shape complementarily, metal–anion complex formation, and hydrophobicity, that chemists are currently employing in their efforts to achieve anion recognition. In the remainder of this chapter, we provide an historical overview that is designed both to trace the origins of the field and introduce many of the molecular recognition motifs that are continuing to play a critical role in terms of the design and synthesis of current state-of-the-art synthetic receptor systems. But first we will start with an overview of the challenges that anion complexation presents to the supramolecular chemist.

1.2 The Challenges of Anion Complexation

The design of anion receptors (and receptors for ion-pairs) is particularly challenging when compared to the design of receptors for cations. There are a number of reasons for this. Anions are larger than the equivalent isoelectronic cations (see Table 1.1)[1] and hence have a lower charge to radius ratio. The more diffuse nature of anions means that electrostatic binding interactions are less effective than they would be for the corresponding isoelectronic cation.

Anions may be pH sensitive (becoming protonated at low pH and so losing their negative charge). Thus, receptors must function within the pH window of their target anion. This is a particular problem when designing protonated receptors for anions (*e.g.*, ammonium containing receptors) as the protonation window of the receptor (and the anion) must also be considered. It is, of course, less of a problem for neutral receptors, or those containing permanent built-in charges, designed to operate in aprotic media.

Anionic species have a wide range of geometries (Figure 1.1) and therefore generally a higher degree of design and complementarity is required to make receptors that are selective for a particular anionic guest than for most simple cations.

Table 1.1 *The difference in radii for typical isoelectronic cations and anions (in octahedral environments) serves to underscore the more diffuse nature of the anionic species (taken from R.D. Shannon, Acta Cryst., 1976,* ***A32****, 751)*

Group 1 (cations)		*Group 17 (anions)*		Δr
Na^+	1.16 Å	F^-	1.19 Å	0.03 Å
K^+	1.52 Å	Cl^-	1.67 Å	0.15 Å
Rb^+	1.66 Å	Br^-	1.82 Å	0.16 Å
Cs^+	1.81 Å	I^-	2.06 Å	0.25 Å

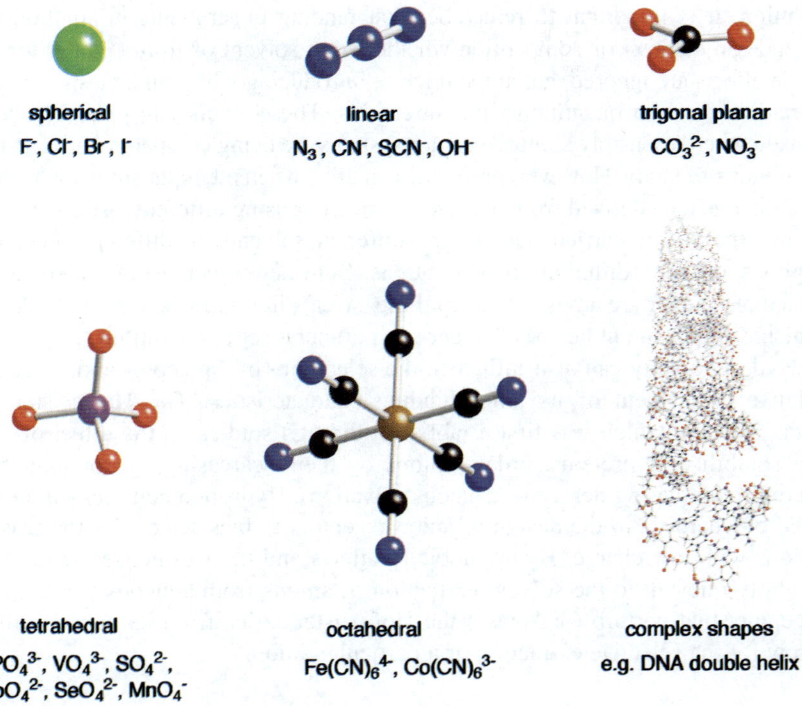

Figure 1.1 *Anions come in many shapes and sizes!*

The nature of the solvent in which the anion-binding event occurs plays a crucial role in controlling anion-binding strength and selectivity. Electrostatic interactions generally dominate over other recognition forces and are particularly important in stabilizing anions in solution. However, hydroxylic solvents are also noted for their ability to form strong hydrogen bonds with anions. A potential anion receptor must, therefore, compete effectively with the solvent environment in which the anion recognition event is to take place. For example, a neutral receptor that binds anions solely through hydrogen-bonding interactions is less likely to be capable of competing with the polar protic solvation shell surrounding the target anion in a hydroxylic solvent and hence may only function as an anion receptor in aprotic organic solvents (in which the anion interacts more weakly with the solvent). A charged receptor, on the other hand, can benefit from electrostatic effects and thus may compete more effectively with polar protic solvents. For example, protonated polyammonium macrocycles are capable of binding anions in water. Of course, the anion receptor must not just compete with the solvent but also with the counter cation that is necessarily paired with the targeted anion.

Ion-pairing can be very significant, particularly in non-polar solvents.[2] Therefore, when studying anion complexation there is a necessary trade-off associated with the choice of solvent. In non-polar solvents, the anion may be weakly solvated but there may be significant ion-pairing. In more polar solvents, the solvation may be stronger – but solvation of both the cation and anion will reduce the strength of ion-pairs in solution. It is important to remember that binding experiments in solution always include an element of competition whether from solvent or from counter ion. Often these effects are ignored, but at the price of introducing what can be a significant systematic error into quantitative measurements. These errors can generally be overlooked when ostensibly similar receptor systems are being compared under analogous conditions of study. However, problems can arise when attempts are made to contrast quantitative data derived from analyses carried out using different instrumentation, or from experiments carried out in, *e.g.*, different solvents, at different concentration regimes, or using different counter cations. Both newcomers to the field and established researcher are advised to keep these caveats in mind when trying to determine which receptor might be "best" under any particular set of conditions.

Hydrophobicity can also influence the selectivity of a receptor and, as such, any relative assessment of its anion-binding characteristics. The Hofmeister series[3] (Scheme 1.1), which was first established through studies on the effect of salts on the solubility of proteins, orders anions by their decreasing hydrophobicity (and therefore increasing degree of aqueous solvation). Hydrophobicity may therefore be used by chemists in the design of anion receptors to bias selectivity towards larger anions with low charge. Hydrophobicity effects and the Hofmeister series are particularly relevant to the solvent extraction of anions from aqueous solution. Anion receptors that perturb (or "bias") the Hofmeister series from its normal order can allow for the selective extraction of a particular anion.

organic anions > ClO_4^- > SCN^- > I^- > salicylate$^-$ > NO_3^- > Br^- > Cl^- > HCO_3^- > $H_2PO_4^-$ > F^-, SO_4^{2-} > HPO_4^{2-}

Scheme 1.1 *The Hofmeister series*

1.3 Anions in Biological Systems

Anions are ubiquitous in biology. They are present in roughly 70% of all enzymatic sites, play essential structural roles in many proteins, and are critical for the manipulation and storage of genetic information (DNA and RNA are polyanions). Anions are also involved in regulating osmotic pressure, activating signal transduction pathways, maintaining cell volume, and in the production of electrical signals. Not surprisingly, therefore, the disruption of anion flux across cell membranes (especially chloride, present in cells at the 5–15 mM concentration level[4]) is increasingly recognized as being the primary determinant of many diseases, including cystic fibrosis,[5] Bartter's syndrome,[6] Dent's disease,[7] Pendred's syndrome,[8,9] and osteopetrosis.[10] In fact, the transport of anions through cell phospholipid bilayers is known to be mediated by a variety of channels and anion transport proteins with at least 14 mitochondrial anion transport systems having been identified so far.[11] These include (among others) systems responsible for the trafficking of ADP, ATP, phosphate, citrate, maleate, oxaloacetate, sulfate, glutamate, fumarate, and halide anions.

Recently, several X-ray crystal structures have been solved that have allowed the direct visualization of enzyme–anionic substrate complexes that are stabilized via multiple hydrogen-bonding interactions. In particular, the structure of the DNA helicase

RepA sulfate complex, solved to 1.95 Å resolution, shows six hydrogen-bonding interactions between the sulfate anion and the RepA protein scaffold. The sulfate anion is also hydrogen bonded to Asp140 via an intervening water molecule (Figure 1.2).[12]

The sulfate-binding protein (SBP) found in *Salmonella typhimurium* is involved in sulfate transport and consists of two globular domains that are linked by a flexible hinge. The structure of the sulfate complex of this protein, elucidated in 1985 by Pflugrath and Quiocho, reveals that the sulfate anion is bound 7 Å below the surface of the protein in a cleft between the two globular domains. When bound in this site, the sulfate anion is inaccessible to solvent. There are no charged residues present in the sulfate-binding site (Figure 1.3). Instead, the sulfate anion is bound by seven hydrogen bonds from neutral residues, five from peptide NH groups, one from a serine OH residue and one from a tryptophan NH group (Figure 1.3). The charge on the

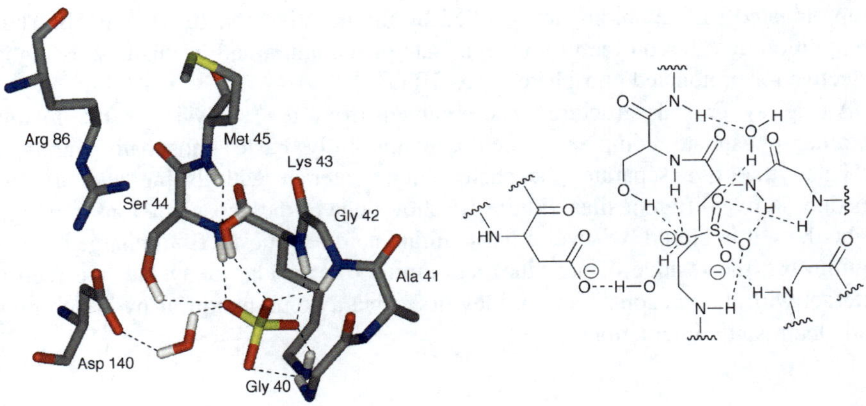

Figure 1.2 *ATPase active site in DNA helicase Rep A showing the interaction of the bound sulfate anion with various P-loop residues*

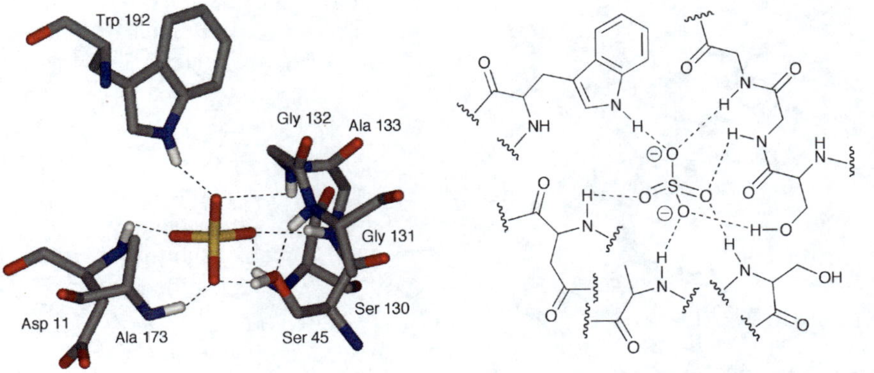

Figure 1.3 *The X-ray crystal structure and schematic of the sulfate-binding site in the sulfate-binding protein. The anion is bound by seven hydrogen bonds from neutral NH and OH hydrogen bond donor groups*

sulfate dianion is stabilized through several hydrogen bond relay systems that presumably serve to spread well into the protein matrix.[13]

The phosphate-binding protein (PBP) is a polypeptide chain with a molecular mass of 34,400 Da. Like SBP, this protein consists of two globular domains. The phosphate-binding site is located 8 Å below the protein surface in a cleft between the two domains. The protein is involved in the transport of phosphate into bacterial cells and shows high selectivity for phosphate, binding this anion at least five orders of magnitude more strongly than sulfate. The phosphate-binding site is shown schematically in Figure 1.4.[14] In contradistinction to SBP, the phosphate anion in PBP is bound by 12 hydrogen bonds from a combination of charged and neutral peptide residue interactions that include salt bridge involving the guanidinium group of Arg135. An aspartate residue (Asp56 shown in blue in Figure 1.4) is thought to be responsible for the high phosphate selectivity of this protein. This residue acts as a hydrogen bond acceptor and can form a hydrogen bond with partially protonated phosphate anions. Unprotonated sulfate anions are repelled by the negative charge on Asp156. Thus, recognition depends on the protonation state of the anion and in this case is highly selective for protonated phosphates (*e.g.*, HPO_4^{2-}).

An X-ray crystal structure was obtained from the analysis of the histone octamer–phosphate complex, which contains fully basic phosphate anions; it revealed that five separate phosphate anions interact with lysine and arginine residues at five different sites. Figure 1.5 shows one of the phosphate ions and highlights how it is bound by several basic amino acid residues.[15] Both charged guanidinium (cf. the residue of arginine) and amine (the residue of lysine) are seen to interact with the phosphate anion; they do so via a combination of hydrogen bond and electrostatic interactions.

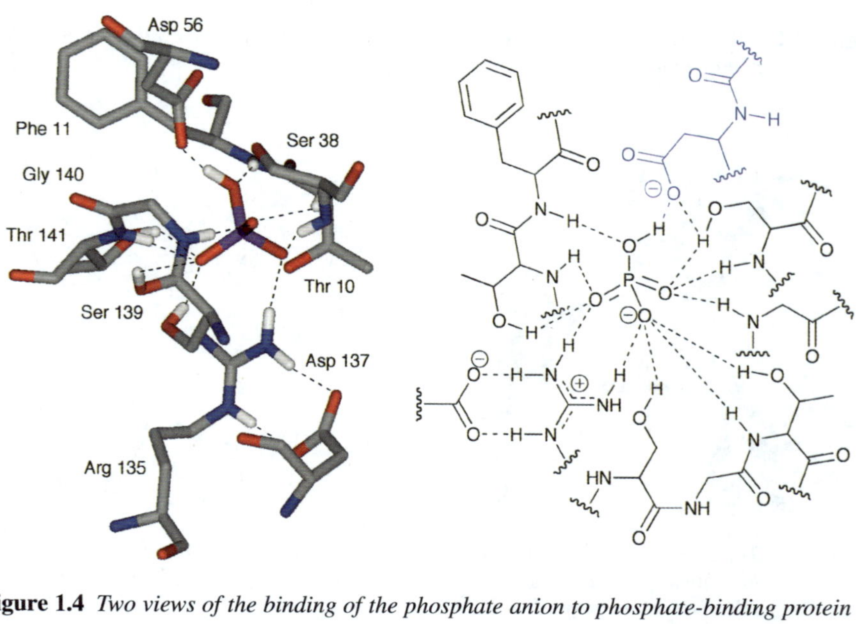

Figure 1.4 *Two views of the binding of the phosphate anion to phosphate-binding protein*

Acyclic polyamines such as spermine, $NH_2(CH_2)_3NH(CH_2)_4NH(CH_2)_3NH_2$, are involved in promoting cell growth, inducing the biosynthesis of DNA, RNA, and proteins[16] and regulating various enzymatic activities.[17] It is also well known that polyamines display high phosphate anion affinities and bind well, for example, to phenylalanine transfer RNA.[18] Recently, Ohishi and co-workers reported the X-ray crystal structure of a spermine–phosphate anion complex.[19] This structure revealed that the protonated spermine interacts with the bound phosphate anion via hydrogen bonds involving intervening water molecules. In other new work, Steed and co-workers reported an X-ray crystal structure that confirmed the ability of tetraprotonated spermine binds two hydrogen phosphate anions in the solid state (Figure 1.6).[20]

We have seen that the guanidinium group present in the active site of PBP can form salt bridges with phosphate anions. Arginine residues play a very important role in stabilising a wide range of protein polyphosphate complexes including ones

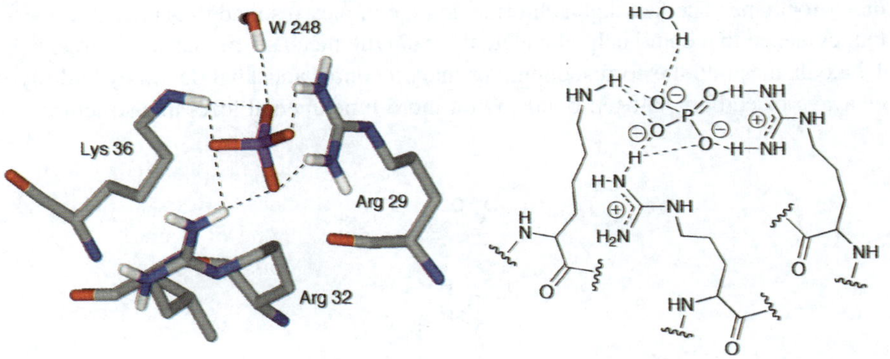

Figure 1.5 *The amino acid residues that interact with the phosphate anion in a structurally characterized histone octamer–phosphate complex. Only one of five bound phosphates is shown*

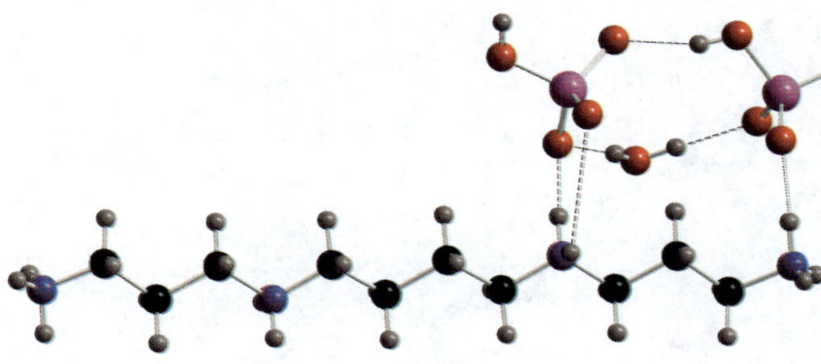

Figure 1.6 *The complex between tetraprotonated spermine and two hydrogen phosphate anions. Five water molecules have been removed to improve the clarity of the representation. In this and subsequent figures depicting structures from X-ray diffraction analysis, the carbon atoms will be shown in black, hydrogen atoms in grey, nitrogen in blue, and oxygen in red. Other colours will be used to designate other atoms (e.g., magenta for phosphorus in this case)*

involving DNA and RNA. One mode of binding that has been proposed for arginine allows it to recognize loops and bulges in RNA. RNA-binding motifs in proteins such as the human immuno-deficiency virus tat protein consist of arginine rich regions. It has been proposed that the arginine group recognizes bulges in RNA by bridging between phosphate groups that are close together in space. This binding mode, shown in Figure 1.7, has been termed "the Arginine Fork."

In contrast to the phosphate and sulfate anions, the chloride anion is spherical. It is, however, no less involved in important biological process and, not surprisingly, has been extensively studied in this context. For instance, Dijkstra and co-workers reported a 2.4 Å-resolution X-ray structure showing chloride anion bound to the active site of haloalkane dehalogenase from *Xanthobacter autotrophicus* GJ10. In this structure, the bound chloride is seen to interact with Trp 125 and 175 via hydrogen bond interactions with the indole NH protons (Cl···N = 3.6 and 3.2 Å, respectively). These interactions are shown in Figure 1.8.[21] It is noteworthy that the crystals used to provide this X-ray structure were grown from a medium containing 1,2-dichloroethane. The fact that a chloride anion complex resulted thus provides important evidence that could help elucidate the catalytic mechanism that serves to convert 1-haloalkanes into the corresponding primary alcohols and a halide ion by hydrolytic cleavage of carbon–halogen bonds. On a more fundamental level this structure also

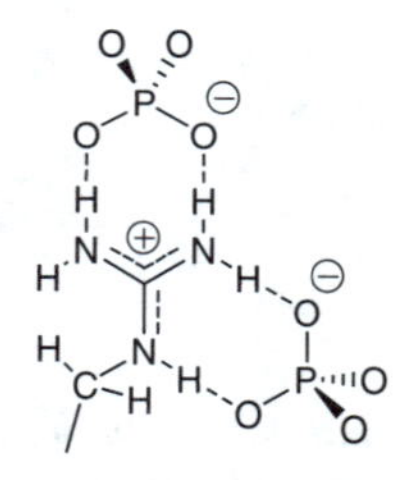

Figure 1.7 *The arginine fork motif*

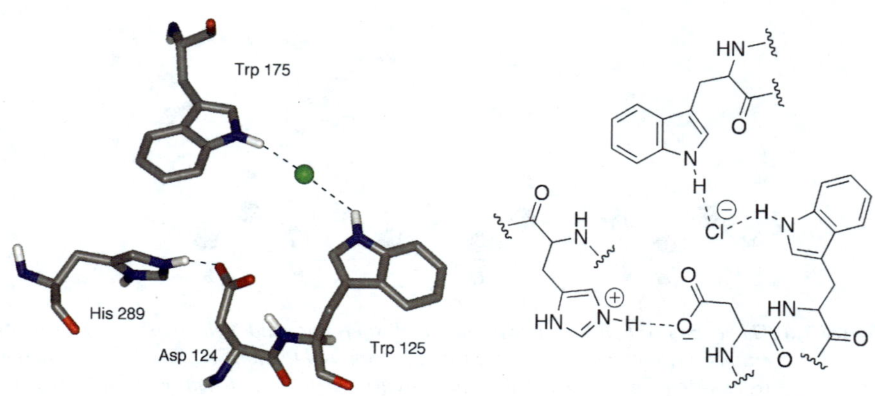

Figure 1.8 *Enzymatic active site of haloalkane dehalogenase revealing the presence of a bound chloride anion*

provides an unequivocal proof this enzyme can interact with chloride anion. Such information is of potential practical importance since it is well known that the nitrogen-fixing hydrogen bacterium, *X. autotrophicus* GJ10, containing haloalkane dehalogenase degrade harmful halogenated compounds to primary alcohol with water as a co-substrate and without the need for either oxygen or ancillary co-factors.[22]

In 2002, the 3 Å-resolution X-ray structure of the StClC chloride ion channel was solved.[23] This long-awaited crystal structure revealed a homodimer-derived channel that is notable for its hour glass shape and the complete absence of positively charged amino acid side chains anywhere near where the chloride anions would pass (Figure 1.9). On the other hand, the structure does indicate the presence of a negatively charged glutamate side chain just above the channel entrance. This residue is thought to act as an anion-regulating gate. By swinging out to open the channel, it allows Cl^- ions to enter the channel pore from whence they are pulled (presumably) towards the constricted, neutral centre of the channel by surfaces rich in positively polarized (but not charged) residues. In spite of the elegance of this structure, and a considerable body of work devoted to understanding biological ion transport in general, our understanding of through-membrane anion transport remains extremely limited. One way of addressing this deficiency is through the synthesis and study of synthetic anion receptors and artificial anion channels, and this, in turn, is providing an important motivation to produce such systems.[24]

The year 2002 was also marked by the exciting discovery of a new, genetically encoded amino acid, L-pyrrolysine, whose proposed biological function relies in part on pyrrole-like NH-anion recognition (cf. Figure 1.10).[25] Pyrrole NH-chloride anion recognition and transport have also been proposed recently as an explanation for the anticancer and immunosuppressive activity displayed by prodigiosin, a naturally

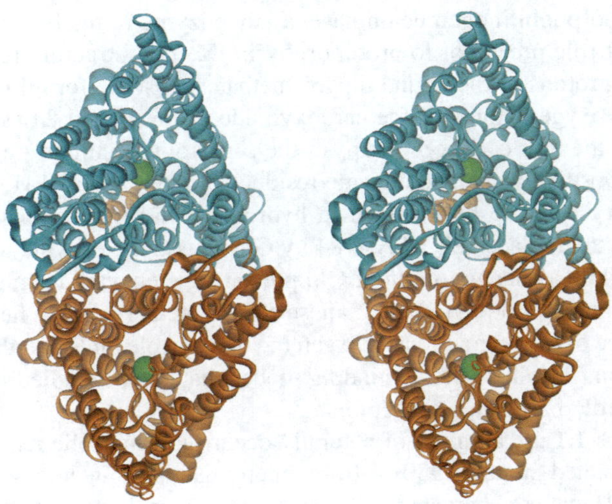

Figure 1.9 *Stereo view of a ribbon representation of the StClC dimer visualized from the extracellular side. The two subunits are brown and cyan. A chloride anion in the selectivity filter is represented as a green sphere*

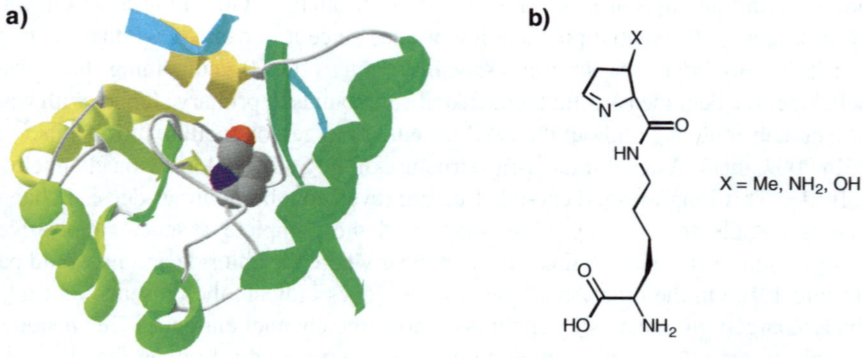

Figure 1.10 *(a) Ribbon diagram of Methanosarcina barketi momomethylamine methyltransterase subunit. The L-pyrrolysine is shown using a space-filling model. (b) Stick-diagram of proposed L-pyrrolysine amino acid*

occurring tripyrrolic pigment known since the 1930s (*vide infra*). However, an alternative explanation for the observed activity has also been put forward that is based on copper complex formation and DNA cleavage.[26] A desire to distinguish between these two limiting mechanisms and to test whether synthetic oligopyrrole species can act as anion carriers thus provides a further incentive to make and study pyrrole-based anion receptors.

In 1992, Jordan and co-workers reported a most exciting discovery that has remained essentially overlooked by the anion recognition community but which provides a further important incentive for the design of pyrrole-based anion receptors. Specifically, these workers reported the 1.9 Å-resolution X-ray crystal structure of porphobilinogen deaminase, a key enzyme in the biosynthesis of the linear tetrapyrrole precursor to protoporphyrin IX. This structure reveals that the pyrrolic NH protons of the bound dipyrromethane substrate form hydrogen bonds with the two oxygen atoms of the carboxyl side chain of Asp 84 (Figure 1.11).[27] Furthermore, the four carboxyl groups of the β-pyrrolic positions are seen to interact with the positively charged enzyme residues of Arg 11, Arg 131, Arg 132, Arg 155, and Lys 83 and to be involved in hydrogen bond interactions with Ser 13. Interestingly, the replacement of Asp 84 by Glu causes the enzyme to lose 99% of its activity, while a change to Asn or Glu prevents the enzyme from catalyzing the formation of preeuroporphyringen.[28] These complementary results help underscore the importance of the interactions between the two pyrrole units and the Asp 84 carboxylate anion. They also define binding modes that can be studied in detail using synthetic pyrrole-based anion receptors.

Prodigiosins **1.1** are a family of naturally occurring tripyrrolic red pigments that were first isolated in the 1930s[29] from microorganisms including *Serratia* and *Streptomyces*[30] that are characterized by a common pyrrolylpyrromethene skeleton. These molecules, especially prodigiosin 25-C, **1.1b**, have been studied extensively for their promising immunosuppressive[31] and anticancer activity.[32] Various prodigiosins have also been found to induce apoptosis in dozens of human cancer cell

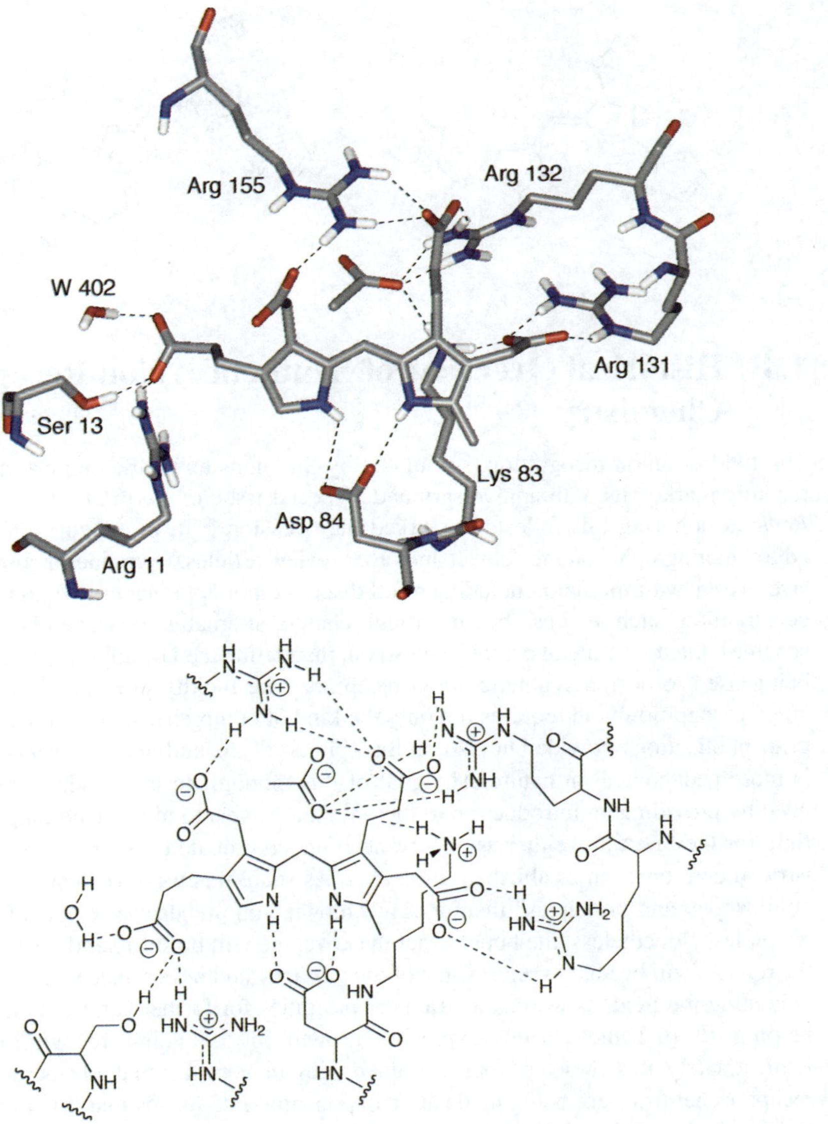

Figure 1.11 *Partial view of the X-ray structure of the enzyme porphobilinogen deaminase showing the binding of a dipyrromethane co-factor through pyrrolic NH to Asp 84 and pyrrolic side chains (acetate and propionate) to Arg 132, Arg 131, Arg 155, and Lys 83 hydrogen-bonding interactions*

lines, including liver cancer,[33] human breast cancer,[34] human colon cancer,[32] gastric cancer,[35] and haematopoietic cancer cell lines.[36] Taken in concert, this combination of properties has made the prodigiosin an attractive target for use in various combination of drug therapies. It is also inspiring the synthesis of new pyrrole-based anion carriers as discussed further in Chapters 3 and 5.

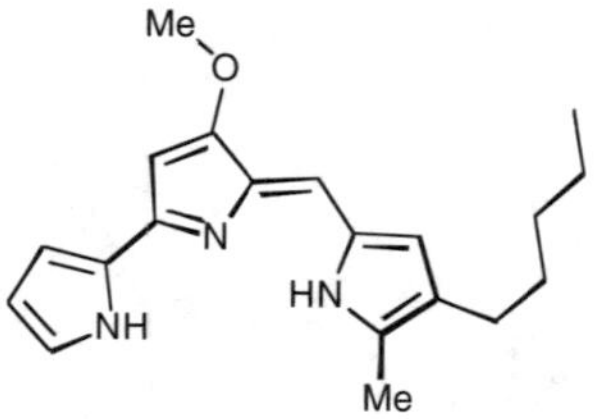

1.1a

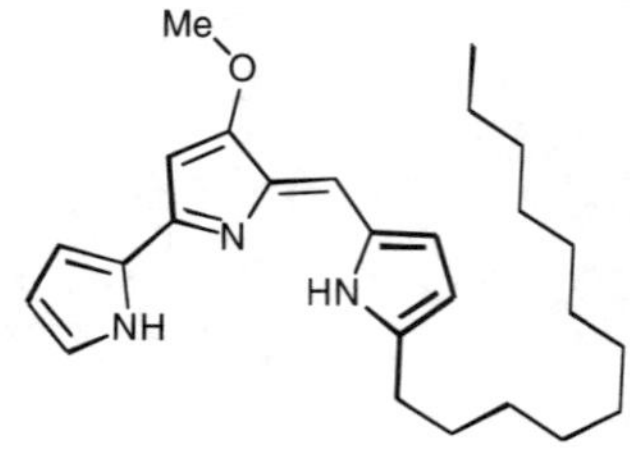

1.1b

1.4 Historical Overview of Synthetic Anion Receptor Chemistry

The field of anion recognition, stimulated by Simmons and Park's report in 1968,[37] recently marked its 35th anniversary and a special issue of *Coordination Chemistry Reviews* honoring this milestone marked the occasion.[38] It, in conjunction with an edited monograph[39] on the subject and other review articles,[40] provides a comprehensive overview of the field, including a full discussion of approaches to anion receptor construction, such as those based on high charge, alternative hydrogen bond donor schemes, the use of metal centres, and so on, that define the key approaches currently being used to design synthetic anion receptors. The rapidly growing nature of the anion recognition field requires not only the kind of comprehensive review that these prior publications provide but also, in the opinion of the authors, a presentation that is more pedagogical in nature. The goal of this monograph is to address this latter need by providing an introduction to the field that is geared to those coming into the field for the first time, either as an advanced undergraduate or graduate-level chemistry student or as an established researcher. As such, this work is focused more on "philosophy and principles" than on a full-blown, comprehensive review of specific examples. Nonetheless, the hope is that the coverage will be sufficiently detailed that the readers will be able to sense some of the vibrancy and importance that is currently animating the field, as well as the rich opportunities for further contribution that are helping to fuel the current explosive growth in the anion recognition area. Unfortunately, this choice of focus, coupled with the very fact that advances in anion receptor chemistry are being made at an exponential rate means that it is impossible to include the work, or all of the work, of many colleagues. The authors necessarily apologize for this but trust that our fellow practitioners will appreciate the importance of our goals and forgive us for any real or perceived oversights.

Traditionally, as implied above, the field of synthetic anion receptor chemistry traces its origins back to the 1968 communication by Simmons and Park from DuPont Central Research in Delaware. In this seminal work, the halide binding properties of several macrobicyclic receptors, consisting of two ammonium bridgehead centres spanned by three alkyl linkers, were reported. The authors noted that the proton on the ammonium group could either point out of or into the cavity of the receptor, and it was found that the different conformations could be observed by ^{1}H NMR spectroscopy. From the various compounds studied, with differently sized linkers, varying from 7 to 10 methylene groups in length, compound **1.2** was found to have the highest affinity for chloride with $K_a = 4\ M^{-1}$ in 50% TFA solution. Under these conditions the only conformer observed, in the presence of the chloride anion, had both ammonium hydrogen atoms pointed in towards the centre of the cavity, leading to the suggestion that the anion was bound within the cavity of this cavity. The crystal structure of this complex was reported by Marsh and co-workers in 1975. It was found that the macrobicyclic chloride anion complex crystallized together with an $H_{13}O_6^+$ cation (which was the focus of the paper) and eight chloride counteranions.[41] However, importantly, the crystal structure did serve to confirm that the chloride anion was bound within the central cavity of the positively charged receptor in the solid state.

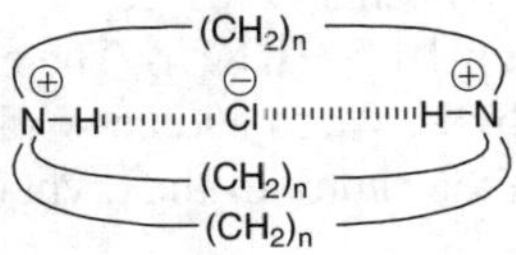

1.2-Cl⁻ n = 9

Simmons and Park's landmark contribution foreshadowed the extensive work that was to be carried out on macrocyclic ammonium-based anion receptors in the ensuing decades. Consequently, this work is generally regarded as the first example of what has come to be known as synthetic anion receptor chemistry. However, 14 years earlier (*i.e.*, in 1954), Tanford inferred through changes in the effective pK_a values that a complex is formed between the anionic conjugate base of acetic acid and guanidinium moieties. In fact, quantitative association constants were obtained. However, it was also found that under these conditions, the association constant (log K_a) was low, being less than roughly 0.5.[42] Using this approach, the association constants for guanidinium complexes formed from a number of other carboxylate anions and from phosphates were measured (*viz.* log $K_a = 0.37, 0.32, 0.43$, and $1.37\ M^{-1}$ for acetate, formate, chloroacetate, and dihydrogen phosphate, respectively, in 1.02 M tetramethylammonium chloride aqueous solution).[43]

Subsequent to Tanford's report but earlier than Simmons and Park's paper, the formation of a chelated complex between methoxide anion and a bidentate Lewis acid, namely the 1,2 ethane derivative **1.3**, was reported by Shriver and Biallas.[44] In the abstract of their communication the authors noted: "*This is the converse of the usual situation where a central metal ion serves as an acceptor toward a difunctional base.*" Clearly Shriver and Biallas appreciated the *umpolung* nature of their chelated system with respect to traditional coordination chemistry. Today, Lewis acidic boron centres continue to be used in anion complexation agents with a plethora of systems having been reported in the literature along with other receptors containing mercury,[45] tin,[46] germanium,[47] and silicon.[48]

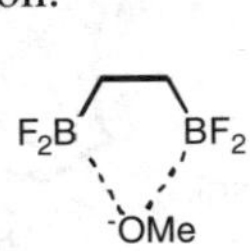

1.3-OMe⁻

Following Simmons and Park's report, the next step forward came in the mid-1970s, when Lehn and co-workers described the anion-binding properties of a variety of macrobicyclic and macrotricyclic ammonium-based receptors. This research clearly demonstrated how optimizing the fit of an anion for a given charged cavity could lead to strong binding. For example, the synthesis and binding of halide anions by compounds **1.4** and **1.5**, cryptand-like receptors containing four amine centres, was shown to be selective by size.[49] Upon conversion to their corresponding tetraprotonated forms, these receptors were found to bind chloride anions selectively with an association constant (log K_a) of more than 4 in aqueous solution. Iodide is too large to fit into the cavity and is therefore bound considerably less strongly. The model compound **1.6** was also studied and displayed a much lower affinity for anions than the macrobicyclic systems (a log K_a of 1.7 ± 0.1 was observed for the binding of chloride in water at pH 1.5 (HNO_3)). The crystal structure of the chloride complex of **1.4** was reported by R. Weiss and co-workers in 1976; it confirmed that the chloride anion is bound in the centre of the cryptand.[50]

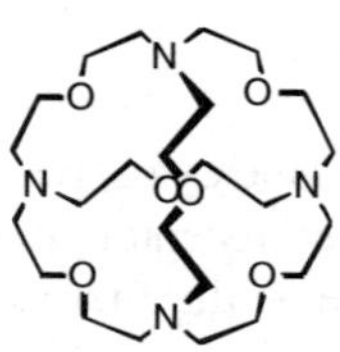

1.4

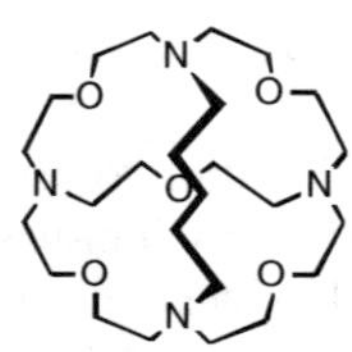

1.5

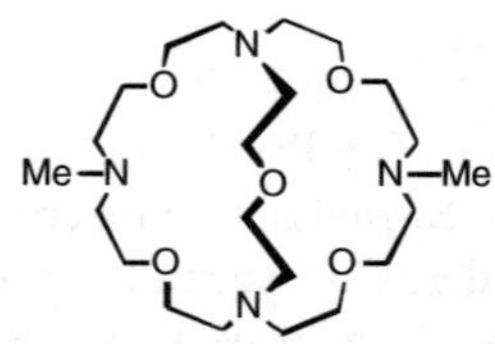

1.6

In contradistinction to receptors **1.4** and **1.5**, receptor **1.7**, an ellipsoidal hexaprotonated cryptand also synthesized and studied by Lehn and co-workers, was found to be selective for linear anions such as azide N_3^- (added as the sodium salt). This anion is complementary to the *shape* of the cavity and was found to be bound with high affinity in aqueous media (log K_a = 4.6 in aqueous solution at 25 °C as determined by ^{13}C NMR spectroscopic titration methods). The spherical anion, chloride, displays a much weaker interaction (log K_a < 1.0) under identical conditions.[51] A crystal structure of the chloride complex revealed that the anion is bound in an octahedral fashion via six ammonium groups (Figure 1.12a). The receptor in this structure is considerably distorted, a finding that was thought to reflect the reduced stability of the complex. An X-ray structure of the complex formed with the azide anion, however, reveals that this more complementary anion is bound by six hydrogen bonds arranged in two trigonal arrays of hydrogen bonds, each binding a terminal nitrogen atom (Figure 1.12b).[52]

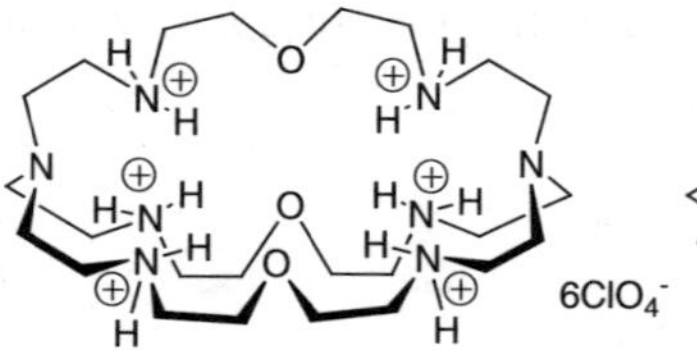

1.7

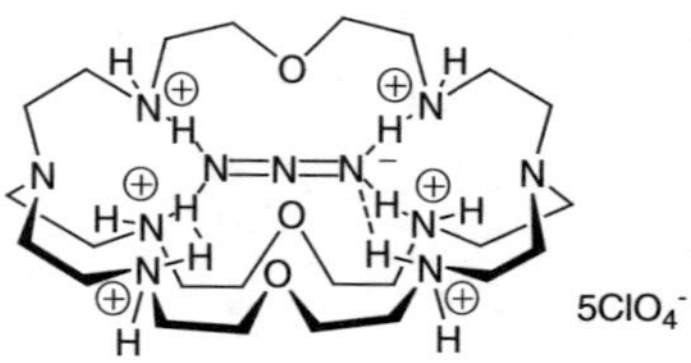

1.7–N_3^-

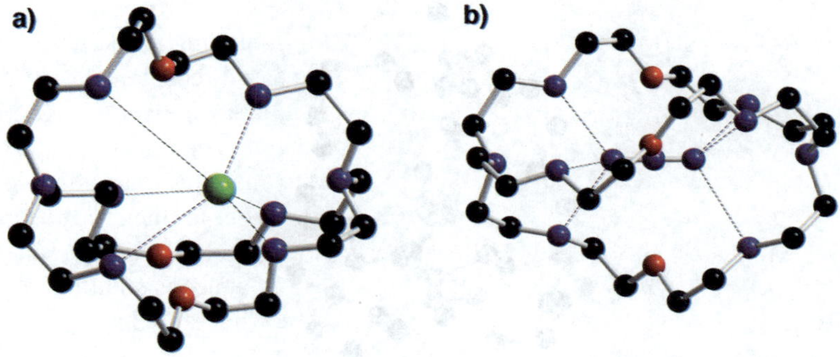

Figure 1.12 *The X-ray crystal structures of (a) the chloride complex and (b) the azide complex of receptor 1.7. The same colour codes are used in this figure as in Figure 1.6, with the addition that the chloride anion is shown in green*

Another early pioneer in the anion recognition arena is Schmidtchen from the Technische Universität München. Schmidtchen produced a series of receptors that do not rely on hydrogen-bonding interactions to bind anions.[53] Instead, these receptors employ quaternary ammonium groups arranged in a tetrahedral manner. Hence, the anion is bound by these cage-like receptors (*e.g.*, receptors **1.8** and **1.9**) solely by electrostatic interactions. By altering the length of the alkyl chain between the ammonium centres, Schmidtchen was able to "tune" the selectivity of the receptors for particular halides. In fact, the cavity in receptor **1.8** has an internal diameter of approximately 4.6 Å that provides a good size match for iodide anion (diameter 4.12 Å). A crystal structure of the iodide salt of **1.8** (Figure 1.13) revealed that one of the iodide counteranions is encapsulated within the macrotricycle. The larger receptor **1.9** is able to form complexes with anions such as *p*-nitrophenolate that are too large to form complexes with receptor **1.8**.

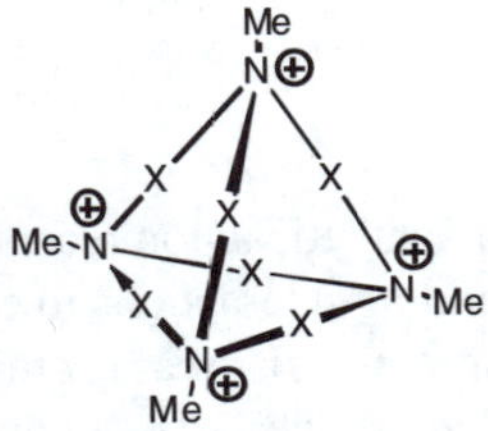

1.8 X = $-(CH_2)_6-$
1.9 X = $-(CH_2)_8-$

Receptors **1.8** and **1.9** are positively charged and are therefore associated with counteranions that may compete for the anion-binding site. To overcome this limitation, Schmidtchen produced zwitterionic receptors such as **1.10** and **1.11** that are neutral.[54] In this case, ^{1}H NMR spectroscopic studies carried out in D_2O served to confirm that receptor **1.11** forms stronger complexes with chloride, bromide, and iodide anions than receptor **1.8**. Taken in concert, this body of work was

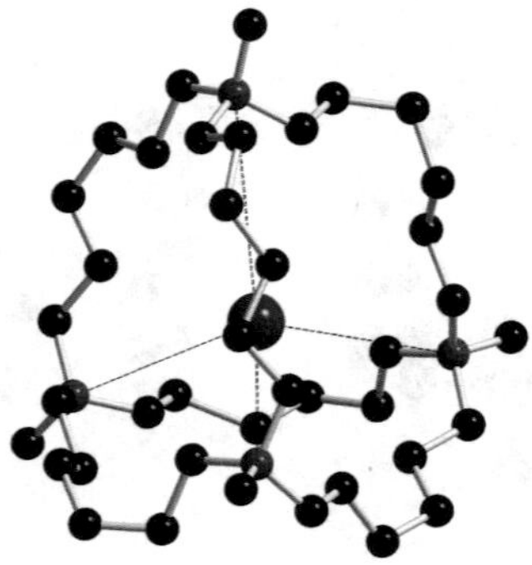

Figure 1.13 *The X-ray crystal structure of the iodide salt of receptor **1.8**. Only the internally bound iodide (purple) is shown*

seminal in that it established the viability of receptor designs predicated on purely electrostatic interactions. Nonetheless, perhaps because of perceived synthetic challenges, very few other systems have been produced that rely solely on this recognition motif.

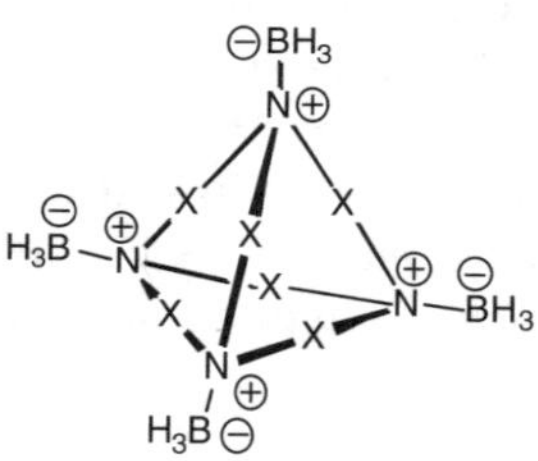

1.10 X = –(CH$_2$)$_6$–

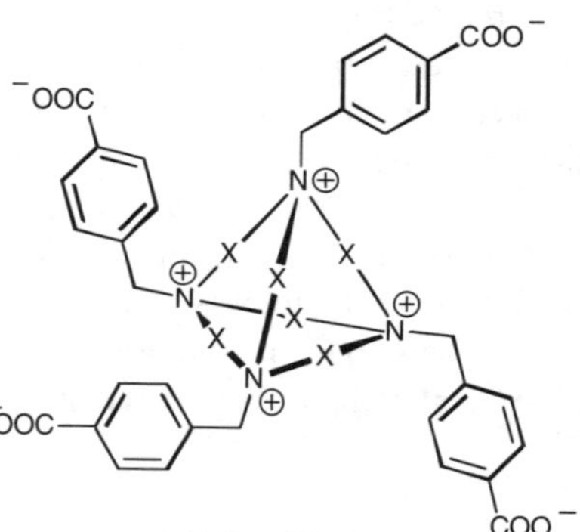

1.11 X = –(CH$_2$)$_6$–

Conversely, a wide variety of synthetic anion receptors are known that are neutral and which rely on hydrogen-bonding interactions to effect anion recognition. These receptors include amides (or thio-amides), ureas (or thio-ureas), or pyrroles. The first synthetic anion receptor to utilize amide N-H···anion interactions was reported by Pascal and co-workers in 1986.[55] The system in question, receptor **1.12** contains three amide NH groups that can orientate to form a convergent binding site within the cyclophane host. The X-ray crystal structure of **1.12** is shown in Figure 1.14. It highlights the fact that in the solid state the NH groups do not point into the cavity of the host but rather, they are inclined by between 47° and 68° relative to radii drawn from the central axis through the nitrogen atoms. In spite of the fact that the NH donor groups are not apparently oriented in an ideal arrangement, this receptor was found to bind fluoride anions in DMSO-d_6 solution, as judged from ^{1}H and ^{19}F NMR spectroscopic studies.

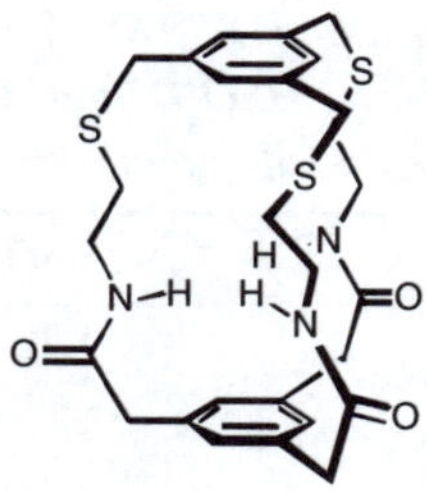

1.12

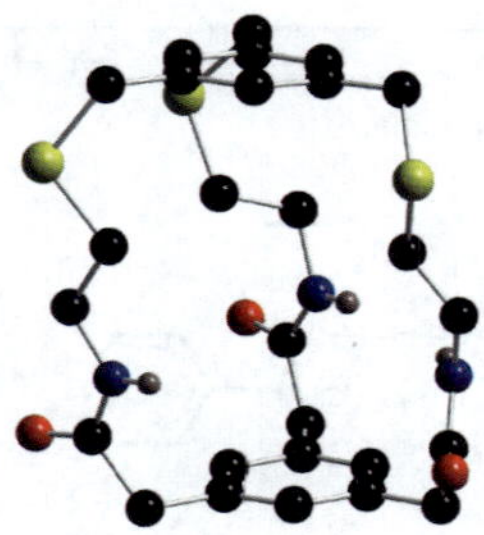

Figure 1.14 *Pascal's amide-based cyclophane anion receptor **1.12**. The sulfur atoms are shown in gold*

In 1993, Reinhoudt and co-workers[56] produced a series of acyclic tripodal receptors containing amide groups (**1.13–1.18**). The association constants of these receptors with $H_2PO_4^-$, HSO_4^-, and Cl^- were measured by conductivity experiments in acetonitrile. The receptors were found to bind dihydrogen phosphate anion selectively over chloride anion, a substrate that in turn is bound more strongly than hydrogen sulfate (Table 1.2). Receptor **1.18** shows the highest affinity for the dihydrogen phosphate anion, presumably due to the electrophilicity of the sulfonamide groups and preorganization of the binding site via π–π stacking of the naphthyl groups.

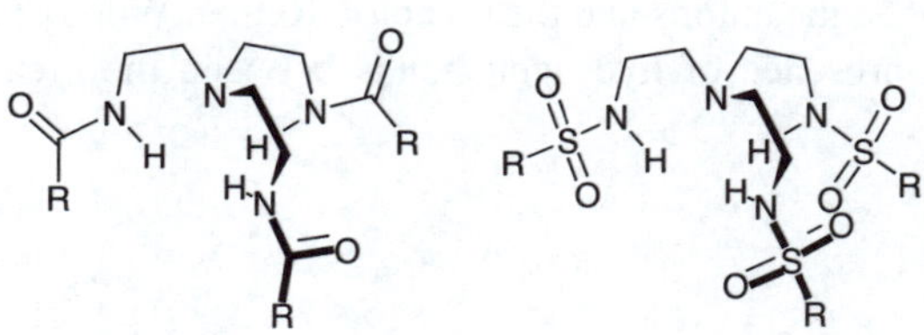

1.13 R = CH_2Cl
1.14 R = $(CH_2)_4CH_3$
1.15 R = C_6H_5
1.16 R = 4-$MeOC_6H_4$

1.17 R = 4-MeC_6H_5
1.18 R = 2-naphthyl

Urea and thio-urea groups have been used to construct a considerable number of receptors for anionic species. These groups have particularly high affinities for oxo-anions, as they are capable of forming two hydrogen bonds to the oxo-anion (Figure 1.15). One of the earliest examples of oxo-anion complexation by a

Table 1.2 *Stability constants K_a (M^{-1}) for **1.13–1.18** with the anions $H_2PO_4^-$, HSO_4^-, and Cl^- in acetonitrile, as determined by conductometry*[a]

	$H_2PO_4^-$	HSO_4^-	Cl^-
1.13	6100	170	1740
1.14	280	31	290
1.15	870	56	100
1.16	510	73	190
1.17	3500	79	540
1.18	14200	38	1600

[a]Error is 5% for $K_a > 10^2$ M^{-1} and 10% for $K_a < 10^2$ M^{-1}. The anions were studied in the form of their n-Bu_4N^+ salts.

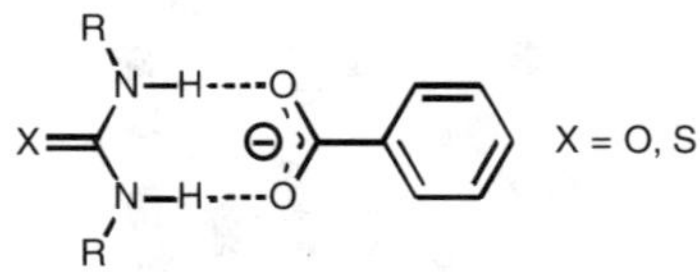

Figure 1.15 *Schematic representation designed to illustrate why urea and thio-urea groups are excellent receptors for oxo-anions such as carboxylates*

synthetic receptor containing a built-in urea subunit (*e.g.*, **1.19** and **1.20**) was reported by Wilcox and co-workers in 1992.[57] In this seminal study, UV/Vis titrations were carried out in chloroform solution at 298 K using receptor **1.19** and a variety of oxo-anions (in the form of their tetrabutylammonium salts). Association constants of $2.7 \pm 0.8 \times 10^4$, $9.0 \pm 2.0 \times 10^3$, $6.1 \pm 1.2 \times 10^3$, and $6.9 \pm 1.4 \times 10^3$ M^{-1} were found for the complexes formed between **1.19** and anions A–D, respectively. The fact that similar association constants were observed for the interaction between **1.19** and both tosylate and camphorsulfonate led the authors to conclude that the critical receptor–anion recognition process does not involve π-stacking interactions between the anions and the receptor. Rather, Wilcox attributed complex formation to the presence of hydrogen bonds between the urea and oxo-anionic guests.

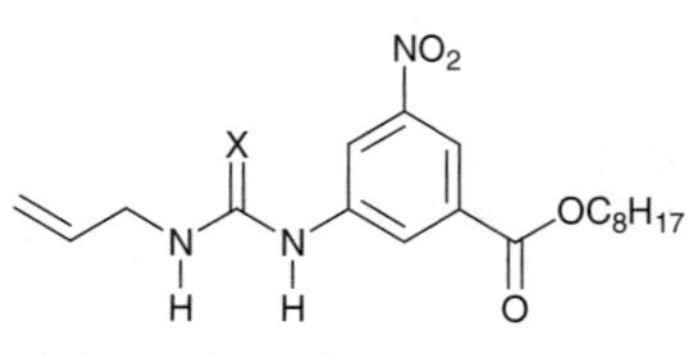

1.19 X = O
1.20 X = S

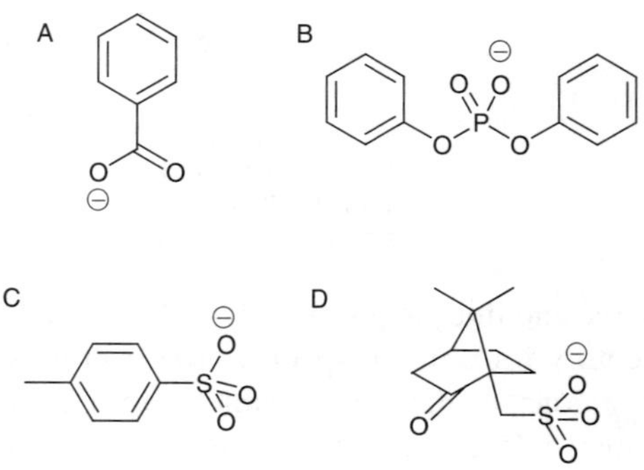

As noted in the introductory portions of this chapter, care must be taken with protonated polyammonium receptors. On the one hand, it is necessary to insure that the environment is sufficiently acidic for the receptors themselves to remain protonated, while, on the other, making certain that the proton concentration is not so high as to protonate any putative anionic guest. One way of overcoming this dilemma is to use guanidinium subunits as the binding motif. The guanidinium cation is stabilized by resonance and charge delocalization. It has a pK_a of 13.6 and is approximately three orders of magnitude more stable in its cationic form than a protonated secondary amine ($pK_a \approx 10.5$). Therefore, guanidinium subunits generally remain protonated at relatively high pH values, a feature that is ideal for extending the pH range over which positively charged, protonated anion receptors can operate.

Nature appears well aware of these benefits. The positively charged guanidinium group is widely distributed in biological systems as a side chain of arginine and, not surprisingly given its charge, this moiety plays a major role in the binding of anionic substrates (*vide supra*). Synthetic guanidinium-based receptors are also well known and have played an important role in the development of anion recognition chemistry and many examples will be discussed later on in this monograph (see Chapter 2). Like ureas, these systems show strong affinities for carboxylates, phosphates, sulfates, and nitrates, binding these substrates through hydrogen bonding but unlike urea-based systems, there is a strong electrostatic component to the binding, allowing for substantial anion–receptor recognition in aqueous or partially aqueous environments.

Lehn's group reported the synthesis and binding properties of the first guanidinium-based anion receptors in 1978.[58] However, even though macrocyclic, these systems showed relatively poor anion affinities. A breakthrough advance came when Schmidtchen incorporated the guanidinium motif into fused bicyclic ring systems, thereby pre-organising the NH hydrogen-bonding array.[59] These receptors, *e.g.*, **1.21–1.23**, are characterized by hydrogen-bonding arrays that are similar to those present in ureas. They also show high affinities for complementary carboxylate or phosphate guests. For example, the tetraphenylborate salt of receptor **1.21** forms a very stable complex ($K_a = 1.4 \times 10^5\ M^{-1}$) with *p*-nitrobenzoate in chloroform.[59] This has led to extensive use of guanidinium-based receptors for the binding of this kind of anionic substrate.

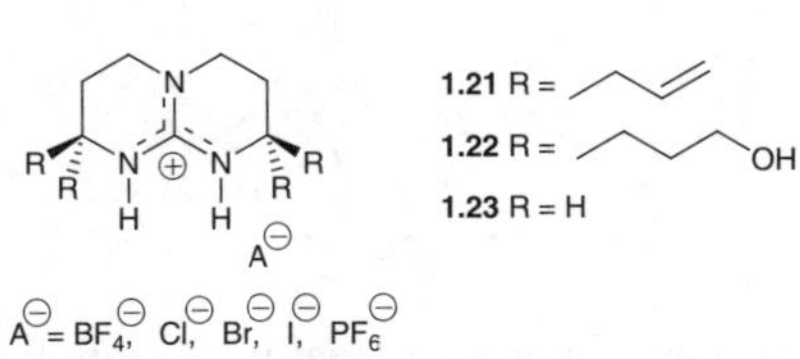

Cyclic and acyclic receptors based on pyrrole also show high affinity for anions. In the late 1980s, Sessler, one of the authors of this monograph, was interested in the metal complexation chemistry of expanded porphyrins such as sapphyrin (*e.g.*, **1.24**), a pentapyrrolic macrocycle first synthesized by Woodward and co-workers.[60] Sapphyrin is much more readily protonated than porphyrin, resulting in the facile generation of a mono- or dipositive macrocycle. In early work, Sessler and co-workers attempted to crystallize the bis-hexafluorophosphate salt of sapphyrin. Instead of obtaining crystals of the expected salt, a mixed fluoride/hexafluorophosphate salt

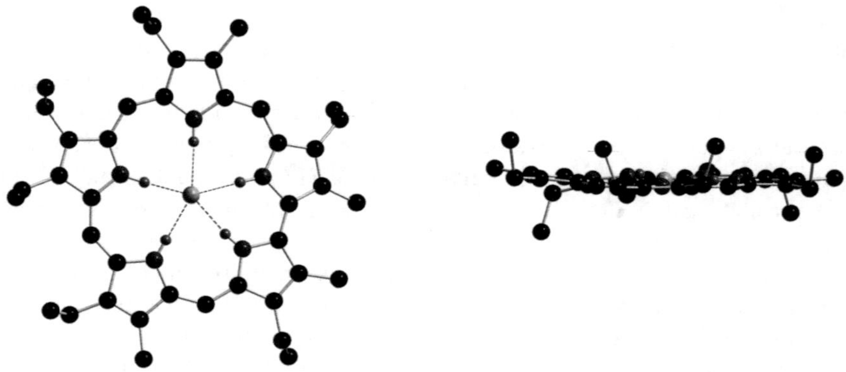

Figure 1.16 *X-ray crystal structure of the mixed fluoride/hexafluorophosphate salt of sapphyrin. The bound fluoride anion is shown in yellow, while the unbound hexafluorophosphate counteranion is not shown*

crystallized (Figure 1.16). The crystal structure was elucidated by Ibers and it was found that the fluoride anion was bound within the macrocyclic core of the doubly protonated sapphyrin, being held three by five hydrogen bonds, as well as, presumably, the dipositive charge.[61] Solution phase experiments in several organic solvents revealed that fluoride anion is bound over 10^3 times more strongly to diprotonated sapphyrin than either bromide or chloride.[62] This discovery has led to an exploration of the rich anion complexation chemistry of pyrrole-containing systems by Sessler, Gale, and others (see Chapters 3 and 5).

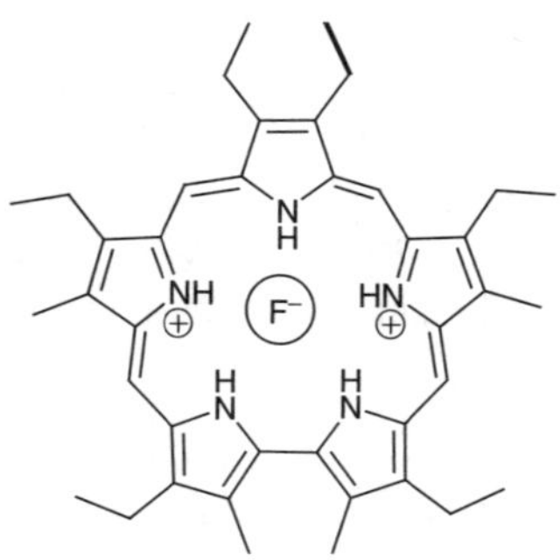

$[2H·\mathbf{1.24}]^{2+}–F^-$

A final type of anion–receptor interaction that is receiving considerable attention in the literature involves the use of anions as templating agents for the formation of self-assembled molecular architectures. Perhaps the most spectacular example of this is Lehn's pentameric circular helicate (Figure 1.17). The pentametallic circular helicate **1.25** only forms in the presence of chloride anions.[63] The complex may be produced by mixing tris-bipyridine ligands with equimolar amounts of $FeCl_2$ in ethylene glycol at 170 °C. The chloride anion bound in the centre of the helicate is locked in place and cannot be exchanged for other anions such as hexafluorophosphate or triflate. If iron(II) salts, such as tetrafluoroborate or sulfate are used, a hexameric complex **1.25** is obtained. The chloride anion, therefore, plays a role in the assembly of this remarkable

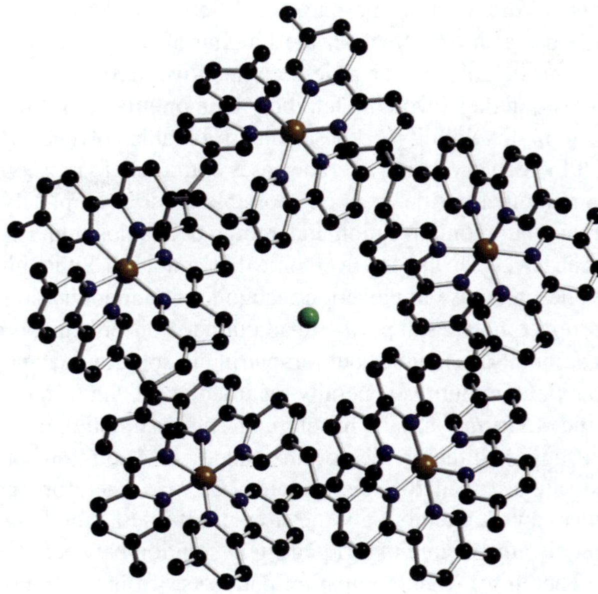

Figure 1.17 *Lehn's chloride templated circular helicate complex **1.25**. The iron atoms are shown in brown*

complex (Figure 1.17). The chemistry of other self-assembling systems will be discussed in Chapter 9.

1.5 Measurement Methods: Caveats and Limitations

Of course, it is not only the role of the supramolecular chemist to design and synthesize receptors, but also to evaluate their binding properties and selectivity. In the field of anion receptor chemistry, a variety of techniques have been employed to measure the stability constants of hosts with guests. These usually involve titrating a guest into a solution of the host. The titration may be followed by using one or more of a variety of spectroscopic or calorimetric tools including ^{1}H NMR spectroscopy (or spectroscopy involving another NMR detectable nuclei), UV/Vis absorption spectroscopy, fluorescence emission spectroscopy, or isothermal titration calorimetry (ITC). Each of these techniques looks at a different part of the binding process and/or overall equilibrium. For instance, titrations involving ^{1}H NMR spectroscopy and monitoring NH proton shifts in, *e.g.*, amide or pyrrole type receptors, provide insights into the direct interaction of the anion with the hydrogen bond donor subunits of the receptor. By contrast, UV/Vis and fluorescence spectroscopy reflect changes in the optical properties of the light absorbing/emitting portions of the receptor (including any appended chromophore), whereas ITC provides information about changes in energy for the system as a whole. These techniques often operate over different sensitivity ranges, typically 10^{-3} M for NMR spectroscopy, 10^{-4} for ITC, and 10^{-5} or lower for UV/Vis or fluorescence spectroscopy.

Measurements are often made in a range of different solvents and the polarity of the solvent often has a direct effect on the binding affinities, with, in general, the affinities being considerably higher in less competitive aprotic organic solvents. In these latter solvents, studies involving tetrabutylammonium anion salts are common. This is due to the high solubility of these salts in organic solvents. However, these salts are difficult to keep dry, being in some cases extremely hygroscopic. In the case of tetrabutylammonium fluoride, it is not possible to dry completely this material without the tetrabutylammonium cation undergoing decomposition. Cryptand salts of potassium fluoride have been used to overcome this limitation, although this approach has yet to see widespread use in the anion recognition community, in part because it is difficult to assure complete salt purity. In addition to concerns involving salt purity, many common assumptions made about ion-pairing in solution may not be valid. For example, tetrabutylammonium is generally regarded as an "innocent" counter cation with little tendency to form ion-pairs in solution. This assumption is incorrect and, in fact, it has been suggested that in dichloromethane 1 mM solutions of tetrabutylammonium chloride are less than 20% dissociated at 22 °C.[2] Therefore, going from one solvent to another can change not only the strength of interaction between the anion and receptor, but also the degree of ion-pairing in solution between the anion and its counter cation. Therefore, when comparing data sets of binding constants across a range of receptors, it is essential that the binding studies be conducted under *identical* conditions (*e.g.*, temperature, solvent, concentration, and even measurement method); otherwise, any conclusions drawn may well be invalid.

1.6 Summary Remarks

This book will examine the different strategies synthetic chemists have used to bind anions and ion-pairs. The context of this discussion will examine the roles anion complexation can play in various technologies, namely the production of sensors, in extraction, and in the generation of supports for chromatographic applications. We will also point out opportunities for further work in the area of anion complexation, as well as highlight the role anion receptors could play in the area of biomedicine.

References

1. R.D. Shannon, *Acta Cryst.*, 1976, **A32**, 751.
2. S. Alunni, A. Pero and G. Reichenback, *J. Chem. Soc. Perkin Trans. 2*, 1998, 1747.
3. F. Hofmeister, *Arch. Exp. Pathol. Pharmacol.*, 1888, **24**, 247.
4. B. Alberts, D. Bray, J. Lewis, M. Raff, K. Roberts and J.D. Watson, in *Molecular Biology of the Cell*, 3rd edn, 1994, Garland Science, New York.
5. M.P. Anderson, R.J. Gregory, S. Thompson, D.W. Souza, S. Paul, R.C. Mulligan, A.E. Smith and M.J. Welsh, *Science*, 1991, **253**, 202.
6. D.B. Simon, R.S. Bindra, T.A. Mansfield, C. Nelson-Williams, E. Mendonca, R. Stone, S. Schurman, A. Nayir, H. Alpay, A. Bakkaloglu, J. Rodriguez-Soriano,

J.M. Morales, S.A. Sanjad, C.M. Taylor, D. Pilz, A. Brem, H. Trachtman, W. Griswold, G.A. Richard, E. John and R.P. Lifton, *Nat. Genet.*, 1997, **17**, 171.

7. O. Devuyst, P.T. Christie, P.J. Courtoy, R. Beauwens and R.V. Thakker, *Hum. Mol. Genet.*, 1999, **8**, 247.
8. D.A. Scott, R. Wang, T.M. Kreman, V.C. Sheffield and L.P. Karniski, *Nat. Genet.*, 1999, **21**, 440.
9. A. Yoshida, S. Taniguchi, I. Hisatome, I.E. Royaux, E.D. Green, L.D. Kohn and K. Suzuki, *J. Clin. Endo. Metab.*, 2002, **87**, 3356.
10. U. Kornak, D. Kasper, M.R. Bosl, E. Kaiser, M. Schweizer, A. Schulz, W. Friedrich, G. Delling and T.J. Jentsch, *Cell*, 2001, **104**, 205.
11. R.S. Kaplan, *J. Membr. Biol.*, 2001, **179**, 165.
12. H. Xu, N. Sträter, W. Schröder, C. Böttcher, K. Ludwing and W. Saenger, *Acta Cryst.*, 2003, **D59**, 815.
13. J.W. Pflugrath and F.A. Quiocho, *Nature*, 1985, **314**, 257; J.W. Pflugrath and F.A. Quiocho, *J. Mol. Biol.*, 1988, **200**, 163; J.J. He and F.A. Quiocho, *Science*, 1991, **251**, 1497.
14. H. Luecke and F.A. Quiocho, *Nature*, 1990, **347**, 402.
15. L. Chantalat, J.M. Nicholson, S.J. Lambert, A.J. Reid, M.J. Donovan, C.D. Reynolds, C.M. Wood and J.P. Baldwin, *Acta Cryst.*, 2003, **D59**, 1395.
16. E. Rowatt and R.J. Williams, *Biochem. J.*, 1987, **245**, 641.
17. H. Alexandre and M. Geuskens, *Arch. Biol.*, 1984, **95**, 55.
18. G.J. Quigley, M.M. Teeter and A. Rich, *Proc. Natl. Acad. Sci. USA*, 1978, **75**, 64; I. Labadi, E. Jenei, R. Lahti and H. Lonnberg, *Acta Chem. Scand.*, 1991, **45**, 1055; E. Lahti and H. Lonnberg, *Biochem. J.*, 1989, **259**, 55.
19. H. Ohishi, I. Nakanishi, K. Inubushi, G.v.d. Marel, J.H.v. Boom, A. Rich, A.H.J. Wang, T. Hakoshima and K.-I. Tomita, *FEBS Lett.*, 1996, **391**, 153; H. Ohishi, N. Terasoma, I. Nakanishi, G. van der Marel, J.H. van Boom, A. Rich, A.H.J. Wang, T. Hakoshima and K.-I. Tomita, *FEBS Lett.*, 1996, **398**, 291.
20. C.A. Ilioudis, D.G. Georganopoulou and J.W. Steed, *CrystEngComm*, 2002, **4**, 26.
21. K.H.G. Verschueren, F. Seljee, H.J. Rozeboom, K.H. Kalk and B.W. Dijkstra, *Nature*, 1993, **363**, 693; K.H.G. Verschueren, S.M. Franken, H.J. Rozeboom, K.H. Kalk and B.W. Dijkstra, *J. Mol. Biol.*, 1993, **232**, 856.
22. D.B. Janssen, A. Scheper, L. Dijkhuizen and B. Witholt, *Appl. Environ. Microbiol.*, 1985, **49**, 673; S. Keuning, D.B. Janssen and B. Witholt, *J. Bacteriol.*, 1985, **163**, 635; C. Wischnak and R. Muller, *Biotechnology*, 2000, **11b**, 241.
23. R. Dutzler, B. Campbell Ernest, M. Cadene, T. Chait Brian and R. MacKinnon, *Nature*, 2002, **415**, 287.
24. P.H. Schlesinger, R. Ferdani, J. Liu, J. Pajewska, R. Pajewski, M. Saito, H. Shabany and G.W. Gokel, *J. Am. Chem. Soc.*, 2002, **124**, 1848; P.H. Schlesinger, R. Ferdani, R. Pajewski, J. Pajewska and G.W. Gokel, *Chem. Commun.*, 2002, 840.
25. G. Srinivasan, C.M. James and J.A. Krzycki, *Science*, 2002, **296**, 1459; B. Hao, W. Gong, T.K. Ferguson, C.M. James, J.A. Krzycki and M.K. Chan, *Science*, 2002, **296**, 1462.

26. A. Fürstner, *Angew. Chem. Int. Ed.*, 2003, **42**, 3582; M.S. Melvin, K.E. Wooton, C.C. Rich, G.R. Saluta, G.L. Kucera, N. Lindquist and R.A. Manderville, *J. Inorg. Biochem.*, 2001, **87**, 129; M.S. Melvin, J.T. Tomlinson, G.R. Saluta, G.L. Kucera, N. Lindquist and R.A. Manderville, *J. Am. Chem. Soc.*, 2000, **122**, 6333; A. Fürstner and E.J. Grabowski, *ChemBioChem*, 2001, **2**, 706.
27. G.V. Louie, P.D. Brownlie, R. Lambert, J.B. Cooper, T.L. Blundell, S.P. Wood, M.J. Warren, S.C. Woodcock and P.M. Jordan, *Nature*, 1992, **359**, 33.
28. S.C. Woodcock and P.M. Jordan, *Biochemistry*, 1994, **33**, 2688.
29. F. Wrede and A. Rothhaas, *Z. Physiol. Chem.*, 1934, **226**, 95; F. Wrede, *Z. Physiol. Chem.*, 1932, **210**, 125; V.F. Wrede and A. Rothhaas, *Z. Physiol. Chem.*, 1933, **222**, 203; F. Wrede and A. Rothhass, *Z. Physiol. Chem.*, 1933, **219**, 267; F. Wrede and A. Rothhaas, *Z. Physiol. Chem.*, 1933, **215**, 67.
30. H.H. Wasserman, G.C. Rodgers and D.D. Keith, *Tetrahedron*, 1976, **32**, 1851; K. Harashima, T. Tanaka and J. Nagatsu, *Agr. Biol. Chem.*, 1967, **31**, 481.
31. S.B. Han, H.M. Kim, Y.H. Kim, C.W. Lee, E.-S. Jang, K.H. Son, S.U. Kim and Y.K. Kim, *Int. J. Immunopharmacol.*, 1998, **20**, 1.
32. B. Montaner and R. Perez-Tomas, *Life Sci.*, 2001, **68**, 2025.
33. C. Yamamoto, H. Takemoto, K. Kuno, D. Yamamoto, A. Tsubura, K. Kamata, H. Hirata, A. Yamamoto, H. Kano, T. Seki and K. Inoue, *Hepatology*, 1999, **30**, 894.
34. D. Yamamoto, Y. Uemura, K. Tanaka, K. Nakai, C. Yamamoto, H. Takemoto, K. Kamata, H. Hirata and K. Hioki, *Int. J. Cancer*, 2000, **88**, 121.
35. C. Diaz-Ruiz, B. Montaner and R. Perez-Tomas, *Histol. Histopathol.*, 2001, **16**, 415.
36. B. Montaner, S. Navarro, M. Pique, M. Vilaseca, M. Martinell, E. Giralt, J. Gil and R. Perez-Tomas, *Br. J. Pharmacol.*, 2000, **131**, 585.
37. C.H. Park and H.E. Simmons, *J. Am. Chem. Soc.*, 1968, **90**, 2431.
38. P.A. Gale, *Coord. Chem. Rev.*, 2003, **240**, 1; M.D. Best, S.L. Tobey and E.V. Anslyn, *Coord. Chem. Rev.*, 2003, **240**, 3; J.L. Sessler, S. Camiolo and P.A. Gale, *Coord. Chem. Rev.*, 2003, **240**, 17; J.M. Llinares, D. Powell and K. Bowman-James, *Coord. Chem. Rev.*, 2003, **240**, 57; C.R. Bondy and S.J. Leob, *Coord. Chem. Rev.*, 2003, **240**, 77; K. Choi and A.D. Hamilton, *Coord. Chem. Rev.*, 2003, **240**, 101; T.J. Wedge and F.M. Hawthorne, *Coord. Chem. Rev.*, 2003, **240**, 111; T.N. Lambert and B.D. Smith, *Coord. Chem. Rev.*, 2003, **240**, 129; A.P. Davis and J.-B. Joos, *Coord. Chem. Rev.*, 2003, **240**, 143; M.W. Hosseini, *Coord. Chem. Rev.*, 2003, **240**, 157; P.D. Beer and E.J. Hayes, *Coord. Chem. Rev.*, 2003, **240**, 167; P.A. Gale, *Coord. Chem. Rev.*, 2003, **240**, 191.
39. A. Bianchi, K. Bowman-James and E. Garcia-España (eds), *Supramolecular Chemistry of Anions*, Wiley–VCH, New York, 1997.
40. F.P. Schmidtchen and M. Berger, *Chem. Rev.*, 1997, **97**, 1609; M.C.T. Fyfe, P.T. Glink, S.Menzer, J.F. Stoddart, A.J.P. White and D.J. Williams, *Angew. Chem. Int. Ed.*, 1997, **36**, 2068; P.A. Gale, *Coord. Chem. Rev.*, 1999, **199**, 181; P.A. Gale, *Coord. Chem. Rev.*, 2000, **199**, 181; P.D. Beer and P.A. Gale, *Angew. Chem. Int. Ed. Engl.*, 2001, **40**, 486; P.A. Gale, P. Anzenbacher Jr. and J.L. Sessler, *Coord. Chem. Rev.*, 2001, **222**, 57; S.L. Wiskur, H. Ait-Haddou, J.J. Lavigne and E.V. Anslyn, *Acc. Chem. Res.*, 2001, **34**, 963; V. McKee, J. Nelson

and R.M. Town, *Chem. Soc. Rev.*, 2003, **32**, 309; R. Martínez-Máñez and F. Sancenón, *Chem. Rev.*, 2003, **103**, 4419; R. Vilar, *Angew. Chem. Int. Ed. Engl.*, 2003, **42**, 1460. S. Kubik, C. Reyheller and S. Stüwe, *J. Incl. Phenom.*, 2005, **52**, 137.

41. R.A. Bell, G.G. Christoph, F.R. Fronczek and R.E. Marsh, *Science*, 1975, **190**, 151.
42. C. Tanford, *J. Am. Chem. Soc.*, 1954, **76**, 945.
43. J.I. Watters and S. Matsumoto, *J. Am. Chem. Soc.*, 1964, **86**, 3961; B. Springs and P. Haake, *Bioorg. Chem.*, 1977, **6**, 181.
44. D.F. Shriver and M.J. Biallas, *J. Am. Chem. Soc.*, 1967, **89**, 1078.
45. X. Yang, C.B. Knobler and M.F. Hawthorne, *Angew. Chem. Int. Ed.*, 1991, **30**, 1507; X. Yang, C.B. Knobler and M.F. Hawthorne, *J. Am. Chem. Soc.*, 1992, **114**, 380; Z. Zheng, X. Yang, C. B. Knobler and M.F. Hawthorne, *J. Am. Chem. Soc.*, 1993, **115**, 5320.
46. Y. Azuma and M. Newcomb, *Organometallics*, 1984, **3**, 9; M. Newcomb, M.T. Blanda, Y. Azuma and T.J. Delord, *J. Chem. Soc. Chem. Commun.*, 1984, 1159; M. Newcomb, A.M. Madonik, M.T. Blanda and J.K. Judice, *Organometallics*, 1987, **6**, 145; M. Newcomb, J.H. Horner and M.T. Blanda, *J. Am. Chem. Soc.*, 1987, **109**, 7878; M. Newcomb and M.T. Blanda, *Tetrahedron Lett.*, 1988, **29**, 4261; M. Newcomb, J.H. Horner, M.T. Blanda and P.J. Squattrito, *J. Am. Chem. Soc.*, 1989, **111**, 6294.
47. K. Ogawa, S. Aoyaki and Y. Takeuchi, *J. Chem. Soc. Perkin Trans.*, 1993, **2**, 2389; S. Aoyagi, K. Tanaka, I. Zicmane and Y. Takeuchi, *J. Chem. Soc. Perkin Trans.*, 1992, **2**, 2217; S. Aoyagi, K. Tanaka and Y. Takeuchi, *J. Chem. Soc. Perkin Trans. 2*, 1994, 1549.
48. K. Tamao, T. Hayashi and Y. Ito, *J. Organomet. Chem.*, 1996, **506**, 85.
49. E. Graf and J.-M. Lehn, *J. Am. Chem. Soc.*, 1975, **97**, 5022; E. Graf and J.-M. Lehn, *J. Am. Chem. Soc.*, 1976, **98**, 6403.
50. B. Metz, J. M. Rosalky and R. Weiss, *J. Chem. Soc. Chem. Commun.*, 1976, 533.
51. J.-M. Lehn, E. Sonveaux and A.K. Willard, *J. Am. Chem. Soc.*, 1978, **100**, 4914; J.-P. Kintzinger, J.-M. Lehn, E. Kauffmann, J.L. Dye and A.I. Popov, *J. Am. Chem. Soc.*, 1983, **105**, 7549.
52. B. Dietrich, J. Guilhem, J.-M. Lehn, C. Pascard and E. Sonveaux, *Helv. Chim. Acta*, 1984, **67**, 91.
53. F.P. Schmidtchen, *Angew. Chem. Int. Ed.*, 1977, **16**, 720; F.P. Schmidtchen, *Chem. Ber.*, 1980, **113**, 864; F.P. Schmidtchen, *Chem. Ber.*, 1981, **114**, 597; F.P. Schmidtchen and G. Muller, *J. Chem. Soc. Chem. Commun.*, 1984, 1115.
54. K. Worm, F.P. Schmidtchen, A. Schier, A. Schafer and M. Hesse, *Angew. Chem. Int. Ed.*, 1994, **33**, 327; K. Worm and F.P. Schmidtchen, *Angew. Chem. Int. Ed.*, 1995, **34**, 65.
55. R.A. Pascal, J. Spergel and D.V. Engbersen, *Tetrahedron Lett.*, 1986, **27**, 4099.
56. S. Valiyaveettil, J.F.J. Engbersen, W. Verboom and D.N. Reinhoudt, *Angew. Chem. Int. Ed.*, 1993, **32**, 900.
57. P.J. Smith, M.V. Reddington and C.S. Wilcox, *Tetrahedron Lett.*, 1992, **35**, 6085.

58. B. Dietrich, T.M. Fyles, J.-M. Lehn, L.G. Pease and D.L. Fyles, *J. Chem. Soc. Chem. Commun.*, 1978, 934.
59. G. Müller, J. Riede and F.P. Schmidtchen, *Angew. Chem. Int. Ed.*, 1988, **27**, 1516.
60. V.J. Bauer, D.L.J. Clive, D. Dolphin, J.B. Paine III, F.L. Harris, M.M. King, J. Loder, S.-W.C. Wang and R.B. Woodward, *J. Am. Chem. Soc.*, 1983, **105**, 6429.
61. M. Shionoya, H. Furuta, V. Lynch, A. Harriman and J.L. Sessler, *J. Am. Chem. Soc.*, 1992, **114**, 5714.
62. J.L. Sessler, M.J. Cyr, V. Lynch, E. McGhee and J.A. Ibers, *J. Am. Chem. Soc.*, 1990, **112**, 2810; J.L. Sessler, M. Cyr, H. Furuta, V. Král, T. Mody, T. Morishima, M. Shionoya and S.J. Weghorn, *Pure Appl. Chem.*, 1993, **65**, 393.
63. B. Hasenknopf, J.-M. Lehn, B.O. Kneisel, G. Baum and D. Fenske, *Angew. Chem. Int. Ed.*, 1996, **35**, 1838; B. Hasenknopf, J.-M. Lehn, N. Boumediene, A. Dupont-Gervais, A. van Dorsselaer, B. Kneisel and D. Fenske, *J. Am. Chem. Soc.*, 1997, **119**, 10956.

CHAPTER 2

Classic Charged Non-Metallic Systems

2.1 Polyammoniums

Since the days of Park and Simmons[1] (*cf.* Chapter 1), polyammonium receptors have emerged as true workhorses in the area of anion recognition chemistry. Polyammonium receptors can be easily generated via the protonation of polyamines and, as a rule, most polyamines are multiply protonated at pH 7. The resulting protonated receptors generally display strong anion-binding tendencies in both organic and aqueous solvents as a result of both strong electrostatic and hydrogen-bonding interactions. These receptors also have the advantage that they can be accessed using a range of synthetic pathways, allowing for a considerable diversity in terms of size, charge, and shape.

In this chapter, the chemistry of polyammonium and other classic charged-anion receptor systems, namely those based on quarternary ammonium, guanidinium, amidinium, imidazolium, and thiouronium motifs, respectively, is reviewed. The presentation is organized according to each of these five major receptor types, giving rise to six major sections. Most of these sections are further subdivided according to the structure, leading to individual subsections devoted to "acyclic receptors" or "linear systems", "monocyclic receptors", "bicyclic receptors", and "polycyclic receptors", respectively. Because of this geometry-based organization, this chapter tends to be front-loaded so as to introduce in some detail the basic structural elements that have been commonly used to link a variety of different binding motifs.

2.1.1 Acyclic Systems

A big advance in the development of cationic anion receptor systems came in 1977 when the X-ray crystal structure of **2.1a**, in the form of its citrate complex, was reported by Glusker and co-workers[2] (*cf.* Figure 2.1a). This seminal result made it clear that even simple systems, such as the ethylenediamonium cation, are capable of binding anions, at least solid state. Two years after the Glusker report, the binding of carboxylate anions in solution was studied by Lehn and co-workers[3]; from an

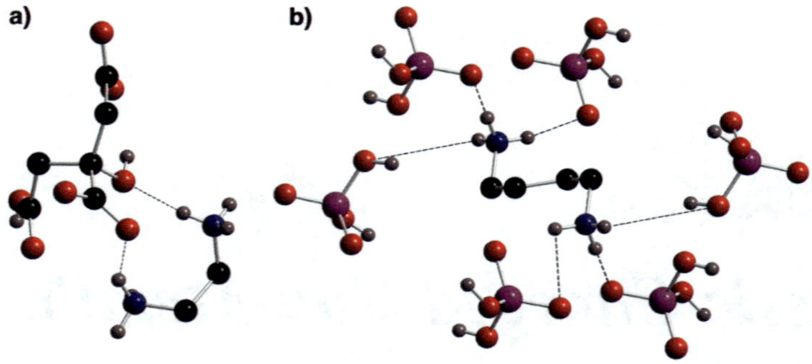

Figure 2.1 *Single crystal X-ray structures of (a) the citrate complex of diammonium cation **2.1a** and (b) the hexakis-dihydrogen phosphate complex of congener **2.1c***

analysis of pH-metric titration curves, a log K_a of 3.4 was deduced for the binding of maleate monoanion to **2.1a** in MeOH/H_2O 9:1. In independent work, the structure of the salt formed between protonated putrescine, **2.1c**, and dihydrogen phosphate was determined in the solid state from a single crystal X-ray diffraction analysis carried out at 85 K (Figure 2.1b).[4] The structure obtained in this way was considered as a simple model for amine–nucleic acid interactions.[5] In 1977, Nakai and Glinsmann[6] measured the binding constant between putrescine **2.1c** and adenosine phosphates at pH 7.5 and at 303 K (K_a = 82, 200, and 290 M^{-1} for AMP, ADP, and ATP sodium salts, respectively). Quantitative analyses of the interactions of this (**2.1c**) and other diammonium receptors, namely **2.1b** and **2.1d**, with tetrasodium pyrophosphate were carried out by Lönnberg and co-workers[7] using potentiometric titation methods; the log K_a values calculated in this way were 2.6, 2.1, and 1.9 for **2.1b**, **2.1c**, and **2.1d**, respectively.

2.1a n = 0
2.1b n = 1
2.1c n = 2
2.1d n = 3

Spermidine **2.2** and spermine **2.3** are among the most important polyamines in biology. They are widely distributed in living cells.[8] These two polyamines, present in their respective protonated forms at physiological pH, affect the rate and extent of nucleic acid biosynthesis with spermine, in particular, being known to bind the phosphate groups in yeast phenylalanine transfer RNA.[9] Their strong affinity for nucleic acids is also thought to have a stabilizing effect on DNA.[10] In 1969, Huse and Iitaka[11] reported the X-ray crystal structure of the complex formed between spermidine **2.2** and hydrogen phosphate (*cf.* Figure 2.2a), although the hydrogen atoms were not specifically identified. Far more recently, Steed and co-workers[12] succeeded in solving the single crystal X-ray structures of this complex (*i.e.*, hydrogen

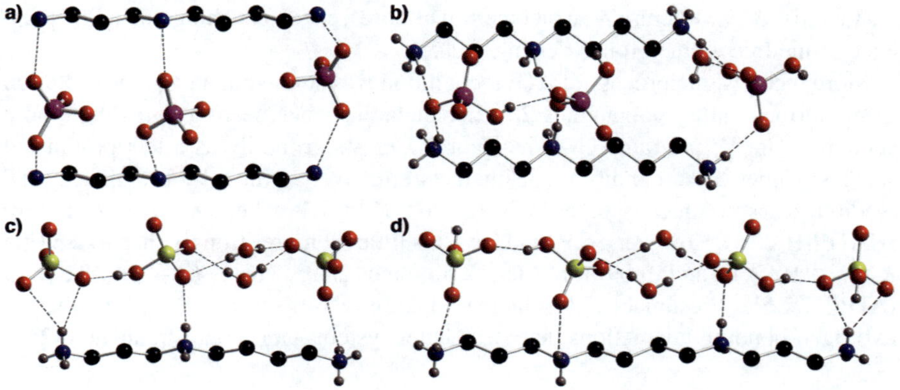

Figure 2.2 *(a) and (b) Single crystal X-ray structures (separate determinations) of the tris-hydrogen phosphate complex of **2.2**. Six water molecules are removed for clarity in both (a) and (b). (c) Single crystal X-ray structure of the diaquo tris-hydrogen sulfate complex of **2.2**; (d) diaquo tetrakis-hydrogen sulfate complex of **2.3**. The sulfur atom is shown in yellow*

phosphate–**2.2**) and two related complexes, namely hydrogen sulfate–**2.2** and hydrogen sulfate–**2.3** (*cf.* Figures 2.2b–d). The association constants (K_a or $K_{a1} \times K_{a2}$) for the binding of AMP, ADP, ATP, and pyrophosphate anion to **2.2** (**2.3**) were determined at pH 7.5 and found to be 230 × 22 (360 × 28), 330 × 82 (1300 × 120), 900 × 280 (9500 × 2400), and 640 (2700) $M^{-2\,(-1)}$, respectively.[6,13]

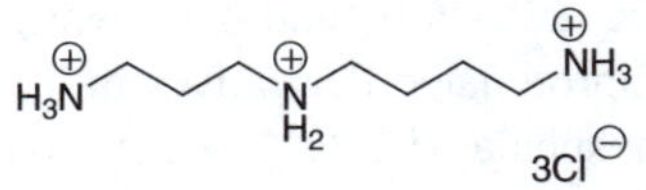

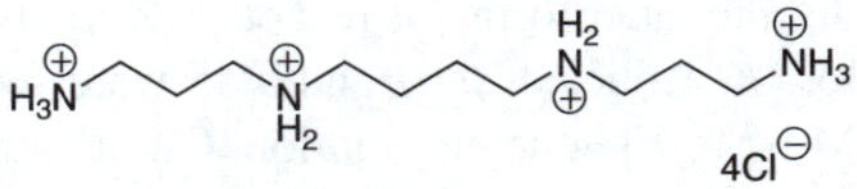

In order to convert the process of anion binding into a detectable signalling event, Czarnik and co-workers synthesized the anthrylpolyamine **2.4**. This species exists as the trication at pH 6. Under these conditions, the addition of oxyanions, such as hydrogen phosphate, sulfate, and acetate, lead to changes in the fluorescence emission intensity.[14] The most dramatic spectral change was observed in the presence of hydrogen phosphate, with an increase in intensity of more than 145% being observed. On the other hand, the addition of ATP, acetate, and dimethylphosphate anions induced a decrease in the fluorescence intensity. These changes allowed the relevant association constants to be calculated quantitatively (log K_a = 4.2, ≤ 0.6, and ≤ 0.5 for ATP, acetate, and dimethyl phosphate anions, respectively, and 0.82 for hydrogen phosphate). Taken in concert, these results support the hypothesis that complex hydrogen phosphate–**2.4** exists in the form of Structure **B**, which precludes the intermolecular quenching process, while the other anion complexes are best characterized by Structure **A**, which permits fluorescence quenching due to the presence of the free amine group (Scheme 2.1).

More recently, the process of ATP recognition was studied in an aqueous solution using various anthrylpolyamines **2.5–2.6**, including ones bearing two fluorophore subunits.[15] In the pH range 3–6, receptor **2.5** exists primarily as a tetraprotonated species. Under these conditions, it interacts effectively with monoprotonated ATP (sodium salt) in aqueous 0.15 M NaCl (log K_a = 7.1, where K_a = [H_5ATP·**2.5**]/[HATP][H_4**2.5**]). However, ATP binds to **2.6** a little bit more strongly than to **2.5** (log K_a = 9.9, 8.1, and 9.09 for **2.6a**, **2.6b**, and **2.6c**, where K_a = [H_5ATP·**2.6**]/[ATP][H_5**2.6**]). Presumably, this increased affinity reflects the larger number of hydrogen-bonding interactions that these latter systems can support relative to **2.5**.

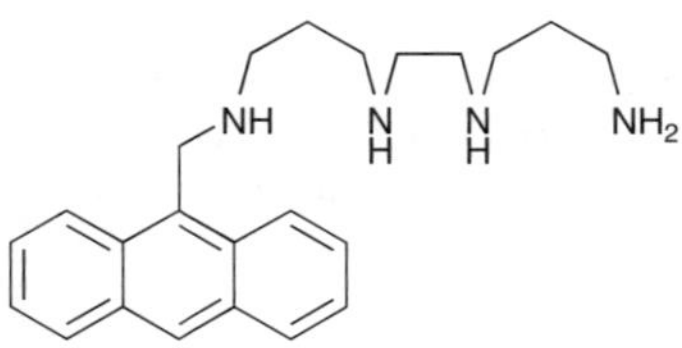

2.5

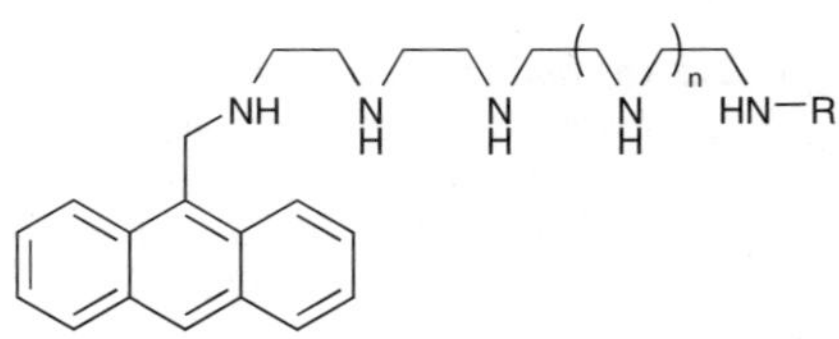

2.6a n = 1, R = H
2.6b n = 2, R = H
2.6c n = 2, R = CH_2(9-naphthyl)

Bianchi and co-workers[16] prepared the cleft-like hexamine ligands **2.7a** and **2.7b** containing hetro-aromatic moieties and studied them as linear polyammonium-type anion receptors. The association constants between these receptors and ATP were found to be strictly pH dependent. For example, mono protonated ATP binds to the diprotonated forms of receptors **2.7a** and **2.7b** with log K_a = 4.08 and 4.46, while log K_a values of 5.01 and 5.45 were seen for the tetraprotonated derivatives in 0.1 M NMe_4Cl aqueous solution. On the other hand, triphosphate, $H_2P_3O_{10}^{3-}$, was found to bind to receptor **2.7a** less well than ATP, even though both species contain the same triphosphate unit (the log K_a for the interaction of H_4**2.7a**$^{4+}$ with $H_2P_3O_{10}^{3-}$ was 3.25). Presumably, the rigid aromatic unit helps provide a binding site that is better

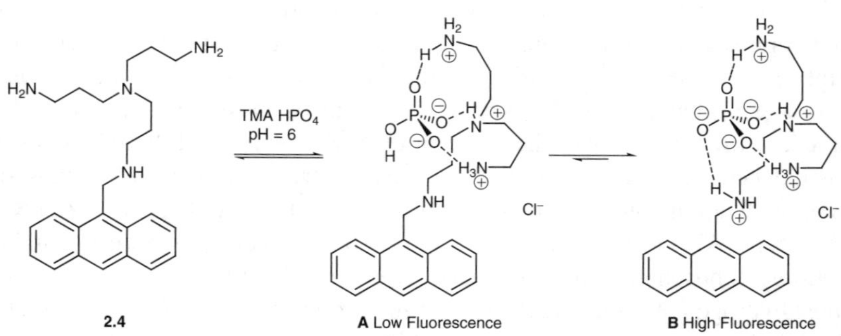

Scheme 2.1 *Protonation of **2.4** by hydrogen phosphate*

optimized for the nucleotide-containing anions as the result of, *e.g.*, allowing for π–π stacking and hydrophobic interactions that are less beneficial in the case of $H_2P_3O_{10}^{3-}$.

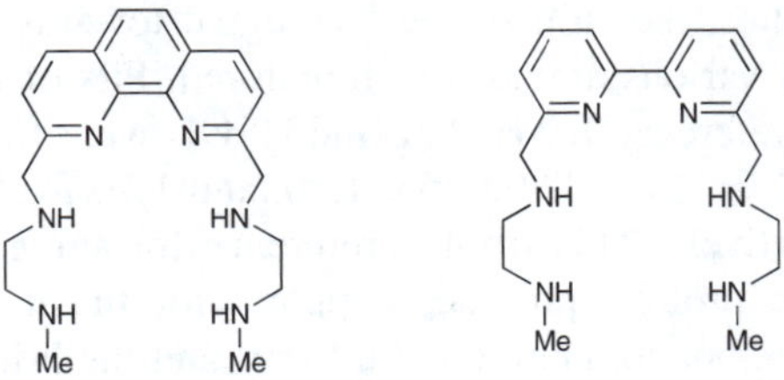

2.7a **2.7b**

Tren-based tripodal polyammonium species have played an important role in the development of anion recognition chemistry, seeing application, for instance, in such practical areas as pertechnetate and perrhenate anion extraction.[17] This utility reflects in large measure the fact that such species are relatively easy to make and modify.

Among the important studies in this area is the 2004 report from Bowman-James and co-workers[18] who examined the anion-binding ability of the tren-based tripodal polyammonium receptor **2.8** in both the solid sate and in solution. For instance, these workers reported the crystal structure of the dihydrogen phosphate-anion complex shown in Figure 2.3a. In this case, it was found that one phosphate, presumably in the form of dihydrogen phosphate, is located in the middle of the tren-derived "backbone", being held in place via three N–H···O hydrogen-bonding interactions. The remaining three of the four phosphates bound per receptor are found in between each of the tren "arms" again, presumably, as dihydrogen phosphate. However, the corresponding structure (*cf.* Figure 2.3b) reveals the presence of only three bound bromide anions per equivalent of **2.8**, which is thus presumably

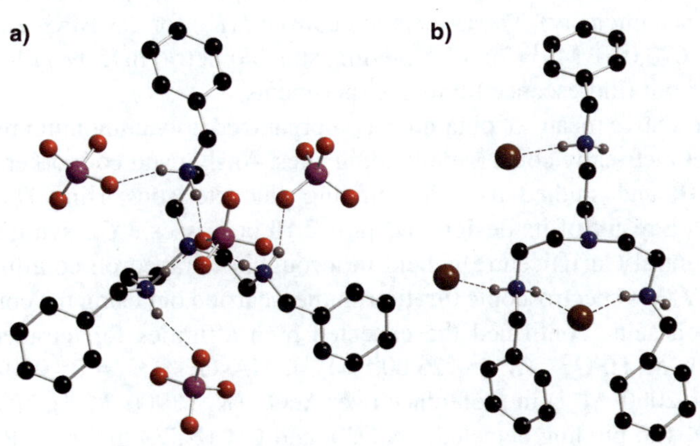

Figure 2.3 *Single crystal X-ray structures of (a) the tetrakis-phosphate and (b) tris-bromide anion complexes stabilized by the protonated forms of* **2.8***. Bromide anions are shown in brown. The other colour codes are as defined previously*

triprotonated, not tetraprotonated, in this case. Quantitative assessments of the interaction between **2.8** and anions (studied as the corresponding tetrabutylammonium (TBA) salts) came from ^{1}H NMR spectroscopic titrations carried out in $CDCl_3$. In these studies, the shift of the NH signal towards lower field observed upon the addition of various anions was monitored. Fits of the resulting binding profiles revealed a preference for $H_2PO_4^-$ and HSO_4^- over other anions included in the study, *i.e.*, log K_a = 3.25, 3.20, 1.55, 1.80, and 1.70 for $H_2PO_4^-$, HSO_4^-, NO_3^-, Cl^-, and Br^-, respectively. This strong preference for anions that contain one or more residual acidic protons provides supports for the hypothesis that proton exchange between the bound anion (acidic form) and the bridgehead amine nitrogen atom could be contributing to the binding process.

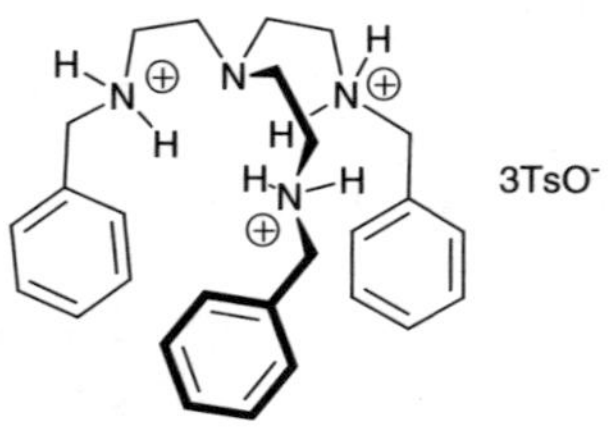

2.8

Receptor **2.9**, which relies on triethylbenzene as the backbone, was designed by Anslyn and Best[19] to act as a fluorescent sensor for 2,3-bisphosphoglycerate (2,3-BPG). Here, it was noted that the three ethyl group present in the backbone would force the other three appended recognition moieties to point in the same direction, thus creating a well-defined binding cavity. The fact that one of these motifs was a europium tetra-*N*-oxide bipyridine complex was expected not only to enhance the binding process, but allow for it to be followed via spectroscopic means (*e.g.*, change in fluorescence intensity). The association constant (K_a) for 2,3-BPG–**2.9** was determined to be 670,000 M^{-1} (for a 1:1 binding stoichiometry) in (50%) MeOH/MeCN, as inferred from fluorescence titration experiments.

As an alternative means of obtaining a preorganized polyammonium receptor with multiple, geometrically convergent binding sites, Anslyn and co-workers[20] prepared receptor **2.10** and studied its anion-binding characteristics using UV–Vis spectroscopy. As a result of its design, receptor **2.10** possesses a C_{3v} symmetric cavity, while the central Cu(II) centre is held in a roughly tetrahedron coordination environment. UV–Vis spectroscopic titration studies, carried out using the corresponding sodium anion salts, confirmed the expected high affinities for tetrahedral-shaped anions, such as HPO_4^{2-} (K_a = 25,000 M^{-1}), $HAsO_4^{2-}$ (K_a = 25,000 M^{-1}), and ReO_4^- (K_a = 2000 M^{-1}) in preference over AcO^- (K_a < 900 M^{-1}), NO_3^- (K_a < 20 M^{-1}), HCO_3^- (no binding detected; "N.D."), and Cl^- (N.D.) in the HEPES solution (pH 7.4). The selectivity displayed by system **2.10** was attributed to a combination of receptor size, shape, and charge complementarity effects.

Anslyn and Tobey[21] have also reported the carboxylate anion-binding properties of **2.10**. In this work, the association constants for the interaction with the 1,2,3, 4-butanetetracarboxylate, tricarballate, glutarate, and acetate anions were measured by UV–Vis and isothermal calorimetric (ITC) titration methods in HEPES buffer (pH 7.4). It was found that the tetra- and tricarboxylate anions were bound to receptor **2.10** with similar affinity (K_a = 220,000 (18,000) and 90,000 (19,000) M^{-1} by UV–Vis (ITC)), from which it was inferred that both anions bind to the receptor via one primary Cu(II)-carboxylate and two ancillary ammonium-carboxylate interactions. Consistent with such thinking, the association constants for the di- and monocarboxylate substrates were found to be almost two orders of magnitude lower (K_a = 2000 (400) and 900 (300) M^{-1} by UV–Vis (ITC)) than the more complex tri- and tetracarboxylate targets.

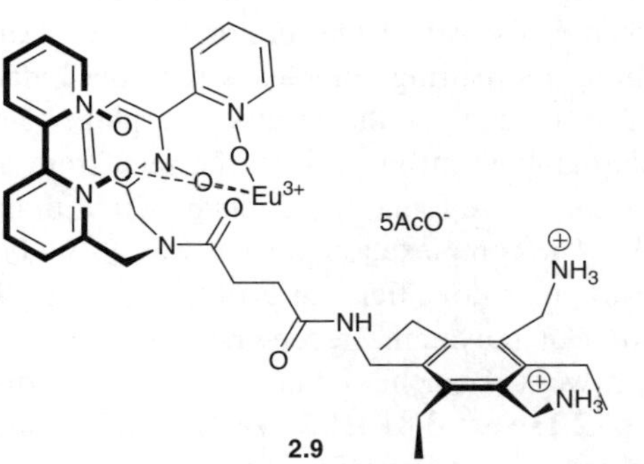

2.9

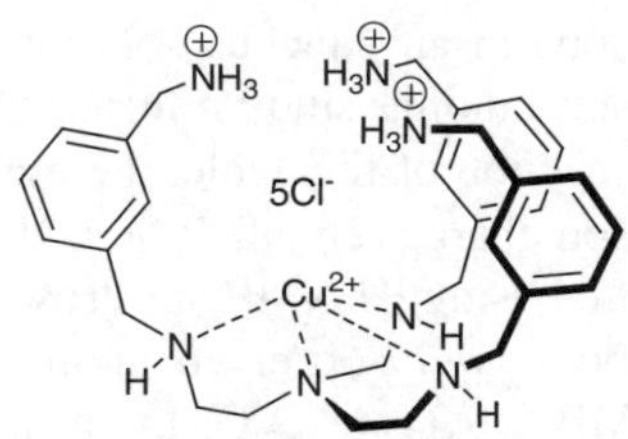

2.10

2.1.2 Monocyclic Systems

In 1981, Kimura and co-workers[22] indirectly observed that the protonated cyclic tetra- and pentaamine receptors, **2.11a**, **2.11b**, and **2.14**, are capable of binding carboxylate-type anions. This finding came about while these researchers were seeking to develop simple and reliable analytical methods for identifying and separating various cyclic polyamines. Specifically, in the course of testing the electrophoretic properties of polyamines on a cellulose citrate membrane at pH 6, it was found that protonated polyamines, **2.11a**, **2.11b**, and **2.14**, migrated towards the anode, *i.e.*, in exactly the opposite direction expected for a cationic species. This unexpected observation was rationalized by assuming that a citrate-anion complex was being formed whose overall charge was negative, rather than positive. Figure 2.4 illustrates the citrate complex that is thought to form with **2.14**.

In follow-up work, the carboxylate-binding properties of pentamines **2.12** and heaxamine **2.14a** were studied at neutral pH using polarographic methods.[23] Under these conditions, the amines are triprotonated and are able to bind to citrate (K_a = 55, 250, 1000, and 240 M^{-1} for **2.12a**, **2.12b**, **2.12c**, and **2.14**, respectively) more tightly than various simple dicarboxylate anions (*e.g.*, succinate, malonate, malate, maleate, and fumarate). This result was taken as evidence that electrostatic interactions play a major role in the complexation process. Nonetheless, it was found that the 15-membered cyclic

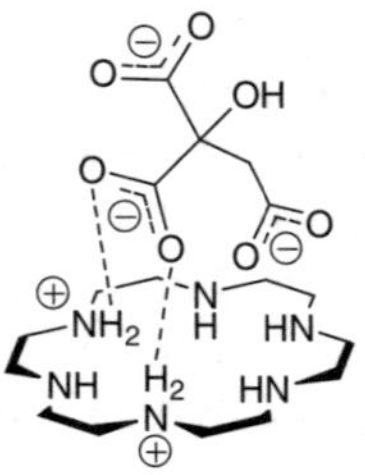

Figure 2.4 *Proposed model for the 1:1 complex formed between protonated macrocycle **2.14** and citrate anion*

receptor **2.12a** displays a rather low affinity, at least in relative terms, reflecting the fact that unfavourable steric and conformation effects can modulate the anion affinities.

One year later, Kimura and co-workers[24] discovered that protonated tetra- and pentaamine receptors can also form strong ion-pairing complexes with phosphate anions in an aqueous solution. Table 2.1 summarizes the results of a phosphate-anion-binding study effected using polarographic means. As can be seen from an inspection of this table, the binding affinities were found to correlate well with net anion charge (*i.e.*, ATP > ADP > AMP). The complexation process was also monitored using ^{1}H NMR spectroscopy. In particular, downfield shifts of the adenine CH protons were observed upon the addition of polyamine receptors to solutions of AMP or ATP in D_2O. In an independent work, Bianchi and co-workers[25] reported that the association constants (log K_a) of **2.11e** are 3.81 (H_4**2.11e**$^{4+}$ + ATP^{4-}) and 3.04 (H_4**2.11e**$^{4+}$ + $HATP^{3-}$), as derived from potentiometric titration studies.

In 1982, Suet and Handel[26] demonstrated the interaction between fluoride anion and the tetramines **2.11c**, **2.11d**, and **2.11e**. Fluoride anion was found to bind more strongly to the tetraprotonated species **2.11e** (log K_a = 2.8) than to its tetraprotonated analogues **2.11b** (log K_a = 2.0) or **2.11c** (log K_a = 1.9) in an aqueous solution (pH 1). Based on an analysis of CPK models, it was proposed that the cavity of **2.11e** (radius 1.4 Å) provides a good match for the fluoride anion (ionic radius taken as 1.36 Å), whereas **2.11c** and **2.11d** have cavities that are too small (radius = *ca.* 0.7 and 1.0 Å, respectively).

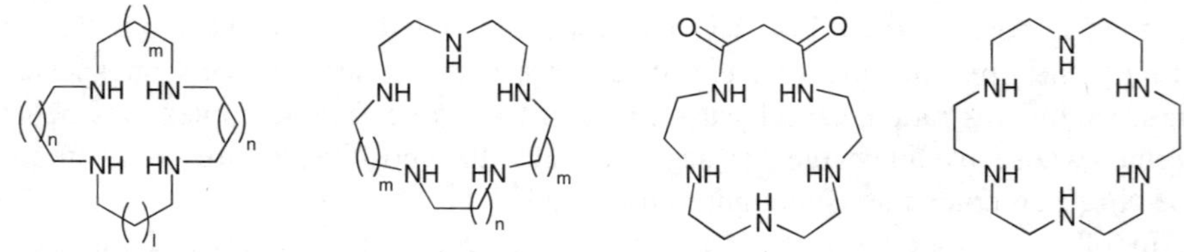

2.11a n = 1, m = 2, l = 0
2.11b n = l = 1, m = 2
2.11c n = m = l = 1
2.11d n = 1, m = l = 2
2.11e n = m = l = 2

2.12a n = m = 1
2.12b n = 2, m = 1
2.12c n = 1, m = 2

2.13

2.14

Subsequent to Kimura's studies on the anion binding ability of **2.14**, the first single X-ray crystal structure of a chloride anion complex involving the tetraprotonated form of **2.14** was reported by Gelb and co-workers.[27] As can be inferred from this structure, reproduced in Figure 2.5a, direct, and presumably strong, $^{+}NH \cdots Cl^{-}$ hydrogen-bonding interactions help stabilize this complex. Independent pH-potentiometric and conductometric titration experiments confirmed that receptor **2.14** exists in its tetraprotonated form at $2 < pH < 3$ and displays a relatively weak affinity for nitrate and chloride anions ($K_a \approx 10^2\ M^{-1}$ in both cases). Subsequently, Spiccia and co-workers[28] also reported structures of the chloride-, bromide-, and iodide-anion complexes of this receptor. These structures are quite similar to that shown in Figure 2.5a.

Table 2.1 *Association constants (K_a) for the formation of nucleotide phosphate complexes with various protonated polyamines in 0.2 M aqueous $NaClO_4$ at 25 °C*

	H_4**2.11d**$^{4+}$	H_3**2.12a**$^{3+}$	H_3**2.12b**$^{3+}$	H_3**2.12c**$^{3+}$	H**2.13**$^{1+}$	H_3**2.14**$^{3+}$
AMP^{2-}	6.9×10^3	1.56×10^3	1.30×10^3	6.86×10^2	5.03×10^2	1.77×10^3
ADP^{3-}	3.2×10^4	8.72×10^3	1.47×10^3	1.00×10^3	1.21×10^2	4.47×10^5
ATP^{4-}	4.5×10^6	1.03×10^4	4.23×10^3	5.12×10^3	4.77×10^2	2.50×10^6

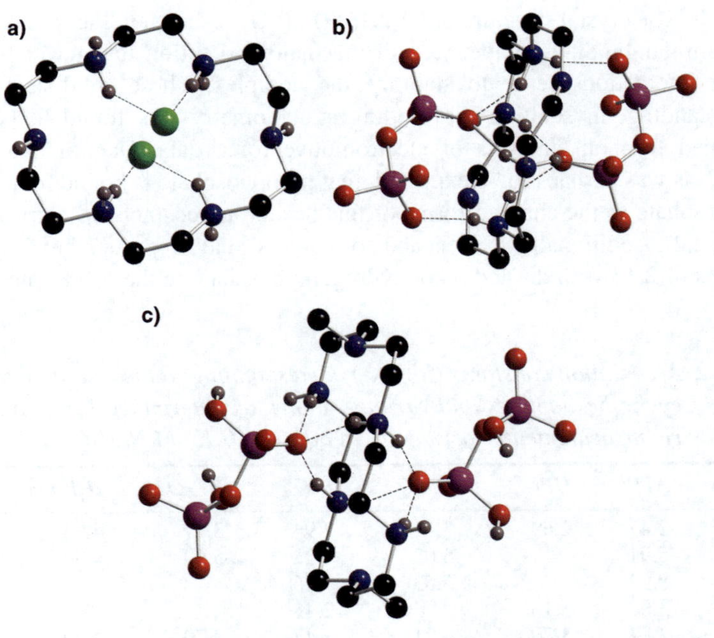

Figure 2.5 *Single crystal X-ray structures of (a) the bis-chloride anion complex of tetraprotonated **2.14**, (b) the 2:1 complex formed between $H_2P_2O_7^{2-}$ and the tetraprotonated form of **2.14** in the solid state, and (c) the 2:1 complex formed between $H_2P_2O_7^{2-}$ and the tetraprotonated form of **2.15**. In all cases, further counter anions not proximate to the receptor were found in the crystal lattice. These anions are not shown for the sake of clarity*

Table 2.2 summarizes many of the anion-binding constants that Bianchi and co-workers[29–32] derived for the cyclic hexaamines, **2.14**, **2.15**, and **2.16**, as the result of work that first commenced in 1992. Generally, the association constants for these systems were found to correlate with the degree of protonation. Also, stronger complexes were observed to form with highly negatively charged anion such as ATP and $P_2O_7^{4-}$. At a given protonation level, **2.14** and the methylated systems **2.15** and **2.16** were found to display similar affinities for phosphate-type anions. However, the tetraprotonated forms of receptors **2.15** and **2.16** were seen to bind ATP with lower affinity than the non-methylated analogue, **2.14**.

In addition to the binding studies, kinetic experiments designed to test receptor-promoted ATP cleavage using macrocycles **2.14–2.16** were also carried out. These were effected by monitoring the rate of ATP loss using ^{31}P NMR.[29,33] While tetramethylated hexamine **2.16** was found to produce a large rate enhancement (*e.g.*, $k = 43 \times 10^3$ min^{-1} pH 3, 80 °C), the analogous dimethylated receptor **2.15** (*e.g.*, $k = 5.6 \times 10^3$ min^{-1} pH 3, 80 °C) proved to be much less effective, showing a rate that was slower even than that displayed by **2.14** (*e.g.*, $k = 13 \times 10^3$ min^{-1} pH 3, 80 °C).

In 1999 two more crystal structures, namely of (H_4**2.14**)·$(H_2P_2O_7)_2$ and (H_4**2.15**)·$(H_2P_2O_7)_2$, were eludicated in conjunction with these studies.[29,31] As can be seen from Figure 2.5b, in the first of these structures, the two $H_2P_2O_7^{2-}$ anions are bound above and below the pseudo-plane defined by the protonated macrocycle; they are held in place by a combination of ion-pairing interactions and eight hydrogen bonds. The crystal structure of (H_4**2.15**)·$(H_2P_2O_7)_2$, shown in Figure 2.5c, reveals a very similar binding geometry, with the major exception that fewer hydrogen-bonding interactions serve to stabilize the complex. These solid-state findings notwithstanding, in solution the binding stoichiometry was found to be 1:1, as determined from an analysis of electromotive force data. Quantitative analyses revealed, as was perhaps to be expected, that pyrophosphate is bound more strongly than phosphate to the charged forms of this hexamine receptor (*cf.* entries for **2.14** in Table 2.2). Additionally, Spiccia and co-workers[34] have reported the X-ray crystal structures of **2.14** with di- and monohydrogen phosphate in the solid state.

Table 2.2 *Association constants (log K_a) corresponding to the interaction between cyclic hexaamines and various anions, as determined from potentiometric measurements carried out in aqueous 0.15 M $NaClO_4$ at 298 K*

	ATP^{4-}	HPO_4^{2-}	$H_2PO_4^-$	$P_2O_7^{4-}$	$HP_2O_7^{3-}$	$H_2P_2O_7^{2-}$	SO_4^{2-}
H_3**2.14**$^{3+}$	2.47	2.69	3.71	2.94[a]	3.27	4.04	
H_4**2.14**$^{4+}$	5.91		5.02	7.22	5.69	4.37	3.84
H_5**2.14**$^{5+}$	8.92		5.53	11.60			4.44
H_2**2.15**$^{2+}$	3.00	3.16		2.45			2.82
H_3**2.15**$^{3+}$	3.82	3.70	4.21	3.47	3.79	4.29	3.34
H_4**2.15**$^{4+}$	7.39		5.20		6.33	4.53	4.89
H_2**2.16**$^{2+}$		2.74					2.93
H_3**2.16**$^{3+}$	3.30	3.24	3.40	3.49[a]	2.52		3.38
H_4**2.16**$^{4+}$	7.48		3.83	7.48			4.48

[a]In reference 31 these values are reported as 3.27 and 3.88 instead of 2.94 and 3.49, respectively.

Sulfate represents the last anion whose binding interactions with **2.14–2.16** were studied in detail. As above, these analyses were carried out in an aqueous solution by means of potentiometric techniques.[32] It was found that the affinity for sulfate increases with increasing positive charge on the receptors. However, little differentiation among the three receptors was observed. In other words, the various protonated forms of **2.14**, **2.15**, and **2.16** that were studied were found to display similar binding behaviour, even though two of the receptors, namely **2.15** and **2.16**, bear *N*-methyl substituents.

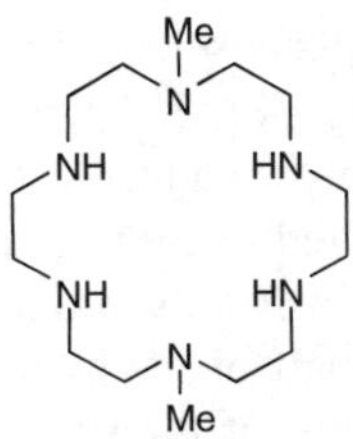

2.15

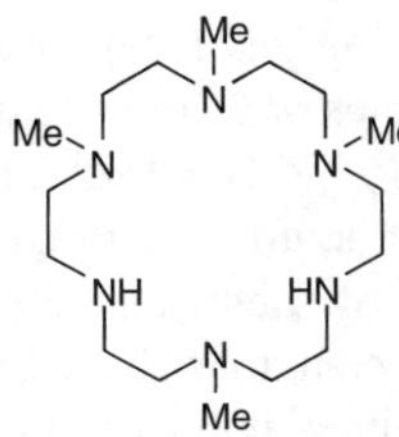

2.16

In an independent work carried out in the same year, but reported slightly after Kimura's original study, Lehn and co-workers[35] were able to demonstrate that various anions were bound to the protonated form of a slightly different hexamine receptor, namely **2.17**. In this case, computer analyses of the pH-metric titration data revealed that the fully protonated form of **2.17** (*i.e.*, H_6**2.17**$^{6+}\cdot 6Cl^-$) forms strong complexes with both organic and inorganic di- and polyanions (*e.g.*, log K_a = 4.0, 3.8, 3.3, 2.4, 4.7, 3.9, 3.4, 6.5, and 8.9 for sulfate^{2-}, oxalate^{2-}, malonate^{2-}, succinate^{2-}, citrate^{3-}, $Co(CN)_6^{3-}$, AMP^{2-}, ADP^{3-}, and ATP^{4-}, respectively, in aqueous 0.1 M tetramethylammonium chloride (TMACl)). However, no evidence of appreciable binding was seen in the case of various monoanions, including acetate, nitrate, and tetrafluoroborate anions, presumably reflecting competition from chloride anion. The fact that the strongest complexes were formed with small and highly charged anions led to the conclusion that electrostatic interactions play a major role in defining the anion recognition event in receptors of this type. Such a conclusion was supported by data from follow-up studies in which anion binding was analyzed as a function of pH.[36,37]

In order to improve the selectivity towards dicarboxylates, Lehn and Hosseini[36] designed and prepared several isomeric hexamine receptors, namely **2.18a**, **2.18b**, and **2.19c**, wherein bridges of different length are used to link the tripod-like subunits. While receptor **2.17** displays a preference towards oxalate dianion (log K_a = 3.8, 3.2, and 2.6 for H_6**2.17**$^{6+}$, H_5**2.17**$^{5+}$, and H_4**2.17**$^{4+}$, respectively, in aqueous 0.01 M TMACl) over anionic malonate, succinate, glutarate, adipate, and pimelate, the corresponding hexamine receptor **2.18a** shows a selectivity towards succinate dianion (log K_a = 3.4, 2.85, and 2.45 for H_6**2.18a**$^{6+}$, H_5**2.18a**$^{5+}$, and H_4**2.18a**$^{4+}$, respectively, in aqueous 0.01 M TMACl) and glutarate dianion (log K_a = 3.4, 2.9, and 2.5 for H_6**2.18a**$^{6+}$, H_5**2.18a**$^{5+}$, and H_4**2.18a**$^{4+}$, respectively, in aqueous 0.01 M TMACl) over anionic oxalate, malonate, adipate, and pimelate.

Receptor **2.18b** is of special interest. Not only does this system display the highest affinity for oxalate (log K_a = 6.3, 4.7, and 2.85 for H_6**2.18b**$^{6+}$, H_5**2.18b**$^{5+}$, and H_4**2.18b**$^{4+}$, respectively, in aqueous 0.01 M TMA-Cl), it also shows appreciable selectivity towards pimelate (log K_a = 3.4, 2.85, and 2.70 for H_6**2.18b**$^{6+}$, H_5**2.18b**$^{5+}$, and H_4**2.18b**$^{4+}$, respectively, in aqueous 0.01 M TMA-Cl) and suberate (log K_a = 3.45, 3.0, and 2.65 for H_6**2.18b**$^{6+}$, H_5**2.18b**$^{5+}$, and H_4**2.18b**$^{4+}$, respectively, in aqueous 0.01 M TMA-Cl). The authors did not provide a detailed rationale for the high oxalate-binding affinity. Nonetheless, these results do serve to underscore the fact that the complexation selectivity can be controlled in receptors of this type via suitable structural modifications. Consistent with this generalized conclusion is the finding that receptor **2.19c** displays selectivity for glutarate, binding to this latter species with rather high affinity (log K_a = 6.1, 5.5, and 2.95 for H_6**2.19c**$^{6+}$, H_5**2.19c**$^{5+}$, and H_4**2.19c**$^{4+}$, respectively, in aqueous 0.01 M TMA-Cl).

Later, the crystal structure of **2.19a**, characterized in the form of its hydrochloride salt, was elucidated by Lehn and co-workers.[38] It was found that, in the solid state, H_6**2.19a**$^{6+}$ is rectangular shaped and that half of the hydrogen atoms present on the protonated amines point towards the centre of the ring and are involved in hydrogen-bond interactions with the two bound chloride anions (Figure 2.6).

Recently, Bowman-James and co-workers[39] reported the crystal structure of the phosphoric acid/dihydrogen phosphate complex of hexaprotonated **2.19b**. This crystal structure is noteworthy because it provides evidence that phosphate can exist in the form of $H_2PO_4^-$ (as well as H_3PO_4) in complexes isolated at low pH. Solution-phase potentiometric titrations revealed that the pentaprotonated and hexaprotonated forms of **2.19b** bind two $H_2PO_4^-$ anions at pH values between 2 and 4 (log K_a = 6.16 and 6.44 for (H_5**2.19b**)·$(H_2PO_4)_2$ and (H_6**2.19b**)·$(H_2PO_4)_2$, respectively).

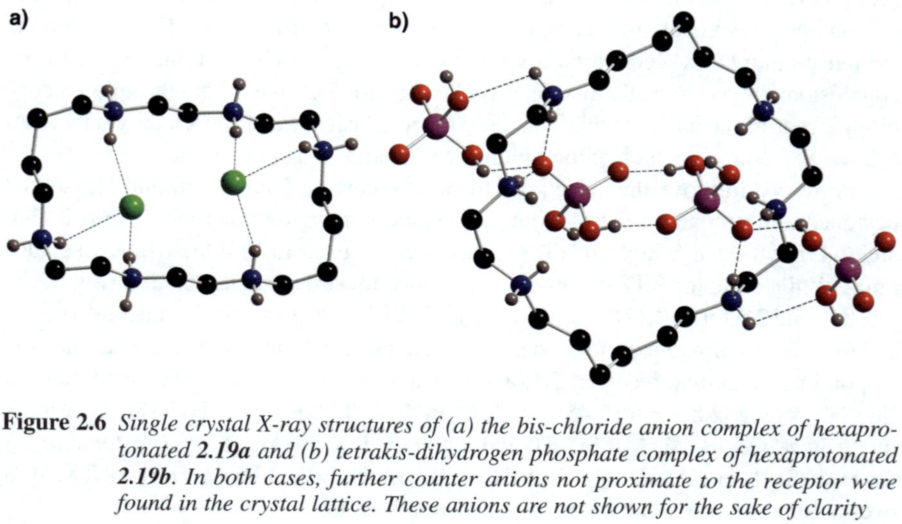

Figure 2.6 *Single crystal X-ray structures of (a) the bis-chloride anion complex of hexaprotonated **2.19a** and (b) tetrakis-dihydrogen phosphate complex of hexaprotonated **2.19b**. In both cases, further counter anions not proximate to the receptor were found in the crystal lattice. These anions are not shown for the sake of clarity*

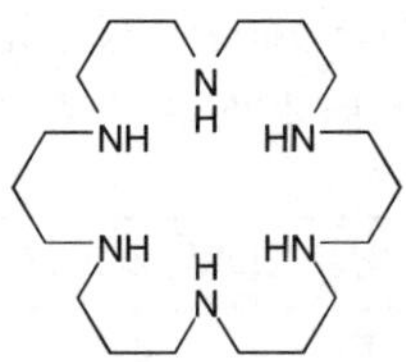

2.17

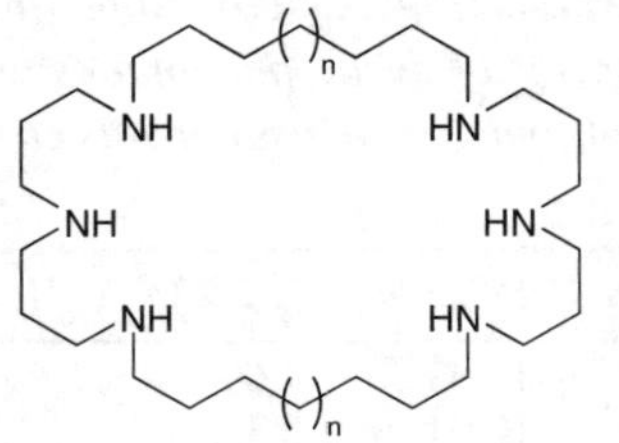

2.18a n = 1
2.18b n = 4

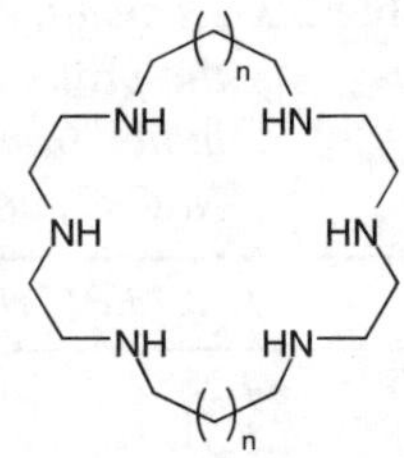

2.19a n = 2
2.19b n = 4
2.19c n = 7

The first optically active cyclic polyamine **2.20** was synthesized by Burrows and Marecek[40] in 1986. In this case, a solution-phase binding study, carried out using ^{31}P NMR spectroscopy, provided evidence that the protonated form of **2.20** binds ATP with a 1:1 binding stoichiometry. Unfortunately, no further analyses of this system have apparently been carried out.

Recently, Lehn and co-workers[41] generated the two optically active receptors **2.21** and **2.21** using a chemoenzymatic approach. The enantioselectivity of these two systems was found to depend strongly on the structure of the anion being targeted for binding and the degree of protonation. Hexamine **2.21** was found to display a moderate preference for D-tartarate (log K_a = 4.10 (3.49) and 3.34 (2.71) for H_6**2.21**$^{6+}$ and H_5**2.21**$^{5+}$, respectively, with D-tartarate (L-tartarate) in aqueous 0.1 M TMA-Cl). However, the hexaprotonated receptor **2.22** containing a second chiral moiety was found to display a rather respectable enantiomeric preference for *N*-Ac-D-asparate (log K_a = 5.34) over its enantiomer (log K_a = 4.57). Interestingly, the binding behaviour of receptor **2.21** towards *N*-Ac-asparate was found to change as a function of protonation. For example, the hexaprotonated form **2.21** is selective for the D-antipode (log K_a = 4.52 and 4.14 for D- and L-*N*-Ac-asparate), whereas the converse is true for the corresponding tetraprotonated form (log K_a = 1.76 and 1.99 for D- and L-*N*-Ac-asparate).

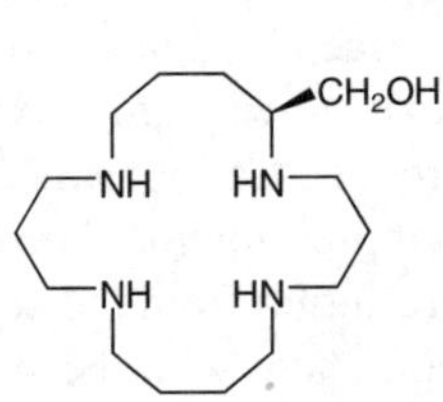

2.20

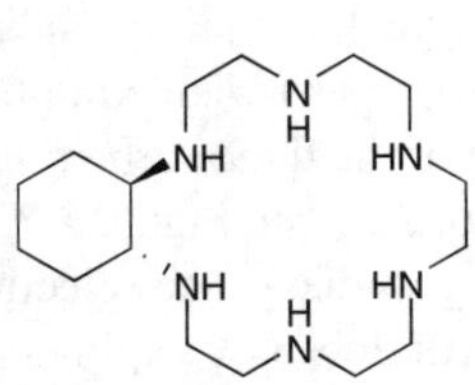

2.21

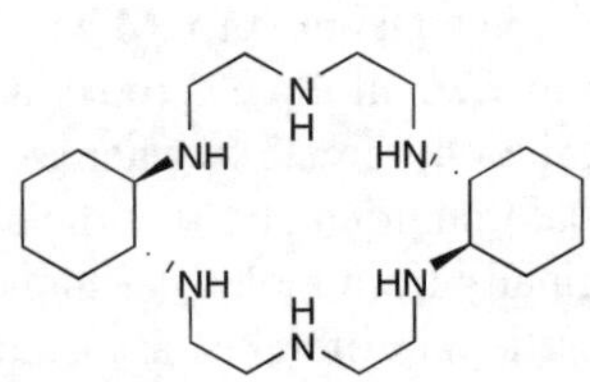

2.22

Table 2.3 summarizes the anion-binding affinities of the cyclic hepta- and octaamine-based receptors, **2.23a**, **2.23b**, **2.24**, and **2.25**.[30–33,35,42,43] From a quick inspection of this table, it is clear that the anion-binding affinities of these receptors increase as (1) the degree of protonation increases and (2) the size of the ring is decreased at any given level of protonation (*e.g.*, log K_a = 8.92, 7.28, and 6.93 for the binding of ATP^{4-} by H_5**2.14**$^{5+}$, H_5**2.23a**$^{5+}$, and H_5**2.24**$^{5+}$, respectively). Also

Table 2.3 *Association constants (log K_a) corresponding to the interaction between the protonated forms of cyclic polyamines and various anions, as determined from potentiometric measurements carried out in aqueous 0.15 M $NaClO_4$ at 298 K*

	H_5**2.23a**$^{5+}$	H_6**2.23a**$^{6+}$	H_7**2.23a**$^{7+}$	H_4**2.23b**$^{4+}$	H_5**2.23b**$^{5+}$	H_6**2.23b**$^{6+}$	H_8**2.25**$^{8+a}$
ATP^{4-}	7.28	10.52	12.93	4.69	7.54	10.04	8.5
ADP^{3-}	5.31	7.96	10.04	5.33	8.23	11.09	7.5
AMP^{2-}	4.99	7.00	8.87	4.91	7.59	10.18	4.1
$H_2P_2O_7^{2-}$	6.53	8.29					
SO_4^{2-}	4.05	5.42		4.09	5.12	6.94	4.0

[a]0.1 M TMA-Cl aqueous solution; for APT^{4-}, log K_a = 9.79, 11.89, and 13.81 for H_6**2.24**$^{6+}$, H_7**2.24**$^{7+}$, and H_8**2.24**$^{8+}$, respectively.

noteworthy is the finding that binding affinities for nucleotide phosphates decrease according to the order ATP > ADP > AMP.[29,30,42]

Recently, the interactions of **2.23a** and dinucleotides NAD and NADP in aqueous 0.15 M $NaClO_4$ were studied using potentiometric titration, cyclic voltametry, and ^{31}P NMR spectroscopic methods.[44] It was found that NADP forms complexes with the charged forms of **2.23a** that are stronger than those formed with NAD. Presumably, this reflects the presence of its "extra" phosphate unit that helps stabilize the complex (log K_a = 4.27 and 5.02 for H_4**2.23a**$^{4+}$ and H_5**2.23a**$^{5+}$, respectively, with NAD; log K_a = 4.74 and 7.19 for H_4**2.23a**$^{4+}$ and H_5**2.23a**$^{5+}$, respectively, with NADP).

In 1994, Bianchi and co-workers[43] studied the selectivity of receptor **2.23a** towards Y-shape-carboxylate anions using potentiometric and electrochemical techniques. This receptor was found to bind tricarboxylate anions in preference to dicarboxylate anions. In the case of the 1,3,5-cyclohexane tricarboxylate anions, better binding was seen for **2.26** (log K_a = 5.2, 8.6, 12.9, and 15.7 for H_4**2.23a**$^{4+}$, H_5**2.23a**$^{5+}$, H_6**2.23a**$^{6+}$, and H_7**2.23a**$^{7+}$, respectively, in aqueous 0.15 M $NaClO_4$) than for **2.27** (log K_a = 4.0, 6.6, 9.7, and 11.7 for H_4**2.23a**$^{4+}$, H_5**2.23a**$^{5+}$, H_6**2.23a**$^{6+}$, and H_7**2.23a**$^{7+}$, respectively, in aqueous 0.15 M $NaClO_4$). Such a finding can be explained by appreciating that in **2.26** all three carboxylate groups are oriented in the same "downward" (axial orientation) direction, whereas, one of the three carboxylate groups faces "outwards" (equatorial orientation) in the case of **2.27** (*cf.* Figure 2.7 for a representation derived originally from molecular modelling studies). The selectivity resulting from the configurational difference associated with these two epimers is consistent with the suggestion that receptor **2.23a** provides a flat, positively charged surface that is well matched to the negative surface present in **2.26**.

Additional evidence that protonated polyamine receptors can bind Y-shape anions, comes from a structural analysis of the nitrate complex of protonated **2.24** reported by Bowman-James and co-workers.[45] In the solid state, **2.24** is fully protonated and adopts an almost flat conformation. It interacts with eight nitrate counter anions via a combination of hydrogen-bond interactions and ion-pairing (*cf.* Figure 2.8). Two of the nitrate anions are encapsulated inside the cavity in almost perfect fashion, while six nitrate anions are bound outside the macrocyclic ring.

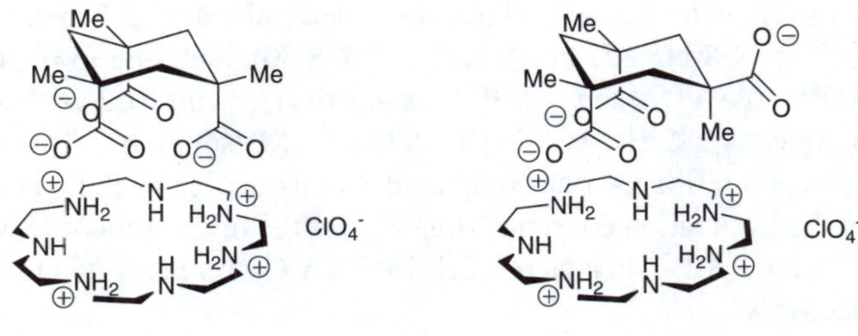

2.23a-2.26 **2.23a-2.27**

Figure 2.7 *Proposed molecular model for the 1:1 complexes formed between tetraprotonated* ***2.23a*** *and trianions* ***2.26*** *and* ***2.27****, respectively*

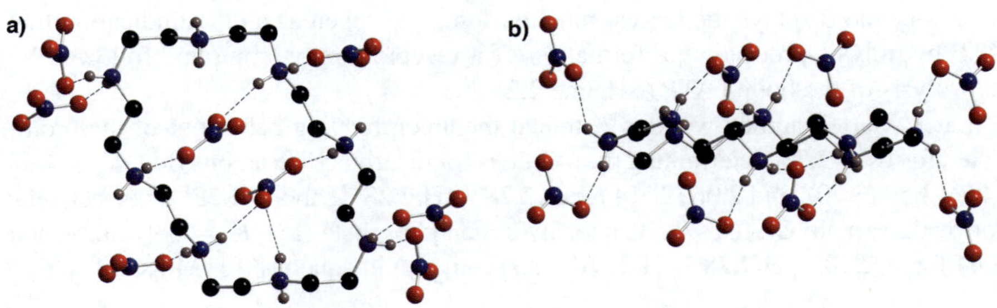

Figure 2.8 *Single crystal X-ray structure of the complex formed between fully (i.e., eight-fold) protonated* ***2.24*** *and its eight nitrate counter anions. Shown are front (a) and side (b) views*

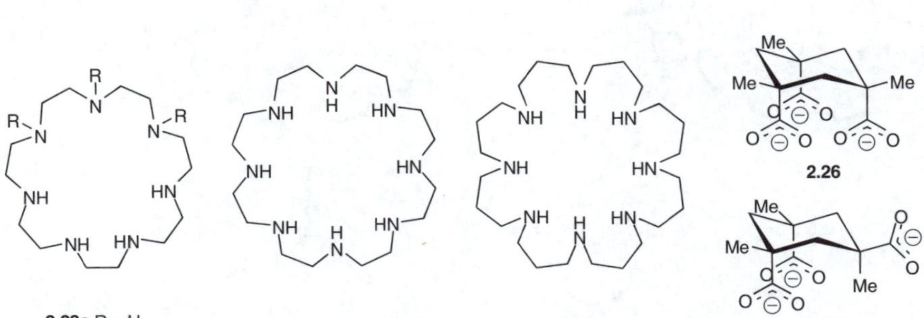

2.23a R = H
2.23b R = Me

2.24 **2.25** **2.26** **2.27**

In one of the historically most important studies involving polyammonium receptors, it was found that macrocycle **2.28** not only interacts strongly with nucleotide phosphates, but also acts as a catalyst for such biomimetic reactions as phosphoryl group transfer and hydrolysis.[46–48] In important predicative work, dating to 1983, the association constants corresponding to the interaction between the protonated forms of **2.28** and various nucleotide phosphates were determined by pH-metric titrations; values of log K_a = 4.80 (8.15, 11.0), 3.40 (6.20, 8.30), and 2.85 (5.00, 6.95) were found for ATP^{4-}, ADP^{3-}, and AMP^{2-}, respectively, with H_4**2.28**$^{4+}$ (H_5**2.28**$^{5+}$, H_6**2.28**$^{6+}$) in aqueous TsOH (0.01 M)/TsONa (0.1 M) solution.[47] This study thus served to extend and refine the one completed 2 years earlier by Lehn and co-workers,[35] wherein the association constants (log K_a) for the interaction of H_6**2.31**$^{6+}$ with ATP^{4-}, ADP^{3-}, and AMP^{2-} in aqueous 0.1 M TMA-Cl were reported to be 9.1, 7.7, and 4.7, respectively.

Noting that the pentaethylenehexamine[49] produced a significant enhancement of the ATP-hydrolysis rate, the first-order hydrolysis rate constants of polyammonium **2.28** and **2.31** were determined quantitatively from plots of log[ATP] *vs.* time using ^{31}P NMR spectroscopy.[47] At pH 3.5, receptor **2.28** ($k = 70 \times 10^3$ min^{-1}) displays the fastest hydrolysis rate with **2.31** ($k = 22 \times 10^3$ min^{-1}) and **2.14** ($k = 22 \times 10^3$ min^{-1}) being more effective than **2.17** ($k = 6.2 \times 10^3$ min^{-1}; all rates in D_2O/H_2O (1:9) at 60 °C). The fact that receptor **2.28** binds ATP more strongly than the other receptors and displays the largest rate constant, was taken as a clear indication that ATP hydrolysis proceeds via formation of a receptor–anion complex, followed by hydrolysis of the bound ATP (Scheme 2.2).

Later, Martell and co-workers[50] studied the anion-binding behaviour of the protonated forms of **2.28**, determining the affinities for dicarboxylate anions (log K_a = 4.68 (4.2), 3.59 (3.70), and 2.06 (2.38) for H_6**2.28**$^{6+}$, H_5**2.28**$^{5+}$, and H_4**2.28**$^{4+}$, respectively, for oxalate (malonate)), as well as for hydrogen phosphate (log K_a = 6.97, 5.29, and 2.64 for H_6**2.28**$^{6+}$, H_5**2.28**$^{5+}$, H_4**2.28**$^{4+}$, respectively) in aqueous 0.1 M KCl.

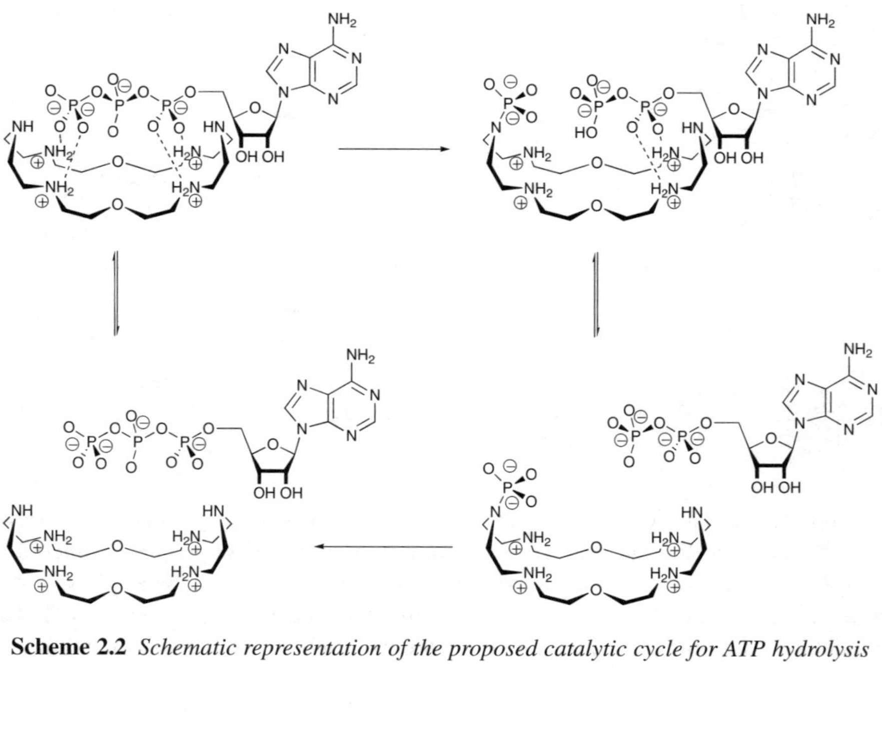

Scheme 2.2 *Schematic representation of the proposed catalytic cycle for ATP hydrolysis*

In an effort to improve the binding affinities towards nucleotide phosphates, one or two pendant acridine units were attached to the macrocycle.[48,51] In aqueous solution at neutral pH, the order of nucleotide-anion binding was found to decrease in the order **2.30** > **2.29** > **2.28**, presumably reflecting the importance of π-stacking interactions between the appended acridine and the bound nucleotide. Specifically, system **2.30** was found to form a stronger complex with NADPH ($K_a \geq 3 \times 10^8$ M^{-1}) than with NADP ($K_a = 5 \times 10^5$ M^{-1}), ATP ($K_a = 7 \times 10^7$ M^{-1}), or triphosphate ($K_a \geq 7 \times 10^7$ M^{-1}).

In 1991, Lehn and co-workers[38] reported the X-ray crystal structure of the chloride-anion complex of the hexaprotonated receptor **2.28** (Figure 2.9a). Five years later, Bowman-James and co-workers[52] elucidated the crystal structure of the nitrate complex generated from the tetraprotonated form of **2.28** (Figure 2.9b). In both cases, the receptor **2.28** (charged forms) adopts a conformation that serves to define a "pocket-like" binding site. In the case of the nitrate complex, molecular-dynamic simulations provided support for the notion that solvation effects in solution are important in terms of defining what is presumed to be a folded, flexible structure.[38,45,52]

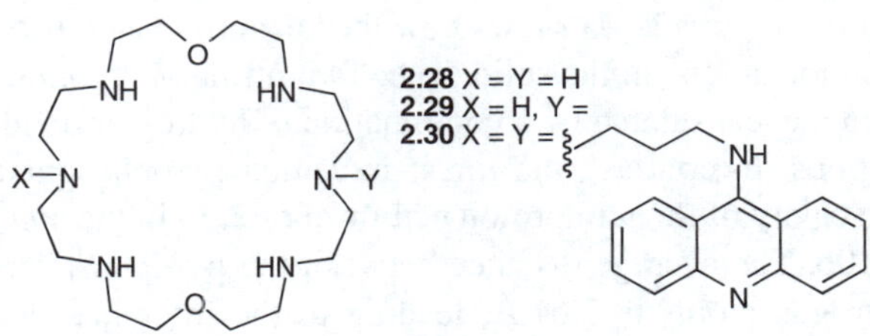

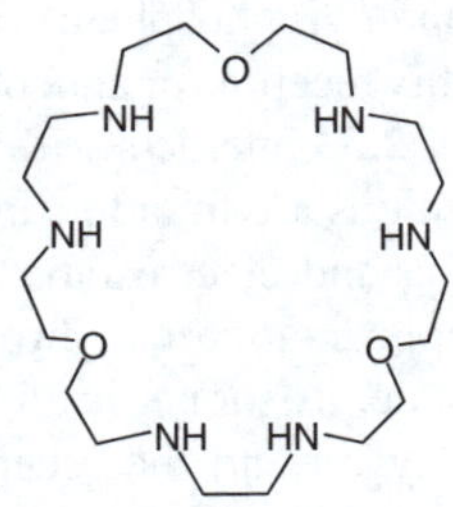

2.31

In what represents an alternative multifunctional approach to molecular recognition, Lehn and co-workers[53] designed and prepared the bis-intercaland-type receptor **2.32**. The protonated form of **2.32** contains two flat naphthalene subunits that are designed to provide a π-stacking unit along with two charged diethylenetriamine subunits,

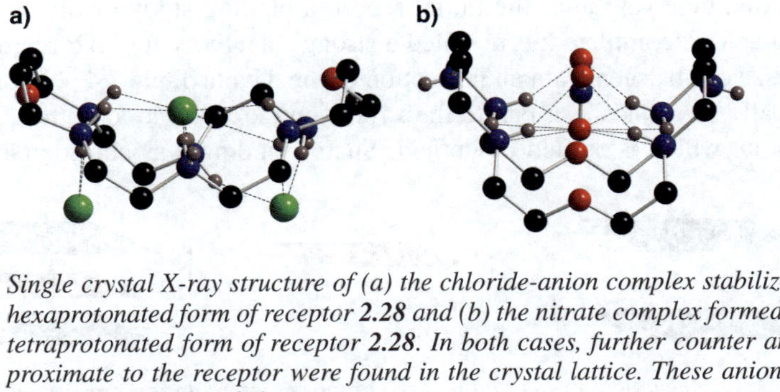

Figure 2.9 *Single crystal X-ray structure of (a) the chloride-anion complex stabilized by the hexaprotonated form of receptor* ***2.28*** *and (b) the nitrate complex formed from the tetraprotonated form of receptor* ***2.28****. In both cases, further counter anions not proximate to the receptor were found in the crystal lattice. These anions are not shown for the sake of clarity*

which were expected to act as the primary anion binding sites. In an aqueous solution, a complex with a dianion could thus be formed via a combination of hydrophobic (π-stacking), electrostatic, and hydrogen bonding interactions (Scheme 2.3).

Quantitative analyses of the anion binding behaviour of **2.32** were made using standard ^{1}H NMR spectroscopic titration methods. In some cases, corroborating studies were carried out using fluorescence titrations and were found to be fully consistent with the results from the NMR spectroscopic analyses. According to the NMR spectroscopic titration results, tetraprotonated **2.32** displays a preference for ATP (log K_a = 5.2), ADP (log K_a = 5.1), isophthalate (log K_a = 5), and terephthalate (log K_a = 5.2) over other anions, such as maleate (log K_a = 3.5), fumarate (log K_a = 4.4), AMP (log K_a = 4.3), cytidine-5′-phosphate (CMP) (log K_a = 3.7), and uridine monophosphate (UMP) (log K_a = 4.1) in aqueous solution at pH 6. The strong binding of benzene dicarboxylate anions over simple alkane dicarboxylate anions was taken as an indication that the extra stabilization provided by the π-stacking interactions between the receptor naphthalene subunits and the benzene rings of the bound guests are energetically significant.

In 1996, Martell and co-workers[54] reported the crystal structures of the bromide and pyrophosphate complexes of protonated **2.33** and studied binding behaviour of this receptor in aqueous solution. Figure 2.10a shows how the hexaprotonated form of **2.33** interacts with six bromide anions in the solid state. Two of the six bromide anions are found inside the ring, encapsulated by a combination of hydrogen-bonding and electrostatic interactions. In contrast, the linear oxyanion pyrophosphate appears to thread through the cavity of the tetraprotonated form of **2.33** in the solid state, as shown in Figure 2.10b. The average distance between the pyrophosphate oxygen and the receptor nitrogen atoms is 2.67 Å, leading to the inference that hydrogen-bond interactions help stabilize this host–guest complex. Anion binding studies, carried out in an aqueous solution using potentiometric methods, revealed that hydrogen phosphate is bound more strongly as the degree of protonation on **2.33** increases. Presumably, this reflects the larger number of hydrogen bonds and greater Coulombic attraction such protonation provides (Table 2.4). Receptor **2.33** was also found to be a slightly better receptor than its aryl-free analogue **2.28**. A similar binding trend was also found in the case of pyrophosphate and triphosphate. However, the charged receptor **2.33** was found to bind to simple triphosphate anion more strongly than either pyrophosphate or hydrogen phosphate at a given level of protonation. One year after the initial report, a binding study involving nucleotide phosphates was completed; it revealed a strong selectivity for ATP relative to ADP or AMP, as can be seen from an inspection of the data in Table 2.4.[55] In the presence of ATP, all CH signals ascribed to the xylene unit are shifted to higher field except one proton, which is shifted downfield. Such a finding was considered consistent

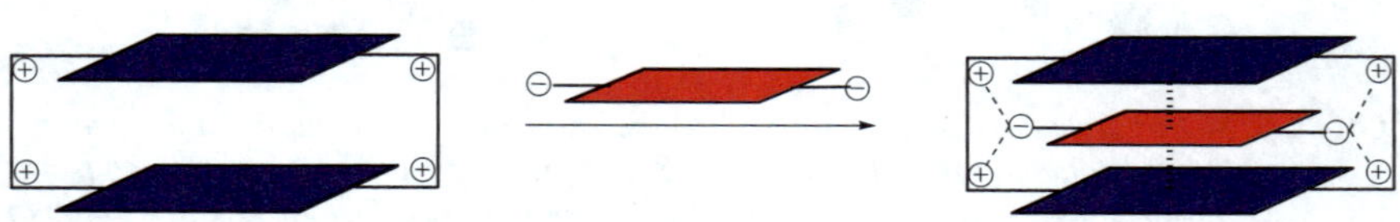

Scheme 2.3 *Schematic representation of the proposed intercalation process involved in the binding of dianionic aromatic substrates by* ***2.32***

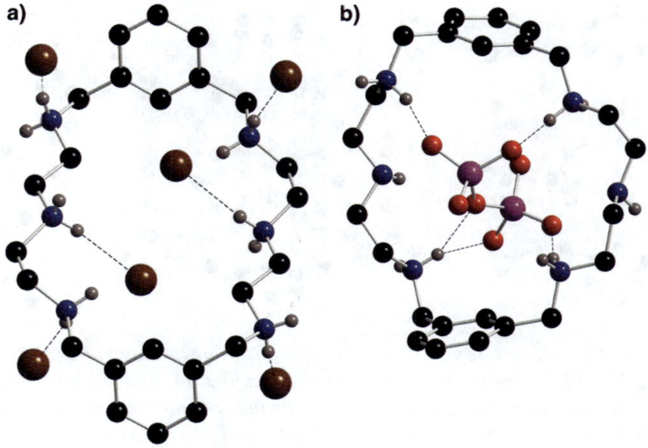

Figure 2.10 *Single crystal X-ray structures of the (a) hexakis-bromide complex of hexaprotonated **2.33** and (b) pyrophosphate complex of tetraprotonated **2.33**. In all cases, further counter anions not proximate to the receptor were found in the crystal lattice. These anions are not shown for the sake of clarity*

Table 2.4 *Association constants (log K_a) for the interaction between cyclic polyamines **2.33** and various phosphorylated substrates as determined from potentiometric titrations carried out in aqueous 0.1 M KCl at 298 K*

	HPO_4^{2-}	$P_2O_7^{4-}$	$P_3O_{10}^{5-}$	ATP^{4-}	ADP^{3-}	AMP^{2-}
H_6**2.33**$^{6+}$	7.36	13.07	14.19	11.16	9.47	7.12
H_5**2.33**$^{5+}$	5.47	9.94	10.85	8.69	7.42	5.6
H_4**2.33**$^{4+}$	2.87	5.73	6.47	5.27	4.37	3.33
H_3**2.33**$^{3+}$			4.71	3.35	3.07	2.7

with a stacking interaction involving the nucleotide and benzene subunits. Based on this and other observations, a non-planar conformation was inferred wherein the system folds over to generate a well-defined sandwich-like binding pocket (*cf.* Scheme 2.3).

Recently, Bowman-James and co-workers[56,57] reported X-ray crystal structures of the nitrate- and sulfate-anion complexes of the hexaprotonated form of **2.33**. As shown in Figures 2.11a and b, both structures are relatively planar with similar hydrogen-bonding patterns being observed in the two cases.

Solution-phase sulfate anion binding studies, carried out in aqueous 0.1 M sodium toslyate (NaOTs), were performed using two different techniques, namely potentiometry and NMR spectroscopic titrations. The results from these independent studies gave rise to concordant results, namely log K_a = 3.53 (3.04) and 4.36 (4.07) for H_5**2.33**$^{5+}$ and H_6**2.33**$^{6+}$, as determined from the potentiometric measurement (NMR titration). Recently, the fluoride anion affinities were also measured and found to be considerably weaker (log K_a ~ 2) than those for sulfate.[57]

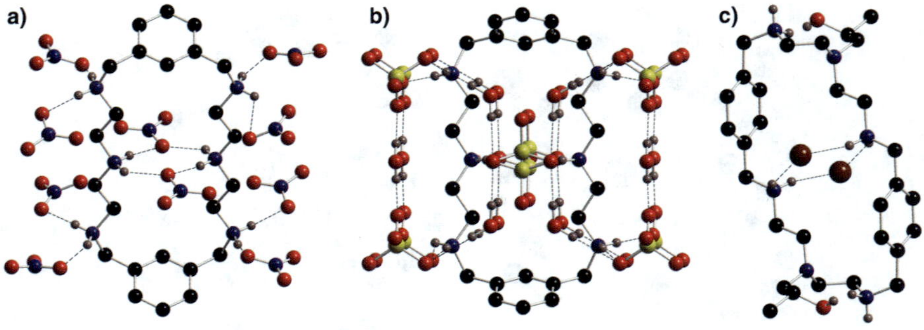

Figure 2.11 *Single crystal X-ray structures of (a) of the nitrate complex of hexaprotonated **2.33**, (b) sulfate complex of hexaprotonated **2.33**, and (c) bis-bromide complex of tetraprotonated **2.34**. In all cases, further counter anions not proximate to the receptor were found in the crystal lattice. These anions are not shown for the sake of clarity*

In 2004, receptor **2.34**, containing two pendant hydroxylpropyl groups, was prepared by Sun and co-workers.[58] Its ability to bind to amino acids (*e.g.*, glycine, aspartic acid, and lysine) when protonated was reported, along with the single crystal X-ray structure of its bromide complex. This latter structure, involving the tetraprotonated form, revealed that, in the solid state, two bromide anions are encapsulated within the central cavity, as depicted in Figure 2.11c. In solution, much of the driving force for the formation of the proposed amino acid complexes is thought to involve interactions between the negative carboxylic oxygen of the substrate and the charged nitrogen atoms of the receptor.

A quantitative analysis of the binding properties of **2.34** revealed a strong selectivity for aspartate (log K_a = 8.53, 7.95, and 7.06 for H_6**2.34**$^{6+}$, H_5**2.34**$^{5+}$, and H_4**2.34**$^{4+}$, respectively) over glycinate (log K_a = 7.60, 6.91, and 6.33 for H_6**2.34**$^{6+}$, H_5**2.34**$^{5+}$, and H_4**2.34**$^{4+}$, respectively) and monohydrogen lysinate (log K_a = 5.25, 5.03, and 4.52 for H_5**2.34**$^{5+}$, H_4**2.34**$^{4+}$, and H_3**2.34**$^{3+}$, respectively) in aqueous 0.1 M $NaNO_3$. Presumably, this selectivity reflects the increased negative charge present on aspartate due to the additional carboxyl group.

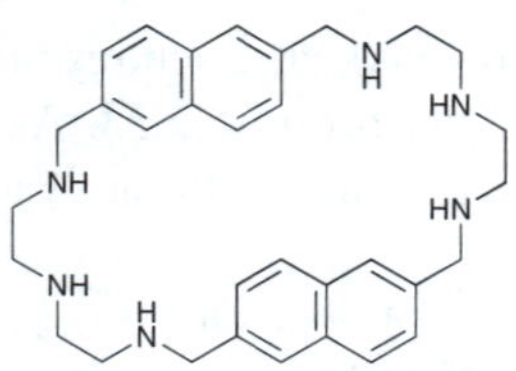

2.32

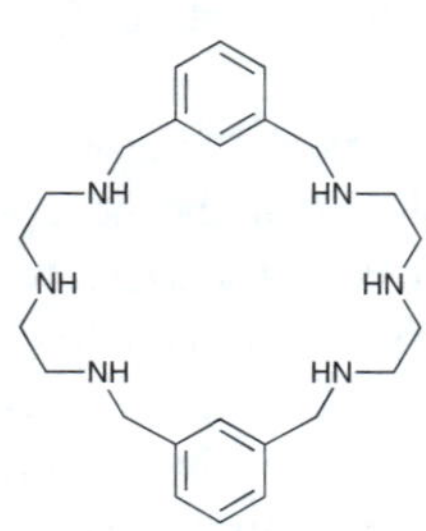

2.33

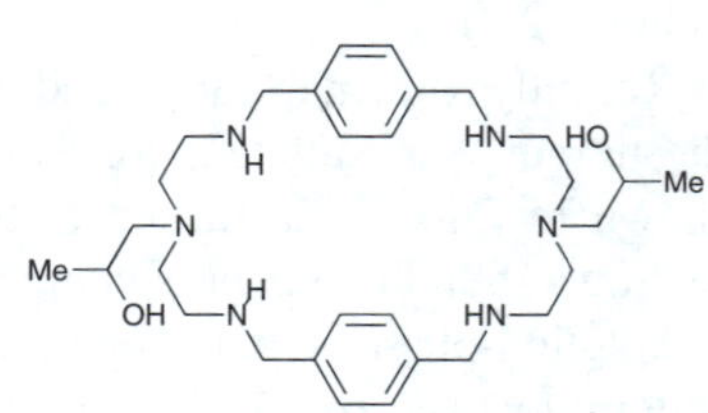

2.34

In 1995, Martell and co-workers[59] reported furan-bridged receptor **2.35** along with the crystal structures of two of its anion complexes. This receptor has a structure similar to that of receptor **2.28**. However, it is characterized by a greater degree of rigidity due to the two aromatic furan rings. In the case of the oxalate anion complex of the hexaprotonated form (*cf.* Figure 2.12a), receptor **2.35** adopts a chair-like conformation wherein three nitrogen atoms lie in a plane and the two furan rings point "up" and "down" in opposite directions. The oxalate counter anion is located out of the ring (above or below depending on perspective) instead of being encapsulated inside the macrocyclic cavity. However, as normal in complexes of this type, this anionic guest is held in place by a combination of hydrogen-bonding and electrostatic interactions.

In contrast to the above, the pyrophosphate anion complex of hexaprotonated **2.35** (*cf.* Figure 2.12b) is characterized by a binding mode in which the anionic substrate is encapsulated within the receptor cavity, again via a combination of multiple ion-pairing and hydrogen-bonding interactions. Consistent with this presumed more favourable binding seen in the solid state, titration studies carried out in an aqueous solution revealed that pyrophosphate is bound more strongly than oxalate by receptor **2.35**. This is true at all levels of protonation (*cf.* Table 2.5). The solution-phase studies were extended to include a range of other anionic substrates and the trends observed were seen to match those seen for other analogous receptors.[60,61] In fact, receptor **2.35** exhibits a slightly higher affinity for nucleotide phosphates than does its structural analogue **2.28**. For instance, log K_a values of 11.4 and 11.0 were recorded for the binding of ATP to the hexaprotonated forms of **2.35** and **2.28**, respectively, under essentially analogous conditions (*cf.* Table 2.5 and discussion above).

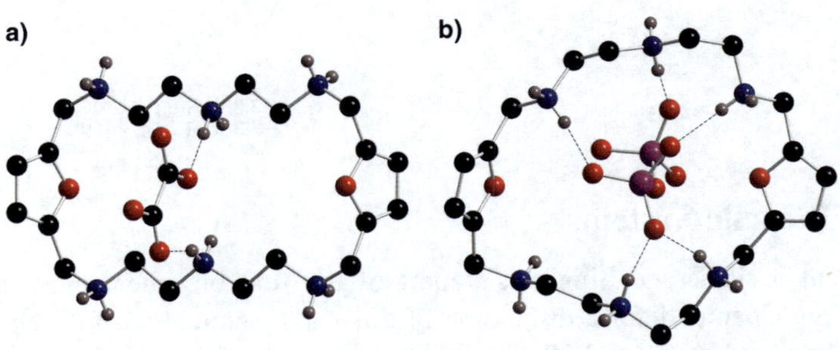

Figure 2.12 *Single crystal X-ray structures of (a) oxalate complex of tetraprotonated* ***2.35*** *(four chloride and four water molecules per macrocycle have been removed for clarity) and (b) pyrophosphate complex of tetraprotonated* ***2.35***

Table 2.5 *Association constants (log K_a) corresponding to the interaction between cyclic polyamines* ***2.35*** *and various anionic substrates, as determined via potentiometric measurements carried out in aqueous 0.1 M KCl at 298 K*

	$Oxalate^{2-}$	HPO_4^{2-}	$P_2O_7^{4-}$	$P_3O_{10}^{5-}$	ATP^{4-}	ADP^{3-}	AMP^{2-}
H_6**2.35**$^{6+}$	4.97	7.29	12.55	13.05	11.43	9.21	6.48
H_5**2.35**$^{5+}$	4.12	5.83	9.53	9.79	8.83	6.82	4.69
H_4**2.35**$^{4+}$	2.97	3.51	5.70	5.78	5.49	4.00	2.66

In spite of displaying nucleotide-binding affinities that were as good or better than those of macrocycle **2.28**, receptor **2.35** was found to be less effective at promoting ATP hydrolysis ($k = 7.24 \times 10^3$ and 0.34×10^3 min^{-1} at pH 4.4 and 7.4, respectively, for **2.35** *vs.* $k = 20.91 \times 10^3$ and 6.05×10^3 min^{-1} at pH 4.4 and 7.4, respectively, for **2.28**; all measurements in 0.1 M KCl aqueous solution).[61]

Recently, Navarro, García-España and co-workers[62] designed and prepared macrocycles **2.36a** and **2.36b** containing two pyrazole units as a receptor for L-glutamate. The pyrazole ring can behave as a hydrogen bond donor or acceptor.[63] In contrast to what one might predict *a priori*, the stability of the complex formed between **2.36a** and L-glutamate ($HGlu^-$) was found to increase with increasing pH (log K_a = 3.49, 3.44, 3.33, 3.02, and 3.11 for H**2.36a**$^+$, H_2**2.36a**$^{2+}$, H_3**2.36a**$^{3+}$, H_4**2.36a**$^{4+}$, and H_5**2.36a**$^{5+}$, respectively). On the other hand, the tetramine receptor **2.36b** displays affinities for L-glutamate that show the expected inverse correlation with pH (*i.e.*, log K_a = 3.71, 4.14, 4.65, 4.77, and 6.76 for H**2.36b**$^+$, H_2**2.36b**$^{2+}$, H_3**2.36b**$^{3+}$, H_4**2.36b**$^{4+}$, H_5**2.36b**$^{5+}$, and H_6**2.36b**$^{6+}$). These results were rationalized in terms of the appended benzyl groups playing a role in modulating the binding behaviour of receptor **2.36b**. Further examples of monocyclic systems are listed in the bibliography.[64]

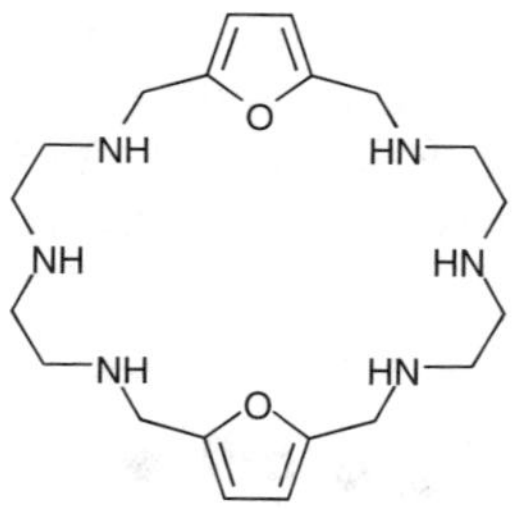

2.35

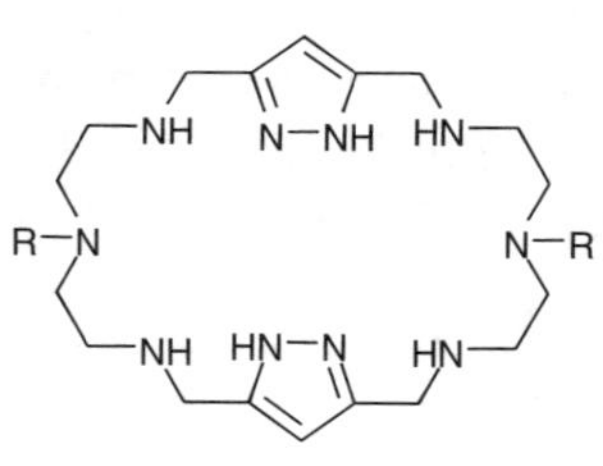

2.36a R = H
2.36b R = Bz

2.1.3 Bicyclic Systems

Subsequent to Park and Simmons' report of the first bicyclic ammonium-anion receptor (*cf.* Chapter 1 for a discussion of this early, seminal work), Lehn and co-workers prepared receptor **2.37**. This system was specifically designed to provide more anion-binding sites within the cavity. After initial studies involving the use of NMR spectroscopy, more detailed binding experiments were performed using pH-metric titrations carried out in aqueous 0.1 M perchlorate. From these latter analyses, the association constants (log K_a) for the binding of nitrate and chloride anions to H_6**2.37**$^{6+}$ were found to be 2.93 and 2.26, respectively.[65]

In 1984, the crystal structures of several anion complexes of protonated **2.37** were reported, along with the results of anion binding studies. Attempts to crystallize a complex of the linear F–H–F anion yielded a structure wherein a single fluoride anion is held somewhat along one edge of the cavity (Figure 2.13a). However, the corresponding bromide anion complex, shown in Figure 2.13b, reveals the anion interacting with each of the potential binding sites; it is thus fully coordinated in an octahedral fashion by six hydrogen-bonding interactions.[66]

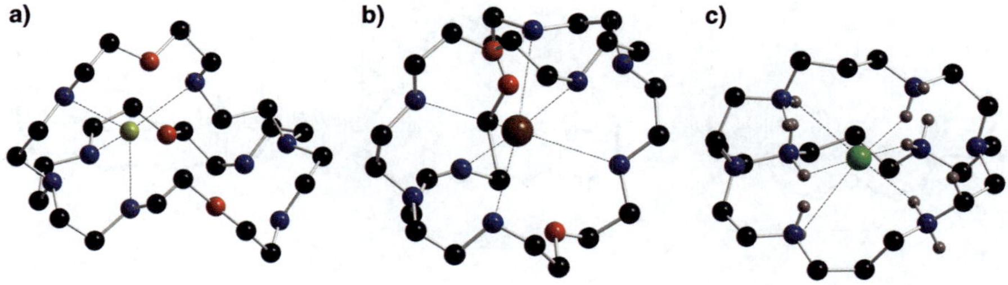

Figure 2.13 *Single crystal X-ray structures of (a) fluoride and (b) bromide complexes of hexaprotonated* ***2.37****. (c) X-ray structure of the chloride-anion complex of hexaprotonated* ***2.39***

Solution-phase potentiometric studies, carried out in aqueous 0.1 M NaOTs solution, revealed that the most stable complexes were obtained with charge dense anions such as F^- (log $K_a = 4.10$), $P_2O_7^{4-}$ (log $K_a = 10.30$), ATP^{4-} (log $K_a = 8.00$), ADP^{3-} (log $K_a = 5.85$), AMP^{2-}(log $K_a = 3.85$), and SO_4^{2-} (log $K_a = 4.90$).[66] This is consistent with binding interactions that are largely electrostatic in nature. However, the fact that the association constants of hexaprotonated **2.37** towards nucleotide phosphates are lower than those of the simple macrocycle **2.28**, at the same level of protonation, provides support for the notion that larger anions are only partially bound within the cavity defined by receptor **2.37**. Consistent with this conclusion is the finding that **2.37** is less effective than **2.28** in terms of promoting ATP hydrolysis (*i.e.*, $k = 85 \times 10^3$ and 1.4×10^3 min^{-1} for **2.28** and **2.37**, respectively, at pH 5.5).[47] In the case of mononegative-charged anions, the binding selectivity of the hexaprotonated form of **2.37** was found to follow the order $I^- < Br^-, Cl^-, NO_3^- \ll N_3^-$.

In contrast to what was seen for **2.37**, a related receptor, **2.38**, was found to form a rather stable complex with I^- and, indeed, within the halide series contrasting anion-binding behaviour was seen (log K_a = 2.40 (2.10), 2.95 (2.65), and 3.40 (3.00) for H_8**2.38**$^{8+}$ (H_7**2.38**$^{7+}$) with Cl^-, Br^-, and I^-, respectively, in aqueous 0.1 M NaOTs at 25 °C).[37] Presumably, these findings reflect the fact that in the case of receptor **2.38**, the two $N(CH_2CH_2CH_2NH_2)_3$ tripod subunits linked by the $CH_2CH_2CH_2$ chains provide a larger and more spherical cavity than does **2.37**. It was also found that the hexaprotonated form of **2.38** binds oxalate more effectively than does the corresponding hexaprotonated species **2.17**. By contrast, receptor H_6**2.17**$^{6+}$ forms a more stable malonate complex than the analogous receptor H_6**2.38**$^{6+}$. The selectivity of **2.38** towards oxalate can be explained by the size of the inside cavity, which is a near-perfect match for oxalate. Thus, taken in concert, these results clearly demonstrate the importance of the bicyclic effect, at least in the case of appropriately matched receptor–substrate pairs.

Another analogous receptor, **2.39**, was also prepared; this system displays a preference for fluoride and chloride over bromide and iodide anions in aqueous solution (log K_a = 6.10, 5.75, 4.40, and 2.25 for H_6**2.39**$^{6+}$ with F^-, Cl^-, Br^-, and I^-, respectively).[67] The single crystal X-ray structure of the chloride anion complex was solved (*cf.* Figure 2.13c). It revealed that the bound chloride anion is completely encapsulated within the hexaprotonated receptor **2.39**.

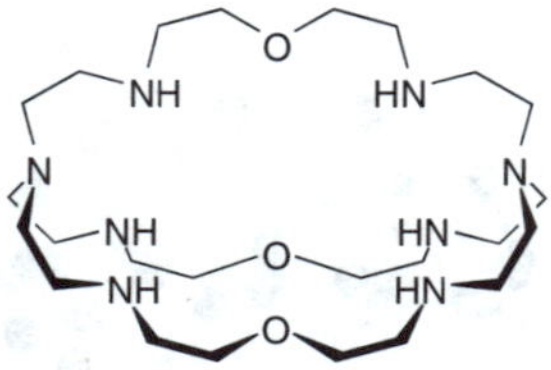

2.37

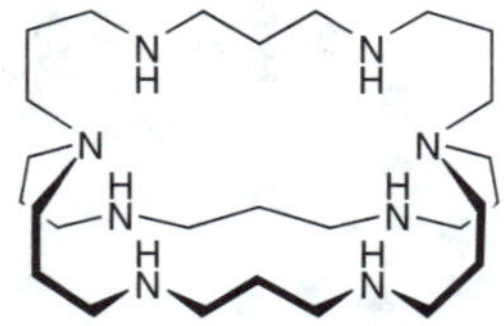

2.38

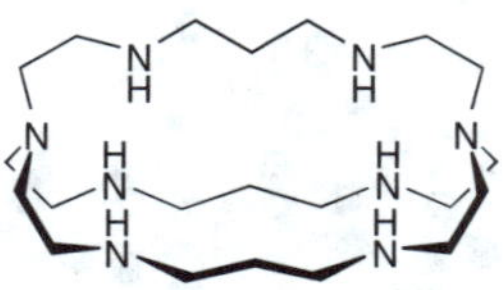

2.39

Among the series of azacryptand receptors produced by the Lehn laboratory, the bicyclic ligand **2.40** is noteworthy. It has the smallest cavity and was considered well-tailored for the binding of fluoride anion. The crystal structure of the fluoride-anion complex of hexaprotonated **2.40** was reported in 1989 (Figure 2.14a). It served to confirm the proposed anion recognition, at least in the solid state, revealing, in particular, the presence of a single fluoride anion located in the middle of the cavity. The N–H···F distances were found to be in the range of 2.76 and 2.86 Å.[68] Eleven years later, Bowman-James and co-workers[69] reported the structure of the corresponding chloride-anion complex, wherein the anion is once again encapsulated within the three-dimensional binding cavity (*cf.* Figure 2.14b). While similar hydrogen-bonding interactions are seen in both the fluoride and chloride complexes, longer average hydrogen bond distances of 2.99–3.18 Å are seen in the case of the latter anion.

In aqueous solution, a remarkable selectivity for fluoride over chloride was observed. For instance, the association constants (log K_a) for fluoride and chloride (0.1 M KNO_3 for fluoride and 0.1 M KCl for chloride) were found to be ≥ 8.8 and ≥ 1.2 for H_5**2.40**·F and H_5**2.40**·Cl, respectively, as determined by Smith and co-workers.[70] Later, Lehn and co-workers[67] also obtained similar results in a slightly different aqueous environment (log $K_a = 10.55$ and < 2 respectively, for the binding of fluoride and chloride to H_6**2.40**$^{6+}$ in the presence of 0.1 M TMA-TsO).

As an extension of this work, proton NMR spectroscopic methods were used to investigate the anion-binding behaviour of **2.40** at low pH. While a significant downfield shift in the peak corresponding to the CH proton adjacent to the amines was seen in the presence of fluoride anion over a pD range from 5.5 to 2.5, the corresponding

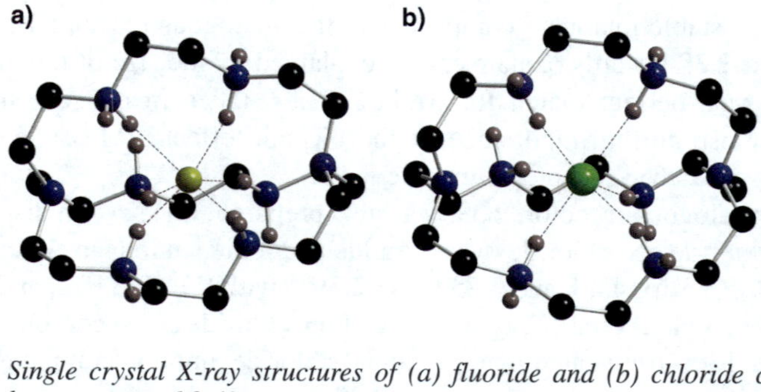

Figure 2.14 *Single crystal X-ray structures of (a) fluoride and (b) chloride complexes of hexaprotonated* ***2.40***

chemical shifts recorded in the presence of chloride, bromide, nitrate, and iodide were not found to be substantially displaced from their original position in anion-free **2.40** as determined at the same pD.[69,71]

At a pD of 2.0, the presence of both fluoride and chloride produced a similar downfield shift in this key CH signal. By contrast, NMR spectra recorded in the presence of other test anions, such as nitrate, bromide, and iodide, revealed little change. These observations were considered indicative of chloride anion encapsulation at low pD, in analogy to what is seen in the solid state (*cf.* Figure 2.12b). Concurrent with this experimental work, Wipff and co-workers[72] carried out molecular dynamics simulations, considering the question of fluoride and chloride-anion selectivity.

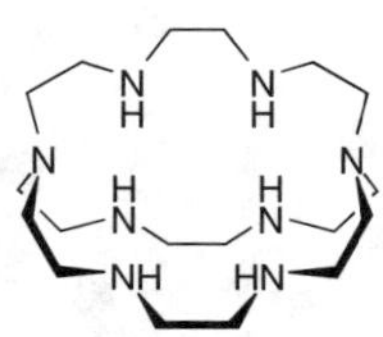

2.40

The more rigid cyclophane-type macrobicyclic anion receptors, **2.41a**, **2.41b**, and **2.42b**, were prepared by Lehn and Heyer[73] using a tripod–tripod coupling strategy. Again, proton NMR spectroscopy was used to obtain qualitative insights into the anion-binding properties of the hexaprotonated forms of these receptors. In the presence of nitrate, sulfate, and chloride anions, large shifts were seen for the CH signals in H_6**2.41a**$^{6+}$. Conversely, only small peak shifts were observed with H_6**2.41b**$^{6+}$ and H_6**2.42b**$^{6+}$. Quantitative pH-metric studies revealed that the log K_a values lie in the range of 2.5–4.0 for monovalent anions, such as Cl^-, NO_3^-, and N_3^-, binding to H_6**2.41a**$^{6+}$, H_6**2.41b**$^{6+}$, and H_6**2.42b**$^{6+}$; by contrast log K_a values of 5.0–6.5 were recorded for dianions such as SO_4^{2-}, $S_2O_6^{2-}$, and $^-O_2CCO_2^-$.

Quite recently, Steed and co-workers[74] reported the synthesis and halide anion-binding properties of a similar pre-organized macrobicyclic receptor, namely **2.42a**, along with crystal structures of their halide anion complexes. The resulting structures, shown in Figures 2.15a–d, reveal that the hexaprotonated receptor **2.42a** binds to most halide anions in a way that involves near-complete encapsulation. Interestingly, the bromide anion was observed to be bound by the heptaprotonated form of **2.42a** in the solid state. The average N···X distances are 2.65, 3.18, 3.31, and 3.45 Å for the fluoride, chloride, bromide, and iodide complexes, respectively. Surprisingly, the fluoride complex, shown in Figure 2.15a, displays the longest distance from the centre of the benzene core to the anion (4.506 Å), while the other three complexes were characterized by shorter distances (*ca.* 3.6 Å). Presumably, the longer distance seen in the fluoride complex reflects repulsion between the electron-rich π-surface of the benzene ring and the highly charge π dense anion.

Potentiometric titrations carried out in aqueous 0.1 M NaOTs revealed a high selectivity for fluoride over chloride (log K_F/log K_{Cl} > 5; log K_a = 9.54, 7.84, 5.65, and 4.86, respectively, for H_6**2.42a**$^{6+}$, H_5**2.42a**$^{5+}$, H_4**2.42a**$^{4+}$, and H_3**2.42a**$^{3+}$ and fluoride anion; log K_a = 4.19, 3.88, and 2.06, respectively, for H_6**2.42a**$^{6+}$,

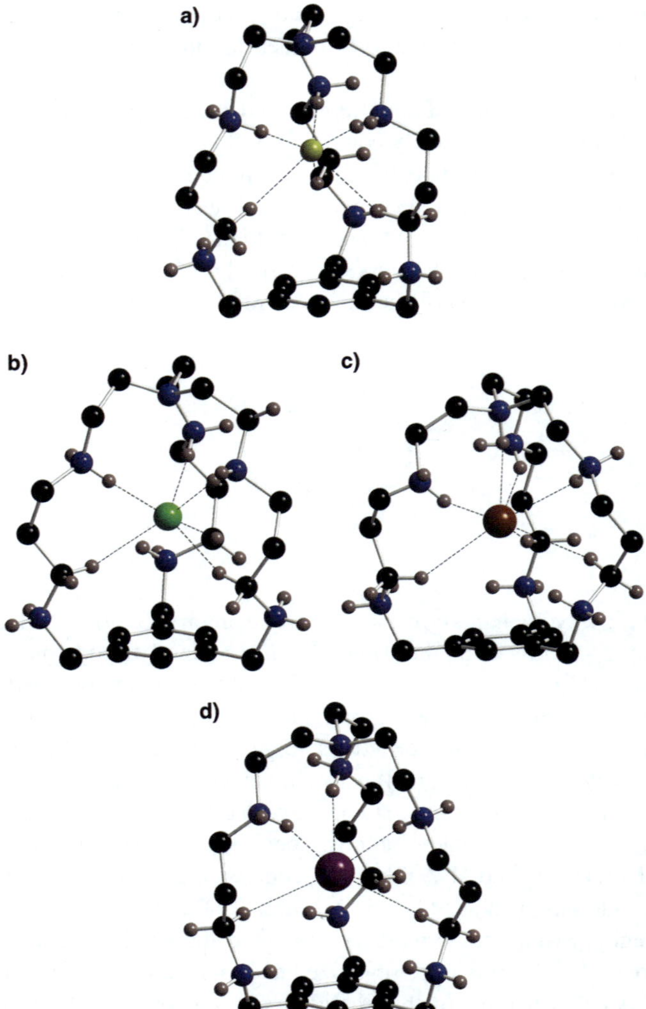

Figure 2.15 *Single crystal X-ray structures of the (a) fluoride, (b) chloride, (c) bromide, and (d) iodide complexes of protonated* ***2.42a***

H_5**2.42a**$^{5+}$, and H_4**2.42a**$^{4+}$ and chloride anion). However, no appreciable binding of bromide or iodide was observed.

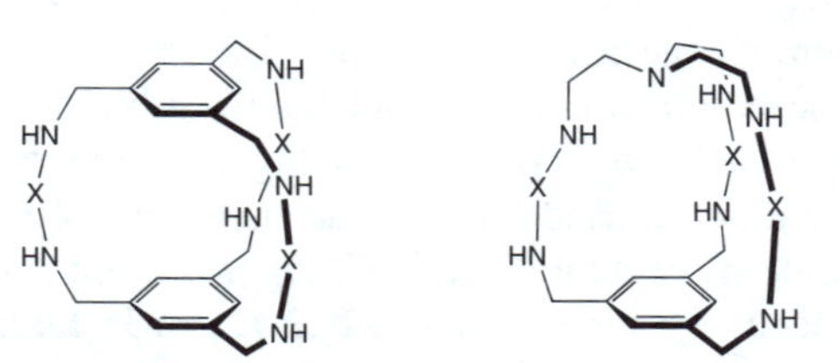

2.41a X = $-CH_2CH_2CH_2-$
2.41b X = $-CH_2CH_2OCH_2CH_2-$

2.42a X = $-CH_2CH_2CH_2-$
2.42b X = $-CH_2CH_2OCH_2CH_2-$

In 1991, Lehn and co-workers[75] reported the synthesis of a more extended macrobicyclic receptor, **2.43**, that was designed to bind to dicarboxylate anions in aqueous solution. Solid-state structural analyses of the terephthalate complex of the hexaprotonated form of **2.43** (*cf.* Figure 2.16) confirmed that this dianionic guest is indeed bound within the cylindrical cavity, being held in place, at least in part, via four well-oriented $^{+}$N–H···O^{-} hydrogen bonds (N–O bond lengths ranging between 2.77 and 2.96 Å). As determined from NMR spectral titrations carried out in aqueous solution, adipate anion is bound more strongly than other shorter or longer α,ω-dicarboxylate anions, $^{-}O_2C–(CH_2)_n–CO_2^{-}$ (K_a = 2600 M^{-1} for adipate). Presumably, this reflects the fact that adipate anion provides the best fit within this series of putative substrates. The strongest binding was, however, actually observed in the case of the more rigid terephthalate anion (K_a = 25,000 M^{-1}), a finding that was also interpreted in terms of structural complementarity.

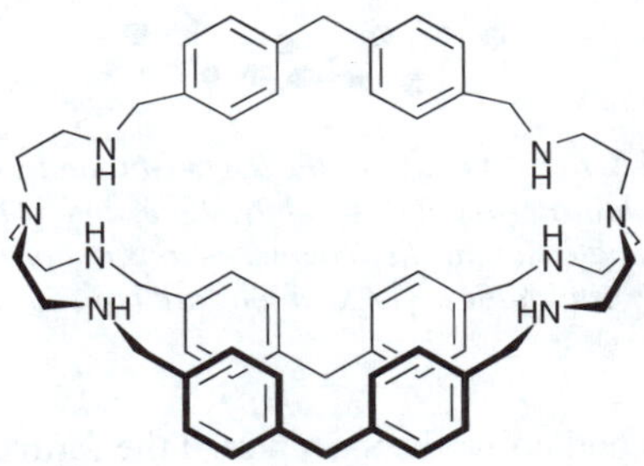

2.43

Essentially concurrent with this work, Martell and co-workers[76] reported a two-step synthesis of receptor **2.44**, along with an X-ray structure of its bromide anion complex. This complex, involving the "Y"-shaped octaprotonated form of **2.44**, is characterized by the presence of three bromide anions held in place by hydrogen-bonding interactions (Figure 2.17a).

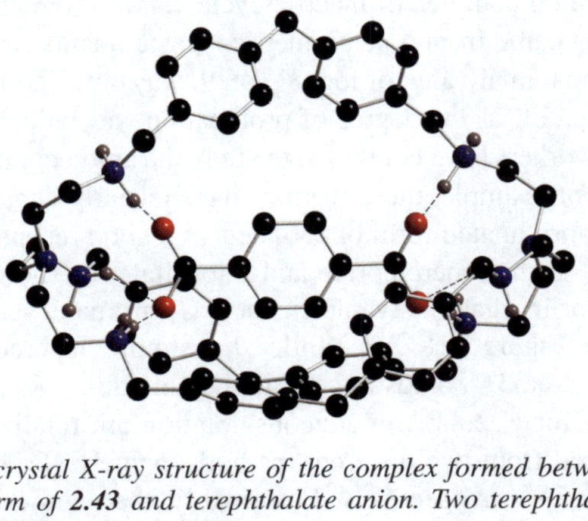

Figure 2.16 *Single crystal X-ray structure of the complex formed between the hexaprotonated form of* **2.43** *and terephthalate anion. Two terephthalate anions are not shown for the sake of clarity*

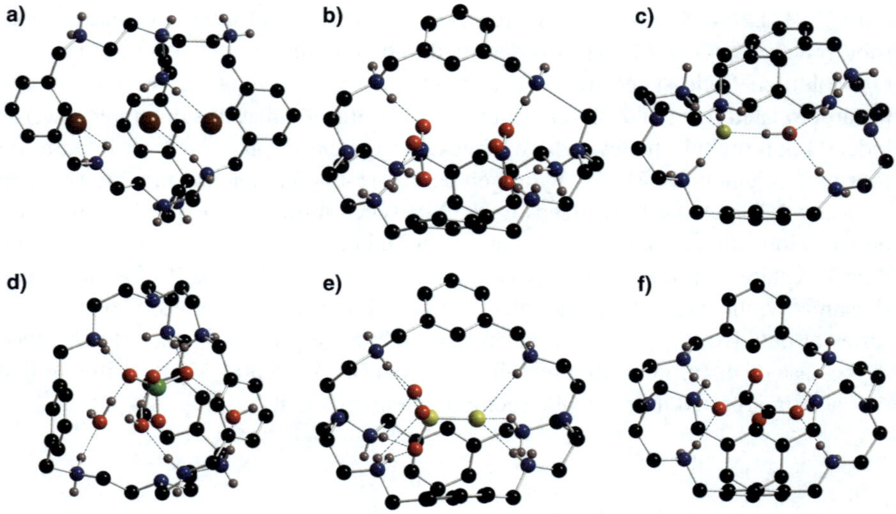

Figure 2.17 *Single crystal X-ray structures of the (a) tris-bromide, (b) bis-nitrate, (c) fluoride-aquo, (d) perchlorate-trisaquo, (e) thiosulfate, and (f) oxalate complexes of protonated **2.44**. In all cases, further counter anions not proximate to the receptor were found in the crystal lattice. These anions are not shown for the sake of clarity*

Later, Bowman-James and co-workers[77] reported the formation and characterization of the corresponding nitrate and fluoride anion complexes. In the case of nitrate, the complex was obtained from an anion-exchange reaction involving the addition of $AgNO_3$ to the hydrochloride salt. X-ray diffraction analysis revealed the presence of two nitrate anions nestled within the cavity, being held there in an eclipse-like arrangement (Figure 2.17b). The association constants (log K_a) for the formation of the first nitrate and second nitrate complex were found to be 3.02 ± 0.03 and 2.38 ± 0.08, respectively.

In the case of the fluoride anion complex, one fluoride anion and one water molecule are bound within the cavity in the solid state (*cf.* Figure 2.17c).[57,78] Solution-phase ^{19}F NMR studies confirmed an NH–F peak shift over a broad pH range for the putative fluoride anion complex of macrobicycle **2.44**. More detailed studies of fluoride anion binding came from potentiometric measurements carried out in aqueous 0.1 M NaOTs. A maximal value of log K_a, 4.29, was obtained for H_7**2.44**$^{7+}$,[57] with the affinity dropping off as the degree of protonation was reduced.

Nelson and co-workers have continued to study the anion complexation properties of receptor **2.44**. For example, these workers have recently demonstrated that in the solid state the hexaprotonated form of **2.44** acts as a good receptor for organic tetrahedral oxo-anions, such as perchlorate and thiosulfate, as shown in Figures 2.17d and e, as well as for inorganic oxyanions, such as chromate, selenate, and perrhenate, as shown in Figure 2.18.[79,80] While the simple macrocyclic analogue of H_6**2.44**$^{6+}$, namely H_6**2.33**$^{6+}$, binds to perchlorate with a log $K_a \leq 1$, the association constants (log K_a) for H_6**2.44**$^{6+}$ in aqueous solution are relatively high, *e.g.*, 3.53 and 3.25, as inferred from two different methods, namely NMR spectroscopic and potentiometric titrations, respectively. Presumably, the geometric match (size and shape complementarity) between the trigonal host and the bound tetrahedral anion accounts for this relative enhancement in binding affinity.

Doubly negativly charged oxyanions are potentially capable of forming a larger number of hydrogen bonds with protonated receptor **2.44** than can the singly charged species, perchlorate. Such anionic guests were thus expected to be bound well in solution and in the solid state. As can be seen from an inspection of Figure 2.17e, an X-ray structural analysis revealed that thiosulfate is held within receptor **2.44** (hexaprotonated form) as the result of hydrogen-bonding interactions involving all three oxygen atoms of the substrate and all six NH protons of the macrobicycle. NMR spectroscopic titration studies of hexaprotonated **2.44** confirmed the expected higher binding affinities (*e.g.*, log $K_a = 4.0$ for both SeO_4^{2-} and $S_2O_3^{2-}$). More detailed studies of the solution-phase anion-binding properties came from potentiometry studies (Table 2.6). These latter analyses revealed that, as a general rule, double negative anions, such as thiosulfate, are held more strongly than mononegative anions, such as nitrate and perchlorate, at least at a comparable level of receptor protonation.[81]

Recently, Nelson and co-workers found that protonated receptor **2.44** also binds dicarboxylate anions with stability constants in selected cases that are quite high. In the context of these studies, the crystal structure of the oxalate anion complex was solved (Figure 2.17f). As can be seen from an inspection of this figure, the oxalate anion exists in a twisted conformation and is completely encapsulated within the receptor, being held in place by a combination of electrostatic, hydrogen bonding (*e.g.*, NH···O), and

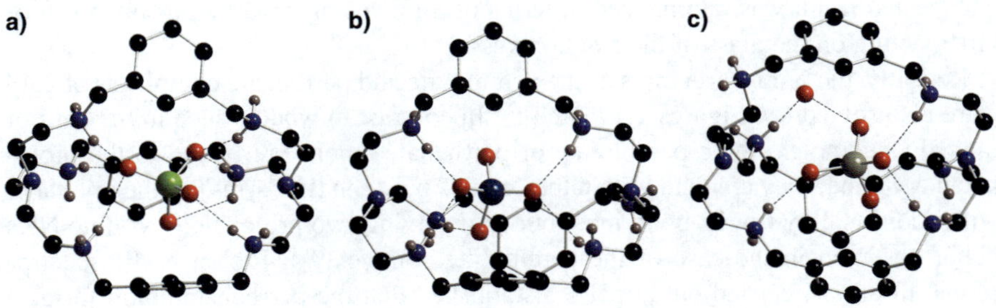

Figure 2.18 *Single crystal X-ray structures of inorganic anion complexes involving protonated receptor **2.44**. (a) Selenate, (b) chromate, and (c) perrhenate complexes. In all cases, further counter anions not proximate to the receptor were found in the crystal lattice. These anions are not shown for the sake of clarity*

Table 2.6 *Association constants (log K_a) corresponding to the interaction between cyclic polyamine **2.44** (various protonation forms) and representative oxo-anions, as determined from potentiometry measurements carried out in aqueous 0.1 M NaOTs at 298 K*

	ReO_4^-	SO_4^{2-}	SeO_4^{2-}	$S_2O_3^{2-}$	*Oxalate*$^{2-}$	*Malonate*$^{2-}$	*Acetate*$^-$	*Lactate*$^-$
H_6**2.44**$^{6+}$	3.71	6.57	7.24	8.51	10.71	7.22	4.00	3.86
H_5**2.44**$^{5+}$	3.45	4.72	5.39	6.40	8.64	6.20	4.21	4.06
H_4**2.44**$^{4+}$	3.06	3.70	4.77	5.49	7.15	5.79	4.27	3.61
H_3**2.44**$^{3+}$	2.81	3.47	4.18	4.74	5.49	4.87	4.05	3.45

π-stacking (*i.e.*, C=O···aromatic benzene) interactions.[82] As revealed by the values included in Table 2.6, receptor **2.44** (charged forms) displays an exceptionally high affinity for oxalate relative to other carboxylate anions, a finding that is easily rationalized in terms of good size and shape complementarity.[83]

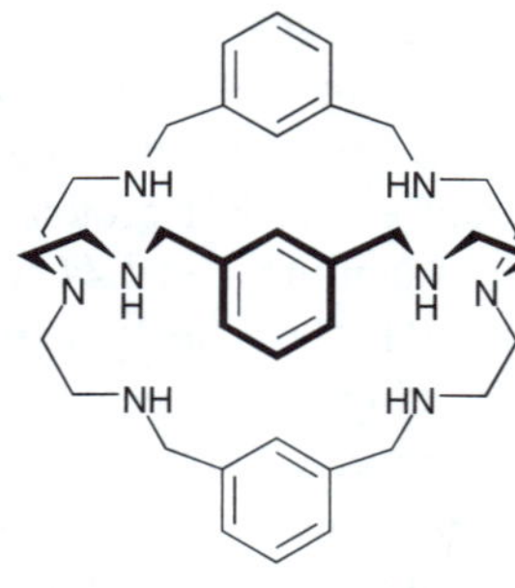

2.44

The tris-pyridine analogue of cryptand **2.44**, macrobicycle **2.45**, has been extensively studied by McKee and co-workers.[84] In the course of trying to grow diffraction-grade crystals of the complex presumed to be formed from the reaction of **2.45** with HBF_4 these workers isolated crystals of the SiF_6^{2-} complex, whose structure was elucidated via single crystal X-ray diffraction analysis (*cf.* Figure 2.19a). This unexpected result was rationalized in terms of SiF_6^{2-} being produced as the result of HBF_4 acting on the glass of the reaction vessel.

Recently, the X-ray structures of the perchlorate and perrhenate complexes of **2.45** were also solved (*cf.* Figures 2.19b and c). In contrast to what is seen in the case of the SiF_6^{2-} complex, three perchlorate or perrhenate anions are bound to the protonated cryptand; they are held within the "clefts" between the "arms" of the cryptand, with additional hydrogen-bond interactions involving two protonated pyridine NHs being observed in the case of the perchlorate complex.[80,85] Proton NMR spectroscopic titrations, carried out at pH 3, established that the perrhenate anion forms a

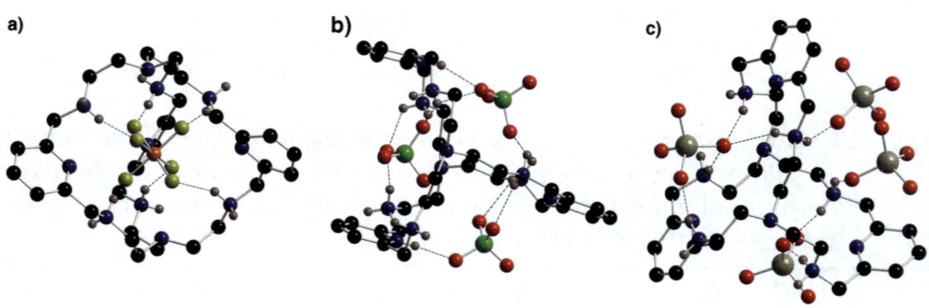

Figure 2.19 *Single crystal X-ray structures of the (a) SiF_6^{2-}, (b) ClO_4^-, and (c) ReO_4^- complexes of protonated receptor **2.45**. In all cases, further counter anions not proximate to the receptor were found in the crystal lattice. These anions are not shown for the sake of clarity*

stronger complex than does perchlorate (*e.g.*, for H_6**2.45**$^{6+}$, log K_a = 3.20 and 2.56 for ReO_4^- and ClO_4^-, respectively).[80]

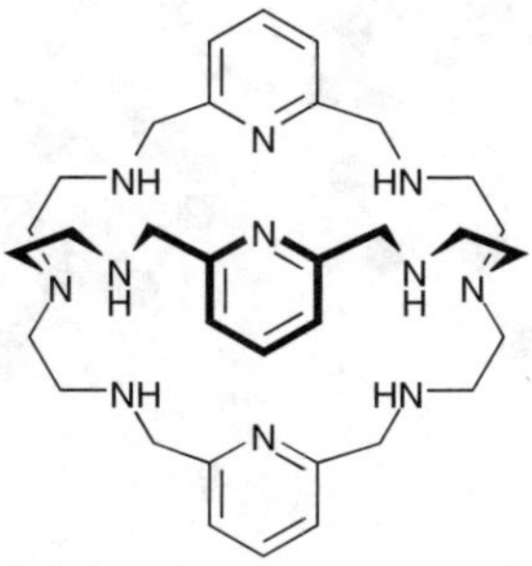

2.45

In 1995, the structure of the perchlorate complex of **2.46** was reported. As can be seen from an inspection of Figure 2.20a, this structure, apparently the first involving an oxyanion and this receptor, provided unambiguous proof for the encapsulation of the perchlorate anion within the cavity.[84] Single crystal X-ray structures of the oxalate, thiosulfate, sulfate, and chromate complexes were also solved (shown in Figures 2.20b, 2.20c, 2.20d, and 2.20e, respectively).[81,82] Across the board, the structures obtained were found to resemble the corresponding complexes derived from receptor **2.44**. As can be seen from the data collected in Table 2.7, the solution-phase binding behaviour of **2.46** is also highly analogous to that of **2.44**.

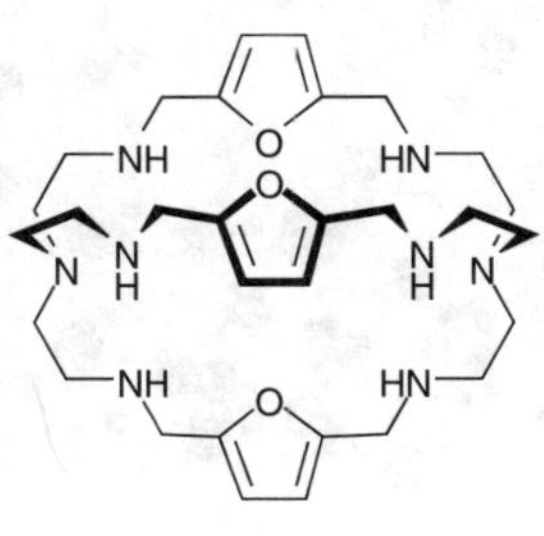

2.46

Replacing the *m*-xylene spacer present in **2.44** with a *p*-xylene spacer generated a slightly larger cavity in the form of receptor **2.47**. This "expansion" allowed for the in-cavity binding of two discrete fluoride anions and one bridging water molecule as determined from a single crystal X-ray-diffraction analysis. As can be seen from an inspection of Figure 2.21, the two fluoride anions reside almost in the centre of the cavity.[86] However, the separation between these fluoride anions and the surrounding amine nitrogen atoms is slightly longer than what one would expect for strong hydrogen bonds (*i.e.*, these distances range from 2.6 to 2.7 Å). Nonetheless, strong binding is seen in solution, as inferred from potentiometric measurements carried out in aqueous solution (*i.e.*, for fluoride anion log K_a = 3.16, 3.24, 3.96, and 6.54 for H_3**2.47**$^{3+}$, H_4**2.47**$^{4+}$, H_5**2.47**$^{5+}$, and H_6**2.47**$^{6+}$,

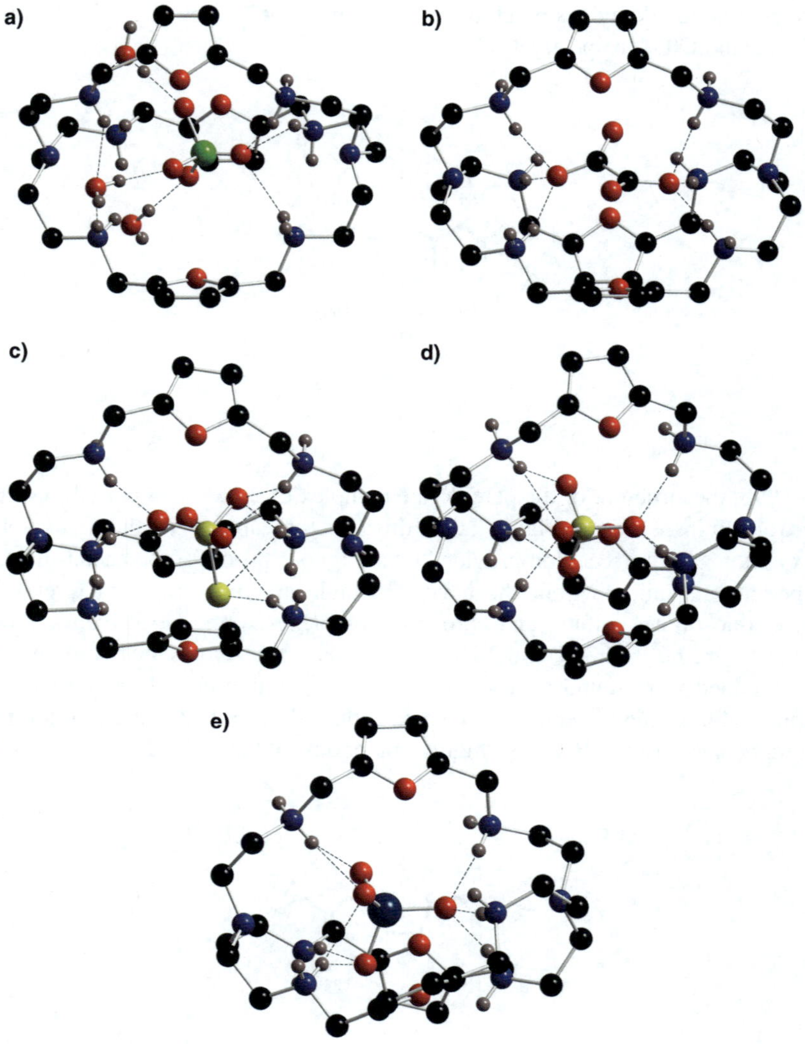

Figure 2.20 *Single crystal X-ray structures of (a) perchlorate, (b) oxalate, (c) thiosulfate, (d) sulfate, (e) and chromate complexes of protonated receptor* ***2.46****. In all cases, further counter anions not proximate to the receptor were found in the crystal lattice. These anions are not shown for the sake of clarity*

Table 2.7 *Association constants (log K_a) corresponding to the interaction between cyclic polyamines* ***2.46*** *and various oxyanions as determined via potentiometry in aqueous 0.1 M NaOTs at 298 K*

	Oxalate^{2-}	*Malonate*$^{2-}$	*Acetate*$^{-}$	*Lactate*$^{-}$	SO_4^{2-}	SeO_4^{2-}	$S_2O_3^{2-}$
H_6**2.46**$^{6+}$	8.30	6.62	3.73	3.50	7.21	7.27	7.65
H_5**2.46**$^{5+}$	7.12	5.55	3.99	3.75	5.21	5.38	5.11
H_4**2.46**$^{4+}$	5.53	5.00	4.01	3.30	4.32	4.53	5.09
H_3**2.46**$^{3+}$	4.82	4.42	3.93	3.13	4.02	4.15	4.56

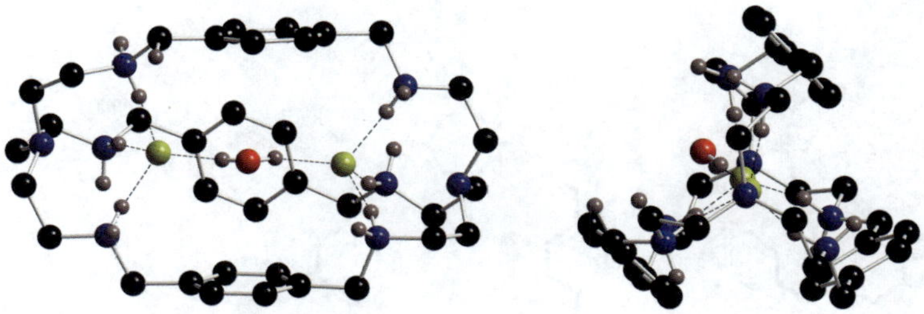

Figure 2.21 *Single crystal X-ray structure of the bis-fluoride aquo complex of protonated* **2.47** *pictured in front (left) and side (right) views*

respectively). Recently, further X-ray crystal structures of halide (*i.e.*, chloride and bromide) complexes of protonated **2.47** were reported by Bowman-James and co-workers.[87]

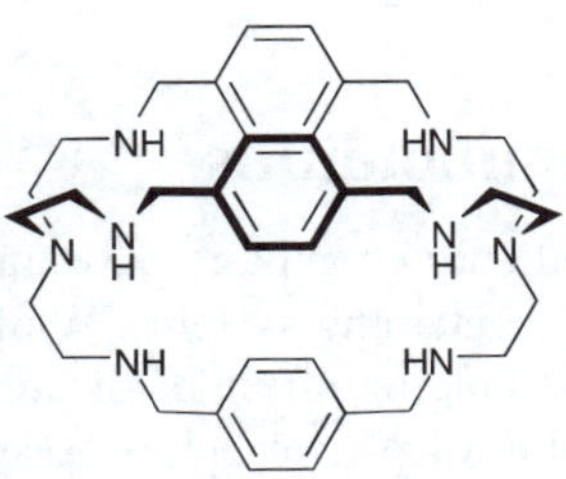

2.47

2.1.4 Polycyclic Systems

Lehn and Fujita prepared the dome-shaped cyclophane-type macrotricyclic receptors **2.48a** and **2.48b** via a three-step synthesis. The advantage of this procedure lies not only in its simplicity, but also in the fact that the size of the cavity can be tuned by varying the lengths of the methylene bridges. Preliminary results indicate that **2.48**, when hexaprotonated, possesses three-fold symmetry and forms a stable complex with nitrate anion.[88]

An even more complicated and extended system, the so-called "kyuphane" (Kyushu + cyclophane) **2.49**, was prepared by Murakami and Snyder in 1991. This cubical system has six faces and was expected to act as a rather rigid cage-type receptor. Kyuphane **2.49** was found to bind to both non-ionic and anionic guest through hydrophobic and electrostatic interactions. At pH 4.0, it exhibits a high affinity for magnesium bis(8-anilinonaphthalene-1-sulfonate) [$Mg(ANS)_2$], Orange G (disodisodium 7-hydroxy-8-phenylzaonaphthalene-1,3-disulfonate) [$Na_2(OG)$], and potassium 6-*p*-toluidinonaphthalene-2-sulfonate [K(TNS)] (*i.e.*, association constants, K_a, of over 10^5 M^{-1}), with selectivity over disodium naphthalene disulfonate (*i.e.*, disodium naphthalene-1,5-disulfonate and disodium naphthalene-2,7-disulfonate) and monosulfonate (*i.e.*, sodium naphthalene-1-sulfonate and sodium naphthalene-2-sulfonate).[89,90]

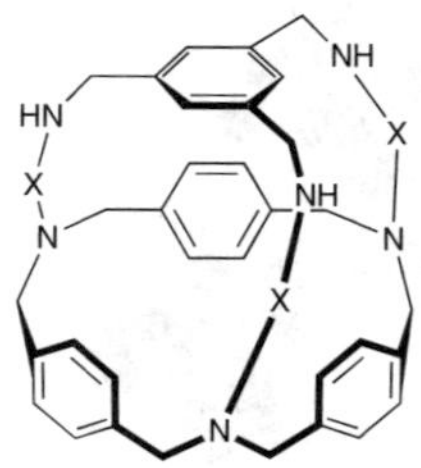

2.48a X = $(CH_2)_3$
2.48b X = $(CH_2)_4$

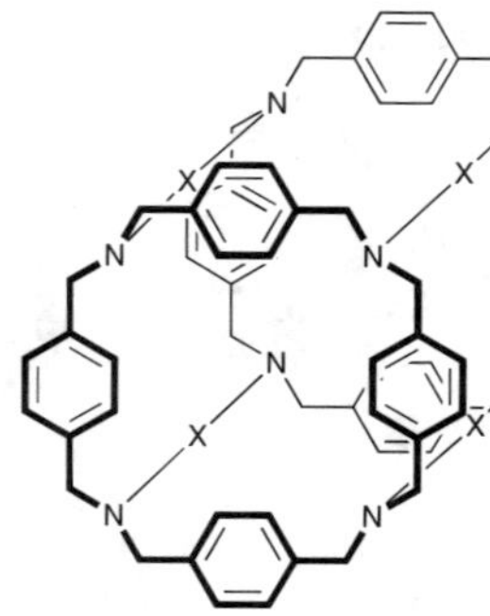

2.49 X =

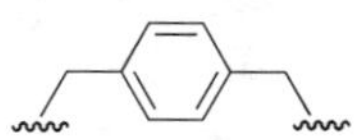

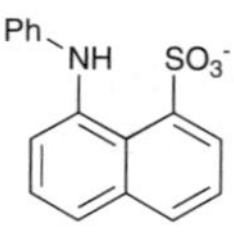

ANS

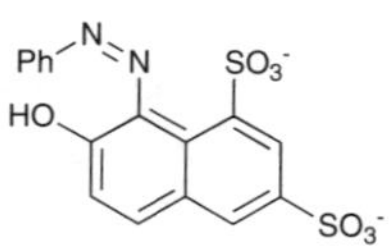

OG

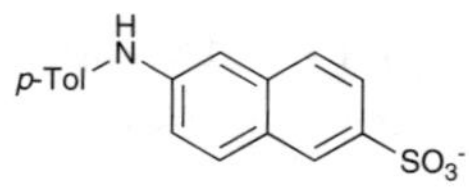

TNS

2.2 Quaternary Ammoniums

Quaternary ammonium-based anion receptors have unique advantages when compared with other ammonium-containing systems. While, the extent of protonation, net charge, and overall anion-binding affinities of ammonium-based receptors are generally a significant function of pH, receptors based on quaternary ammonium groups usually display binding behaviour that is pH independent as a consequence of their intrinsic positive charge. On the other hand, these systems do suffer from the disadvantage that they are unable to interact with bound substrates via hydrogen bonds; they simply lack the NH donor sites.

2.2.1 Linear Systems

In 1981, Breslow and co-workers[91] demonstrated the selective functionalization of flexible biscarboxylate chains ion-paired to the doubly functionalized quaternary ammonium salt (producing the pre-organized ensemble **2.50**). In this case, photolysis led to the production of **2.51** with good selectivity, a result that was thought to reflect the favourable geometrical matching of the two reactants via complementary electrostatic interactions (Scheme 2.4).

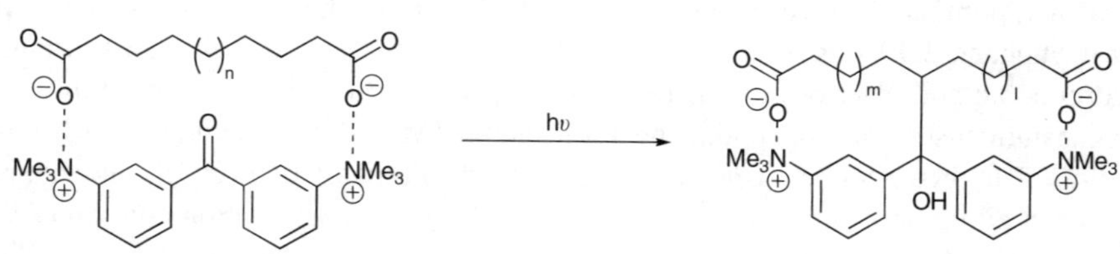

2.50 **2.51**

Scheme 2.4 *Photolysis of compound **2.50***

Almost at the same time, Tabushi and co-workers reported that the lipophilic diammonium cation **2.52** acted as an efficient and selective phase-transfer agent, as well as an extractant for nucleotides. In both cases, the anions were thought to form complexes with the two cationic sites present in **2.52** producing an overall neutral salt that was then able to move from the aqueous layer to the organic phase. In the case of the adenosine phosphates, it was found that the extent of extraction generally increased in the order of AMP $<$ ADP $<$ ATP.[92] With chloroform as the organic phase, maximal selectivity for ATP/AMP and ADP/AMP was seen at pH 3 (extraction ratios of 7500 and 45, respectively). Furthermore, transport experiments, involving a so-called Pressman cell (a system described in greater detail in Chapter 3), revealed that the rate of transport was correlated with the number of phosphate units present on the adenosine core. For instance, under typical conditions, selectivity ratios of 60 for ATP/AMP and 51 for ADP/AMP were observed.

In an effort to improve the transport efficiency and to increase its selectivity, the di- and tricationic systems **2.53** and **2.54** were prepared by Diederich and Li[93] in 1992. Their ability to effect the transport of ATP, cytidine 5′-triphosphate (CTP), 2′,3′-dideoxythymidine 5′-triphosphate (ddTTP), and 3′-azido-2′-deoxythymidine 5′-triphosphate (AZTTP) was studied using a U-tube-type Pressman cell with a chloroform "membrane". Based on these studies, compound **2.53** was found to be the more efficient carrier for all anionic substrates tested ($k = 7.1 \times 10^{-9}$, 8.3×10^{-9}, 10.0×10^{-9}, and 3.7×10^{-9} mol cm^{-2} h^{-1} for ATP, CTP, ddTTP, and AZTTP, respectively). Concentration-dependent extraction studies provided support for the notion that **2.53** permits the formation of a 2:1 carrier-ATP complex in chloroform. While generally inferior to **2.53** as a carrier, system **2.54** did allow for the rather selective transport of ddTTP ($k = 11.8 \times 10^{-9}$ mol cm^{-2} h^{-1}) under these same model conditions.

The tetracationic compound **2.55**, containing two diquaternary 1,4-diaza[2.2.2]-bicyclooocatane (DABCO) subunits, was also studied. This system forms a 1:1 complex with most nucleotide phosphates in dichloromethane. However, it proved rather ineffective as a carrier (the transport rates for **2.55** were approximately 1 magnitude slower than those for carrier **2.53** under identical U-tube model membrane test conditions).[94] Interestingly, studies involving liposomes, rather than a Pressman-type cell, revealed that both **2.53** and **2.55** induced through-membrane leakage of [2,8-^{3}H]ATP not by acting as specific nucleotide triphosphate carriers, but rather by acting as detergents and breaking up the actual liposomal structure. Both **2.53** and **2.55** proved much more effective for this latter, undesirable purpose than the well-known detergent Triton X-100.

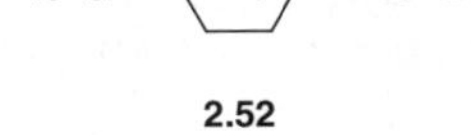

2.52

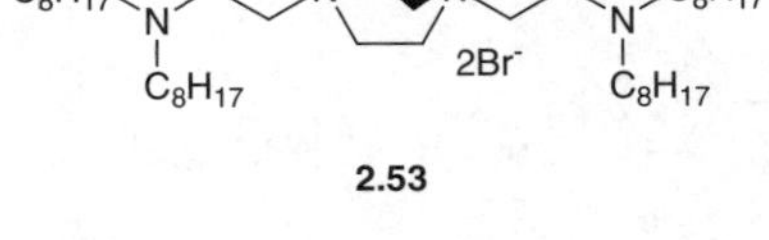

2.53

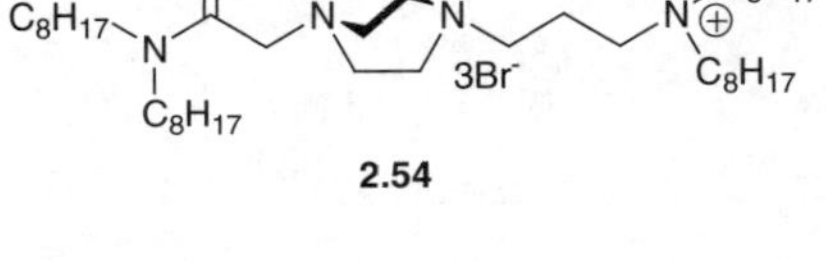

2.54

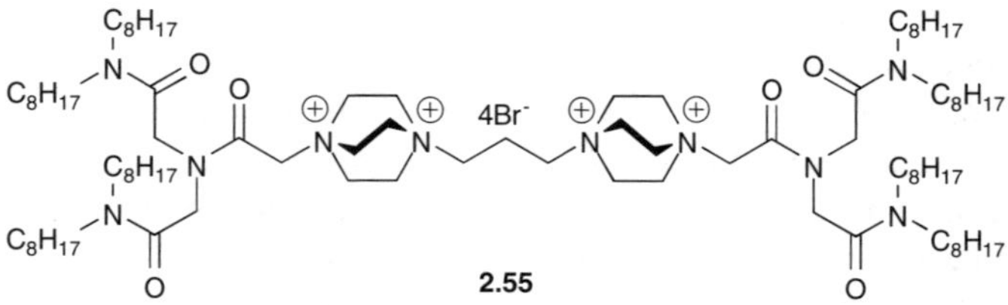

With the stated goal of quantifying the energetic determinants of non-covalent interactions, including both Coulombic and van der Waals terms, Schneider and co-workers[95,96] prepared the quaternary ammonium receptors **2.56–2.60** as their chloride salts and studied their interactions with several anionic guests, namely **A–I** as their sodium salts. The association constants corresponding to the formation of ion-pairs in water were determined by ^{1}H NMR spectroscopic methods and via conductivity experiments. It was found that the dicationic receptors **2.57**, **2.59a**, and **2.59b** interacted well with dianionic guest **F** ($-\Delta G$ values on the order of 12.4–14.3 kJ mol^{-1}). The pyridinium-based dicationic compound **2.58** was also found to form relatively strong ion-pairs with guests **D**, **F**, and **G**, but with little selectivity ($-\Delta G$ values on the order of 15.3–16.3 kJ mol^{-1}). In the case of the tricationic receptor **2.60**, strong ion-pairs were formed with **H** and **I** that are stabilized mostly due to electrostatic effects ($-\Delta G$ = 17.1 and 22.1 kJ mol^{-1} for **H** and **I**, respectively). From these and other systematic analyses, it was concluded that each additional ion-pairing site increases the Coulomb effect by roughly 5 ± 1 kJ mol^{-1} (*i.e.*, per individual salt bridge) in aqueous media.

Recently, Hossain and Schneider[97] studied the effect of flexibility in organic ion-pairs by carrying out NMR spectroscopic titrations in D_2O. It was found that complexes derived from the rigid hosts **2.56** and **2.57** and rigid guests **B** and **A** were slightly more stable than those produced from **2.57** and the flexible guest **C**. Further, in the case of guest **E**, complexes formed from **2.59c** bearing a longer alkyl bridge were found to be less stable than those of **2.59b** with a shorter alkyl bridge; this was taken as support for the notion that good geometrical matching is important in optimizing these kinds of electrostatic interactions. From quantitative studies (*i.e.*, the full set of titration results), it was concluded that a reasonably linear correlation exists between ΔG and the number of flexible single bonds, with the addition of the latter providing an energetic disadvantage of approximately $\Delta\Delta G$ = 0.5 kJ mol^{-1} per single bond. Ionic strength effects were also tested but not found to be of significance under the aqueous conditions of these experiments.

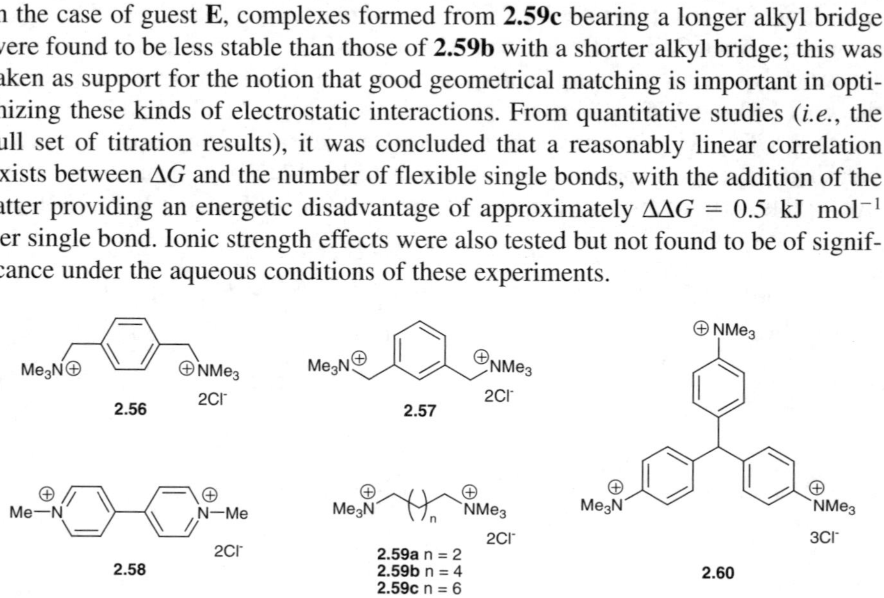

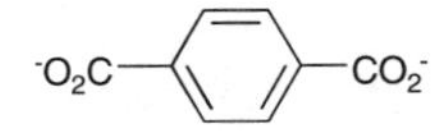

A

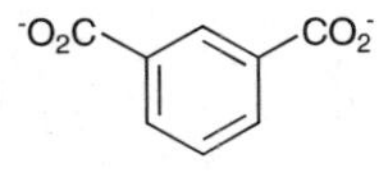

B

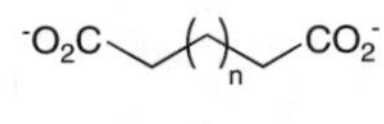

C n = 2

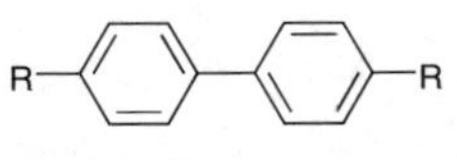

D R = SO_3^-
E R = CO_2^-

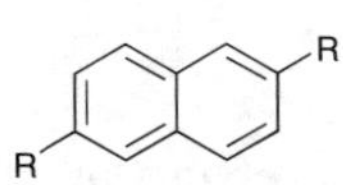

F R = SO_3^-
G R = CO_2^-

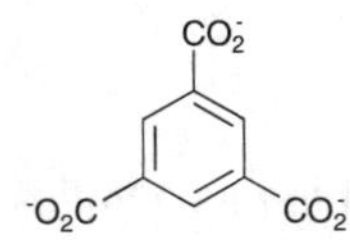

H

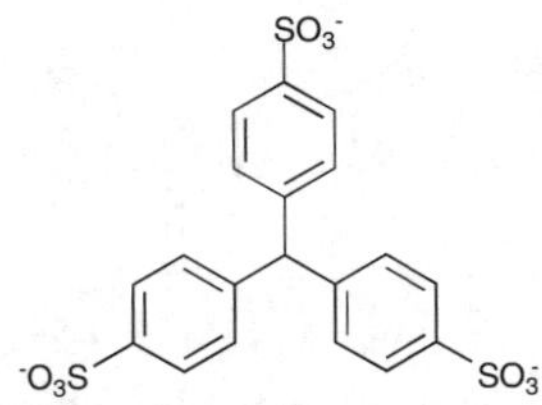

I

The new acyclic quaternary polypyridinium receptors **2.61–2.63** were prepared by Beer and co-workers[98] in 1992. A preliminary assessment of their anion-binding behaviour was made via ^{1}H NMR and ^{13}C NMR spectroscopy. The addition of tetra-butylammonium chloride, TBA-Cl, to these three receptors in DMSO-d_6 induced a large downfield shift in the CH proton resonance due to a change in the electrostatic environment of the receptors, as well as induced changes in the conformation. Significant shifts in the carbon resonances, up to 1.11 ppm, were also observed, which were put forward as additional evidence in support of anion binding.

Four years later, the results of quantitative anion-binding analyses involving **2.61–2.63** and the new receptors **2.64** and **2.65** were reported.[99] Here, both NMR spectroscopic and electrochemical methods were used. From the NMR spectroscopic studies, the pyridinium-based receptors, **2.62**, **2.63**, and **2.65**, were found to display low chloride-anion affinities (association constants, K_a, in DMSO-d_6 of 30–40 M^{-1} for a 1:1 binding process). In contrast, receptor **2.64**, which includes three extra three amide subunits, was found to bind to chloride anion reasonably well (K_a = 110 M^{-1} in DMSO-d_6), being the best in this series of receptors. From the electrochemical analyses, carried out in acetonitrile using cyclovoltammetry (tetra-butylammonium hexafluorophosphate, TBAPF$_6$, as the supporting electrode), it was found that receptors **2.62** and **2.65** undergo a reversible two-wave reduction process, whereas receptor **2.63** exhibits two irreversible reduction waves. In the presence of 10 equiv. chloride anion, the first reduction wave undergoes a significant cathodic shift (up to ΔE = 50, 50, and 130 mV in the case of **2.62**, **2.65**, and **2.63**, respectively), as would be expected for a system engaged in anion binding.

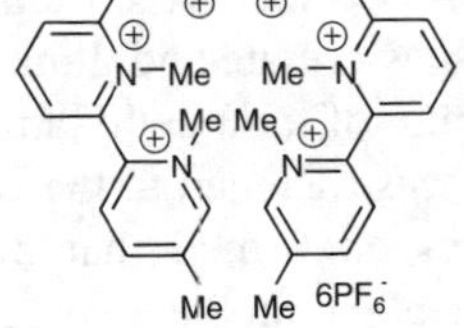

2.61

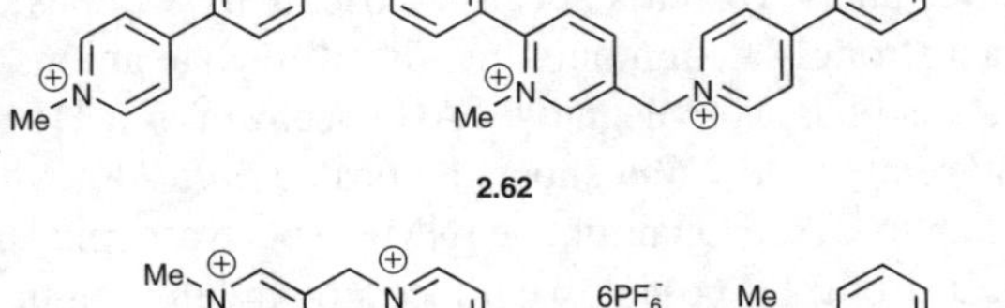

2.62

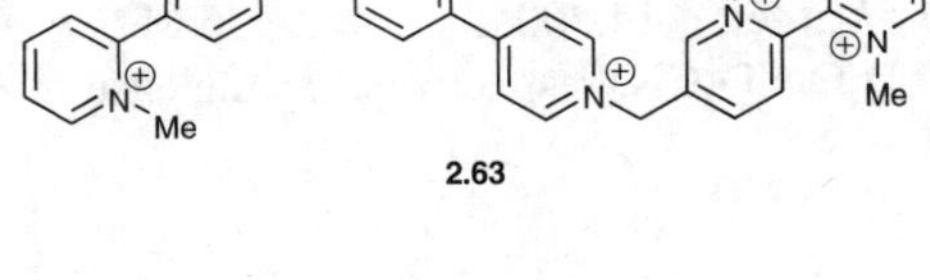

2.63

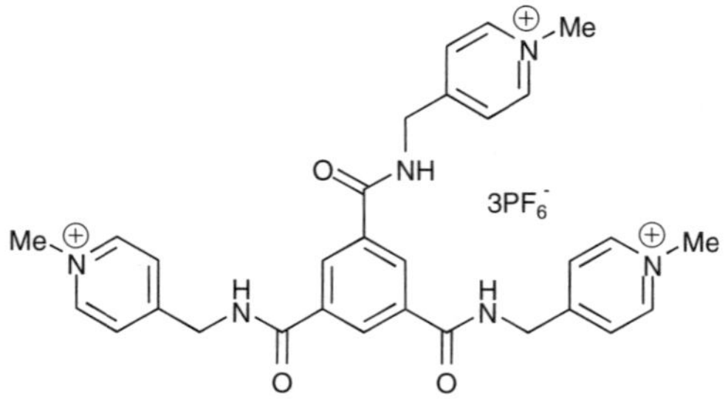

2.64

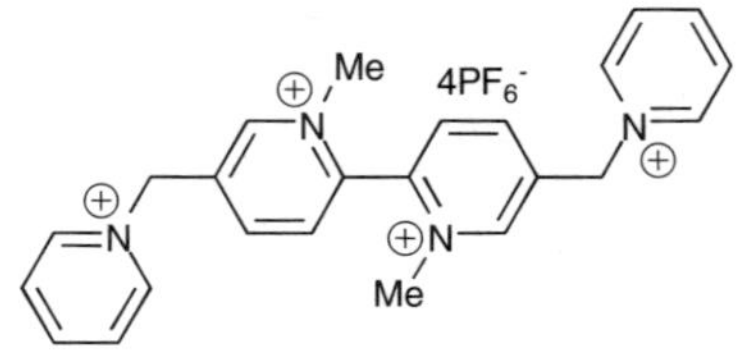

2.65

More systematic anion-binding studies involving pyridinium-based receptors **2.66–2.68** were reported by Jeong and Cho[100] in 1997. From ^{1}H NMR spectroscopic titrations carried out in DMSO-d_6, it was concluded that the association constant (K_a) for the binding of benzoate anion by **2.66** is *ca.* 300 M^{-1}. This value is much higher than that recorded in the case of the corresponding neutral 3-(acetylamino)pyridine control compound (*i.e.*, K_a = 16 M^{-1}). Presumably, this increased affinity reflects the presence of electrostatic interactions between the pyridinium receptor and the anionic carboxylate substrate, as well as the increased hydrogen donor ability of the amide NH proton that arises as the result of pyridine *N*-methylation. Not surprisingly, the bispyridinium receptors, **2.67a**, **2.67b**, and **2.68**, were found to form strong complexes ($K_a > 10^3$ M^{-1}) with dicarboxylate anions, such as adipate, in 10% D_2O/DMSO-d_6. Under these conditions, the monomeric receptor **2.66** was found to bind to butyrate, the monoanionic "analogue" of adipate, much less well (K_a = 30 M^{-1}). Based on these results, it was suggested that the two pyridinium subunits interact cooperatively in binding adipate. Further examples of pyridinium-based anion receptors are listed in the References.[101]

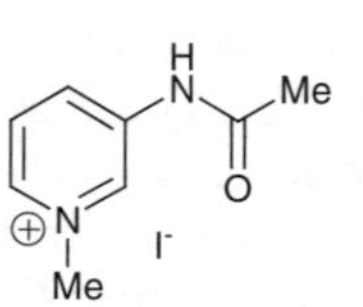

2.66

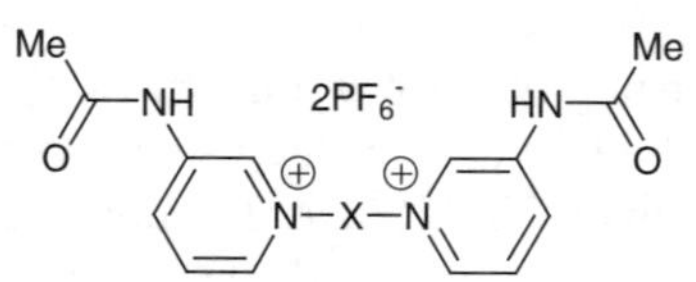

2.67a X = -$(CH_2)_5$-
2.67b X = -CH_2(*p*-C_6H_4)CH_2-

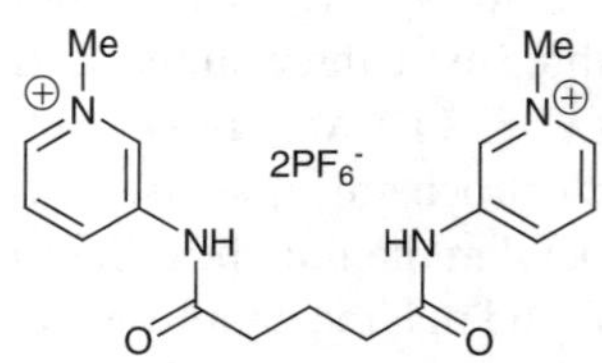

2.68

In separate work by Costa and co-workers,[102] the two squaramido-based receptors **2.69a** and **2.70** were prepared. These systems, reported in 1998, were designed to have a rigid square-shaped framework. As determined from NMR spectroscopic titrations and FAB mass spectrometric analyses, these two receptors were found to interact strongly with mono- and dicarboxylate anions such as acetate and glutarate. In the gas phase, the negative FAB spectrum of a 1:1 mixture of tetramethylammonium acetate with **2.69a** showed a peak at *m*/*z* 346, which was assigned to the complex **2.69a**·OAc. Quantitative NMR spectroscopic studies confirmed that **2.69a** forms a strong 1:1 complex with tetramethylammonium acetate, with the association constants, K_a, being 14,200, 311, and 396 M^{-1} in DMSO-d_6, 10% D_2O/DMSO-d_6, and 10% D_2O/CD_3CN, respectively. In the case of **2.70** containing two squaramido units linked by an *m*-xylene spacer, selective binding of dicarboxylate anions was observed. In fact, the association constant corresponding to the interaction between **2.70** and $(TBA)_2$ glutarate proved too large to determine accurately in 10% D_2O/DMSO-d_6 ($K_a \approx 3.6 \times 10^5\ M^{-1}$). Therefore, a more polar mixture consisting of 30% D_2O/CD_3CN was used; under these conditions a K_a value of $5.6 \times 10^2\ M^{-1}$ was recorded.

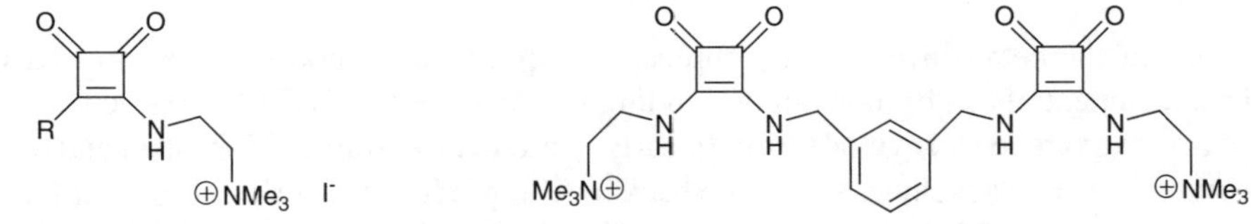

2.69a R = NHBn
2.69b R = NHBu

2.70

Four years later, the interactions between the squaramide receptors **2.69b**, **2.70**, and **2.71a** and representative carboxylate anions were characterized by isothermal titration calorimetry (ITC).[103] In pure DMSO, such analyses indicated that in all cases anion complexation was both enthalpically and entropically favoured. On the other hand, a corresponding thermodynamic analysis of **2.70** and **2.71a**, carried out in methanol, revealed a carboxylate-anion-binding process that was enthalpically unfavoured but entropically favoured, with the later term predominating. This disparate binding behaviour was taken as a clear indication that the solvent is involved in the equilibrium process. In DMSO, **2.69b** was found to bind to tetramethylammonium acetate strongly (K_a = 14,000 M^{-1}) but to TBA *p*-nitrobenzoate weakly (K_a = 498 M^{-1}). Under these same conditions, receptor **2.70** shows a preference for isophthalate over oxalate (K_a = 75,000 and 31,000 M^{-1} for isophthalate and oxalate, respectively). In the case of **2.71a**, the following stability order (K_a) was found: squarate (250,000 M^{-1}) > oxalate (19,000 M^{-1}) > isophthalate (15,000 M^{-1}) (TBA salts) in DMSO. However, in methanol a preference for oxalate (39,000 M^{-1}) over squarate (26,000 M^{-1}) was observed. Again, these findings serve to highlight the fact that the effect of solvent can be significant.

Quite recently, Costa and co-workers[104] used a colorimetic complexation-induced displacement approach to evaluate the relative selectivities of **2.71b** and **2.72** towards SO_4^{2-} and HPO_4^{2-}. In this study, Cresol Red was chosen as the signalling agent (*i.e.*, the dye being displaced), with the result that EtOH–H_2O (9:1; v/v) solutions of **2.71b** and **2.72** containing Cresol Red, presumably tied up in the form of a complex with these receptors, changed from purple (λ_{max} = 580 nm) to yellow (λ_{max} = 428 nm) upon the addition of sulfate and hydrogen phosphate anions. However, the addition of carbonate, nitrate, nitrite, fluoride, chloride, bromide, and iodide anions failed to produce any detectable spectral change. Based on competitive colorimetric titration studies, it was concluded that receptor **2.72** exhibits a moderate selectivity for sulfate over hydrogen phosphate ($K_{rel}(SO_4^{2-}/HPO_4^{2-})$ = 2.28), while **2.71b** is not particularly selective ($K_{rel}(SO_4^{2-}/HPO_4^{2-})$ = 1.06).

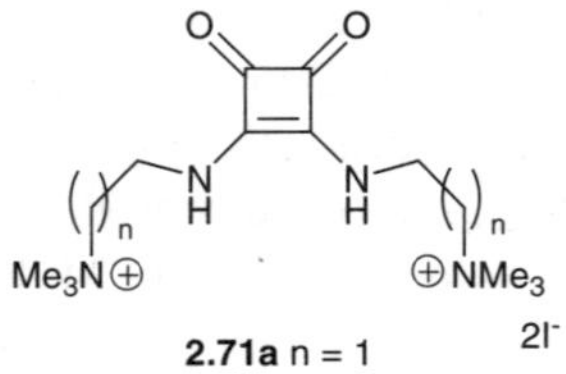

2.71a n = 1
2.71b n = 2

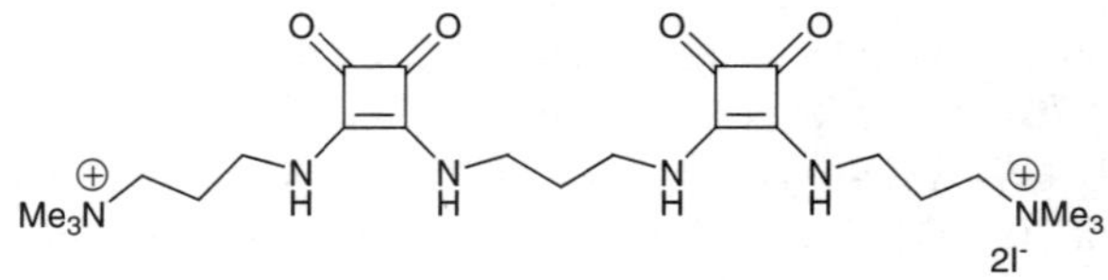

2.72

One of the better-investigated approaches to producing anion receptors involves the attachment of one or more anion-binding site(s) to well-known rigid frameworks, such as calixarene and cyclodextrin. In early work, dating from 1989, Matsui and co-workers[105] used α-cyclodextrin as the backbone to produce the series of pyridinium-based receptors **2.73**. In aqueous solution, the HCO_3^- salts of receptors **2.73a–2.73d** were found to form strong 1:1 charge-transfer complexes with bromide, iodide, and isothiocyanate anions via what was presumed to be a combination of hydrophobic, van der Waals, and electrostatic interactions. Quantitative UV–Vis titrations were carried out. Based on such analyses, also carried out in aqueous media, it was concluded that (1) the selectivity order for all receptors was $I^- > SCN^- > Br^-$ and (2) the anion affinities increase as the number of positively charged pyridinium subunits is increased (*e.g.*, for KI, $K_a = 1.36 \times 10^2$, 1.32×10^3, 1.11×10^4, and 9.0×10^5 M^{-1} for **2.73a**, **2.73b**, **2.73c**, and **2.73d**, respectively).

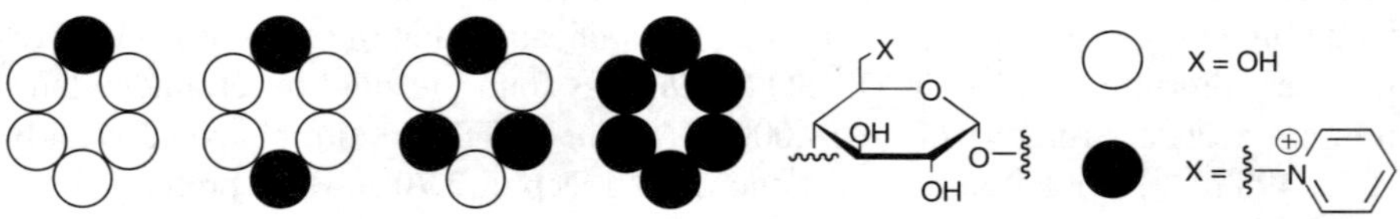

2.73a **2.73b** **2.73c** **2.73d**

Two years later, Hong and co-workers[106] reported a rather unique water-soluble receptor **2.74** that was found to complex anions in aqueous media. This receptor was synthesized by reacting bromoresorcin[4]arene with hexamethylenetetramine. The resulting bowl-shaped cavitand receptor **2.74** was found to exhibit a high affinity for the aromatic carboxylate **J** (K_a = 12,600 M^{-1} in D_2O), demonstrating a selectivity for this species relative to the other anions studied.[106] Based on the ^1H NMR spectroscopic analyses, it was concluded that the methoxy group of **J** is located within the internal cavity of **2.74**, as shown in Scheme 2.5.

2.2.2 Monocyclic Systems

The tetrasulfonium-paracyclophane **2.75** is a classic model systems for ammoninum-based macrocycles. It was prepared by the late Iwao Tabushi and co-workers via the methylation of the corresponding neutral tetrasulfur-bridged cyclophane with Me_3OBF_4. Space-filling models indicated that receptor **2.75** has a square hydrophobic cavity that was expected to allow for the inclusion of hydrophobic guests. In fact, the addition of **2.75** to an aqueous solution of 8-anilinonaphthalene-1-sulfonate (1,8-ANS) led to a

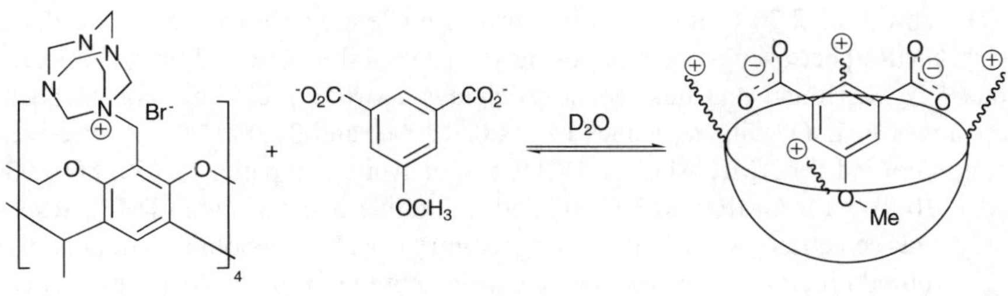

2.74 J 2.74-J

Scheme 2.5 *Proposed binding mode of* ***2.74*** *with* ***J***

strong enhancement in the fluorescence intensity of the ANS. Benesi-Hildebrand analysis (*i.e.*, a plot of fluorescence intensity *vs*. concentration of **2.75**) produced a straight line, from which an association constant of $K_a = 1.6 \times 10^3\ M^{-1}$ was derived.[107]

Schneider and co-workers studied the corresponding quaternary ammonium receptor **2.76** using NMR, fluorescence, and UV–Vis spectroscopic titrations. The use of different spectroscopic methods was attributed to their limited sensitivity; NMR spectroscopy was used for $K_a < 10^4\ M^{-1}$, UV–Vis spectroscopy for $10^2\ M^{-1} < K_a < 10^5\ M^{-1}$, and fluorescence spectroscopy for $10^5\ M^{-1} < K_a < 10^7\ M^{-1}$. Increasing the water content in organic solvents such as ethanol and methanol was found to increase the affinity of **2.76** for both ANS ($K_a = 3.54 \times 10^5$, 5.5×10^4, and $1.94 \times 10^4\ M^{-1}$ in H_2O, 20% EtOH, and 60% EtOH, respectively) and 2,4-dinitronaphthalen-1-olate (DNNO) ($K_a = 2.11 \times 10^4$, 1.52×10^3, and $2.6 \times 10^2\ M^{-1}$ in H_2O, 50% MeOH, and 80% MeOH, respectively).[108] In light of the fact that the effect of electrostatic binding was expected to decrease as the percentage of water was raised, these findings were considered to support the assumption that van der Waals forces play an important role in regulating the binding process, even if the latter is often dominated by electrostatic interactions.

One year later, complexation studies involving **2.76** and aliphatic (*e.g.*, cyclohexane carboxylate and 2-cyclohexylacetate) and aromatic (*e.g.*, benzoate and 2-phenylacetate) anions, chosen to be similar in size and shape, were carried out. Taken in concert, these analyses served to demonstrate that the binding of aromatic anions is favoured by a factor of up to 60.[109] Schneider[95,110] also demonstrated that neutral aromatic naphthalene derivatives are bound with greater affinity than analogous non-aromatic guests. Again, these results were taken as a further indication that van der Waals interactions are important in stabilizing various host–guest complexes derived from **2.76**.

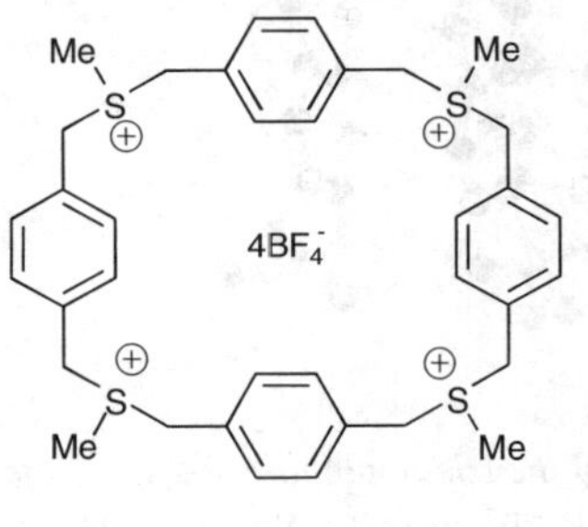

2.75

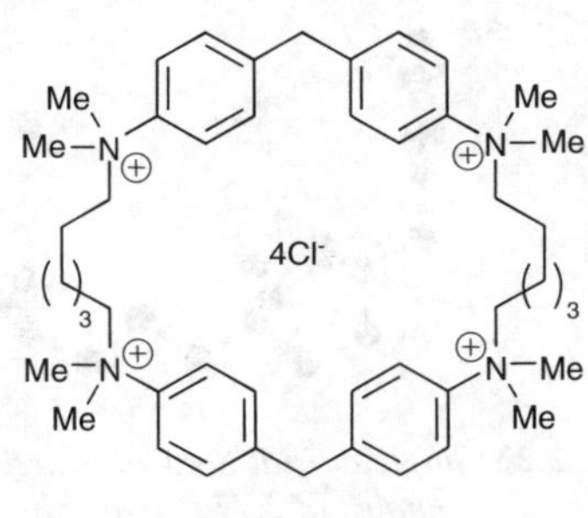

2.76

The ability of **2.76** to interact with various nucleotide phosphates was assessed using NMR spectroscopy by monitoring the peak shifts observed upon complexation.[111] It was found that this receptor displays a strong preference for adenosine derivatives in D_2O, with K_a values of 1900, 13000, and 37,000 M^{-1}, respectively, being recorded for AMP, ADP, and ATP, respectively, compared to 450, 800, 930, and 1210 M^{-1} for GMP, UMP, CMP, and thymidine-5′-phosphate (TMP), respectively. This selectivity was ascribed to a geometry (*i.e.*, hydrophobic-binding pocket) that favoured binding of the aromatic adenosine core via van der Waals interactions, thus fine-tuning a recognition process that was otherwise expected to be driven mostly by electrostatics.

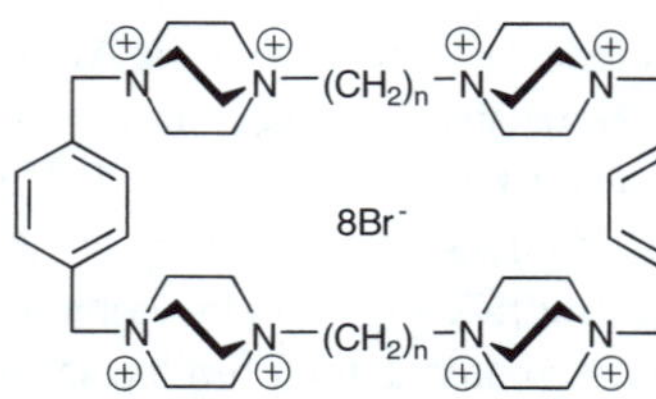

2.77a n = 3
2.77b n = 5
2.77c n = 7

In order to generate a DABCO-based cyclophane with an enhanced propensity to bind to substrates via electrostatic means, the octacationic cyclophanes **2.77** were prepared by Menger and Catlin[112] and reported in 1995. A single crystal X-ray structural analysis of **2.77a**, crystallized in the presence of naphthalene-2,7-disulfonate (NDS), reveals the presence of three large NDS anions that do not bind inside cavity, but rather interact with the exterior of the host through a combination of π–π stacking and/or electrostatic interactions (Figure 2.22). Interestingly, two chloride counterions are found within the cyclophane core. In aqueous solution, quantitative studies of **2.77a** and test substrates benzenesulfonate, naphthalene-2-sulfonate, and NDS (K_a = 61, 447, 3390 M^{-1}, respectively) led to the suggestion that the binding process was

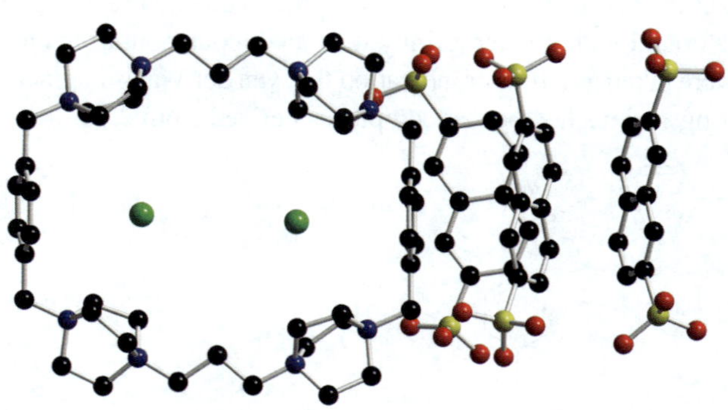

Figure 2.22 *Single crystal X-ray diffraction structure of the tris-naphthalene-2,7-disulfonate complex of **2.77a**. Two chloride counterions are present inside the cavity*

dominated by Coulombic effects. Receptors **2.77a–2.77c** were also found to bind to ATP well at pH 8.0–8.5 with little effect of cavity size being seen on the affinities (K_a = 13,300, 15,900, and 12,100 M^{-1} for **2.77a**, **2.77b**, and **2.77c**, respectively).

Slightly after the above study appeared, a new cyclophane **2.78**, a synthetic mimic of the Vancomycin carboxylate-binding site, was prepared by Diederich and co-workers.[113] This tricationic receptor was found to bind dicarboxylate anions reasonably well in aqueous media (K_a = 230 and 1800 M^{-1} for the disodium salts of malonic acid and maleic acid, respectively, in D_2O) with good selectivity over various monocarboxylate anions, for which the corresponding K_a values were all less than 70 M^{-1} in D_2O.[113] In all cases, ion-pairing interactions between the cationic site of the receptor and the anionic guest were considered to provide the major driving force for host–guest complex formation.

The semi-synthetic alkaloid-based cyclophane **2.79** was prepared by Eliseev and co-workers[114] via the dibenzylation of the natural-occurring compound (*S*,*S*)-(+)-tetrandrine. Within the series of alkyl α,ω-dicarboxylate anions, this receptor shows the greatest selectivity for succinate (K_a = 58 M^{-1} in aqueous 0.05 M NaCl). Higher across-the-board affinities were seen in the case of aromatic dicarboxylates, presumably reflecting the presence of additional π-stacking or hydrophobic interactions. Interestingly, in aqueous media, receptor **2.79** was found to bind *o*-phthalate (K_{a1} = 135 M^{-1} and K_{a2} = 12 M^{-1}) with a 1:2 host:guest ratio but terephthalate (K_a = 110 M^{-1}) in a 1:1 manner. Both species were bound more strongly than isophthalate (K_{a1} = 49 M^{-1}).

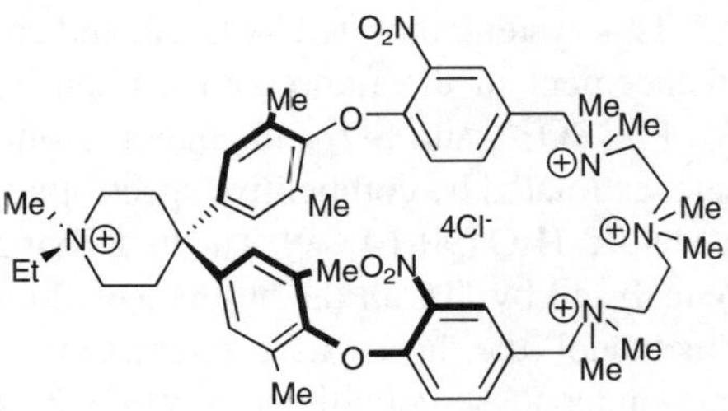

2.78

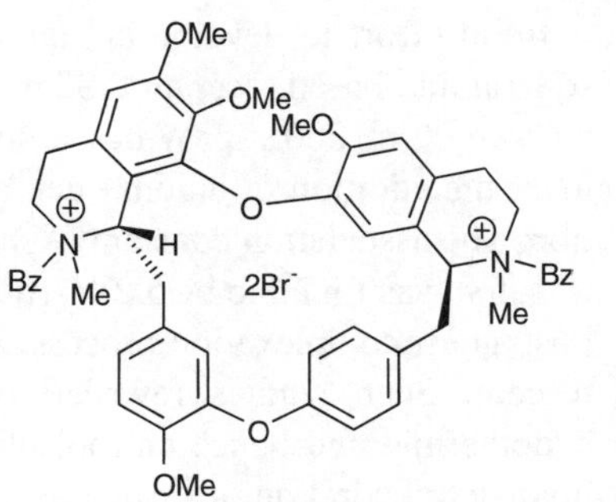

2.79

In 1998, Shinoda and co-workers[115] reported the one-step synthesis and crystal structure of a new type of quaternary tetrapyridinium macrocyle, **2.80**, as well as its anion-binding properties. In the solid state, two bromide counterions are located inside the ring and two are found outside the cavity (Figure 2.23). It was found that the distance between the bromide anions and the pyridine ring were relatively short (*i.e.*, 2.65–2.79 Å) and that **2.80** exists in a 1,2-alternate conformation. In D_2O at pH 7–8, receptor **2.80** was found to bind to rigid tricaboxylate anions (present in their trianionic forms at this pH) with high affinity. Within the test series defined by structures **2.81**, selectivity towards **2.81b** was seen (log K_a = 5.1, 4.5, 4.4, and 4.1 for **2.81b**, **2.81c**, **2.81a**, and **2.81d**, respectively).

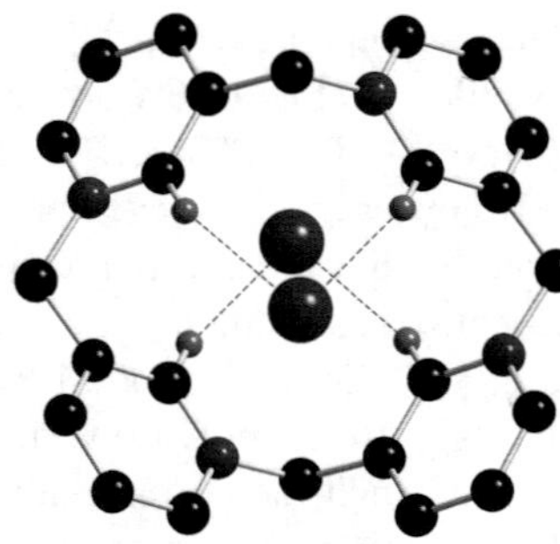

Figure 2.23 *X-ray crystal structures of the bromide complex of* ***2.80****. Two additional bromide anions are found outside the ring; they have been removed for clarity*

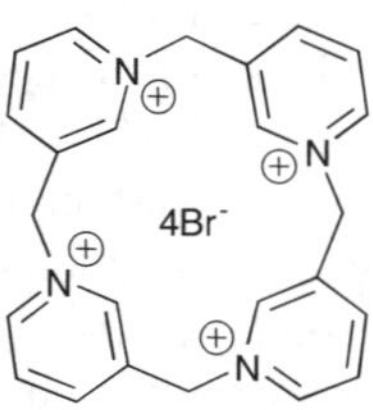

2.80

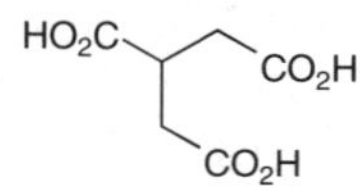

2.81a

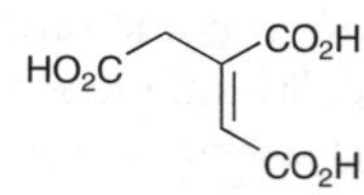

2.81b

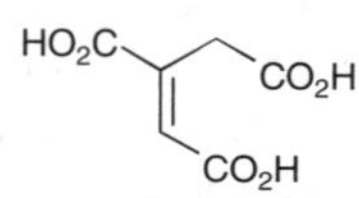

2.81c

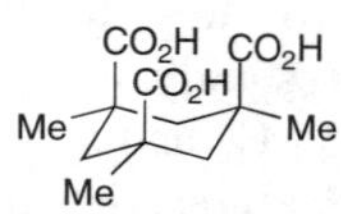

2.81d

In an effort to develop an anion-sensing agent, the fluorescein complex of the squaramide-based receptor **2.82** was studied. This system, reported by Costa and co-workers[116] in 2001, provides a strong enhancement in the fluorescence intensity upon the addition of anions such as $C_2O_4^{2-}$, $PhOPO_3^{2-}$, and SO_4^{2-} in aqueous solution. The association constant (K_a) for sulfate, estimated by competitive spectrophotometry, was found to be $5.2 \times 10^6\ M^{-1}$ in MeOH–H_2O (90:10 v/v). The underlying host–guest complexation process was also analyzed by ITC in the absence of fluorescein. Such studies revealed that in methanol the host–guest interaction is endothermic and hence entropically driven. Under these conditions, receptor **2.82** binds tetrahedral anions, such as sulfate ($K_a = 4.6 \times 10^6\ M^{-1}$), in preference over divalent anions, such as $C_2O_4^{2-}$ ($K_a = 3.2 \times 10^5\ M^{-1}$) and $PhOPO_3^{2-}$ ($K_a = 1.5 \times 10^4\ M^{-1}$). Various test monoanions, such as halides, nitrate, and acetate, were found to be bound with little or no affinity.

In 2003, Bowman-James and co-workers[117] reported a new class of amide/quaternized amine macrocycle, **2.83a** and **2.83b**. In this case, the use of a pyridine or phenyl linkage translates into significantly different anion-binding behaviour and structural features in the solid state. For instance, a single crystal X-ray structure of the chloride-anion adduct of **2.83a** revealed, as shown in Figure 2.24a, that the two cationic sites are located far from each other, presumably as the result of electrostatic repulsion (N···N = 12.744 Å). All chloride counterions are outside the cavity. By contrast, single crystals of **2.83b** grown in the presence of iodide-anions (Figure 2.24b) gave rise to a structure wherein interactions between the amide NH protons

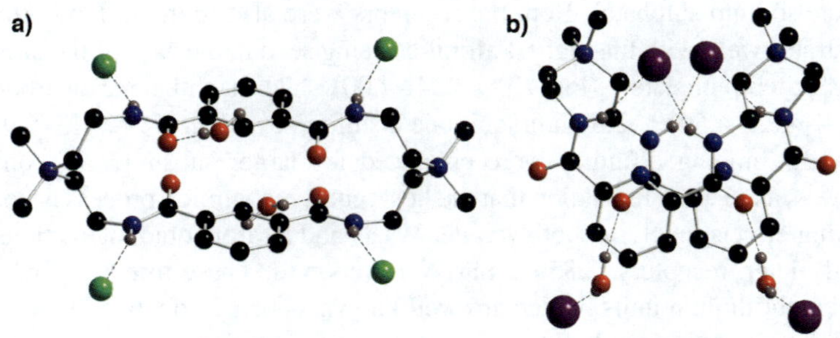

Figure 2.24 *Single crystal X-ray structures of (a) the tetrakis chloride–diaquo adduct of* ***2.83a*** *and (b) tetrakis-iodide–diaquo adduct of* ***2.83b***

and the pyridine nitrogen atoms help stabilize a folded conformation that brings the two cationic sites closer to one another (N···N = 6.527 Å).

The binding properties of **2.83a** and **2.83b** were analyzed in DMSO-d_6 solution using standard NMR spectroscopic titration methods. It was found that, with the exception of fluoride, **2.83b** is a better anion receptor than its phenyl-linked congener **2.83a** for all anions subject to study. Both the receptors proved highly selective for dihydrogen phosphate (log K_a = 5.32 and 4.06 for **2.83b** and **2.83a**, respectively). In addition, receptor **2.83b** displays a higher affinity for chloride anion (log K_a = 4.75) than for bromide (log K_a = 4.38), iodide (log K_a = 2.21), hydrogen sulfate (log K_a = 3.90), nitrate (log K_a = 2.32), or perchlorate (log K_a = 2.40).

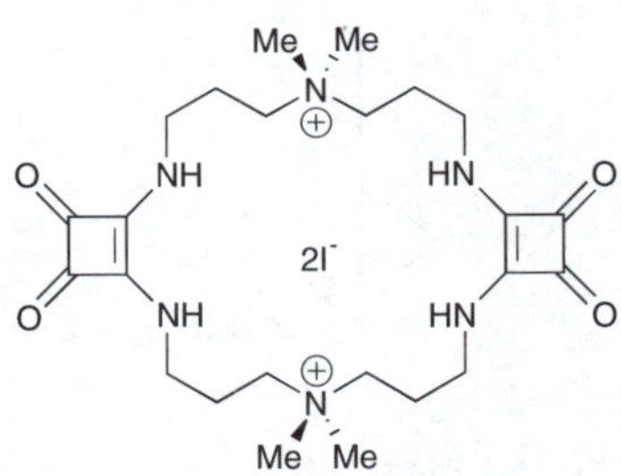

2.82

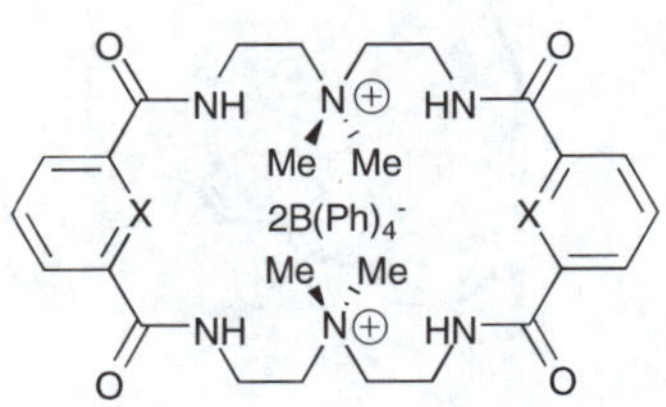

2.83a X = CH
2.83b X = N

2.2.3 Polycyclic Systems

The tricyclic receptors **2.84a–2.84d** were designed by Lehn and co-workers[118] to effect the recognition of planar anionic, as well as neutural guests, such as aromatic carboxylates, nucleosides, and nucleotides through a combination of stacking and/or electrostatic interactions. These rigid receptors were found to form remarkably stable complexes with the neutral nucleoside adenosine (log K_a = 4.00, 3.87, 3.86, and 3.99 for **2.84a**, **2.84b**, **2.84c**, and **2.84d**, respectively) and the corresponding doubly charged monophosphate, AMP (log K_a = 4.08, 3.79, 3.92, and 4.08 for **2.84a**, **2.84b**, **2.84c**, and **2.84d**, respectively) in an aqueous buffer (pH 7.8; 0.01 M/0.01 M sodium cacodylate/sodium sulphate). Both the receptors were also found to bind other large, flat substrates well, with the highest affinities being seen in the case of the largest substrates (*e.g.*, terephthalate^{2-} (log K_a = 3.54–4.00) < 2,6-naphthalene dicarboxylate^{2-} (log K_a = 4.25–4.59) < 2,6-anthraquinone disulfonate^{2-} (log K_a = 5.13–5.95)). The fact that the binding affinities were enhanced for larger substrates of comparable charge, was taken as an indication that the host–guest recognition process is dominated by stacking effects involving both van der Waals and hydrophobic interactions.

Slightly later, receptors, **2.85a–2.85c** were reported. These three systems contain two phenanthridinium units, which are well known as being effective DNA intercalators.[119] The presence of the highly fluorescent phenanthridinium unit also allowed the determination of various association constants via fluorimetric titrations. The results of these binding studies, carried out in an aqueous buffer at pH 6, revealed K_a values on the order of 10^5–10^6 M^{-1} for various nucleotides (*e.g.*, ATP, ADP, AMP, GMP, CMP, UMP, TMP, and adenosine). As in the case of **2.84** above, the neutral substrate adenosine (log K_a = 5.62, 5.12, and 5.35 for **2.85a**, **2.85b**, and **2.85c**, respectively) was found to bind to these three receptors with affinities that are similar to those seen for their charged nucleotide analogues (*e.g.*, log K_a = 5.80 (5.65), 5.98 (5.68), and 5.95 (5.78), respectively, for the interaction of **2.85a**, **2.85b**, and **2.85c** with ATP (ADP)). However, across the board higher association constants were seen for series **2.85** than for **2.84**, a finding that was considered to reflect the superiority of phenanthridinium as a subunit for the construction of water-soluble receptors of this generalized type.

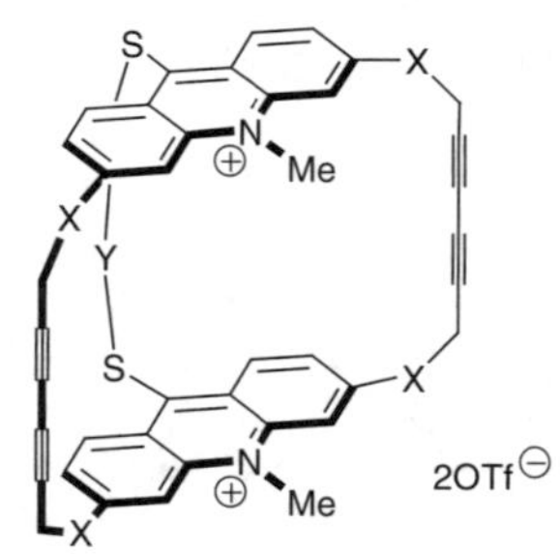

2.84a X = O, Y = -$(CH_2)_6$-
2.84b X = O, Y = -$CH_2CH_2OCH_2CH_2$-
2.84c X = O, Y = -$CH_2CH_2OCH_2CH_2OCH_2CH_2$-
2.84d X = NH, Y = -$CH_2CH_2OCH_2CH_2$-

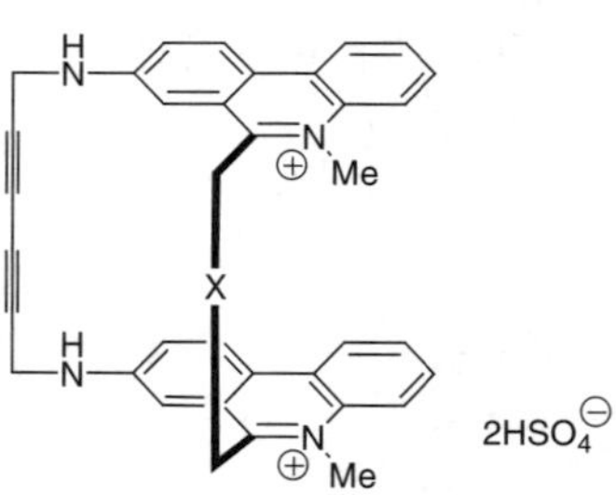

2.85a X = -$(CH_2)_4$-
2.85b X = -$(CH_2)_6$-
2.85c X = *p*-C_6H_4

One of the early pioneers in the area of ammonium-based receptors, Schmidtchen[120] first reported the synthesis and anion-binding properties of a series of macrotricyclic quaternary ammonium-based receptors, namely **2.86a–2.86c**, in 1977. Based on CPK modelling studies, it was proposed that receptor **2.86a**, containing the smallest cavity (diameter = *ca.* 3.9 Å), would be able to accommodate only smaller anions such as chloride (ionic diameter = 3.62 Å) and bromide (ionic diameter = 3.92 Å). By contrast, receptors **2.86b** (diameter = 4.6 Å) and **2.86c** (diameter = 7.6 Å) were expected to have a cavity that was suitable for the uptake of iodide anion (ionic diameter = 4.56 Å). NMR spectroscopic titration studies, carried out in aqueous solution, revealed that these three receptors exhibit considerable selectivity towards bromide anion (log K_a = 1.8, 2.45, and 2.45 for **2.86a**, **2.86b**, and **2.86c**, respectively) and iodide (log K_a = 2.2 and 2.4 for **2.86b** and **2.86c**, respectively) over chloride anion (log K_a = 1.0, 1.3, and < 0.5 for **2.86a**, **2.86b**, and **2.86c**, respectively). Based on ^{35}Cl and ^{19}F NMR spectroscopic studies, it was proposed that these halide anions are bound within the central cavity, a conclusion supported by a single crystal X-ray structure of the iodide complex (*cf.* Figure 1.11).[121,122] Later on, further synthetic details were reported, along with the results of in-depth anion-binding studies involving halides, carboxylates, and phosphates.[123,124]

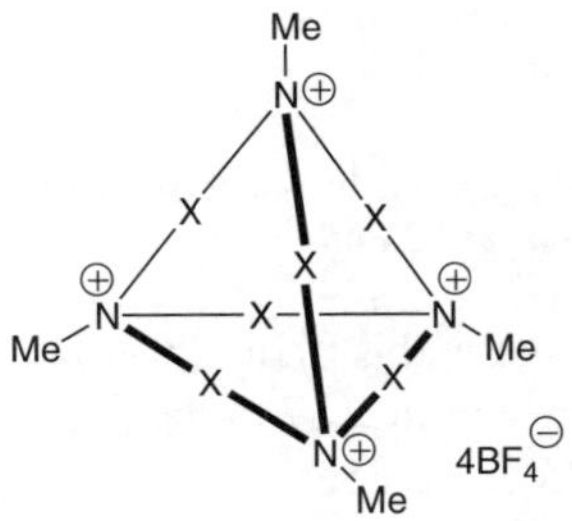

2.86a X = -$(CH_2)_2O(CH_2)_2$-
2.86b X = -$(CH_2)_6$-
2.86c X = -$(CH_2)_8$-

Inspired by natural enzymes that achieve extraordinary specificity and catalytic activity by binding substrates within what are often very sophisticated binding pockets, Schmidtchen[125] applied two synthetic macrotricyclic receptors, namely **2.86a** and **2.86b**, to the anion-mediated reactions shown in Scheme 2.6a. In the presence of receptor **2.86a**, the reaction of fluoro(dinitro)benzene with azide in an aqueous environment consisting of 25% methanol in water, was found to be slower in the absence of the receptor. This enhancement effect was even more significant in the case of compound **2.86b**; here, k_{cat}/k_{un} = 48.[126] The fact that **2.86b** proved superior, led to the hypothesis that while receptor **2.86a** forms a stronger complex with azide (K_a = 500 M^{-1}), receptor **2.86b** (K_a = 83 M^{-1}) is better able to stabilize polarizable species such as those presumably involved in the transition state of the reaction. Follow-up studies involving other similar substrates confirmed that **2.86b** is a better enzyme mimic than **2.86a**, at least for these kinds of reactions.[126,127] In related work, it was found that the rate of decarboxylation of 6-nitrobenisoxazole-3-carboxylate (*cf.* Scheme 2.6b) was found to be augmented in the presence of macrobicycle **2.86b** (k_{cat}/k_{un} = 110).[128]

Connecting two macrotricyclic ammonium receptors of different sizes by a *p*-xylene bridge produced the ditopic receptor **2.87**.[129] The ability of this receptor to bind dianionic guests of structure **2.88a–2.88e** was probed through UV spectroscopic titrations carried out in water. Such studies revealed that the affinities for typical ditopic anions increased sharply as the size of the guests increased (K_a = 714, 322, 2041, 5265, and 10,000 M^{-1} for **2.88a**, **2.88b**, **2.88c**, **2.88d**, and **2.88e**, respectively). On this basis, it was inferred that guests, **2.88d** and **2.88e**, provide the best geometrical match for this ditopic receptor (*i.e.*, **2.87**) and that both of the binding subunits participate in guest recognition. A comparison of the association constants with those for the analogous monotopic receptor, system **2.86b**, revealed that the ditopic receptor

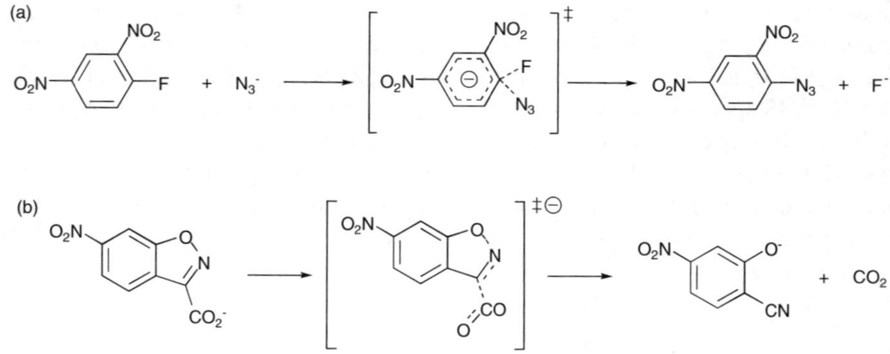

Scheme 2.6 *This substitution reaction is accelerated by quaternary ammonium-based receptors **2.86a** and **2.86b***

2.87 is roughly three times more effective. Again, this finding was taken as support for the notion that both recognition subunits present in **2.87** participate in the binding of appropriately sized substrates, such as **2.88d** and **2.88e**.

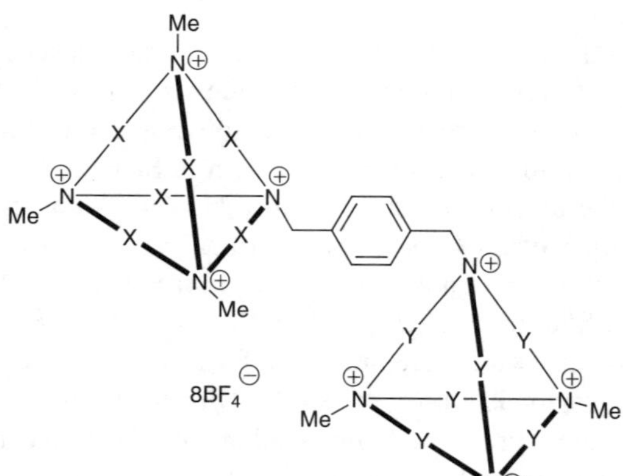

2.87 X = $-(CH_2)_6-$, Y = $-(CH_2)_8-$

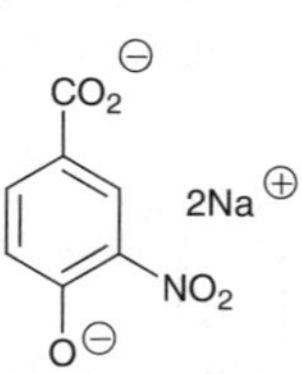

2.88a

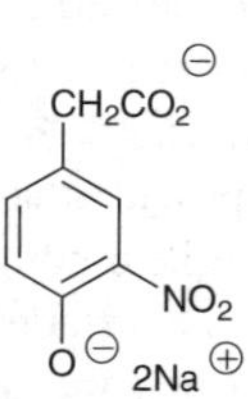

2.88b

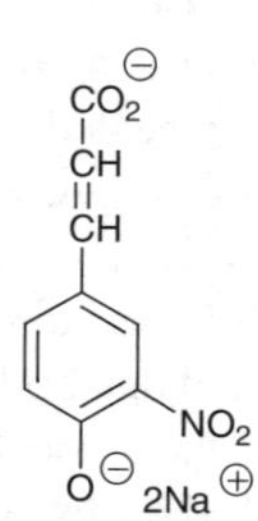

2.88c

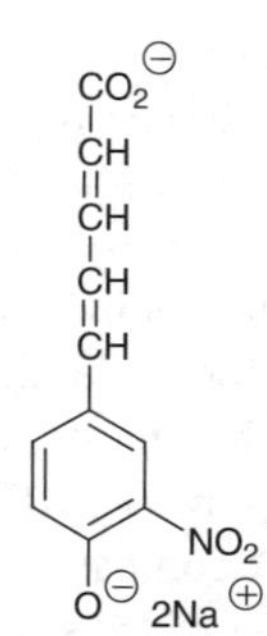

2.88d

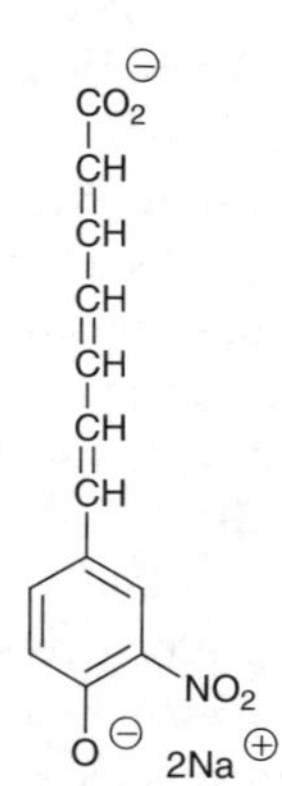

2.88e

Ichikawa and Hossain[122] have also studied a range of polyammonium receptors, including **2.89–2.90**. In the case of receptor **2.89a**, having the smallest cavity, fluoride was found to bind with a $K_a = 1.5 \times 10^4$ M^{-1} in water, while no apparent chloride anion binding was observed.[122] By contrast, the slightly larger system, **2.89b**, yielded X-ray quality crystals of the corresponding chloride anion complex. As seen in Figure 2.25a, in this case, the chloride anion is held in the middle of cavity, presumably, as the result of electrostatic interactions involving the four positive ammonium centres (the Cl···N bond distances range from 3.84 to 4.45 Å).[130]

Three years later, the crystal structure of the bromide complex of **2.90b** was solved (Figure 2.25b). It was found that the receptor–halide bond distances were between 0.01 and 0.20 Å longer than in the case of the chloride complex of **2.89b**.[131] Quantitative NMR spectroscopic studies, carried out in D_2O using analogue **2.90a**, revealed a preference for bromide (K_a = 990 M^{-1}) over chloride (K_a = 110 M^{-1}), as well as over fluoride and iodide (no evidence of interaction in the latter two cases). Quantitative NMR spectroscopic titration results, carried out in acetonitrile-d_3 using analogue **2.90c** containing a methyl group, revealed a selectivity for chloride ($K_a > 10^4$ M^{-1}) over bromide ($K_a \approx 340$ M^{-1}).[132]

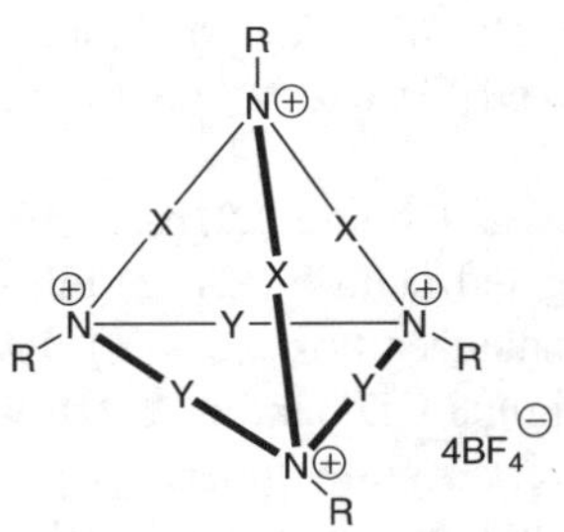

2.89a X = Y = -$(CH_2)_5$-, R = Me
2.89b X = -$(CH_2)_5$-, Y = -$(CH_2)_6$-, R = Bz

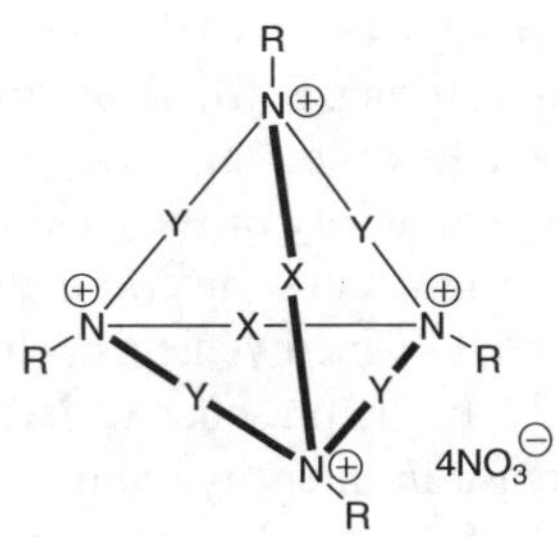

2.90a X = -$(CH_2)_5$-, Y = -$(CH_2)_6$-, R = Me
2.90b X = -$(CH_2)_5$-, Y = -$(CH_2)_6$-, R = Bz
2.90c X = -$(CH_2)_6$-, Y = -$(CH_2)_5$-, R = Me

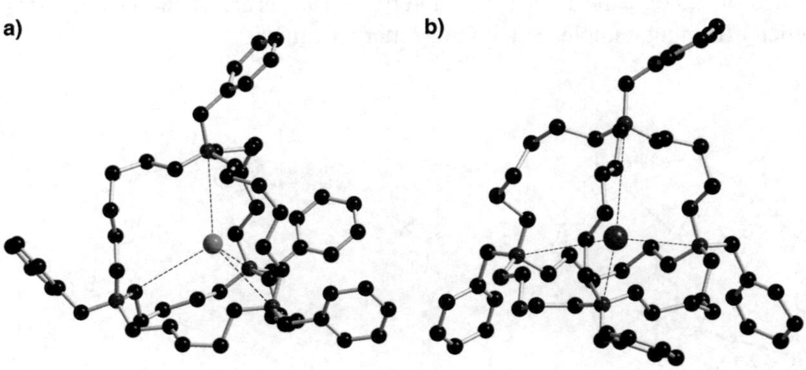

Figure 2.25 *Single crystal X-ray structures of (a) the chloride anion complex of **2.89b** and (b) the bromide anion complex of **2.90b**. In both cases, further counter anions not proximate to the receptor were found in the crystal lattice. These anions are not shown for the sake of clarity*

An alternative approach to building cage-type receptors was introduced by Murakami and co-workers.[90,133–136] These researchers took advantage of two well-known, rigid "scaffolds", namely tetraaza[6.1.6.1]paracyclophane and tetra[3.3.3.3]paracyclophane, to generate azaparacyclophanes, **2.91a** and **2.91b**, which are characterized by the presence of chiral-binding sites. Computer modelling experiments suggested a cavity size with a maximal inner diameter of *ca.* 10 Å. In addition, because it was expected that the four bridges would twist in the same direction, it was proposed that the receptor as a whole would provide a chiral-binding environment. This latter expectation was confirmed by circular dichroism (CD) studies. For instance, in an aqueous carbonate buffer, values of $[\theta]$ equal to 1.2×10^5 and -1.3×10^5 deg cm^2 dmol^{-1} were recorded for **2.91a** and **2.91b** at 237 and 236 nm, respectively.

In 1991, the first anion-binding studies were performed with **2.91a** using ANS as the substrate. Based on the NMR and fluorescence spectroscopic titrations, association constants, K_a, of 4.3×10^3 M^{-1} in D_2O–DMSO-d_6 (7:3 v/v) and 2.8×10^4 M^{-1} were determined in an aqueous buffer at pH 4.1.[133] It was also found that **2.91a** forms complexes with a number of anionic dyes in a 1:1 host–guest ratio. The association constants, K_a, evaluated on the basis of a Benesi-Hildebrand analysis, were 3.7×10^5, 3.4×10^5, 2.5×10^5, and 1.2×10^5 M^{-1} for Naphthol Yellow S, Dimethylsulfonazo III, Bromopyrogallol Red, and Orange G, respectively, in an aqueous HEPES buffer.[134]

Later, the ability of the cage-type cyclophanes, **2.91b** and **2.91c**, to effect chirality-based molecular discrimination was investigated using (4*Z*,15*Z*)-bilirubin IXα (**BR**) and 4,4′-methylenebis(3-hydroxy-2-naphthalenecarboxylic acid) (**PA**) as substrates.[136] In the presence of **BR** and **PA**, the original CD band of **2.91b** was found to decrease in intensity, while two new bands were seen to appear ($[\theta] = -2.3 \times 10^4$ and 1.8×10^4 deg cm^2 dmol^{-1} at 464 and 404 nm for **BR** and $[\theta] = -2.4 \times 10^3$ and 1.2×10^3 deg cm^2 dmol^{-1} at 379 and 339 nm for **PA**). Based on an analysis of Cotton effects, only *S*-**BR** and *S*-**PA** were thought to form complexes with **2.91b**. However, inverted bisignate CD bands were observed for **BR** and **PA** in the presence of **2.91c**, a result that was interpreted in terms of the enantiomers with *R*-helicity forming complexes with this latter receptor.

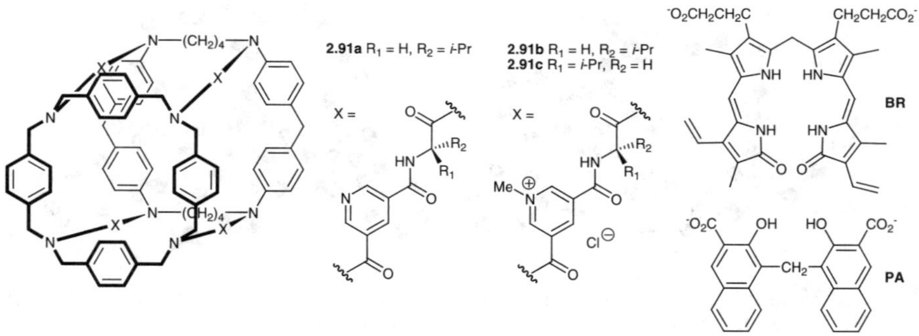

2.3 Guanidiniums

Since the early days of supramolecular chemistry, the guanidinium group has attracted the attention of those interested in anion receptor design. As the arginyl residues of proteins, the guanidinium group plays an important role in maintaining protein tertiary structure through, *e.g.*, interactions with carboxylate-bearing side chains, as well as in the targeted recognition of many anionic enzymatic substrates. A key feature of the guanidinium moiety is its high basicity, which has the consequence that this group remains protonated over a wide pH range ($pK_a = 13.5$).[137] The ability to provide two protons that point in roughly the same direction is also critical; it allows for the stabilization of two parallel hydrogen bonds, as seen in the X-ray crystal structures of many guanidinium salts (Figure 2.26).[138] As a consequence of these two near-unique features, guanidinium-based receptors often show high affinities and selectivities for oxyanion substrates (*e.g.*, carboxylates, phosphates, sulfates, and nitrates), even in polar organic solvents and in aqueous environments.

In work from 1954 that antedates the rise of supramolecular and anion recognition chemistry as we now know it, Tanford[139] used changes in the pK_a of acetic acid to establish the formation of a complex between the anionic conjugate base (*i.e.*, acetate) and guanidinium. Under these conditions, it was found that the association constant ($\log K_a$) was low, being less than roughly 0.5. This method was subsequently used by other researchers to measure the stability constants for guanidinium complexes formed from a number of carboxylates and from phosphates (*e.g.*, $\log K_a = 0.37$,

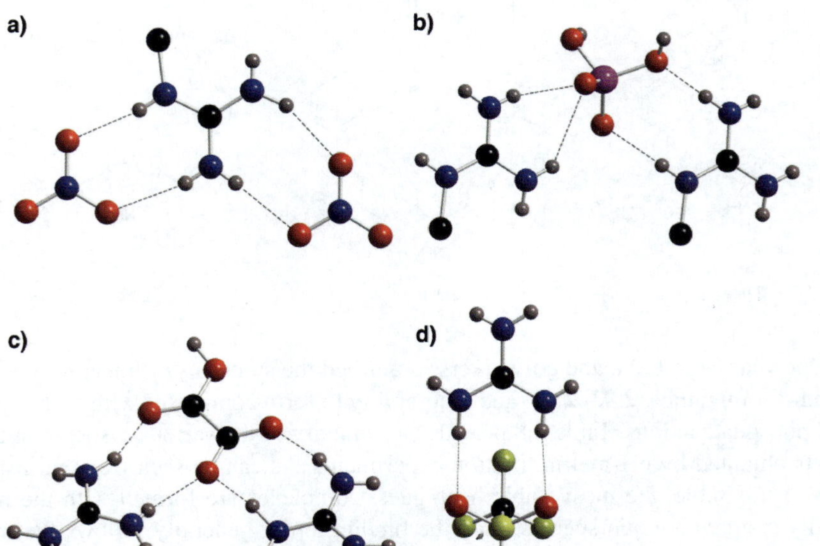

Figure 2.26 *Single crystal X-ray structures of (a) nitrate-anion complex of methylguanidinium, (b) dihydrogen phosphate-anion complex of methylguanidinium, (c) oxalate complex of guanidinium, and (d) trifluoromethylsulfonate complex of guanidinium*

0.32, 0.43, and 1.37 for acetate, formate, chloroacetate, and dihydrogen phosphate, respectively, in aqueous 1.02 M TMACl).[140] Much more recently, Hamilton and Linton[141] used isothermal titration calorimetry to analyze the thermodynamics of the interaction between guanidinium and acetate anion; they found $K_a = 7900\ M^{-1}$, $\Delta H = -14.6\ kJ\ mol^{-1}$, and $\Delta S = 25.1\ J\ mol^{-1}\ K^{-1}$ in DMSO at 25 °C.

In 1978, Lehn and co-workers[142] reported the synthesis and anion-binding properties of the flexible guanidinium-based macrocycles **2.92**. In the context of this work, three receptors were synthesized. They were prepared via the conversion of the corresponding thiourea-containing systems into first their *S*-ethyl-thiouronium derivatives and then, after reaction with ammonia, into the desired target systems **2.92a–2.92c**.

The stability constants of macrocycles **2.92** were determined by analysis of the pH titration curves recorded in the presence and absence of various putative anionic substrates. Taken in concert, these studies revealed that macrocycles **2.92a**, **2.92b**, and **2.92c** bind to trianionic phosphate (PO_4^{3-}) with a 1:1 ratio and with affinity constants, log K_a = 3.1, 3.4, and 4.3, respectively, in methanol–water (9:1) solution at 20 °C.[142] The fact that receptor **2.92c** demonstrated the highest stability constant, relative to the other receptors, **2.92a** and **2.92b**, was considered consistent with the appealing notion that, at least within this series of ostensibly related macrocycles, the PO_4^{3-} anion-binding affinity is related to the number of positively charged guanidinium units available for interaction.

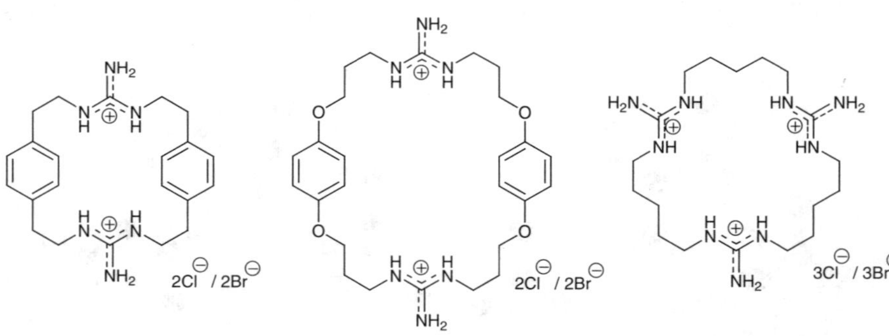

2.92a **2.92b** **2.92c**

One year later, Lehn and co-workers[3] described the synthesis of linear polyguanidinium compounds, **2.93–2.96**, and their ability to form complexes with carboxylate and phosphate anions. Table 2.8 provides a summary of the various association constants obtained by pH-metric titration experiments. As can be seen from an inspection of this table, the most stable host–guest complexes are formed with the most highly charged anionic species (*e.g.*, the binding order generally follows the trend $P_2O_7^{4-} > HP_2O_7^{3-}$ and $CH_3CO_2^- > {}^-O_2CCH_2CO_2^-$). Likewise, for a given anionic substrate, the highest affinities are seen for those receptors bearing the greatest positive charge. Such trends are consistent with electrostatic effects dominating the binding process. Nonetheless, the fact that a high selectivity for malonate over fumarate is seen in the case of **2.95** and **2.96** provides an indication that geometrical effects could play a role in regulating the anion-binding process.

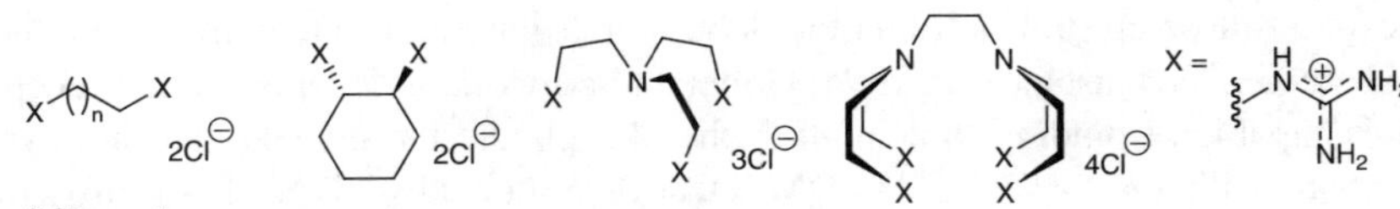

2.93a n = 1
2.93b n = 2 **2.94** **2.95** **2.96**

Some years after Lehn's extensive studies, Hamilton and co-workers[143–145] prepared the bis-acylguanidinium-based receptors **2.97** and **2.98**, open chain bis-guanidinium systems linked by iso- and terephthalate spacers, respectively.

An intramolecular hydrogen-bond interaction between the NH and CO moieties, established by proton NMR studies, is thought to endow receptor **2.97** with considerable rigidity. The higher degree of preorganization provides four hydrogen bond donors on the central cavity of the receptor, which was designed to favour the binding of a single phosphate substrate. The association constant for diphenyl phosphate binding, studied in the form of its TBA salt, was obtained by recording the change in the UV absorption intensity at 266 nm observed upon diluting a 1:1 mixture of the host and guest in CH_3CN; from an analysis of the resulting binding profile, a K_a of 4.6×10^4 M^{-1} was determined.[143]

In contrast to the above, the relatively flexible receptor **2.98** was found to bind glutarate strongly (K_a = 8500 and 480 M^{-1} in 12% and 25% D_2O/DMSO-d_6, respectively) as inferred from similar spectroscopic studies.[144] Based on more detailed temperature-dependent NMR spectroscopic titrations, it was concluded that that the glutarate-binding affinities increase with increasing temperature (K_a = 600, 720, 1010, and 1430 M^{-1} at 10, 20, 40, and 50 °C, respectively, in 25% D_2O/CD_3OD). The thermodynamic parameters derived from these studies (ΔH = 15.9 kJ mol^{-1} and ΔS = 108.8 J mol^{-1} K^{-1}) were found to match well with those obtained from independent ITC experiments carried out in pure methanol at 25 °C (K_a = 2700 M^{-1}, ΔH = 15.5 kJ mol^{-1}, and ΔS = 117 J mol^{-1} K^{-1}). Based on this, it was concluded that the complexation process in pure methanol and in various methanol/water mixtures is endothermic and thus entropy driven.[145]

Table 2.8 *Association constants (log K_a) corresponding to the formation of phosphate and carboxylate anion complexes of receptors **2.93–2.96**, as determined in aqueous 0.1 M tetramethylammonium chloride (TMACl) (phosphate anions) and methanol/water (9:1) (carboxylate anions) at 25 °C*

Anion	***2.93a***	***2.93b***	***2.94***	***2.95***	***2.96***
PO_4^{3-}	2.3	1.8	2.4	3.5	3.1
HPO_4^{2-}	—	—	1.2	1.9	1.9
$P_2O_7^{4-}$	2.5	2.2	2.7	4.3	4.1
$HP_2O_7^{3-}$	1.3	1.4	2.0	3.0	2.6
$CH_3CO_2^-$	2.2	1.8	—	2.4	2.7
$^-O_2CCH_2CO_2^-$	4.2	—	3.9	4.4	5.1
$^-O_2CC_2H_2CO_2^-$	—	1.4	3.0	2.5	3.5

Several follow-up studies from other laboratories have served to complement these initial studies. For instance, in 1994, Göbel and co-workers[146] reported that receptor **2.99** is capable of binding to a small cyclic phosphodiester, namely catechol cyclic phosphate, with a $K_a = 95\ M^{-1}$ in DMSO, as determined by ^{31}P NMR spectroscopy. Further, Grossel and co-workers[147] succeeded in obtaining the X-ray crystal structures of the sulfate-anion complexes of **2.97** and **2.99**. As shown in Figure 2.27, in the case of receptor **2.97**, the sulfate anion is held within the rigid cavity by hydrogen-bond interactions involving all four oxygen atoms. By contrast, outside binding is seen in the case of **2.99**.

The strong anion-binding affinities seen for receptors **2.97** and **2.99** led three different groups to apply these systems to the problem of phosphodiester cleavage; it was found that these systems do indeed act as effective "artificial catalysts" or, in some cases, "anticatalysts" for RNA hydrolysis.[146,148]

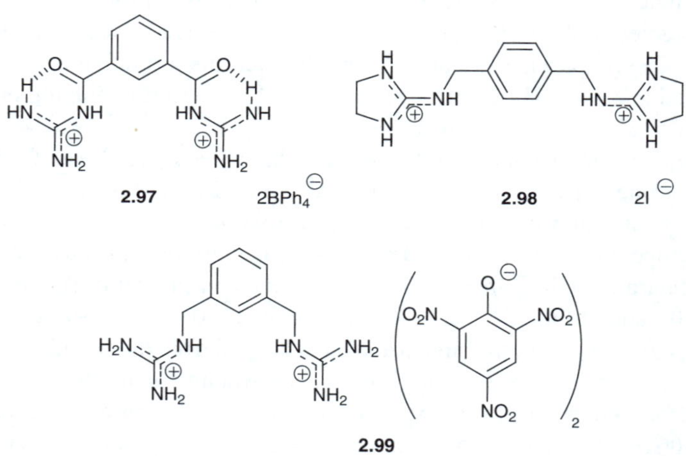

In independent work, Anslyn and Ariga designed receptors **2.100a** and **2.100b** as mimics for the active site of staphylococcal nuclease (SNase), an enzyme capable of effecting a 10^{16}-fold rate enhancement for DNA hydrolysis.[149] This particular work was carried out within the cadre of trying to develop metal-free catalysts capable of

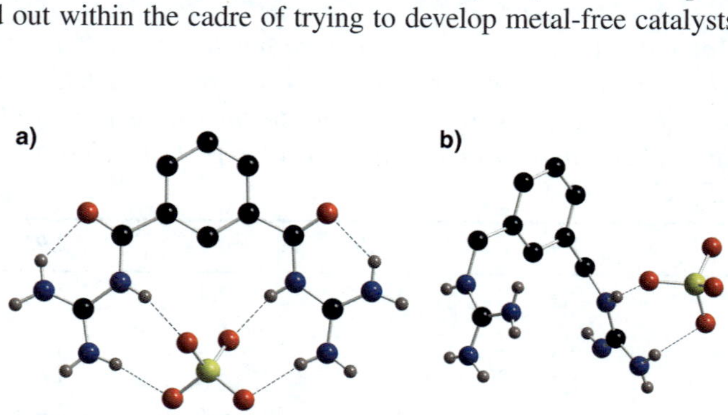

Figure 2.27 *Single crystal X-ray structures of the sulfate-anion complexes of (a) sulfate* ***2.97*** *and (b)* ***2.99***

enhancing the hydrolysis of phosphodiesters.[150] These pre-organized receptors contain a V-shape cavity that was designed to bind the targeted phosphodiester substrate strongly via a combination of electrostatic interactions and four oriented hydrogen bonds. Two X-ray crystal structures, shown in Figure 2.28, provided evidence of binding in the solid state. Solution-phase anion-binding studies, carried out using dibenzylphosphate as a model substrate and ^{31}P NMR spectroscopic titrations as the analysis method, revealed that receptors **2.100a** and **2.100b** both exhibit good binding in DMSO (K_a = 2200 and 1400 M^{-1} for these two receptors, respectively). However, the binding affinities for both receptors were seen to drop dramatically in pure water (*i.e.*, to below 30 M^{-1}).[150] In spite of this latter drop off, receptor **2.100b** was found to bind the phosphomonoester, UMP, well in pure water (*i.e.*, K_a = 960 M^{-1}).[151] Eventually, it was found that compound **2.100a** catalyzes mRNA cleavage with a rate that is enhanced by over 20-fold relative to the uncatalyzed reaction.[152]

Another bis-guanidinium system, **2.101**, linked through a *trans*-decalin spacer, was introduced by Göbel and co-workers[153] in 1994 as a potential phosphodiester hydrolysis catalyst. This receptor was found to bind the anionic form of a small cyclic phosphodiester (**CP**), with a K_a = 110 M^{-1} in DMSO, as determined by ^{31}P NMR spectroscopy. However, only weak catalytic activity was observed.

In 1995, Hamilton and co-workers[154] reported the synthetic bis-guanidinium receptor **2.102**. This system was designed to bind two carboxylate side chains in α-helical peptides. It thus relies on a rigid scaffold to orient two guanidinium groups in a parallel fashion and at a well-defined distance. To test the efficacy of the design, three 16-mer model peptides, containing two aspartate units located at different positions (i+3, i+4, and i+11) along the peptide chain, were studied. In a solution consisting of 10% D_2O/CD_3OD, receptor **2.102** was found to form a strong complex with the peptide containing two aspartic acid moieties three residues apart (K_a = 2200 M^{-1} for i+3). Moreover, a significant enhancement in the helicity of this peptide and its i+4 congener (up to 9%) was seen, as inferred from CD measurements. These results were taken as support for the proposal that the host–guest interactions occur preferentially in the case of appropriately chosen helical peptides. Several analogues of **2.102** were also studied; however, these systems displayed lower affinites.[155]

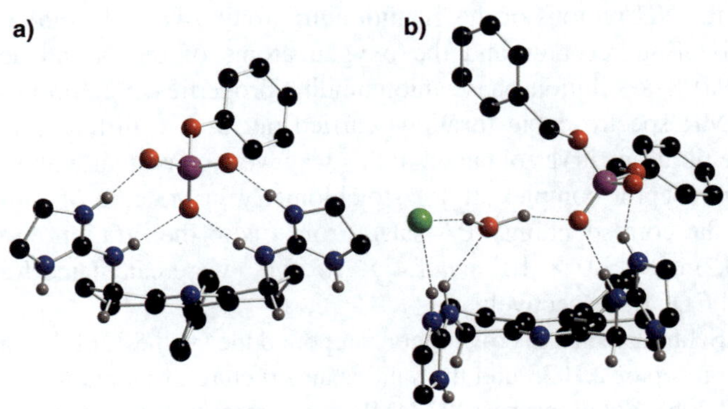

Figure 2.28 *Single crystal X-ray structures of (a) phenyl phosphate-anion complex of* ***2.100b*** *and (b) mixed dibenzyl phosphate, chloride anion, and aquo complex of* ***2.100b***

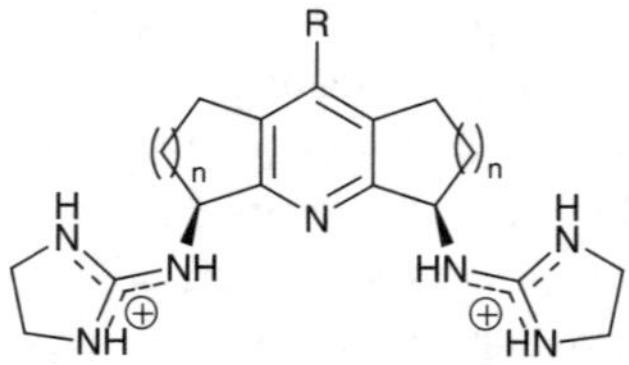

2.100a n = 2, R = CO_2Et, $2Cl^{\ominus}$

2.100b n = 1, R = Ph, $2BPh_4^{\ominus}$

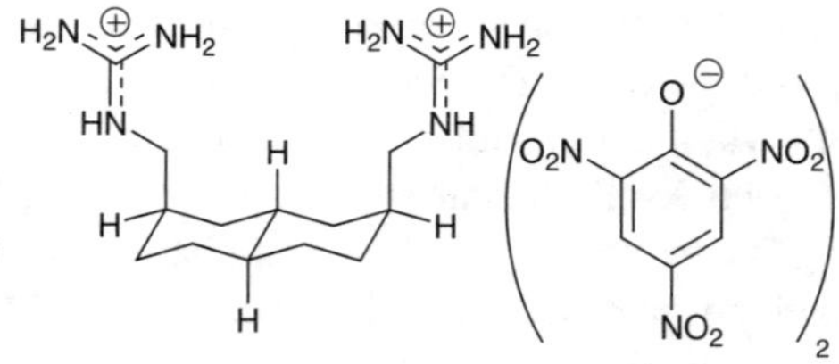

2.101

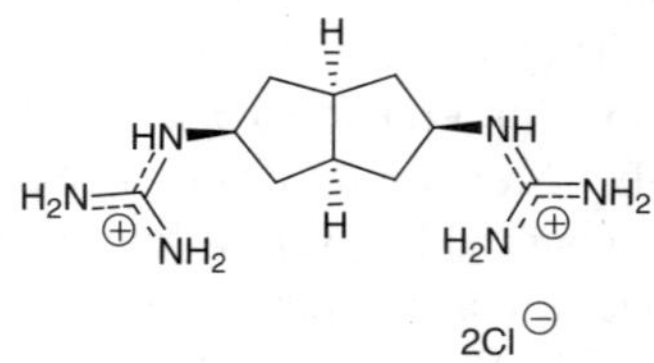

2.102

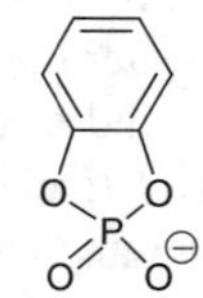

CP

As part of a generalized effort to improve the binding characteristics of guanidinium receptors, the guanidinium group has been incorporated into a number of bicyclic systems. The advantage of such structures is that they provide two pre-organized protons that point in the same direction. This allows receptors of this type to form complexes easily with various anionic guests. The first bicyclic compounds containing a guanidinium group, systems with generalized structure **2.103**, were actually reported by McKay and Kreling[156] in the late 1950s. However, it was not until the 1980s, when Schmidtchen[123] developed synthetic procedures that allowed him to access compounds of general structure **2.103** easily and, accordingly, to study their anion-binding properties in detail,[157] that bicyclic guanidinium systems came to be appreciated as potentially powerful anion receptors.

A single crystal X-ray diffraction analysis of **2.103c** was carried out. It revealed a structure wherein two acetate anions are bound by the embedded guanidinium motifs of two separate molecules of **2.103c** via a combination of electrostatic interactions and oriented hydrogen bonds. These latter involve the oxygen atoms of the acetate anions and the NH protons of the guanidinium groups, as well as the OH moieties of a neighbouring receptor and the oxygen atoms of the bound acetate anion (Figure 2.29a). The solution-phase anion binding properties of **2.103b** were deduced from ^1H NMR spectroscopic titrations carried out in two different solvents. The shape of the titration curves obtained in this way were consistent with the formation of an anion–receptor complex of 1:1 stoichiometry in the case of *p*-nitrobenzoate (studied as the corresponding TBA salt). From fits to the titration curves, affinity constants (K_a) of $> 1.0 \times 10^4$ and 1.4×10^5 M^{-1} were calculated for binding in $CDCN_3$ and $CDCl_3$, respectively.

Later on, Schmidtchen and co-workers[158] reported the synthesis of the chiral bicyclic guanidinium receptor **2.103d** and the solid-state structure of its nitrate-anion complex (*cf.* Figure 2.29b). Solution-phase ^1H NMR spectroscopic experiments carried out in CD_3CN provided evidence that diastereomeric host–guest complexes with racemic carboxylates, such as *N*-acetyl-D,L-alanine or D,L-2-methylbutyrate, were being formed.

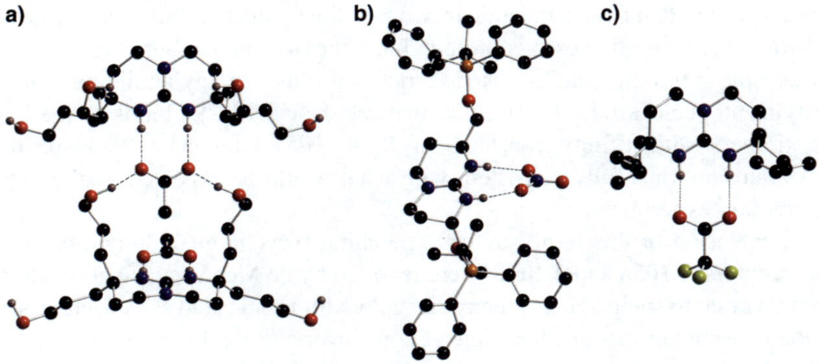

Figure 2.29 *Single crystal X-ray structures of (a) the 2:2 acetate-anion complex formed from receptor **2.103c**, (b) the nitrate-anion complex of **2.103d**, and (c) the trifluoroacetate-anion complex of **2.103b***

More recently, the X-ray crystal structure of the trifluoroacetate complex of **2.103b** was reported by Schmidtchen and co-workers.[159] In the solid state (*cf.* Figure 2.29c), this complex is stabilized by two well-oriented hydrogen bonds involving the two parallel NH subunits, as well as what are presumed to be more isotropic electrostatic interactions. The interaction between **2.103b** and benzoate was studied in acetonitrile using ITC. It was found to depend strongly on the counter anion of the host. For example, the use of BF_4^- as the counter anion produced the highest association constant K_a = 414,000 M^{-1}, whereas the corresponding Cl^- salt generated a value that was considerably lower 38,000 M^{-1}. It was confirmed that the iodide salts of **2.103b** and **2.103d** bind to benzoate with K_a = 280,000 and 203,000 M^{-1}, respectively.

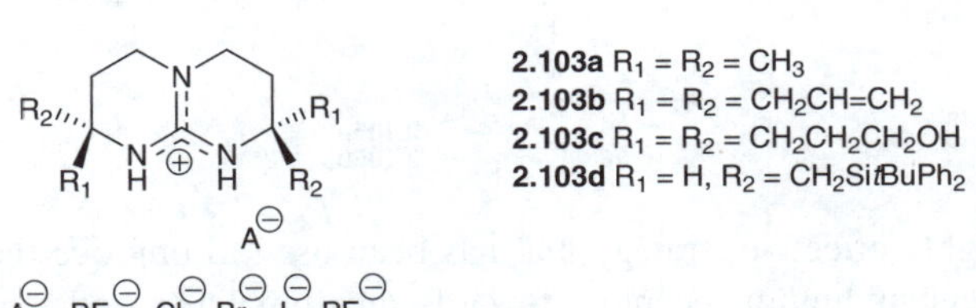

In work that appeared one year after Schmidtchen's initial 1987 report involving the anion-recognition properties of **2.103**, Lehn and co-workers[160] described the first chiral bicyclic guanidinium receptor, namely the bis-naphthoate **2.104**. It was found that this receptor is capable of extracting *p*-nitrobenzoate anion from an aqueous media into organic solvents, lending credence to the notion that it can bind oxyanions. Quantitative analysis of the binding constants, carried out using standard ^1H NMR spectroscopic titration protocols, gave a K_a of 1609 M^{-1} for the binding of *p*-nitrobenzoate (studied as the corresponding triethylammonium salt) in $CDCl_3$. As judged from extraction experiments, the asymmetric nature of **2.104**-*SS* allowed for the enantioselective recognition of chiral carboxylate anions. For instance, in one study it was shown that receptors **2.104**-*SS* could be used to extract preferentially

L-*N*-acetyltryptophan from a racemic mixture of the L- and D-enantiomers. Likewise, **2.104**-*RR* could be used to extract preferentially the D-antipode. Supporting ^{1}H NMR spectroscopic titration studies, carried out with the triethylammonium salt of *N*-acetyltryptophan in $CDCl_3$, revealed that receptor **2.104**-*SS* binds to the L- and D-enantiomers with affinity constants of K_a = 1051 and 534 M^{-1}, respectively. These values are thus fully consistent with what would be expected on the basis of the extraction experiments.[160]

In an extension of this basic strategy, the chiral bicyclic guanidinium-based synthetic receptors, **2.105a** and **2.105b**, were reported by de Mendoza and co-workers.[161] These two systems, prepared as putative receptors for uronic acid salts, were designed with the expectation that ancillary interactions involving the hydroxyl groups of the targeted glycopyranosyl guests and the convergent hydroxyl groups of the curved deoxychlolic acid-derived "arms" of the "tweezer-like" receptors would serve to complement those provided by guanidinium-carboxylate ion-pairing. It was also appreciated that the rigid steroid subunit would provide a lipophilic outer surface, thus permitting good solubility in organic solvents. As it transpired, receptor **2.105a**, the system with the shorter spacer, was found to display higher carbohydrate affinities than **2.105b** (K_a = 7000 and 2800 M^{-1} for the binding of D-glucuronate by **2.105a** and **2.105b**, respectively), as determined from ^{1}H NMR titrations carried out in $CD_3CN/CDCl_3$. Also, a slight preference for glucuronate over galacturonate was observed in the case of receptor **2.105a** (K_{rel} = 1.2 for D-glucuronate/D-galacturonate).

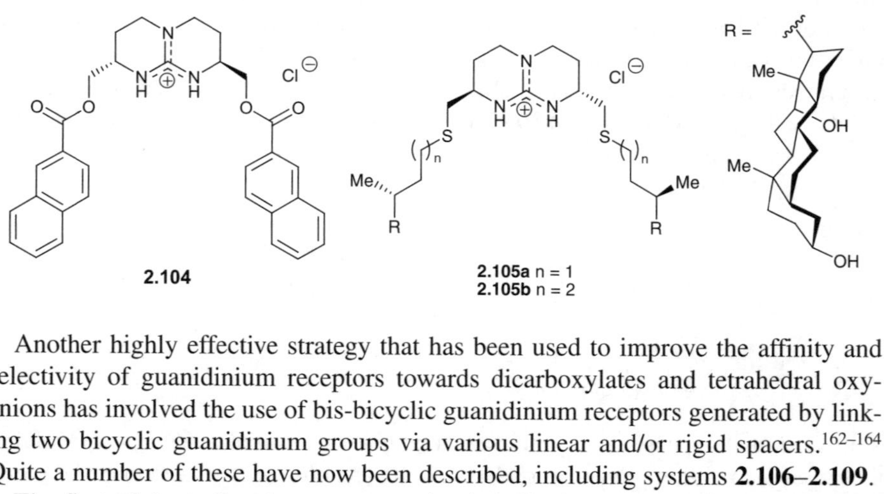

2.104

2.105a n = 1
2.105b n = 2

Another highly effective strategy that has been used to improve the affinity and selectivity of guanidinium receptors towards dicarboxylates and tetrahedral oxyanions has involved the use of bis-bicyclic guanidinium receptors generated by linking two bicyclic guanidinium groups via various linear and/or rigid spacers.[162–164] Quite a number of these have now been described, including systems **2.106–2.109**.

The first of these ditopic receptors to be described, namely **2.106a** and **2.106b**, consist of two bicyclic guanidinium units attached via a urethane linker.[162] Data from ^{1}H NMR spectroscopic titration experiments carried out in CD_3OD and D_2O confirmed that receptor **2.106a** is capable of forming 1:1 complexes with biologically relevant phosphates, such as *p*-nitrophenyl phosphate, CMP, and TMP, although the affinities were not particularly high (*e.g.*, K_a = 10.6 M^{-1} for the binding of TMP in D_2O). Receptor **2.106b** was also found to be capable of effecting the extraction of dicarboxylate anions (*e.g.*, succinate, fumarate, folate, and

N-acetylaspartate), but not monocarboxylate anions, from an aqueous layer into a chloroform phase.

The foldable bis-guanidinium receptor **2.107a**, wherein the bicyclic guanidiniums are linked via ether linkages, was also found to be a good receptor for dicarboxylate anions. For instance, K_a values of 2540, 16,500, 6060, and 14,500 M^{-1} were obtained for oxylate, malonate, isophthalate, and *p*-nitroisophthalate anions, respectively (1:1 binding), as determined from ^1H NMR titration analyses in CD_3OD. In contrast, under identical conditions of study, no evidence of binding was seen in the case of simpler monoanions, such acetate and iodide. In follow-up studies, it was found that receptor **2.107a**, as well as congener **2.107b**, allowed for the recognition of nucleotide phosphates in polar solvents, as inferred from ^1H NMR spectroscopic analyses.[164] For instance, in the case of receptor **2.107a**, affinities (K_a) of between 1.8×10^4 and 3.8×10^4 M^{-1} were recorded in CD_3CD for various phosphate anions, including nucleotide monophosphates. Likewise, receptor **2.107b** was found to bind HPO_4^{2-} with a $K_a = 970$ M^{-1} in D_2O.

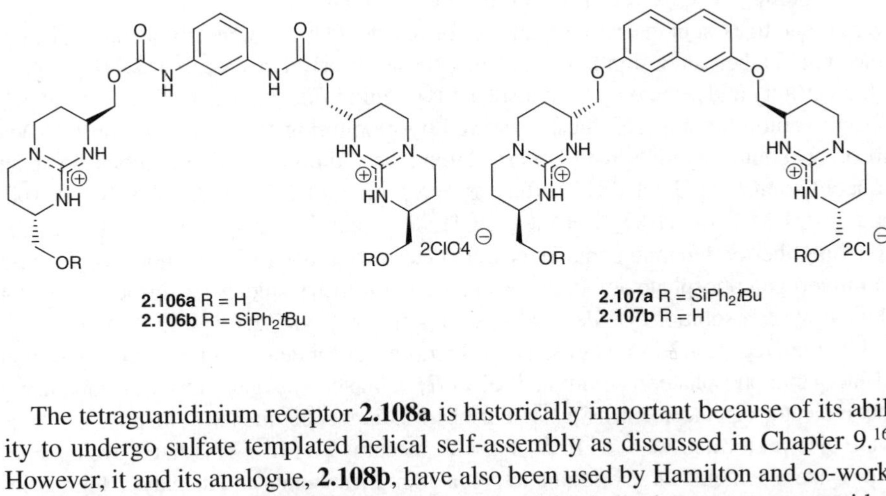

2.106a R = H
2.106b R = SiPh$_2$*t*Bu

2.107a R = SiPh$_2$*t*Bu
2.107b R = H

The tetraguanidinium receptor **2.108a** is historically important because of its ability to undergo sulfate templated helical self-assembly as discussed in Chapter 9.[165] However, it and its analogue, **2.108b**, have also been used by Hamilton and co-workers[166–169] to effect the recognition and stabilization of α-helical tetraasparate peptides. Within the series of test peptides defined by **2.109a–2.109f**, the strongest interaction was seen in the case of **2.108a–2.109a**, which contains four aspartate (**D**) units ($K_a = 1.56 \times 10^5$ M^{-1}, as determined from CD titrations carried out in 9:1 CH_3OH/H_2O).[166] Based on the ITC experiments performed with **2.108b** and peptides containing both Asp (**2.109a**) and Glu (**2.109d**) subunits, it was concluded that the recognition of the shorter peptide **2.109a** is enthalpy driven ($\Delta H = -23.6$ kJ mol^{-1} and $T\Delta S = 5.40$ kJ mol^{-1}), whereas that of the longer target (*i.e.*, **2.109d**) is entropy driven ($\Delta H = 15.4$ kJ mol^{-1} and $T\Delta S = 44.9$ kJ mol^{-1}).[168] The relative location of Asp (**D**) and Trp (**W**) residues within a peptide sequence was also found to play an important role in defining the molecular recognition process.[169] For instance, in trifluoroethanol:water (1:1) the affinity of **2.108b** for peptide **2.109f** ($K_a = 4.2 \times 10^6$ M^{-1}) is around 26-fold lower than that for **2.109g** ($K_a = 1.1 \times 10^8$ M^{-1}).

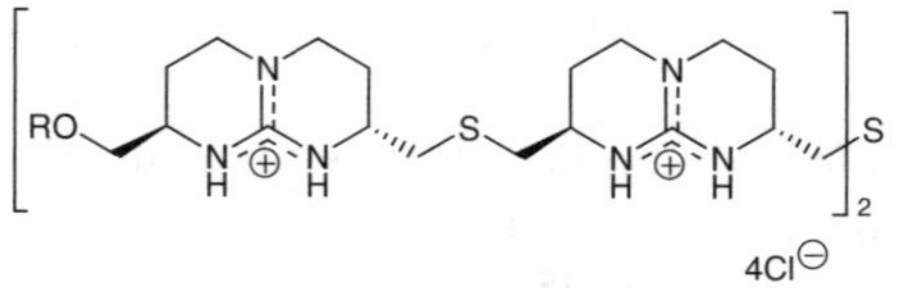

2.108a R = $SiPh_2tBu$
2.108b R = H

2.109a Ac-A-A-A-**D**-Q-L-**D**-A-L-**D**-A-Q-**D**-A-A-Y-NH_2
2.109b Ac-A-A-A-**D**-Q-L-**D**-A-L-**D**-A-Q-**N**-A-A-Y-NH_2
2.109c Ac-A-A-A-**D**-Q-L-**D**-A-L-**N**-A-Q-**N**-A-A-Y-NH_2
2.109d Ac-A-A-A-**E**-Q-L-**E**-A-L-**E**-A-Q-**E**-A-A-Y-NH_2
2.109e Ac-A-A-A-**D**-Q-L-**D**-A-L-**D**-A-Q-**D**-A-A-Y-$CONH_2$
2.109f Ac-A-A-A-**D**-Q-L-**D**-**W**-L-**D**-**W**-A-**D**-**W**-A-Q-**W**-A-A-A-$CONH_2$
2.109g Ac-A-A-A-**W**-Q-L-**W**-**D**-L-**W**-**D**-A-**W**-**D**-A-Q-**D**-A-A-A-$CONH_2$

In an effort to design synthetic receptors displaying strong and selective anion binding in aqueous media and which would potentially permit sensing applications using an indicator-displacement assay (*cf.* Chapter 8), anslyn and co-workers[170] have synthesized a number of guanidinium-based systems based on the triethylbenzene scaffold. For instance, the tri- and tetraguanidinium-based receptors **2.110** and **2.111** containing a Cu^{2+}-binding site were recently reported by the Anslyn group.[20,171,172] Both receptors were found to exist in their pre-organized forms after Cu^{2+} complexation and to display selectivity in their anion-binding behaviour. It was observed that receptor **2.110** displays a high affinity and selectivity for phosphate-type anions in aqueous media at neutral pH. Specific anion-binding affinities, measured by monitoring the change in the UV–Vis absorbance intensity at 790 nm observed upon the addition of sodium-anion salts in an aqueous solution (pH 7.4 TRIS buffer), gave K_a values of 1.5×10^4, 1.7×10^4, < 100, and <100 M^{-1} for HPO_4^{2-}, $HAsO_4^{2-}$, $CH_3CO_2^-$, and Cl^-, respectively.[20] The corresponding thermodynamic parameters were determined using ITC methods. In the case of dihydrogen phosphate anion, for examples, calorimetry studies performed at pH 7.4 (HEPES buffer solution) yielded $\Delta H = -15.9$ kJ mol^{-1} and $\Delta G = -22.2$ kJ mol^{-1}.[171]

Later on, receptor **2.110** was used as a chemosensor for determining the concentration of inorganic phosphate in serum and saliva (*cf.* Chapter 8). Again, this was done using the indicator-displacement assay.[173] Further binding experiments with phosphoesters were also performed with **2.110**[174] and with receptor **2.111**.[172] This latter cyclophane-type system contains three bridging tetraguanidinium subunits and displays a high affinity ($K_a = 8 \times 10^8$ M^{-1} in water/methanol 1:1 at pH 6.8 solution) and also a good selectivity for 2,3-bisphosphoglycerate (2,3-BPG) among the various anions tested.[172]

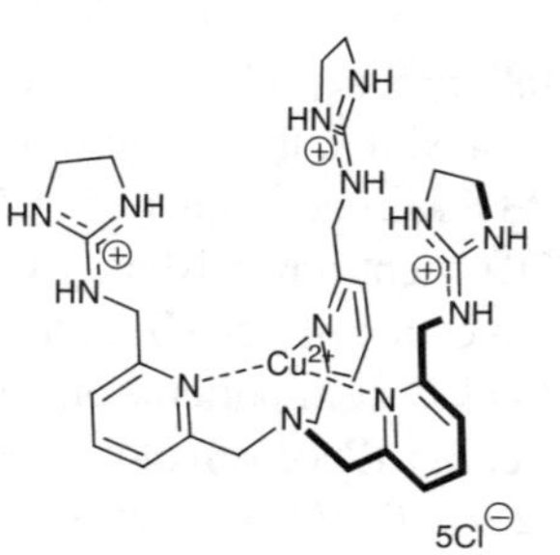

2.110

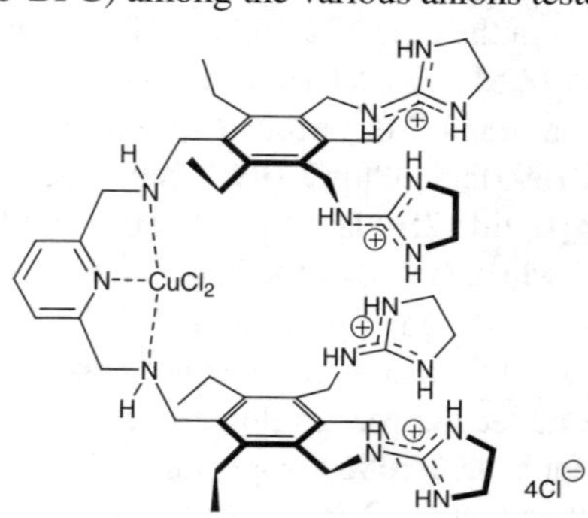

2.111

The steroidal guanidinium receptors **2.112–2.115** were recently introduced by A.P. Davis and co-workers.[175] These systems, designed to effect the enantioselective recognition of *N*-acyl-α-amino acids, showed remarkably consistent selectivity towards the L-forms of *N*-acetyl-α-amino acids, as judged from extraction experiments. Support for this finding came from NMR spectroscopic analysis and Monte Carlo molecular mechanics (MCMM) analyses, which revealed a better substrate–receptor fit in the case of the L-enantiomers.[175]

Davis, de Mendoza, and de Vries have employed steroidal guanidinium receptors to effect the enantioselective transport of *N*-Ac-phenylalanine through dichloromethane and dichloroethane bulk liquid membranes in a standard U-tube apparatus. Transport through octanol/hexane was also achieved using hollow fiber membrane contactors. In both sets of experiments, good enantioselectivities were observed.[176]

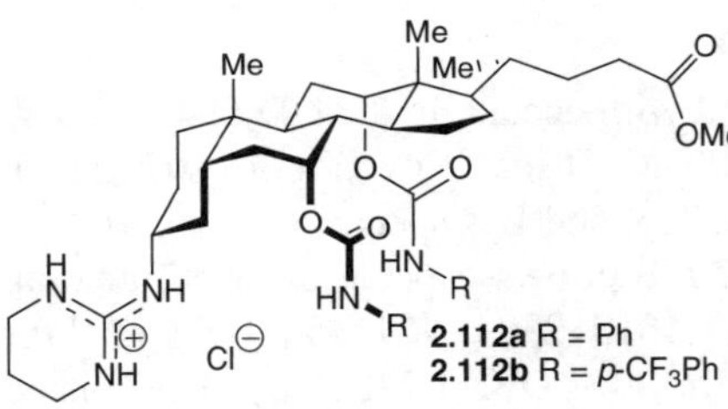

2.112a R = Ph
2.112b R = *p*-CF$_3$Ph

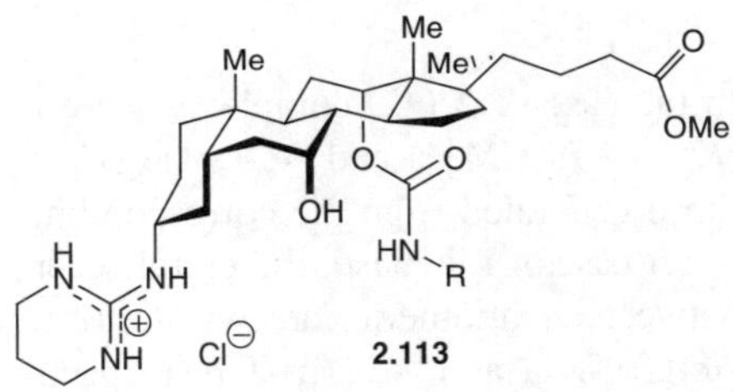

2.113

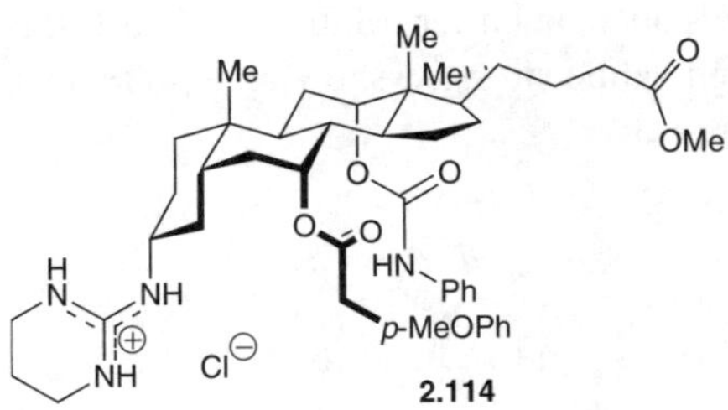

2.114

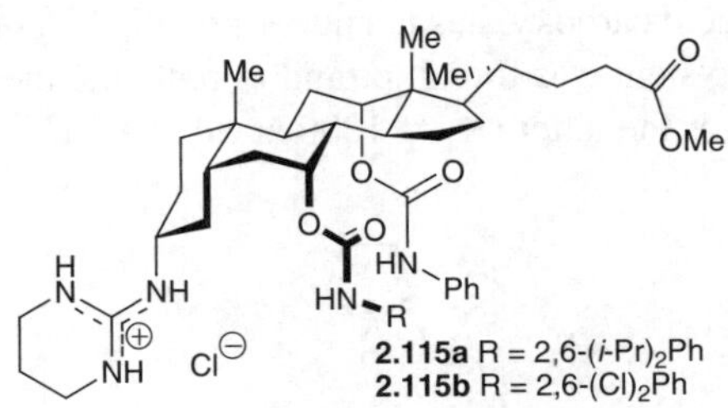

2.115a R = 2,6-(*i*-Pr)$_2$Ph
2.115b R = 2,6-(Cl)$_2$Ph

Over the last several years, Schmuck's group has shown how the binding affinities of guanidinium salts towards carboxylate anions can be enhanced via the incorporation of additional hydrogen bond motifs.[177–180] Early on in the course of this work, the crystal structure of the acetate complex of a simple guanidiniocarbonyl pyrrole receptor (**2.116**) was solved. This structure, shown in Figure 2.30a, established that the anionic acetate guest is bound not only via the expected bidentate guanidinium NH-to-oxygen hydrogen bonds, but also through an additional intermolecular pyrrole NH–oxygen interaction. These findings, not withstanding a computer-based molecular dynamics calculation, led to the suggestion that a tridentate-binding mode is the one favoured in solution (Figure 2.30b). Analysis of the ^{1}H NMR spectrum recorded in DMSO-d_6, revealed large, complexation-induced downfield shifts for all of the NHs signals, a finding that was considered consistent with the proposed tridentate-binding mode.[177]

Modification at the pyrrolic α-positions led to receptors **2.117** and **2.118**, which were found to bind *N*-Ac-amino acid carboxylate anions in highly polar media.[178]

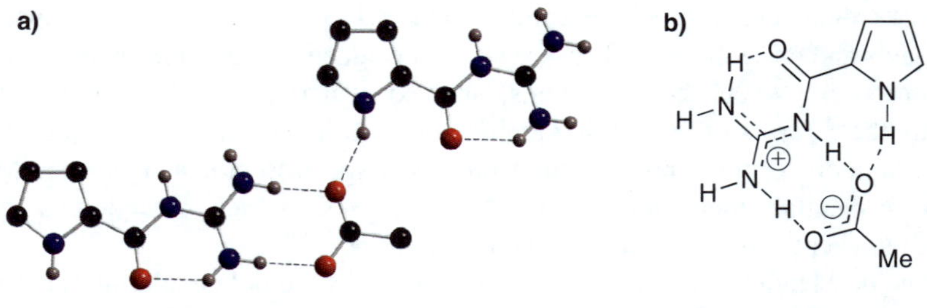

Figure 2.30 *(a) Single crystal X-ray structure of the acetate-anion complex of two guanidiniocarbonyl pyrroles* ***2.116*** *as observed in the solid state via X-ray diffraction analysis and (b) schematic representation of the tridentate complex expected to dominate under most solution-phase conditions*

In the case of **2.117**, complexes were formed with acetate in 40% D_2O–DMSO-d_6 (K_a = 2790 M^{-1}) and Ac-L-Phe (K_a = 1700 M^{-1}) that proved more stable than those generated from the corresponding Ala, Trp, and Lys species (K_a < 900 M^{-1} in all cases). Likewise, the chiral receptor **2.118** proved capable of discriminating between enantiomeric carboxylates (*e.g.*, K_a = 1610 (930) and 1145 (1005) M^{-1} for Ac-L(D)-Ala and Ac-L(D)-Trp, respectively, in 40% D_2O–DMSO-d_6). The related valine-derived receptor, **2.119**, was also found to be an efficient receptor for amino acid carboxylates in aqueous media.[179] For instance, in buffer solution at pH 6.1, this system was found to bind to both alanine and valine strongly, showing a preference for the latter (K_a = 1000 and 1750 M^{-1}, respectively).

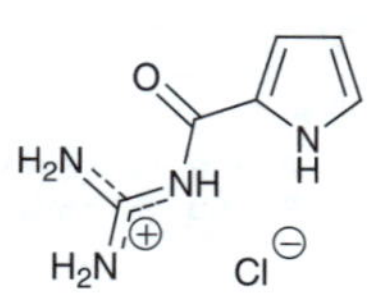

2.116

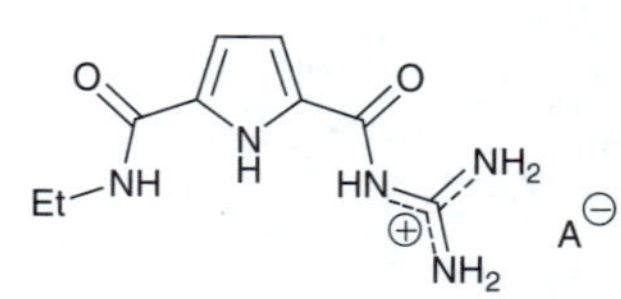

2.117

$A^{\ominus}$ = picrate or $Cl^{\ominus}$

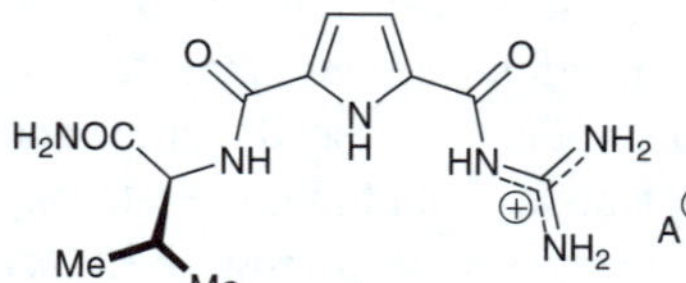

2.118

$A^{\ominus}$ = picrate

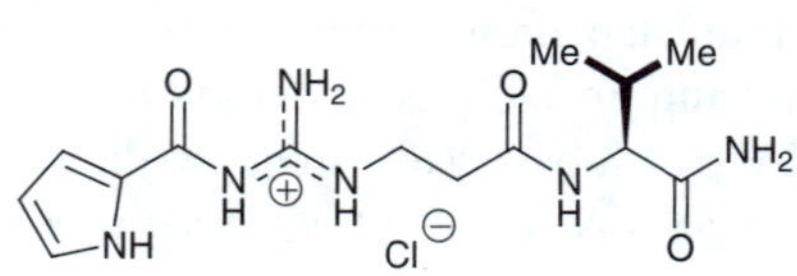

2.119

2.4 Amidiniums

Amidinium cation (*e.g.*, **2.121**), a member of the generalized guanidinium family, represents an important motif that has proved extremely useful in the design of receptors for oxyanions, such as carboxylates and phosphates. Several X-ray crystal structures of amidinium–oxyanion complexes have been analyzed and all have served to confirm a binding mode that includes two linear hydrogen bonds between the oxygen atoms of the anion and the amidinium NH protons (Figure 2.31).[181,182] Although related to guanidinium cations, amidiniums are slightly less basic ($pK_a > 11$).[183] Therefore, amidiniums generally act as better hydrogen donors after protonation (Scheme 2.7).

Amidinium–carboxylate salts are of particular interest as model systems for the arginine–aspartate salt bridges found in many biological structures including zinc finger/DNA complexes,[184] RNA stem loops,[185] and the active site of dihydrofolate reductase.[186] Amidinium and benzamidinium compounds are also important anti-trypsin agents, as discussed further below.

Trypsin-like serine proteinases play a central role in the regulation of many biological processes.[187] Generally, they cleave the peptide bonds between the carboxyterminal side of positively charged arginine and lysine residues that are bound to an aspartate residue (Asp 189) within a specific binding pocket. Therefore, many natural and synthetic proteinase inhibitors contain an arginine or lysine residue.[188] In 1965, Mares-Guia and Shaw[189] reported that amidiniums are good model systems for the arginine side chain of trypsin substrates. Subsequently, the crystal structure of a benzamidinium–trypsin complex was determined by X-ray diffraction analysis.[190] As can be seen from Figure 2.32, which shows the relevant active site, the benzamidinium is bound near the so-called catalytic triad thought responsible for catalytic activity, being held in place through a combination of electrostatic interactions and five hydrogen bonds involving Gly 219, Asp 189, Ser 190, and an internal water molecule. This binding is thought to shield the active site by preventing *inter alia* binding of normal oligopeptide substrates. Recently, the interaction between

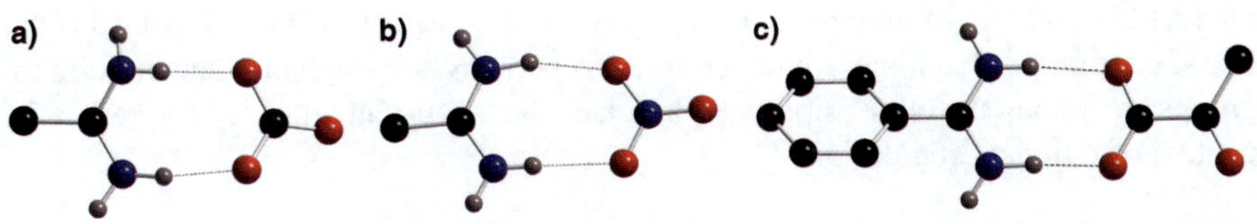

Figure 2.31 *Single crystal X-ray structures of (a) carbonate complex of amidinium **2.121** (R = CH_3), (b) nitrate complex of amidinium **2.121** (R = CH_3), and (c) pyruvate complex of benzamidinium **2.121** (R = C_6H_5)*

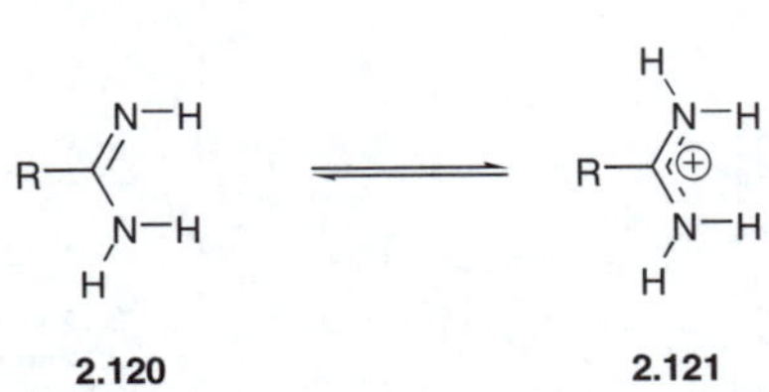

Scheme 2.7 *Amidinium protonation equilibrium*

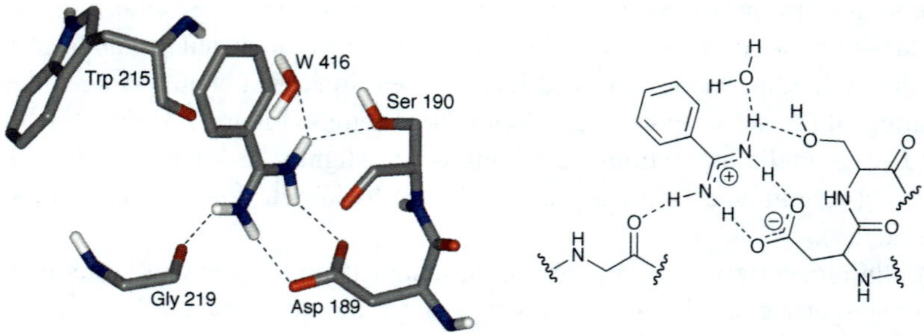

Figure 2.32 *Partial view showing the active site portion of the complex formed between benzamidinium and trypsin, as determined by X-ray structural analysis (left). The corresponding structural formula is shown on the right*

benzamidinium and trypsin was studied by ITC, yielding a K_a of 4.5×10^4 M^{-1} in Tris buffer at pH 8.0.[191]

Separate from the above, a number of aromatic diamidinium cations have been found to possess antiprotozoal activity and modest antiviral activity.[192] For example, pentamidinium **2.123** and furamidinium **2.125a** have been extensively used for the treatment of *Pneumocystis carinii* pneumonia (PCP), which occurs in over 70% of AIDS-patients.[193] This anti-PCP activity is thought to reflect the fact that aromatic bis-amidiniums can bind to the minor groove of duplex DNA, particularly in AT-rich regions.[194] This binding is clearly seen in a number of crystal structures obtained to date, including that between the antitrypanosomal drug berenil **2.122** and d(CGCGAATTCGCG)$_2$,[195] as well those between this model duplex DNA sequence and several other bis-amidinium agents (*e.g.*, **2.123–2.125**), as shown in Figure 2.33.[196] This structural work has been complemented by solution-phase binding studies, from which K_a values of 6.6×10^5 and 6.7×10^6 M^{-1} were derived for the binding of berenil (**2.122**) and furamidinium (**2.125a**), respectively, to d(CGCGAATTCGCG)$_2$ in MES20 buffer.[197] Given these interactions, it is perhaps not surprising that amidinium groups are among the more important pharmacophores currently being explored in the context of rational drug design.[198]

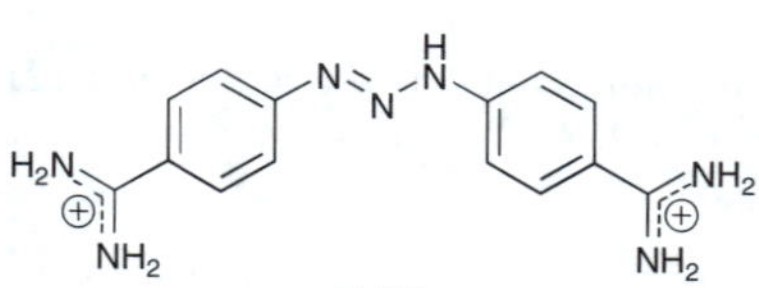

2.122

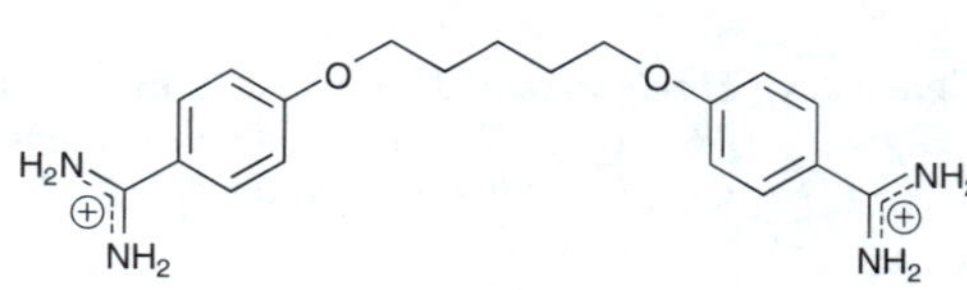

2.123

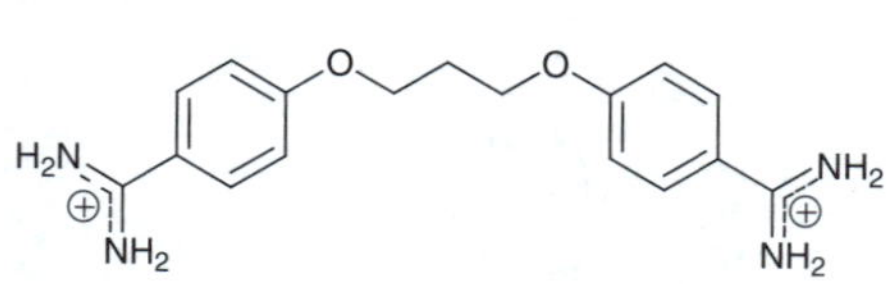

2Cl⁻

2.124

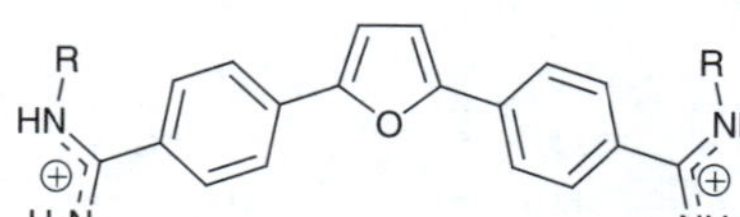

2.125a R = H
2.125b R = *i*-Pr
2.125c R = cyclopropyl
2.125d R = cyclobutyl

2Cl⁻

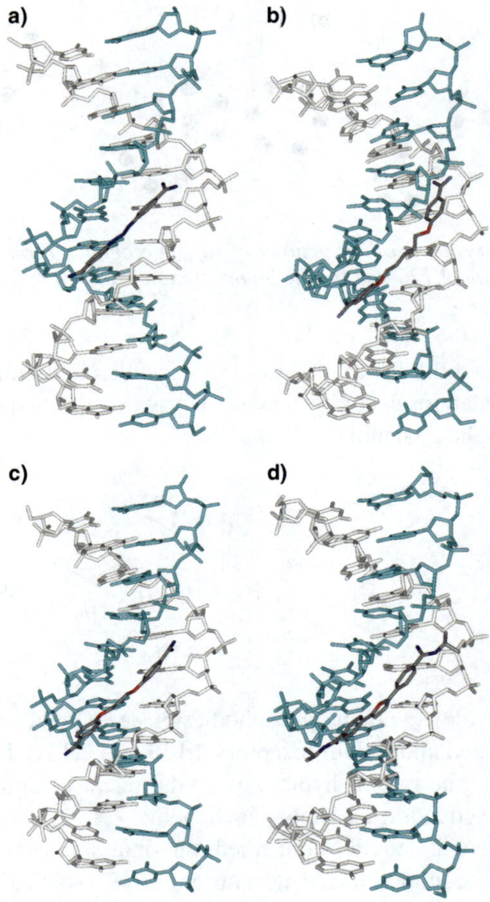

Figure 2.33 *Single crystal X-ray structures of the complexes formed between d(CGC-GAATTCGCG)$_2$ and (a) berenil (**2.122**), (b) pentaamidinium (**2.123**), (c) propamidinium (**2.124**), and (d) the furan derivative (**2.125b**)*

Eschenmoser and co-workers[199] were the first to provide evidence that the amidinium group could be used to effect anion recognition; these researchers showed that compound **2.126** was capable of extracting hydrogen sulfate from an aqueous media into an organic phase. Subsequently, A.P. Davis and co-workers[200] reported the X-ray crystal structure of the complex formed between the bicyclic amidinium form of **2.126** and nitronate **2.129b** as shown in Figure 2.34a. Based on supporting NMR spectroscopic analyses involving **2.129a**, the formation of this complex was thought to reflect proton transfer from the nitroalkane to the neutral form of amidine **2.126**. Two years later, the X-ray crystal structure of the complex formed between the protonated form of **2.127** and *S*-naproxenate was obtained (*cf.* Figure 2.34b).[201] It confirmed the ability of amidinium units to interact with oxyanions via a combination of electrostatic and hydrogen-bond interactions, as reflected in the NH···O distance of 1.76 Å seen in this particular structure.

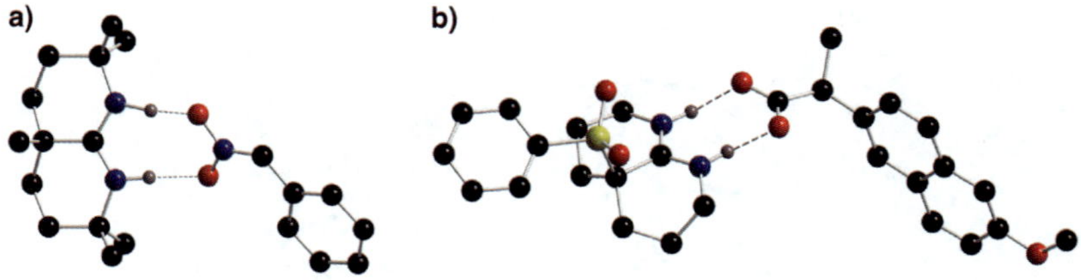

Figure 2.34 *Single crystal X-ray structures of (a) the complex formed between nitronate **2.129b** and **2.126** and (b) the S-naproxenate complex of **2.127***

Davis also prepared the chiral, non-racemic bicyclic amidinium **2.128**. However, little evidence of enantioselectivity was seen in extraction experiments involving L- or D,L-*N*-(BOC)-phenylalaninate.

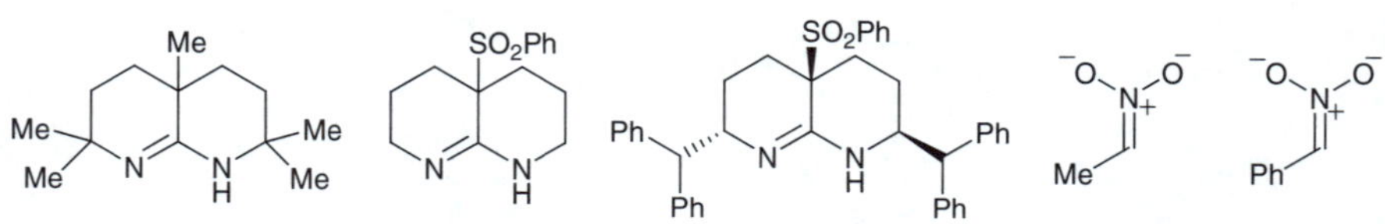

In an effort to develop synthetic phosphodiesterases, Göbel and co-workers synthesized the elaborated amidinium receptors **2.130** and **2.131**. These systems were expected to enhance the rate of hydrolysis as the result of amidinium–phosphate complexation and activation as shown in Scheme 2.8. Concentration-dependent ^{1}H NMR spectroscopic studies confirmed the interaction between **2.130** and dimethylphosphate sodium salt and acetate anion (K_a = 190 and 1300 M^{-1} in DMSO-d_6, respectively), as well as the formation of various control complexes such as **2.133** in this and other solvents.[202] In addition, evidence for the formation of a complex between **2.130** and dimethyl phosphate in the solid state was obtained from X-ray structural analyses (*cf.* Figure 2.35a). X-ray diffraction analysis (*cf.* Figure 2.35b) was also used to provide an unambiguous identification of the putative reaction product, allowing NMR spectroscopic methods to be used to follow the course of the reaction. Once the above predicative studies were complete, the authors analyzed the ability of **2.130** and **2.131** to enhance the hydrolysis of dimethylphosphate. Rate enhancements of *ca.* 3000 and 9000 were observed relative to **2.134**.

As a part of these studies, the chiral amidinium receptor **2.132** was tested. This species offers the advantage that it allows the host–guest interaction to be monitored by following the substrate-derived changes (if any) observed in the CD spectrum. For instance, the negative peak around 420 nm seen in the CD spectrum of **2.132** disappears in the presence of **2.133**. Complementary ^{1}H NMR spectroscopic titrations, carried out in DMSO-d_6, confirmed that **2.132** binds dimethyl phosphate sodium salt with an association constant (K_a) of 250 M^{-1}.

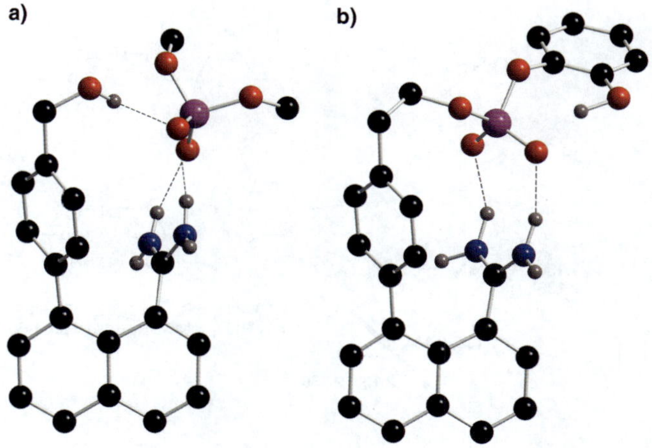

Figure 2.35 *Single crystal X-ray diffraction structures of (a) the dimethyl phosphate complex of **2.130** and (b) the product from the reaction of **2.131** and **2.133***

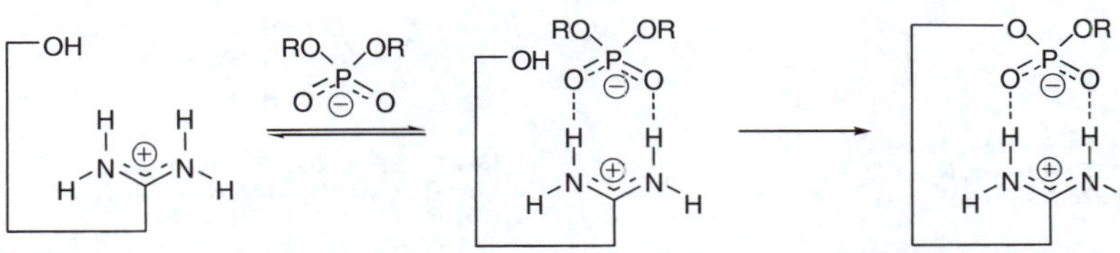

Scheme 2.8 *Proposed mechanism of supramolecular transesterification effected within an ion-pair complex*

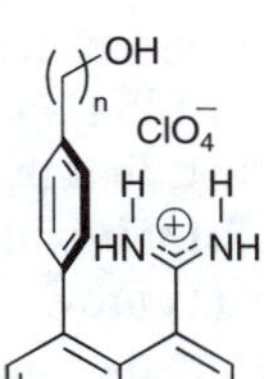

2.130 n = 1
2.131 n = 2

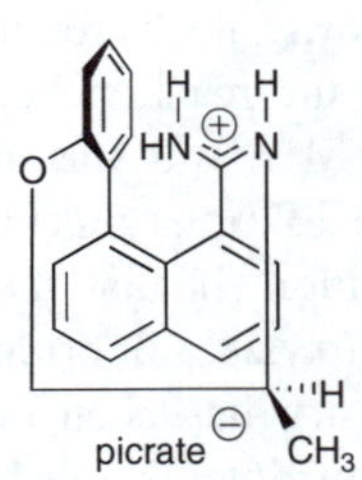

2.132

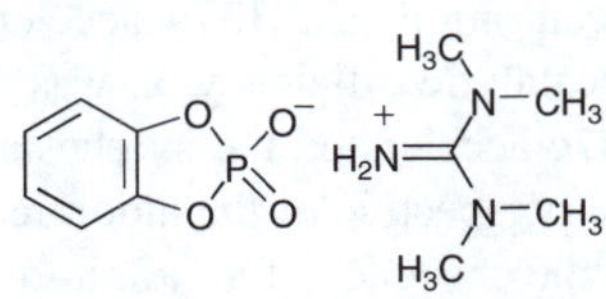

2.133

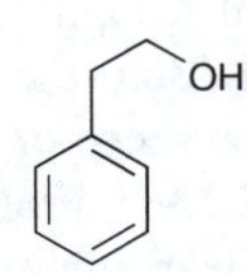

2.134

In work of a very different nature, an artificial self-replicating system that relied on the use of an amidinium–carboxylate salt bridge was put forward by von Kiedrowski and Terfort[203] (Scheme 2.9). Based on the ^{1}H NMR spectroscopic titration studies, using benzamidinium tetrafluoroborate and cesium acetate, it was proposed that the amidinium–carboxylate complexation would be sufficiently strong so as to allow the self-replication process to be viable (K_a = 350 M^{-1} in DMSO-d_6 at 35 °C). The reaction between **2.135** and **2.136** yielded **2.137**, a "template" that is capable of acting as an autocatalyst for the further formation of **2.138** (*cf.* Scheme 2.9). As evidence in

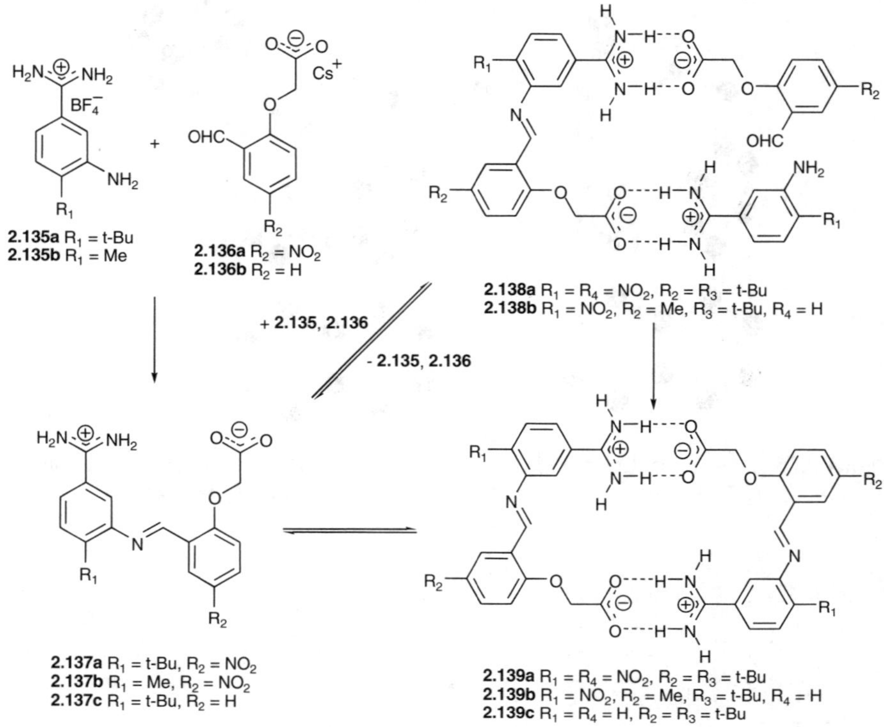

Scheme 2.9 *Reactions associated with the generation of a proposed autocatalytic, self-replicating system*

support of the proposed autocatalytic reaction pathway, it was found that the presence of a higher concentration of **2.137a** accelerated the reaction between **2.135a** and **2.136a** (the autocatalytic efficiency, ε, was 16.4 $M^{-1/2}$). It was also found that the presence of **2.137c** accelerated the synthesis of **2.137b** via reaction of **2.135b** with **2.136a**. From the perspective of the anion recognition chemist, these results provide clear support for the existence of amidinium–carboxylate salt bridges during the various reactions in question. Other related chemistry has been described by these researchers, including scenarios that are based on the interactions between water-soluble amidinium moieties and carboxylate anions.[204]

In a unique use of the interaction between carboxylate anions and amidinium cations, the Nocera group generated the functionalized porphyrin systems, **2.140–2.143**. These systems were used to probe the mechanism of proton-coupled electron transfer (PCET), a process which is important not only in terms of understanding the basics of energy conversion in photosynthesis but also the function of many proteins and enzymes.[205] Specifically, these systems were designed to model the arginine–aspartate salt bridge, which plays an important role in stabilizing many biological structures.[184–186,206]

Initial studies were performed with porphyrin **2.140** as the electron donor and 3,4-dinitrobenzoic acid as the electron acceptor. The strength of the carboxylate-anion complex formed between these two species was determined from a dilution experiment carried out in DMSO-d_6. In particular, by following the change in the chemical shift of the NH peak as a function of the overall concentration of the constituents using ^{1}H NMR spectroscopy, a K_a value of 3500 M^{-1} could be calculated.[207] In the presence of benzoic acid, the luminescence of the neutral form of **2.140b** in CH_2Cl_2 was seen to decrease, with the excited singlet state lifetime of the resulting complex being 434 ps (λ_{exc} = 580 nm, λ_{det} = 760 nm, where "exc" and "det" refer to excitation and detection, respectively). Replacing the benzoic acid by 3,4-dinitrobenzoic acid did not change the luminescence intensity but did serve to reduce the excited state lifetime of the complex to 327 ps. From this latter result, a unimolecular rate constant for PCET of 7.5×10^8 s^{-1} was calculated. It was also concluded that electron transfer from the singlet excited state of **2.140b** to the bound 3,4-dinitrobenzoate anion occurs through the amidinium–carboxylate salt bridge.

Two years later, similar experiments were performed using a slightly different porphyrin donor, **2.141**, which bears an amidinium subunit directly linked to a β-pyrrolic position.[208] In this case, a single crystal X-ray structure of the complex formed between **2.141a** and benzoate anion could be obtained. As seen in Figure 2.36, the expected salt bridge is apparent in the solid state (the NH···O distances are 1.80 and 1.87 Å). This salt bridge also proved to be very stable in aprotic solutions, such as DMSO and CH_2Cl_2. This was confirmed by NMR and UV–Vis spectroscopic titrations, from which association constants (K_a) for the complexes formed between **2.141a** and benzoate (3,5-dinitrobenzoate) anion of 1550 (267) M^{-1} and 5.7×10^5 (7.7×10^5) M^{-1} were obtained in DMSO-d_6 and CH_2Cl_2, respectively. The luminescence of porphyrin **2.141b** was seen to decay monoexponentially with a lifetime of 1.5 ns following excitation in the presence of excess benzoate. By contrast, in the case of the more electron deficient acceptor, 3,5-dinitrobenzoate anion, the rate of excited state decay was substantially increased (k_{PCET} = 6.4×10^7 s^{-1}).

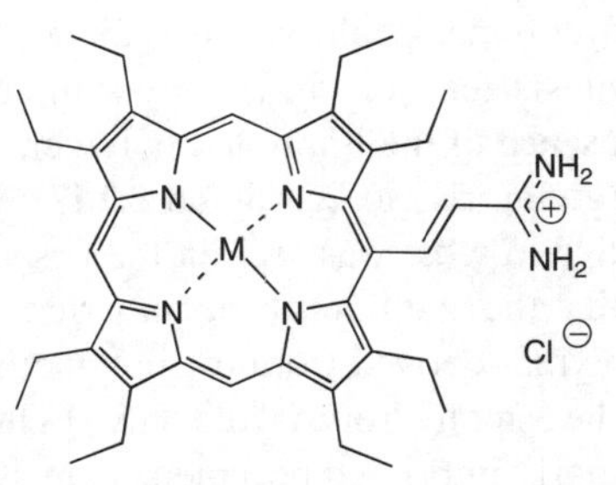

2.140a M = Ni
2.140b M = Zn

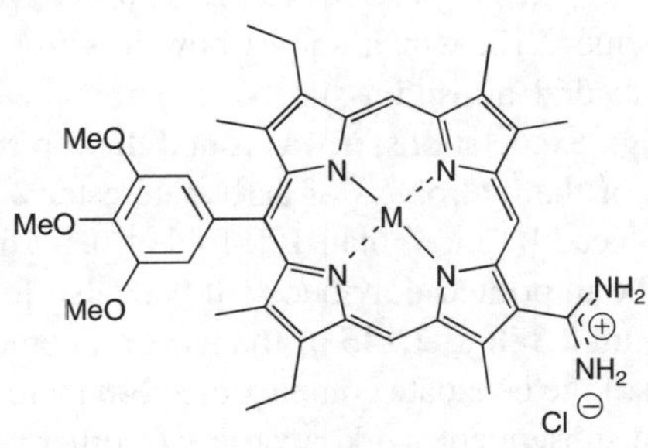

2.141a M = Ni
2.141b M = Zn

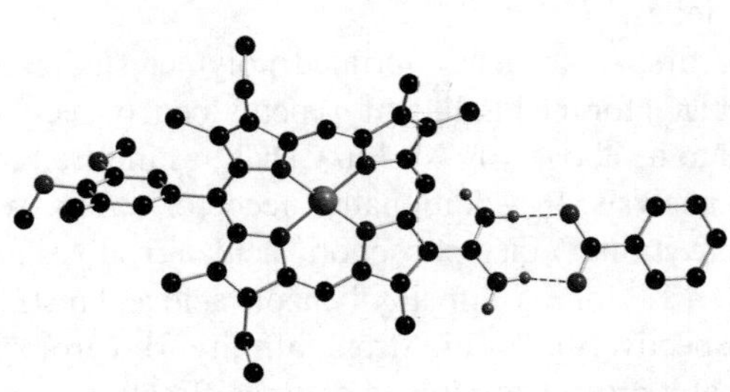

Figure 2.36 *Single crystal X-ray structure of the benzoate anion complex of* ***2.141a***

In follow-up studies, the structurally homologous β- and *meso*-amidinium porphyrins, **2.142a–2.143b**, were tested as the photodonors in PCET models analogous to those described above.[209] Based on these studies, it was concluded that an amidinium group attached to a β-pyrrolic position is more strongly coupled electronically to the porphyrin ring than one that is linked to the meso position. Extensions of this approach have been reported by the Nocera group, including the analysis of a multicomponent systems containing Ru(II)bipyridine, as well as porphyrin, photoactive groups. The interested reader is referred to the primary literature.[210]

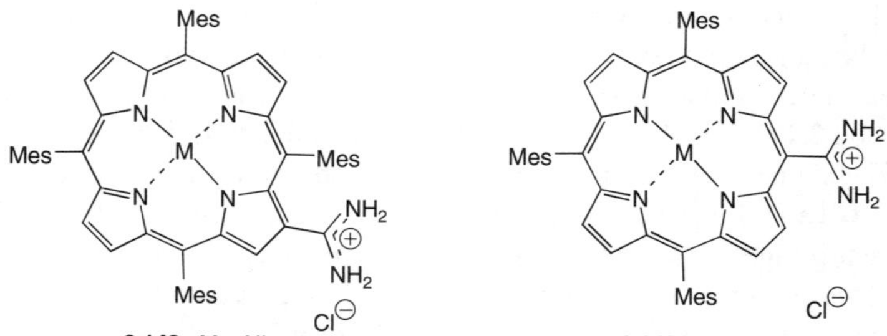

2.142a M = Ni
2.142b M = Zn

2.143a M = Ni
2.143b M = Zn

In 1997, Wulff and co-workers[211] prepared a molecularly imprinted polymer obtained via the polymerization of **2.148** using 80% ethylene dimethylacrylate as a cross-linker and the phosphonic monoester **2.144** as a template. Following polymerization and work-up, removal of the template produced a polymer with two amidinium units inside the cavity. These embedded groups, it was proposed, could act as binding sites or serve as catalytically active groups. This latter point is illustrated in Scheme 2.10, which shows how this imprinted polymer could act to stabilize the tetrahedral transition state of a carbonate hydrolysis reaction. In accord with such design expectations, it was found that, in the presence of the imprinted polymer, the rate of the hydrolysis of carbonate ester **2.145** (giving rise to **2.146** and **2.147**) was enhanced by more than 100-fold at pH 7.6 relative to what was seen in the absence of the imprinted polymer.[211] It was also found that the addition of monomeric amidinium **2.148** to **2.145** or the use of a control polymer derived from the polymerization of the benzoate complex of **2.148** increased the ester hydrolysis rate only slightly.

In subsequent work, a variety of other molecularly imprinted polymers were synthesized via the polymerization of **2.148** with various templates; many showed high catalytic activity and good selectivity for the hydrolysis of carbonate- and carbamate-containing moieties.[212,213]

Slightly after the first molecular imprinted polymer studies were reported, the association constant (K_a) for the binding of *p*-methylbenzoic acid to **2.148** was determined; it was found to be above 10^6 M^{-1} in $CDCl_3$, as inferred from an NMR spectroscopic titration analysis.[212] Additionally, receptor **2.148** was found to bind carboxylic acids more strongly than phosphoric acid derivatives in CD_3CN ($K_a = 1.2 \times 10^4$ and 4.6×10^3 M^{-1} for 3,5-dimethylbenzoic acid and bis(3,5-dimethylphenyl) phosphoric acid, respectively).[214] This strong affinity for carboxylic acids was supported by the fact that it proved possible to obtain diffraction grade single crystals of

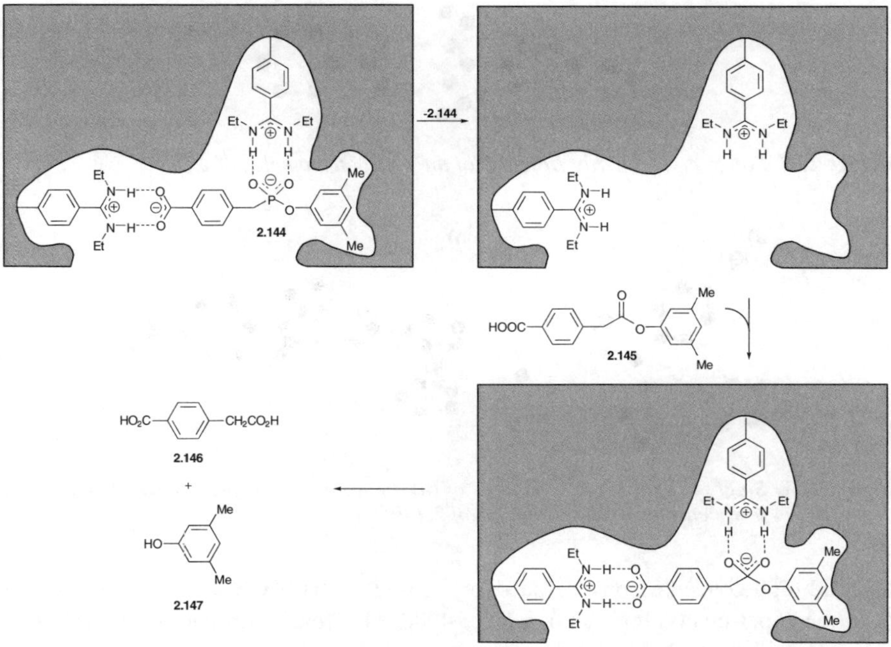

Scheme 2.10 *Schematic representation of the molecular imprinting process and proposed ester hydrolysis mechanism*

the benzoate complex of the protonated form of **2.149**, a saturated analogue of **2.148**. The resulting structure, shown in Figure 2.37, is characterized by a high degree of symmetry and is thus considered consistent with a recognition process that is stabilized via a combination of electrostatic and hydrogen-bonding interactions (the NH···O distance is 1.85 Å).[215]

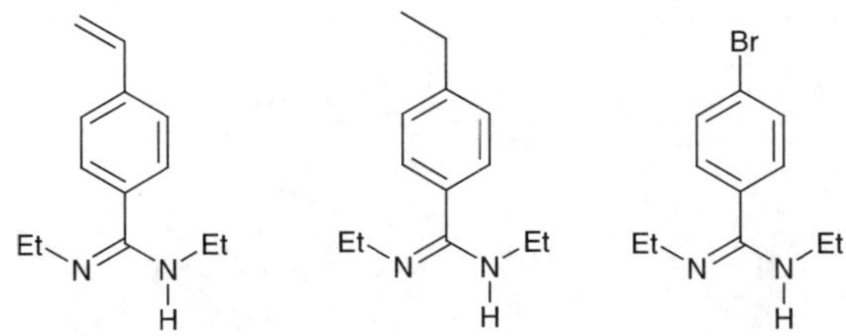

2.148 **2.149** **2.150**

In 2001 Kraft and co-workers reported that the *N,N′*-substituted benzamidinium **2.150** binds to tetrazoles. While strictly speaking work that falls outside the purview of this book, this finding is important since tetrazoles can be considered as bioisosteric replacements for carboxylic acids.[216] An X-ray crystal analysis revealed that **2.150** exists in an *E,Z*-type conformation when complexed to tetrazole and that an infinite chain, held together by hydrogen bonds and salt bridges, is stabilized in the solid state (Figure 2.38a). In $CDCl_3$ solution, however, the binding mode proved to be a function of concentration, with **2.150** adopting *E,E*- and *E,Z*-like conformations at

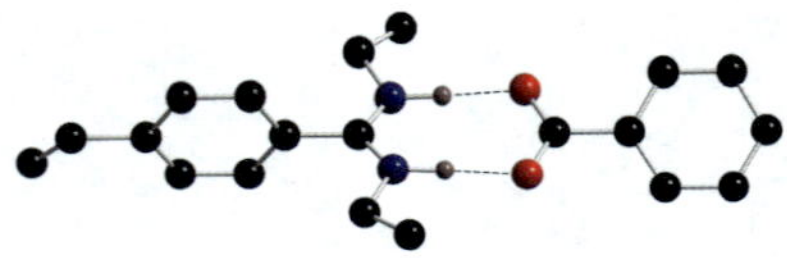

Figure 2.37 *Single crystal X-ray structure of the benzoate complex of* ***2.149***

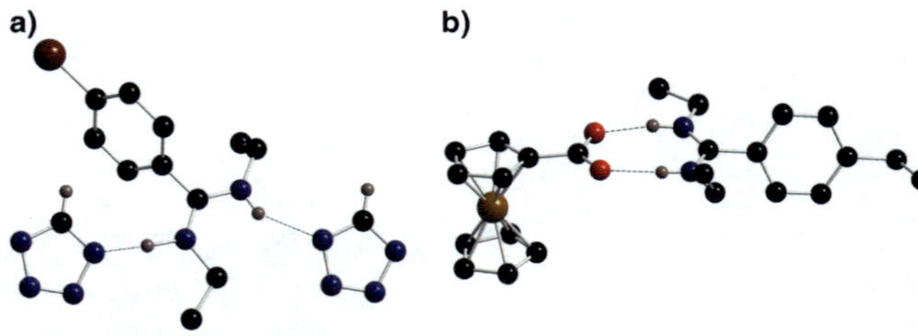

Figure 2.38 *Single crystal X-ray structures of (a) the tetrazole complex of* ***2.150*** *and (b) the ferrocenecarboxylate complex of* ***2.149***

low and high concentrations, respectively. Based on ^{1}H NMR spectroscopic studies (dilution experiments; Job plots), a K_a of 4000 M^{-1} for the formation of a 1:1 complex in $CDCl_3/CD_3CN$ (6:1) was inferred.

An interesting feature of amidinium–carboxylate salt bridges has recently been described by Cooke and Rotello. These workers noted that in the presence of excess **2.149** in CH_2Cl_2, the half-wave potential of ferrocenecarboxylic acid was shifted from 0.81 to 0.54 V.[217] This remarkably large negative shift (26 kJ mol^{-1}) led to the suggestion that the salt bridge stabilizes the ferrocenium state of ferrocenecarboxylic acid. The formation of the proposed salt bridge between **2.149** and ferrocenecarboxylate was confirmed by X-ray diffraction analysis, as shown in Figure 2.38b.

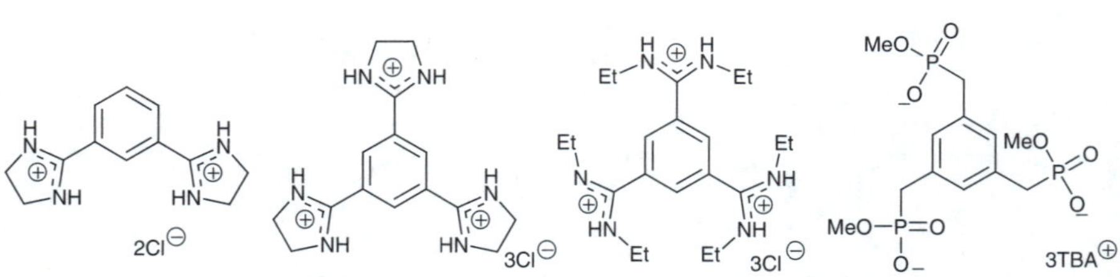

2.151 **2.152** **2.153** **2.154**

In an extension of this generalized approach to molecular recognition, Kraft and Fröhlich[218] prepared the doubly- and triply-substituted amidinium receptors, **2.151–2.153**. In the solid state, the complex formed between **2.152** and the trifluoroacetate counter anions is stabilized by a combination of electrostatic interactions and six well-defined hydrogen bonds. The net result is a complex with C_3-type symmetry, as shown in Figure 2.39. By contrast, the use of a symmetrical trisphosphate (**2.154**) led to the production of a simple 1:1 complexes in the case of the triply-substituted receptors **2.152** and **2.153**, as inferred from ^{1}H NMR spectroscopic studies carried out in CD_3OD solution (*e.g.*, for the interaction of **2.153** with **2.154**, $K_a = 1.1 \times 10^6\ M^{-1}$).[219]

In 1998, two bis-amidinium calixarenes, **2.155** and **2.156**, prepared in the form of their chloride-anion salts, were synthesized by Gale[220] as templates for the formation of self-assembled ditopic receptors. Based on UV–Vis spectroscopic titrations, both receptors were found to bind carboxylate anions, such as 4-nitrobenzoate, **2.157a**, and **2.157b**, in DMSO solution in a 1:2 (amidinium:carboxylate) stoichiometric ratio. Three years later, the ability of receptors **2.155** and **2.156** to bind bis-carboxylate anions was probed using NMR spectroscopic methods and single crystal X-ray structural analysis.[221] While the crystal structure of the picrate complex (*cf.* Figure 2.40a) reveals a 1:2 host–guest stoichiometry, the corresponding malonate complex in Figure 2.40b is characterized by three different binding modes. The crystal structure of the complex between **2.155** and difluorophosphinate anion was also solved; interestingly, it confirmed the presence of hydrogen bonds between the amidinium NH protons and both the fluorine and oxygen atoms present in the anionic guest (Figure 2.40c). The occurrence of difluorophosphinate in this complex was unexpected and resulted from the use of an impure batch of silver hexafluorophosphate.

The above finding highlights a potential problem that warrants special mention. Hexafluorophosphate anion is commonly used as counter anion in charged anion receptor systems. Care must always be taken to ensure that $PF_2O_2^-$ is not present. This can easily be achieved by a ^{31}P NMR analysis. A similar phenomenon was observed in the first anion complex of sapphyrin, which although crystallized in the presence of HPF_6, formed a mixed fluoride/hexafluorophosphate salt (see Chapter 3).

In solution, the association constants for the step-wise interaction of **2.155** with acetate in DMSO-d_6 were also studied; the resulting association constants were

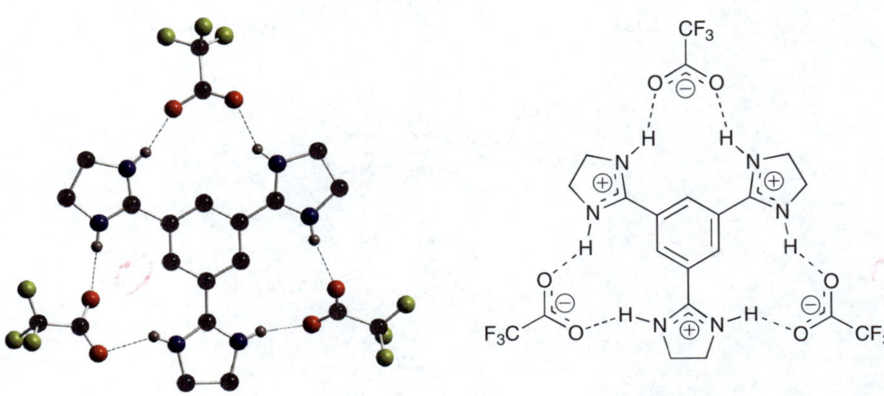

2.152-$3CF_3CO_2^-$

Figure 2.39 *Single crystal X-ray structure and schematic representation of the tris-trifluoroacetate-anion complex of* ***2.152***

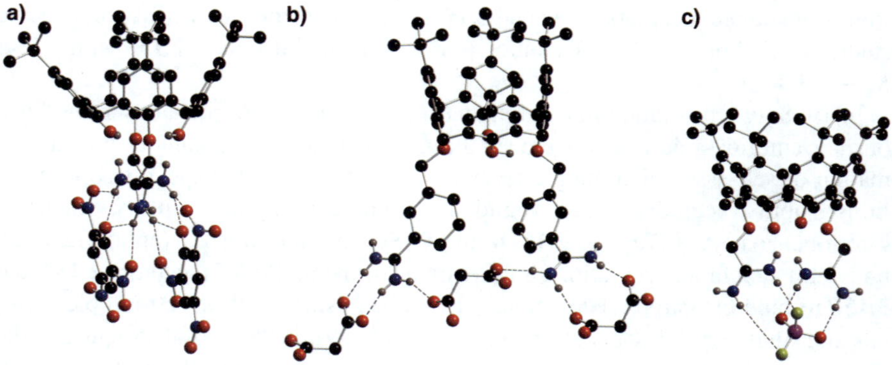

Figure 2.40 *Single crystal X-ray structure of (a) the picrate complex of **2.155**, (b) the malonate complex of **2.156**, and (c) the difluorophosphinate complex of **2.155***

found to be $K_{a1} = 990\ M^{-1}$ and $K_{a2} = 960\ M^{-1}$, as inferred from ^1H NMR spectroscopic titrations.[221]

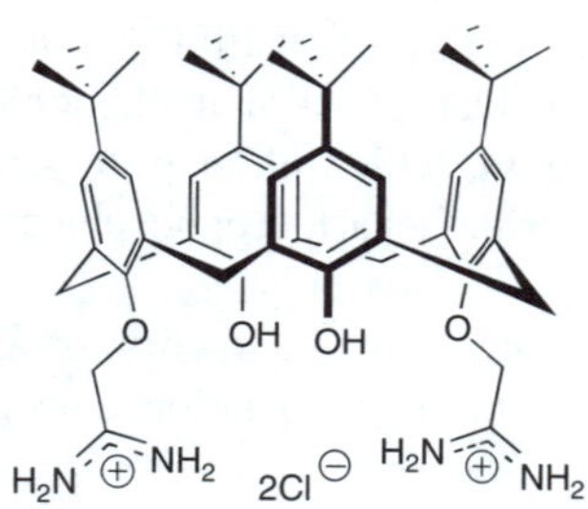

2.155

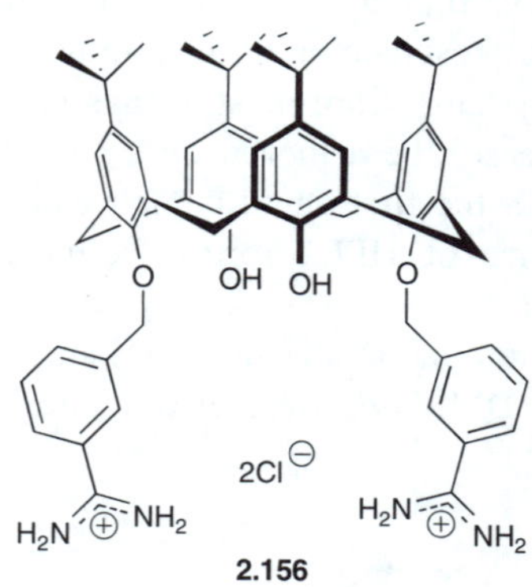

2.156

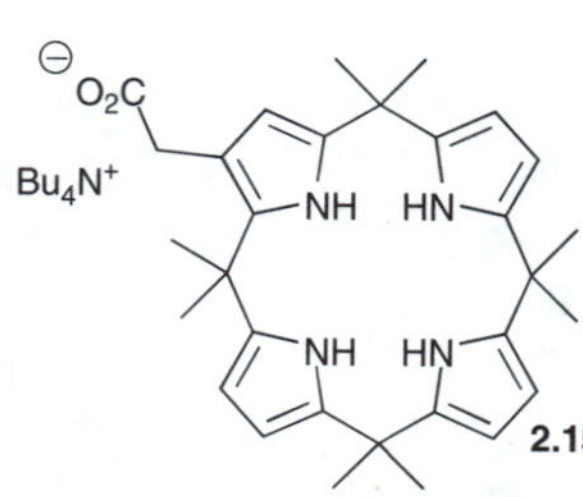

2.157a

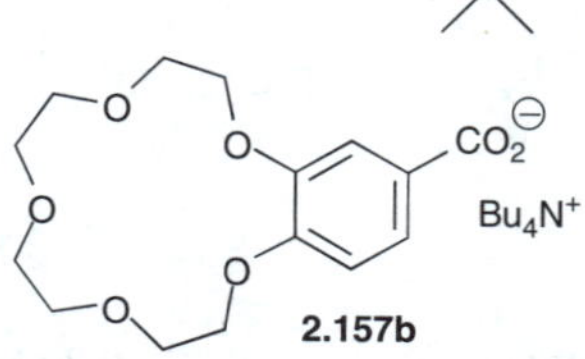

2.157b

In an effort to develop agents that are able to achieve the selective recognition of dicarboxylate substrates, Diederich and co-workers[222] prepared the cleft-type diamidinium receptor (±)-**2.158**; this was done by attaching two phenylamidinium motifs to a 1,1′-binaphthalene scaffold. NMR spectroscopic titrations confirmed that receptor (±)-**2.158** forms strong complexes with dicarboxylate anions in competitive protic solvents (*e.g.*, K_a = 8200, 5100, 8300, and 10,000 M^{-1} for the binding of the TBA salts of glutarate, **2.159a**, **2.159b**, and **2.159c**, respectively, in CD_3OD). However, no appreciable selectivity for the series of rigid-isophthalate anions was seen relative to the more flexible glutarate guest. The finding was rationalized in terms of non-directional coulombic charge–charge interactions, rather than oriented hydrogen bonds, dominating the molecular recognition process. The thermodynamics of the various host–guest-binding events were obtained from a van't Hoff analysis of the variable-temperature ^{1}H NMR spectroscopic titrations and from ITC analyses. Taken together, these studies revealed that complexation in MeOH is strongly entropically driven with an unfavourable enthalpic change. For example, the interaction between (±)-**2.158** and **2.159a** yield values of ΔH = 7.7 kJ mol^{-1} and $T\Delta S$ = 29.0 kJ mol^{-1} from an NMR base van't Hoff analysis and ΔH = 11.6 kJ mol^{-1}, $T\Delta S$ = 35.0 kJ mol^{-1} from ITC measurements, respectively.

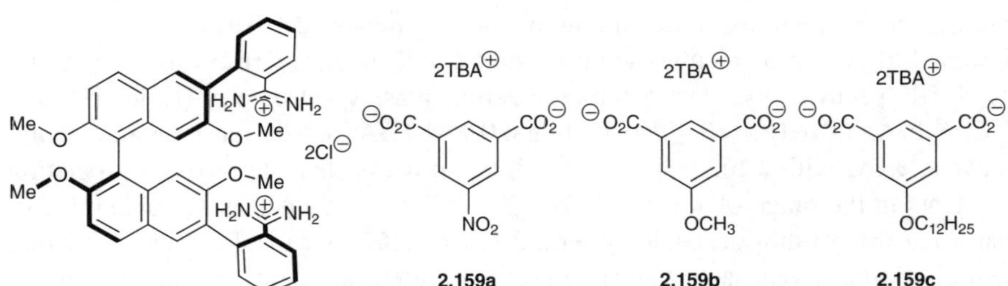

2.158 **2.159a** **2.159b** **2.159c**

Diederich and Sebo also reported the synthesis of three bowl-type cavitand receptors **2.160a–2.160c**. However, anion-binding studies were performed with only **2.160c** due to the poor solubility of **2.160a** and **2.160b**. This tetraamidinium functionalized resorcin[4]arene (*i.e.*, **2.160c**) was found to bind the isophthalate anions **2.159a** and **2.159b** with a 1:2 binding stoichiometry in both CD_3OD and D_2O, as revealed by standard Job's analysis. Significant selectivity towards the sodium salt of **2.159a** (K_{a1} = 14,800 M^{-1} and K_{a2} = 3800 M^{-1}) over the sodium salt of **2.159b** (K_{a1} = 86,000 M^{-1} and K_{a2} = 7700 M^{-1}) was observed in D_2O.

Receptor **2.160c** was also found to form strong 1:1 complexes with various nucleotide phosphates in D_2O. In the case of the adenosine phosphates, the association constants increased as a function of nucleotide charge (*i.e.*, the affinity order (K_a, M^{-1}) was cAMP (1400) < AMP (10,000) < ADP (48,700) < ATP (660,000) in pure D_2O). This receptor also shows moderate selectivity towards AMP (K_a = 10,000 M^{-1}) over other nucleotide monophosphate such as GMP (K_a = 5200 M^{-1}), CMP (K_a = 3500 M^{-1}), TMP (K_a = 5900 M^{-1}), and UMP (K_a = 3800 M^{-1}) in D_2O containing Tris buffer (2.5 mM, pH 8.3).

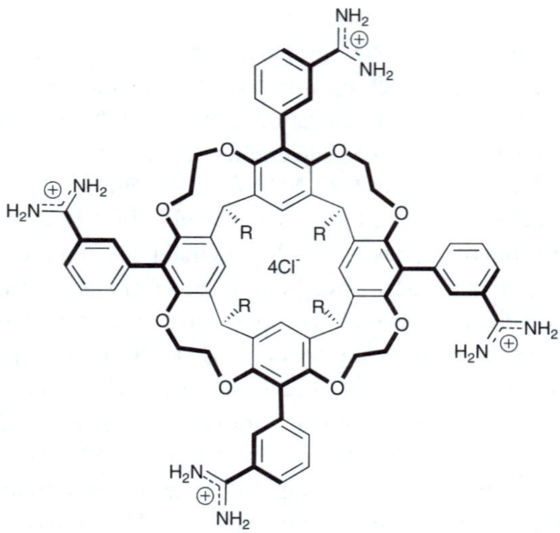

2.160a R = $(CH_2)_{10}CH_3$
2.160b R = CH_2CH_2Ph
2.160c R = $(CH_2)_3(OOCH_2CH_2)_3OCH_3$

In 2002, Reinhoudt and co-workers described an elegant way to build molecular capsules based on the use of amidinium-sulfate-anion salt bridges. In particular, the interaction between the tetraamidinium calix[4]arenes, **2.161a–2.161d**, and the designed-to-be-complementary tetrasulfonato calix[4]arene (**2.162**) was analyzed by ^1H NMR spectroscopy, electrospray ionization–mass spectrometry (ESI–MS), and ITC.[223] In MeOH/H_2O (60/40 containing 0.01 M TBA–perchlorate), the interaction between **2.161** with **2.162** is an entropy-driven process characterized by association constants in the range of 1.1×10^6 to 8.5×10^6 M^{-1}. Evidence was also put forward that the resulting capsule system, **2.161a–2.162**, is capable of encapsulating neutral guests. A year later, the same group reported the X-ray crystal structure of **2.161a–2.162** (*cf.* Figure 2.41), as well as a slightly different capsule system based on amidinium–carboxylate salt bridges, namely **2.161a–2.163**.[224]

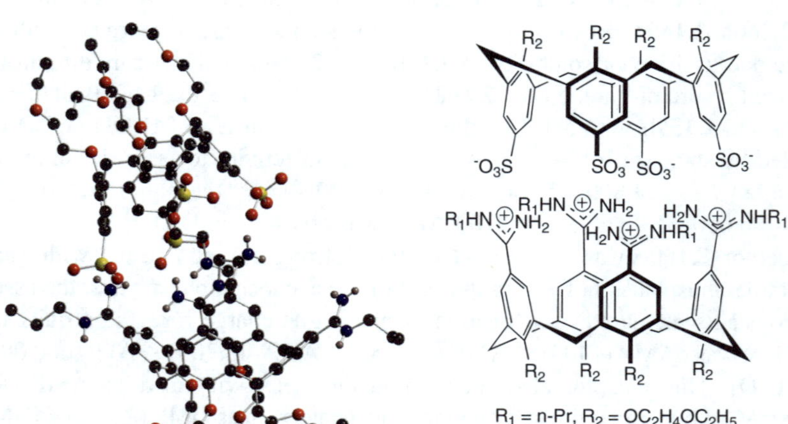

2.161a-2.162

Figure 2.41 *Single crystal X-ray structure and schematic representation of the capsule formed between **2.161a** and **2.162** (referred to in the text as **2.161a–2.162**)*

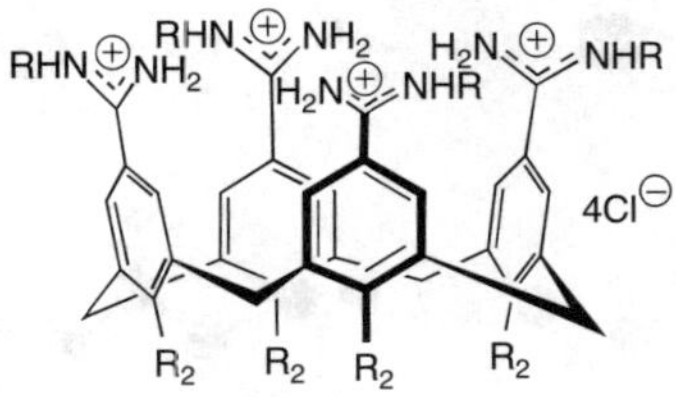

2.161a R = n-Pr
2.161b R = i-Pr
2.161c R = *p*-tolyl
2.161d R = H

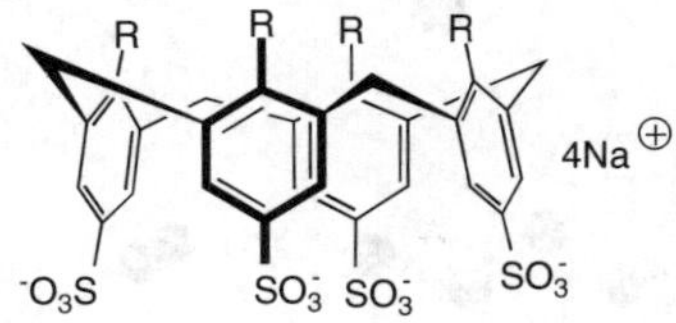

2.162 R = $OC_2H_4OC_2H_5$

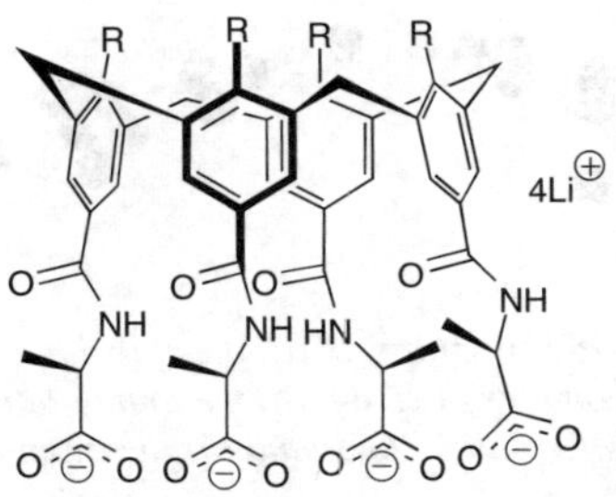

2.163 R = $OC_2H_4OC_2H_5$

2.5 Imidazoliums

In recent years, a relatively new cationic subunit for anion recognition has been introduced, namely the 1,3-disubstituted imidazolium motif. While most other cationic anion complexing systems (*e.g.*, protonated polyammonium, guanidinium, amidinium, and thiouronium) rely on a combination of electrostatic interactions and $^+NH \cdots A^-$ hydrogen bonds to achieve anion (A^-) recognition, receptors based on the imidazolium cation stabilize the corresponding anion complexes via a combination of electrostatic interactions and $^+CH \cdots A^-$ type hydrogen bonds. Thus, compared to systems based on NH hydrogen bond donors, imidazolium-based receptors can offer a significant advantage, namely pH-independent binding.

2.5.1 Acyclic Systems

Over the past decade, a considerable number of studies involving the use of 1,3-dialkylimidazolium chlorides have appeared. Many of these studies have been inspired by the fact that dialkylimidazolium chlorides often display noteworthy ionic liquid properties which, in turn, has made them useful as electrolytes in batteries and photoelectrochemical cells, as well as in electroplating applications.[225] Although once a point of controversy, the fact that systems such as **2.164** can engage in CH-anion hydrogen-bonding interactions is now widely accepted.[226,227] For instance, such hydrogen bonds have been observed in a number of imidazolium halide salt structures determined by single crystal X-ray diffraction analysis.[227,228] Figure 2.42 shows several selected crystal structures of anion salts of imidazolium cations **2.164–2.166**.

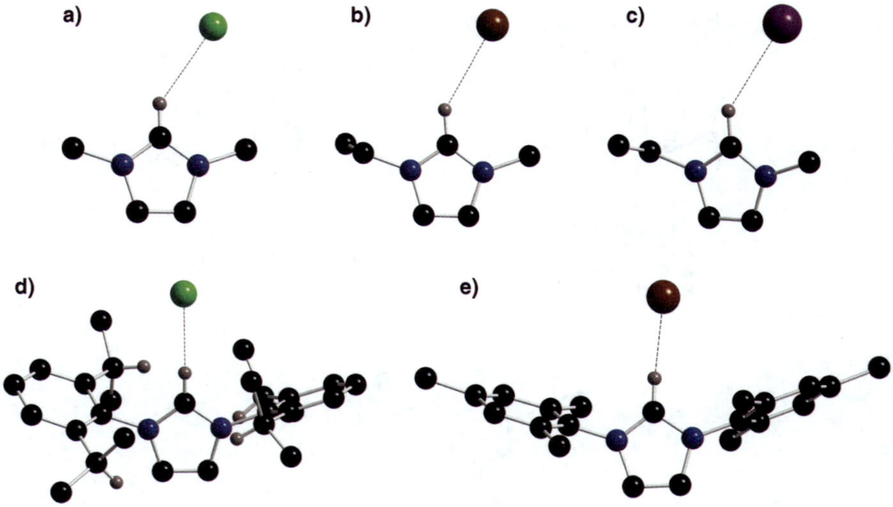

Figure 2.42 *Single crystal X-ray structures of (a) the chloride-anion complex of* ***2.165****, (b) the bromide-anion complex of* ***2.164****, (c) the iodide-anion complex of* ***2.164****, (d) chloride-anion complex with* ***2.166a****, and (e) the bromide-anion complex of* ***2.166b***

Such solid-state analyses have been complemented by ESI–MS studies, which have allowed the relative strength of the putative hydrogen-bond interactions between various imidazolium cations and anions to be determined in the gas phase.[229]

In 1994, using multinuclear NMR spectroscopy, Welton and co-workers[230] provided evidence that **2.164** can interact with halide anions in solution. In particular, it was found that the series of complexes **2.164**-X (X = Cl^-, Br^-, and I^-) displayed distinct (and different) CH resonances in their respective ^1H and ^{13}C NMR spectra as recorded in $CDCl_3$. The greatest differences were seen in the case of the signals for the C(2)H group located between the two imidazolium nitrogen atoms. However, differences were also seen for the C(4)H and C(5)H resonances. Such findings provide strong support for the notion that both cation–anion electrostatic and CH-anion hydrogen-bond interactions contribute to anion binding under these solution-phase conditions. Later on, quantitative studies, involving analysis of the structural analogue **2.165** (hexafluorophosphate salt), were carried out by Sato and co-workers[231]; this yielded association constants (K_a) of 78, 59, and 28 M^{-1} for chloride, bromide, and iodide, respectively, in CD_3CN.

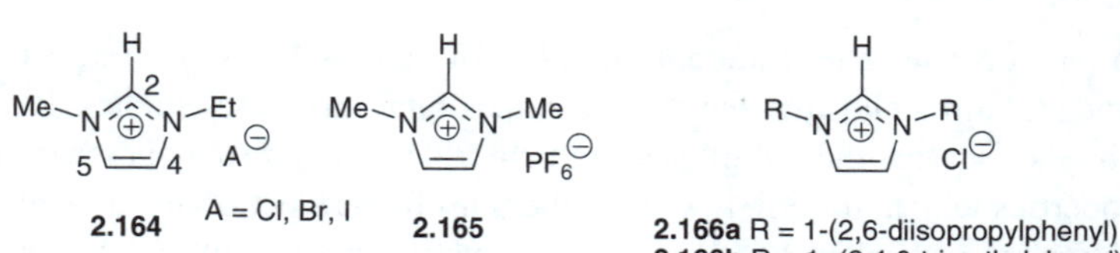

Two bis-imidazolium receptors, **2.167** and **2.168**, that contain a pair of imidazolium subunits connected through a benzene spacer, were also prepared by the Sato group; not surprisingly, these systems proved to be better anion-binding agents than the corresponding monoimidazolium analogues.[231,232] For instance, ^1H NMR titration studies revealed that the association constant (K_a) corresponding to the formation of **2.167**-Cl^- in CD_3CN was 1300 M^{-1}. Similarly, K_a values of 140 and 60 M^{-1} were determined for the chloride and bromide complexes of **2.168**, respectively, in DMSO-d_6.[231,232] Subsequently, the fluoride anion affinity of **2.167** (K_a = 990 M^{-1} in CD_3CN) was determined on the basis of competitive ^{1}H NMR spectroscopic titration studies.[233]

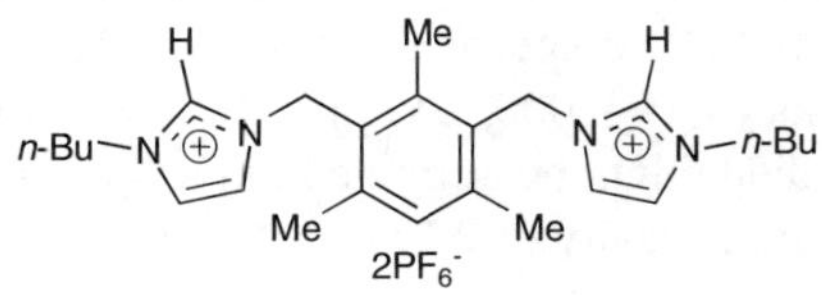

2.167

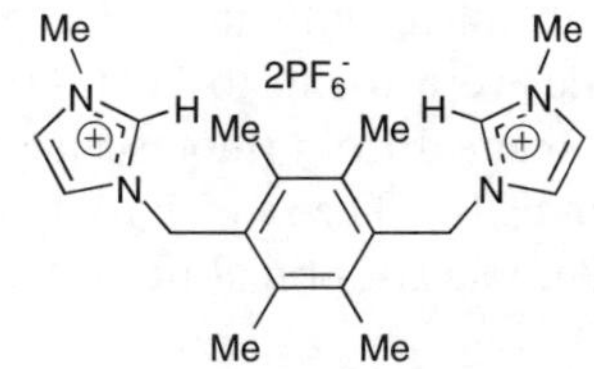

2.168

Recently, K.S. Kim and J. Yoon designed and prepared two new imidazolium-based fluorescent chemosensors, namely **2.169** and **2.170**. These systems rely on a bridging anthracene fluorophore and were designed to target various biologically important phosphate anions.[234]

In the case of **2.169**, fluorescence spectral titrations revealed K_a values of ~1,300,000, 7900, 4500, and 600 M^{-1} for $H_2PO_4^-$, Cl^-, Br^-, and I^-, respectively, in acetonitrile.[234] The high-K_a value and selectivity for dihydrogen phosphate within this set of anions (the corresponding K_a for F^- binding could not be obtained due to inconsistencies in the titration curve) were rationalized in terms of the directional hydrogen-bond interactions ($^+CH \cdots X^-$) provided by the pre-organized and relatively rigid anion-binding sites present in **2.169**. However, the inherent basicity of the anions in question is surely also a contributing factor, especially to the selectivity.

The other imidazolium anthracene sensor studied by this group, namely **2.170**, was found capable of differentiating the structurally similar substrates, GTP and ATP. For instance, a complexation-induced quenching of the fluorescence intensity was observed in the presence of GTP in aqueous media, while it was enhanced upon the addition of ATP, ADP, and AMP in aqueous solution. No appreciable fluorescence changes were observed with pyrophosphate, dihydrogen phosphate, fluoride, or chloride anions. Quantitative fluorescence titration studies carried out in an aqueous solution (pH 7.4) revealed that receptor **2.170** forms a complex with GTP (K_a = 87,000 M^{-1}) that is about six times more stable than that formed with ATP (K_a = 15,000 M^{-1}) and over 100 times more stable than those formed from ADP (K_a = 614 M^{-1}) or AMP (K_a = 121 M^{-1}). This selectivity for GTP over ATP was ascribed to the presence of π–H interactions, which were calculated to be more important in the case of GTP than ATP.

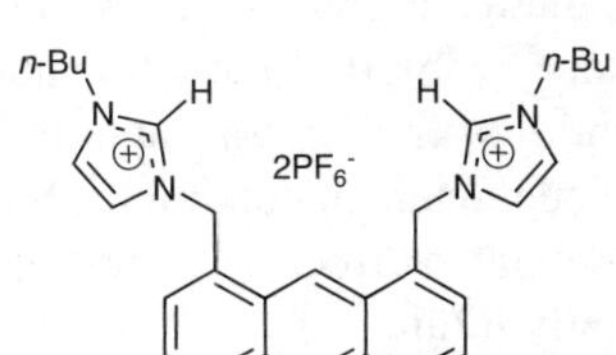

2.169

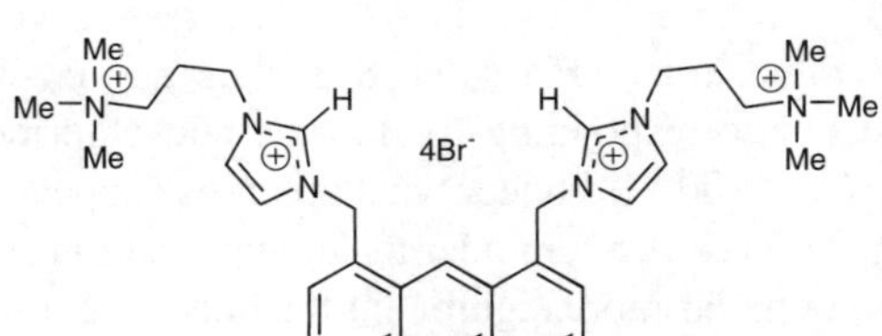

2.170

Quite recently, H. Kim and J. Kang[235] prepared an iodide-selective imidazolium receptor **2.171** consisting of two methylene bridged bis-imidazolium rings attached to a naphthalene. The halide-anion-binding properties of this receptor were investigated by ^{1}H NMR spectroscopic and fluorescence titrations. The addition of TBA halide salts to a solution of **2.171** in CD_3CN containing 10% DMSO-d_6 induced detectable downfield shift of the imidazolium CH resonance. By monitoring changes in the chemical shifts in the CH group, the association constants (K_a) for iodide and bromide were found to be 1600 and 140 M^{-1}, respectively. Also, similar values were obtained from fluorescence titrations (K_a = 5000, 243, and 185 M^{-1} for I^-, Br^-, and Cl^-). The selectivity towards iodide might be attributed to the special, and apparently near-optimal fit of this anion in the cavity of **2.171**.

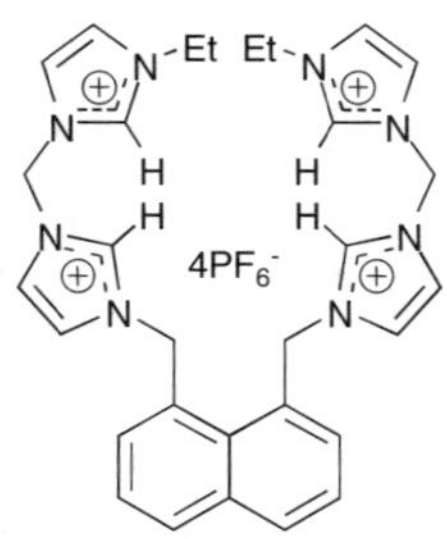

2.171

Inspired by what has been achieved using other anion-binding motifs, several groups have exploited the 1,3,5-triply-substituted benzene framework to produce tripodal imidazolium-anion receptors, such as **2.172–2.175**. As a general rule, these systems have proved effective as halide-anion-binding agents.[231–233,236,237]

In the case of **2.173**, quantitative determinations of the association constants (K_a) were made by Sato and co-workers[231] using ^{1}H NMR spectroscopic titrations carried out in CD_3CN; these studies revealed, as expected, that this pre-organized receptor forms rather stable complexes with chloride (K_a = 75,000 M^{-1}) and bromide (K_a = 46,000 M^{-1}) but binds to iodide anion only weakly (K_a = 7200 M^{-1}). Independently, Kim and co-workers determined the affinity for fluoride via a competitive ^{1}H NMR titration experiment and found that this anion is also tightly bound in CD_3CN (K_a = 210,000 M^{-1}).[231,233] In contrast, the more flexible system **2.172** was found to bind to chloride anion only weakly (K_a = 1500 M^{-1} in CD_3CN).

Slightly later, the association constants for **2.173** with various anions were reported by the Sato group;[232] in DMSO-d_6 selectivity for $H_2PO_4^-$ was observed (K_a = 760, 200, 94, 200, 290, and >10^3 M^{-1} for TEA-Cl, TEA-Br, TEA-I, TBA-HSO_4, TBA-ClO_4, and TBA-H_2PO_4, respectively). Interestingly, using the TBA salts, rather than TEA salts, K.S. Kim and co-workers[233] reported the association constants corresponding to fluoride and chloride-binding binding constants to be 2400 and 1500 M^{-1} under what otherwise appeared to be indentical conditions. Such findings serve as a reminder that counter cation and associated effects (*e.g.*, ion-pairing) can be far more significant than one might normally intuit.

In an effort to improve the anion-binding ability of these kinds of tripodal receptors, the nitro-substituted system, **2.175**, was synthesized by K.S. Kim and co-workers,[236] along with the control system, **2.174**. As inferred from ^{1}H NMR spectral studies, both receptors form 1:1 complexes with halide anions, showing good selectivity for chloride over bromide or iodide. In DMSO-d_6, it was found that the association constants of **2.175** (K_a = 4800, 490, and 90 M^{-1} for Cl^-, Br^-, and I^-, respectively) are around four times higher than those of **2.174** (K_a = 1100, 180, and 60 M^{-1} for Cl^-, Br^-, and I^-, respectively). Receptor **2.175** was also found to bind $H_2PO_4^-$ in preference over HSO_4^- (the relevant association constants are 2500 and 350 M^{-1}, respectively, as determined in DMSO-d_6).

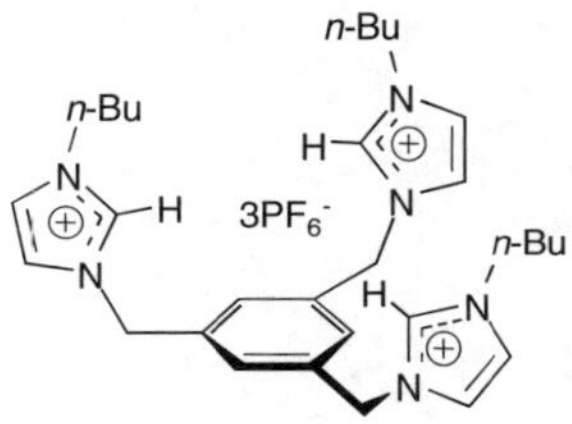

2.172

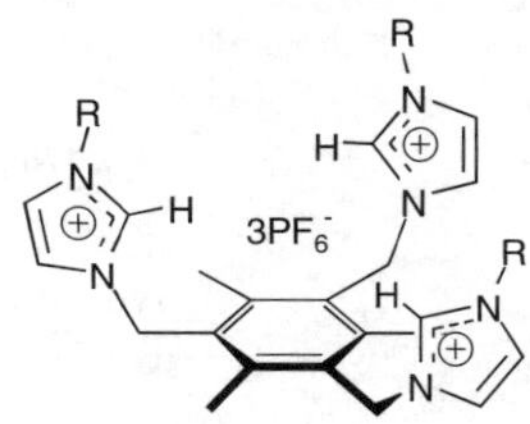

2.173 R = n-Bu
2.174 R = Me

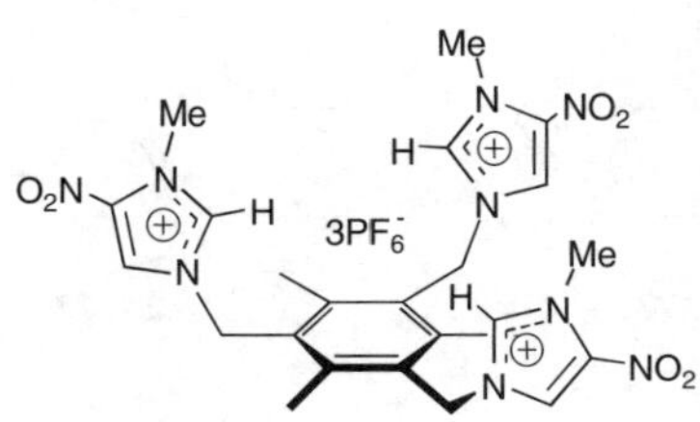

2.175

As in the case of other anion-binding agents, the cavitand scaffold has been exploited to produce a new kind of imidazolium receptor, namely the quadruple-functionalized system **2.176**.[238] This system forms a 1:1 host–guest complex with dicarboxylate anion **2.177**; however, 1:2 complex formation was observed in the case of halide anions and acetate anion, as inferred from Job plots. From ^{1}H NMR spectroscopic titrations carried out in DMSO-d_6, association constants, K_a, of 200, 2100, 8100, 7300, and 16,200 M^{-1} were determined for succinate dianion, dicarboxylates **2.177a**, **2.177b**, **2.177c**, and **2.177d**, respectively. However, only weak interactions were observed with chloride, bromide, and acetate anions ($K_a \leq 400$ M^{-1}). These observations led to the suggestion that the anion affinities are correlated with anion size and structure, in particular the length of the spacer connecting the individual carboxylate entities in anions such as **2.177**.

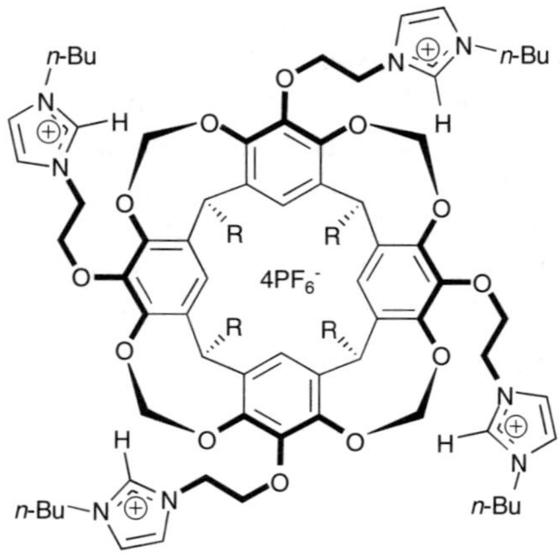

2.176 R = CH_2CH_2Ph

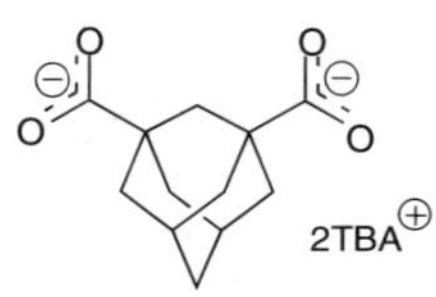

2.177a

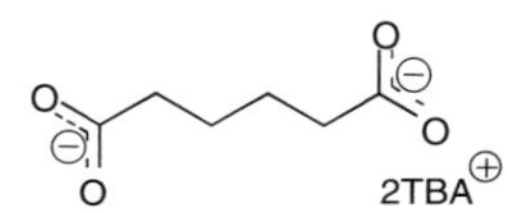

2.177b

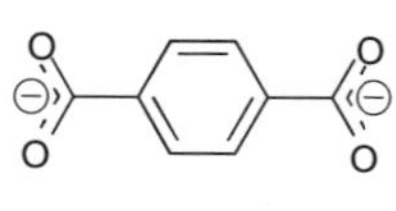

2.177c 2TBA$^{\oplus}$

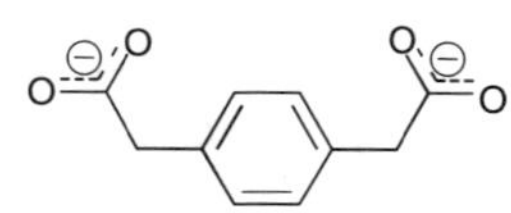

2.177d 2TBA$^{\oplus}$

2.5.2 Cyclic Systems

Even though a number of imidazolium-based cyclic systems have been synthesized in the last few years using anion-templated strategies (see Chapter 9), it was only in 1999 that the first report of their anion-binding properties appeared from the group of Alcalde.[239] These studies involved the dicationic imidazoliophanes **2.178a–2.178c**, species prepared using a "3+1" synthetic approach. The structures of these macrocycles were determined using general spectroscopic methods and were confirmed in the case of **2.178a** and **2.178b** by X-ray crystallographic analysis. In the solid state, **2.178a** adopts a 1,2-alternative chair-like conformation and is bound to one chloride counter anion via a highly directional C–H···Cl interaction (*cf.* Figure 2.43a). However, the conformation of the corresponding complex formed between **2.178b** and chloride anion, shown in Figure 2.43b, is cone-like. While quantitative information was not forthcoming, significant CH shifts in the ^{1}H NMR resonances were observed upon the addition of various anions to **2.178c**·$2PF_6$ in DMSO-d_6, with the order of peak shifts being $H_2PO_4^- > F^- > CH_3CO_2^- > CN^- > Cl^-$.

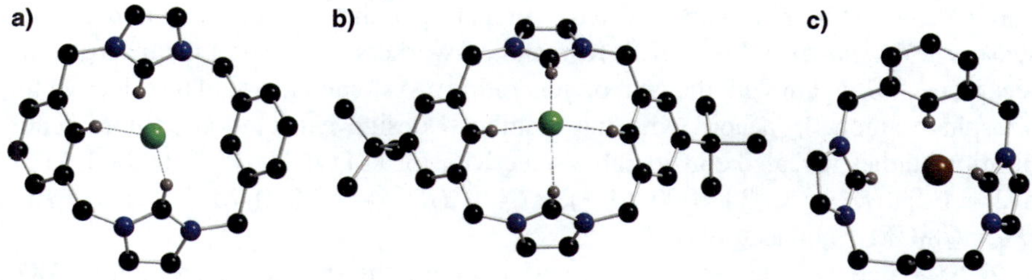

Figure 2.43 *Single crystal X-ray structures of (a) the chloride-anion complex of* ***2.178a****, (b) the chloride-anion complex of* ***2.178b****, and (c) the bromide-anion complex of* ***2.179a***

In 2002, Xie and co-workers[240] reported the synthesis of a couple of new imidazolium cyclophanes, namely **2.179a–2.180b**, along with their anion-binding affinities. The single crystal X-ray analysis of **2.179a**-Br, shown in Figure 2.43c, serves to highlight the hydrogen-bond interaction between the CH protons of the dicationic receptor and the bound bromide anion (the hydrogen bond distance is 2.66 Å). This complexation was confirmed by ^{1}H NMR spectroscopic studies carried out in DMSO-d_6, which revealed a downfield shift in the CH imidazolium resonances in the presence of added bromide anion. Quantitative determinations of the association constants (K_a) were then made by using UV–Vis titrations performed in acetonitrile. While receptor **2.179a** displays a strong affinity for chloride and bromide, it is not particularly selective (K_a = 2210, 15,700, 17,900, and 4700 M^{-1} for F^-, Cl^-, Br^-, and I^-, respectively). In contrast, cyclophane **2.179b** exhibits good selectivity towards chloride anion, as well as the highest overall affinities among this series of receptors (K_a = 40,600, 7760, 18,400, and 20 M^{-1} for Cl^-, F^-, Br^-, and I^-, respectively, in acetonitrile). Considering the electronegativity order of the halide anions, these results provide clear evidence that the inherent anion-binding selectivity of imidazolium receptors may be modulated via appropriate structural design. Consistent with this hypothesis, the pyridine receptors **2.180a** and **2.180b** were found to display bromide anion affinities that were about one-third of those of **2.179a** and **2.179b** (K_a = 6100 and 5850 M^{-1}, respectively, for **2.180a** and **2.180b** binding bromide anion in acetonitrile).

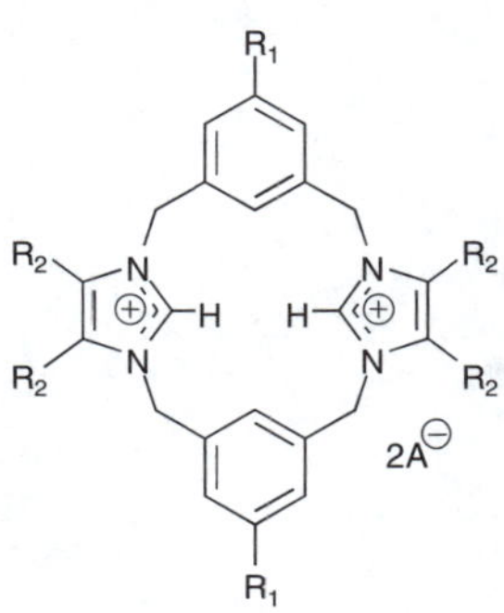

2.178a R_1 = H, R_2 = H
2.178b R_1 = *t*-Bu, R_2 = H
2.178c R_1 = *t*-Bu, R_2 = *n*-Bu

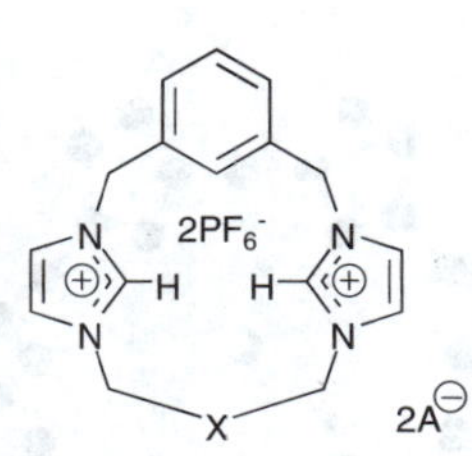

2.179a X = -$(CH_2CH_2)_2$-
2.179b X =

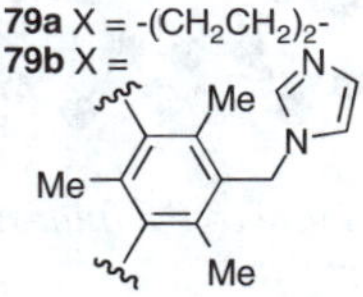

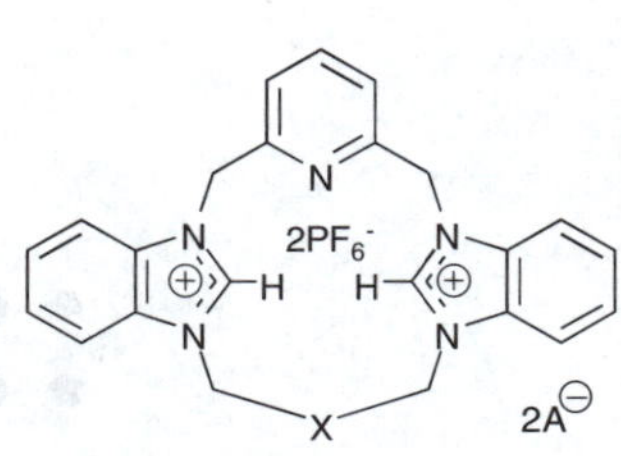

2.180a X = -$(CH_2CH_2)_2$-
2.180b X =

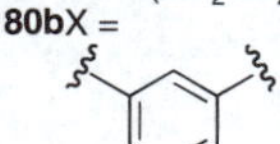

In the context of their work devoted to preparing anthracene-based diimidazolium sensors (*vide supra*), J. Yoon, K.S. Kim and co-workers[241] prepared the macrocyclic receptor **2.181**, again with the goal of generating a system that would be selective for phosphate anion. In accord with these authors' design expectations, fluorescence titration studies revealed considerable selectivity for $H_2PO_4^-$ ($K_a > 1{,}300{,}000\ M^{-1}$) over F^- ($K_a = 340{,}000\ M^{-1}$), Cl^- ($K_a = 2000\ M^{-1}$), and Br^- ($K_a = 780\ M^{-1}$) in acetonitrile.

In 2003, one year earlier than the report detailing the synthesis of receptor **2.181**, Sato and co-workers[232] reported the synthesis of the tetracationic imidazoliophane **2.182**. In DMSO-d_6, receptor **2.182** shows a preference for bromide anion among the three halide anions tested (K_a = 720, 2230, and 460 M^{-1} for Cl^-, Br^-, and I^-, respectively). However, the highest affinity was seen for hydrogen sulfate, with little affinity being seen for dihydrogen phosphate or perchlorate (K_a = 8500, 1350, and 290 M^{-1} for HSO_4^-, $H_2PO_4^-$, and ClO_4^-, respectively, in DMSO-d_6).

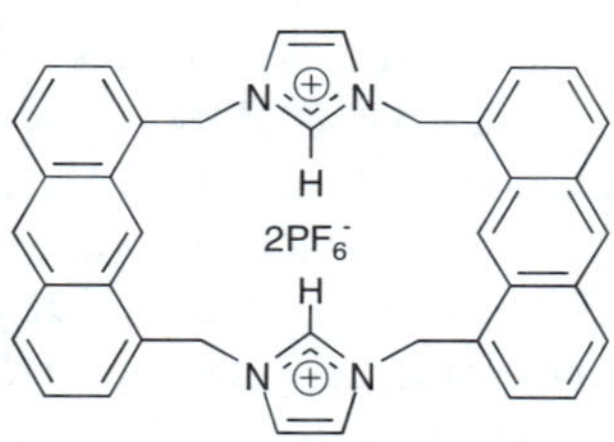

2.181

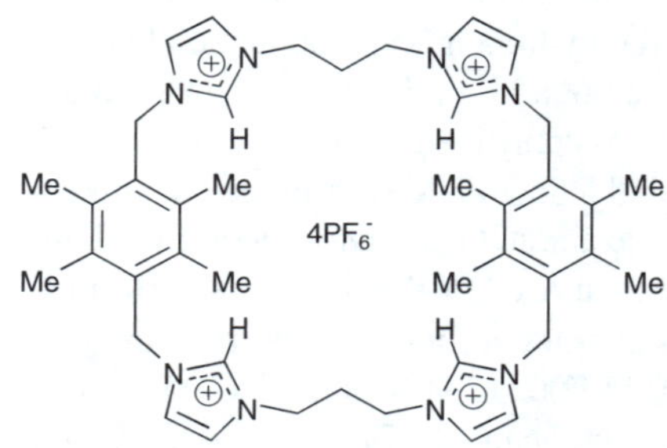

2.182

I.-C. Hwang, K.S. Kim and co-workers[242] have reported the synthesis and anion-binding properties of calix[4]imidazolium[2]pyridine (**2.183**). The single crystal X-ray structure of **2.183** with fluoride, shown in Figure 2.44, reveals that the anion is held in the centre of the macrocyclic core via strong H···F interactions (1.991 and 2.082 Å). In solution, quantitative association constants (K_a) were obtained using ^{1}H NMR spectroscopic titrations in DMSO-d_6. These studies showed that the receptor binds to fluoride with a stability constant of 28,900 M^{-1} in 1:1 stoichiometry.

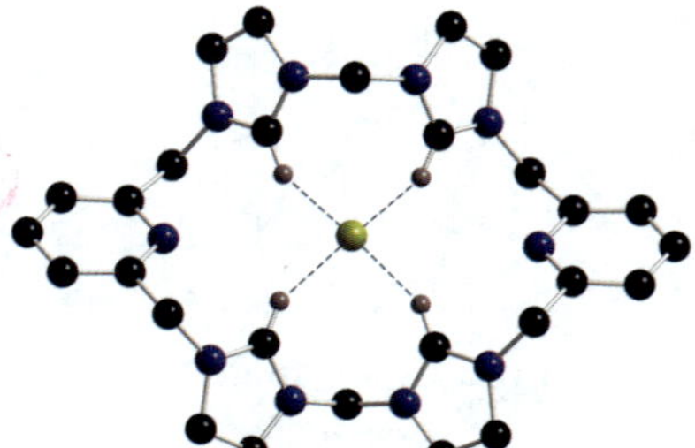

Figure 2.44 *Single crystal X-ray structure of the fluoride anion complex of* ***2.183*** *(three* PF_6^- *anions are removed for clarity)*

However, 1:2 complexes of **2.183** with chloride, bromide, iodide, acetate, and hydrogen sulfate in DMSO-d_6 were observed with K_{a1} (K_{a2}) = 2030 (2790), 100 (10,700), 130 (3330), 5040 (1940), and 40 (1120) M^{-1}.

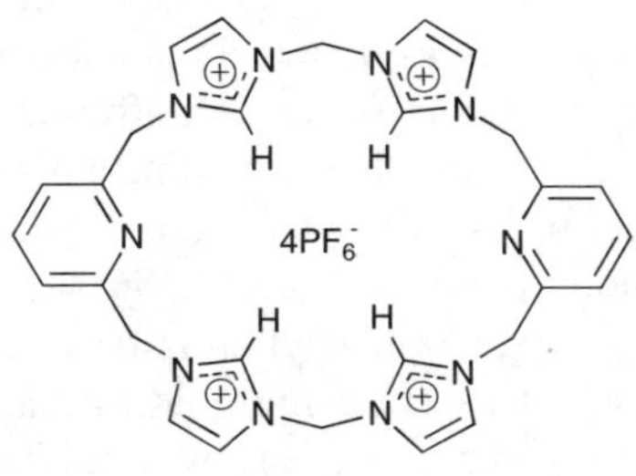

2.183

As should be apparent from the earlier sections of this chapter, the use of three-dimensional charged systems often gives rise to receptors with considerable selectivity and generally enhanced anion recognition ability. Appreciating this, Inove prepared the imidazolinium-based cryptand-type receptor **2.185**; it was prepared from **2.184**, a well-known fluoride-anion receptor,[67] by treating with triethylorthofomate at 120 °C.[243] A single crystal X-ray diffraction analysis of **2.185** confirmed the tricationic cryptand-like structure and revealed a small inside cavity with an approximate diameter of 4.6 Å, which was thought suitable for accommodating small anions such as fluoride. Another crystal structure, shown in Figure 2.45, confirmed that fluoride anion can indeed be encapsulated inside this cavity. In this latter structure, two hydrogen-bond interactions were observed between the two protonated bridgehead nitrogen atoms and the bound fluoride anion. However, no evidence of interaction with the imidazolium CH protons was seen. In aqueous solution (pD = 1.0), however, a significant downfield shift of the imidazolium CH resonance, from 6.20 to 7.15, was observed in the presence of F^-; this was taken as evidence that these protons participate in anion binding. Under these conditions, no corresponding peak shifts were seen in the presence of chloride anion. Based on competitive ^{1}H NMR spectroscopic titrations, an exceptionally high affinity for fluoride anion was inferred (log K_a = 12.5).

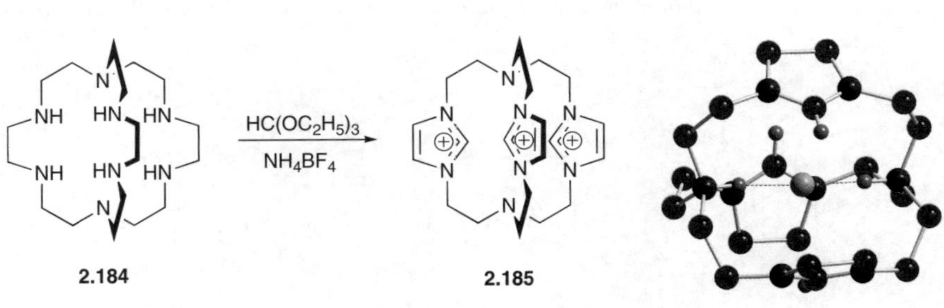

Figure 2.45 *Single crystal X-ray structure of the fluoride-anion complex of* ***2.185***

2.6 Thiouronium

The final positively charged anion-binding motif to be considered in this chapter is the thiouronium group. Formally a thiourea derivative, thiouronium subunits provide two parallel hydrogen bond donors in analogy to the similarly charged guanidinium unit and neutral donor groups such as urea and thiourea. Not surprisingly, therefore, cationic thiouronium-based receptors show high affinities for oxyanions. However, even simple thiouronium units are capable of binding anions, at least in the solid state, as documented by a number of X-ray crystallographic studies (*cf.* Figure 2.46).[182,244]

The first synthesis of thiouronium salts (*e.g.*, **2.186**) was realized by Bernthsen and Klinger[245] in 1878 via the reaction of benzyl bromide with thioureas (Scheme 2.11), and by quite early on these salts were being used as reagents for the isolation and characterization of various organic acids.[246] However, effective as this latter process was, it is not appreciated as being anion recognition *per se*.

The first thiouronium-based anion receptors to be recognized as such were systems **2.187–2.190**, species designed by Yeo and Hong[247] to function as 5′-AMP extractants and carriers. The thiouronium units in these receptors were expected to interact with the phosphate group of this and other mononucleotides through a combination of hydrogen bond and electrostatic interactions. In addition, on the basis of precedent established in the case of sapphyrin-based carriers (*cf.* Chapter 3), it was further

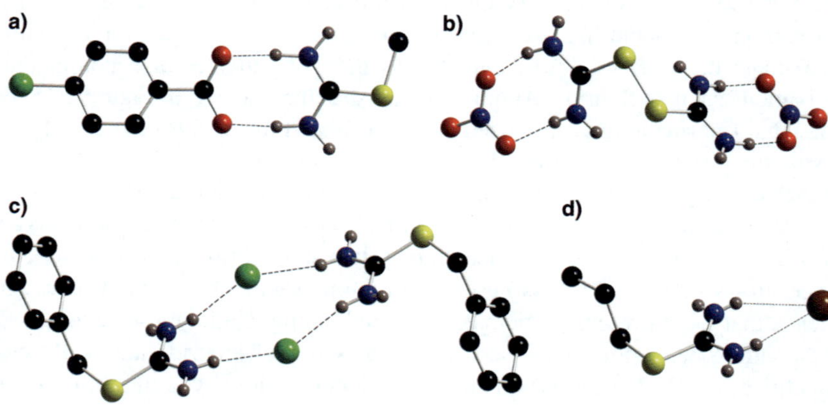

Figure 2.46 *Single crystal X-ray structures of (a) p-chlorobenzoate complex of S-methylthiouronium, (b) nitrate complex of dithiouronium, (c) chloride-anion complex of S-benzylthiouronium, and (d) bromide-anion complex of S-ethylthiouronium*

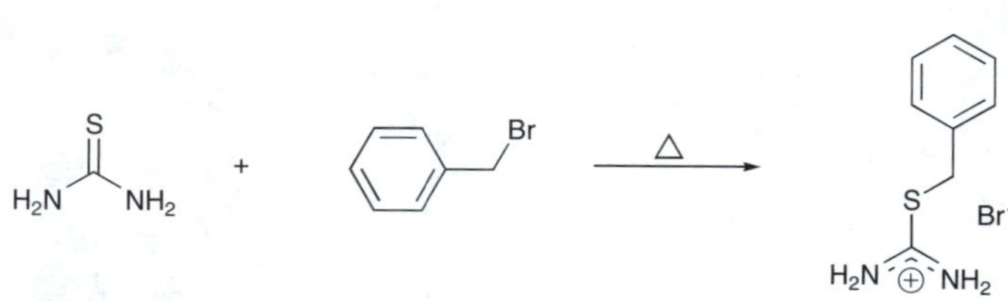

2.186

Scheme 2.11 *Synthesis of S-benzyl thiouronium **2.186***

expected that ancillary hydrogen-bond interactions, involving the thymine residues present in receptors **2.187** and **2.188** and the adenosine functionality present in the 5′-AMP target, would improve the binding and transport selectivity. As shown in Table 2.9, both **2.187** and **2.188** do indeed display higher 5′-AMP transport rates as compared to their structural analogues, **2.189** and **2.190**, presumably reflecting the benefit of these extra Watson–Crick interactions. The thiouronium–thymine carriers **2.187** and **2.188** also showed selectivity for 5′-AMP over 5′-GMP in the transport experiments, providing support for the notion that selective base-pairing plays a role in defining the underlying recognition chemistry. In addition to these effects, the bis(thiouronium) receptors **2.187** and **2.189** generally proved to be better carriers than their mono(thiouronium) analogues, **2.188** and **2.190**. The best extractions and the fastest transport rates were obtained with receptor **2.187**, as would be expected for a system that combines both bis(thiouronium) sites for phosphate-anion binding and bis(thymine) groups for adenosine recognition (Figure 2.47).

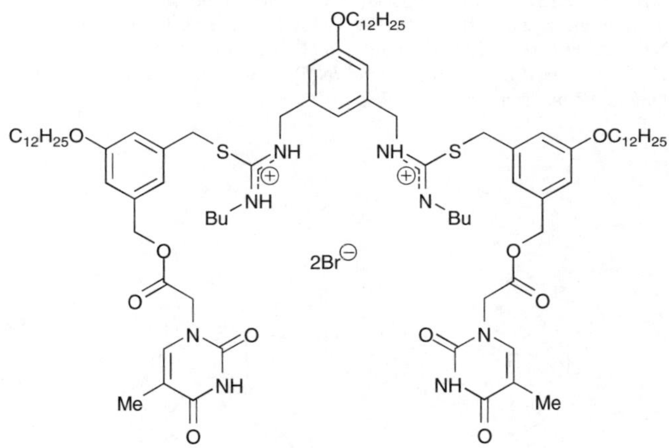

2.187

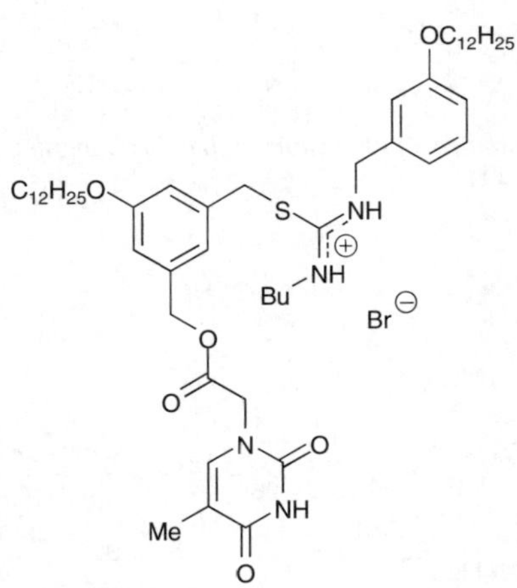

2.188

Table 2.9 *Extraction and transport of 5′-AMP and 5′-GMP from an initial aqueous layer into or through an organic phase as effected by receptors **2.187–2.190***

		Extraction (%)[a]		Transport rate (10^{-8} M h^{-1} cm^{-1})[b]	
Carrier	*pH*	*5′-AMP*	*5′-GMP*	*5′-AMP*	*5′-GMP*
2.187	5.0	9.9	9.1	15.1	2.4
	7.0	11.7	8.8	5.9	~2[d]
2.188	5.0	7.3	2.4	0.9	0.7
	7.0	9.3	7.5	1.3	N.D.[c]
2.189	5.0	7.1	4.3	1.2	0.6
	7.0	8.3	3.7	2.7	0.7
2.190	5.0	5.3	4.2	0.5	~1
	7.0	7.0	5.8	~0.2[d]	N.D.[c]

[a]In the case of **2.187** and **2.189** = 5 mM, [Guest] = 0.1 mM; in the case of **2.188** and **2.190** = 10 mM, [Guest] = 0.1 mM. [b]In the case of **2.187** and **2.189**, Guest 0.1 M H_2O, 4 mL/Carrier 1.0×10^{-3} = 0.1 M, $CHCl_3$, 8 mL/NaBr 2.5×10^{-2} M, H_2O, 4 mL; in the case of **2.188** and **2.190**, Guest 0.1 M H_2O, 4 mL/Carrier 2.0×10^{-3} = 0.1 M, $CHCl_3$, 8 mL/NaBr 2.5×10^{-2} M, H_2O. [c]N.D. = not determined. [d] Detectable but could not be determined accurately.

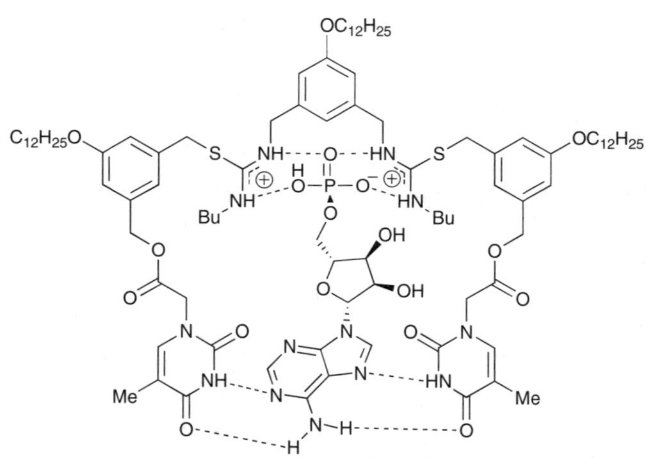

Figure 2.47 *Proposed structure of the supramolecular complex formed between receptor **2.187** and 5′-AMP*

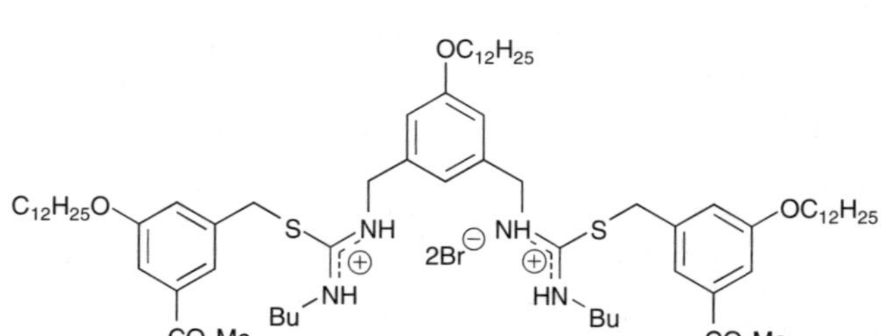

2.189

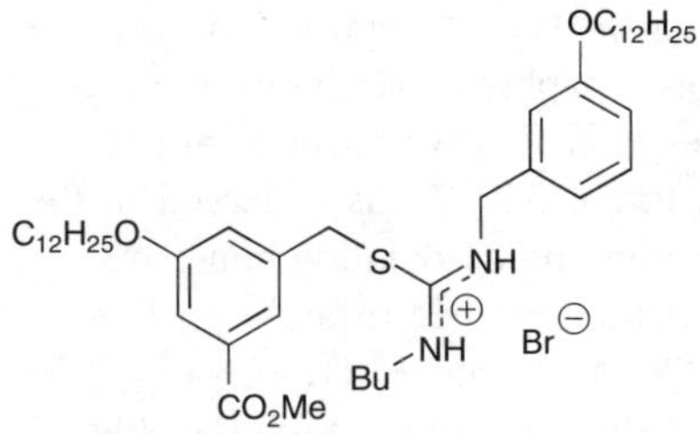

2.190

In the context of these efforts, the association constants of simple thiouronium receptors **2.191a** and **2.192a** were measured using ^{1}H NMR and UV–Vis spectroscopic titration methods.[247,248] In the case of receptor **2.191a**, studied in the form of its bromide salt, the binding affinities (K_a) for various oxyanions were found to follow the trend $PhPO_3^{2-}$ (4350 M^{-1}) > $PhOPO_3^{2-}$ (3700 M^{-1}) > $H_2PO_4^-$ (1080 M^{-1}) > $PhCO_2^-$ (590 M^{-1}) > $PhP(OH)O_2^-$ (150 M^{-1}) > $PhOP(OH)O_2^-$ (50 M^{-1}) > $PhSO_3^-$ (~5 M^{-1}) in DMSO-d_6, as determined from ^{1}H NMR spectroscopic titrations. As a general rule, these binding affinities were found to correlate well with the Brönsted basicity of the anionic guests. The selectivity for $H_2PO_4^-$ over $PhCO_2^-$ is an exception; it was rationalized in terms of differences in binding geometry, as shown in Figure 2.48. In the UV–Vis spectrum, the addition of $H_2PO_4^-$ was found to increase the absorbance intensity of receptor **2.191a**. From these spectral changes, an association constant 34,000 M^{-1} for $H_2PO_4^-$ binding was obtained in 1,2-dichloroethane. While this bis(thiouronium) receptor bound oxyanions strongly, the corresponding mono(thiouronium) receptor **2.192a** displayed only relatively weak affinities for $H_2PO_4^-$ (K_a = 340 M^{-1}) and $CH_3CO_2^-$ (K_a = 800 M^{-1}) in DMSO-d_6. Appreciating these anion-binding properties, the thiouronium derivatives **2.191b**, **2.192b**, and **2.192c** were used by Suzuki and co-workers[249] to generate new ion-selective electrodes. The interested reader is referred to the original publication.

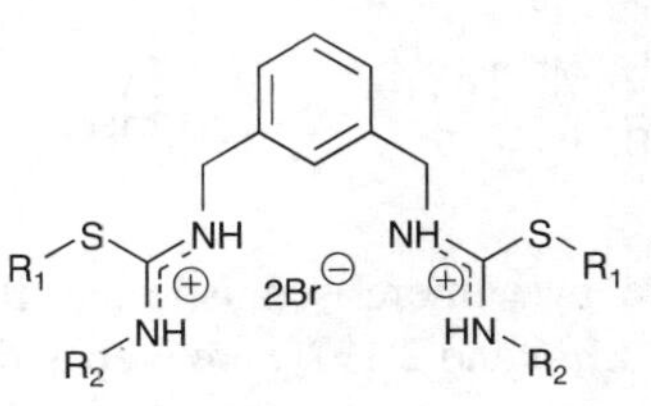

2.191a R_1 = *p*-NO_2Bz, R_2 = *n*-Bu
2.191b R_1 = *n*-Bu, R_2 = *n*-$C_{12}H_{25}$

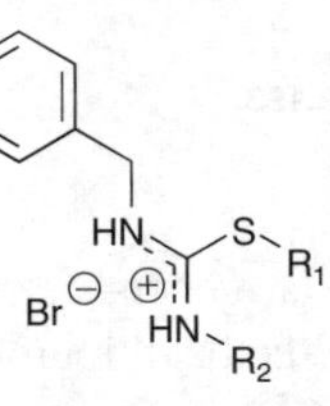

2.192a R_1 = *p*-NO_2Bz, R_2 = *n*-Bu
2.192b R_1 = Me. R_2 = *n*-Bu
2.192c R_1 = *n*-Bu, R_2 = *n*-$C_{12}H_{25}$

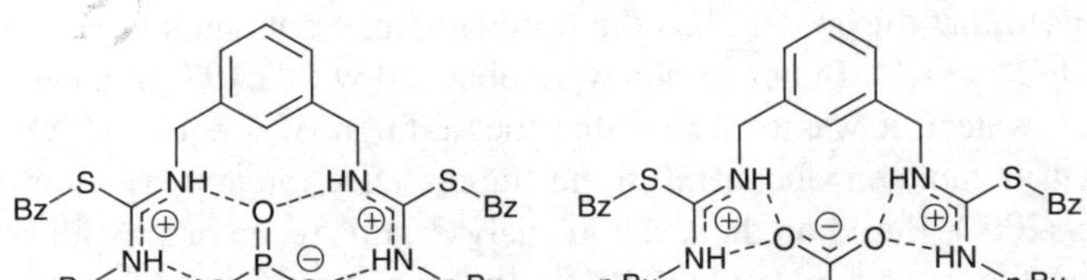

Figure 2.48 *Proposed structures for complexes formed between receptor* ***2.191a*** *and dihydrogen phosphate (left) and benzoate (right)*

In 2000, Kubo and co-workers[250] reported the acetate selective chemosensor **2.193**, which is based on a naphthalene–thiouronium dyad. While receptor **2.193** shows weak fluorescence when dissolved in acetonitrile, an enhancement in the fluorescence intensity of up to *ca.* 350% was observed in the presence of 1 equiv. of acetate anion. The corresponding association constants, calculated from non-linear curve-fits of the observed fluorescence titration profiles, confirmed the high selectivity for acetate ($K_a > 10^6$ M^{-1}) over $(BuO)_2PO_2^-$ ($K_a = 5.6 \times 10^4$ M^{-1}) in CH_3CN at 25 °C. In the case of chloride anion, no appreciable change in the fluorescence intensity was observed.

In independent work reported in 2002, Teramae and co-workers[251] detailed the synthesis of the naphthalene- and anthracene-functionalized mono(thiouronium) receptors, **2.194a–2.195b**, which were also designed to function as photoinduced electron transfer sensors. Unfortunately, only receptor **2.194b** proved stable in methanol under conditions of photoillumination. However, this system was found to display a significant enhancement in fluorescence intensity upon treatment with certain oxyanions. In the case of HPO_4^{2-} and $CH_3CO_2^-$, the relevant association constants were calculated to be 1.1×10^4 and 1.7×10^3 M^{-1} for 1:1 binding in CH_3OH ([K^+-18-crown-6] salts were used as the anion source). In contrast, no association constants could be calculated in the case of $H_2PO_4^-$ and Cl^- because the spectral changes proved too small.

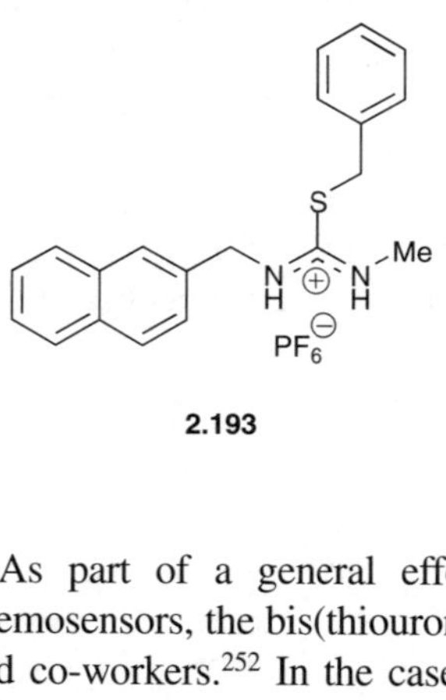

2.193

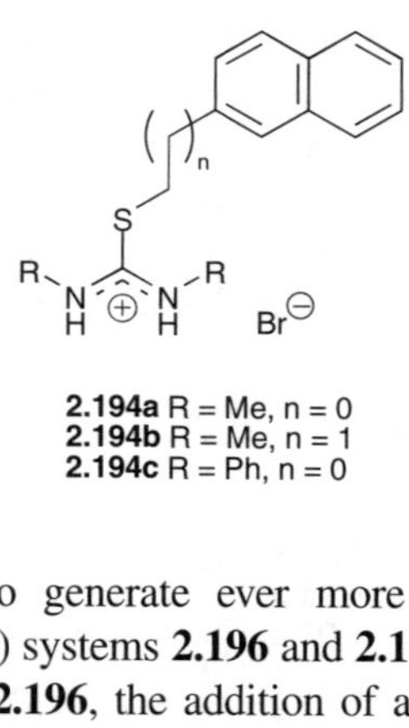

2.194a R = Me, n = 0
2.194b R = Me, n = 1
2.194c R = Ph, n = 0

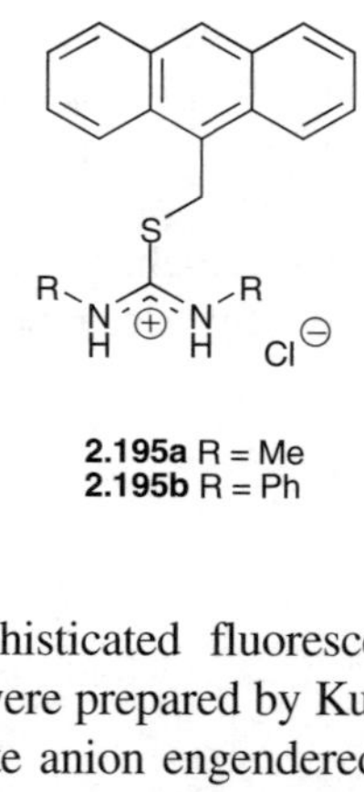

2.195a R = Me
2.195b R = Ph

As part of a general effort to generate ever more sophisticated fluorescent chemosensors, the bis(thiouronium) systems **2.196** and **2.197** were prepared by Kubo and co-workers.[252] In the case of **2.196**, the addition of acetate anion engendered a maximal enhancement in the fluorescence intensity of *ca.* 1600% in acetonitrile. Unfortunately, the host–guest-binding stoichiometry was not clean. Thus, association constants could not be measured. Nonetheless, based on the observed optical changes, it was proposed that this system had the following inherent anion selectivities: $AcO^- > (n\text{-}BuO)_2PO_2^- > Cl^-$. Better results were obtained with **2.197**. In aqueous acetonitrile (6% v/v water), it was observed that the addition of 4 equiv. HPO_4^{2-} to **2.197** caused a considerable enhancement in the fluorescence intensity (*i.e.*, up to a maximum of *ca.* 520%). Based on the titration analyses, this receptor was found to form a strong 2:1 host–guest complex with HPO_4^{2-} in this same solvent mixture ($K_{a21} = 1.6 \times 10^{11}$ M^{-1}). In contrast, 1:2 host–guest-binding was observed in the case of acetate anion under identical conditions ($K_{a11} = 5.4 \times 10^3$ M^{-1} and $K_{a12} = 7.3 \times 10^5$ M^{-1}).

While difficult to rationalize fully, these findings do serve to demonstrate, at the very least, the fact that this receptor is able to interact in different ways with anionic guests of different sizes and shapes.

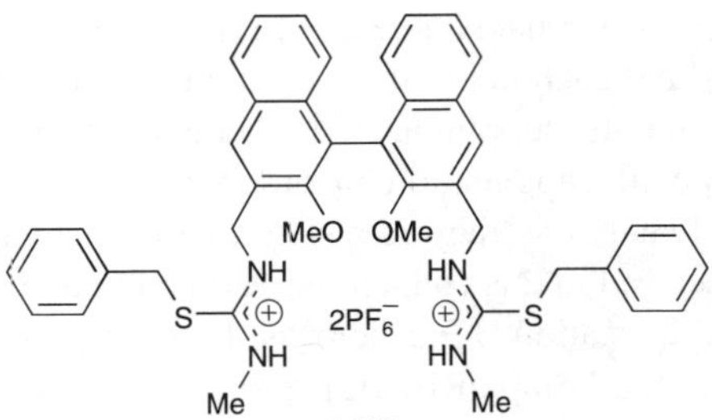

2.196

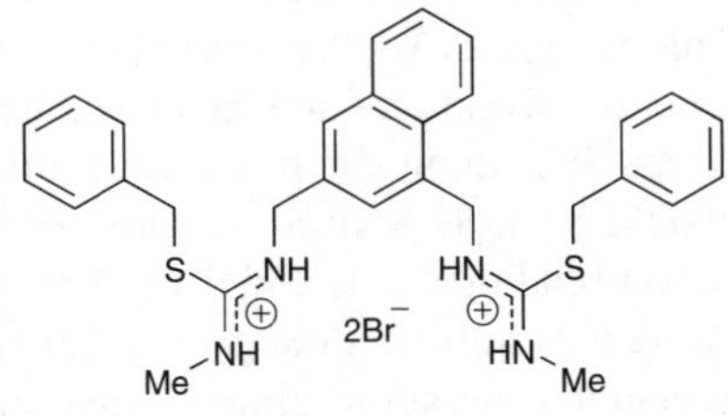

2.197

In analogy to what has been done with other anion recognition motifs, the triply-substituted 1,3,5-benzene platform was used by Ahn and co-workers[253] to prepare the tripodal tri(thiouronium) receptors **2.199a** and **2.199b**, as well as the simpler system **2.198** as a control. In this case, anion-binding studies were performed using ITC analyses carried out in methanol at 303 K. Both tripodal receptors exhibit large $\Delta G°$ values, although in general the binding process proved to be entropy driven. Based on these calorimetric studies, it was concluded that the tripodal receptors **2.199a** and **2.199b** bind to sulfate anion much more effectively than does the bis(tetramethylammonium) control system **2.198** (*i.e.*, for **2.199a**$\cdot SO_4^{2-}$: $\Delta G° = -34.7$ kJ mol^{-1}, $\Delta H° = 26.8$ kJ mol^{-1}, $T\Delta S° = 61.5$ kJ mol^{-1}; for **2.199b**$\cdot SO_4^{2-}$: $\Delta G° = -35.1$ kJ mol^{-1}, $\Delta H° = 30.1$ kJ mol^{-1}, $T\Delta S° = 65.3$ kJ mol^{-1}; for **2.198**$\cdot SO_4^{2-}$: $\Delta G° = -25.1$ kJ mol^{-1}, $\Delta H° = 36.8$ kJ mol^{-1}, $T\Delta S° = 61.9$ kJ mol^{-1}). Similar ITC experiments involving PO_4^{3-}, but carried out under conditions of "reverse titration" wherein the receptor was added to a solution of PO_4^{3-}, provided a $\Delta G°$ value of –31.8 kJ mol^{-1} for **2.199a**. Presumably, the preference towards sulfate reflects a subtle, but effective degree of structural complementarity that favours this particular tetrahedral anion.

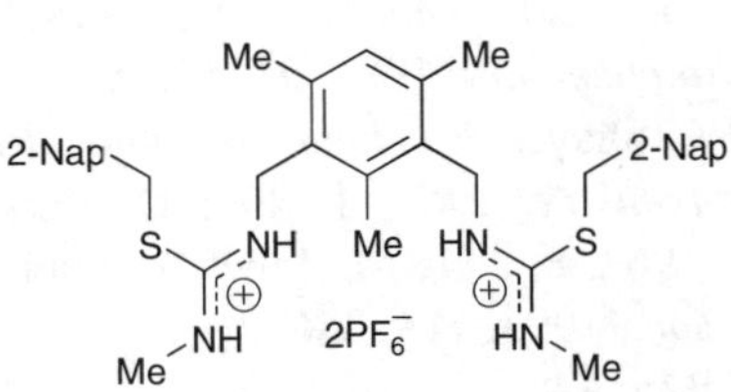

2.198

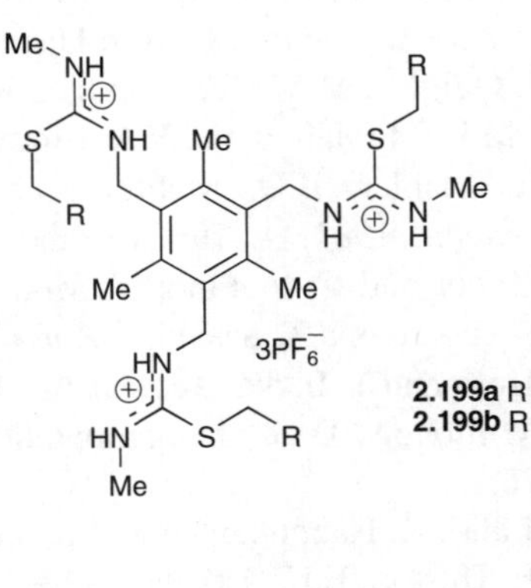

2.199a R = Ph
2.199b R = 2-Nap

2.7 Summary Remarks

In this chapter, we have looked at the most important charged motifs that have been used to effect anion recognition. While many of the systems in question are now considered classics in the field, others are rather new and serve to illustrate the power of charged-based approaches to anion recognition in such disparate application areas as separations, transport, and sensing. A recurring theme in all cases is that control of binding stoichiometry via the construction of receptors of commensurate size, and to a lesser extent charge, can produce systems of considerable selectivity for a given anionic guest. Another important theme is that each class of charged binding motif imparts advantages and disadvantages and that the supramolecular chemist targeting a specific anion or, more generally, a specific application of anion recognition, would be well advised to consider these features as they relate to his or her own anion-binding needs. Finally, the work described in this chapter has served to underscore how similar strategies (*e.g.*, cryptand formation, 1,3,5-benzene-based tripodal receptor generation, dimer formation, etc.) can be applied with benefit using very different anion binding "building blocks", such as guanidinium, imidazolium, etc. Again, this serves to highlight the importance of receptor design and the fundamental utility associated with not only finding new binding motifs but also new ways of arranging them in space. In the ensuing chapters, other classes of recognition units will be introduced. It is hoped that the reader will appreciate the new opportunities these systems offer in the context of both these themes.

References

1. C.H. Park and H.E. Simmons, *J. Am. Chem. Soc.*, 1968, **90**, 2431.
2. N. Gavrushenko, H.L. Carrell, W.C. Stallings and J.P. Glusker, *Acta Cryst.*, 1977, **B33**, 3936.
3. B. Dietrich, D.L. Fyles, T.M. Fyles and J.-M. Lehn, *Helv. Chim. Acta*, 1979, **62**, 2763.
4. F. Takusagawa and T.F. Koetzle, *Acta Cryst.*, 1979, **B35**, 867.
5. N.H. Woo, N.C. Seeman and A. Rich, *Biopolymers*, 1979, **18**, 539.
6. C. Nakai and W. Glinsmann, *Biochemistry*, 1977, **16**, 5636.
7. I. Labadi, E. Jenei, R. Lahti and H. Lönnberg, *Acta Chem. Scand.*, 1991, **45**, 1055.
8. C.W. Tabor and H. Tabor, *Annu. Rev. Biochem.*, 1984, **53**, 749; S.S. Cohen, *A Guide to Polyamines*, Oxford University Press, 1998.
9. G.J. Quigley, M.M. Teeter and A. Rich, *Proc. Natl. Acad. Sci. USA*, 1978, **75**, 64; M.E. McMahon and V.A. Erdmann, *Biochemistry*, 1982, **21**, 5280.
10. H.R. Mahler, B.D. Mehrotra and C.W. Shays, *Biochem. Biophys. Res. Commun.*, 1961, **4**, 79; H. Tabor, *Biochemistry*, 1962, **1**, 496; D. Kaiser, H. Tabor and C.W. Tabor, *J. Mol. Biol.*, 1963, **6**, 141; M. Tabet, V. Labroo, P. Sheppard and T. Sasaki, *J. Am. Chem. Soc.*, 1993, **115**, 3866.
11. Y. Huse and Y. Iitaka, *Acta Cryst.*, 1969, **B25**, 498.
12. C.A. Ilioudis, D.G. Georganopoulou and J.W. Steed, *CrystEngComm*, 2002, **4**, 26.
13. R. Lahti, R. Hannukainen and H. Lönnberg, *Biochem. J.*, 1989, **259**, 55.
14. M.E. Huston, E.U. Akkaya and A.W. Czarnik, *J. Am. Chem. Soc.*, 1989, **111**, 8735.
15. M.T. Albelda, M.A. Bernardo, E. García-España, M.L. Godino-Salido, S.V. Luis, M.J. Melo, F. Pina and C. Soriano, *J. Chem. Soc. Perkin Trans. 2*, 1999, 2545.

16. C. Bazzicalupi, A. Bencini, E. Berni, A. Bianchi, P. Fornasari, C. Giorgi, A. Masotti, P. Paoletti and B. Valtancoli, *J. Phys. Org. Chem.*, 2001, **14**, 432.
17. V. McKee, J. Nelson and R.M. Town, *Chem. Soc. Rev.*, 2003, **32**, 309.
18. M.A. Hossain, J.A. Liljegren, D. Powell and K. Bowman-James, *Inorg. Chem.*, 2004, **43**, 3751.
19. M.D. Best and E.V. Anslyn, *Chem. Eur. J.*, 2003, **9**, 51.
20. S.L. Tobey, B.D. Jones and E.V. Anslyn, *J. Am. Chem. Soc.*, 2003, **125**, 4026.
21. S.L. Tobey and E.V. Anslyn, *J. Am. Chem. Soc.*, 2003, **125**, 10963.
22. T. Yatsunami, A. Sakonaka and E. Kimura, *Anal. Chem.*, 1981, **53**, 477.
23. E. Kimura, A. Sakonaka, T. Yatsunami and M. Kodama, *J. Am. Chem. Soc.*, 1981, **103**, 3041.
24. E. Kimura, M. Kodama and T. Yatsunami, *J. Am. Chem. Soc.*, 1982, **104**, 3182.
25. A. Bianchi, M. Micheloni and P. Paoletti, *Inorg. Chim. Acta*, 1988, **151**, 269.
26. E. Suet and H. Handel, *Tetrahedron Lett.*, 1984, **25**, 645.
27. J. Cullinane, R.I. Gelb, T.N. Margulis and L.J. Zompa, *J. Am. Chem. Soc.*, 1982, **104**, 3048.
28. A.C. Warden, M. Warren, M.T.W. Hearn and L. Spiccia, *New J. Chem.*, 2004, **28**, 1160.
29. A. Andrés, J. Aragó, A. Bencini, A. Bianchi, A. Doménech, V. Fusi, E. García-España, P. Paoletti and J.A. Ramírez, *Inorg. Chem.*, 1993, **32**, 3418.
30. A. Andrés, C. Bazzicalupi, A. Bencini, A. Bianchi, V. Fusi, E. García-España, C. Giorgi, N. Nardi, P. Paoletti, J.A. Ramírez and B. Valtancoli, *J. Chem. Soc. Perkin Trans. 2*, 1994, 2367.
31. C. Bazzicalupi, A. Bencini, A. Bianchi, M. Cecchi, B. Escuder, V. Fusi, E. García-España, C. Giorgi, S.V. Luis, G. Maccagni, V. Marcelino, P. Paoletti and B. Valtancoli, *J. Am. Chem. Soc.*, 1999, **121**, 6807.
32. P. Arranz, A. Bencini, A. Bianchi, P. Diaz, E. García-España, C. Giorgi, S.V. Luis, M. Querol and B. Valtancoli, *J. Chem. Soc. Perkin Trans. 2*, 2001, 1765.
33. A. Bencini, A. Bianchi, C. Giorgi, P. Paoletti, B. Valtancoli, V. Fusi, E. García-España, J.M. Llinares and J.A. Ramírez, *Inorg. Chem.*, 1996, **35**, 1114.
34. A.C. Warden, M. Warren, M.T.W. Hearn and L. Spiccia, *Inorg. Chem.*, 2004, **43**, 6936.
35. B. Dietrich, M.W. Hosseini, J.-M. Lehn and R.B. Sessions, *J. Am. Chem. Soc.*, 1981, **103**, 1282.
36. M.W. Hosseini and J.-M. Lehn, *Helv. Chim. Acta*, 1986, **69**, 587.
37. M.W. Hosseini and J.-M. Lehn, *Helv. Chim. Acta*, 1988, **71**, 749.
38. S. Boudon, A. Decian, J. Fischer, M.W. Hosseini, J.-M. Lehn and G. Wipff, *J. Coord. Chem.*, 1991, **23**, 113.
39. O.A. Gerasimchuk, S. Mason, J.M. Llinares, M. Song, N.W. Alcock and K. Bowman-James, *Inorg. Chem.*, 2000, **39**, 1371.
40. J.F. Marecek and C.J. Burrows, *Tetrahedron Lett.*, 1986, **27**, 5943.
41. I. Alfonso, B. Dietrich, F. Rebolledo, V. Gotor and J.-M. Lehn, *Helv. Chim. Acta*, 2001, **84**, 280.
42. A. Bencini, A. Bianchi, E. García-España, E.C. Scott, L. Morales, B. Wang, T. Deffo, F. Takusagawa, M.P. Mertes, K.B. Mertes and P. Paoletti, *Bioorg. Chem.*, 1992, **20**, 8.

43. A. Bencini, A. Bianchi, M.I. Buruete, P. Dapporto, A. Doménech, E. García-España, S.V. Luis, P. Paoli and J.A. Ramírez, *J. Chem. Soc. Perkin Trans. 2*, 1994, 569.
44. A. Doménech, E. García-España, J.A. Ramírez, B. Celda, M.C. Martínez, D. Monleón, R. Tejero, A. Bencini and A. Bianchi, *J. Chem. Soc. Perkin Trans. 2*, 1999, 23.
45. J. Wiórkiewicz-Kuczera, K. Kuczera, C. Bazzicalupi, A. Bencini, B. Valtancoli, A. Bianchi and K. Bowman-James, *New J. Chem.*, 1999, **23**, 1007.
46. Z. Jiang, P. Chalabi, K.B. Mertes, H. Jahansouz, R.H. Himes and M.P. Mertes, *Bioorg. Chem.*, 1989, **17**, 313; M.P. Mertes and K.B. Mertes, *Acc. Chem. Res.*, 1990, **23**, 413; M.A. Hossain, J.-M. Lehn, K.C. Jones, K.E. Plute, K.B. Mertes and M.P. Mertes, *J. Am. Chem. Soc.*, 1989, **111**, 6330; M.W. Hosseini and J.-M. Lehn, *J. Chem. Soc. Chem. Commun.*, 1985, 1155; P.G. Yohannes, K.E. Plute, M.P. Mertes and K.B. Mertes, *Inorg. Chem.*, 1987, **26**, 1751; P.G. Yohannes, M.P. Mertes and K.B. Mertes, *J. Am. Chem. Soc.*, 1985, **107**, 8288; M.W. Hosseini and J.-M. Lehn, *J. Am. Chem. Soc.*, 1987, **109**, 7047; M.A. Hossain, J.-M. Lehn, L. Maggiora, K.B. Mertes and M.P. Mertes, *J. Am. Chem. Soc.*, 1987, **109**, 537.
47. M.W. Hosseini, J.-M. Lehn and M.P. Mertes, *Helv. Chim. Acta*, 1983, **66**, 2454.
48. M.W. Hosseini, A.J. Blacker and J.-M. Lehn, *J. Am. Chem. Soc.*, 1990, **112**, 3896.
49. S. Suzuki, T. Higashiyama and A. Nakahara, *Bioorg. Chem.*, 1973, **2**, 145; S. Suzuki and A. Nakahara, *Bioorg. Chem.*, 1975, **4**, 250.
50. A.E. Martell and R.J. Motekaitis, *J. Am. Chem. Soc.*, 1988, **110**, 8059; R.J. Motekaitis and A.E. Martell, *Inorg. Chem.*, 1992, **31**, 5534; R.J. Motekaitis and A.E. Martell, *Inorg. Chem.*, 1994, **33**, 1032.
51. H. Fenniri, M.W. Hosseini and J.-M. Lehn, *Helv. Chim. Acta*, 1997, **80**, 786.
52. G. Papoyan, K.-J. Gu, J. Wiórkiewicz-Kuczera, K. Kuczera and K. Bowman-James, *J. Am. Chem. Soc.*, 1996, **118**, 1354.
53. M. Dhaenens, J.-M. Lehn and J.-P. Vigneron, *J. Chem. Soc. Perkin Trans. 2*, 1993, 1379.
54. D.A. Nation, J. Reibenspies and A.E. Martell, *Inorg. Chem.*, 1996, **35**, 4597.
55. D.A. Nation, Q. Lu and A.E. Martell, *Inorg. Chim. Acta*, 1997, **263**, 209.
56. T. Clifford, A. Danby, J.M. Llinares, S. Mason, N.W. Alcock, D. Powell, J.A. Aguilar, E. García-España and K. Bowman-James, *Inorg. Chem.*, 2001, **40**, 4710.
57. J.A. Aguilar, T. Clifford, A. Danby, J.M. Llinares, S. Mason, E. García-España and K. Bowman-James, *Supramol. Chem.*, 2001, **13**, 405.
58. J. Huang, S.-A. Li, D.-F. Li, D.-X. Yang, W.-Y. Sun and W.-X. Tang, *Bioorg. Med. Chem.*, 2004, **12**, 529.
59. Q. Lu, R.J. Motekaitis, J.J. Reibenspies and A.E. Martell, *Inorg. Chem.*, 1995, **34**, 4958.
60. Q. Lu, J.H. Reibenspies, R.I. Carroll, A.E. Martell and A. Clearfiled, *Inorg. Chim. Acta*, 1998, **270**, 207.
61. Q. Lu, A.E. Martell and R.J. Motekaitis, *Inorg. Chim. Acta*, 1996, **251**, 365.

62. C. Miranda, F. Escartí, L. Lamarque, M.J.R. Yunta, P. Navarro, E. García-España and M.L. Jimeno, *J. Am. Chem. Soc.*, 2004, **126**, 823.
63. L. Lamarque, P. Navarro, C. Miranda, V.J. Arán, C. Ochoa, F. Escartí, E. García-España, J. Latorre, S.V. Luis and J.F. Miravet, *J. Am. Chem. Soc.*, 2001, **123**, 10560; C. Foces-Foces, A. Echevarría, N. Jagerovic, I. Alkorta, J. Elguero, U. Langer, O. Klein, M. Minguet-Bonvehí and H.-H. Limbach, *J. Am. Chem. Soc.*, 2001, **123**, 7898.
64. M. Cesario, J. Guilhem and C. Pascard, *Supramol. Chem.*, 1993, **2**, 331; A. Slama-Schwok, M.-P. Teulade-Fichou, J.-P. Vigneron, E. Taillandier and J.-M. Lehn, *J. Am. Chem. Soc.*, 1995, **117**, 6822; M.-P. Teulade-Fichou, J.-P. Vigneron and J.-M. Lehn, *Supramol. Chem.*, 1995, **5**, 139; C. Bazzicalupi, A. Bencini, A. Bianchi, L. Borsari, C. Giorgi, B. Valtancoli, C. Anda and A. Llobet, *J. Org. Chem.*, 2005, **70**, 4257; C. Anda, M.A. Martínez and A. Llobet, *Supramol. Chem.*, 2005, **17**, 257; C. Anda, A. Llobet and V. Salvado, *Inorg. Chem.*, 2000, **39**, 2986; C. Anda, A. Llobet, V. Salvado, A.E. Martell and R.J. Motekaitis, *Inorg. Chem.*, 2000, **39**, 3000; M.A. Bernardo, J.A. Guerrero, E. García-España, S.V. Luis, J.M. Llinares, F. Pina, J.A. Ramírez and C. Soriano, *J. Chem. Soc. Perkin Trans. 2*, 1996, 2335; N. Marcotte and A. Taglietti, *Supramol. Chem.*, 2003, **15**, 617.
65. R.J. Motekaitis, A.E. Martell, J.-M. Lehn and E.-I. Watanabe, *Inorg. Chem.*, 1982, **21**, 4253.
66. B. Dietrich, J. Guilhem, J.-M. Lehn, C. Pascard and E. Sonveaux, *Helv. Chim. Acta*, 1984, **67**, 91.
67. B. Dietrich, B. Dilworth, J.-M. Lehn and J.-P. Souchez, *Helv. Chim. Acta*, 1996, **79**, 569.
68. B. Dietrich, J.-M. Lehn, J. Guilhem and C. Pascard, *Tetrahedron Lett.*, 1989, **30**, 4125.
69. M.A. Hossain, J.M. Llinares, C.A. Miller, L. Seib and K. Bowman-James, *Chem. Commun.*, 2000, 2269.
70. S.D. Reilly, G.R.K. Khalsa, D.K. Ford, J.R. Brainard, B.P. Hay and P.H. Smith, *Inorg. Chem.*, 1995, **34**, 569.
71. M.A. Hossain, J.M. Llinares, N.W. Alcock, D. Powell and K. Bowman-James, *J. Supramol. Chem.*, 2002, **2**, 143.
72. P. Jost, R. Schurhammer and G. Wipff, *Chem. Eur. J.*, 2000, **6**, 4257.
73. D. Heyer and J.-M. Lehn, *Tetrahedron Lett.*, 1986, **27**, 5869.
74. C.A. Llioudis, D.A. Tocher and J.W. Steed, *J. Am. Chem. Soc.*, 2004, **126**, 12395.
75. J.-M. Lehn, R. Méric, J.-P. Vigneron, I. Bkouche-Waksman and C. Pascard, *J. Chem. Soc. Chem. Commun.*, 1991, 62.
76. R. Menif, J. Reibenspies and A.E. Martell, *Inorg. Chem.*, 1991, **30**, 3446.
77. S. Mason, T. Clifford, L. Seib, K. Kuczera and K. Bowman-James, *J. Am. Chem. Soc.*, 1998, **120**, 8899.
78. S. Mason, J.M. Llinares, M. Morton, T. Clifford and K. Bowman-James, *J. Am. Chem. Soc.*, 2000, **122**, 1814.
79. F. Arnaud-Neu, S. Fuangswasdi, B. Maubert, J. Nelson and V. McKee, *Inorg. Chem.*, 2000, **39**, 573; M.J. Hynes, B. Maubert, V. McKee, R.M. Town and J. Nelson, *J. Chem. Soc. Dalton Trans.*, 2000, 2853; B. Maubert, J. Nelson,

V. McKee, R.M. Town and I. Pál, *J. Chem. Soc. Dalton Trans.*, 2001, 1395.
80. D. Farrell, K. Gloe, K. Gloe, G. Goretzki, V. McKee, J. Nelson, M. Nieuwenhuyzen, I. Pál, H. Stephan, R.M. Town and K. Wichmann, *J. Chem. Soc. Dalton Trans.*, 2003, 1961.
81. J. Nelson, M. Nieuwenhuyzen, I. Pál and R.M. Town, *J. Chem. Soc. Dalton Trans.*, 2004, 2303.
82. J. Nelson, M. Nieuwenhuyzen, I. Pál and R.M. Town, *Chem. Commun.*, 2002, 2266.
83. J. Nelson, M. Nieuwenhuyzen, I. Pál and R.M. Town, *J. Chem. Soc. Dalton Trans.*, 2004, 229.
84. G.G. Morgan, V. McKee and J. Nelson, *J. Chem. Soc. Chem. Commun.*, 1995, 1649.
85. V. McKee and G.G. Morgan, *Acta Cryst.*, 2003, **C59**, O150.
86. M.A. Hossain, J.M. Llinares, S. Mason, P. Morehouse, D. Powell and K. Bowman-James, *Angew. Chem. Int. Ed.*, 2002, **41**, 2335.
87. M.A. Hossain, P. Morehouse, D. Powell and K. Bowman-James, *Inorg. Chem.*, 2005, **44**, 2143.
88. T. Fujita and J.-M. Lehn, *Tetrahedron Lett.*, 1988, **29**, 1709.
89. Y. Murakami, J.-I. Kikuchi and T. Hirayama, *Chem. Lett.*, 1987, **16**, 161; Y. Murakami, J.-I. Kikuchi, T. Ohno and T. Hirayama, *Chem. Lett.*, 1989, **18**, 881; Y. Murakami, J.-I. Kikuchi, T. Ohno, T. Hirayama and H. Nishimura, *Chem. Lett.*, 1989, **18**, 1199.
90. Y. Murakami, J.-I. Kikuchi, T. Ohno, T. Hirayama, Y. Hisaeda, H. Nishimura, J.P. Snyder and K. Steliou, *J. Am. Chem. Soc.*, 1991, **113**, 8229.
91. R. Breslow, R. Rajagopalan and J. Schwarz, *J. Am. Chem. Soc.*, 1981, **103**, 2905.
92. I. Tabushi, Y. Kobuke and J.-I. Imuta, *J. Am. Chem. Soc.*, 1981, **103**, 6152.
93. T. Li and F. Diederich, *J. Org. Chem.*, 1992, **57**, 3449.
94. T. Li, S.J. Krasne, B. Persson, H.R. Kaback and F. Diederich, *J. Org. Chem.*, 1993, **58**, 380.
95. H.-J. Schneider, *Angew. Chem. Int. Ed.*, 1991, **30**, 1417.
96. H.-J. Schneider, T. Schiestel and P. Zimmermann, *J. Am. Chem. Soc.*, 1992, **114**, 7698.
97. M.A. Hossain and H.-J. Schneider, *Chem. Eur. J.*, 1999, **5**, 1284.
98. P.D. Beer, J.W. Wheeler, A. Grieve, C.P. Moore and T. Wear, *J. Chem. Soc. Chem. Commun.*, 1992, 1225.
99. P.D. Beer, N.C. Fletcher, A. Grieve, J.W. Wheeler, C.P. Moore and T. Wear, *J. Chem. Soc. Perkin Trans. 2*, 1996, 1545.
100. K.-S. Jeong and Y.L. Cho, *Tetrahedron Lett.*, 1997, **38**, 3279.
101. L.O. Abouderbala, W.J. Belcher, M.G. Boutelle, P.J. Cragg, J. Dhaliwal, M. Fabre, J.W. Steed, D.R. Turner and K.J. Wallace, *Chem. Commun.*, 2002, 358; S. Atilgan and E.U. Akkaya, *Tetrahedron Lett.*, 2004, **45**, 9269.
102. R. Prohens, S. Tomàs, J. Morey, P.M. Deyà, P. Ballester and A. Costa, *Tetrahedron Lett.*, 1998, **39**, 1063.
103. R. Prohens, M.C. Rotger, M.N. Piña, P.M. Deyà, J. Morey, P. Ballester and A. Costa, *Tetrahedron Lett.*, 2001, **42**, 4933.

104. M.N. Piña, M.C. Rotger, A. Costa, P. Ballester and P.M. Deyà, *Tetrahedron Lett.*, 2004, **45**, 3749.
105. Y. Matsui, M. Fujie and H. Sakate, *Carbohyd. Res.*, 1989, **192**, 91.
106. D.-R. Ahn, T.W. Kim and J.-I. Hong, *Tetrahedron Lett.*, 1999, **40**, 6045.
107. I. Tabushi, H. Sasaki and Y. Kuroda, *J. Am. Chem. Soc.*, 1976, **98**, 5727.
108. H.-J. Schneider, R. Kramer, S. Simova and U. Schneider, *J. Am. Chem. Soc.*, 1988, **110**, 6442.
109. H.-J. Schneider, T. Blatter, S. Simova and I. Theis, *J. Chem. Soc. Chem. Commun.*, 1989, 580.
110. H.-J. Schneider, *Recl. Trav. Chim. Pays-Bas*, 1993, **112**, 412.
111. H.-J. Schneider, T. Blatter, B. Palm, U. Pfingstag, V. Rüdiger and I. Theis, *J. Am. Chem. Soc.*, 1992, **114**, 7704; H.-J. Schneider, T. Blatter, A. Eliseev, V. Rüdiger and O.A. Raevsky, *Pure Appl. Chem.*, 1993, **65**, 2329.
112. F.M. Menger and K.K. Catlin, *Angew. Chem. Int. Ed.*, 1995, **34**, 2147.
113. B. Hinzen, P. Seiler and F. Diederich, *Helv. Chim. Acta*, 1996, **79**, 942.
114. K.O. Lara, C. Godoy-Alcántar, I.L. Rivera, A.V. Eliseev and A.K. Yatsimirsky, *J. Phys. Org. Chem.*, 2001, **14**, 453.
115. S. Shinoda, M. Tadokoro, H. Tsukube and R. Arakawa, *Chem. Commun.*, 1998, 181.
116. R. Prohens, G. Martorell, P. Ballester and A. Costa, *Chem. Commun.*, 2001, 1456.
117. M.A. Hossain, S.O. Kang, D. Powell and K. Bowman-James, *Inorg. Chem.*, 2003, **42**, 1397.
118. S. Claude, J.-M. Lehn and J.-P. Vigneron, *Tetrahedron Lett.*, 1989, **30**, 941; S. Claude, J.-M. Lehn, F. Schmidt and J.-P. Vigneron, *J. Chem. Soc. Chem. Commun.*, 1991, 1182.
119. P. Čudić, M. Žinić, V. Tomišić, V. Simeon, J.-P. Vigneron and J.-M. Lehn, *J. Chem. Soc. Chem. Commun.*, 1995, 1073.
120. F.P. Schmidtchen, *Angew. Chem. Int. Ed.*, 1977, **16**, 720.
121. F.P. Schmidtchen and G. Müller, *J. Chem. Soc. Chem. Commun.*, 1984, 1115; K. Ichikawa, A. Yamamoto and M.A. Hossain, *Chem. Lett.*, 1993, **22**, 2175.
122. M.A. Hossain and K. Ichikawa, *Tetrahedron Lett.*, 1994, **35**, 8393.
123. F.P. Schmidtchen, *Chem. Ber.*, 1980, **113**, 864.
124. F.P. Schmidtchen, *Chem. Ber.*, 1981, **114**, 597.
125. F.P. Schmidtchen, *Angew. Chem. Int. Ed.*, 1981, **20**, 466.
126. F.P. Schmidtchen, *Chem. Ber.*, 1984, **117**, 725.
127. F.P. Schmidtchen, *Chem. Ber.*, 1984, **117**, 1287.
128. F.P. Schmidtchen, *J. Chem. Soc. Perkin Trans. 2*, 1986, 135.
129. F.P. Schmidtchen, *Tetrahedron Lett.*, 1986, **27**, 1987; F.P. Schmidtchen, *J. Am. Chem. Soc.*, 1986, **108**, 8249; F.P. Schmidtchen, *J. Incl. Phenom.*, 1987, **5**, 161.
130. K. Ichikawa and M.A. Hossain, *Chem. Commun.*, 1996, 1721.
131. N. Ito, M. Izumi, K. Ichikawa and M. Shiro, *Chem. Lett.*, 1999, **28**, 1001.
132. K. Ichikawa, M. Izumi, D. Goto and N. Ito, *Chem. Eur. J.*, 2001, **7**, 5094.
133. Y. Murakami, T. Ohno, O. Hayashida and Y. Hisaeda, *J. Chem. Soc. Chem. Commun.*, 1991, 950.
134. Y. Murakami, T. Ohno, O. Hayashida and Y. Hisaeda, *Chem. Lett.*, 1991, **20**, 1595.

135. Y. Murakami, O. Hayashida, T. Ito and Y. Hisaeda, *Chem. Lett.*, 1992, **21**, 497; Y. Murakami, O. Hayashida, T. Ito and Y. Hisaeda, *Pure Appl. Chem.*, 1993, **65**, 551.
136. Y. Murakami, O. Hayashida and Y. Nagai, *J. Am. Chem. Soc.*, 1994, **116**, 2611; O. Hayashida, K. Ono, Y. Hisaeda and Y. Murakami, *Tetrahedron*, 1995, **51**, 8423.
137. T.H. Wirth and N. Davidson, *J. Am. Chem. Soc.*, 1964, **86**, 4325; R. Schwesinger, *Chimia*, 1985, **39**, 269.
138. R.M. Curtis and R.A. Pasternak, *Acta Cryst.*, 1955, **8**, 675; F.A. Cotton, V.W. Day, E.E.J. Hazen and S. Larsen, *J. Am. Chem. Soc.*, 1973, **95**, 4834; J.M. Adams and R.W.H. Small, *Acta Cryst.*, 1974, **B30**, 2191; J.M. Adams and R.W.H. Small, *Acta Cryst.*, 1976, **B32**, 832; J.M. Adams and R.G. Pritchard, *Acta Cryst.*, 1976, **B32**, 2438; M. Cygler, M.J. Grabowski, A. Stepien and E. Wajsman, *Acta Cryst.*, 1976, **B32**, 2391; E. Wajsman, M. Cygler, M.J. Grabowski and A. Stepien, *Roczniki Chemii*, 1976, **50**, 1587; A. Stepien and M.J. Grabowski, *Acta Cryst.*, 1977, **B33**, 2924; J.M. Adams and V. Ramdas, *Acta Cryst.*, 1978, **B34**, 2150; J.M. Adams, *Acta Cryst.*, 1978, **B34**, 1218; T. Kolev, H. Preut, P. Bleckmann and V. Radomirska, *Acta Cryst.*, 1997, **C53**, 805; A. Waskowska, *Acta Cryst.*, 1997, **C53**, 128; D.A. Baldwin, L. Denner, T.J. Egan and A.J. Markwell, *Acta Cryst.*, 1986, **C42**, 1197; B. Abrahams, M.G. Haywood, T.A. Hudson and R. Robson, *Angew. Chem. Int. Ed.*, 2004, **43**, 6157.
139. C. Tanford, *J. Am. Chem. Soc.*, 1954, **76**, 945.
140. B. Springs and P. Haake, *Bioorg. Chem.*, 1977, **6**, 181; J.I. Watters and S. Matsumoto, *J. Am. Chem. Soc.*, 1964, **86**, 3961.
141. B. Linton and A.D. Hamilton, *Tetrahedron*, 1999, **55**, 6027.
142. B. Dietrich, T.M. Fyles, J.-M. Lehn, L.G. Pease and D.L. Fyles, *J. Chem. Soc. Chem. Commun.*, 1978, 934.
143. R.P. Dixon, S.J. Geib and A.D. Hamilton, *J. Am. Chem. Soc.*, 1992, **114**, 365.
144. E. Fan, S.A. Van Arman, S. Kincaid and A.D. Hamilton, *J. Am. Chem. Soc.*, 1993, **115**, 369.
145. B. Linton, M.S. Goodman, E. Fan, S.A. Van Arman and A.D. Hamilton, *J. Org. Chem.*, 2001, **66**, 7313.
146. R. Gross, G. Dürner and M.W. Göbel, *Liebigs Ann. Chem.*, 1994, 49.
147. M.G. Hutchings, M.C. Grossel, D.A.S. Merckel, A.M. Chippendale, M. Kenworthy and G. McGeorge, *Cryst. Growth Des.*, 2001, **1**, 339; M.C. Grossel, D.A.S. Merckel and M.G. Hutchings, *CrystEngComm*, 2003, **5**, 77.
148. V. Jubian, R.P. Dixon and A.D. Hamilton, *J. Am. Chem. Soc.*, 1992, **114**, 1120; V. Jubian, A. Veronese, R.P. Dixon and A.D. Hamilton, *Angew. Chem. Int. Ed.*, 1995, **34**, 1237; H.H. Zepik and S.A. Benner, *J. Org. Chem.*, 1999, **64**, 8080; U. Scheffer, A. Strick, V. Ludwig, S. Peter, E. Kalden and M.W. Göbel, *J. Am. Chem. Soc.*, 2005, **127**, 2211.
149. J. Åqvist and A. Warshel, *Biochemistry*, 1989, **28**, 4680; D.J. Weber, A.K. Meeker and A.S. Mildvan, *Biochemistry*, 1991, **30**, 6103.
150. K. Ariga and E.V. Anslyn, *J. Org. Chem.*, 1992, **57**, 417; D.M. Kneeland, K. Ariga, V.M. Lynch, C.-Y. Huang and E.V. Anslyn, *J. Am. Chem. Soc.*, 1993, **115**, 10042.

151. D.M. Perreault, X. Chen and E.V. Anslyn, *Tetrahedron*, 1995, **51**, 353.
152. J. Smith, K. Ariga and E.V. Anslyn, *J. Am. Chem. Soc.*, 1993, **115**, 362; D.M. Perreault, L.A. Cabell and E.V. Anslyn, *Bioorg. Med. Chem.*, 1997, **5**, 1209.
153. R. Gross, J.W. Bats and M.W. Göbel, *Liebigs Ann. Chem.*, 1994, 205.
154. J.S. Albert, M.S. Goodman and A.D. Hamilton, *J. Am. Chem. Soc.*, 1995, **117**, 1143.
155. J.S. Albert, M.W. Peczuh and A.D. Hamilton, *Bioorg. Med. Chem.*, 1997, **5**, 1455.
156. A.F. McKay and M.E. Kreling, *Can. J. Chem.*, 1957, **35**, 1438.
157. G. Müller, J. Riede and F.P. Schmidtchen, *Angew. Chem. Int. Ed.*, 1988, **27**, 1516.
158. A. Gleich, F.P. Schmidtchen, P. Mikulcik and G. Müller, *J. Chem. Soc. Chem. Commun.*, 1990, 55; H. Kurzmeier and F.P. Schmidtchen, *J. Org. Chem.*, 1990, **55**, 3749.
159. M. Haj-Zaroubi, N.W. Mitzel and F.P. Schmidtchen, *Angew. Chem. Int. Ed.*, 2002, **41**, 104.
160. A. Echavarren, A. Galán, J.-M. Lehn and J. de Mendoza, *J. Am. Chem. Soc.*, 1989, **111**, 4994.
161. M. Segura, V. Alcázar, P. Prados and J. de Mendoza, *Tetrahedron*, 1997, **53**, 13119.
162. F.P. Schmidtchen, *Tetrahedron Lett.*, 1989, **30**, 4493.
163. P. Schieβl and F.P. Schmidtchen, *Tetrahedron Lett.*, 1993, **34**, 2449; H. Stephan, K. Gloe, P. Schieβl and F.P. Schmidtchen, *Supramol. Chem.*, 1995, **5**, 273; W. Peschke and F.P. Schmidtchen, *Tetrahedron Lett.*, 1995, **36**, 5155; M. Berger and F.P. Schmidtchen, *J. Am. Chem. Soc.*, 1996, **118**, 8947; M. Fibbioli, M. Berger, F.P. Schmidtchen and E. Pretsch, *Anal. Chem.*, 2000, **72**, 156.
164. P. Schieβl and F.P. Schmidtchen, *J. Org. Chem.*, 1994, **59**, 509.
165. J. Sánchez-Quesada, C. Seel, P. Prados and J. de Mendoza, *J. Am. Chem. Soc.*, 1996, **118**, 277.
166. M.W. Peczuh, A.D. Hamilton, J. Sánchez-Quesada, J. de Mendoza, T. Haack and E. Giralt, *J. Am. Chem. Soc.*, 1997, **119**, 9327.
167. T. Haack, M.W. Peczuh, X. Salvatella, J. Sánchez-Quesada, J. de Mendoza, A.D. Hamilton and E. Giralt, *J. Am. Chem. Soc.*, 1999, **121**, 11813.
168. X. Salvatella, M.W. Peczuh, M. Gairí, R.K. Jain, J. Sánchez-Quesada, J. de Mendoza, A.D. Hamilton and E. Giralt, *Chem. Commun.*, 2000, 1399.
169. B.P. Orner, X. Salvatella, J. Sánchez-Quesada, J. de Mendoza, E. Giralt and A.D. Hamilton, *Angew. Chem. Int. Ed.*, 2002, **41**, 117.
170. A. Metzger, V.M. Lynch and E.V. Anslyn, *Angew. Chem. Int. Ed.*, 1997, **36**, 862; A. Metzger and E.V. Anslyn, *Angew. Chem. Int. Ed.*, 1998, **37**, 649; S.C. McCleskey, A. Metzger, C.S. Simmons and E.V. Anslyn, *Tetrahedron*, 2002, **58**, 621; S.C. McCleskey, P.N. Floriano, S.L. Wiskur, E.V. Anslyn and J.T. McDevitt, *Tetrahedron*, 2003, **59**, 10089; M. Rekharsky, Y. Inoue, S.L. Tobey, A. Metzger and E.V. Anslyn, *J. Am. Chem. Soc.*, 2002, **124**, 14959; K. Niikura, A. Metzger and E.V. Anslyn, *J. Am. Chem. Soc.*, 1998, **120**, 8533; J.P. Lorand and J.O. Edwards, *J. Org. Chem.*, 1959, **24**, 769; J.J. Lavigne and E.V. Anslyn,

Angew. Chem. Int. Ed., 1999, **38**, 3666; S.L. Wiskur, P.N. Floriano, E.V. Anslyn and J.T. McDevitt, *Angew. Chem. Int. Ed.*, 2003, **42**, 2070; A.M. Piatek, Y.J. Bomble, S.L. Wiskur and E.V. Anslyn, *J. Am. Chem. Soc.*, 2004, **126**, 6072; B.T. Nguyen, S.L. Wiskur and E.V. Anslyn, *Org. Lett.*, 2004, **6**, 2499; S.L. Wiskur and E.V. Anslyn, *J. Am. Chem. Soc.*, 2001, **123**, 10109; S.L. Wiskur, J.J. Lavigne, A. Metzger, S.L. Tobey, V. Lynch and E.V. Anslyn, *Chem. Eur. J.*, 2004, **10**, 3792.

171. S.L. Tobey and E.V. Anslyn, *J. Am. Chem. Soc.*, 2003, **125**, 14807.
172. Z. Zhong and E.V. Anslyn, *Angew. Chem. Int. Ed.*, 2003, **42**, 3005.
173. S.L. Tobey and E.V. Anslyn, *Org. Lett.*, 2003, **5**, 2029.
174. T. Zhang and E.V. Anslyn, *Tetrahedron*, 2004, **60**, 11117.
175. A.P. Davis and L.J. Lawless, *Chem. Commun.*, 1999, 9; L.J. Lawless, A.G. Blackburn, A.J. Ayling, M.N. Pérez-Payán and A.P. Davis, *J. Chem. Soc. Perkin Trans. 1*, 2001, 1329.
176. B. Baragaña, A.G. Blackburn, P. Breccia, A.P. Davis, J. de Mendoza, J.M. Padrón-Carrillo, P. Prados, J. Riedner and J.G. de Vries, *Chem. Eur. J.*, 2002, **8**, 2931.
177. C. Schmuck and J. Lex, *Org. Lett.*, 1999, **1**, 1779.
178. C. Schmuck, *Chem. Commun.*, 1999, 843; C. Schmuck, *Chem. Eur. J.*, 2000, **6**, 709.
179. C. Schmuck and V. Bickert, *Org. Lett.*, 2003, **5**, 4579.
180. C. Schmuck and M. Heil, *Org. Biomol. Chem.*, 2003, **1**, 633; C. Schmuck and L. Geiger, *Curr. Org. Chem.*, 2003, **7**, 1485; C. Schmuck and L. Geiger, *J. Am. Chem. Soc.*, 2004, **126**, 8898; C. Schmuck and U. Machon, *Chem. Eur. J.*, 2005, **11**, 1109; C. Schmuck and L. Geiger, *Chem. Commun.*, 2005, 772; C. Schmuck and M. Schwegmann, *J. Am. Chem. Soc.*, 2005, **127**, 3373; C. Schmuck and S. Graupner, *Tetrahedron Lett.*, 2005, **46**, 1295; C. Schmuck and M. Schwegmann, *Org. Lett.*, 2005, **7**, 3517; C. Schmuck and L. Geiger, *J. Am. Chem. Soc.*, 2005, **127**, 10486.
181. O. Kennard and J. Walker, *Acta Cryst.*, 1961, **14**, 91; R. Norrestam, *Acta Cryst.*, 1984, **C40**, 297; I.K. Larsen, *Acta Cryst.*, 1985, **C41**, 749; B. Kratochvíl, J. Ondrácek, J. Krechl and J. Hasek, *Acta Cryst.*, 1987, **C43**, 2182; M. Hjorth and R. Norrestam, *Acta Cryst.*, 1987, **C43**, 1589; K.C. Joshi, R. Bohra, R. Jain and S. Arora, *Acta Cryst.*, 1994, **C50**, 1792; J. Ondrácek, S. Pakhomova, B. Kratochvíl and S. Smrckova, *Collect. Czech. Chem. Commun.*, 1995, **60**, 576; J.C. Barnes and T.J.R. Weakley, *Acta Cryst.*, 1998, **C54**, 1170; J. Barker, W. Errington and M.G.H. Wallbridge, *Acta Cryst.*, 1999, **C55**, 1583; M.N. Kopylovich, V.Y. Kukushkin, M.F.C.G. da Silva, M. Haukka, J.J.R.F. da Silva and A.J.L. Pombeiro, *J. Chem. Soc. Perkin Trans. 1*, 2001, 1569.
182. O. Kennard and J. Walker, *J. Chem. Soc.*, 1963, 5513.
183. G. Hafelinger, in *The Chemistry of Amidines and Imidates*, S. Patai (ed), Wiley, New York, 1975.
184. N.P. Pavletich and C.O. Pabo, *Science*, 1991, **252**, 809; J.M. Berg, *Acc. Chem. Res.*, 1995, **28**, 14.
185. J.D. Puglisi, L. Chen, A.D. Frankel and J.R. Williamson, *Proc. Natl. Acad. Sci. USA*, 1993, **90**, 3680.
186. E.H. Howell, J.E. Villafranca, M.S. Warren, S.J. Oatley and J. Kraut, *Science*, 1986, **231**, 1125.

187. E. Reich, D.B. Rifkin and E. Shaw, *Proteases and Biological Control*, Cold Spring Harbor Lab., New York, 1975; H. Hörn and A. Heidland, *Proteases: Potential Role in Health and Diseases*, Plenum Press, New York, 1992.
188. A.J. Barrett and G. Salvesen, *Proteinase Inhibitors*, Elsevier, New York, 1986.
189. M. Mares-Guia and E. Shaw, *J. Biol. Chem.*, 1965, **240**, 1579.
190. W. Bode and P. Schwager, *J. Mol. Biol.*, 1975, **98**, 693; M. Marquart, J. Walter, J. Deisenhofer, W. Bode and R. Huber, *Acta Cryst.*, 1983, **B39**, 480.
191. R. Talhout and J.B.F.N. Engberts, *Eur. J. Biochem.*, 2001, **268**, 1554; R. Talhout and J.B.F.N. Engberts, *Org. Biomol. Chem.*, 2004, **2**, 3071.
192. E. de Clercq and O. Dann, *J. Med. Chem.*, 1980, **23**, 787.
193. M. Sands, M.A. Kron and R.B. Brown, *Rev. Infect. Dis.*, 1985, **7**, 625; A.B. Montgomery, J.M. Luce, J. Turner, E.T. Lin, R.J. Debs, K.J. Corkery, E.N. Brunette and P.C. Hopewell, *Lancet ii*, 1987, 480; B.G. Gazzard, *J. Antimicrob. Chemother*, 1989, **23**, 67; J.A. Golden, D. Chernoff, H. Hollander, D. Feigal and J.E. Conte, *Lancet i*, 1989, 654; B. Wispelwey and R.D. Pearson, *Infect. Control Hosp. Epidemiol.*, 1991, **12**, 375; D.W. Boykin, A. Kumar, J. Spychala, M. Zhou, R.J. Lombardy, W.D. Wilson, C.C. Dykstra, S.K. Jones, J.E. Hall, R.R. Tidwell, C.A. Laughton, C.M. Nunn and S. Neidle, *J. Med. Chem.*, 1995, **38**, 912.
194. R.R. Tidwell, S.K. Jones, J.D. Geratz, K.A. Ohemeng, M. Cory and J.E. Hall, *J. Med. Chem.*, 1990, **33**, 1252; M. Cory, R.R. Tidwell and T.A. Fairley, *J. Med. Chem.*, 1992, **35**, 431; I.O. Donker, R.R. Tidwell and S.K. Jones, *J. Med. Chem.*, 1994, **37**, 4554.
195. D.G. Brown, M.R. Sanderson, J.V. Skelly, T.C. Jenkins, T. Brown, E. Garman, D.I. Stuart and S. Neidle, *EMBO J.*, 1990, **9**, 1329.
196. K.J. Edwards, T.C. Jenkins and S. Neidle, *Biochemistry*, 1992, **31**, 7104; D.G. Brown, M.R. Sanderson, E. Garman and S. Neidle, *J. Mol. Biol.*, 1992, **226**, 481; C.M. Nunn, T.C. Jenkins and S. Neidle, *Biochemistry*, 1993, **32**, 13838; J.O. Trent, G.R. Clark, A. Kumar, W.D. Wilson, D.W. Boykin, J.E. Hall, R.R. Tidwell, B.L. Blagburn and S. Neidle, *J. Med. Chem.*, 1996, **39**, 4554; I.J. Simpson, M. Lee, A. Kumar, D.W. Boykin and S. Neidle, *Bioorg. Med. Chem. Lett.*, 2000, **10**, 2593; B. Nguyen, C. Tardy, C. Bailly, P. Colson, C. Houssier, A. Kumar, D.W. Boykin and W.D. Wilson, *Biopolymers*, 2002, **63**, 281; B. Nguyen, M.P.H. Lee, D. Hamelberg, A. Joubert, C. Bailly, R. Brun, S. Neidle and W.D. Wilson, *J. Am. Chem. Soc.*, 2002, **124**, 13680.
197. C.A. Laughton, F. Tanious, C.M. Nunn, D.W. Boykin, W.D. Wilson and S. Neidle, *Biochemistry*, 1996, **35**, 5655.
198. A.T. Fuller, I.M. Tonkin and J. Walker, *J. Chem. Soc.*, 1945, 633; H.S. Forrest, A.T. Fuller and J. Walker, *J. Chem. Soc.*, 1948, 1501; H.S. Forrest and J. Walker, *J. Chem. Soc.*, 1948, 1939; K.J. Shaw, W.J. Guilford, J.L. Dallas, S.K. Koovakkaat, M.A. McCarrick, A. Liang, D.R. Light and M.M. Morrissey, *J. Med. Chem.*, 1998, **41**, 3551; M. Renatus, W. Bode, R. Huber, J. Stürzebecher and M.T. Stubbs, *J. Med. Chem.*, 1998, **41**, 5445; B. Garbriel, M.T. Stubbs, A. Bergner, J. Hauptmann, W. Bode, J. Stürzebecher and L. Moroder, *J. Med. Chem.*, 1998, **41**, 4240; G. Phillips, D.D. Davey, K.A. Eagen, S.K. Koovakkat, A. Liang, H.P. Ng, M. Pinkerton, L. Trinh, M. Whitlow, A.M. Beatty and M.M. Morrissey, *J. Med. Chem.*, 1999, **42**, 1749; W.J. Guilford, K.J. Shaw,

J.L. Dallas, S.K. Koovakkat, W. Lee, A. Liang, D.R. Light, M.A. McCarrick, M. Whitlow, B. Ye and M.M. Morrissey, *J. Med. Chem.*, 1999, **42**, 5415.
199. F. Heinzer, M. Soukup and A. Eschenmoser, *Helv. Chim. Acta*, 1978, **61**, 2851.
200. P.H. Boyle, M.A. Convery, A.P. Davis, G.D. Hosken and B.A. Murray, *J. Chem. Soc. Chem. Commun.*, 1992, 239.
201. M.A. Convery, A.P. Davis, C.J. Dunne and J.W. MacKinnon, *J. Chem. Soc. Chem. Commun.*, 1994, 2557.
202. M.W. Göbel, J.W. Bats and G. Dürner, *Angew. Chem. Int. Ed.*, 1992, **31**, 207; G. Müller, G. Dürner, J.W. Bats and M.W. Göbel, *Liebigs Ann. Chem.*, 1994, 1075; S. Lehr, K. Schütz, M. Bauch and M.W. Göbel, *Angew. Chem. Int. Ed.*, 1994, **33**, 984.
203. A. Terfort and G. von Kiedrowski, *Angew. Chem. Int. Ed.*, 1992, **31**, 654.
204. I. Kuzmenko, M. Kindermann, K. Kjaer, P.B. Howes, J. Als-Nielsen, R. Granek, G.V. Kiedrowski, L. Leiserowitz and M. Lahav, *J. Am. Chem. Soc.*, 2001, **123**, 3771.
205. M.Y. Okamura and G. Feher, *Annu. Rev. Biochem.*, 1992, **61**, 861; A. Müller, H. Ratajczaks, W. Junge and E. Diemann, *Electron and Proton Transfer in Chemistry and Biology*, Elsevier, New York, 1992; S. Ferguson-Miller and G.T. Babcock, *Chem. Rev.*, 1996, **96**, 2889.
206. B.R. Crane, L.M. Siegel and E.D. Getzoff, *Science*, 1995, **270**, 59.
207. J.P. Kirby, N.A. van Dantzig, C.K. Chang and D.G. Nocera, *Tetrahedron Lett.*, 1995, **36**, 3477.
208. Y. Deng, J.A. Roberts, S.-M. Peng, C.K. Chang and D.G. Nocera, *Angew. Chem. Int. Ed.*, 1997, **36**, 2124.
209. C.-Y. Yeh, S.E. Miller, S.D. Carpenter and D.G. Nocera, *Inorg. Chem.*, 2001, **40**, 3643.
210. J.P. Kirby, J.A. Roberts and D.G. Nocera, *J. Am. Chem. Soc.*, 1997, **119**, 9230; J. Otsuki, M. Takatsuki, M. Kaneko, H. Miwa, T. Takido, M. Seno, K. Okamoto, H. Imahori, M. Fujitsuka, Y. Araki, O. Ito and S. Fukuzumi, *J. Phys. Chem. A*, 2003, **107**, 379; J. Otsuki, K. Iwasaki, Y. Nakano, M. Itou, Y. Araki and O. Ito, *Chem. Eur. J.*, 2004, **10**, 3461.
211. G. Wulff, T. Gross and R. Schönfeld, *Angew. Chem. Int. Ed.*, 1997, **36**, 1962.
212. G. Wulff and R. Schönfeld, *Adv. Mater.*, 1998, **10**, 957.
213. A.G. Strikovsky, D. Kasper, M. Grün, B.S. Green, J. Hradil and G. Wulff, *J. Am. Chem. Soc.*, 2000, **122**, 6295; J.-Q. Liu and G. Wulff, *J. Am. Chem. Soc.*, 2004, **126**, 7452.
214. G. Wulff and A. Biffis, in *Molecular Imprinted Polymers-Manmade Mimics of Antibodies and their Application in Analytical Chemistry*, B. Sellergren (ed), Elsevier Amsterdam, 2001; G. Wulff, *Chem. Rev.*, 2002, **102**, 1.
215. A. Kraft, L. Peters and H.R. Powell, *Tetrahedron*, 2002, **58**, 3499.
216. L. Peters, R. Fröhlich, A.S.F. Boyd and A. Kraft, *J. Org. Chem.*, 2001, **66**, 3291.
217. G. Cooke, F.M.A. Duclairoir, A. Kraft, G. Rosair and V.M. Rotello, *Tetrahedron Lett.*, 2004, **45**, 557.
218. A. Kraft and R. Fröhlich, *Chem. Commun.*, 1998, 1085; A. Kraft, *J. Chem. Soc. Perkin Trans. 1*, 1999, 705.

219. T. Grawe, T. Schrader, M. Gurrath, A. Kraft and F. Osterod, *Org. Lett.*, 2000, **2**, 29.
220. P.A. Gale, *Tetrahedron Lett.*, 1998, **39**, 3873.
221. S. Camiolo, P.A. Gale, M.I. Ogden, B.W. Skelton and A.H. White, *J. Chem. Soc. Perkin Trans. 2*, 2001, 1294; S. Camiolo, P.A. Gale, M.E. Light and M.B. Hursthouse, *Supramol. Chem.*, 2001, **13**, 613.
222. L. Sebo, B. Schweizer and F. Diederich, *Helv. Chim. Acta*, 2000, **83**, 80; L. Sebo and F. Diederich, *Helv. Chim. Acta*, 2000, **83**, 93.
223. F. Corbellini, R. Fiammengo, P. Timmerman, M. Crego-Calama, K. Versluis, A.J. Heck, I. Luyten and D.N. Reinhoudt, *J. Am. Chem. Soc.*, 2002, **124**, 6569.
224. F. Corbellini, L.D. Costanzo, M. Crego-Calama, S. Geremia and D.N. Reinhoudt, *J. Am. Chem. Soc.*, 2003, **125**, 9946; F. Corbellini, R.M.A. Knegtel, P.D.J. Grootenhuis, M. Crego-Calama and D.N. Reinhoudt, *Chem. Eur. J.*, 2005, **11**, 298.
225. J.S. Wilkes, J.A. Levisky, R.A. Wilson and C.L. Hussey, *Inorg. Chem.*, 1982, **21**, 1263; C.L. Hussey, *Adv. Molten Salt Chem.*, 1983, **5**, 185; Z.-B. Zhou, H. Matsumoto and K. Tatsumi, *Chem. Eur. J.*, 2004, **10**, 6581; V. Cimpeanu, V.I. Parvulescu, P. Amorós, D. Beltrán, J.M. Thompson and C. Hardacre, *Chem. Eur. J.*, 2004, **10**, 4640.
226. A.A. Fannin Jr., L.A. King, J.A. Levisky and J.S. Wilkes, *J. Phys. Chem.*, 1984, **88**, 2609; K.M. Dieter, C.J. Dymek Jr., N.E. Heimer, J.W. Rovang and J.S. Wilkes, *J. Am. Chem. Soc.*, 1988, **110**, 2722; C.J. Dymek Jr. and J.J.P. Stewart, *Inorg. Chem.*, 1989, **28**, 1472.
227. A.K. Abdul-Sada, A.M. Greenway, P.B. Hitchcock, T.J. Mohammed, K.R. Seddon and J.A. Zora, *J. Chem. Soc. Chem. Commun.*, 1986, 1753; C.J. Dymek Jr., D.A. Grossie, A.V. Fratini and W.W. Adams, *J. Mol. Struct.*, 1989, **213**, 25.
228. A.K. Abdul-Sada, S. Al-Juaid, A.M. Greenway, P.B. Hitchcock, M.J. Howells, K.R. Seddon and T. Welton, *Struct. Chem.*, 1990, **1**, 391; J.S. Wilkes and M.J. Zaworotko, *J. Chem. Soc. Chem. Commun.*, 1992, 965; A.J. Arduengo III, H.V.R. Dias, R.L. Harlow and M. Kline, *J. Am. Chem. Soc.*, 1992, **114**, 5530; P.B. Hitchcock, K.R. Seddon and T. Welton, *J. Chem. Soc. Dalton Trans.*, 1993, 2639; J. Fuller, R.T. Carlin, H.C. de Long and D. Haworth, *J. Chem. Soc. Chem. Commun.*, 1994, 299; A. Elaiwi, P.B. Hitchcock, K.R. Seddon, N. Srinivasan, Y.-M. Tan, T. Welton and J.A. Zora, *J. Chem. Soc. Dalton Trans.*, 1995, 3467; A.J. Arduengo, III, R. Krafczyk, R. Schmutzler, H.A. Craig, J.R. Goerlich, W.J. Marshall and M. Unverzagt, *Tetrahedron*, 1999, **55**, 14523; M.L. Cole, C. Jones and P.C. Junk, *New J. Chem.*, 2002, **262**, 1296; J.D. Holbrey, W.M. Reichert and R.D. Rogers, *J. Chem. Soc. Dalton Trans.*, 2004, 2267.
229. F.C. Gozzo, L.S. Santos, R. Augusti, C.S. Consorti, J. Dupont and M.N. Eberlin, *Chem. Eur. J.*, 2004, **10**, 6187.
230. A.G. Avent, P.A. Chaloner, M.P. Day, K.R. Seddon and T. Welton, *J. Chem. Soc. Dalton Trans.*, 1994, 3405.
231. K. Sato, S. Arai and T. Yamagishi, *Tetrahedron Lett.*, 1999, **40**, 5219.
232. K. Sato, T. Onitake, S. Arai and T. Yamagishi, *Heterocycles*, 2003, **60**, 779.

233. S. Yun, H. Ihm, H.G. Kim, C.-W. Lee, B. Indrajit, K.S. Oh, Y.J. Gong, J.W. Lee, J. Yoon, H.C. Lee and K.S. Kim, *J. Org. Chem.*, 2003, **68**, 2467.
234. S.K. Kim, N.J. Singh, S.J. Kim, H.G. Kim, J.K. Kim, J.W. Lee, K.S. Kim and J. Yoon, *Org. Lett.*, 2003, **5**, 2083; J.Y. Kwon, N.J. Singh, H.N. Kim, S.K. Kim, K.S. Kim and J. Yoon, *J. Am. Chem. Soc.*, 2004, **126**, 8892.
235. H. Kim and J. Kang, *Tetrahedron Lett.*, 2005, **46**, 5443.
236. H. Ihm, S. Yun, H.G. Kim, J.K. Kim and K.S. Kim, *Org. Lett.*, 2002, **4**, 2897.
237. J. Howarth and N.A. Al-Hashimy, *Tetrahedron Lett.*, 2001, **42**, 5777; Y. Bai, B.-G. Zhang, J. Xu, C.-Y. Duan, D.-B. Dang, D.-J. Liu and Q.-J. Meng, *New J. Chem.*, 2005, **29**, 777.
238. S.K. Kim, B.-G. Kang, H.S. Koh, Y.J. Yoon, S.J. Jung, B. Jeong, K.-D. Lee and J. Yoon, *Org. Lett.*, 2004, **6**, 4655.
239. E. Alcalde, C. Alvarez-Rúa, S. García-Granda, E. García-Rodriguez, N. Mesquida and L. Pérez-García, *Chem. Commun.*, 1999, 295.
240. Y.Y. Yuan, G. Gao, Z.-L. Jiang, J.-S. You, Z.-Y. Zhou, D.-Q. Yuan and R.-G. Xie, *Tetrahedron*, 2002, **58**, 8993.
241. J. Yoon, S.K. Kim, N.J. Singh, J.W. Lee, Y.J. Yang, K. Chellappan and K.S. Kim, *J. Org. Chem.*, 2004, **69**, 581.
242. K. Chellappan, N.J. Singh, I.-C. Hwang, J.W. Lee and K.S. Kim, *Angew. Chem. Int. Ed.*, 2005, **44**, 2899.
243. B.-G. Zhang, P. Cai, C.-Y. Doan, R. Miao, L.-G. Zho, T. Niitsu and H. Inove, *Chem. Common.,* 2004, 2206.
244. G.B. Jameson, E. Blazsò, N. Seferiadis and H.R. Oswald, *Acta Cryst.*, 1982, **B38**, 2272; V.K. Beískii, F.V. Babilev, T.P. Tryapitsyna and E.A. Mukhin, *Proc. Natl. Acad. Sci. USSR*, 1985, **282**, 605; A.N. Chekholv, *Crystallogr. Rep.*, 1992, **37**, 696; S.W. Ng, *Acta Cryst.*, 1995, **C51**, 1143; J. Barker and H.R. Powell, *Acta Cryst.*, 1998, **C54**, 2019.
245. A. Bernthsen and H. Klinger, *Chem. Ber.*, 1878, **11**, 492.
246. J.J. Donleavy, *J. Am. Chem. Soc.*, 1936, **58**, 1004; B.T. Dewey and R.B. Sperry, *J. Am. Chem. Soc.*, 1939, **61**, 3251; J.P. Kass, J. Nichols and G.O. Burr, *J. Am. Chem. Soc.*, 1942, **64**, 1061; W.A. Bonner, *J. Am. Chem. Soc.*, 1948, **70**, 3508.
247. W.-S. Yeo and J.-I. Hong, *Tetrahedron Lett.*, 1998, **39**, 3769.
248. W.-S. Yeo and J.-I. Hong, *Tetrahedron Lett.*, 1998, **39**, 8137.
249. S.-I. Sasaki, A. Hashizume, S. Ozawa, D. Citterio, N. Iwasawa and K. Suzuki, *Chem. Lett.*, 2001, 382.
250. Y. Kubo, M. Tsukahara, S. Ishihara and S. Tokita, *Chem. Commun.*, 2000, 653.
251. S. Nishizawa, Y.-Y. Cui, M. Minagawa, K. Morita, Y. Kato, S. Taniguchi, R. Kato and N. Teramae, *J. Chem. Soc. Perkin Trans. 2*, 2002, 866.
252. Y. Kubo, S. Ishihara, M. Tsukahara and S. Tokita, *J. Chem. Soc. Perkin Trans. 2*, 2002, 1455; Y. Kubo, M. Kato, Y. Misawa and S. Tokita, *Tetrahedron Lett.*, 2004, **45**, 3769.
253. H.R. Seong, D.-S. Kim, S.-G. Kim, H.-J. Choi and K.H. Ahn, *Tetrahedron Lett.*, 2004, **45**, 723.

CHAPTER 3

Protonated Expanded Porphyrins and Linear Analogues

3.1 Introduction

As detailed in the previous chapter, it has long been appreciated that protonated polyamines and other highly charged species (*e.g.*, guanidiniums) can bind anions well as the result of strong electrostatic and hydrogen-bonding interactions. Over the past decade or so, it has also been appreciated that protonated oligopyrrolic macrocycles and their open-chain analogues can also act as highly efficient anion receptors. Conjugated oligopyrrolic species often contain sp^2 hybridized "pyridine-like" pyrrolic nitrogens that are partially or completely protonated in neutral media. They also generally contain normal pyrrolic NH subunits, and this unique combination of features (hydrogen-bond donor and electrostatic character) makes them attractive as potential anion-binding or transporting agents. The fact that the biological activity of prodigiosin **3.1**, a pyrrolylbipyrrole, could be due to H^+ and Cl^- symport arising from protonation and anion recognition, as presented briefly in Chapter 1, has served to heighten interest in this area. In this chapter, we review the chemistry of various oligopyrrolic systems whose documented anion-binding ability is predicated on the prior protonation of one or more basic imine-like nitrogens. In fact, this means the primary focus will be on the so-called "expanded porphyrins", synthetic analogues of the well-known blood pigment porphyrin.[1] However, some discussion of open-chain systems is included, although work with such receptors, in spite of their potential relevance to prodigiosin, is far less advanced. Neutral pyrrole-based anion receptors are not included in this chapter but instead are discussed in Chapter 5.

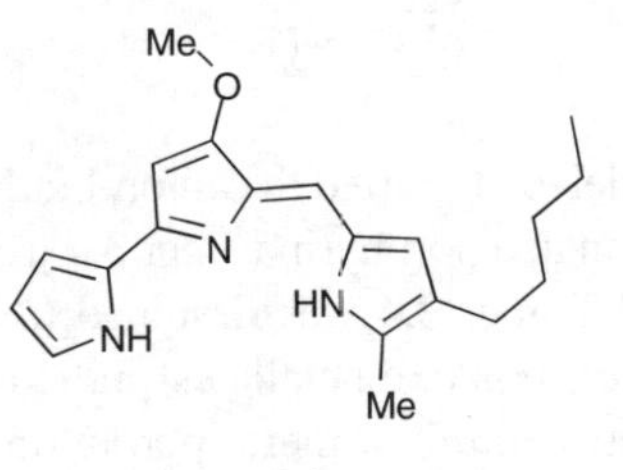

3.1

3.2 Cyclic Systems

3.2.1 Tetrapyrrolic Systems

Porphyrins are tetrapyrrolic, aromatic macrocycles containing an 18 π-electron periphery. They have been extensively studied both because of their recognized biological importance and their long-appreciated ability to form strong metal complexes with a range of cations, including most metalloids in the periodic table. In their free-base forms, porphyrins contain two pyrrole (NH) and two pyridine-like (N) nitrogen atoms within their central cores. In the absence of extensive *meso*- and beta-pyrrolic substitution, most neutral and metal-coordinated porphyrins adopt conformations that are relatively planar. So far, there is no evidence that such normal, planar porphyrins are capable of binding anions in aqueous solution at physiological pH, or even in organic media under most normally accessible conditions. Presumably, this reflects the fact that most porphyrins exist as neutral species due to the low pK_a values of their corresponding mono- and diprotonated (cationic) forms. However, there is ample evidence that the latter forms, although difficult to access in solution, will bind anions in the solid state. In particular, single-crystal X-ray diffraction analyses of the salt forms of porphyrins **3.2a–3.2d** have served to establish that these diprotonated species often interact well with their counteranions in the solid state as the result of electrostatic effects and pyrrole NH-anion hydrogen-bonding interactions (Figure 3.1). As the result of these interactions, distortions from planarity are seen, even in those systems where such distortions are not expected on the basis of substituent effects. To date, anion complexes of carboxylate,[2–4] perchlorate,[2,5,6] sulfate,[4,7] and chloride[4,8] have been characterized in this way. This leads to the suggestion that these and other anions might be bound in solution, at least under conditions of high proton activity. However, as yet, little work along the latter lines has been carried out.

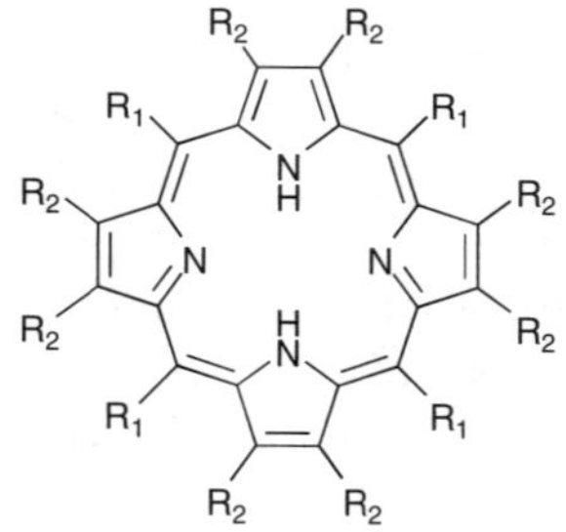

3.2a R_1 = Ph, R_2 = Br
3.2b R_1 = H, R_2 = Et
3.2c R_1 = Ph, R_2 = H
3.2d R_1 = Ph, R_2 = Et

In 1992, Sanders and Anderson reported the anion-binding properties of diprotonated porphyrin **3.3** and hexaprotonated porphyrin trimer **3.4** using Fast-Atom Bombardment (FAB) mass spectrometry.[9] Trimer **3.4** was designed to act as a receptor for "giant" anions, such as Keggin-type polyoxometallates and high-nuclearity metal-carbonyl clusters. The interaction between this trimeric porphyrin and the targeted anions was confirmed via the detection of peaks at *m/z* values corresponding to complexes formed

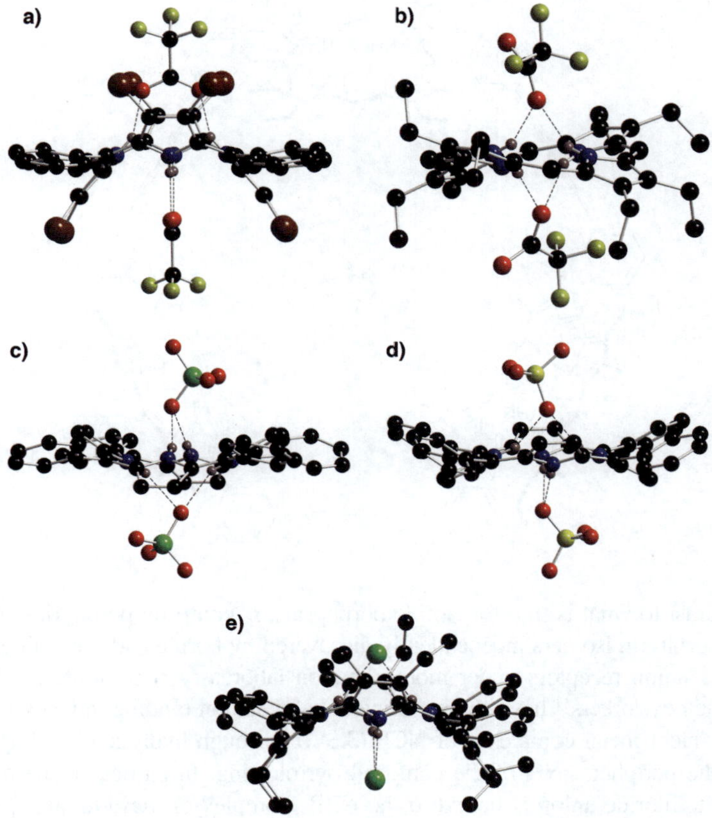

Figure 3.1 *X-ray crystal structures of (a) bis-TFA salt of **3.2a**, (b) bis-TFA salt of **3.2b**, (c) perchloric acid salt of **3.2c**, (d) bis-sulfuric acid salt of **3.2c**, and (e) bis-hydrochloric acid salt of **3.2d***

between $H_6\cdot\mathbf{3.4}^{6+}$ and $PW_{12}O_{40}^{3-}$, $SiW_{12}O_{40}^{4-}$, and $Os_{10}C(CO)_{24}^{2-}$ in the mass spectra of **3.4** recorded in the presence of an excess of the corresponding acids. Corresponding peaks were not seen in the case of the monomer **3.3**. Moreover, no peaks corresponding either to $H_6\cdot\mathbf{3.4}^{6+}-6X^-$ or $H_2\cdot\mathbf{3.3}^{2+}-2X^-$ ($X = Cl^-$, $CF_3CO_2^-$, and 1/2 SO_4^{2-}) were observed, presumably reflecting the fact that anion complexation, were it occurring, would give rise to species with no net charge (*i.e.*, not detectable by MS). Thus, at least to the extent monitoring was carried out solely by mass spectrometric means, **3.4** can be viewed as being a selective "sensor" for the desired "giant" anions.

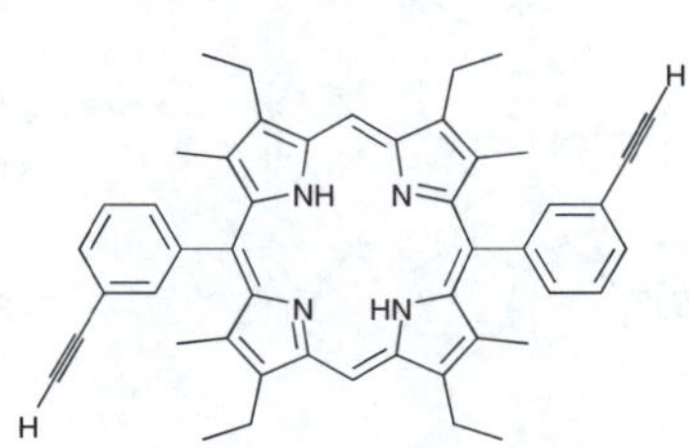

3.3

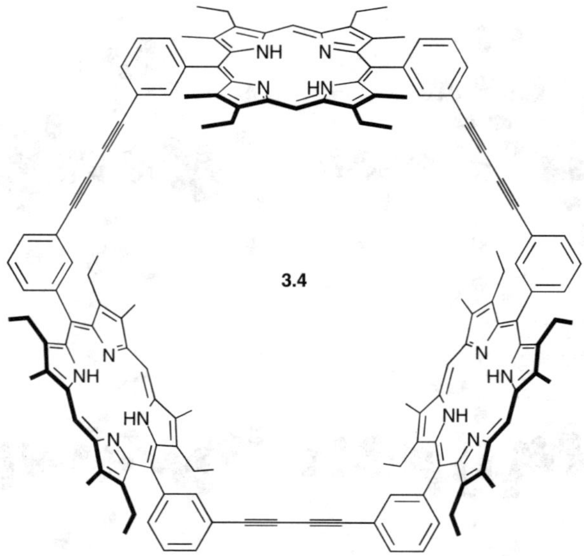

In contrast to what is true for simple porphyrins, *N*-confused porphyrins (NCPs), a class of porphyrin isomers independently discovered by Furuta and Latos-Grażyński,[10] can act as anion receptors under more common laboratory conditions. For instance, Furuta and co-workers[11] have shown efficient halide anion binding in CH_2Cl_2 effected by the divalent metal complexes of NCP (**3.5-M**) through hydrogen-bonding interactions at the peripheral NH of the confused pyrrole ring. In particular, these workers found that chloride anion is bound to the Ni(II) complex of *meso*-tetrakis(pentafluorophenyl)-NCP (**3.5-Ni**) with an affinity constant of $5.7 \times 10^4\ M^{-1}$ in CH_2Cl_2. The zwitterionic resonance form of NCP and the acidity of the *outer* NH proton of the confused pyrrole subunit are thought to play integral roles in the anion-binding process (Scheme 3.1). In a continuation of this theme, Furuta and co-workers[12] subsequently reported that the Cu(II) complex of the *trans*-doubly NCP (*trans*-N_2CP, **3.6-Cu**) was a better Cl^- receptor ($K_a = 9.0 \times 10^4\ M^{-1}$ in CH_2Cl_2) than its singly confused counterpart, **3.5-Cu**.

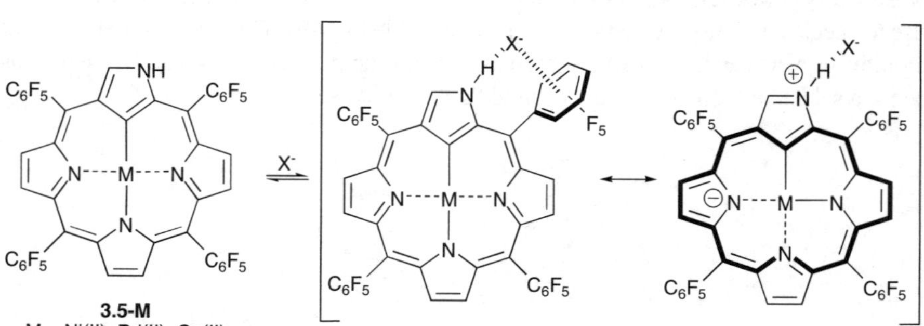

Scheme 3.1 *Anion binding of N-confused porphyrin **3.5-M** and resonance forms*

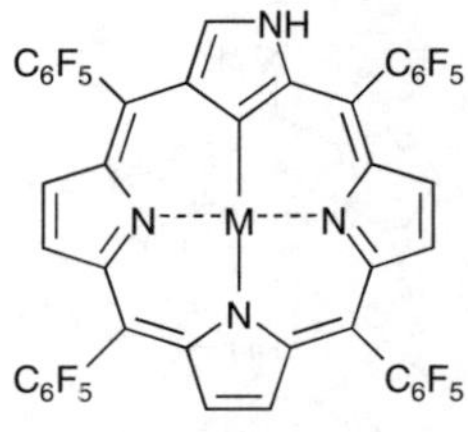

3.5-M
M = Ni(II), Pd(II), Cu(II)

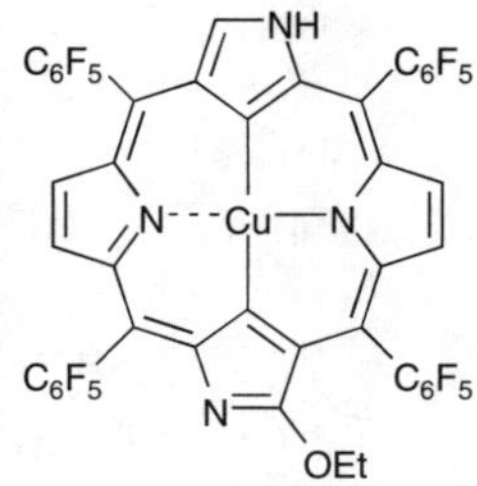

3.6

Prior to describing the anion-binding behaviour of NCP, Furuta and co-workers reported the anion-binding properties of the *N*-fused system, **3.8**. This product was prepared from the corrin isomer **3.7** as shown in Scheme 3.2. Compound **3.7**, in turn, was isolated as a significant by-product during the reaction between an *N*-confused dipyrromethane and an aldehyde. It was found to display good affinity and selectivity towards fluoride over other halide anions. Specifically, the addition of fluoride anion was found to cause the Soret band to shift from 425 to 431 nm, while reducing the intensity of the emission band at 631 nm. Titrations carried out in $CHCl_3$ and monitored by spectroscopic methods (both absorbance and fluorescence) gave rise to fluoride anion-affinity values on the order of $K_a \approx 10^4\ M^{-1}$ when the data were fit to a 1:1 equilibrium expression.[13]

In analogy to what is seen in the case of regular porphyrins, several other porphyrin isomers and analogues, including corrole (**3.9**), porphycene (**3.10**), corrphycene (**3.11**), the corrole derivatives **3.12** and **3.13**, and benziporphryin (**3.14** and **3.15**), were found to form complexes with anions in the solid state as shown in Figures 3.2–3.4.[6,14] As yet, however, no evidence has been forwarded in support of anion binding taking place in solution under neutral conditions.

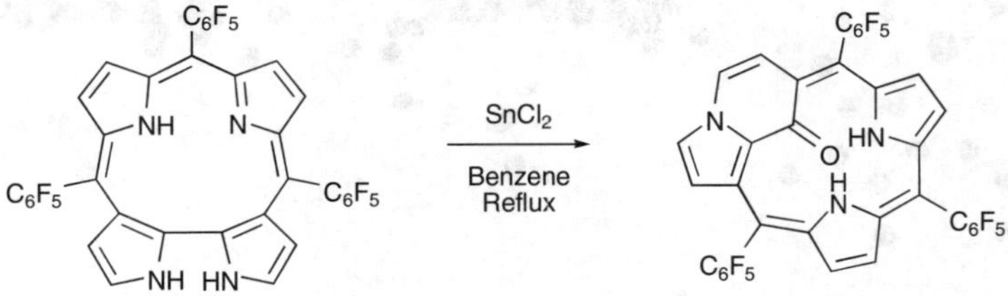

3.7 **3.8**

Scheme 3.2 *Synthesis of N-fused oxyindolphyrin 3.8*

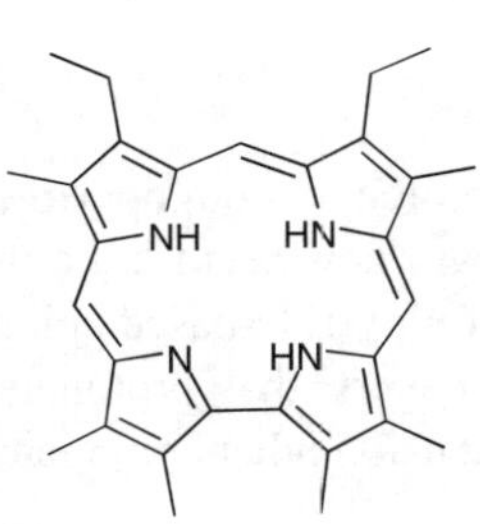

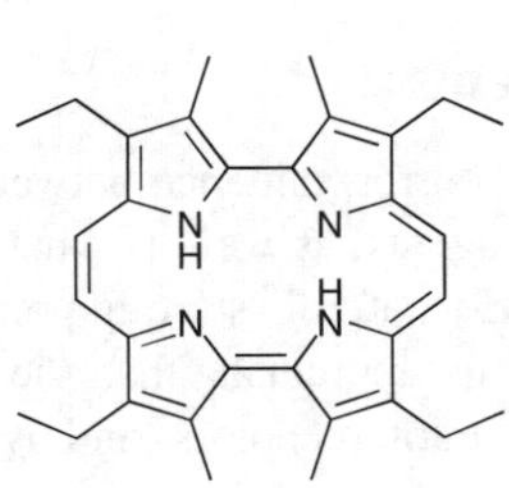

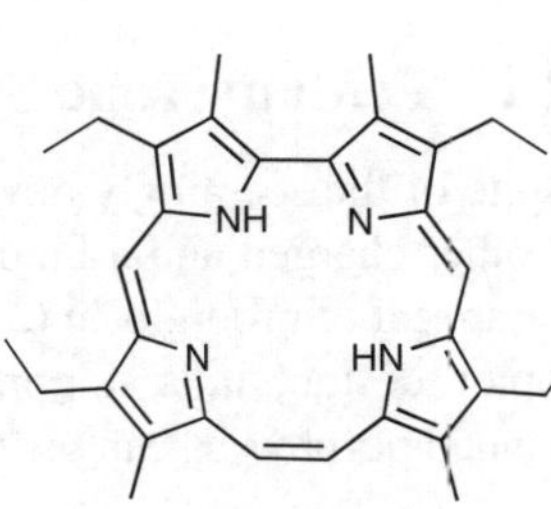

3.9 **3.10** **3.11**

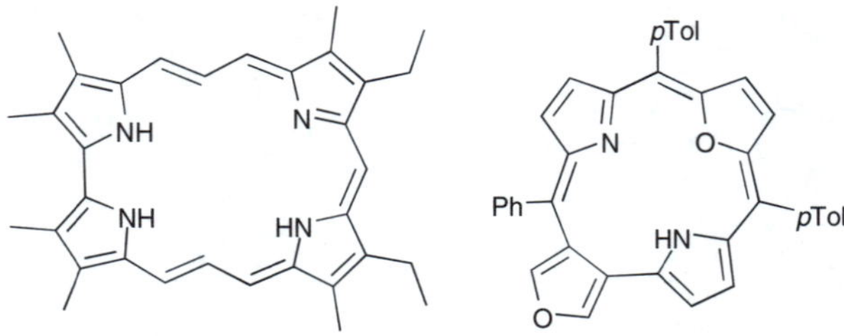

3.12 **3.13**

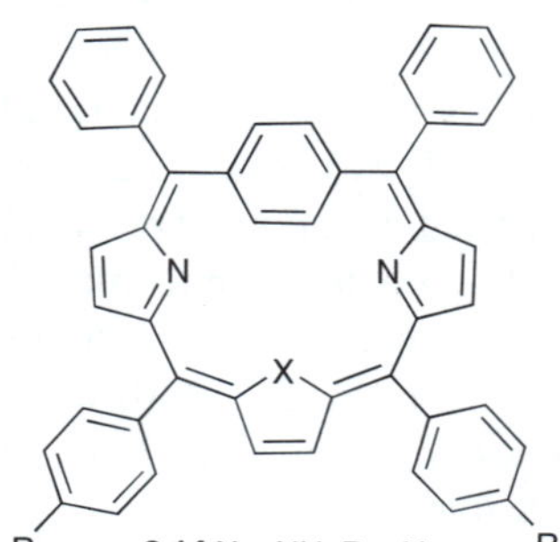

3.14 X = NH, R = H
3.15 X = S, R = Me

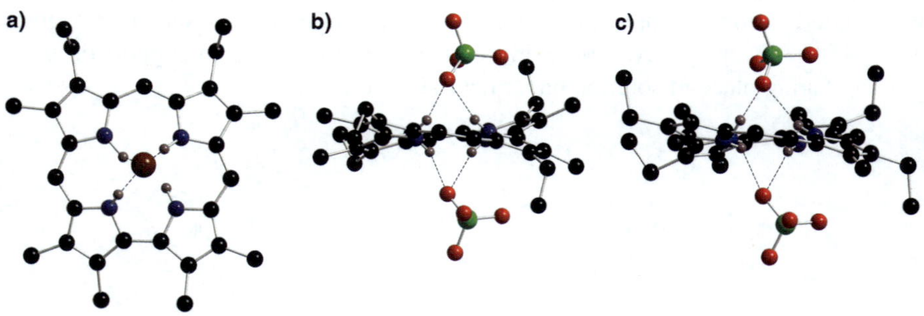

Figure 3.2 *X-ray crystal structures of (a) the bromide-anion complex of* $H{\cdot}\mathbf{3.9}^{+}$*, (b) the bis-perchlorate-anion complex of* $H_2{\cdot}\mathbf{3.10}^{2+}$*, and (c) the bis-perchlorate-anion complex of* $H_2{\cdot}\mathbf{3.11}^{2+}$

3.2.2 Pentapyrrolic Systems

In spite of the seemingly obvious correspondence between protonated oligopyrroles and other charged anion-binding agents, as well as what would now seem to be the clear appeal of prodigiosin (**3.1**) as a natural "starting point" for pyrrole-based anion-receptor design, it is important to appreciate that the discovery that protonated expanded porphyrins can serve as anion receptors (and by extension other conjugated

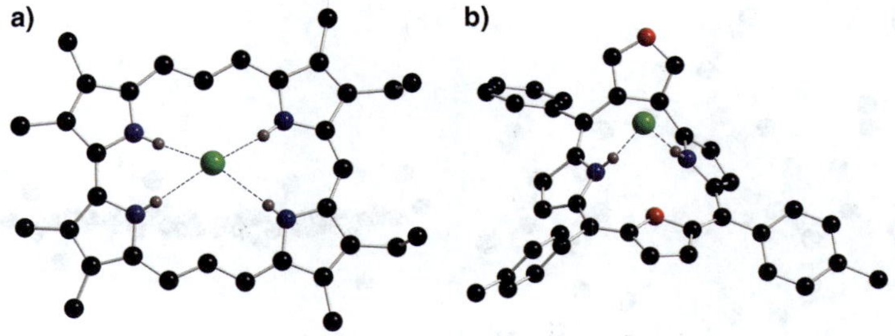

Figure 3.3 *X-ray crystal structures of (a) the chloride-anion complex of H·**3.12**$^+$ and (b) the chloride-anion complex of H·**3.13**$^+$*

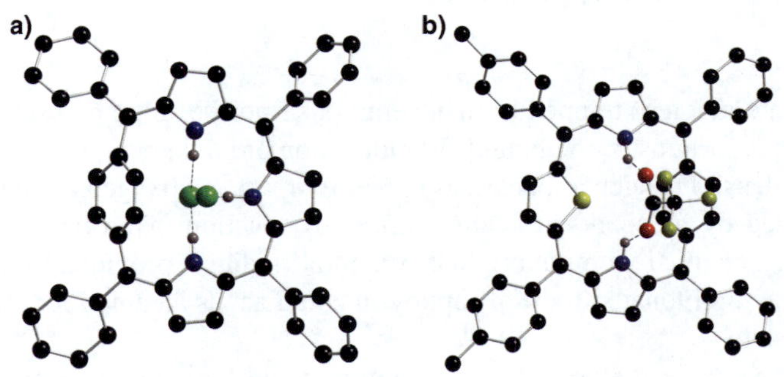

Figure 3.4 *X-ray crystal structures of (a) the chloride-anion complex of H_2·**3.14**$^{2+}$ and (b) the TFA-anion complex of H_2·**3.15**$^{2+}$*

oligopyrroles) was entirely fortuitous. It came about during the course of synthetic studies directed towards streamlining Woodward's original synthesis of sapphyrin, **3.16**.[15] Sapphyrin is a pentapyrrolic porphyrin analogue that contains four bridging *meso*-like carbon atoms and a 22 π-electron periphery. Although first reported by Woodward in 1966,[16] this prototypical expanded porphyrin remained largely unstudied until the late 1980s.[17] For instance, no high-yielding syntheses existed, few well-characterized metal complexes had been reported, and no X-ray diffraction-based structural information was available. Thus, once an improved synthesis of sapphyrin was developed by Sessler and co-workers, efforts were made to obtain X-ray-diffraction-quality single crystals. Appreciating that the diprotonated form would be more symmetrical (as well as potentially more stable), crystals of what was thought to be the bis-HPF_6 salt were grown. The resulting structure, reported by Sessler and Ibers and co-workers in 1990,[15] revealed the presence of only one PF_6^- counteranion. However, the same analysis indicated that a fluoride atom was bound inside the sapphyrin core being held there via a regular array of five N–H···F hydrogen bonds (*ca.* 2.7 Å) (Figure 3.5). Presumably, the bound fluoride anion was either extracted from

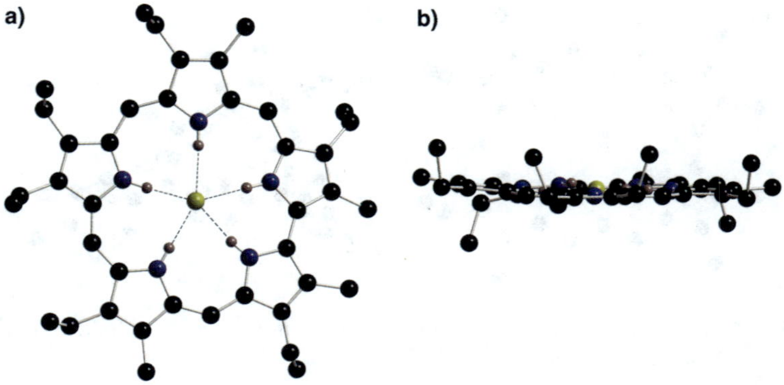

Figure 3.5 *Single crystal X-ray structure (top and side views) of the fluoride-anion complex of diprotonated sapphyrin* H_2·***3.16***$^{2+}$

the PF_6^- counteranion or trapped from adventitious fluoride anion present in solution. While the exact origins of the bound fluoride anion are unknown, the assignment of the electron density to such a species, as opposed to, *e.g.*, an oxygen atom from water, was confirmed by independent synthesis, *i.e.*, by reaction of protonated sapphyrin with F^-. This, coupled with the original structural findings, provided irrefutable evidence that the diprotonated form of sapphyrin could act as an anion receptor, at least in the solid state.

Further support for the notion that sapphyrin could bind anions in the solid state came 2 years after the original Sessler and Ibers report when the X-ray structure of the bis-hydrochloride salt of sapphyrin **3.16** was elucidated.[18] In contrast to what was seen in the case of the fluoride-anion complex, neither of the chloride counter-anions was located in the plane of the diprotonated sapphyrin (Figure 3.6). Rather, they were found to sit above and below the sapphyrin ring, being held there via a combination of hydrogen-bonding and electrostatic interactions. This out-of-plane binding mode reflects the fact that anionic chloride is larger than anionic fluoride and that only the latter is of appropriate size to fit within the relatively small sapphyrin core.

The observation that chloride was bound in an out-of-plane fashion, whereas fluoride anion was accommodated within the core, provided a strong "hint" that chloride anion would be bound less well than fluoride anion in solution. This expectation was confirmed in initial solution-phase studies carried out using fluorescence and UV–Vis spectroscopy. In dichloromethane solution, the most intense absorbance feature seen in the UV–Vis spectrum of **3.16**·2HF is a Soret-like band at $\lambda_{max} = 446$ nm. This band was found to be shifted to the red in the case of **3.16**·2HCl and **3.16**·2HBr ($\lambda_{max} = 456$ and 458 nm, respectively).[18] These differences were found to be enhanced in the corresponding fluorescence emission spectra. From associated spectral titrations, the fluoride-anion binding constant of **3.16** in dichloromethane was calculated to be $>10^8$ M^{-1}. By contrast, those for chloride and bromide were found to be lower (1.8×10^7 M^{-1} and 1.5×10^6 M^{-1}, respectively).[18,19]

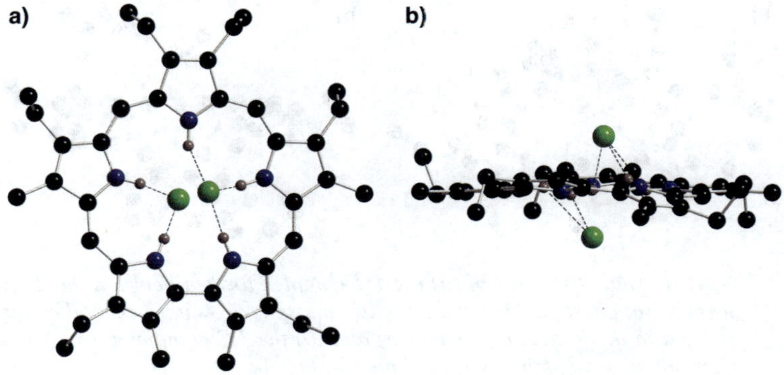

Figure 3.6 *Single crystal X-ray structure (top and side views) of the bis-hydrochloride salt of sapphyrin **3.16***

Unfortunately, the association between the diprotonated form of sapphyrin **3.16** and fluoride was found to be too large to measure accurately in dichloromethane. Thus, the corresponding values were determined in methanol, again for sapphyrin **3.16**, and found to be 2.8×10^5, *ca.* 10^2, and $<10^2$ M^{-1} for fluoride, chloride, and bromide anions, respectively.[18]

A key difference between sapphyrin and porphyrin is that in the case of the larger macrocycle the two sp^2-hydridized pyridine-like core nitrogen, atoms are relatively basic with pK_a values of *ca.* 4.8 and 8.8 being recorded for the corresponding conjugate acids in water.[20,21] As a consequence, water-solubilized sapphyrins were expected to exist in their monoprotonated forms at neutral pH and, in contrast to what is true for porphyrins, act as anion receptors under such conditions. Here, a particularly intriguing question was whether sapphyrins could be used to bind phosphate and related anions of biological importance in water or in other highly polar media.

Initial support for the proposal that the protonated forms of sapphyrin could bind phosphate-type anions came from single-crystal X-ray-diffraction analyses. These analyses proved particularly revealing and served to show that various "pure" and "mixed" anion complexes could be formed in the presence of phosphates, including 1:1 and 1:2 (sapphyrin:phosphate) species. In the case of the structurally characterized 1:1 complex involving $[H_2{\cdot}\mathbf{3.17}]^{2+}$, a dihydrogen phosphate counteranion was found bound to the diprotonated core via five hydrogen-bonding interactions involving the five pyrrolic protons and the bound phosphate oxyanion (Figure 3.7a). Presumably, this mode of binding, which was termed "phosphate chelation" is further stabilized by electrostatic interactions.

A second kind of interaction was revealed by the structure of the 1:2 complex formed between $[H_2{\cdot}\mathbf{3.16}]^{2+}$ and monobasic phenylphosphate (*cf.* Figure 3.7b).[20] In this case, the two counteranions are found bound above and below the minimally distorted sapphyrin plane, in analogy to what was seen in the structure of the bis-hydrochloride salt discussed above.

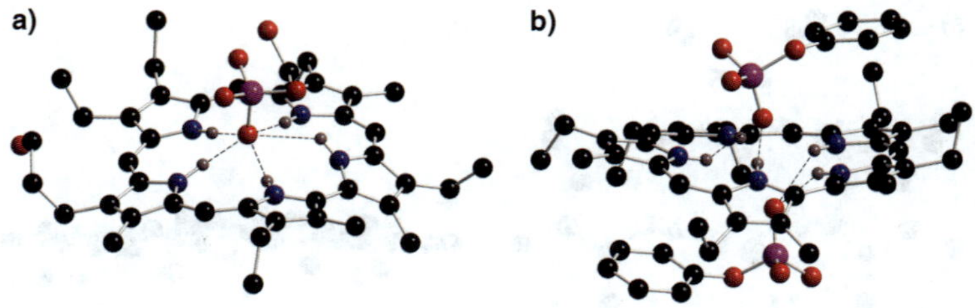

Figure 3.7 *X-ray crystal structures of (a) the 1:1 complex formed between the diprotonated form of sapphyrin **3.17** and monobasic phosphoric acid (a second counteranion, present in the lattice, is not shown) and (b) the 1:2 complex formed between the diprotonated sapphyrin **3.16** and phenylphosphate*

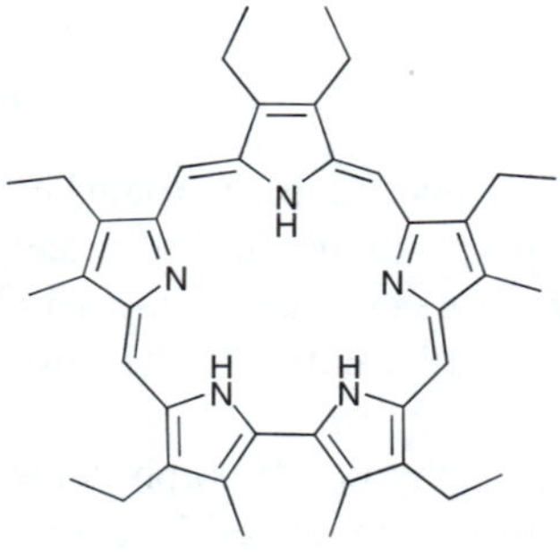

3.16

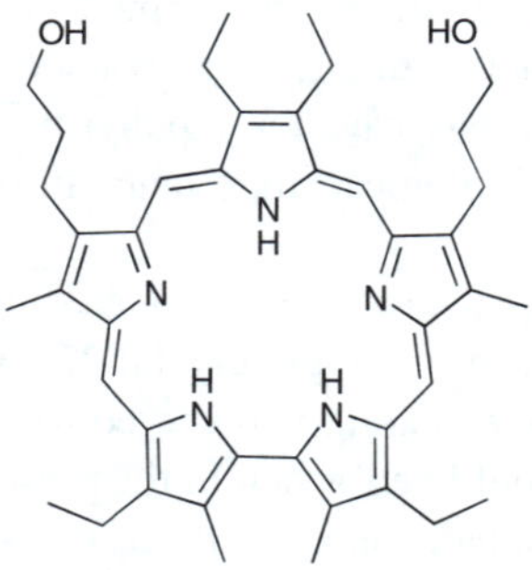

3.17

One of the reasons the various phosphate complexes of protonated sapphyrins attracted attention is that inorganic phosphate (P_i) and various derivatives are ubiquitous in biology. The ability to bind, transport, and modulate the chemical behaviour of such species could prove useful in a number of application areas, including those that could benefit the medical field. For instance, hyperphosphatemia is an ailment caused by a buildup of excess phosphate in the body. It is prevalent among haemodialysis patients and is characterized by heightened calcium phosphate levels that, over time, lead to an increased incidence of cardiovascular mortality.[22] Removing the excess phosphate, perhaps using a carrier-mediated out-of-cell transport-based approach, could serve to reduce this risk. The "other side" of this problem would involve the development of selective carriers for phosphorylated species, including the active forms of known antiviral agents and antisense nucleotides. In this case, through-membrane into cell-anion transport would allow for potentially improved therapeutic treatments. In both cases, what is needed are agents that are capable of binding to a phosphate (or phosphate-like) anion, neutralizing the charge, and producing what would be net hydrophobic species that can pass through the targeted biological membrane. Effective release of the phosphorylated entity after transport is also important.

By virtue of being the first pyrrole-based anion-binding systems to be discovered, sapphyrin derivatives have been extensively studied as possible phosphate-anion carriers. Most of these studies involved transport experiments carried out using a so-called Pressman-type U-tube model membrane apparatus (Figure 3.8). With this arrangement, the anion-transport ability of various putative sapphyrin-type anion carriers could be "screened" by monitoring the movement of a targeted anion from a first aqueous phase (Aq I), through a layer of CH_2Cl_2 (Org), to a second receiving aqueous phase (Aq II). In the case of fluoride anion, it was observed that the presence of sapphyrin **3.16** in the organic layer served to enhance the rate of fluoride anion transport from Aq I to Aq II over a wide range of initial Aq I pH. Further, the transport rate under conditions where sapphyrin **3.16** existed in its monoprotonated form was found to be nearly 100 times greater than the background chloride-anion diffusion rate observed in its absence.[23]

In similar model membrane experiments, the transport of adenosine 5′-monophosphate (5′-AMP) and guanosine 5′-monophosphate (5′-GMP) was achieved using sapphyrin **3.16**. In this study, it was found that decreasing the pH of Aq I below four (to provide a high concentration of the diprotonated sapphyrin) did not serve to enhance the anion-transport rate appreciably.[24] In subsequent work, it was found that the addition of triisopropylsilyl protected cytidine (C-Tips) enhanced the rate of sapphyrin-mediated 5′-GMP transport.[25] This latter finding led Sessler and co-workers to attach a nucleobase to the sapphyrin skeleton. The resulting combined systems, conjugates **3.18** and **3.19**, contain two putative recognition motifs, namely a sapphyrin-derived phosphate binding site and a Watson–Crick base-pairing element.

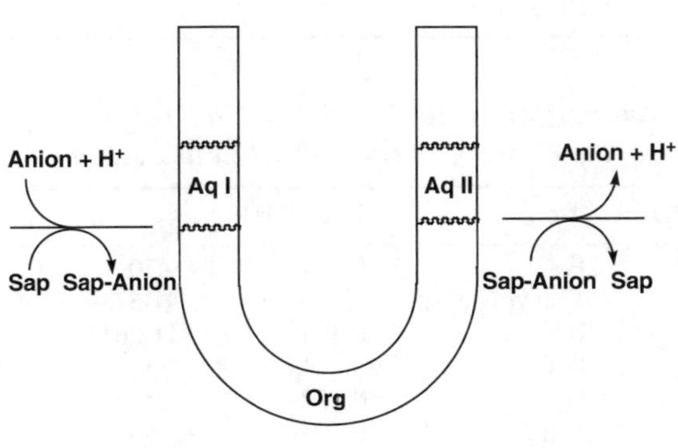

Figure 3.8 *Schematic representation of the Pressman-type model membrane used to measure anion transport rates*

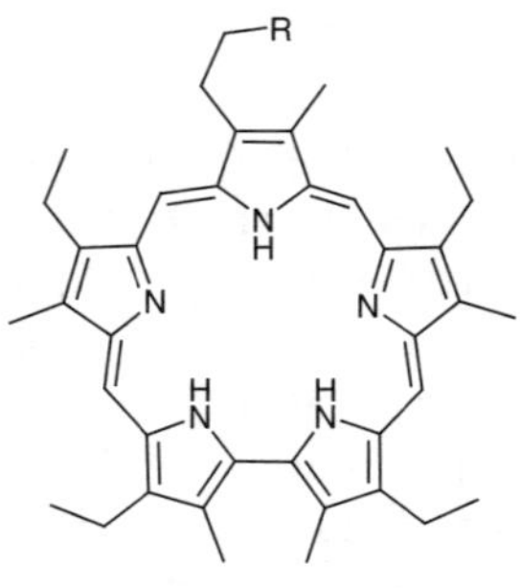

3.18a R = A
3.18b R = B

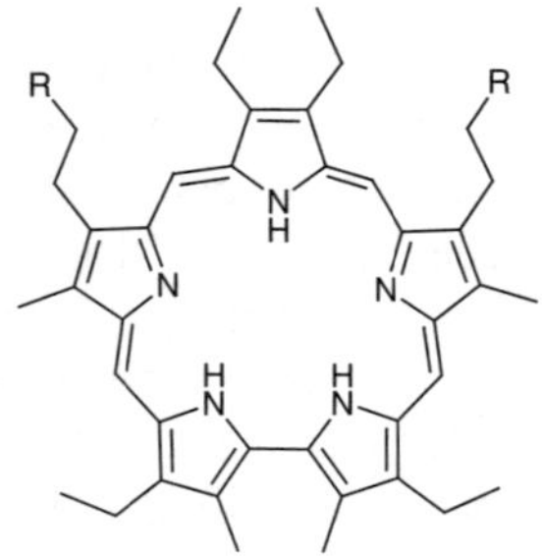

3.19a R = A
3.19b R = B

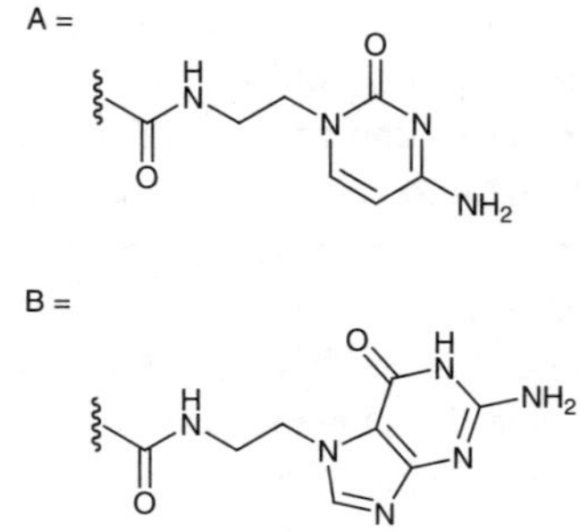

Tables 3.1–3.3 summarize the nucleotide phosphate transport rates observed for sapphyrins **3.18** and **3.19** under several different initial pH conditions (involving both Aq I and Aq II). The cytosine-bearing carriers **3.18a** and **3.19a** were found to exhibit selective transport of 5′-GMP over cytosine 5′-monophosphate (5′-CMP) and 5′-AMP (by a factor of 8–100). These studies also revealed that better transport rates were always observed when the receiving phase (Aq II) is kept highly basic, as

Table 3.1 *Initial nucleotide-5′-monophosphate transport rates (10^{-8} mol cm^2 h^{-1}) observed in the presence of carriers **3.18a** and **3.19a***

Carrier	*Aq I (pH)*	*Aq II*	k_T *5′-GMP*	$k_{5'\text{-GMP}}/k_{5'\text{-CMP}}$	$k_{5'\text{-GMP}}/k_{5'\text{-AMP}}$
3.18a	6.15	H_2O	1.201	101.7	7.66
3.18a	6.70	H_2O	0.287	42.9	8.87
3.18a	7.05	H_2O	0.001	20.1	9.49
3.18a	6.15	10 mM NaOH	1.423	26.3	2.73
3.18a	6.70	10 mM NaOH	1.228	40.8	4.36
3.18a	7.05	10 mM NaOH	0.708	43.3	9.60
3.19a	6.15	H_2O	0.101	6.2	1.38
3.19a	7.05	10 mM NaOH	0.115	23.7	3.18
None	7.00	H_2O	$<10^{-5}$		
3.16	7.00	H_2O	$<10^{-5}$		
3.16	7.00	10 mM NaOH	$<10^{-5}$		

Table 3.2 *Initial transport rates (10^{-8} mol cm^{-2} h^{-1}) observed for different GMP isomers in the presence of carriers **3.18a** and **3.19a***

Carrier	*Aq I (pH)*	*Aq II*	k_T *2′-GMP*	$k_{2'\text{-GMP}}/k_{5'\text{-GMP}}$	$k_{2'\text{-GMP}}/k_{3'\text{-GMP}}$
3.18a	6.70	H_2O	0.767	9.70	7.30
3.18a	6.70	1 mM NaOH	2.989	9.55	5.30
3.18a	7.00	H_2O	0.594	11.06	7.82
3.18a	7.20	H_2O	0.421	$>10^2$	$>10^2$
3.18a	7.35	H_2O	0.352	$>10^2$	$>10^2$
3.19a	6.70	1 mM NaOH	0.104	3.33	2.67
None	7.00	H_2O	$<10^{-5}$		
3.16	7.00	H_2O	2×10^{-5}		

Table 3.3 *Initial nucleotide-5′-monophosphate transport rates (10^{-8} mol cm^{-2} h^{-1}) observed in the presence of carriers* ***3.18b*** *and* ***3.19b***

Carrier	*Aq I (pH)*	*Aq II*	k_T *5′-CMP*	$k_{5'\text{-}CMP}/k_{5'\text{-}GMP}$	$k_{5'\text{-}GMP}/k_{5'\text{-}AMP}$
3.18b	6.70	H_2O	0.129	8.96	3.15
3.18b	6.70	1 mM NaOH	0.541	9.17	3.15
3.19b	6.70	1 mM NaOH	0.147	17.5	10.5

would be expected for a mechanism that involves combined anion + proton transport. In all cases, **3.18a**, which bears only a single attached nucleobase, was found to display a higher selectivity for 5′-GMP then its congener **3.19a**, bearing two appended nucleobase side chains.[25] Rationales for these findings have been put forward that invoke a lower number of competing interactions in the case of **3.18a**.

In order to study the regioselective preference of carrier **3.18a**, three different isomeric species, namely 5′-GMP, 3′-GMP, and 2′-GMP, were used. In most of the cases, the 2′-GMP isomer was transported considerably faster than the 5′ and 3′ isomers (*cf.* Table 3.2). On the other hand, it was observed that the association constants for the formation of H·**3.18a**-5′-GMP and H·**3.18a**-2′-GMP were 8×10^3 and 2.2×10^4 M^{-1}, respectively, as inferred from UV–Vis spectroscopic titrations carried out in methanol.[26] Taken in concert, these findings support the conclusion that the rate of transport is dependent on (at least) two important processes, namely binding and release, and can thus be very sensitive to small structural changes. Figure 3.9 illustrates the proposed complex formed between monoprotonated sapphyrin **3.18a** and 2′-GMP.

In the case of the guanosine-bearing sapphyrins **3.18b** and **3.19b**, transport selectivity for 5′-CMP was observed. As can be seen from an inspection of the data summarized in Table 3.3, the relative rates of carrier-induced, through-membrane transport were found to be smaller for the mono-substituted system, **3.18b**, than for the doubly substituted system, **3.19b**.[26]

The interactions between functionalized sapphyrins and various anions, including nucleotides and oligonucleotides, were also probed via the construction of a modified silica gel chromatography support. This support was produced by attaching a cytosine-substituted sapphyrin containing a carboxylic group to aminopropyl silica gel (Figure 3.10). Once packed in columns, the resulting modified silica gel was found to allow for the HPLC-based separation of GMP, GDP, and GTP from a mixture of mono-, di-, and triphosphates of cytidine, uridine, adenosine, and guanosine under isocratic conditions at pH = 7.[27] The observed selectivity was taken as independent support for the proposed two-site recognition model invoked to rationalize the selective transport of nucleotides by functionalized sapphyrins **3.18** and **3.19**. Control experiments, using a column packed with a sapphyrin-substituted silica gel,[28] allowed for separations between mono-, di-, and triphosphates. However, no nucleoside-based selectivity was observed in this latter case.

In an effort to generate receptors for dianions, several sapphyrin dimers, trimer, and tetramers were prepared. In a first study, dimers **3.20** (K_a = ≤ 20 and 4600 M^{-1} for $CF_3CO_2^-$ and terephthalate in methanol, respectively) and **3.21** were

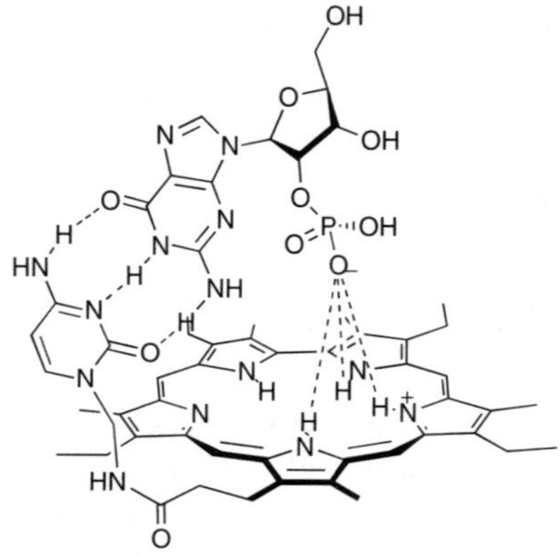

Figure 3.9 *Structure of the proposed supramolecular complex formed between the monoprotonated form of sapphyrin* ***3.18a*** *and 2′-GMP*

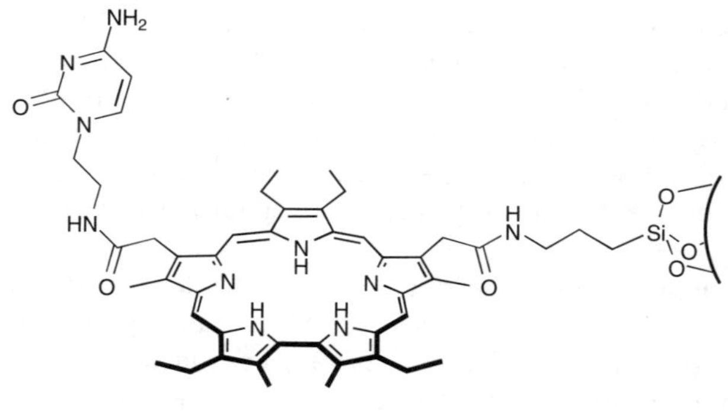

Figure 3.10 *Schematic representation of the functionalized silica gel formed from a cytosine-bearing sapphyrin*

found to be considerably more effective than monomers **3.16**, **3.18**, and **3.19** in enhancing the rate of transport of various dicarboxylate anions through a U-tube membrane model (*cf.* Figure 3.8).[29] Additionally, the sapphyrin dimers linked by chiral linkers (**3.21**) were found to effect the enantioselective recognition of *N*-protected amino acids.[30] The trimeric and tetrameric sapphyrins species **3.22–3.24** were prepared and studied as potential carriers for phosphorylated nucleotides. Using the same U-tube set-up for other studies, it was shown that trimers **3.22** and **3.23** did in fact act as efficient carriers for nucleotide diphosphates. However, they proved ineffective as carriers for nucleotide triphosphates. In contrast, the rate of mono-, di-, and trinucleotides transport was found to be enhanced in the presence of tetramer **3.24**.[31]

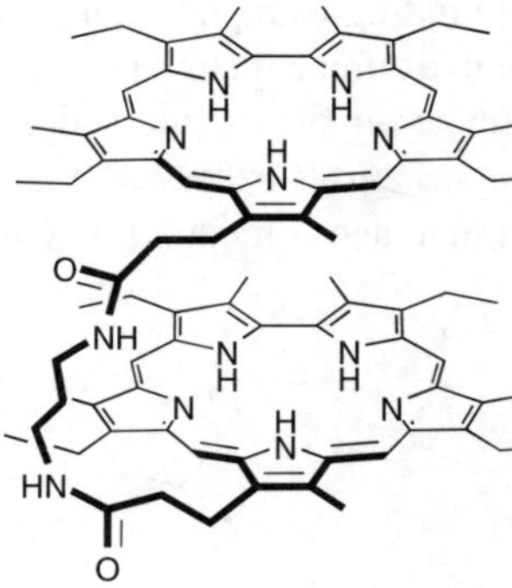

3.20

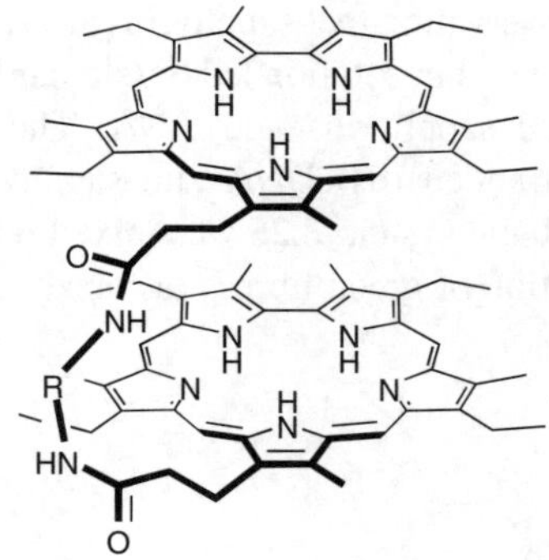

3.21a R =

3.21b R =

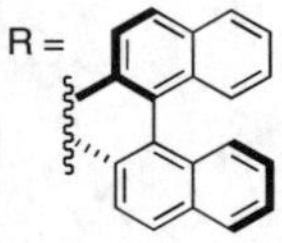

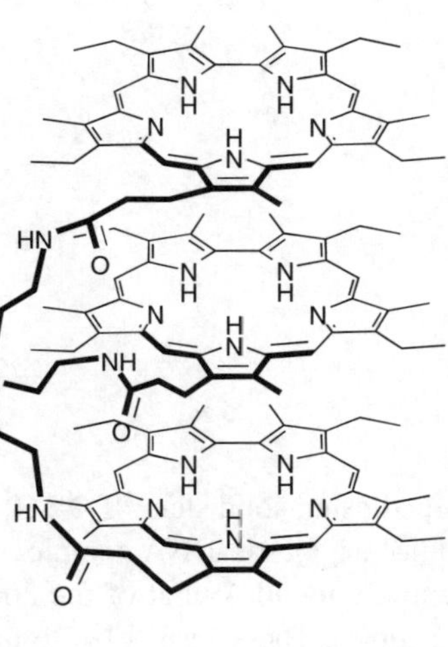

3.22

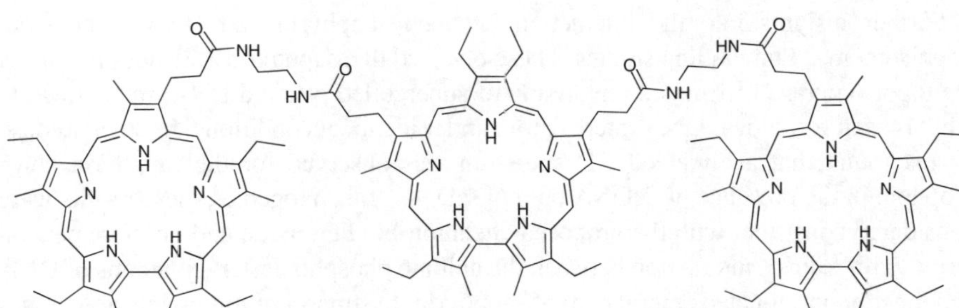

3.23

The observation that sapphyrins are capable of binding with phosphate anion in the solid state and in solution led Sessler and Iverson to examine the interactions between protonated sapphyrins and DNA. The first indication of an interaction between double-stranded (ds) DNA and sapphyrin came when an excess of the sparingly water-soluble system **3.25** was mixed with DNA in aqueous media. Immediately, the precipitation of green fibres was observed.[32]

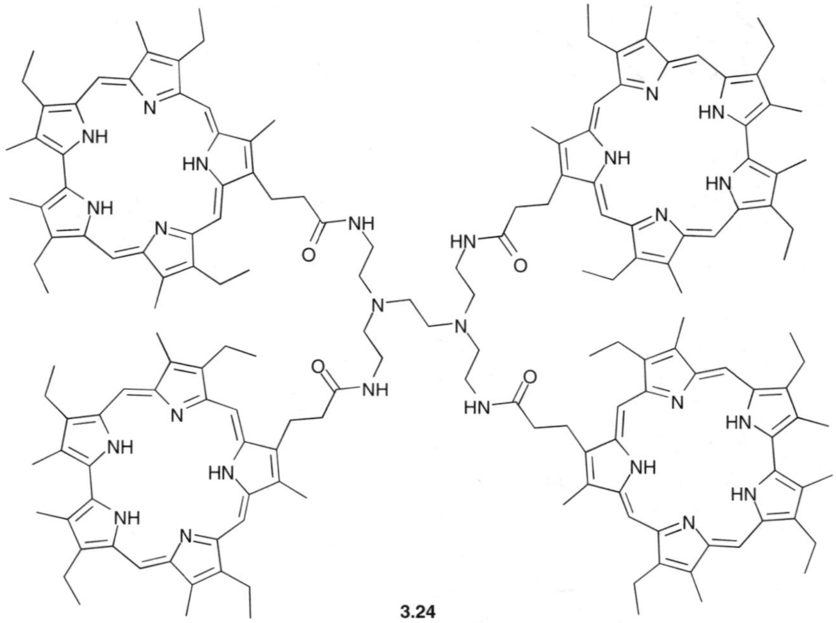

3.24

This precipitate was studied using solid state ^{31}P NMR spectroscopy. The ^{31}P peak corresponding to the presumed sapphyrin-DNA complex was observed at 3.6 ppm, a chemical shift that is essentially identical that of the control sapphyrin–phosphoric acid complex (signal at 3.8 ppm). These results led to the proposal that the interaction between sapphyrin and the phosphate anions present in DNA involves so-called phosphate chelation, in analogy to what is seen in the solid-state structure of $[H_2{\cdot}\mathbf{3.17}]^{2+}{\cdot}H_2PO_4^-$ (*cf.* Figure 3.7 above).

Further insights into the interaction between sapphyrin and DNA came from Topoisomerase I unwinding studies. These revealed that sapphyrin **3.25** does not affect the topoisomerase I-mediated unwinding of supercoiled plasmid DNA, in contrast, to what is seen for known DNA intercalators under identical conditions. In other studies, it was found that an induced CD spectrum was observed for the sapphyrin chromophore in the presence of dsDNA but not ssDNA. Taken together, these results were considered consistent with the proposed chelation-binding mode and, in more general terms, with a direct interaction between the anionic phosphodiester oxyanions of DNA and the monoprotonated central core of sapphyrin. In support of this conclusion, it was recently found that the addition of phosphate anion into aqueous solutions of water soluble, but highly aggregated, sapphyrin **3.25** produced a large enhancement in the fluorescence intensity as the result of anion-induced deaggregation. This latter result is discussed further in Chapter 8.[33]

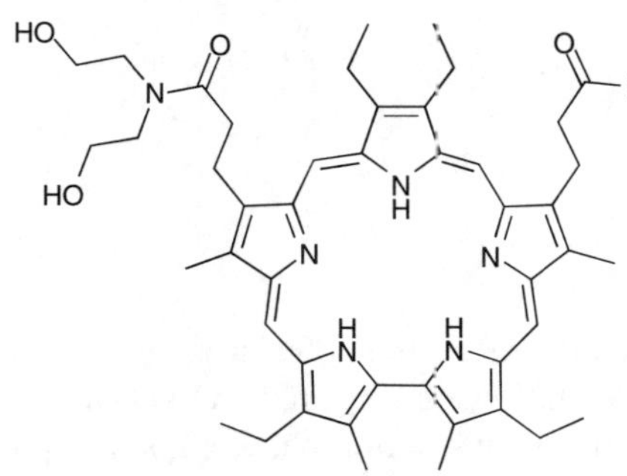

3.25

Photodynamic therapy (PDT) is one of the more appealing new modalities being proposed for use in cancer and cardiovascular therapy.[34] This technique relies on the photochemical production of singlet oxygen via the intermediacy of an appropriate photosensitizer. Currently, Photofrin®, a mixture of haematoporphyrin derivatives, and aminolevulinic acid, an endogenous precursor of protoporphyrin IX, are approved for use in the PDT-based treatments of several cancers. Monoprotonated sapphyrins are capable of generating singlet oxygen upon photoirradiation at neutral pH (albeit less well than a number of so-called second generation photosensitizers); they also bind simple phosphates and DNA under these latter aqueous conditions (see above).[35] It was thus considered of interest to test whether a water-soluble sapphyrin might be able to effect the photocleavage of DNA. In fact, it was found that, upon photoirradiation, sapphyrin **3.25** effected cleavage of pBR322 plasmid DNA with a 17% photoefficiency.[36]

In an effort to improve the DNA-binding affinity of sapphyrin, and with it perhaps, the efficiency of the above photocleavage, a sapphyrin tethered to an oligonucletiode, system **3.26**, was synthesized. This conjugate was found not only to bind to complementary DNA sequences with enhanced affinity, but also to effect sequence-specific modification of DNA when irradiated with >620 nm light; under these conditions, dominant cleavage at the proximate G sites was observed.[37]

As an alternative means of enhancing the DNA-binding affinity, polyamines were also attached to the sapphyrin skeleton. This was expected to produce species that would be positively charged at neutral pH and which would interact with the negatively charged major, or less probable, minor groove of the DNA. Sapphyrins **3.27–3.29** were thus synthesized. They were found to display better water solubility and an improved DNA-binding ability. They were also found to be significantly more effective in terms of DNA cleavage than **3.25**.[38] However, sapphyrins **3.28** and **3.29** were found to show a reduced level of tumour-cell localization, as compared to the sapphyrin **3.25**.[39] This was rationalized in terms of inhibited uptake in the case of the systems containing a large number of positively charged substituents, *i.e.*, **3.27–3.29**.

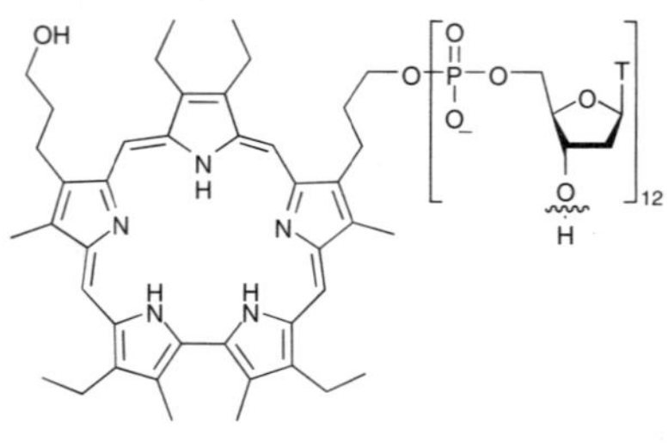

3.26

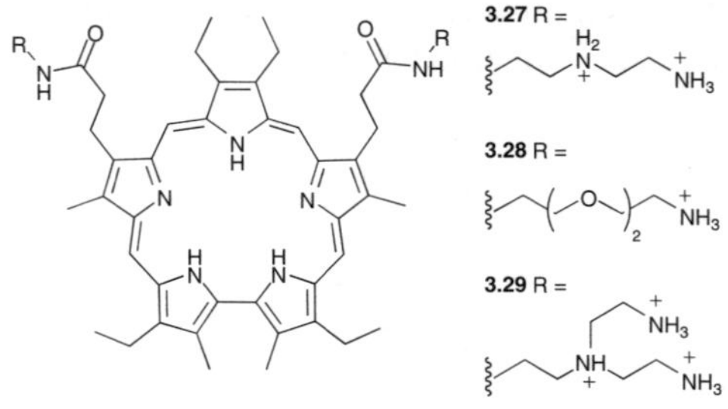

3.27 R =
3.28 R =
3.29 R =

In work that was designed to extend the boundaries of sapphyrin chemistry, the core-modifed sappyrins **3.30–3.32** were prepared. As yet, these systems have not been extensively studied as potential solution-phase anion receptors. However, evidence for anion binding has been seen in the solid state (Figure 3.11).[40] Further studies, involving quantitative comparisons between sapphyrin and its heterosapphyrin analogues, would help probe the determinants of anion recognition and might allow the effect of individual "point mutations" to be better understood.

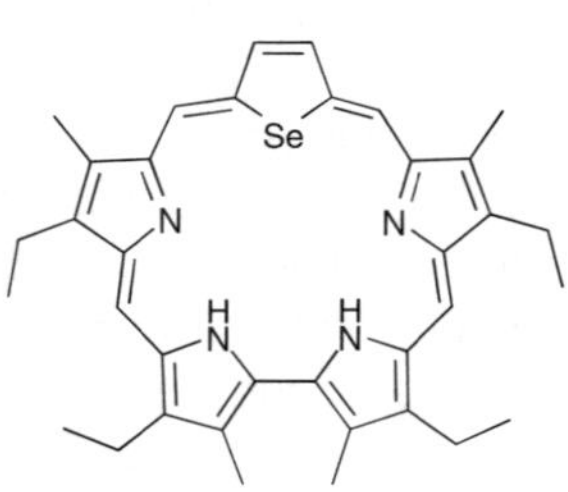

3.30

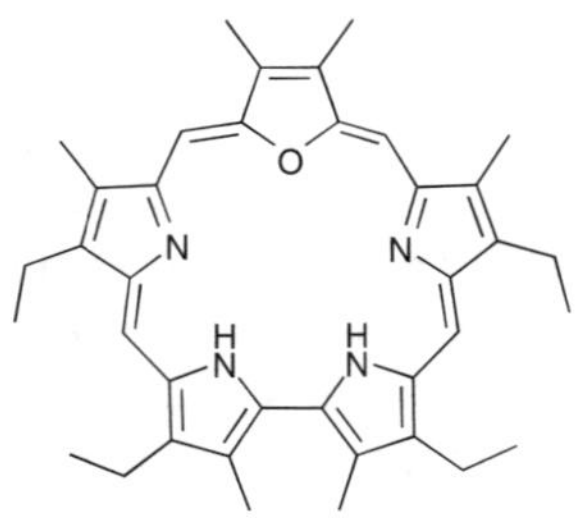

3.31

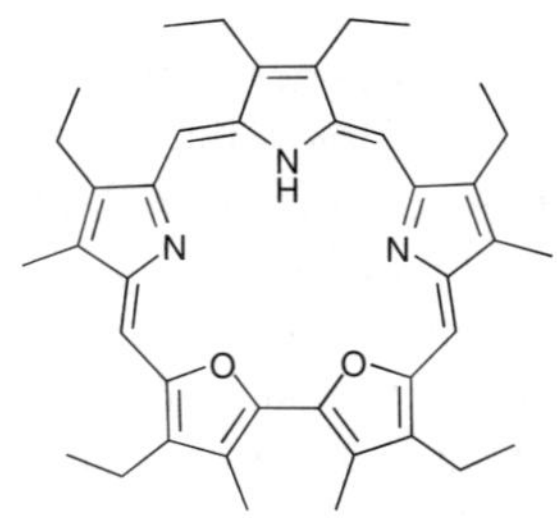

3.32

Studies along these lines have recently been carried out by Chandrashekar and co-workers[41,42] using a series of newly synthesized *meso*-substituted heterosapphyrins **3.33–3.37**. An X-ray diffraction analysis revealed, in analogy to the structures described above, that the diprotonated form of heterosapphyrin **3.33** binds a trifluoroacetate (TFA) counteranion in the solid state via a combination of three directed hydrogen bonds and electrostatic interactions, as shown in Figure 3.12. Quantitative assessments of anion binding in methanol were carried out using UV–Vis spectroscopic titrations. The resulting association constants, summarized in Table 3.4, revealed that the fluoride-anion affinities of **3.33–3.37** are dramatically lowered relative to those of sapphyrin **3.16** (*ca.* $K_a = 2.8 \times 10^5\ M^{-1}$) recorded under comparable conditions. Unfortunately, the corresponding *meso*-substituted all azasapphyrin, although known,[43] has yet to be subject to a comparable, quantitative solution phase analysis. Thus, it is possible that intrinsic electronic effects are playing a role in lowering the anion affinities of **3.33–3.37** in comparison to **3.16**. However, it is likely that the explanation put forward by Chandrashekar and co-workers is the most reasonable, namely that the heteroatom systems contain a decreased number of NH-donor sites relative to the all-aza parent system.

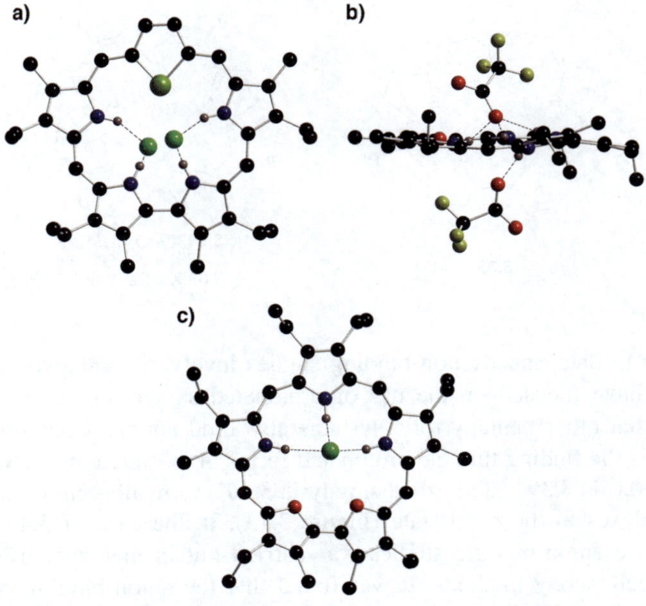

Figure 3.11 *X-ray crystal structures of (a) the bis-chloride anion complex of* $H_2 \cdot \mathbf{3.30}^{2+}$, *(b) the bis-trifluoroacetate (TFA)-anion complex of* $H_2 \cdot \mathbf{3.31}^{2+}$, *and (c) the mono-chloride-anion complex of* $H_2 \cdot \mathbf{3.32}^{2+}$

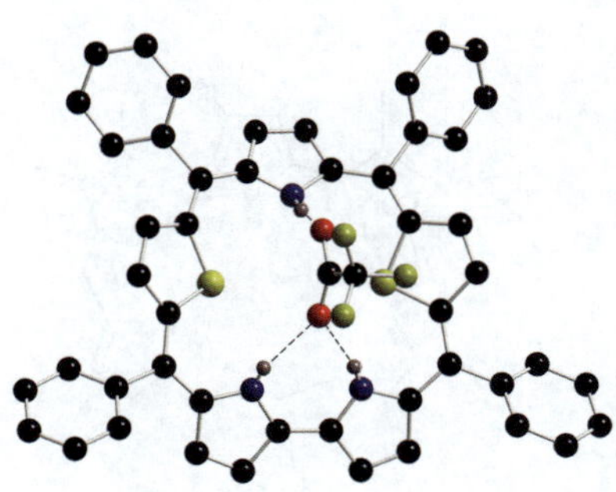

Figure 3.12 *X-ray crystal structure of TFA anion complex with* $H_2 \cdot \mathbf{3.33}^{2+}$

Table 3.4 *Association constants* (M^{-1}) *for heterosapphyrin* ***3.34–3.37*** *with anions* F^-, N_3^-, *and* CO_3^{2-} *measured in methanol by UV–Vis titration method*

	KF	*NaN_3*	*K_2CO_3*
3.34	590	820	1180
3.35	510	540	890
3.36	170	290	830
3.37	210	240	730

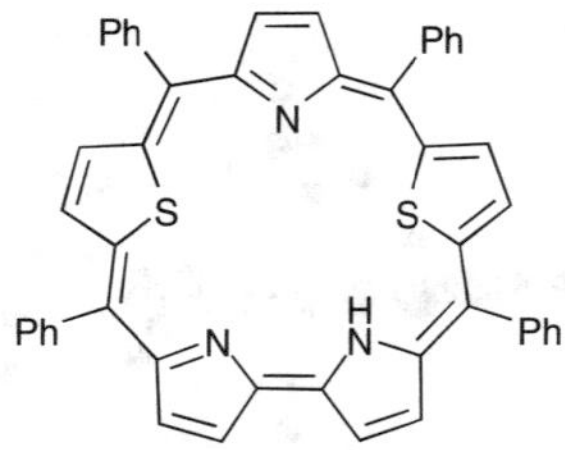

3.33

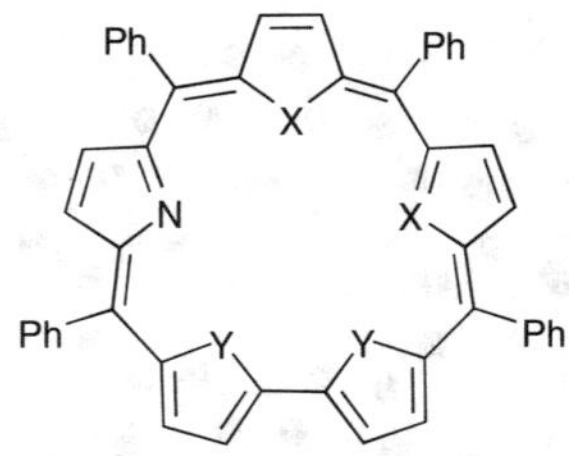

3.34 X = O, Y = S
3.35 X = Y = S
3.36 X = Se, Y = S
3.37 X = O, Y = Se

Although to date, most anion-binding studies involving pentapyrrolic expanded porphyrins have focused on the use of protonated sapphyrins, it is important to appreciate that other pentapyrrolic systems also bind anions. Consistent with this conclusion is the finding that the protonated forms of isosmaragdyrin **3.38**,[44] pentaphyrin-(2.1.0.0.1) **3.39**,[45] and oxasmaragdyrin **3.40**[46] were all seen to form chloride-anion complexes in the solid state (Figure 3.13). In the case of **3.40**, supporting solution-phase anion-binding studies was carried out in methanol using standard UV–Vis spectroscopy methods. It was found that the anion-binding properties of monoprotonated **3.40** differed from those of sapphyrin. While the fluoride-binding affinity of monoprotonated **3.40** is around three times lower than that of diprotonated **3.16**, it shows a dramatically increased affinity for chloride anion ($K_a = 7.6 \times 10^4$, 2.1×10^4, and 2.6×10^4 M^{-1} for F^-, Cl^-, and AMP^-, respectively).

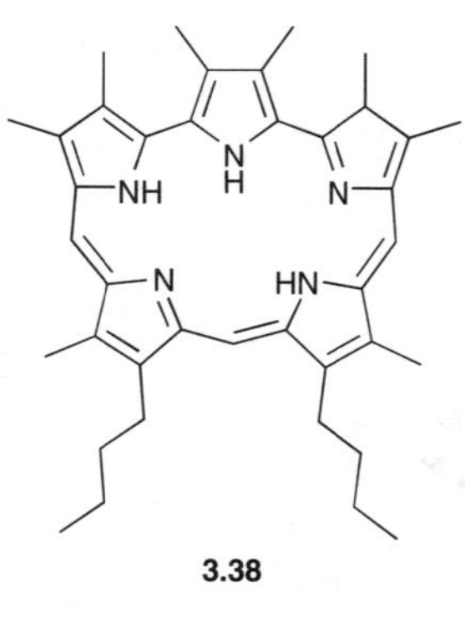

3.38

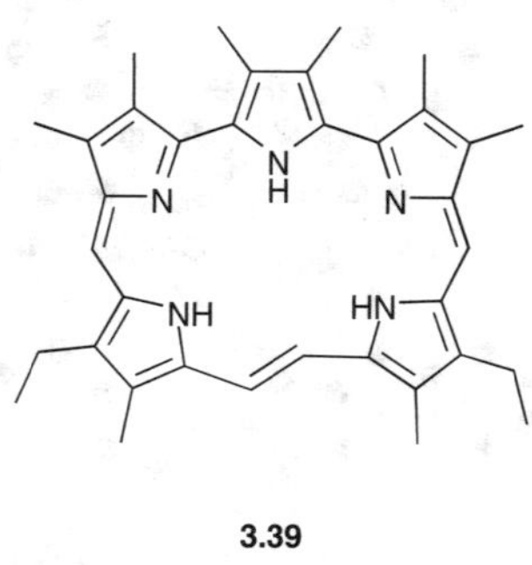

3.39

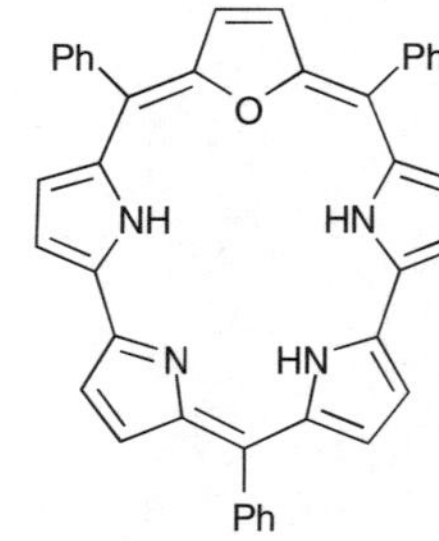

3.40

3.2.3 Hexapyrrolic Systems

The finding that sapphyrin in its protonated forms would bind anions helped stimulate growth in the area of expanded porphyrin chemistry, in particular the synthesis and study of new expanded porphyrins. Among the numerous systems obtained to date, are the hexapyrrolic macrocycles rubyrin **3.41**, core-modified rubyrin **3.42**, rosarin **3.43**, hexaphyrin(1.1.1.1.0.0) **3.44**, and cyclo[6]pyrrole **3.45**.

Rubyrin **3.41a** was prepared early on by the Sessler group and was the first hexapyrrolic system to be characterized by X-ray diffraction analysis. In analogy to sapphyrin, the diprotonated form of rubyrin was found to stabilize a 1:2 chloride anion

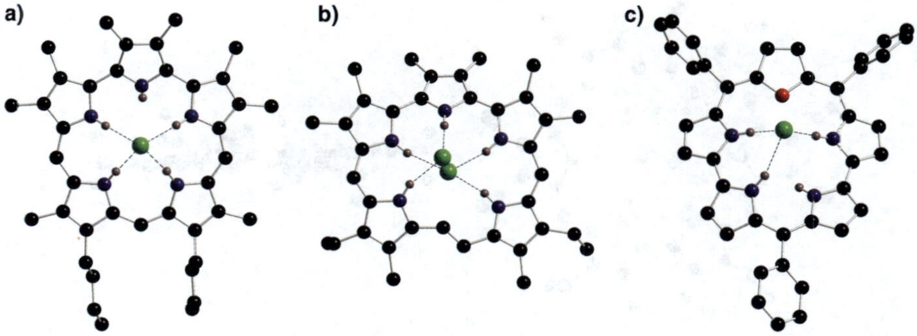

Figure 3.13 *X-ray structures of (a) the hydrochloride salt of isosmaragdyrin **3.38**, (b) the bis-hydrochloride salt of pentaphyrin-(2.1.0.0.1) **3.39**, and (c) the hydrochloride salt of oxasmaragdyrin **3.40***

complex containing two bound chloride anions, as illustrated in Figure 3.14. As true for the corresponding sapphyrin structure, the two chloride anions were found to sit above and below the plane of the relatively planar macrocycle.

In further analogy to sapphyrin, rubyrin was found to be an anion carrier, as inferred from standard U-type anion-transport experiments. Specifically, in the presence of C-Tips (triisopropysilyl cytidine), added as a co-factor, the rate of rubyrin-mediated GMP transport was found to be roughly 30 times faster than that mediated by the smaller sapphyrin system.[47] Complementing these studies is recent work by Umezawa and co-workers on ion-selective electrodes. As detailed in Chapter 8, these researchers found that rubyrin-based membranes exhibit an EMF response towards 3,5-dinitrobenzoate anion.[48]

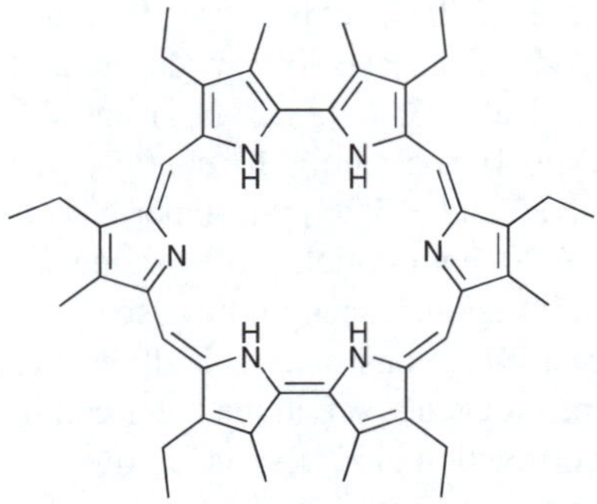

3.41a

Interesting changes in the conformation of rubyrin **3.41b** in the presence of anions were reported in 2005 by Osuka and co-workers. In solution, the addition of HCl to the free-base **3.41b** gave rise to a blue-shift in the Soret-like band from 537 to 518 nm in CH_2Cl_2.[49] Conversely, the addition of TFA led to a red-shift in this same Soret-like band to 555 nm. These differences were attributed to the stabilization of two different conformations upon diprotonation using these two acids. Neither of these conformations was thought to correspond to that of the free-base form. The existence of these

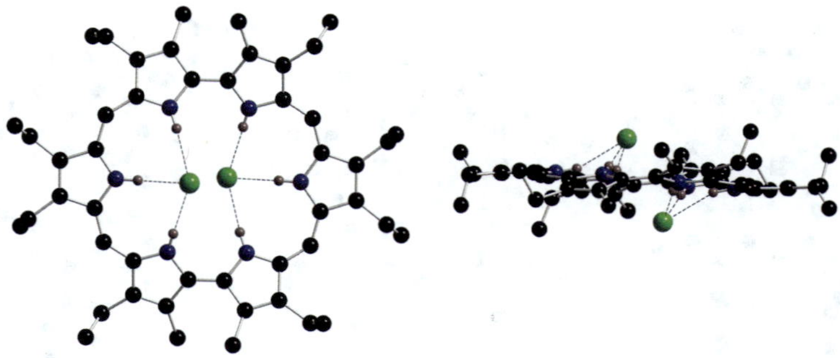

Figure 3.14 *Single crystal X-ray structure of the bis-hydrochloride salt of rubyrin* ***3.41a***

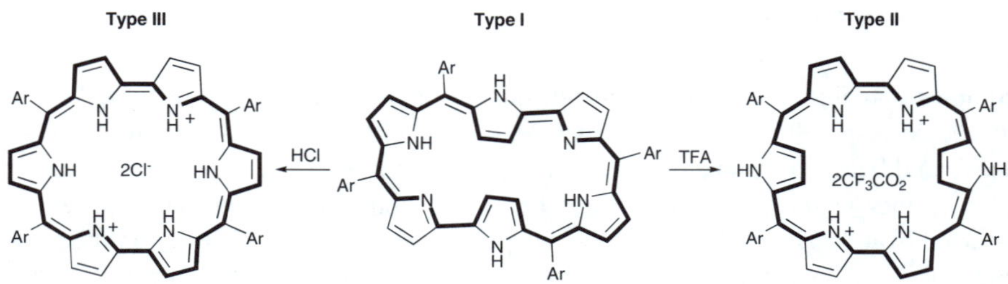

3.41b-$(HCl)_2$ 3.41b Ar = C_6F_5 3.41b-$(TFA)_2$

Scheme 3.3 *Conformational changes observed upon the protonation of* ***3.41b*** *with HCl and TFA, respectively*

three conformational isomers, depicted in Scheme 3.3, were inferred from ^{1}H NMR spectral studies. In the case of the two protonated adducts, the ^{1}H NMR spectra recorded in THF-d_8 revealed the presence of two different NH signals for both salts. These peaks were observed at −3.86 (inner NH) and 16.5 ppm (outer NH) for **3.41b**·2TFA (so-called Type II spectrum). In contrast, the two NH signals were observed at −3.05 (inner NH) and −3.26 ppm (inner NH) for **3.41b**·2HCl (so-called Type III spectrum). The *ca.* 20 ppm upfield shift for one set of NH signals (outer → inner), relative to the bis-TFA adduct, seen in the case of the bis-HCl salt, is particularly noteworthy and is completely consistent with all the NH protons pointing inward towards the centre of the macrocyclic core in the latter complex. Support for this conclusion came from X-ray diffraction analyses, which revealed that in the case of both anions 1:2 complexes were formed in the solid state and that the diprotonated core did indeed adopt very different conformations as shown in Figure 3.15.

Just as is true for the hexa-aza "parent" rubyrin, core-modified rubyrins **3.42a–3.42d** contain two imine-like nitrogens within their central cores. These nitrogen centres are expected to be relatively basic, allowing for the formation of dicationic species that should act as anion-binding agents. Consistent with this hypothesis, a decrease in the absorbance intensity and a shift in the position of the Soret band (by up to *ca.* 30 nm) of **3.42a**·$(HCl)_2$ and **3.42a**·$(HClO_4)_2$ was observed upon the addition of anions. For the protonated form of this macrocycle, an enhancement in the emission

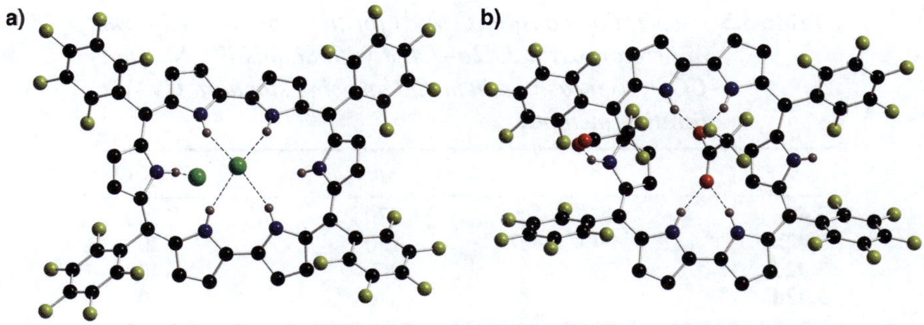

Figure 3.15 *Single crystal X-ray structure of (a) the bis-hydrochloride salt of rubyrin* ***3.41b*** *and (b) the bis-TFA salt of rubyrin* ***3.41b***

intensity was also observed in the presence of selected anions. Association constants were determined by from an analysis of the absorption data, leading to K_a values of 767, 498, and 49 M^{-1} for the binding of N_3^-, AMP, and F^-, respectively, to $[H_2{\cdot}\mathbf{3.42a}]^{2+}$ in methanol.[50] Later, the anion-binding properties of **3.42a–3.42d** were studied in greater detail, but again in methanol and under the same conditions. The results are shown in Table 3.5.[42] The relatively large affinities for carbonate and azide, with selectivity over fluoride anion, was ascribed to the relatively large cavity size of macrocycles such as **3.42a–3.42d**.

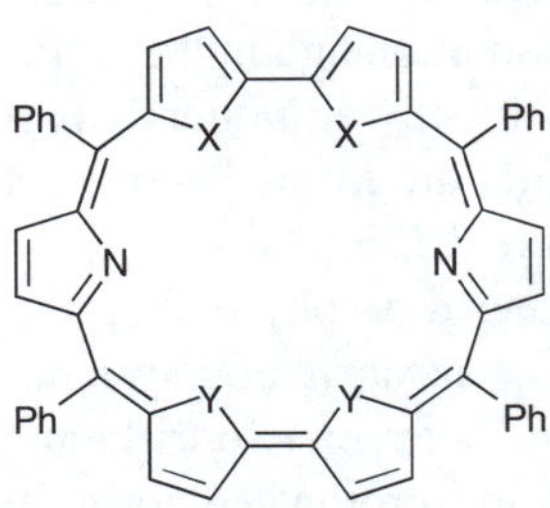

3.42a X = Y = S
3.42b X = Y = Se
3.42c X = O, Y = S
3.42d X = O, Y = Se

The fact that rubyrin proved capable of acting as an anion receptor in its doubly protonated form inspired the synthesis of additional hexapyrrolic systems. Rosarin **3.43**, isorubyrin, **3.44**, and cyclo[6]pyrrole, **3.45**, are examples of such hexapyrrolic macrocycles.

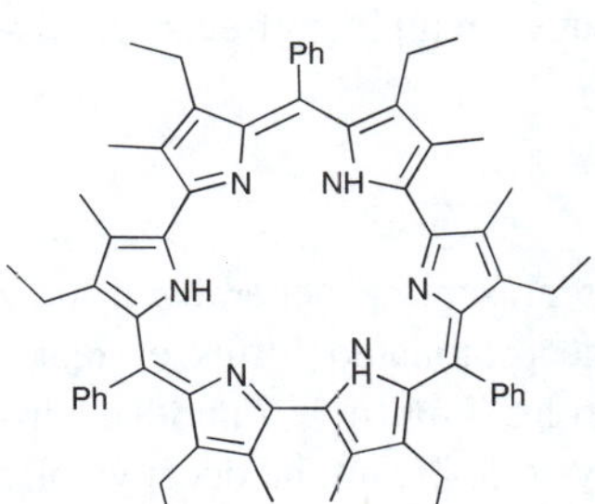

3.43

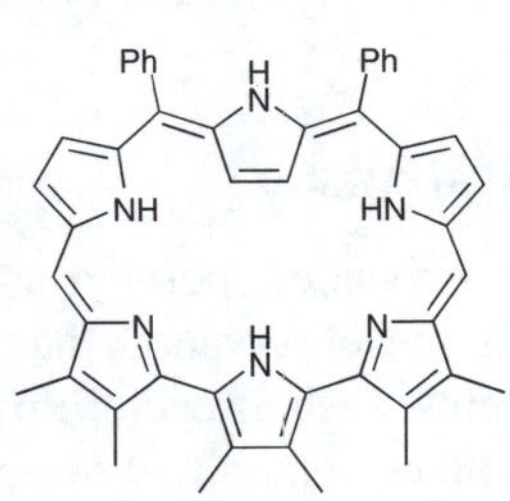

3.44

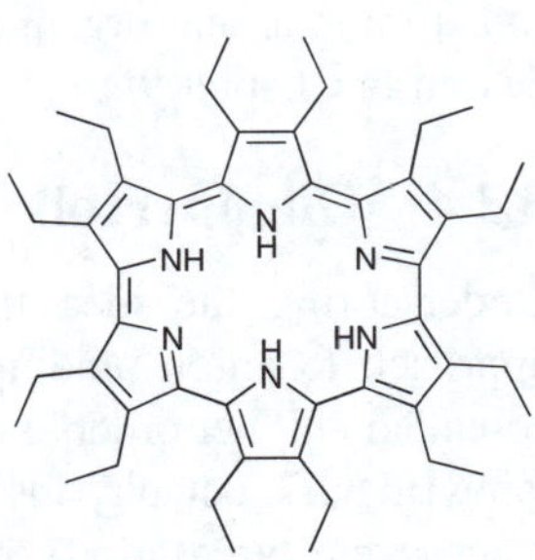

3.45

Table 3.5 *Association constants (M^{-1}) for the diprotonated form of heterorubyrin **3.42a–3.42d** with anions F^-, N_3^-, and CO_3^{2-}, as measured in methanol using standard UV–Vis titration methods*

	KF	*NaN_3*	*K_2CO_3*
3.42a	49	770	1300
3.42b	29	130	220
3.42c	78	900	1400
3.42d	42	370	670

Rosarin **3.43** is a non-aromatic 24 π-electron system that contains three pyrrolic NH and three basic sp^2-hybridized nitrogen atoms. The ability of the fully protonated form to bind chloride anions in the solid state was confirmed by X-ray diffraction analysis.[51] In this case, two of the three chloride anions were found to be bound directly to the triprotonated core with the remaining Cl^- appearing elsewhere in the crystal lattice (*cf.* Figure 3.16a). The presence of the three chloride counteranions is consistent with the proposed 3^+ overall charge and the proposed electronic configuration (*i.e.*, rosarin **3.43**, as prepared, is a 24, rather than a 22 or 26, π-electron system).

Isorubyrin **3.44**, contains the same number of ring-defining atoms as rubyrin. However, in the form of its bis-hydrochloride acid salt, this newer system was found to have one inverted pyrrole unit, both in solution and in the solid state. Thus, in the solid state, only five pyrrolic NH protons directly or indirectly participate in binding the two chloride counteranions. These latter are located above and below the mean plane of the macrocycle as shown in Figure 3.16b.[52]

Cyclo[6]pyrrole **3.45**, which is an expanded porphyrin that was but recently synthesized, represents another example of an hexapyrrolic system that is capable of binding anions. Not only were interactions between two trifluoroacetate counteranions seen in the solid-state structure of the protonated form (Figure 3.16c), but cyclo[6]pyrrole itself was isolated as the result of what appeared to be a specific anion template effect. In particular, it was found to be formed in *ca.* 15% yield when the constituent bipyrrole precursors were subject to Fe(III)-mediated oxidative coupling in the presence of HCl, whereas the use of other acids gave predominantly the corresponding cyclo[8] species (*e.g.*, H_2SO_4) or no isolable macrocyclic product (most mineral and organic acids).[53] These and other anion template effects are discussed in Chapter 9.

3.2.4 Oligopyrrolic Systems

Underscoring the idea that expanded porphyrins represent a generalized new approach to anion binding is the observation that the protonated forms of many so-called "higher order" systems, such as cyclo[*n*]pyrroles **3.46** (n = 7 and 8), heptaphyrin **3.47**, octaphyrins **3.48**, core-modified octaphyrin **3.49** and the decapyrrolic macrocycle turcasarin **3.50** and **3.51** form anion complexes in the solid state (*cf.* Figures 3.17–3.20).[53,54] In the case of cyclo[8]pyrrole **3.46b**, it was further observed

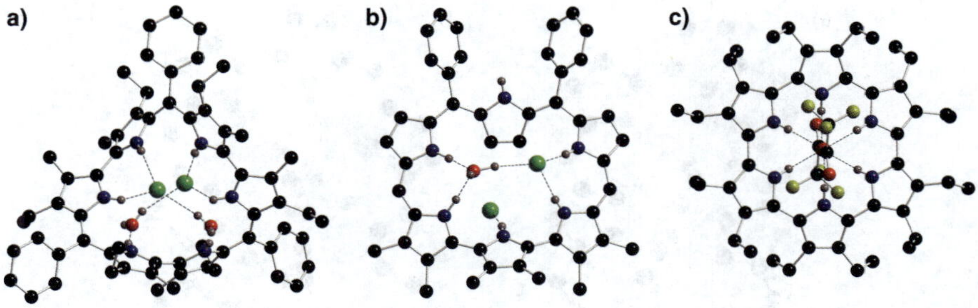

Figure 3.16 *X-ray crystal structures of (a) the trishydrochloride salt of rosarin **3.43**. Only two chloride anions (along with two water molecules) are bound in the core of macrocycle. A third, non-chelated chloride anion (not shown) is found within the crystal lattice; (b) the bis-hydrochloride salt of isorubyrin **3.44** and (c) the bis-TFA salt of cyclo[6]pyrrole **3.45***

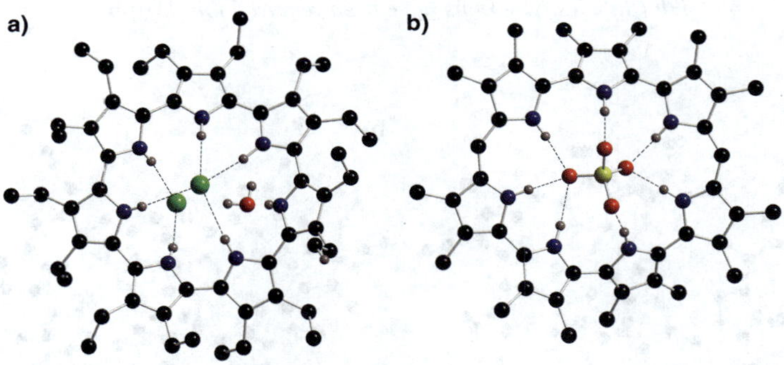

Figure 3.17 *X-ray crystal structures of (a) the bis-hydrochloride salt of cyclo[7]pyrrole **3.46a** and (b) the sulfuric acid salt of heptaphyrin **3.47***

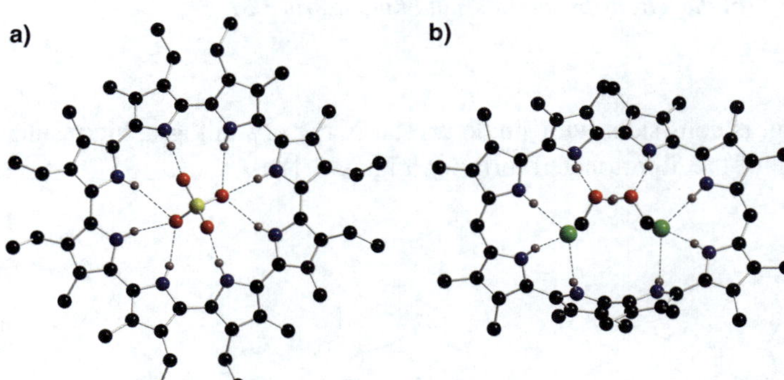

Figure 3.18 *X-ray crystal structures of (a) the dihydrogensulfate salt of cyclo[8]pyrrole **3.46b** and (b) the bis-hydrochloride salt of octaphyrin **3.48***

that the neutral form binds hydrogen sulfate anion with an affinity constant of 5.8×10^3 M^{-1} in DMSO, as judged from isothermal titration calorimetry (ITC) titration analyses.[55] This result, which could reflect partial proton transfer as well as

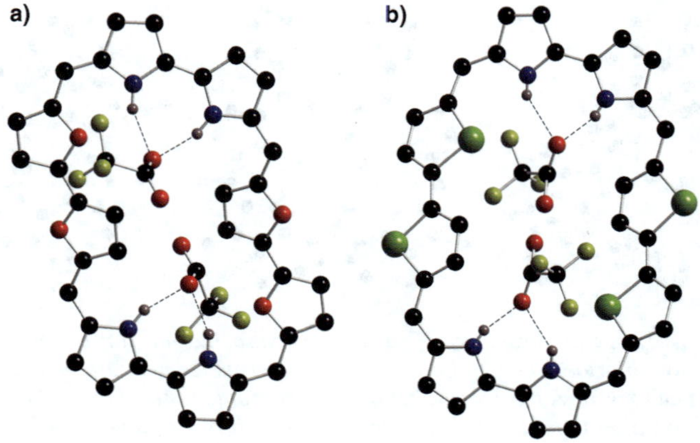

Figure 3.19 *X-ray crystal structures of (a) the bis-TFA salt of **3.49a** and (b) the bis-TFA salt of **3.49b** (the mesityl groups have been removed for clarity)*

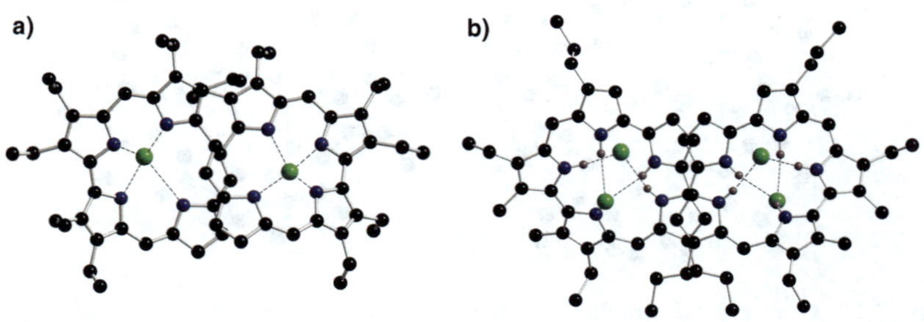

Figure 3.20 *X-ray crystal structures of (a) the bis-hydrochloride salt of octaphyrin **3.50** and (b) the tetrahydrochloride salt of turcasarin **3.51***

H-bonding, is consistent with single crystal X-ray crystallographic results obtained in the case of the diprotonated form (*cf.* Figure 3.18a).

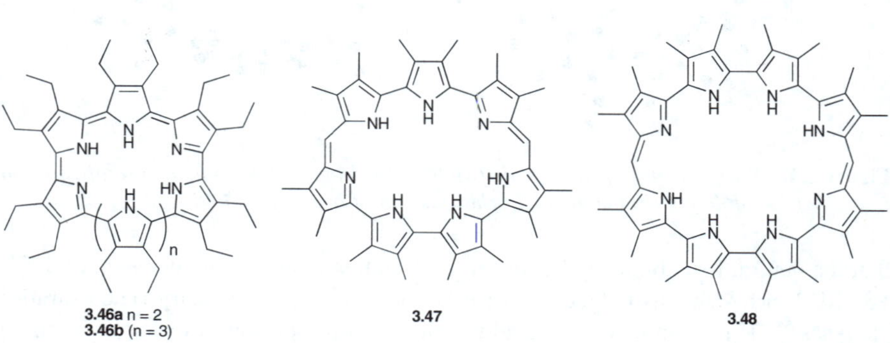

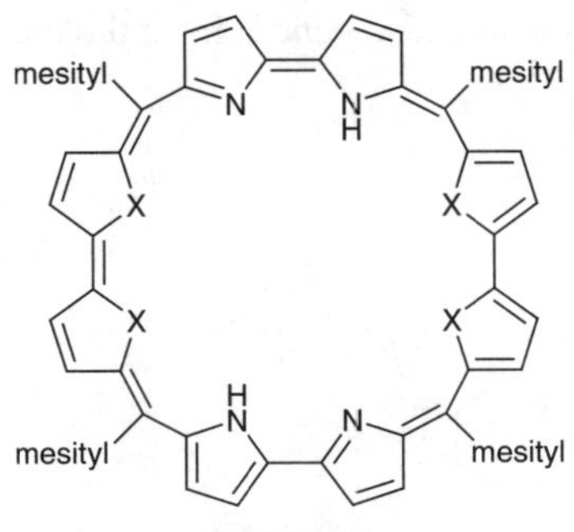

3.49a X = O
3.49b X = Se

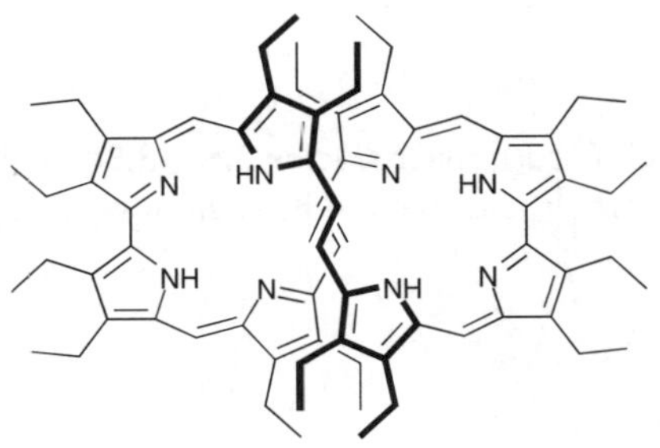

3.50

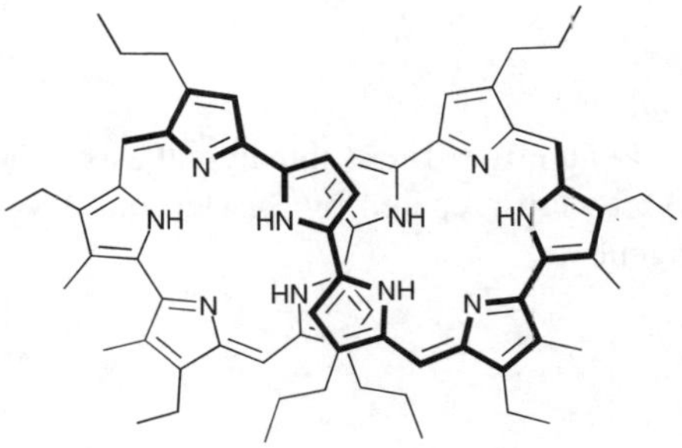

3.51

3.2.5 Imine Linked Receptors and Other Related Systems

The fact that the readily protonated pyridine-like pyrrolic groups resemble imines in a structural sense, led to the consideration that polypyrrolic systems containing true imines could act as anion receptors. The first evidence in support of this hypothesis came from a single-crystal X-ray-diffraction analysis involving the monoprotonated texaphyrinogen **3.52**. Here, a proton on one of the bridging imine subunits was seen to interact with the SCN^- counteranion in the solid state (Figure 3.21).[56] However, the presumed low basicity of this imine nitrogen and the observation of ancillary interactions involving the pyrrole NH protons led to the suggestion that other putative

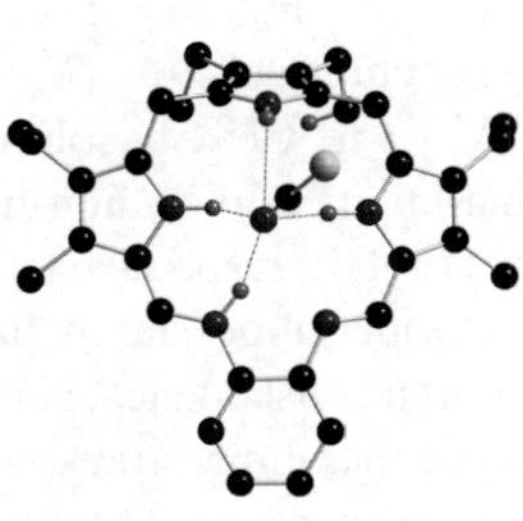

Figure 3.21 *Single crystal X-ray structure of the HSCN salt of **3.52***

imine-type anion receptors would also benefit from the presence of pyrrole NH-anion interactions.

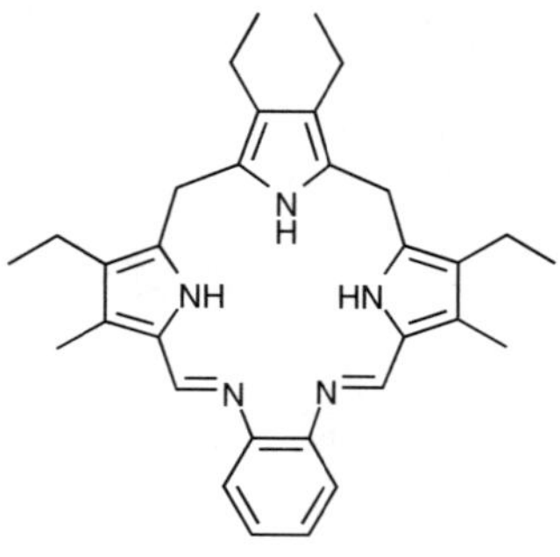

3.52

In an effort to test this hypothesis, other Schiff base macrocycles **3.53**, **3.54**, and **3.55** were prepared by Sessler and co-workers[57–59] and explored as anion-binding agents.

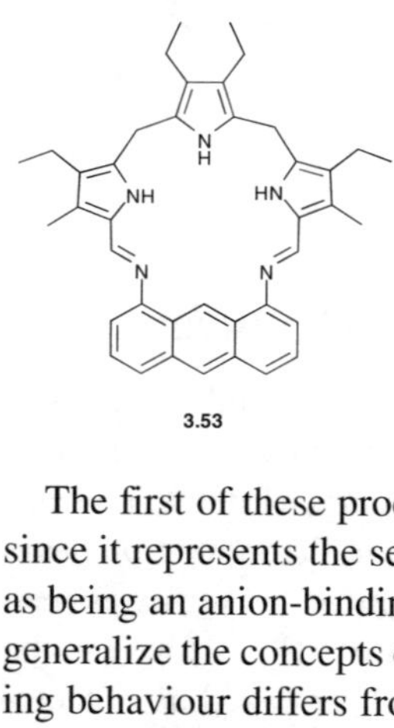

3.53

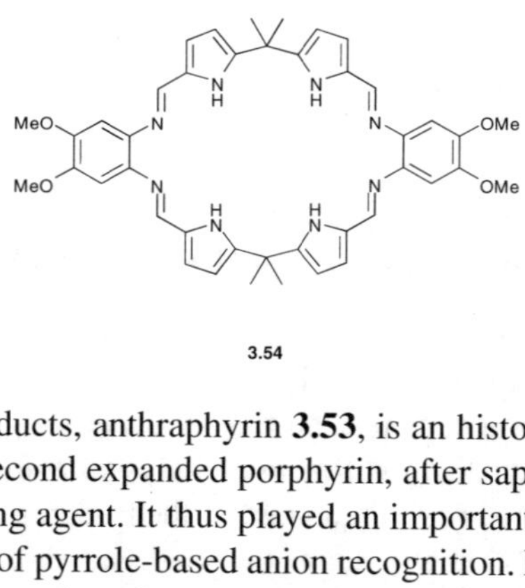

3.54

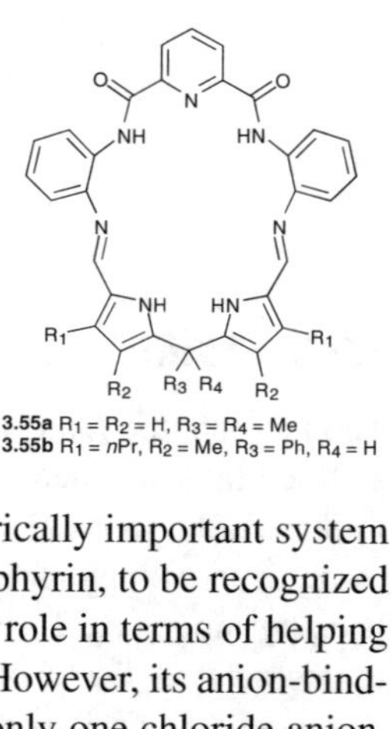

3.55a $R_1 = R_2 = H$, $R_3 = R_4 = Me$
3.55b $R_1 = n\text{Pr}$, $R_2 = Me$, $R_3 = Ph$, $R_4 = H$

The first of these products, anthraphyrin **3.53**, is an historically important system since it represents the second expanded porphyrin, after sapphyrin, to be recognized as being an anion-binding agent. It thus played an important role in terms of helping generalize the concepts of pyrrole-based anion recognition. However, its anion-binding behaviour differs from that of sapphyrin. In particular, only one chloride anion, not two, is bound in the solid state. This anion is found nearly in the centre of the protonated anthraphyrin ring core, being held there via four hydrogen bonds as well as associated electrostatic interactions (Figure 3.22).[57] This Schiff base system also displays a much higher affinity for chloride anion ($K_a = 2 \times 10^5$ M^{-1}) than for fluoride anion ($K_a = 1.4 \times 10^4$ M^{-1}) in CH_2Cl_2 solution, as judged from standard Benesi–Hildebrand and Scatchard plots of the anion-induced UV–Vis spectral shifts (both species studied in the form of their respective tetrabutylammonium salts). This relatively higher affinity for chloride anion stands in contrast to what is seen for diprotonated sapphyrin. Perhaps as a consequence, anthraphyrin acts as a much more effective mediator for the through-membrane transport of fluoride anion than sapphyrin, as judged from model Pressman-type U-tube model experiments.[57] On the other hand, the fluoride-anion transport ability of anthraphyrin is significantly

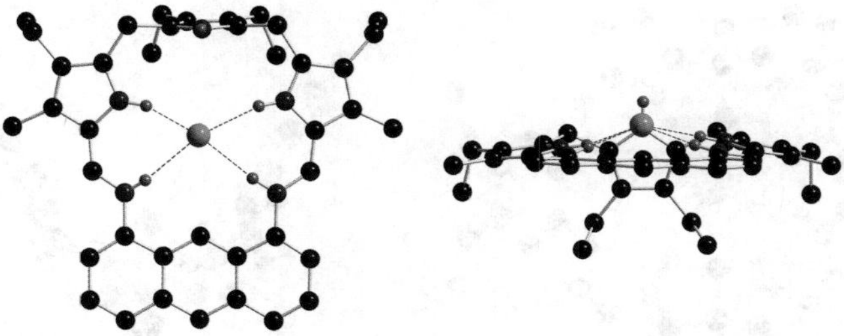

Figure 3.22 *Single crystal X-ray structure of the mixed hydrochloride and hydrotetrafluoroborate salt of anthraphyrin **3.53**. The chloride anion is bound in the diprotonated core (as illustrated). However, the BF_4^- counteranion is not proximate to the core and is not shown*

inhibited by chloride anion. These findings were considered consistent with the proposal that it is the rate of anion release, rather than the overall binding affinity, which determines the efficacy of anion transport.[57]

More recently, another expanded Schiff base macrocycle, **3.54**, which is actually a calix[4]pyrrole-texaphyrin hybrid, was synthesized.[58] This macrocycle has the four pyrrole NH-donor elements characteristic of the calix[4]pyrroles (*cf.* Chapter 5) with the larger cavity size characteristic of texaphyrin, a class of expanded porphyrins that are known *inter alia* for their lanthanide(III)-cation coordination chemistry and potential medical utility.[60] As deduced from an X-ray diffraction analysis of **3.54**·2HCl, in the solid state one chloride anion is bound to the cleft of what is an overall V-shaped Schiff base macrocyclic structure by four pyrrolic NHs and two protonated imine hydrogens (Figure 3.23). This result provided support for the notion that the protonated form would bind chloride anion in solution. In the case of the neutral, non-protonated form, efforts to determine the association constants for chloride and bromide revealed that these species (studied as the corresponding tetrabutylammonium salts) were not bound, at least within the limits of detection in CH_3CN solution. On the other hand, protonation of macrocycle **3.54** serves to increase dramatically its chloride anion-binding affinity ($K_a = 4.6 \times 10^5$ M^{-1} for the monoprotonated form; $K_{a1} = 4.46 \times 10^6$ M^{-1} and $K_{a2} = 2.2 \times 10^5$ M^{-1} for the doubly protonated form in acetonitrile).[58] Such a finding is consistent with the increased positive charge present in the protonated forms serving to strengthen the electrostatic interaction between the nitrogen-rich host and the bound chloride anion.

Another set of hybrid macrocycles, **3.55**, having two imine and two amide functional groups, was prepared from the reaction of a diformyldipyrromethane with a diamine compound. These receptors shows a high selectivity for dihydrogen phosphate and hydrogen sulfate over nitrate and halide anions in acetonitrile solution.[59] Quantitative association constants (K_a) for the binding of anions to **3.55a** were obtained from UV–Vis spectroscopic titrations and were found to be 64,000, 38,000, and 2000 M^{-1} for HSO_4^-, $CH_3CO_2^-$, and Cl^-, respectively, for a 1:1 host–guest binding process. However, this receptor was found to complex $H_2PO_4^-$ with a 1:2 host–guest stoichiometry and give affinity values for the process of $K_{a1} = 342{,}000$ M^{-1} and $K_{a2} = 26{,}000$ M^{-1}, respectively. The observed high selectivity for protic acid anions (*e.g.*, HSO_4^- and

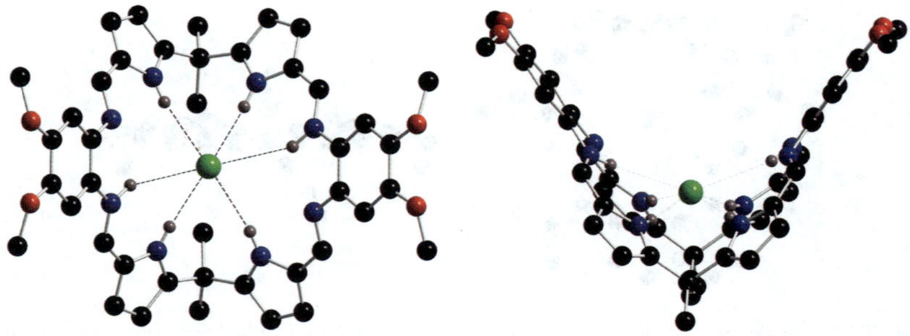

Figure 3.23 *Two views of single crystal X-ray structure of the bis-hydrochloride salt of the Schiff base macrocycle **3.54**. Only one chloride anion is found in the core of macrocycle, while a second nonchelated anion (not shown) is found within the crystal lattice*

$H_2PO_4^-$) is attributed to the presence of extra electrostatic and hydrogen-bond interactions involving the hydrogen atom and the "carbonyl like" oxygen atoms. This hypothesis is consistent with the single crystal X-ray structure of the dihydrogen sulfate salt of analogue **3.55b** shown in Figure 3.24. In particular, the two imine subunits are protonated and participate in hydrogen-bonding interactions with the resulting sulfate anion.[61]

Calixphyrin is a term recently coined by Sessler and co-workers[62] and is meant to designate hybrid systems that are chemical "blends" of calixpyrrole (a class of pyrrolic macrocycles discussed in Chapter 5 that contain sp^3 hybridized pyrrole–pyrrole bridges) and porphyrins (characterized by sp^2 hybrid *meso* carbons). In practice, this definition has been extended to include all porphyrin-like materials containing at least one sp^3 *meso*-carbon bridging atom. In 2001, calix[6]phyrin **3.56** was synthesized and it was observed that the neutral form of the macrocycle did not act as an effective anion receptor, at least for the various anions tested (chloride, bromide, iodide, nitrate, and hydrogensulfate). However, an accompanying X-ray diffraction study revealed that crystals produced in the presence of HCl were monoprotonated with one of the pyrrolic imine units being protonated. As a consequence, one chloride anion was found to be bound within the cavity, being held there via five hydrogen-bonding interactions (Figure 3.25).[63]

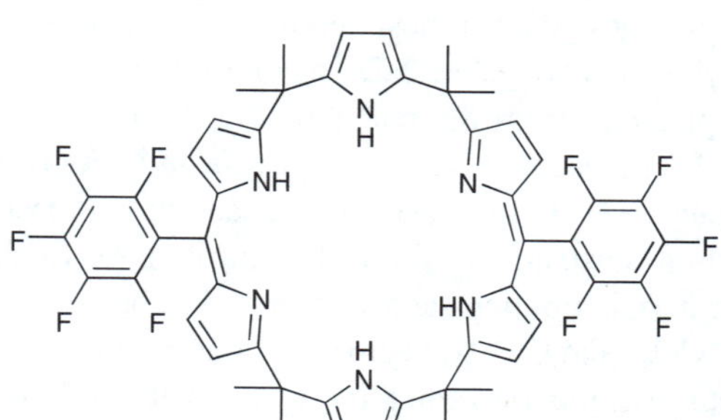

3.56

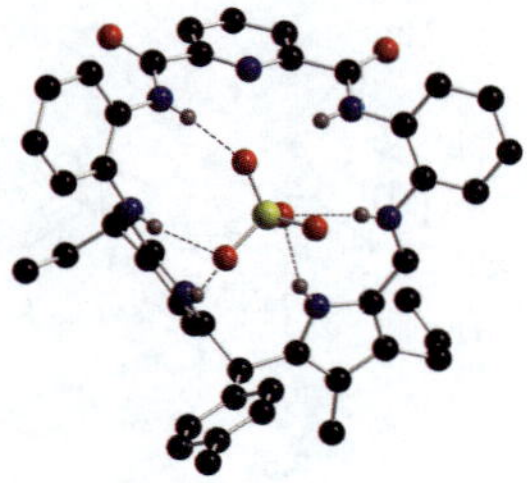

Figure 3.24 *Single crystal X-ray structure of the di-hydrogen sulfate salt of the Schiff base macrocycle **3.55b***

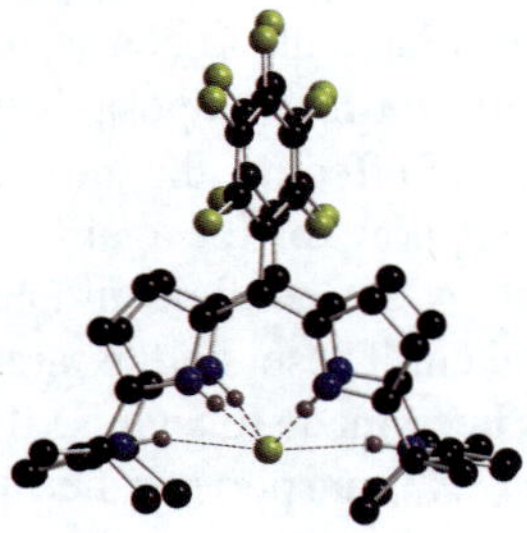

Figure 3.25 *Single crystal X-ray structure of the hydrochloride salt of calix[6]phyrin **3.56***

The fact that the monoprotonated form of **3.56** binds chloride in the solid state led to the proposal that the protonated forms of calix[6]phyrin could function as anion receptors in solution. In acetone solution containing a high concentration of fuming sulfuric acid, added to ensure full protonation, UV–Vis spectroscopic studies revealed that the diprotonated form of calix[6]phyrin, which is dominant under such conditions, binds chloride, bromide, and iodide anions strongly. The affinity constants for I^- and HSO_4^-, which were determined quantitatively, were found to be 2.55×10^4 and 3.5×10^3 M^{-1}, respectively, in this solvent mixture.

3.3 Linear Receptors

Prodigiosins are a family of naturally occurring tripyrrolic red pigments that were first isolated in the 1930s[64] from microorganisms including *Serratia* and *Streptomyces*[65] that are characterized by a common pyrrolylpyrromethene skeleton. These molecules, especially prodigiosin 25-C, **3.57**, have been studied extensively for their promising immunosuppressive[66] and anticancer activity.[67] Various prodigiosins have also been found to induce apoptosis in dozens of human cancer cell lines, including liver cancer,[68] human breast cancer,[69] human colon cancer,[70] gastric cancer,[71] and haematopoietic cancer cell lines.[70] Taken in concert, this combination of properties has made the prodigiosin attractive targets for use in various combination drug therapies. It is also inspiring the synthesis of new pyrrole-based anion carriers.

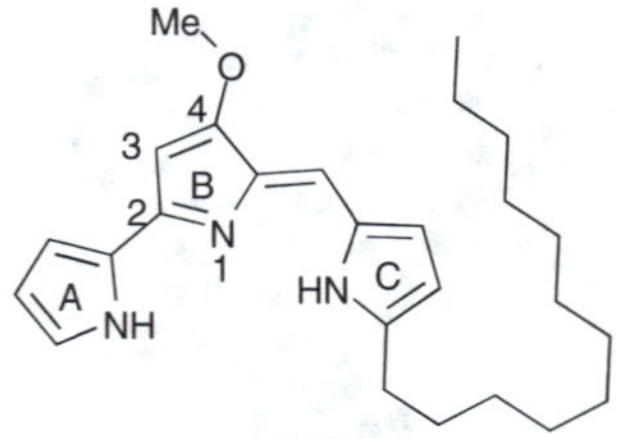

3.57

As implied above, a large number of di-, and, particularly, tripyrrolic species have been made and tested in recent years in the context of studies involving prodigiosins. Other (larger) oligopyrroles have been prepared as potential precursors for polypyrrole macrocycles. Some of these have also been tested as prodigiosin mimics. As mentioned in the introduction, one current proposal is that prodigiosins mediate their physiological effect as the result of effecting the into-cell transport of HCl. As a part of a general effort to provide support for (or against) this proposal, anion-binding studies of linear receptors such as the prodigiosin analogues, tetrapyrrin **3.61**, and hexapyrrin **3.62**, were undertaken. These studies are particularly important since only very recently have efforts been made to analyse the anion-binding behaviour of prodigiosins in either a kinetic (*i.e.*, transport) or thermodynamic (*i.e.*, binding affinity and selectivity) sense.

In early structural work (1978), Treibs and co-workers[72] reported the X-ray crystal structure of the chloride-anion complex of the tripyrrolic species, **3.58**, which contains a dipyrromethane subunit. The HCl complexes of a dipyrromethene **(3.59)** and of a synthetic prodigiosin derivative (**3.60**) have also been elucidated more recently by Manderville and Fürstner (Figure 3.26).[73,74]

In separate work, Manderville and co-workers[75] have shown that protonated prodigiosins can bind to DNA and are capable of complexing copper ions. In the context of this work, Manderville and co-workers[76] reported the crystal structure of a Cu(II) complex of prodigiosin **3.61**, in which the pyrrole ring C is oxidized (Figure 3.26). These findings, considered in concert, led to the suggestion by these workers, as well as by Fürstner,[77] that the biological activity of prodigiosins will more likely prove to be correlated with these chemical features than with $H^+ + Cl^-$ transport ability. This

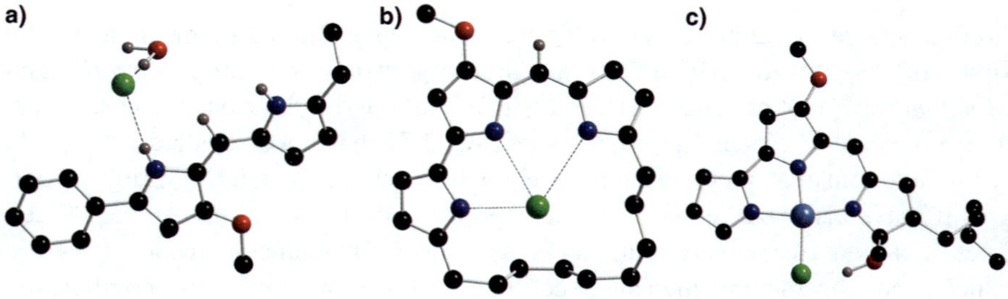

Figure 3.26 *X-ray structure of (a) the hydrochloride salt of the dipyrrolic prodigiosin analogue* ***3.59*** *(b) the hydrochloride salt of the prodigiosin analogue* ***3.60*** *(c) the Cu(II) complex of prodigiosin analogue* ***3.61***

clear disparity in the suggested explanations for the biological activity (transport vs. copper coordination and DNA modification) provides an additional incentive to make and study prodigiosin analogues.

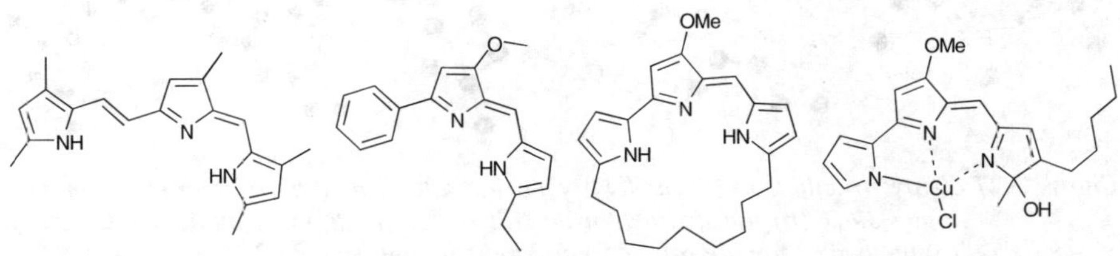

3.58 3.59 3.60 3.61

At present, structure-function studies of prodigiosin analogues are still in their early days.[73,75,78] For the most part, these studies have only involved systems containing the key tripyrrole moieties or its dipyrromethene subunits. Nonetheless, these analyses have been extremely valuable; they have served to highlight the apparent necessity of the 4-methoxy group, the benefit of alkyl substituents on the C ring, and an apparent lack of correlation between pyrrole nitrogen basicity and biological activity.[73]

Recently, the X-ray crystal structure of dipyrromethene **3.62** and prodigiosin analogue **3.63** were solved by Sessler and co-workers[79] as part of an ongoing effort devoted to understanding the anion-binding properties of these latter naturally occurring tripyrrolic species. The structures in question, shown in Figure 3.27, revealed that a bound chloride or bromide anion that is stabilized through a combination of two or three hydrogen-bonding interactions and electrostatic effects. Strong binding was also observed in solution, with the chloride anion being bound with an association constant (K_a) of *ca.* 10^6 and 10^5 M^{-1} for [H·**3.62**]$^+$ and [H·**3.63**]$^+$ as judged from ITC studies carried out in acetonitrile. Such findings are considered important in that they serve to highlight the fact that even very simple protonated pyrrolic systems are capable of binding anions well.

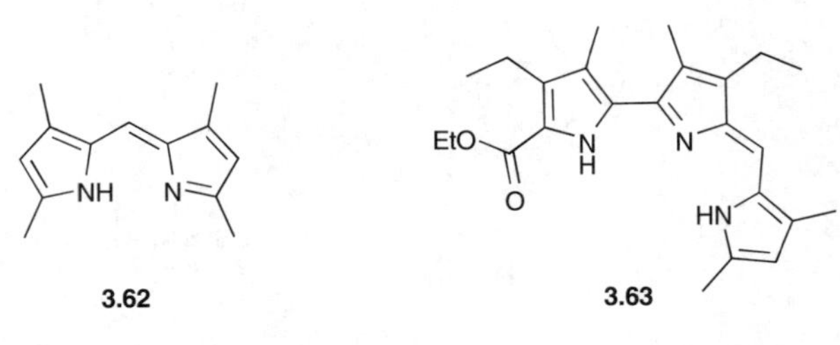

3.62 3.63

In the course of what is ongoing work, Sessler and co-worker have completed the synthesis of three more simple prodigiosins (**3.66a–3.66c**) and dipyrromethene (**3.64**), and determined their chloride anion-binding affinities. Included as part of this study is the known prodigiosin, **3.65a**. As above, the solution phase anion-binding behaviour was probed by carrying out ITC measurements in anhydrous acetonitrile. On the basis of these analyses, the chloride-anion affinities of the protonated forms ([H·**3.64**]$^+$, [H·**3.65**]$^+$, [H·**3.66a**]$^+$, [H·**3.66b**]$^+$, and [H·**3.66c**]$^+$) were found to be

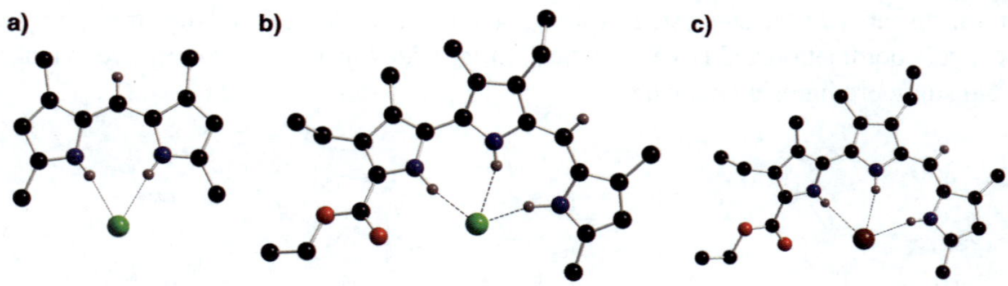

Figure 3.27 *X-ray structures of (a) the hydrochloride salt of the dipyrrolic prodigiosin analogue **3.62**, (b) the hydrochloride salt of the prodigiosin analogue **3.63**, and (c) the hydrobromide salt of the prodigiosin analogue **3.63***

quite high ($K_a = 3.0 \times 10^5$, 5.9×10^5, 1.1×10^5, 4.1×10^4, and 3.8×10^5 M^{-1} in both cases).[79,80] The fact that the protonated forms of [H·**3.65**]$^+$, [H·**3.63**]$^+$, and [H·**3.66a**]$^+$, [H·**3.66b**]$^+$, and [H·**3.66c**]$^+$ were found to bind chloride anion with essentially equal (and high) affinity, coupled with the strong chloride-anion affinities previously recorded for sapphyrin (a species that is also monoprotonated at neutral pH;[20,23] *vide supra*),[21] led to the consideration that other oligomeric pyrrole systems, such as the tetrapyrrole **3.67** and the hexapyrrole **3.68**, might function as HCl receptors or as interesting prodigiosin analogues. In the case of the tetrapyrrin **3.67**, the presence of one more pyrrole unit in relation to the tripyrrolic prodigiosins was expected to lead to an increased binding affinity. Such expectations were realized in the case of the electron-poor dicyano derivative, [H·**3.67b**]$^+$, for which a chloride-anion affinity, K_a, of 4.9×10^5 M^{-1} was recorded for the protonated form; this affinity was found to be enhanced by a factor of roughly 3.9 and 1.7 relative to its α-free analogues [H·**3.67a**]$^+$ and [H·**3.67c**]$^+$.

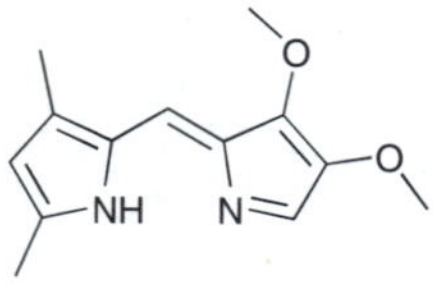

3.64

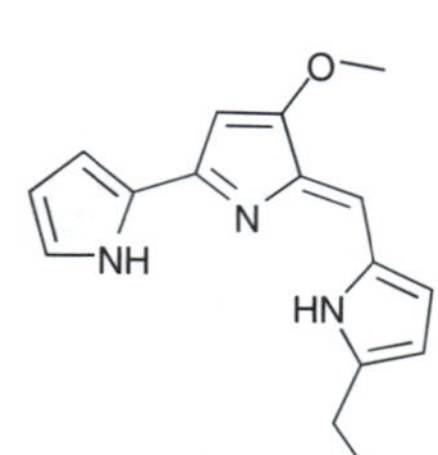

3.65

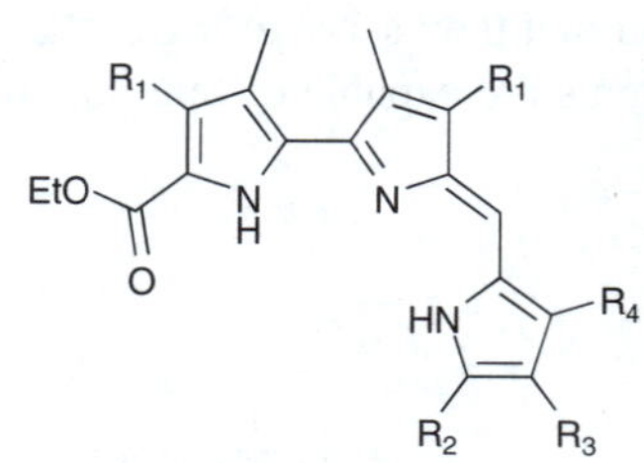

3.66a R_1 = Et, $R_2 = R_3 = R_4$ = H
3.66b R_1 = Et, R_2 = H, $R_3 = R_4$ = OMe
3.66c $R_1 = R_2 = R_4$ = Me, R_3 = H

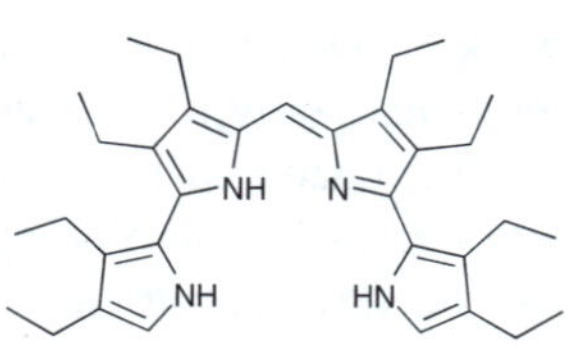

3.67a

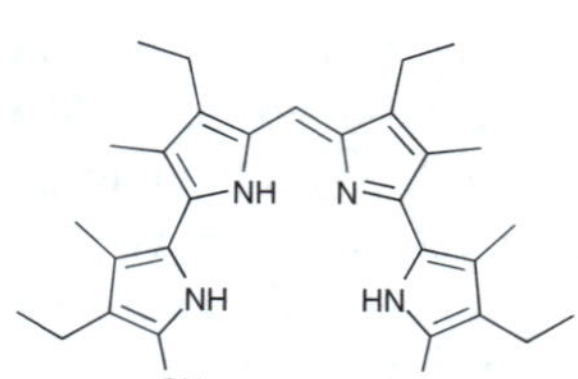

3.67b

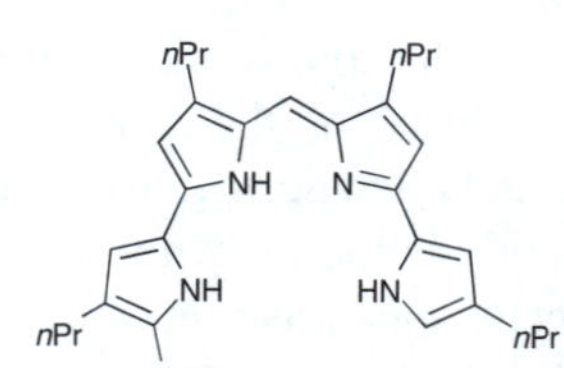

3.67c

The linear hexapyrrolic receptor **3.68**, which is formally an α,α'-linked prodigiosin dimer, also shows promise as an anion receptor. This system contains two potential chloride-anion-binding sites that were expected to allow for the formation of a 2:1 anion-to-receptor complex under appropriate conditions. Consistent with this expectation, the structure of the bis-HCl salt of **3.68a**, determined by X-ray crystal structural analysis (*cf.* Figure 3.28),[81] revealed the presence of two bound chloride anions tethered to the flat diprotonated hexapyrrin via two sets of three-fold hydrogen-bonding interactions. In an accompanying solution phase study that involved the use of a slightly different hexapyrrin (**3.68b**), the diprotonated system was found to bind two chloride anions strongly in acetonitrile ($K_{a1} = 1.2 \times 10^6$, $K_{a2} = 3.2 \times 10^4$ M^{-1}).[80]

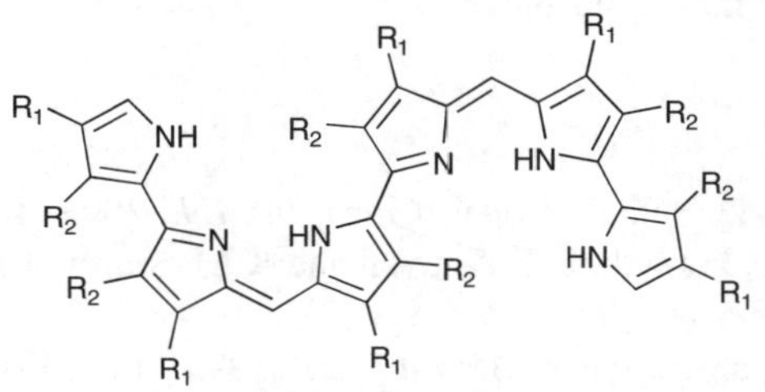

3.68a $R_1 = C_3H_7$, $R_2 = H$
3.68b $R_1 = R_2 = C_2H_5$

Taken together, these results illustrate the promise open-chain oligopyrroles offer as anion receptors and as potential prodigiosin analogues. Consistent with the latter supposition was the finding that tetrapyrroles **3.67a–3.67c** show anticancer activity *in vitro* as judged from cell proliferations studies involving A549 cells (a human lung cancer cell line), although less so than prodigiosin **3.65** and sapphyrin **3.17**.[82] The latter macrocycle was also found to be active *in vitro* and *in vivo* in a Ramos cancer model.[83] However, at the present time it is not known whether its mode of action involves mechanisms related to anion transport. To date, all efforts to produce a stable, non-labile transition metal complexes from sapphyrin have met with failure, a finding that likely rules out copper-based DNA modification as the basis for its activity. In any event, it is clear that the study of prodigiosin analogues and macrocyclic

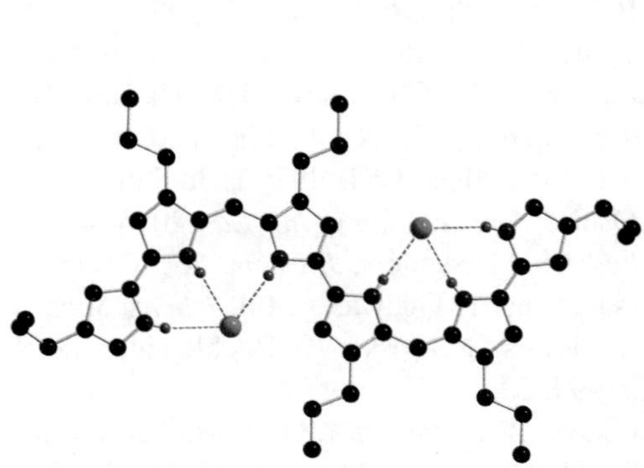

Figure 3.28 *Single crystal X-ray structure of the bis-hydrochloride salt of hexapyrrin **3.68a***

oligopyrroles as potential drug leads could prove quite promising. This is thus seen as a research direction of considerable importance in the near future.

3.4 Summary Remarks

The serendipitous discovery that sapphyrin functions as an excellent fluoride receptor has driven forward research into expanded porphyrins leading to a new family of anion receptor systems over the last 15 years. Research effort is now focussing on linear systems in an effort to produce new prodigiosin-like systems with potentially useful anticancer and ATPase uncoupler properties. Thus, research into protonated pyrrolic systems continues to be driven by the important real-world applications to which these systems may contribute.

References

1. J.L. Sessler and D. Seidel, *Angew. Chem. Int. Ed.*, 2003, **42**, 5134.
2. M.O. Senge, T.P. Forsyth, L.T. Nguyen and K.M. Smith, *Angew. Chem. Int. Ed.*, 1994, **33**, 2485.
3. K.M. Barkigia and J.Fajer, *Acta Cryst.*, 1995, **C51**, 511; M.O. Senge, V. Gerstung, K. Ruhlandt-Senge, S. Runge and I. Lehmann, *J. Chem. Soc. Dalton Trans.*, 1998, 4187; R.G. Khoury, L. Jaquinod, R. Paolesse and K. M. Smith, *Tetrahedron*, 1999, **55**, 6713; R. Paolesse, L. Jaquinod, F.D. Sala, D.J. Nurco, L. Prodi, M. Montalti, C.D. Natale, A. D'Amico, A.D. Carlo, P. Lugli and K.M. Smith, *J. Am. Chem. Soc.*, 2000, **122**, 11295; Y. Mizuno, T. Aida and K. Yamaguchi, *J. Am. Chem. Soc.*, 2000, **122**, 5278; M.O. Senge, *Z. Naturforsch.*, 2000, **55b**, 336.
4. K. Schweiger, H. Hückstädt and H. Homborg, *Z. Naturforsch.*, 1999, **54b**, 963.
5. L.T. Nguyen, M.O. Senge and K.M. Smith, *Tetrahedron Lett.*, 1994, **35**, 7581; B. Cheng, O.Q. Munro, H.M. Marques and W.R. Scheidt, *J. Am. Chem. Soc.*, 1997, **119**, 10732.
6. J.L. Sessler, E.A. Brucker, V.M. Lynch, M. Choe, S. Sorey and E. Vogel, *Chem. Eur. J.*, 1996, **2**, 1527.
7. C.-H. Tsai, J.-Y. Tung, J.-H. Chen, F.-L. Liao, S.-L. Wang, S.-S. Wang, L.-P. Hwang and C.-B. Chen, *Polyhedron*, 2000, **19**, 633.
8. L. Jaquinod, O. Siri, R.G. Khoury and K.M. Smith, *Chem. Commun.*, 1998, 1261; O.S. Finikova, A.V. Cheprakov, P.J. Carroll, S. Dalosto and S.A. Vinogradov, *Inorg. Chem.*, 2002, **41**, 6944; F.B. Larsen, F.G. Hansen and C.J. McKenzie, *Acta Cryst.*, 2004, **E60**, 497; Y. Inokuma, T. Matsunari, N. Ono, H. Uno and A. Osuka, *Angew. Chem., Int. Ed.*, 2005, **44**, 1856.
9. H.L. Anderson and J.K.M. Sanders, *J. Chem. Soc. Chem. Commun.*, 1992, 946.
10. H. Furuta, T. Asano and T. Ogawa, *J. Am. Chem. Soc.*, 1994, **116**, 767; P.J. Chemielewski, L. Latos-Grażyński, K. Rachlewicz and T. Glowiak, *Angew. Chem., Int. Ed.*, 1994, **33**.
11. H. Maeda, Y. Ishikawa, T. Matsuda, A. Osuka and H. Furuta, *J. Am. Chem. Soc.*, 2003, **125**, 11822; H. Maeda, A. Osuka and H. Furuta, *J. Incl. Phenom.*, 2004, **49**, 33.

12. H. Furuta, A. Srinivasan, H. Maeda, T. Ishizuka and T. Morimoto, *Abstracts of Papers, The 3rd International Conference on Porphyrins and Phthalocyanines, New Orleans, United States, July*, 2004, Oral Presentation.
13. H. Furuta, H. Maeda and A. Osuka, *J. Am. Chem. Soc.*, 2001, **123**, 6435.
14. B.F. Anderson, T.J. Bartczak and D.C. Hodgkin, *J. Chem. Soc. Perkin Trans. 2*, 1974, 977; R. Paolesse, R.G. Khoury, F.D. Sala, C.D. Natale, F. Sagone and K.M. Smith, *Angew. Chem., Int. Ed.*, 1999, **38**, 2577; M. Stepien and L. Latos-Grażyński, *J. Am. Chem. Soc.*, 2002, **124**, 3838; C.-H. Hung, C.-Y. Lin, P.-Y. Lin and Y.-J. Chen, *Tetrahedron Lett.*, 2004, **45**, 129; M. Pawlicki, L. Latos-Grażyński and L. Szterenberg, *J. Org. Chem.*, 2002, **67**, 5644.
15. J.L. Sessler, M.J. Cyr, V. Lynch, E. McGhee and J.A. Ibers, *J. Am. Chem. Soc.*, 1990, **112**, 2810.
16. R.B. Woodward, Presented in Post Lecture Discussion during Chemical Society Meeting on Aromaticity, *Sheffield*, 1966. We thank Prof. Emanuel Vogel for calling this to our attention.
17. M.J. Broadhurst, R. Grigg and A.W. Johnson, *J. Chem. Soc. Perkin Trans. 1*, 1972, 2111; V.J. Bauer, D.L.J. Clive, D. Dolphin, J.B. Paine III, F.L. Harris, M.M. King, J. Loder, S.-W.C. Wang and R.B. Woodward, *J. Am. Chem. Soc.*, 1983, **105**, 6429.
18. M. Shionoya, H. Furuta, V. Lynch, A. Harriman and J.L. Sessler, *J. Am. Chem. Soc.*, 1992, **114**, 5714.
19. J.L. Sessler, M. Cyr, H. Furuta, V. Král, T. Mody, T. Morishima, M. Shionoya and S. Weghorn, *Pure Appl. Chem.*, 1993, **65**, 393.
20. V. Král, H. Furuta, K. Shreder, V. Lynch and J.L. Sessler, *J. Am. Chem. Soc.*, 1996, **118**, 1595.
21. B.L. Iverson, K. Shreder, V. Král, P. Sansom, V. Lynch and J.L. Sessler, *J. Am. Chem. Soc.*, 1996, **118**, 1608.
22. B.F. Culleton, M.G. Larson, P.W.F. Wilson, J.C. Evans, P.S. Parfrey and D. Levy, *Kidney Int.*, 1999, **56**, 2214; A. Levin, L. Stevens and P.A. McCullough, *Postgrad. Med.*, 2002, **111**, 53.
23. J.L. Sessler, D.A. Ford, M.J. Cyr and H. Furuta, *J. Chem. Soc. Chem. Commun.*, 1991, 1733.
24. H. Furuta, M.J. Cyr and J.L. Sessler, *J. Am. Chem. Soc.*, 1991, **113**, 6677.
25. V. Král, J.L. Sessler and H. Furuta, *J. Am. Chem. Soc.*, 1992, **114**, 8704.
26. V. Král and J.L. Sessler, *Tetrahedron*, 1995, **51**, 539.
27. J.L. Sessler, J.W. Genge, V. Král and B.L. Iverson, *Supramol. Chem.*, 1996, **8**, 45.
28. B.L. Iverson, R.E. Thomas, V. Král and J.L. Sessler, *J. Am. Chem. Soc.*, 1994, **116**, 2663.
29. V. Král, A. Andrievsky and J.L. Sessler, *J. Am. Chem. Soc.*, 1995, **117**, 2953.
30. J.L. Sessler, A. Andrievsky, V. Král and V. Lynch, *J. Am. Chem. Soc.*, 1997, **119**, 9385.
31. V. Král, A. Andrievsky and J.L. Sessler, *J. Chem. Soc. Chem. Commun.*, 1995, 2349.
32. B.L. Iverson, K. Shreder, V. Král and J.L. Sessler, *J. Am. Chem. Soc.*, 1993, **115**, 11022.

33. J.L. Sessler, J.M. Davis, V. Král, T. Kimbrough and V. Lynch, *Org. Biomol. Chem.*, 2003, **1**, 4113.
34. R.K. Pandey, *J. Porphyrins Phthalocyanines*, 2000, **4**, 368; T.D. Mody, *J. Porphyrins Phthalocyanines*, 2000, **4**, 362; R.K. Pandey and G. Zheng, in *The Porphyrin Handbook*, K.M. Kadish, K.M. Smith and R. Guilard (eds), Academic Press, San Diego, 2000; T. Mody and J.L. Sessler, *J. Porphyrins Phthalocyanines*, 2001, **5**, 134.
35. B.G. Maiya, M. Cyr, A. Harriman and J.L. Sessler, *J. Phys. Chem.*, 1990, **94**, 3597.
36. D. Magda, M. Wright, R.A. Miller, J.L. Sessler and P.I. Sansom, *J. Am. Chem. Soc.*, 1995, **117**, 3629.
37. J.L. Sessler, P.I. Sansom, V. Král, D.O'Connor and B.L. Iverson, *J. Am. Chem. Soc.*, 1996, **118**, 12322.
38. J.L. Sessler, A. Andrievsky, P.I. Sansom, V. Král and B.L. Iverson, *Bioorg. Med. Chem. Lett.*, 1997, **7**, 1433.
39. V. Král, J.M. Davis, A. Andrievsky, J. Kralová, A. Synytsya, P. Poucková and J.L. Sessler, *J. Med. Chem.*, 2002, **45**, 1073.
40. J.L. Sessler, M.C. Hoehner, A. Gebauer, A. Andrievsky and V. Lynch, *J. Org. Chem.*, 1997, **62**, 9251; J. Lisowski, J.L. Sessler and V. Lynch, *Inorg. Chem.*, 1995, **34**, 3567.
41. S.J. Narayanan, B. Sridevi, T.K. Chandrashekar, A. Vij and R. Roy, *J. Am. Chem. Soc.*, 1999, **121**, 9053.
42. A. Srinivasan, V.R.G. Anand, S.K. Pushpan, T.K. Chandrashekar, K.-I. Sugiura and Y. Sakata, *J. Chem. Soc. Perkin Trans. 2*, 2000, 1788.
43. P.J. Chmielewski and L. Latos-Grażyński, *Chem. Eur. J.*, 1995, **1**, 68; K. Rachlewicz, N. Sprutta, L. Latos-Grażyński, P.J. Chmielewski and L. Szterenberg, *J. Chem. Soc. Perkin Trans. 2*, 1998, 959; G.R. Geier III and J.S. Lindsey, *J. Org. Chem.*, 1999, **64**, 1596.
44. J.L. Sessler, J.M. Davis and V. Lynch, *J. Org. Chem.*, 1998, **63**, 7062.
45. S.J. Weghorn, V. Lynch and J.L. Sessler, *Tetrahedron Lett.*, 1995, **36**, 4713.
46. B. Sridevi, S.J. Narayanan, R. Rao and T.K. Chandrashekar, *Inorg. Chem.*, 2000, **39**, 3669.
47. H. Furuta, T. Morishima, V. Král and J.L. Sessler, *Supramol. Chem.*, 1993, **3**, 5.
48. K. Umezawa, K. Tohda, X.M. Lin, J.L. Sessler and Y. Umezawa, *Anal. Chim. Acta*, 2001, **426**, 19.
49. S. Shimizu, R. Taniguchi and A. Osuka, *Angew. Chem. Int. Ed.*, 2005, **44**, 2225.
50. A. Srinivasan, V.M. Reddy, S.J. Narayanan, B. Sridevi, S.K. Pushpan, M. Ravikumar and T.K. Chandrashekar, *Angew. Chem. Int. Ed.*, 1997, **36**, 2598.
51. J.L. Sessler, S.J. Weghorn, Y. Hiseada and V. Lynch, *Chem. Eur. J.*, 1995, **1**, 56.
52. J.L. Sessler, D. Seidel, C. Bucher and V. Lynch, *Chem. Commun.*, 2000, 1473.
53. T. Köhler, D. Seidel, V. Lynch, F.O. Arp, Z. Ou, K.M. Kadish and J.L. Sessler, *J. Am. Chem. Soc.*, 2003, **125**, 6872.
54. J.L. Sessler, S.J. Weghorn, V.M. Lynch and M.R. Johnson, *Angew. Chem. Int. Ed.*, 1994, **33**, 1509; E. Vogel, M. Bröring, J. Fink, D. Rosen, H. Schmickler, J. Lex, K.W.K. Chan, Y.-D. Wu, D.A. Plattner, M. Nendel and K.N. Houk, *Angew. Chem. Int. Ed.*, 1995, **34**, 2511; J.L. Sessler, D. Seidel and V.M. Lynch,

J. Am. Chem. Soc., 1999, **121**, 11257; D. Seidel, V. Lynch, M. and J.L. Sessler, *Angew. Chem. Int. Ed.*, 2002, **41**, 1422; V.G. Anand, S. Venkatraman, H. Rath, T.K. Chandrashekar, W. Teng and K. Ruhland-Senge, *Chem. Eur. J.*, 2003, **9**, 2282.
55. J.L. Sessler and W.-S. Cho, unpublished results.
56. J.L. Sessler, M.R. Johnson and V. Lynch, *J. Org. Chem.*, 1987, **52**, 4394.
57. J.L. Sessler, T.D. Mody, D.A. Ford and V. Lynch, *Angew. Chem. Int. Ed.*, 1992, **31**, 452.
58. J.L. Sessler, W.-S. Cho, S.P. Dudek, L. Hicks, V.M. Lynch and M.T. Huggins, *J. Porphyins Phthalocyanines*, 2003, **7**, 97.
59. J.L. Sessler, E. Katayev, G.D. Pantos and Y.A. Ustynyuk, *Chem. Commun.*, 2004, 1276.
60. G. Givaja, A.J. Blake, C. Wilson, M. Schröder and J.B. Love, *Chem. Commun.*, 2003, 2508; J.M. Veauthier, W.-S. Cho, V.M. Lynch and J.L. Sessler, *Inorg. Chem.*, 2004, **43**, 1220; P.L. Arnold, A.J. Blake, C. Wilson and J.B. Love, *Inorg. Chem.*, 2004, **43**, 8206; G. Givaja, A.J. Blake, C. Wilson, M. Schröder and J.B. Love, *Chem. Commun.*, 2005, 4423; J.L. Sessler, E. Tomat, T.D. Mody, V.M. Lynch, J.M. Veauthier, U. Mirsaidov and J.T. Markert, *Inorg. Chem.*, 2005, **44**, 2125; J.M. Veauthier, E. Tomat, V.M. Lynch, J.L. Sessler, U. Mirsaidov and J.T. Markert, *Inorg. Chem.*, 2005, **44**, 6736.
61. J.L. Sessler, E. Katayev, G.D. Pantos, P. Scherbakov, M.D. Reshetova, V.N. Khrustalev, V.M. Lynch and Y.A. Ustynyuk, *J. Am. Chem. Soc.*, 2005, **127**, 11442.
62. V. Král, J.L. Sessler, R.S. Zimmerman, D. Seidel, V. Lynch and B. Andrioletti, *Angew. Chem. Int. Ed.*, 2000, **39**, 1055; C. Bucher, D. Seidel, V. Lynch, V. Král and J.L. Sessler, *Org. Lett.*, 2000, **2**, 3103.
63. C. Bucher, R.S. Zimmerman, V. Lynch, V. Král and J.L. Sessler, *J. Am. Chem. Soc.*, 2001, **123**, 2099.
64. F. Wrede, *Z. physiol. Chem.*, 1932, **210**, 125; F. Wrede and A. Rothhaas, *Z. Physiol. Chem.*, 1933, **215**, 67; V.F. Wrede and A. Rothhaas, *Z. Physiol. Chem.*, 1933, **222**, 203; F. Wrede and A. Rothhass, *Z. Physiol. Chem.*, 1933, **219**, 267; F. Wrede and A. Rothhaas, *Z. Physiol. Chem.*, 1934, **226**, 95; A. Fürstner, *Angew. Chem. Int. Ed.*, 2003, **42**, 3582.
65. K. Harashima, T. Tanaka and J. Nagatsu, *Agric. Biol. Chem.*, 1967, **31**, 481; H.H. Wasserman, G.C. Rodgers and D.D. Keith, *Tetrahedron*, 1976, **32**, 1851.
66. S.B. Han, H.M. Kim, Y.H. Kim, C.W. Lee, E.-S. Jang, K.H. Son, S.U. Kim and Y.K. Kim, *Int. J. Immunopharmaco.*, 1998, **20**, 1.
67. B. Montaner and R. Perez-Tomas, *Life Sci.*, 2001, **68**, 2025.
68. C. Yamamoto, H. Takemoto, K. Kuno, D. Yamamoto, A. Tsubura, K. Kamata, H. Hirata, A. Yamamoto, H. Kano, T. Seki and K. Inoue, *Hepatology*, 1999, **30**, 894.
69. D. Yamamoto, Y. Uemura, K. Tanaka, K. Nakai, C. Yamamoto, H. Takemoto, K. Kamata, H. Hirata and K. Hioki, *Int. J. Cancer*, 2000, **88**, 121.
70. B. Montaner, S. Navarro, M. Pique, M. Vilaseca, M. Martinell, E. Giralt, J. Gil and R. Perez-Tomas, *Brit. J. Pharmacol.*, 2000, **131**, 585.

71. C. Diaz-Ruiz, B. Montaner and R. Perez-Tomas, *Histol. Histopathol.*, 2001, **16**, 415.
72. A. Treibs, M. Strell, I. Strell and D. Grimn, *Liebigs. Ann. Chem.*, 1978, 289.
73. M.S. Melvin, J.T. Tomlinson, G. Park, C.S. Day, G.R. Saluta, G.L. Kucera and R.A. Manderville, *Chem. Res. Toxicol.*, 2002, **15**, 734.
74. A. Fürstner, J. Grabowski and C.W. Lehmann, *J. Org. Chem.*, 1999, **64**, 8275.
75. M.S. Melvin, M.W. Calcutt, R.E. Noftle and R.A. Manderville, *Chem. Res. Toxicol.*, 2002, **15**, 742.
76. G. Park, J.T. Tomlinson, M.S. Melvin, M.W. Wright, C.S. Day and R.A. Manderville, *Org. Lett.*, 2003, **5**, 113.
77. A. Fürstner and E.J. Grabowski, *ChemBioChem*, 2001, **2**, 706.
78. R.D'Alessio, A. Bargiotti, O. Carlini, F. Colotta, M. Ferrari, P. Gnocchi, A. Isetta, N. Mongelli, P. Motta, A. Rossi, M. Rossi, M. Tibolla and E. Vanotti, *J. Med. Chem.*, 2000, **43**, 2557; A. Fürstner, J. Grabowski, C.W. Lehmann, T. Kataoka and K. Nagai, *ChemBioChem*, 2001, **2**, 60.
79. J.L. Sessler, L.R. Eller, W.-S. Cho, S. Nicolaou, A. Aguilar, J.T. Lee, V.M. Lynch and D.J. Magda, *Angew. Chem. Int. Ed.*, 2005, **44**, 5989.
80. W.-S. Cho, J.L. Sessler, L.R. Eller, G.D. Pantos and B.R. Peterson, *Abstracts of Papers, 225th ACS National Meeting, New Orleans, LA, United States, March 23–27*, 2003, ORGN.
81. J.L. Sessler, S.J. Weghorn, V. Lynch and K. Fransson, *J. Chem. Soc. Chem. Commun.*, 1994, 1289.
82. D. Magda and J.L. Sessler, unpublished results.
83. L. Naumovski, J. Ramos, M. Sirisawad, J. Chen, P. Thieman, P. Lecane, D. Magda, Z. Wang, C. Cortez, G. Boswell, D.G. Cho, J. Sessler and R. Miller, *Mol. Cancer Ther.*, 2005, **4**, 968.

CHAPTER 4

Neutral Non-Metallic Systems

4.1 Amide-Based Anion Receptors

It has long been appreciated that proteins can bind anions not only via interactions involving positively charged residues, but also through the neutral amide linkages that define their very essence. These latter subunits can serve as both hydrogen bond acceptors (oxygen lone pairs) and hydrogen bond donors (NH), with the latter generally being more important for anion recognition. For example, the X-ray crystal structure of the sulfate-binding protein of *Salmonella typhimurium* reveals that five among the seven identified hydrogen-bonding interactions involve amide NH groups of the polypeptide backbone.[1] Given the importance of amide-based binding motifs in Nature, it is not surprising that considerable attention has been devoted to developing synthetic anion receptors that rely on similar kinds of interactions. Highlights of this chemistry are reviewed in this chapter in the following order: "Acyclic Systems", "Cyclic Systems", "Calixarene and Steroid-Based Systems", and "Peptide-Based Receptors".

4.1.1 Acyclic Systems

In what is now recognized as being a seminal contribution to the anion recognition field, in 1997 Crabtree and co-workers[2] reported the synthesis and associated anion-binding studies of the non-preorganized neutral acyclic receptors **4.1a**–**4.1e**. In the context of this work, a single crystal X-ray diffraction analysis of the bromide anion complex of the bis-amide system **4.1a** was carried out; the resulting structure is shown in Figure 4.1.[2] It reveals the presence of two hydrogen bond interactions between two amide NH groups and the bound bromide anion (N···Br = 3.42 and 3.71 Å).

Due to the low solubility of bis-amide **4.1a** in CD_2Cl_2, it was not used for solution phase studies. Rather, such studies were carried out using analogues **4.1b**–**4.1e**. As the result of standard NMR spectroscopic titrations in CD_2Cl_2 and associated Job plots, it was determined that these receptors all bind anions with a 1:1 stoichiometry except for receptor **4.1d**, which displays a mixed 1:1 and 1:2 binding stoichiometry in the case of fluoride and acetate anions. Within the halide anion series, the anion affinities were found to be $Cl^- > Br^- > I^-$. It was also found that the rigid bis-amide receptor **4.1b** exhibits the highest chloride anion selectivity (K_a = 30,000, 61,000, 7100, 460, and

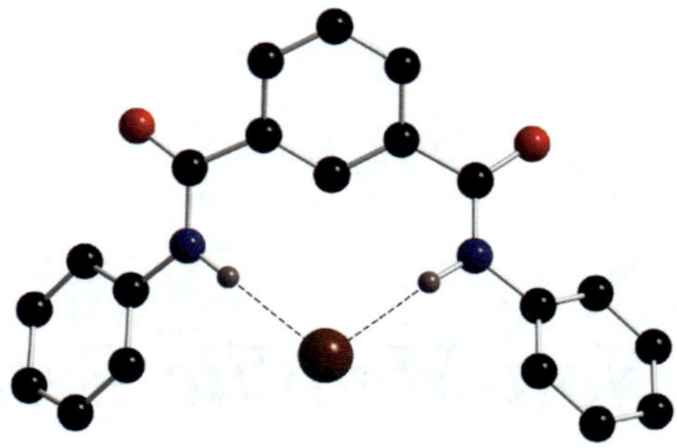

Figure 4.1 *Single crystal X-ray structure of the bromide anion complex of* ***4.1a***

19,800 M^{-1} for F^-, Cl^-, Br^-, I^-, and OAc^-, respectively).[2] By contrast, the more flexible bis-sulfonamide receptor **4.1d** displays a relatively high iodide anionbinding affinity, and hence a reduced selectivity (K_a = 20,000, 4600, and 1200 M^{-1} for Cl^-, Br^-, and I^-, respectively).[3] However, an increased selectivity in favour of smaller halide anions was observed in the case of receptor **4.1e**, presumably reflecting the rigidity of the hydrogen-bonding interactions between the pyridine nitrogen atom and the amide NH proton and the steric effect of the lone pair electron on nitrogen (K_a = 24,000, 1500, 57, and 525 M^{-1} for F^-, Cl^-, Br^-, and OAc^-, respectively).[3]

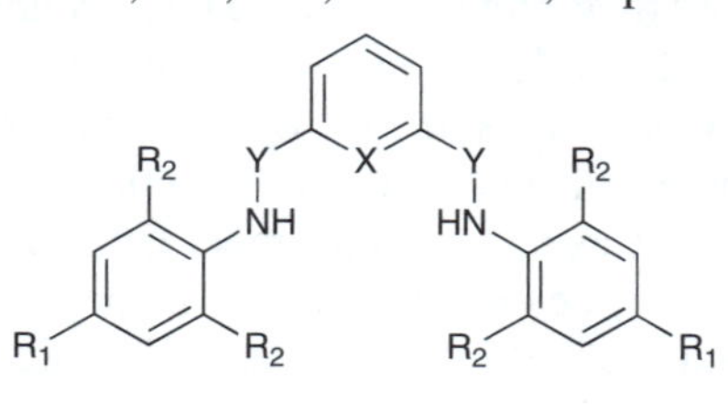

4.1a X = CH, Y = CO, $R_1 = R_2$ = H
4.1b X = CH, Y = CO, R_1 = *n*-Bu, R_2 = H
4.1c X = CH, Y = CO, $R_1 = R_2 = CH_3$
4.1d X = CH, Y = SO_2, $R_1 = R_2$ = H
4.1e X = N, Y = CO, $R_1 = R_2$ = H

Essentially contemporaneous with the above work, the B.D. Smith group[4] reported the synthesis of the structurally modified amide-based receptor **4.2**. This system displayed relatively enhanced carboxylate anion-binding affinity as the result of a cooperative polarization effect arising from the bound boronate group. As indicated by ^{1}H COSY and NOE difference experiments, compound **4.2** adopts a cleft-shaped conformation in DMSO-d_6, as shown in Scheme 4.1. In the presence of acetate anion,

4.2-DMSO **4.2**-OAc⁻

Scheme 4.1 *Proposed equilibrium changes observed in the presence of acetate anion*

receptor **4.2** forms a complex that is roughly 20 times stronger than that formed by **4.1a** (K_a = 2100 and 110 M^{-1} for **4.2** and **4.1a**, respectively, in DMSO-d_6). Recently, similar compounds were also prepared by the Gale group[5,6] in the context of their work on anion-controlled assembly (*cf.* Chapter 9).

The introduction of two extra amide units, as done in the case of receptors of generalized structure **4.3**, represents a different approach to generating improved receptor systems. This simple modification was found to improve the anion-binding affinity as well as the selectivity towards dihydrogen phosphate. All four amide NH groups in receptors **4.3a–4.3c** are thought to combine to form strong complexes with oxyanions, such as acetate and phosphate. The relevant affinity constants were determined from UV–Vis spectroscopic titrations. In acetonitrile containing 0.5% DMSO, both **4.3a** and **4.3b** were found to bind strongly, both acetate (*i.e.*, $K_a = 8.8 \times 10^3$ and 2.2×10^4 M^{-1} for **4.3a** and **4.3b**, respectively) and dihydrogen phosphate (*i.e.*, K_a = 9.6×10^3 and 2.6×10^4 M^{-1} for **4.3a** and **4.3b**, respectively), albeit with little discrimination between these two species.[7] Interestingly, receptors **4.3c** and **4.3d**, bearing one or two pyridyl groups, display a high preference for dihydrogen phosphate anion (*i.e.*, $K_a = 5.6 \times 10^5$ and $> 10^6$ M^{-1} for **4.3c** and **4.3d**, respectively) over acetate (*i.e.*, K_a = 1.8×10^4 and 1.7×10^4 M^{-1} for **4.3c** and **4.3d**, respectively). Presumably, this reflects the presence of ancillary stabilizing interactions between the two OH groups on $H_2PO_4^-$ and one or both of the pyridyl nitrogen atoms.

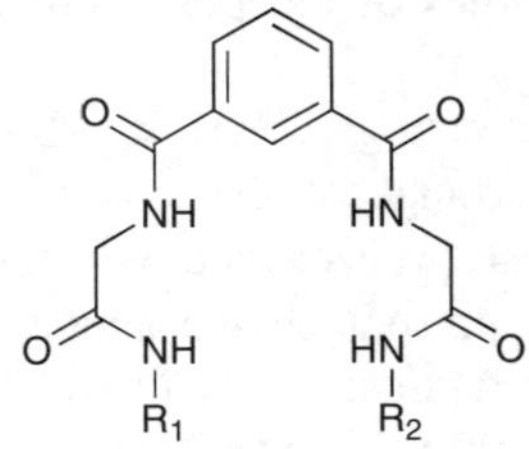

4.3a $R_1 = R_2$ = Bu
4.3b $R_1 = R_2$ = phenyl
4.3c R_1 = phenyl, R_2 = 2-pyridiyl
4.3d $R_1 = R_2$ = 2-pyridiyl

Recently, several other anion receptors that contain spacer-linked amide functional groups have been described, including systems **4.4–4.7**.[8–10] The first of these, the squaramido-based receptors **4.4a** and **4.4b**, contain two parallel hydrogen bonding donors and like various guanidium- and urea-containing systems (*cf.* Chapter 2 and Section 4.3, respectively), show considerable affinity for carboxylate-type anions.[8,11] For example, the association constants for acetate anion binding to **4.4a** and **4.4b** are 217 and 1980 M^{-1}, respectively, in DMSO-d_6.

Receptor **4.5** relies on a biimidazole moiety as its critical bis-amide bridging unit. As a consequence, it contains two additional hydrogen-bonding donor groups (the imidazole NH protons) than do the Crabtree systems **4.1**. Its anion-binding properties were studied by means of fluorescence titrations carried out in dichloromethane. On this basis it was determined that receptors **4.5a** and **4.5b** do not bind nitrate anion well but show strong affinity for dihydrogen phosphate and chloride anions.[9] However, these two systems display slightly different selectivities. For example, in dichloromethane receptor **4.5a** binds Cl^- ($K_a = 1.4 \times 10^5$ M^{-1}) more strongly than $H_2PO_4^-$ (K_a = 6.8×10^4 M^{-1}). On the other hand, receptor **4.5b** binds dihydrogen phosphate anion preferentially ($K_a = 4.6 \times 10^4$ and 3.7×10^4 M^{-1} for $H_2PO_4^-$ and Cl^-, respectively).

Inspired by Crabtree's work, Gale and co-workers[10,12] have developed a series of bis-amide anion receptors where the central bridging element is a pyrrole or pyrrole derivative. These systems are among the simplest and most cost effective of the many pyrrole-based anion receptors currently being developed. As such, they are discussed in Chapter 5, which deals with neutral pyrrole-containing anion receptors. Quite recently, as part of an ongoing effort to generalize this approach to anion recognition, the Gale group has reported the synthesis of the furan and thiophene-linked bis-amide systems **4.6a–4.6d**. In contrast to the corresponding 2,5-diamidopyrrole systems, receptors **4.6a** and **4.6b** do not favour the binding of oxo-anions, such as dihydrogen phosphate and carboxylates, presumably as the result of repulsive interactions between the oxygen atom on the furan and the charge-dense bound anion. On the other hand, both **4.6a** and **4.6b** were found to display fluoride anion selectivity, binding fluoride anion with relatively high affinity. For instance, based on NMR spectroscopic titration studies, the association constants for the interaction of fluoride anion with **4.6a** and **4.6b** were found to be 1140 and 557 M^{-1}, respectively, in DMSO-d_6/0.5% water. An X-ray crystallographic analysis (*cf.* Figure 4.2), reveals that the thiophene-based amide receptor **4.6d** forms a complex with fluoride in the solid state that is characterized by a 2:2 binding stoichiometry. This complex is stabilized via two different hydrogen-bond interactions involving the amide NH and thiophene CH protons. NMR spectroscopic titration experiments, carried out in DMSO-d_6/0.5% water solution by tracking the changes in the chemical shift of the thiophene CH proton, allowed an association constant of 82 M^{-1} to be calculated for the fluoride complex of **4.6c**, while similar studies of receptor **4.6d** revealed the presence of multiple equilibria.

The tetra-amide receptors **4.7a–4.7c** are based on a glycoluril backbone and provide four amide-derived hydrogen-bond donors arranged at the "corners" of the glycoluril unit. The anion-binding properties of these tweezer-type receptors could be, modified by changing the nitrogen substituent present on the amide "tails". For example, as judged by ^{1}H NMR spectroscopic titrations, both **4.7a** and **4.7b** display high affinities for fluoride (K_a = 2,300 and 36,000 M^{-1} for **4.7a** and **4.7b**, respectively, in CD_3CN) and acetate ($K_{a1} \times K_{a2}$ = 180,000 and 2,700,000 M^{-1} for **4.7a** and **4.7b**, respectively, in CD_3CN) over other halide anions and nitrate.[13] However, the highest association constant displayed by the naphthalene substituted receptor **4.7c**,

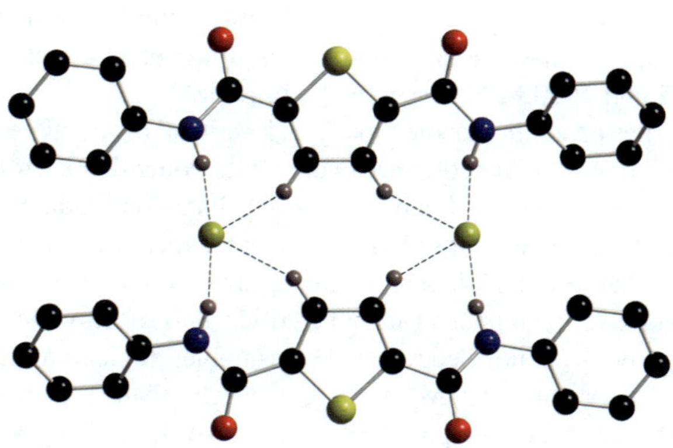

Figure 4.2 *Single crystal X-ray structure of* $(\mathbf{4.6d})_2 \bullet (F^-)_2$

120,000 M^{-1} in acetonitrile as determined using fluorescence, was recorded in the case of bromide anion.[14] The order of selectivity was found to be $Br^- > Cl^- > F^- > I^-$. The differing order of selectivity (as compared to **4.7a** and **4.7b**) and preference for bromide anion are attributed to the cavity size of receptor **4.7c**; presumably, this latter is especially suitable in terms of both size and shape for bromide in a way that is not true for the other anions studied.

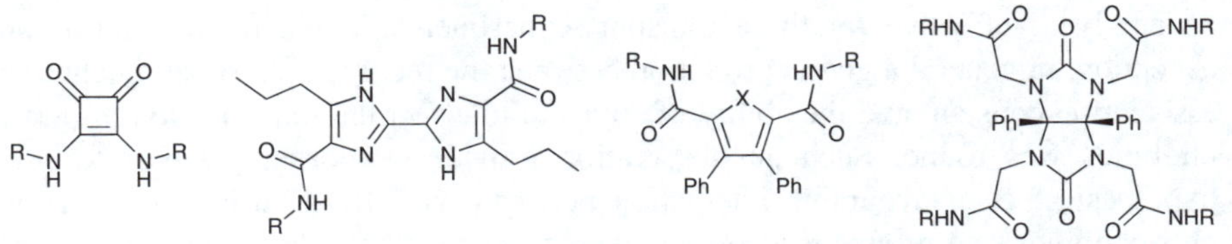

4.4a R = *t*-Bu
4.4b R = Bn

4.5a R = 3,5-difluorobenzyl
4.5b R = 4-methylbenzyl

4.6a X = O, R = *n*-Bu
4.6b X = O, R = Ph
4.6c X = S, R = *n*-Bu
4.6d X = S, R = Ph

4.7a R = *n*-Bu
4.7b R = 4-nitrophenyl
4.7c R = 2-naphthaly

In 1993, Reinhoudt and co-workers[15] first reported a series of neutral amide- and sulfonamide-containing anion receptors, **4.8a–4.8d**, **4.9a**, and **4.9b**, that are based on a tris(2-aminoethyl)amine (TREN) "core" (see Chapter 1). The crystal structure of receptor **4.9b**•H_2O was characterized by X-ray diffraction analysis (Figure 4.3). The C_3-symmetric three-dimensional arrangement seen in the solid state is stabilized by four hydrogen bonds involving three NH···O interactions and one SO···H linkage. This solid state structure provided support for the notion that these TREN-derived receptors would provide "anion-binding pockets" in solution that would lead to *bona fide* anion recognition under equilibrium conditions.

In what is more recent work, Bowman-James and Moyer reported a dual-host strategy for improving the extraction of $CsNO_3$ from aqueous media into an organic solvent; it is based on the use of the functionalized TREN receptor, **4.8b**, for nitrate anion recognition, in conjunction with the known Cs^+ cation complexing agent, tetrabenzo-24-crown-8.[16] This concept is discussed in more detail in Chapter 9.

In a separate line of investigation, Smith and co-workers[17,18] studied the sulfonamide and amide derivtives of TREN, **4.8e** and **4.9b–4.9e**, as new potential translocases that could bind to phosphatidylcholine and effect the through-membrane

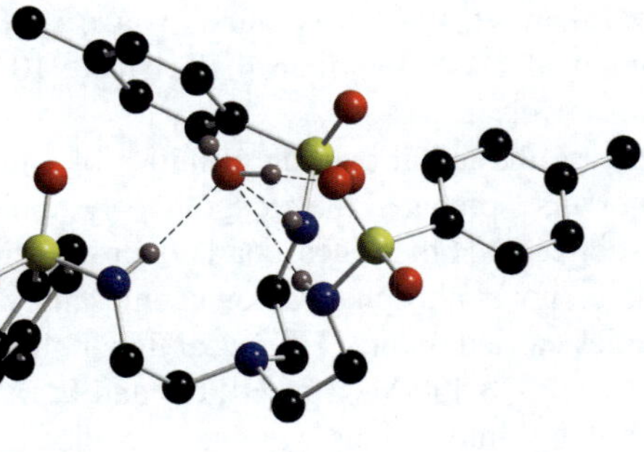

Figure 4.3 *Single crystal X-ray structure of* ***4.8b***•H_2O

transfer of a fluorescent phosphatidylchloline probe. As part of this study, it was found that these receptors bind 1-palmitoyl-2-oleoyl-*sn*-glycero-3-phosphocholine (POPC) in accord with the following affinity sequence, namely **4.9e** > **4.9d** > **4.9b** > **4.9c** > **4.8e**. In the corresponding translocation experiments, TREN **4.9b** was found to be more effective than **4.8e** at translocating the phosphatidylcholine probe across the membrane surface of differentiated vesicles.[18] Interestingly, in spite of displaying very different binding behaviour from **4.9b,** receptor **4.9c** was found to be nearly as effective for these transport experiments. While noteworthy as an exception, in general a good correlation between the recorded phosphatidylcholine association constant and the ability of an agent to effect the translocation of phospholipids was found. Such an observation provides important support for the hypothesis that a favourable interaction between the TREN derivatives and the phospholipid head groups leads to an enhancement in through-membrane translocation ability.

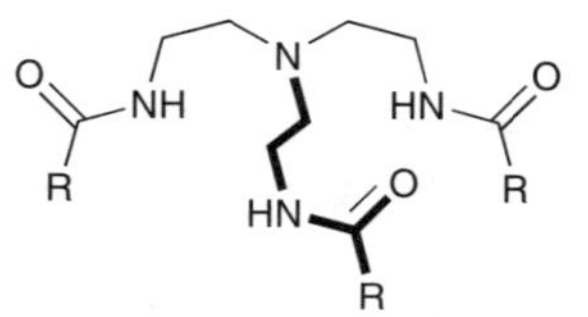

4.8a R = CH_2Cl
4.8b R = $(CH_2)_4CH_3$
4.8c R = C_6H_5
4.8d R = 4-$CH_3OC_6H_4$
4.8e R = 4-$CH_3C_6H_4$

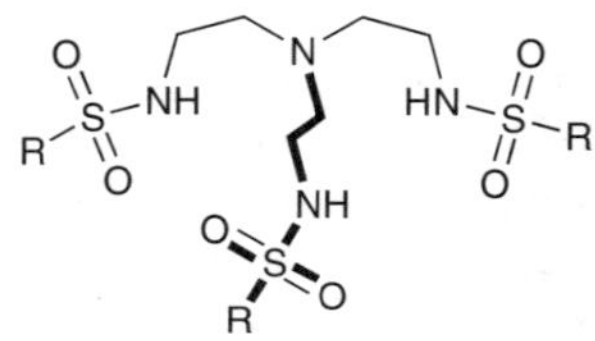

4.9a R = 2-naphthyl
4.9b R = 4-$CH_3C_6H_4$
4.9c R = $(CH_2)_2CH_3$
4.9d R = 4-$CH_3OC_6H_4$
4.9e R = 3-$CF_3C_6H_4$

A different three-fold symmetric core was used to produce the tripod shaped receptors **4.10a** and **4.10b.**[19] These two amide-rich systems were synthesized from the all-*cis* isomer of cyclohexane 1,3,5-tricarboxylic acid (Kemp's triacid) and two different amines. In the case of receptor **4.10b**, the resulting product possesses six potential hydrogen-bond donors and a number of potential hydrogen-bond acceptors. As a result, both phosphate-type anions and phosphonic acids are strongly bound by this receptor. For example, the association constant for propylphosphonic acid binding was found to be $1.6 \times 10^2\ M^{-1}$ in $CDCl_3$/10% CD_3OH, as determined by ^1H NMR spectroscopic titrations. By contrast, the corresponding association constant for receptor **4.10a**, which contains fewer hydrogen-bond donor sites, proved too weak to measure. Likewise, **4.10b** was found to have a substantially higher affinity for phosphate anion in DMSO-d_6 than **4.10a** ($K_a \geq 10^5$ and $1 \times 10^2\ M^{-1}$ for **4.10b** and **4.10a**, respectively).

In an effort to increase the anion-binding affinities of amide-based receptor systems, Fang and co-workers[20] prepared the wedge-like system **4.11**. This system contains multiple amide-derived NH hydrogen-bond donor moieties and was put forward by the authors as being a novel phosphate anion chemosensor. The results from fluorescence titration studies carried out in THF reveal that in this solvent the association constants are 186,300 and 328,400 M^{-1} for $H_2PO_4^-$ and PO_4^{3-}, respectively. By contrast, no significant binding interactions were seen in the case of Br^-, Cl^-, SCN^-,

ClO_4^-, or HSO_4^-, supporting the claim that system **4.11** could have a role to play as a selective anion sensor.

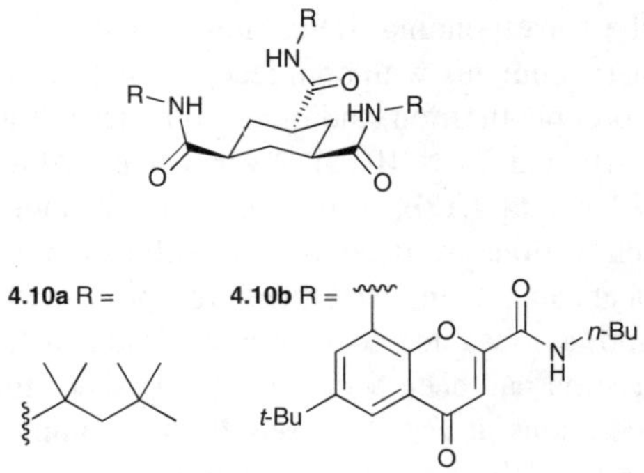

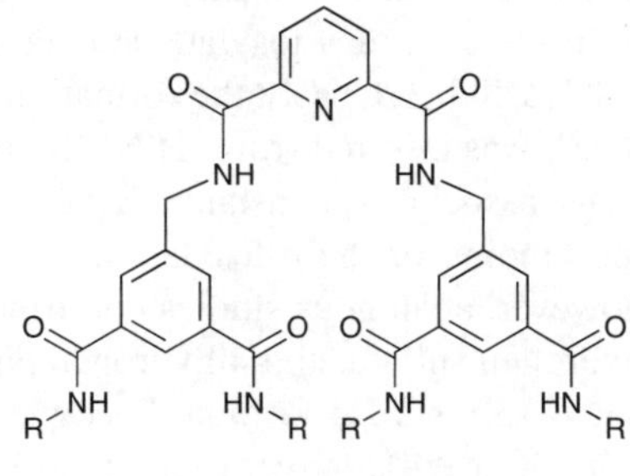

4.11 R = (1-pyrenyl)methyl

4.1.2 Cyclic Systems

Vancomycin, which is discussed in greater detail in Section 4.2 dealing with peptide-type receptor systems, is a naturally occurring cyclic polyamide that has been used extensively as an "antibiotic of last resort". It is believed to play a key role in preventing pathogenic bacteria cell-wall construction as a result of binding to the carboxylate motifs present in the L-Lys-D-Ala-D-Ala tripeptide sequences of the requisite cell-wall precursors. In 2000, Pieters reported the results of anion-binding study involving a synthetic analogue of the "carboxylate-binding pocket" of vancomycin, namely **4.12b** and **4.12c**.[21] In some respects, this study represents a continuation of the work of Hamilton and co-workers[22] who reported the association constant for the interaction of vancomycin **4.12a** with cyanoacetic acid in 1988 ($K_a = 580\ M^{-1}$ in $CDCl_3$). In accord with expectations, **4.12c**, which contains two more amide functional groups than **4.12b**, displays a higher *N*-Ac-D-Ala carboxylate-binding affinity than this latter system (in CH_3CN, $K_a = 9.7 \times 10^3$ and $2.3 \times 10^2\ M^{-1}$ for **4.12b** and **4.12c**, respectively).[21] Interestingly, **4.12b** was also found capable of binding acetate in CH_3CN, although the corresponding K_a value is roughly one order of magnitude lower than in the case of *N*-Ac-D-Ala.

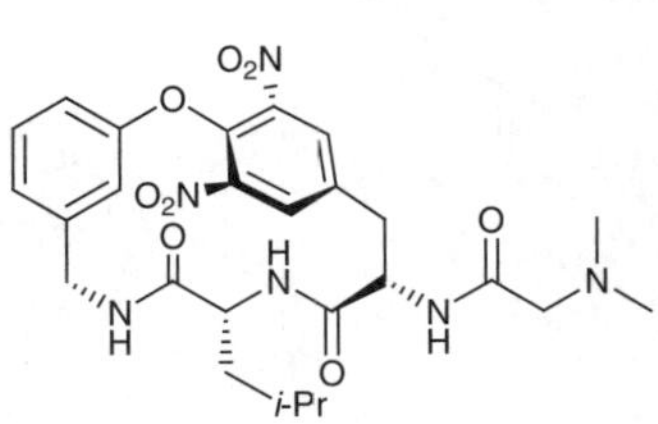

4.12a

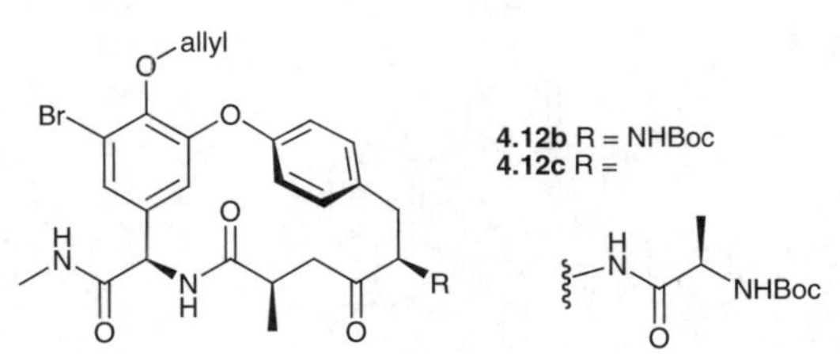

3:1 mixture of atropisomers

Recently, Hamilton and co-workers[23,24] reported the synthesis of a new class of amide-containing receptors, **4.13**. These rigid systems were designed to bind tetrahedral anions strongly. They all contain three amide NH groups constrained within the interior of a macrocyclic core that can act as potential hydrogen-bond donors. MM2 force field calculations provided support for the assumption that the cavity of receptors **4.13** would accommodate tetrahedral oxyanions of appropriate size. Experimental tests of this proposal were made via standard ^{1}H NMR and fluorescence spectral titrations.

In the case of *p*-tosylate, studied as the corresponding triethylammonium salt in 2% DMSO-d_6/$CDCl_3$, the formation of a 1:1 complex with both receptors **4.13a** and **4.13b** was inferred from ^{1}H NMR spectroscopic titration studies.[23] From these titrations, association constants (K_a) of 2.6×10^5 and 2.1×10^5 M^{-1} were calculated for the binding of the *p*-tosylate anion to **4.13a** and **4.13b**, respectively. Interestingly, however, analogous studies confirmed that various halide anions, as well as nitrate, hydrogen sulfate, and dihydrogen phosphate anions, interacted with receptors **4.13a** and **4.13b** in a 2:1 fashion (host:guest ratio) at low anion concentrations and at a 1:1 ratio after ≥ 0.5 eqv. of the anion in question had been added. On the basis of the observed stoichiometry and other considerations, it was proposed that the complex formed at low anion concentrations (relative to the receptor) exists in the form of a sandwich-like structure wherein the anion sits in between two host molecules.

In pure DMSO-d_6, only 1:1 binding behaviour was observed for receptor **4.13b**, while the association constants for various test halide anions, as well as nitrate and *p*-tosylate anions, were found to be dramatically decreased. On the other hand, high binding affinities for hydrogen sulfate and dihydrogen phosphate anions were observed, with the formation of 1:1 complexes being inferred ($K_a = 1.7 \times 10^3$ and 1.5×10^4 M^{-1} for HSO_4^- and $H_2PO_4^-$, respectively).[23]

System **4.13c**, which was designed to act as a fluorescence chemosensor, displayed a similar selectivity to that seen in the case of receptors **4.13a** and **4.13b**. In particular, a high inherent affinity for tetrahedral oxyanions was once again seen.[24] In this case, the association constants were measured by monitoring the changes in the maximum emission intensity observed following the addition of tetrabutylammonium-anion salts in DMSO/1,4-dioxane (1:1 mixture). The results of these fluorescence titrations yielded affinity constants of 2.0×10^6, 6.3×10^5, and 3.8×10^3 M^{-1} for dihydrogen phosphate, phenyl phosphate, and *p*-tosylate, respectively. In this case, similar to that of **4.13a** and **4.13b**, the selectivity for dihydrogen phosphate was considered to reflect the presence of three convergent amide NH hydrogen-bond donor subunits.

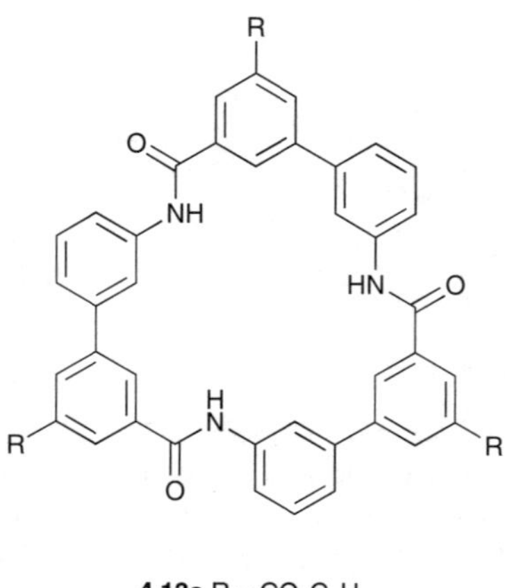

4.13a R = $CO_2C_2H_5$
4.13b R = NHBoc

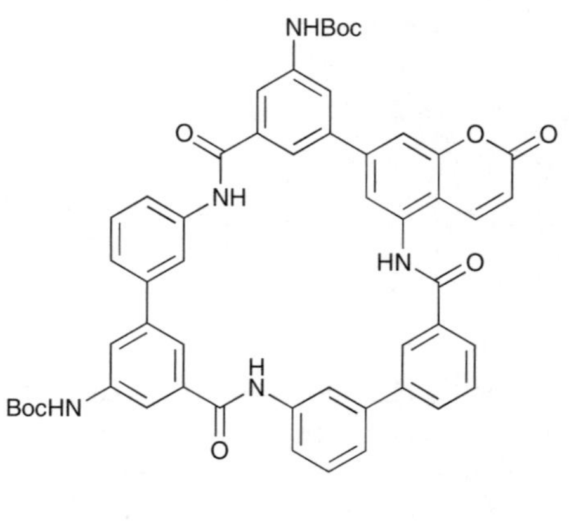

4.13c

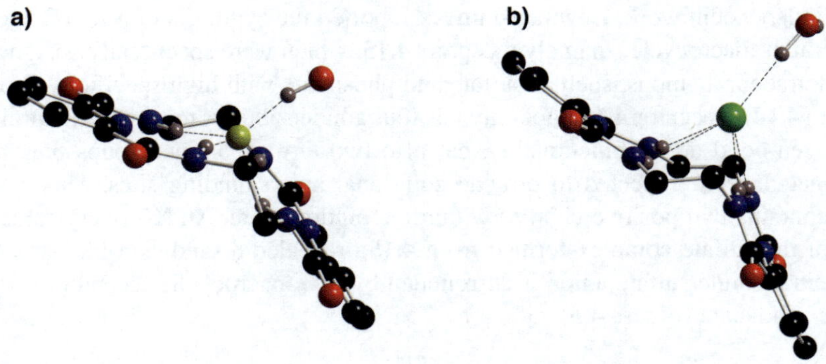

Figure 4.4 *(a) Single crystal X-ray structure of the fluoride anion and water molecule complex of tetralactam **4.14a** showing one-bound water molecule. (b) Single crystal X-ray structure of the chloride anion complex of tetralactam **4.14a***

In 2001, Jurczak[25] reported the synthesis of two neutral macrocyclic polylactam anion receptors, **4.14a** and **4.14b**, that were readily prepared from a 2,6-pyridinedicarboxylate precursor and 1,2-diaminoethane, respectively. The X-ray crystal structure of the fluoride and chloride complexes formed from tetralactam **4.14a** are shown in Figure 4.4; it supports the notion that the size of the cavity is too small to accommodate chloride anion easily, whereas it provides a good fit for the smaller fluoride anion. Specifically, the bound fluoride anion is held in place by four inward-pointing hydrogen bonds.

In order to examine the solution phase anion-binding properties of these receptors, ^{1}H NMR spectroscopic titrations were performed using various tetrabutylammonium-anion salts in DMSO-d_6. The results of these studies served to confirm that the largest anion-binding affinities and highest selectivities were seen for acetate and dihydrogen phosphate anions (K_a = 2640 and 1680 M^{-1}, for $CH_3CO_2^-$ and $H_2PO_4^-$, respectively), but also served to establish that the K_a for fluoride is higher than that of chloride anion (830 and 65 M^{-1} for F^- and Cl^-, respectively).

Quite recently, several tetraamide macrocycle derivatives, containing 2,5-diamidopyridine and 1,3-diamidobenzene subunits bearing but differing in the number of linking methylene "spacers", have been prepared and their anion binding properties have been studied.[26]

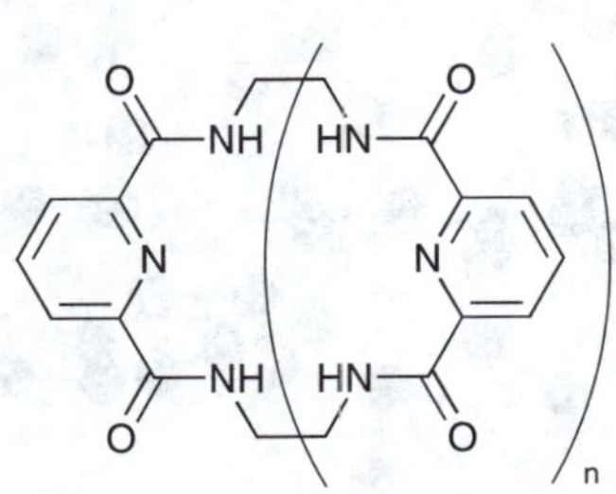

4.14a n = 1
4.14b n = 3

In independent work, Bowman-James[27] reported the synthesis of a set of ostensibly similar macrocycles, namely receptors **4.15**, which were specifically designed to bind tetrahedral anions, such as sulfate and phosphate with high specificity. In contrast to **4.14a**, receptor **4.15a** not only has four amide subunits to provide neutral NH hydrogen-bond donor functionality, but also two tertiary amine groups that, once protonated, were expected to provide additional anion binding sites. This system also contains two positive charges within the binding core. An X-ray crystal structure of the sulfate complex formed from **4.15a** revealed a sandwich-like structure, wherein a single sulfate anion is surrounded by two macrocyclic receptors, at least in the solid state (Figure 4.5).

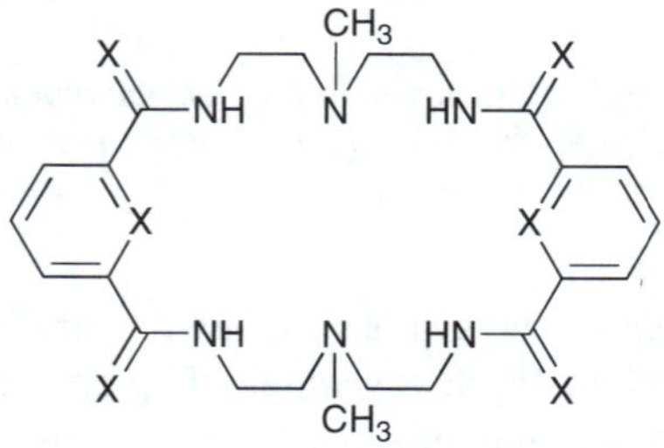

4.15a X = O, Y = CH
4.15b X = O, Y = N
4.15c X = S, Y = CH
4.15d X = S, Y = N

In 1997, Anslyn and co-workers[28] reported the results of studies involving an anion-binding polyamide cage-type receptor, **4.16**. Receptor **4.16**, the first cage-type polyamide receptor to be investigated since Pascal's initial studies of his trisamide cyclophane **1.12** (Chapter 1), contains a rather rigid framework that provides six

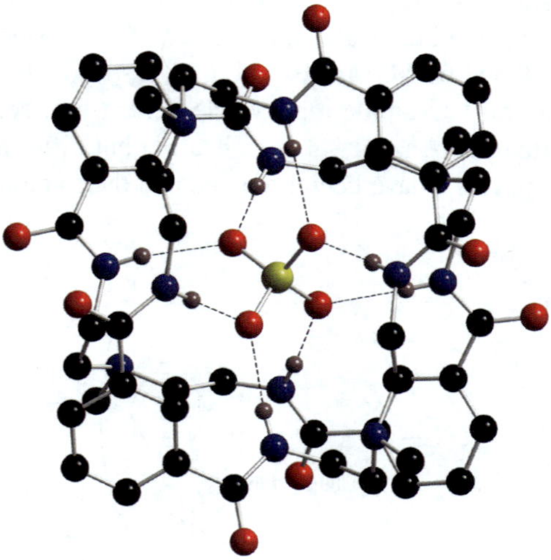

Figure 4.5 *Single crystal X-ray structure of the 1:2 (anion to receptor) solid-state sulfate anion complex formed from receptor* ***4.15a***

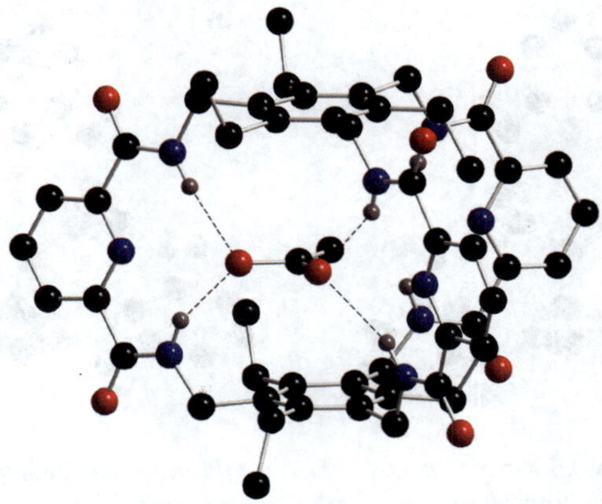

Figure 4.6 *Single crystal X-ray structure of the acetate-anion complex of receptor **4.16***

inward-pointing amide NH hydrogen-bond donor subunits. An X-ray crystal structure of the acetate complex formed from receptor **4.16** reveals a 1:1 binding stoichiometry in the solid state. It also shows that the acetate anion is encapsulated within the cavity via four hydrogen bonds (Figure 4.6). On the basis of standard ^1H NMR titrations, carried out in CD_2Cl_2/CD_3CN (1:3 v/v), affinity constants (K_a) were determined for $CH_3CO_2^-$, NO_3^-, Cl^-, and $H_2PO_4^-$, and were found to be 770, 300, 40, and 25 M^{-1} for these four anionic substrates, respectively.

Taken in concert, these results were considered to support the proposition that the amide functional group, when properly oriented, may be used to bind Y-shaped anions, such as acetate, effectively. Although, the highest binding constant was observed for acetate anion, a relatively high affinity for nitrate was observed, reflecting once again what was thought to be a good geometrical match.

Two years after the original report, the anion binding constants of receptor **4.16** were reinvestigated by monitoring the UV–Vis spectral changes induced upon the addition of anionic substrates under conditions of competitive indicator-displacement.[29] Association constants for NO_3^-, Br^-, and ClO_4^-, calculated in this way, were found to be 380, 220, 130 M^{-1} in CH_3OH/CH_2Cl_2 (1:1) and 500, 190, 70 M^{-1} in CH_3CN/CH_2Cl_2 (3:1), respectively.

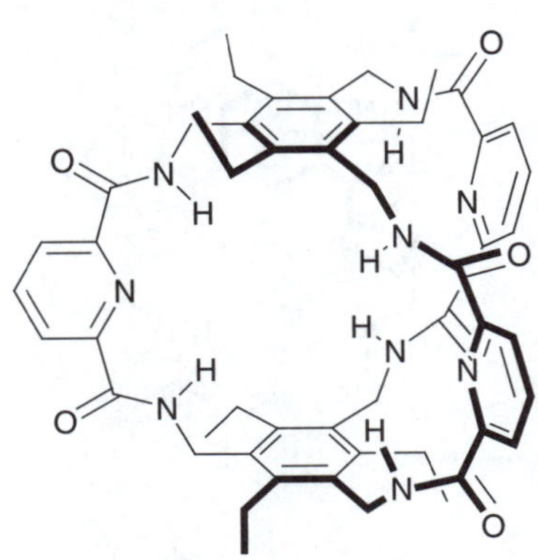

4.16

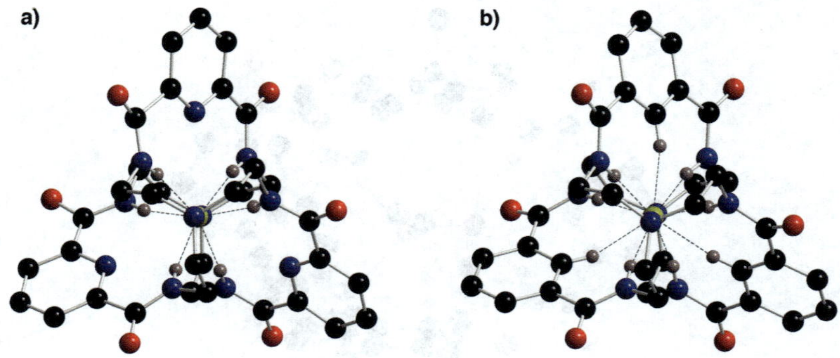

Figure 4.7 *Single crystal X-ray structures of (a) the fluoride-anion complex of receptor* ***4.17a*** *and (b) the fluoride-anion complex of receptor* ***4.17c***

Quite recently, a new type of polyamide cryptand-type receptor, represented by systems, **4.17a–4.17c**, was reported by Bowman-James and co-workers.[30–32] These systems have six amide (thioamide) NH donor subunits and two tertiary aliphatic amine groups. X-ray diffraction analysis revealed that the fluoride anion complex of receptor **4.17a** is stabilized by six hydrogen bonds (N···F = 2.84–2.89 Å) as shown in Figure 4.7a.[31] In the case of **4.17c**, three extra interactions between the central benzene CH proton and the bound fluoride anion are observed (C···F = 3.03–3.08 Å), along with six more normal hydrogen bonds involving the amide subunits and the bound fluoride anion (N···F = 2.94–3.11 Å) (Figure 4.7b).[32] The anion-binding properties of receptors **4.17a** and **4.17b** in DMSO-d_6 solution were measured via standard ^{1}H NMR spectral titrations using various tetrabutylammonium salts. The association constants (log K_a) calculated in this way were found to be: F^- (5.90), Cl^- (3.48), $H_2PO_4^-$ (3.31), HSO_4^- (1.83) Br^- (1.60), and I^- (1.30) for the amide cryptand **4.17a**, and F^- (4.50), Cl^- (1.54), $H_2PO_4^-$ (3.40), HSO_4^- (1.69) Br^- (< 1.0), and I^- (< 1.0) for the thioamide cryptand **4.17b**.[30,33] Both cryptand-type receptors **4.17a** and **4.17b** display higher fluoride anion affinity and lower chloride, acetate, and dihydrogen phosphate anion affinities than their 2-dimensional analogues, receptors **4.15a–4.15d**.

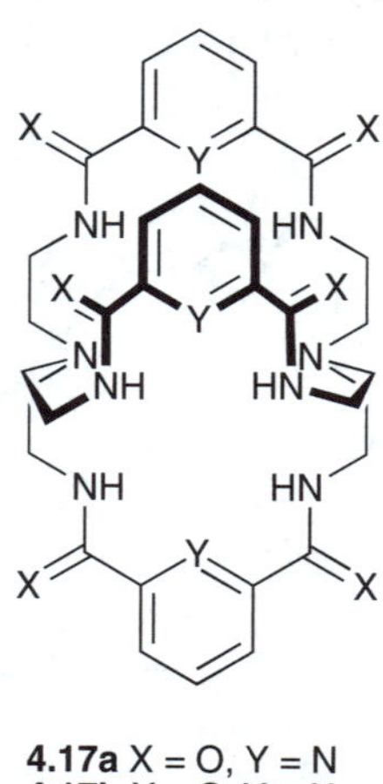

4.17a X = O, Y = N
4.17b X = S, Y = N
4.17c X = O, Y = CH

4.1.3 Calixarene and Steroid-Based Systems

Calixarenes and steroids have been widely used as backbones in the design of supramolecular receptors. This is because these cores can be easily modified and possess well-defined upper and lower faces. In work that builds on early efforts involving TREN-derived sulfonamide anion-binding agents (see above), Reinhoudt and co-workers[34] prepared the sulfonamide functionalized, calixarene-backbone receptor, **4.18**, in 1993. As reflected in the association constants measured via NMR titration methods in $CDCl_3$, receptors **4.18a–4.18c** were found to display considerable selectivity for HSO_4^- over $H_2PO_4^-$, Cl^-, NO_3^-, and ClO_4^-. Of this set, receptor **4.18c** was found to display the highest anion affinities (K_a), presumably as the result of the fact that it contains four "extra" amide groups (K_a = 103,400, 1,250, 513 M^{-1} for HSO_4^-, Cl^-, and NO_3^-, respectively).

In 1997, Cameron and Loeb[35] functionalized the 1,3-positions of the upper rim of a calix[4]arene with amide groups. These modified calixarenes, systems **4.19a** and **4.19b**, act as good receptors for Y-shaped anions, such as carboxylate and dicarboxylate anions. For example, the association constants, K_a, for the binding of $C_6H_5CO_2^-$, $CH_3CO_2^-$, HSO_4^-, $H_2PO_4^-$, and ReO_4^- to **4.19a** in $CDCl_3$ are 107, 88, 27, 22, and < 10 M^{-1}, respectively.

In an effort to increase the anion-binding affinity, a dichloromethyl group was attached to the terminal amide position. Such a modification was expected to increase the acidity of the amide NH protons thus enhancing the anion affinity. In point of fact, receptor **4.19b** shows an affinity for benzoate anion (K_a = 5160 M^{-1}) that is enhanced compared to that of **4.19a** (K_a = 107 M^{-1}, as noted above). These findings support the notion that the affinities and perhaps inherent selectivities of this type of anion receptor can be tuned through the use of inductive effects.

Calixarene **4.19b** also proved effective as a receptor for various simple dicarboxylate anions.[35] In the case of oxalate anion, a 1:1 binding stoichiometry was observed. By contrast, a 1:2 host:guest ratio was found in the case of larger dicarboxylate anions, such as isophthalate, terephthalate, and furamate. The measured association constants (K_a) were determined to be 707, $> 10^6$, $> 10^6$, and 10^6 M^{-1}, for oxalate, isophthalate, terephthalate, and fumarate anions, respectively.

Ungaro and co-workers[36] designed a series of strapped *C*-linked 1,3-dialanyl-calix[4]arenes, **4.20**. These systems contained built-in aromatic linkers to reduce the conformational flexibility. As might be expected for appropriately "preorganized" receptors, both **4.20a** and **4.20b** were found to display enhanced binding affinities towards carboxylate anions (K_a = 44,000 and 7,100 M^{-1}, respectively, for the benzoate and acetate anion complexes of **4.20a** and K_a = 40,100, 10,500, 2,800, and 200 M^{-1} for the benzoate, acetate, chloride, and nitrate anion complexes of **4.20b**, respectively; all in acetone-d_6). The fact that within this class of substrates, a selectivity towards benzoate anion was observed and rationalized in terms of π/π stacking interactions involving the aromatic ring of the strap and/or the aromatic subunits present in the calix[4]arenes themselves. Such ancillary effects would not, of course, be expected to play a role in terms of regulating the anion affinities of the other anions tested. Similar findings were also described by Huang and co-workers.[37]

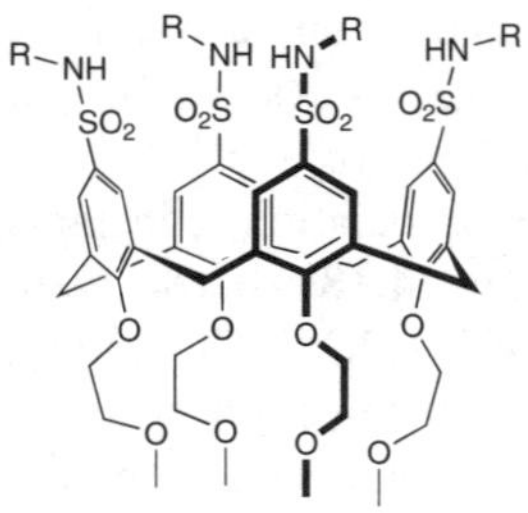

4.18a R = *n*-Pr
4.18b R = *t*-Bu
4.18c R = $CH_2CH_2NHCOCH_3$

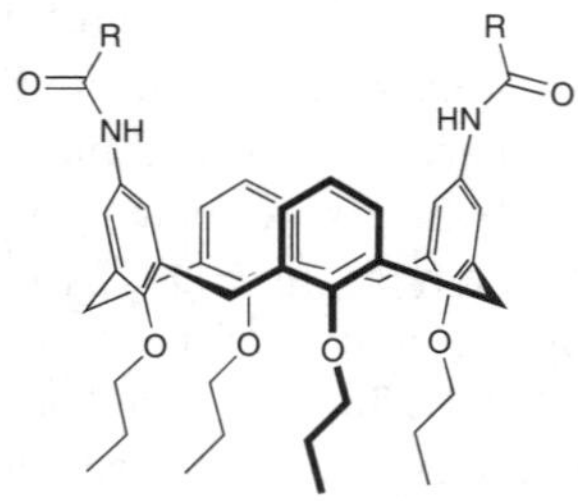

4.19a R = CH_2Cl
4.19b R = $CHCl_2$

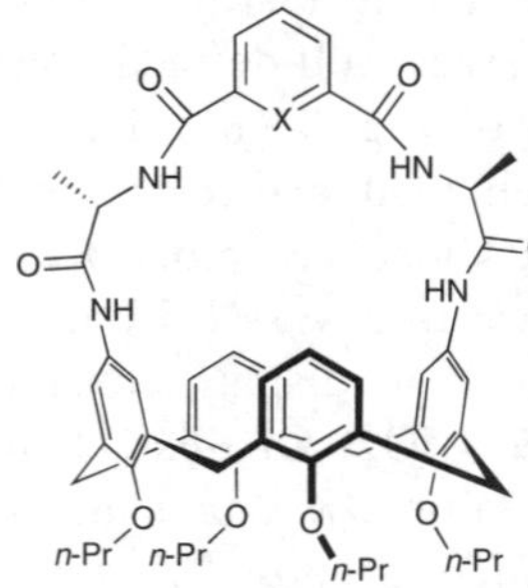

4.20a X = CH
4.20b X = N

Recently, the J.T. Davis' group[38] demonstrated the use of a calix[4]arene (**4.21b**) bearing tetrabutylamide groups to effect the transmembrane transport of H^+ and Cl^- through lipid bilayers. The corresponding methyl-functionalized analogue **4.21a** showed little ion transport activity, presumably due to insufficient hydrophobicity. However, it produced diffraction grade single crystals and the resulting X-ray structure, shown in Figure 4.8, revealed that this system exhibited a 1,3-alternate conformation with the four amide groups self-assembled through water molecules and chloride anions to produce an extended channel. The H^+/Cl^- transport ability of **4.21b**, as well as the open chain control systems **4.22a**–**4.22e**, was evaluated using a

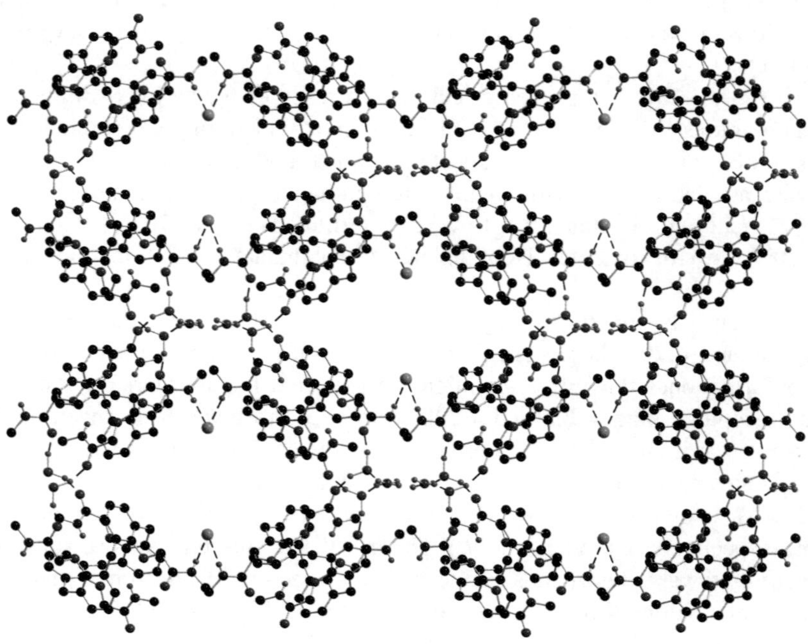

Figure 4.8 *Packing style representation of single crystal X-ray structure of* **4.21a**•HCl•$(H_2O)_3$

pH stat fluorescent assay. Interestingly, these studies revealed that it was trimer **4.22b**, not the calix[4]arene derivative **4.21b**, which proved most efficient for ion transport (5.8×10^{-2} s^{-1} at 5 μM). The overall trend was **4.22b** >> **4.22c** > **4.21b** > **4.22a** > **4.22d** ≈ **4.22e**. While it was proposed that the transport process involved so-called H^+/Cl^- "symport", the mechanism of chloride anion transport was not clearly determined.

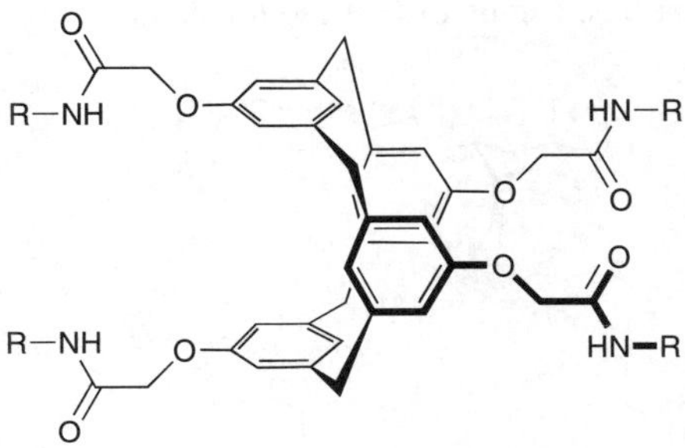

4.21a R = Me
4.21b R = *n*Bu

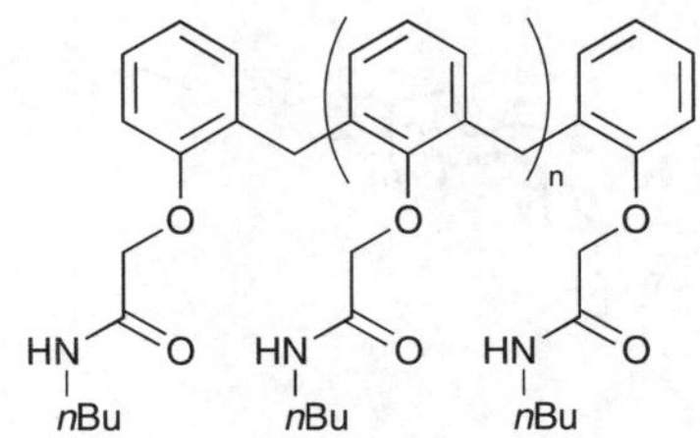

4.22a n = 0
4.22b n = 1
4.22c n = 2
4.22d n = 3
4.22e n = 4

The first steroidal-based tripodal anionophores, **4.23a** and **4.23b**, were developed by A.P. Davis and co-workers.[39] In later work, this group also reported a steroid-based cryptand that binds anions using four OH and two NH hydrogen bonding donor groups (*cf.* Section 4.5). In the case of **4.23a** and **4.23b**, ^{1}H NMR spectroscopic titration studies, carried out in $CDCl_3$, revealed a downfield shift in the NH peak upon the addition of selected anions, including fluoride, chloride, bromide, iodide, and tosylate anions as well as a 1:1 binding stoichiometry. Both tripods **4.23a** and **4.23b** provide well-defined binding sites for small-sized anions such as fluoride anion. For example, the association constants for the sulfonamido bis-carbamate **4.23a** are 15,400, 7200, 7200, 930, and 865 M^{-1} for fluoride, chloride, bromide, iodide, and tosylate anion, respectively. Unfortunately, the fluoride-anion affinity of the tris-sulfonamide **4.23b** proved too high to measure reliably using NMR spectroscopic titration methods. Nonetheless, it was found to show good selectivity for chloride anion relative to other anions included in the study (K_a = 92,000, 9,200, 525, and 950 M^{-1} for chloride, bromide, iodide, and tosylate anion, respectively). Presumably, the higher anion affinities displayed by the tris-sulfonamide **4.23b** reflect its more preorganized structure arising from the presence of internal S=O···HN interactions.

In order to overcome the sensitivity limitations inherent in NMR spectroscopic titrations, A.P. Davis and co-workers[40] adapted their systems such that they could be studied by Cram's extraction method. Towards this end, several new receptors, **4.24a–4.24c** were prepared in 2002. These systems contain eicosyl (C_{20}) side chains (attached via ester linkages) to prevent loss of the receptors into the aqueous phase. They also incorporated additional electron-withdrawing groups (*e.g.*, CF_3 and NO_2), which were expected to improve the hydrogen-bond donor potency and, as a consequence, the effective anion-binding affinity. The results from extraction experiments, carried out using chloride and bromide anions, revealed anion-binding affinities that were significantly improved compared to **4.23**, as would be expected given the modifications in structure. Surprisingly, however, the tris-*p*-nitrosulfonamide **4.24c** was found to bind anions less well (K_a = 3.2 × 10^6 and 1.1 × 10^6 M^{-1} for chloride and bromide anion, respectively) than the bis-carbamate receptors, **4.24a** and **4.24b**, with the latter displaying the highest affinity in the case of chloride and bromide (K_a = 3.2 × 10^7 and 2.9 × 10^7 M^{-1} for these two anions, respectively). These results highlight the fact that the anion-binding properties of ostensibly similar receptors can depend on subtle structural factors that, in the absence of detailed information from X-ray diffraction-based analyses or high-level calculations, can be difficult to understand.

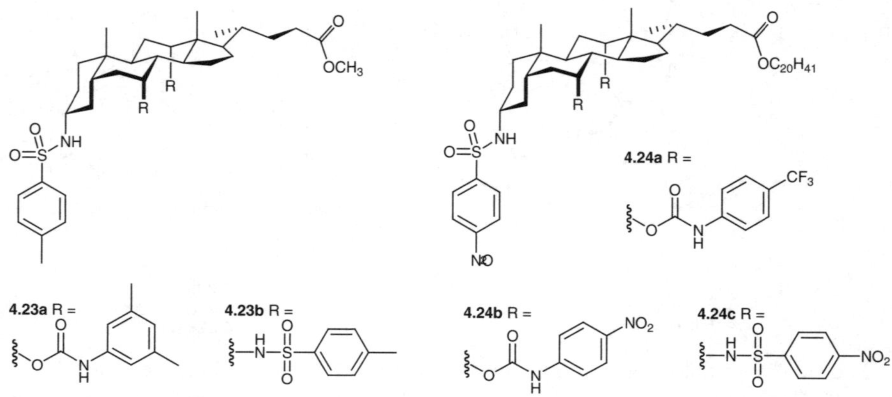

4.2 Peptide-Based Receptors

Recently, peptide-based systems have emerged as an important class of synthetic anion receptor. Much of this interest reflects the enormous importance that anionpeptide interactions play in Nature. In addition to the systems presented in Chapter 1, a nice example of these interactions is provided by vancomycin **4.25**.[41,42] Produced by the microorganism *Amycolatopsis orientalis*, vancomycin is often used as an antibiotic of last resort in the treatment of severe life-threatening infections. It was first isolated from a soil sample in the middle of 1950s, and in 1969, before its structure was established,[43] Perkins[41] suggested that vancomycin mediates its action by binding to bacterial cell walls that terminate in the sequence L-Lys-D-Ala-D-Ala. Once the structure of **4.25** was fully elucidated, the interactions of vancomycin with these bacterial mucopeptide precursor sequences were more fully analyzed via NMR spectroscopy.[44] The results of these studies, carried out with D-Ala-D-Ala-terminating peptides, led to the suggestion that three adjacent amide groups bind to the carboxylate anion of the antibiotic and later on, comprehensive studies of the binding of vancomycin with various peptide analogues were carried out by Nieto and Perkins[45] using UV–Vis spectroscopy. These workers found, for instance, that K_a = 1.5 × 10^6 M^{-1} for the complex of vancomycin with Ac_2-L-Lys-D-Ala-D-Ala in sodium citrate buffer solution at pH 5.1. The Burgen and Williams groups[46] independently investigated vancomycin-peptides complexes using ^{1}H and ^{13}C NMR spectroscopy, finding for instance that K_a = 1.4 × 10^4 M^{-1} for Ac-D-Ala-D-Ala in buffered aqueous solution at pD 5.5. Recently, the thermodynamics of these kinds of interactions have been probed in detail using microcalorimetric titration methods.[47]

Waltho and Williams[48] also discovered that chloroermomycin, **4.25b**, forms an asymmetric dimer in aqueous solution and that this species can bind two Ac_2-Lys-D-Ala-D-Ala groups, with each of these substrates being held in place by four amide-amide

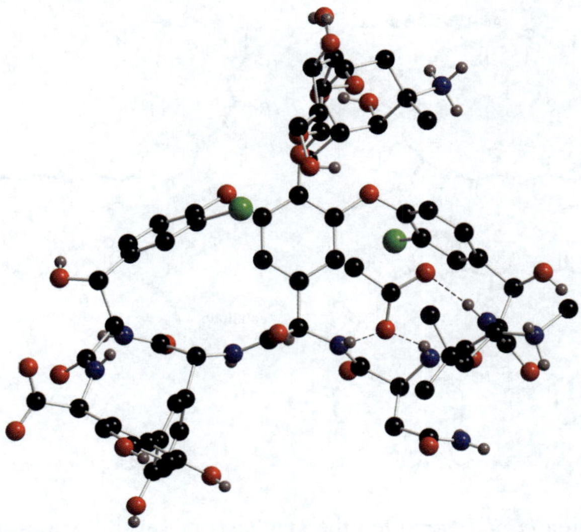

Figure 4.9 *Single crystal X-ray structure of the acetate-anion complex of monomer **4.25a** as observed in the solid state*

hydrogen bonds. Confirmatory support for this proposal was subsequently obtained from an X-ray crystallographic analysis of the acetate complex of **4.25a** (*cf.* Figure 4.9).[49] The resulting structure revealed that the acetate oxygen interacts with one part of the vancomycin homodimer via hydrogen bonds with the three target amide groups (O···H = 1.97, 2.08, and 2.07 Å). Quite recently, Williams' group[50] was able to establish an association constant (K_a) for the binding of two indole-2-carboxylic acid substrates to the dimer-**4.25b**; this was done by monitoring the peak shift of two NH protons (annotated as w and w*) in 9:1 H_2O:D_2O (pH = 6); this gave K_a and K_a^* values of 600 and 100 M^{-1}, respectively.

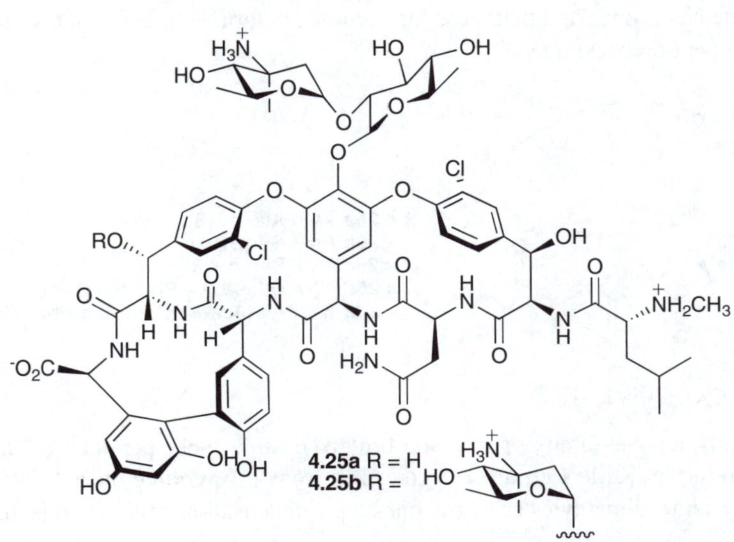

4.25a R = H
4.25b R =

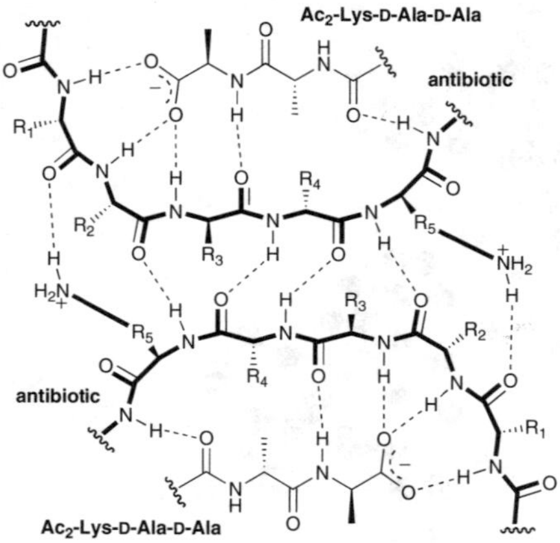

(**4.25b**)$_2$-(Ac$_2$-Lys-D-Ala-D-Ala)$_2$

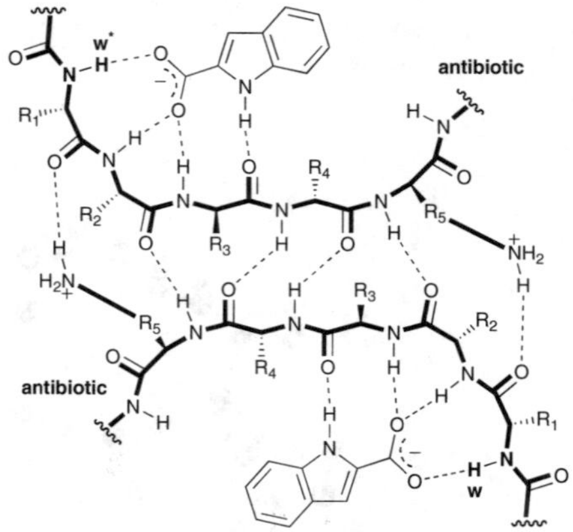

(**4.25b**)$_2$-(indole-carboxylate)$_2$

In 1995, Ishida *et al.*,[51] reported the synthesis of several neutral cyclic peptides *cyclo* (-AA-Aba-)$_n$ (*NB*: AA stands for amino acid and Aba stands for 3-amino-benzoic acid). These receptors, shown as structures **4.26a–4.26e**, include non-natural amino acid in their backbones. Nonetheless, they were found to be highly effective receptors for a 4-nitrophenylphosphate anion, a coloured substrate whose binding to **4.26a–4.26e** could be monitored by UV–Vis spectroscopy. Quantitative analysis revealed, as expected, that the association constants depended on the size of cyclopeptides, with the largest system, **4.26e**, showing the smallest value ($K_a = 6.8 \times 10^3\ M^{-1}$) in DMSO. By contrast, the smallest cyclic peptide, **4.26a**, was found to exhibit the largest binding constant ($K_a = 1.2 \times 10^6\ M^{-1}$) in this same solvent. Accompanying ^{1}H NMR spectroscopic studies revealed that the amide NH peaks are shifted from 8.47 and 10.23 ppm to 10.08 and 12.36 ppm on the addition of disodium 4-nitrophenylphosphate. On this basis, it was concluded that the phosphate oxyanion of 4-nitrophenylphosphate is bound to receptor **4.26a** via interactions with the amide NH groups and that these interactions, in turn, help define the structure of the cyclopeptide backbone.

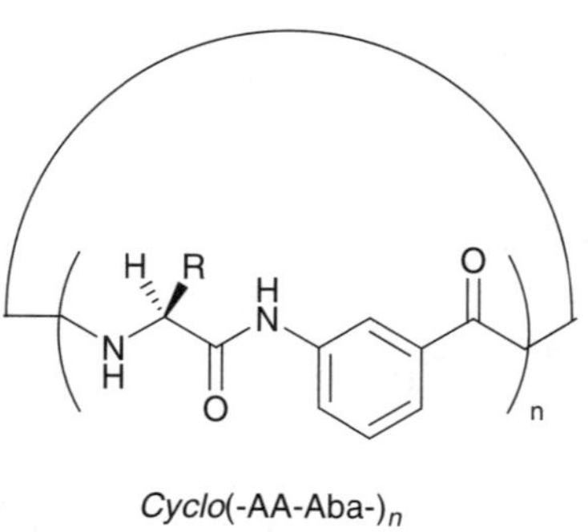

4.26a AA = Ala, n = 3
4.26b AA = Ser(Bzl), n = 3
4.26c AA = Ser, n = 3
4.26d AA = (Ala-Aba)$_2$-Phe-Aba-Ser(Bzl), n = 1
4.26e AA = Ala-Aba-Ser(Bzl)-Aba-Phe-Aba, n = 2

Cyclo(-AA-Aba-)$_n$

Recently, a novel family of aromatic-bridged cystine cyclic peptides **4.27a–4.27c** was reported by Karle's group.[52] Synthesized from 1,3-pyridine dicarbonylchloride and L-cysteine dimethylester via a one-step condensation process, this series of receptors contains a combination of disulfide bridges and aromatic units within its backbone and provides a matched set of well-defined homologues. In the case of the cyclic peptide **4.27a**, ^1H NMR spectroscopic titrations, carried out in $CDCl_3$, were used to assess the anion binding properties. It was found that the NH signals underwent a shift on the addition of ω-alkane dicarboxylate anion salts (*i.e.*, $(CH_2)_n(CO_2^-)_2$; n = 1–4). The maximum NH peak shift was observed in the case of the bis-tetrabutylammonium salt of glutaric acid, a dianionic substrate that was found to be bound with an association constant (K_a) of *ca.* 370 M^{-1}.

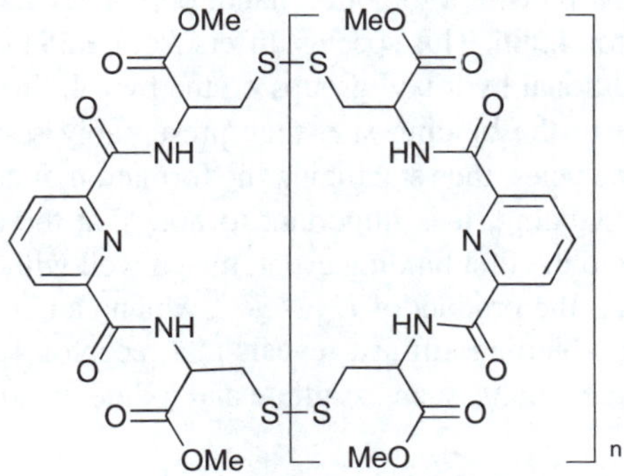

4.27a n = 1
4.27b n = 2
4.27c n = 3

It has long been a goal in the anion recognition community to develop neutral synthetic anion receptors that are effective in polar solvents, such as water. In recent years, Kubik and co-workers[53–57] have made good progress towards attaining this objective. Specifically, they succeeded in preparing a set of cyclic hexapetides, **4.28**, that contain L-proline and 6-aminoicolinoic acid subunits and determining their anion-binding affinities in aqueous media.[53,55] In the context of this work, these workers also determined the solid state structure of the 2:1 (host:guest) iodide complex. As shown in Figure 4.10, in the solid state, this complex exists in the form of

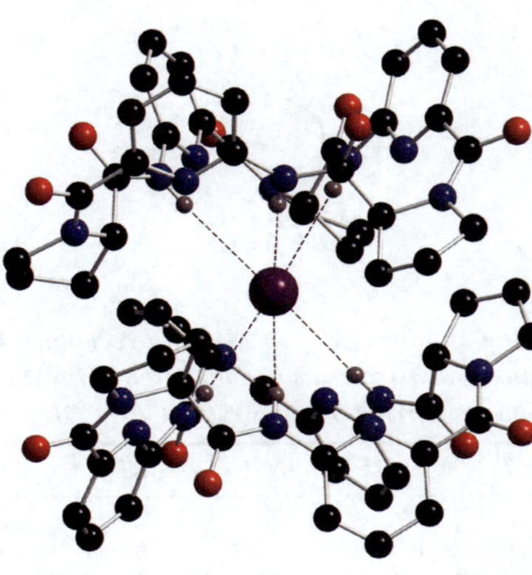

Figure 4.10 *Single crystal X-ray structure of the 1:2 (anion to receptor) iodide-anion complex stabilized by receptor **4.28a** in the solid state*

a sandwich-type structure wherein one iodide anion is bound between two cyclic peptides receptors **4.28a**; it is stabilized in this orientation by six hydrogen bonds involving the NH protons of the macrocyclic polyamide.[55]

The solution phase anion binding properties of **4.28a** and **4.28b** were also studied in aqueous media using ^1H NMR spectroscopic methods, monitoring in particular the shifts in the NH proton signals seen upon the addition of various anion-containing salts; the results are summarized in Table 4.1.[53] In the case of receptor **4.28a**, the 1:2 (anion:host) binding stoichiometry inferred from the X-ray structure of the iodide complex was confirmed by Job plots. On the other hand, only a 1:1 anion:host stoichiometry was seen for receptor **4.28b**. This species differs from **4.28a** only by virtue of the fact that it contains additional hydroxyl groups on the pyrrolidine rings. While these latter are not proximate to the binding site, they presumably serve to prevent aggregation between the macrocycles, thus stabilizing the formation of a 1:1 complex. Such considerations not withstanding, it is important to note that the K_{a1} values of receptor **4.28a**, corresponding to the first binding event, match well with the K_a values of receptor **4.28b**. Furthermore, the product of $K_{a1} \times K_{a2}$, which can be considered a reflection of the absolute anion-binding affinity, reveals that receptor **4.28a** is a very effective anion receptor for large anions, such as sulfate and iodide, in aqueous media.

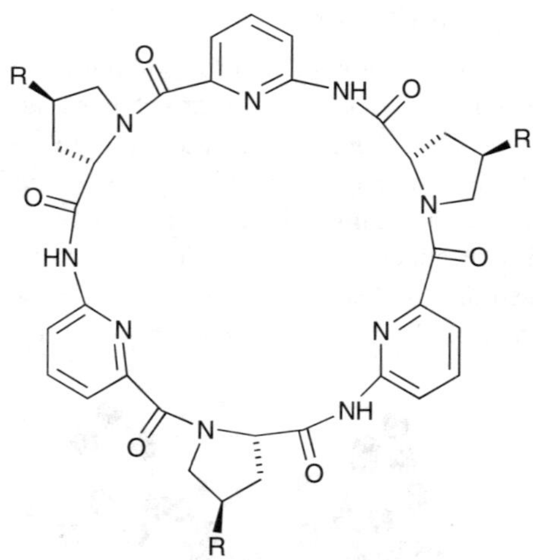

4.28a R = H
4.28b R = OH

Table 4.1 *Anion-binding affinities (M^{-1} or M^{-2}) for receptors **4.28a** and **4.28b** and selected anionic substrates as deduced from ^1H NMR spectroscopic titration experiments carried out in 80% D_2O/CD_3OD*

		NaCl	*NaBr*	*NaI*	*KI*	*$N(CH_3)_4I$*	*Na_2SO_4*
4.28a	K_{a1}	5	16	22	24	24	96
	K_{a2}	6 770	6 820	7 380	6 660	7 470	1 270
	$K_{a1} \times K_{a2}$	3.4×10^4	1.1×10^5	1.6×10^5	1.6×10^5	1.8×10^5	1.2×10^5
4.28b	K_a	8	13	19	–	–	95

Kubik has also prepared a set of dimeric receptors, **4.29**, that contain his basic cyclopeptide core linked by spacers of different lengths.[54,56] These systems were found to bind several large anions, including sulfate, bromide, and iodide. A 1:1 binding stoichiometry was inferred from electrospray ionization mass spectrometric studies in the gas phase and from Job plots constructed from solution phase ^{1}H NMR spectroscopic titrations; these latter provided specific support for the notion that receptor **4.29a** binds iodide anion with a 1:1 stoichiometry in 50% D_2O/CD_3OD solution. This 1:1 binding stoichiometry, in turn, led Kubik to suggest that the two cyclopeptide units present in **4.29a**–**4.29c** fold over to bind these anionic guests like a "molecular oyster" pinching a chemical "pearl".

A first set of quantitative binding studies, carried out using standard ^{1}H NMR titration methods, revealed that receptor **4.29a** has a high selectivity for sulfate anion in 50% D_2O/CD_3OD ($K_a = 3.5 \times 10^5$, 8.9×10^3, and 5.3×10^3 M^{-1} for Na_2SO_4, NaI, and NaBr, respectively).[54] One year after this initial report appeared, a second quantitative anion binding study was performed using ITC in 2:1 (v/v) acetonitrile/ water solution. Under these latter conditions, it was found that receptors **4.29b** and **4.29c** bind sulfate and iodide anions more strongly than receptor **4.29a** (by *ca.* one order magnitude). In the case of sulfate added as the potassium salt, the association constants (K_a) were found to be 2.0×10^5, 5.4×10^6, and 6.7×10^6 M^{-1} for receptors **4.29a**, **4.29b**, and **4.29c**, respectively.[56] The corresponding values for iodide also added as the potassium salt were roughly two orders of magnitude lower across the board but reflected the same trend in favour of the *m*-phenyl linked system, **4.29c** ($K_a = 3.3 \times 10^3$, 2.9×10^4, and 5.6×10^4 M^{-1} for receptors **4.29a**, **4.29b**, and **4.29c**, respectively).

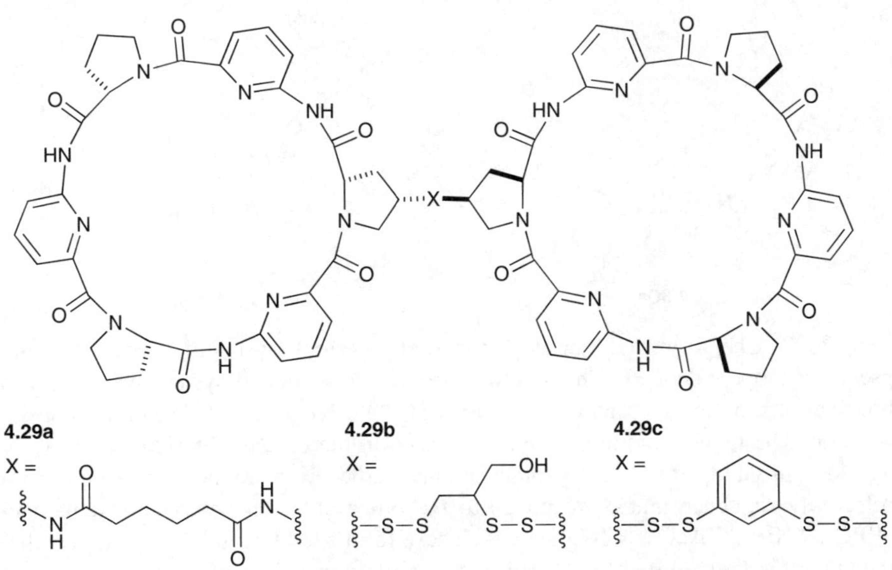

Two similar cyclic hexapeptides, namely **4.30a** and **4.30b** built up from D- and L-α-aminoxy acid building blocks, were also prepared and found to show promise as selective chloride anion receptors.[58,59] In this case, a conformational study, made using a model cyclic hexapeptide bearing methyl side chains, revealed that **4.30a** and

4.30b adopt a "bracelet-like" conformation and thus possess a rather small binding core ($d \leq 1.61$ Å), features that were considered good for binding small ions, such as halide anions.

The association constants for the binding of fluoride and chloride anions to **4.30a** were determined from ^{1}H NMR spectroscopic titration studies carried out in $CDCl_3$ at 223–273 K due to the observation of proton peak broadening at room temperature. The corresponding K_a values at 298 K were then calculated from van't Hoff plots. The results revealed a significant preference for chloride over fluoride anion (K_a = 11,880 and 30 M^{-1}, respectively).[58] Given the higher electronegativity of fluoride anion, this finding is noteworthy; it underscores the power of the basic design strategy, namely that a given anion (*e.g.*, chloride; ionic diameter, $d = 1.67$ Å or fluoride, $d = 1.19$ Å) can be bound with preference by a receptor that is appropriately sized (*e.g.*, **4.30a**; $d = 1.61$ Å), even when a simple consideration of electrostatic effects would argue against the observance of such selectivity. In the case of receptor **4.30b**, the association constants, as judged by ^{1}H NMR spectroscopic titrations carried out in CD_2Cl_2, revealed affinities for the halide anions and nitrate that correlated well with anion size (*i.e.*, K_a = 15,000, 910, 51, and 440 M^{-1} for Cl^-, Br^-, I^-, and NO_3^- at 298 K, respectively).[59]

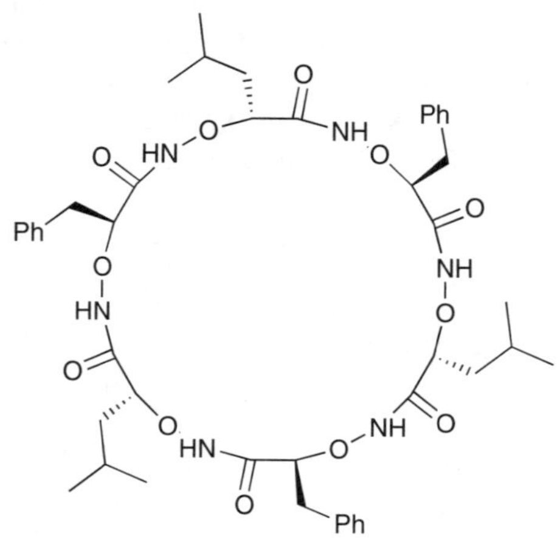

4.30a

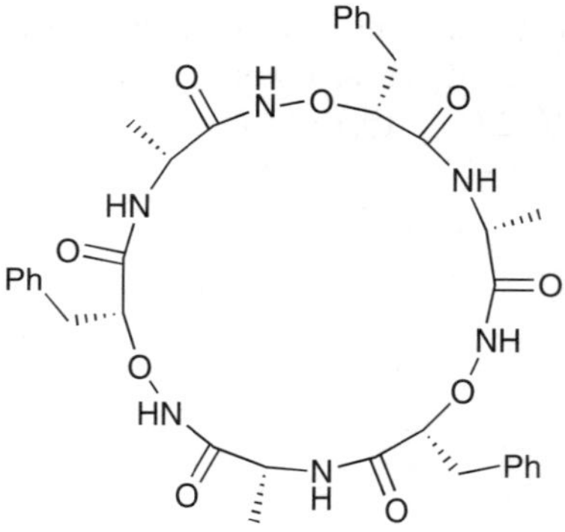

4.30b

In 2002, Cheng and co-workers[60] reported the synthesis of the cystine-bearing pseudo-cyclopeptide **4.31**. This conformationally constrained system was found to bind various anions such as F^-, Cl^-, Br^-, $H_2PO_4^-$, NO_3^-, and AcO^- in acetonitrile solution. The highest binding affinity was seen in the case of fluoride anion (K_a = 418 M^{-1} in CD_3CN), while the other anions studied were found to be bound in a selectivity sequence that corresponds to the order of anion polarizability, namely $H_2PO_4^- > Br^- > AcO^- > NO_3^- > Cl^-$. These results led the authors to suggest that the anion binding ability of **4.31** depends on the polarizability of the anions in question, as well as the strength of the hydrogen bonds these anions are likely to form.

In a continuation of this work, the polymethylene-bridged, cysteine-glycine-containing cyclopeptides **4.32a** and **4.32b** were prepared.[61] Quantitative studies, carried out in $CDCl_3$ using standard NMR spectroscopic methods, revealed that **4.32a** exhibits a preference for chloride anion over fluoride, bromide, or iodide anions (K_a = ca. 1000 M^{-1} for chloride anion). Presumably, the smaller ring size in **4.32a** compared to **4.31** (20- *vs.* 25-membered rings, respectively) plays a key role in establishing its noteworthy chloride anion selectivity.

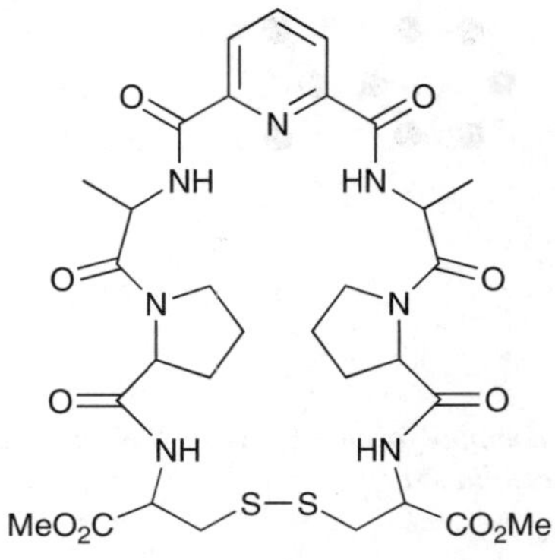

4.31

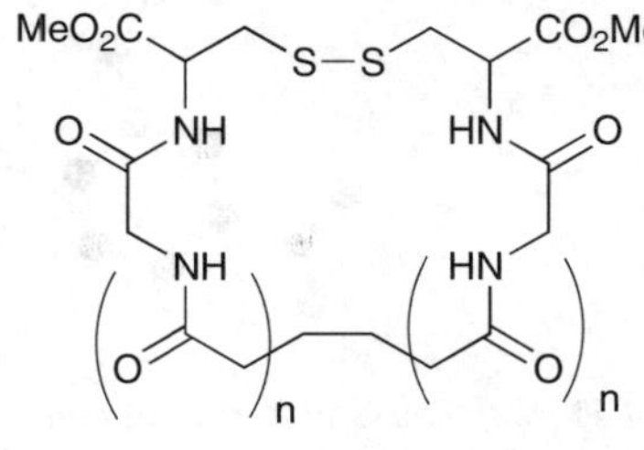

4.32a n = 1
4.32b n = 2

4.3 Urea-Based Anion Receptors

The urea and thiourea functional groups are among the most popular binding motifs currently being used to prepare neutral anion-binding receptors. They provide two hydrogen-bond donor groups that point in the same direction and which are spaced appropriately to interact with a range of anionic substrates, including Y-shaped ones such as carboxylates, nitrates, and phosphates. Further, both groups are relatively easy to prepare and can provide a means of linking together smaller molecular subunits. It is this combination of two-fold hydrogen-bond donor ability and synthetic accessibility that accounts in large measure for their importance and popularity as anion binding subunits.

4.3.1 Acyclic Systems

The hydrogen-bond-mediated-molecular recognition properties of the urea functionality were illustrated early on via solid state structural studies of compounds of type **4.33**.[62] It was found that the urea-containing receptor **4.33b**, bearing electron withdrawing groups on the *meta* position of phenyl ring, forms complexes in the solid state with neutral molecules containing oxygen donor atoms, including such species as butanone, THF, triphenylphosphine, DMSO, and nitroaniline, as judged by, *e.g.*, single crystal X-ray diffraction analysis. As illustrated in Figure 4.11, two hydrogen bond interactions stabilize the complex formed between receptor **4.33b** and *N, N'*-dimethyl-*p*-nitroaniline (N···O = 3.051 Å). This co-crystal structure is important in the context of anion recognition because it provides support for the notion that urea groups might be used to bind other hydrogen-bond acceptor species, including oxy anions in solution.

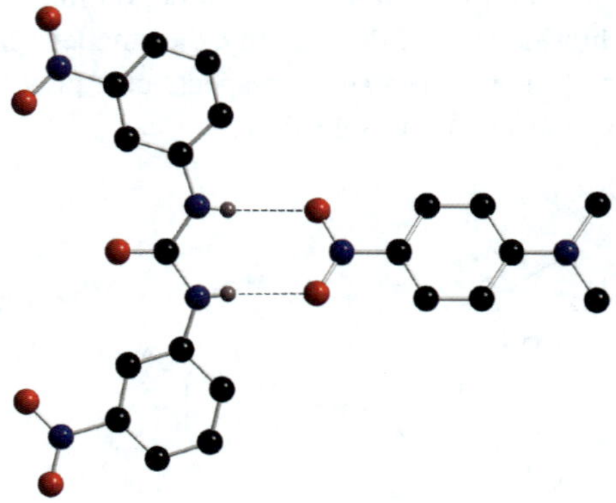

Figure 4.11 *Single crystal X-ray structure of the complex formed between N,N′-dimethyl-p-nitroaniline and receptor* ***4.33b*** *in the solid state*

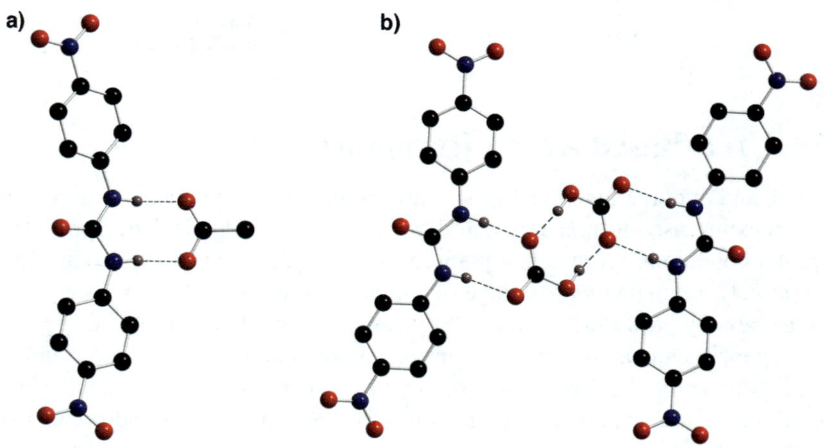

Figure 4.12 *Single crystal X-ray structures of (a) acetate anion complex with* ***4.33d*** *and (b) carbonate anion complex with* ***4.33d***

Quite recently, the Fabbrizzi group studied the interaction of **4.33d** with various anions. The nature of the complex with typical Y-shaped anions (*e.g.*, acetate and carbonate) was inferred from X-ray structural data.[63] In the specific case of acetate (*cf.* Figure 4.12a), the N–O bond lengths were found to be in the range of 2.69 to 2.77 Å. In acetonitrile solution, the addition of oxoanions was found to produce a bright yellow colour with a new band at ~370 nm appearing in the UV–Vis spectrum. Monitoring these spectral changes allowed the association constants to be determined quantitatively. The most stable complexes were formed with fluoride, acetate, and benzoate anions (log K_a = 7.38, 6.61, and 6.42, respectively, in CH_3CN). For the inferred 1:1 binding stiochiometry, the order of selectivity was thus found to be $F^- > CH_3CO_2^- > C_6H_5CO_2^- > H_2PO_4^- > Cl^- > NO_2^- > HSO_4^- > NO_3^-$. Interestingly, in the case of the fluoride anion, hydrogen-bond interactions with **4.33d** serve to stabilize a 1:1 complex at low relative fluoride anion concentrations. However, the addition of more fluoride anion was seen to effect deprotonation of the urea NH proton, resulting in the formation of HF_2^-. The deprotonated form of **4.33d** was characterized by a new band at 475 nm in the UV–Vis spectrum that was clearly present after the addition of 2 equiv. of fluoride anion. The acetonitrile solutions containing this deprotonated urea species were seen to take up CO_2 from air easily to produce a hydrogencarbonate complex, the structure of which was deduced from a single crystal X-ray diffraction analysis (Figure 4.12b). Further studies were reported by Fabbrizzi and Nam.[64] Deprotonation of other hydrogen bond donor receptors by fluoride had previously been observed by Gale[65] and Gunnlaugsson[66] as discussed later in this book. It is thus quite likely that analogous chemistry takes place in the case of other related systems, although it is not clear that it has always been identified as such.

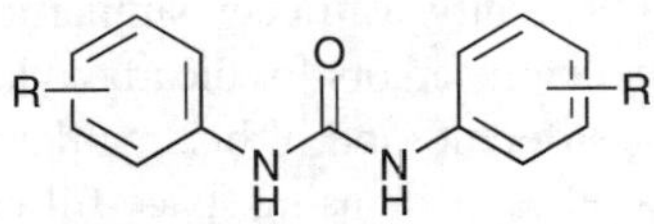

4.33a R = H
4.33b R = 2-NO_2
4.33c R = 3-NO_2
4.33d R = 4-NO_2
4.33e R = 4-OCH_3

In 1997, B.D. Smith and co-workers[4,67] described the preparation of the new boronate-urea receptors **4.35** and **4.36**. As the result of the polarization induced by the internally coordinated Lewis acid centre, these systems were expected to display enhanced carboxylate anion-binding affinities relative to the control, boron-free systems **4.34**. Quantitative studies of acetate-anion binding were carried out using standard NMR spectroscopic titrations in DMSO-d_6. These analyses confirmed that receptors **4.35b** and **4.36** did in fact display significantly enhanced affinity constants as compared to **4.35** (for acetate anion: $K_a = 6 \times 10^4$, $\sim 7 \times 10^3$, and $\sim 3 \times 10^2$ for **4.35b**, **4.36**, and **4.34**, respectively).[4,67] They also revealed that system **4.35b**, containing a coordinated BF_2 subunit, displayed the highest association constant. This is as expected given the strong electron withdrawing nature of the F substituents; it is also consistent with the general notion that boron coordination increases the hydrogen bond donor ability of the urea NH protons.

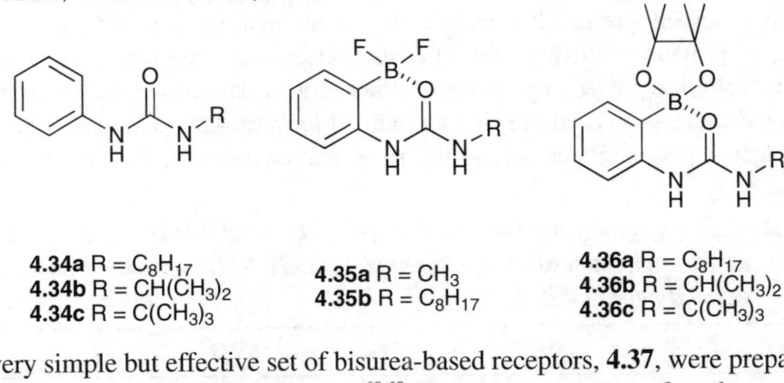

4.34a R = C_8H_{17}
4.34b R = $CH(CH_3)_2$
4.34c R = $C(CH_3)_3$

4.35a R = CH_3
4.35b R = C_8H_{17}

4.36a R = C_8H_{17}
4.36b R = $CH(CH_3)_2$
4.36c R = $C(CH_3)_3$

A very simple but effective set of bisurea-based receptors, **4.37**, were prepared and studied by Hamilton and co-workers.[68,69] These systems were found to exhibit a strong affinity for dicarboxylate anions, even in highly competitive solvents, such as DMSO-d_6. In the context of these studies, it was noted that the thiourea system, **4.37b**, was a more effective receptor than the corresponding urea-containing analogue, **4.37a**, binding glutarate roughly 15 times as well ($K_a = 6.4 \times 10^2$ and 1.0×10^4 M^{-1} for **4.37a** and **4.37b**, respectively, in DMSO-d_6).[68] Later on, more detailed binding studies were performed using standard ITC and NMR spectral titrations.[69] Proton NMR spectroscopic titrations, carried out over the temperature range of 20 to 50 °C, revealed that the anion-binding constants displayed by **4.37** increased as the temperature was lowered (*cf.* Table 4.2), a finding that presumably reflects the extra rotational and translational energy required to stabilize a receptor-anion complex at high temperature. However, slightly different association constants were obtained by calorimetry ($K_a = 1.3 \times 10^3$ and 8.4×10^3 M^{-1} in the case of **4.37a** and **b**, respectively, for glutarate binding in DMSO at 25 °C). Recently, a strategy similar to this one has been used by other research groups to generate receptors for dicarboxylate anions.[70]

In 1994, Kelly reported a different kind of bisurea-derived anion receptor, **4.38**.[71] On the basis of predicative studies made using space-filling molecular models, it was suggested that systems of type **4.38**, particularly system **4.38c**, would prove highly effective as receptors for dicarboxylate-type anions and other electron poor substrates, including *m*-dinitro-substituted benzene derivatives, such as TNT. Quantitative anion binding studies were carried out in DMSO-d_6 using ^{1}H NMR spectral titration methods and are found to be consistent with the author's predictions. In particular, it was found that the association constants (K_a) of **4.38c** for isophthalate, terephthalate, and *m*-dinitrobenzene (studied as the corresponding triethylammonium salts in the former two cases) were 63,000, 745, and 86 M^{-1}, respectively. The fact that receptor **4.38c** displayed a higher affinity towards isophthalate, relative to terephthalate, was taken as a further confirmation that geometry optimization plays a key role in regulating host-guest-binding interactions in systems of this type.

Somewhat prior to Kelly's report, Rebek and co-workers[72] described the synthesis and anion-binding properties of a series of asymmetric receptors of general structure **4.39**. In this case, an initial anion-binding study was performed using system **4.39a** in conjunction with standard NMR titration methods. From these analyses, it was concluded that this receptor binds benzoate anion with an association constant (K_a) of 2×10^5 M^{-1} in $CDCl_3$. The enantioselectivity of the recognition process was also investigated using the *S*-naproxate anion. Unfortunately, no appreciable enantioselectivity was observed, presumably because the chiral environments of **4.39a** (*R,R*) and (*S,S*) did not provide a sufficient level of enantioselective discrimination. However, this class of receptor has proved important in another context. This is because Umezawa and co-workers[73] have succeeded in using a modified form of **4.39** to prepare a new class of ion-selective polymeric membrane electrodes that are selective for chloride anion.

Table 4.2 *Glutarate anion-binding affinities (K_a, M^{-1}; ΔH, $kJmol^{-1}$; ΔS, $Jmol^{-1}K^{-1}$) at various temperatures determined via ^{1}H NMR spectroscopic titrations carried out in DMSO-d_6*

Host	*20 °C*	*30 °C*	*40 °C*	*50 °C*	*ΔH*	*ΔS*
4.37a	790	620	510	420	−16.8	+1.3
4.37b	11 000	8 500	6 300	5 200	−20.1	+8.4

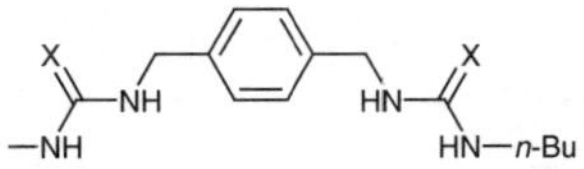

4.37a X = O
4.37b X = S

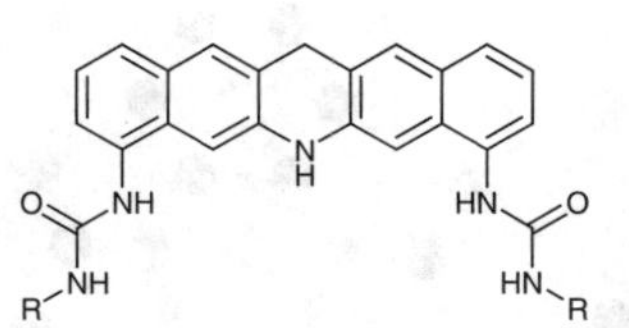

4.38a R = β-tritylethyl
4.38b R = *t*-Bu
4.38c R = *n*-Bu

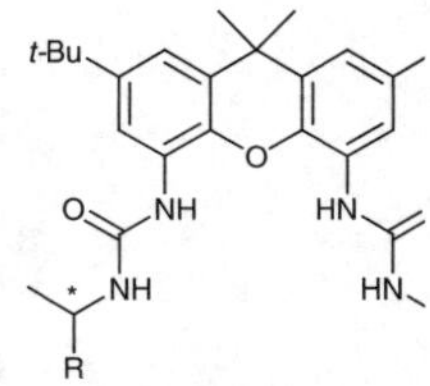

4.39a R = Ph
4.39b R = 1-naphthyl

The *m*-phenyl spaced, bis-urea/thiourea receptors **4.40** represent another important class of anion receptor that displays a strong selectivity for dihydrogen phosphate over other oxy-anions.[74] The selectivity of these receptors **4.40** is attributed to the directionality of the hydrogen-bonding interactions. However, it also reflects the relatively high basicity of the dihydrogen phosphate anion. Quantitative assessments of the binding behaviour were made using NMR spectroscopic titrations carried out in DMSO-d_6. On the basis of these studies, it was concluded that receptor **4.40c** has the highest association constant (K_a = 4600 M^{-1}), presumably as the result of the electron withdrawing phenyl group.[74] It was also found that the bisthiourea system **4.40b** binds dihydrogen phosphate anion more effectively in DMSO-d_6 than its bisurea analogue, **4.40a** (K_a = 820 and 110 M^{-1} for **4.40b** and **4.40a**, respectively).[74]

The anion-binding properties of the *o*-phenylenediamine-based receptor **4.41** were studied by Gale and co-workers.[75] The geometry of this receptor provides a space appropriate for binding carboxylate anions. As evidenced by X-ray crystallographic analysis of the benzoate complex, as shown in Figure 4.13, four hydrogen bond interactions between the urea NH protons, and the oxygen atoms of the bound carboxylate anion serve to stabilize the formation of these kinds of complexes. Consistent with the rather distortion-free binding seen in the solid state, high-association constants, recorded in DMSO-d_6/0.5% water, were obtained with two representative carboxylate anions (K_a = 3,210 and 1,330 M^{-1} for acetate and benzoate, respectively).

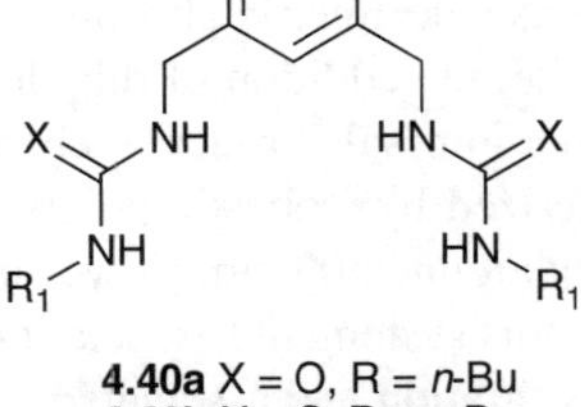

4.40a X = O, R = *n*-Bu
4.40b X = S, R = *n*-Bu
4.40c X = S, R = Ph

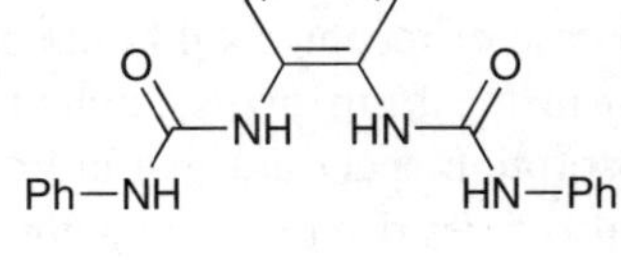

4.41

In order to improve the affinity of urea-based receptors for oxy-anions, systems **4.42** and **4.43**, containing three urea subunits, were synthesized.[19,76] These systems were prepared by condensing various tris-amine spacers with an isocyanate. In the case of receptor **4.42**, the association constant for phosphate-anion binding was found to be 1.1 × 10^3 M^{-1} in DMSO-d_6. However, receptor **4.43a**, containing a trisaminoethyl spacer, proved to be a better phosphate anion receptor than **4.42** (an association constant of 1.1 × 10^4 M^{-1} was recorded for **4.43a** in this same solvent).[19] Presumably, this reflects the fact that the use of the trisaminoethyl spacer provides a cavity that is better suited to accommodate the phosphate anion. More

Figure 4.13 *Single crystal X-ray structure of the benzoate complex of receptor **4.41***

recently, Werner and Schneider[76] studied the halide-anion binding behaviour of receptors **4.43c–4.43h**. Of these, the tripodal systems **4.43f** and **4.43g** were found to exhibit the highest chloride anion association constants DMSO-d_6 (K_a = 1859 and 1979 M^{-1}, respectively). They also exhibited good selectivity for this anion relative to the other anions studied (*e.g.*, bromide and iodide).

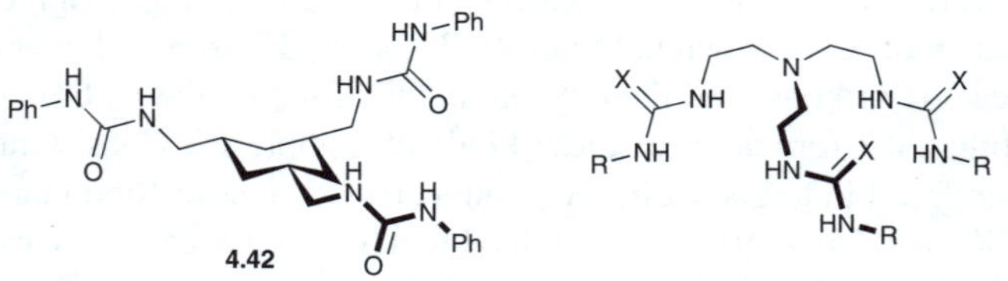

4.43a R = Ph, X = O
4.43b R = Ph, X = S
4.43c R = 4-ClC_6H_4, X = O
4.43d R = 1-naphthyl, X = O
4.43e R = 1-adamantyl, X = O
4.43f R = 4-$NO_2C_6H_4$, X = O
4.43g R = 4-$NO_2C_6H_4$, X = S
4.43h R = *n*-Bu, X = O

Suzuki and co-workers[77] have reported a different type of preorganized tripodal receptor, **4.44**, that draws on the 1,3,5-trisubstituted benzene scaffold popularized by Anslyn. These systems were designed to serve as colorimetric phosphate anion sensors. In the case of receptor **4.44a** in acetonitrile, the addition of dihydrogen phosphate anion in the form of its tetrabutylammonium salt produces changes in the UV–Vis absorption spectrum that are characterized by isobestic points at 310 and 346 nm; it also gives rise to a strong emission at around 500 nm. However, relatively poor selectivity was observed. In particular, for this system, the association constants in CH_3CN, as determined by standard UV–Vis titration and associated curve-fitting methods, were found to be 3.7×10^5 and 1.9×10^5 M^{-1} for dihydrogen phosphate and acetate anion, respectively. In the case of **4.44b**, a different spectral behaviour was seen. Here, addition of phosphate anion was found to produce a quenching of the fluorescence emission intensity, an effect that was attributed to an intra-complex photoinduced electron transfer process.

Recently, another C_{3v}-symmetric tripodal scaffold **4.45** was prepared by Ahn and co-workers[78] and its anion-binding properties were studied in DMSO-d_6 via NMR spectroscopic titrations. It was found that this system displays a higher affinity for dihydrogen phosphate anion (K_a = 521 M^{-1}) relative to the other anions studied (*e.g.*, nitrate, hydrogen sulfate, chloride, and bromide; K_a < 60 M^{-1} in all cases).

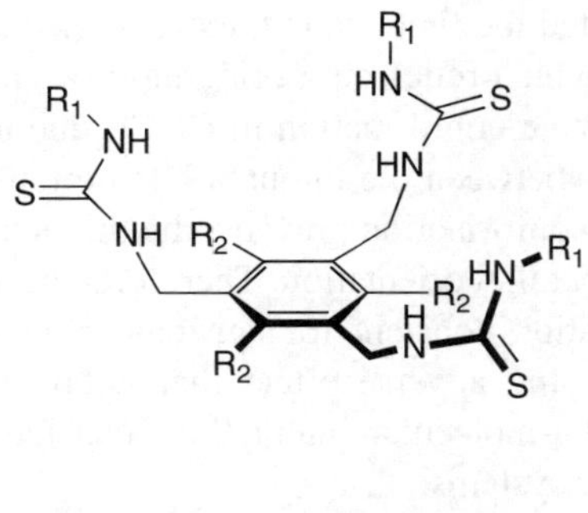

4.44a R_1 = pyren-1-yl, R_2 = *n*-Bu
4.44b R_1 = 9-anthrylmethyl, R_2 = *n*-Bu

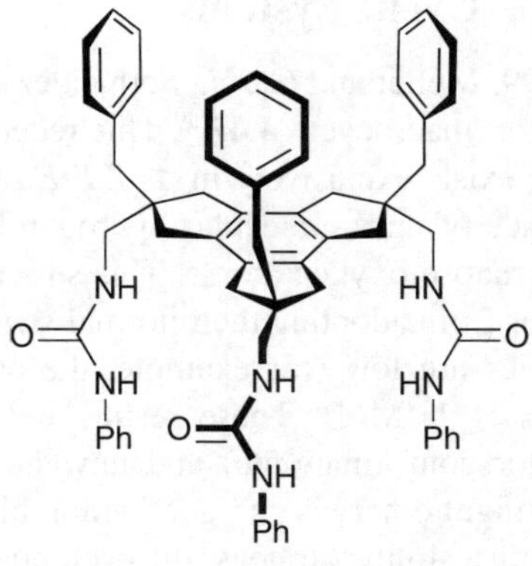

4.45

In another attempt to increase the anion-binding ability of urea-based neutral anion receptors, two open chain systems, **4.46a** and **4.46b**, having four thiourea units linked via two different spacers, namely *p*- and *m*-xylene, were synthesized by Mellet *et al.*[79] In this case, the addition of tetrabutylammonium glutarate caused the urea NH signals seen in the ^{1}H NMR spectrum to shift to the lower field. Such shifts are consistent with the formation of a hydrogen bonded complex involving a bound glutarate anion. In the case of receptor **4.46a**, the binding stoichiometry and association constants were determined from Job plots and from ^{1}H NMR spectroscopic titrations, respectively. Interestingly, it was found that glutarate anion was bound too well to allow the affinity constants to be measured accurately in either $CDCl_3$ or DMSO-d_6. Therefore, analogous titrations were carried out in a more polar medium (10% D_2O/DMSO-d_6). Under these latter conditions, an association constant of 10^3 M^{-1} was deduced for the 1:1 binding of glutarate by **4.46a**.

In contrast to what proved true for **4.46a**, the stoichiometry for glutarate binding by receptor **4.46b** was found to be 1:2 (host:guest), as induced from a Job plot analysis carried out in DMSO-d_6. Consistent with this stoichiometry, a ^{1}H NMR spectroscopic titration analysis revealed a two-step binding process with $K_{a1} = 10^5$ M^{-1} and $K_{a2} = 10^2$ M^{-1}. While the difference between **4.46a** and **4.46b** is notable, the fact that two ostensibly similar receptors display such disparate anion binding patterns providing support for the appealing notion that geometry optimization may be used to fine-tune the anion-recognition behaviour of synthetic receptor systems.

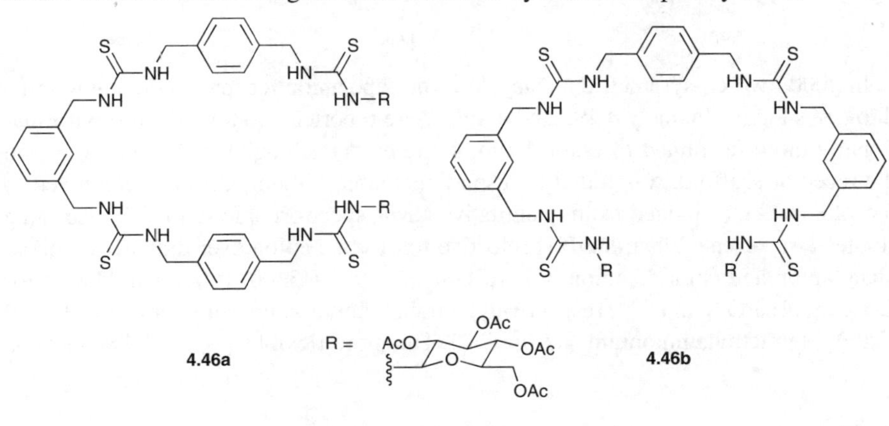

4.46a **4.46b**

4.3.2 Cyclic Systems

In 1999, Mellet and García Fernández reported the first cyclic thiourea-based anion receptor, macrocycle **4.47**.[79] This receptor, which relies on a bridging disaccharide motif, exists exclusively in its *Z,E/E,Z* alternate conformation in $CDCl_3$ due to the presence of intra-molecular hydrogen bonds between the thiourea NH protons and the pyranose oxygen atoms. This structural conformation prevents the thiourea NH protons from adopting their normal *syn* 1,3-parallel orientation. Therefore, the anion affinities are low (for example, the association constant for benzoate binding in $CDCl_3$ is 13 M^{-1}). These results indicated the adverse effect that unfavourable receptor conformation(s) and unwanted intra-molecular interactions can have on what might otherwise be good anion binding systems.

Using a similar strategy, the cyclophane-based cyclic thiourea receptors **4.48a** and **4.48b** were synthesized from the condensation of the appropriate diisothiocynate and diamine precursors under dilute conditions.[80] These cyclic receptors show significant selectivity towards dihydrogen phosphate over other oxyanions, such as acetate and hydrogen sulfate, and the halide anions chloride and bromide in DMSO-d_6. The actual association constants were measured via NMR spectroscopic titrations carried out at 60 °C due to the slow nature of the anion complexation equilibrium at 30 °C. It was found that the *ortho-meta* isomer **4.48a** displays larger association constants (K_a = 12,000, 2,200, 120, 19, and 12 M^{-1} for $H_2PO_4^-$, $CH_3CO_2^-$, Cl^-, HSO_4^-, Br^-, respectively) than does the *meta-meta* isomer **4.48b** (K_a = 2,500, 390, 14, 2, and <1 M^{-1} for $H_2PO_4^-$, $CH_3CO_2^-$, Cl^-, HSO_4^-, Br^-, respectively). In all probability, receptor **4.48a** enforces a distance between the two thiourea subunits that is more suitable for anion binding than does **4.48b**.

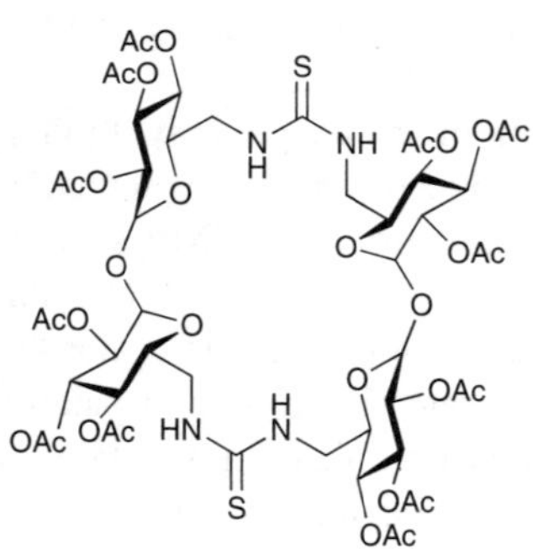

4.47

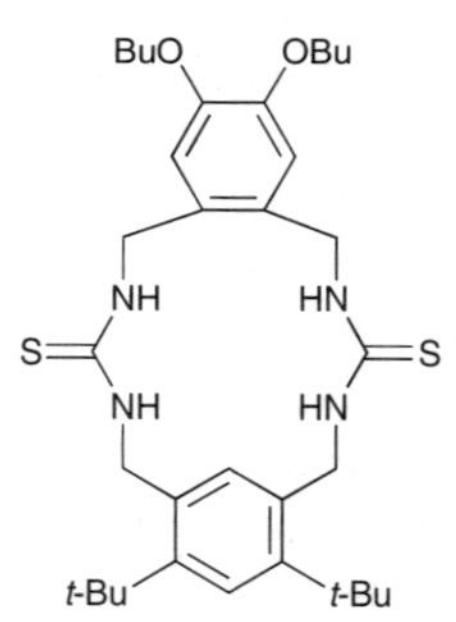

4.48a

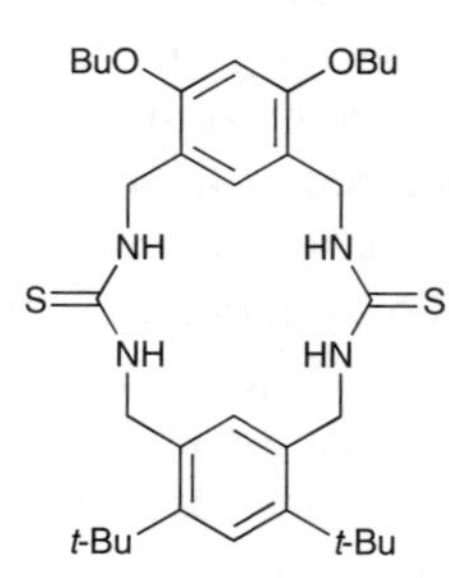

4.48b

In 2000, two C_3-symmetric metacyclophane-type anion receptors containing three thiourea subunits, namely **4.49a** and **4.49b**, were reported by Hong.[81] The conformationally more restrained receptor **4.49b**, made by "attaching" three ethyl groups on the receptor scaffold, was found to display increased binding affinities compared to receptor **4.49a**, as judged from quantitative NMR spectroscopic studies. These same studies also revealed that **4.49b** is selective for acetate anion over dihydrogen phosphate anion and chloride anion in DMSO-d_6 (*i.e.*, K_a = 5300, 1600, and 95 M^{-1} for $CH_3CO_2^-$, $H_2PO_4^-$, and Cl^-, respectively; tetrabutylammonium salts for $CH_3CO_2^-$ and $H_2PO_4^-$; tetraethylammonium salt for Cl^-). The more flexible system **4.49a** shows a different selectivity order and an overall lower affinity for anions with respect to **4.49b** (K_a = 800, 320, and 40 M^{-1} for $H_2PO_4^-$ $CH_3CO_2^-$, and Cl^-, respectively, in DMSO-d_6).

Slightly before the publication describing systems **4.49** appeared, the synthesis and properties of macrocycles **4.50a** and **4.50b**, containing four urea tethering subunits, were described by Reinhoudt and co-workers.[82] The anion-binding affinities of both receptors **4.50a** and **4.50b** were studied using standard ^{1}H NMR spectroscopic titration methods. In these studies, DMSO-d_6 was used as the solvent and the tetrabutylammonium-anion salts were used as the anion source. The more flexible receptor **4.50a** was found to bind dihydrogen phosphate and chloride anions with K_a values of 4000 and < 50 M^{-1}, respectively, and with a 1:1 stoichiometry. This system thus demonstrates a relatively high $H_2PO_4^-/Cl^-$ selectivity (K_{rel} > 100) that stands in contrast to what is seen in the case of the more rigid receptor **4.50b**. Here, K_a values of 2500 and 500 M^{-1} were recorded for dihydrogen phosphate and chloride anion, respectively. Presumably, these findings reflect the fact that **4.50b** is not particularly well optimized for dihydrogen phosphate binding and lacks the flexibility of its congener **4.50a** that would allow it to adjust its size and shape so as to accommodate better this relatively large anion.

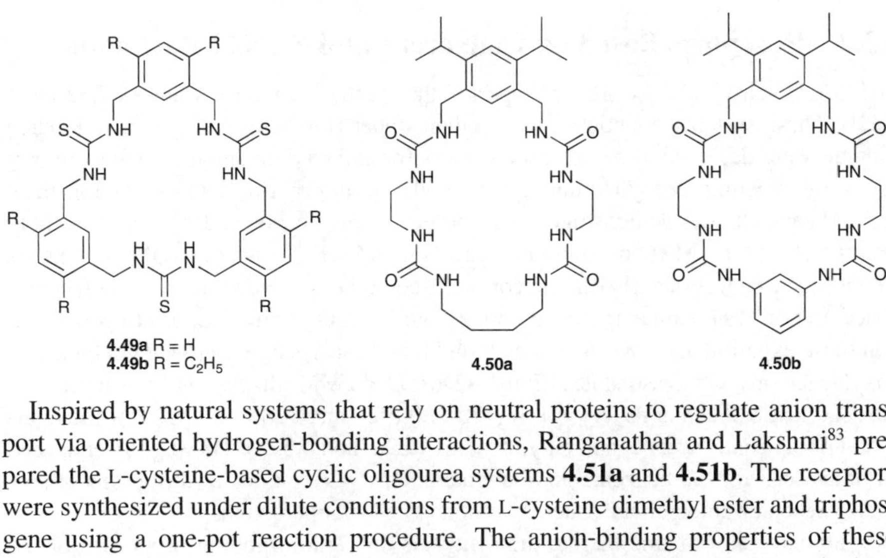

4.49a R = H
4.49b R = C_2H_5 **4.50a** **4.50b**

Inspired by natural systems that rely on neutral proteins to regulate anion transport via oriented hydrogen-bonding interactions, Ranganathan and Lakshmi[83] prepared the L-cysteine-based cyclic oligourea systems **4.51a** and **4.51b**. The receptors were synthesized under dilute conditions from L-cysteine dimethyl ester and triphosgene using a one-pot reaction procedure. The anion-binding properties of these highly symmetric receptors were then studied via standard NMR spectroscopic titrations carried out in $CDCl_3$. The macrocyclic triurea **4.51a** was found to bind chloride anion with high affinity and decent selectivity relative to the larger halide anions, a finding that was rationalized in terms of receptor–anion complementarity. Consistent with this rationale, receptor **4.51a** was also found to complex the nitrate anion, although this latter trigonal planar species was not bound as well as chloride anion (K_a = 2.05 × 10^3, 2.01 × 10^2, and 5.2 × 10^2 M^{-1} for Cl^-, Br^-, and NO_3^-, respectively).

In contrast to what was seen in the case of **4.51a**, the cyclic tetraurea **4.51b** was not found to bind chloride, bromide, or nitrate well. However, it was found to be a good receptor for the squarate dianion; this latter delocalized planar tetraoxyanion was found to be bound with a $K_a = 3.2 \times 10^3$ M^{-1} in $CDCl_3$.

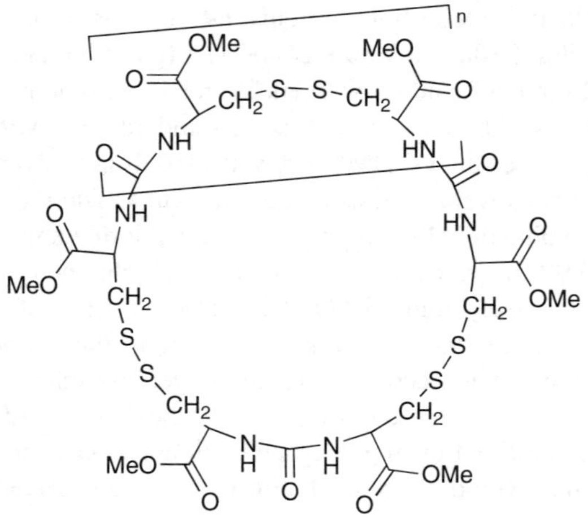

4.51a n = 1
4.51b n = 2

4.3.3 Receptors Based on Calixarene and Steroid Backbones

In 1996, Ungaro and co-workers[84] reported the synthesis of calix[4]arenes **4.52a** and **4.52b**. These systems, functionalized on their upper rims with a single urea/thiourea subunit, were designed to be neutral receptors for carboxylate anions. They were prepared by reacting phenyliso(thio)cynate with an upper rim amino-functionalized calix[4]arene. It was demonstrated that these systems do not undergo self-assembly over the 10^{-1}–10^{-3} M concentration range when DMSO-d_6 was used as the solvent. As a consequence, accurate association constants could be obtained provided the receptor concentration was maintained at an appropriate concentration. Taking into consideration these experimental caveats, it was found that these systems were able to bind carboxylate anions with reasonable affinity in DMSO-d_6, while displaying little affinity for the spherical halide anions (K_a = 118 (170), 80 (92), and 128 (339) M^{-1} for the interaction of benzoate, acetate, and butyrate with **4.52a** (and **4.52b**). The slightly enhanced affinities seen for the benzoate and butyrate anions were rationalized in terms of secondary receptor-anion effects, such as π–π and CH_3–π interactions.

In an extension of this theme, the synthesis and anion binding properties of the bis-urea functionalized calixarenes **4.53a** and **4.53b** were also investigated.[84] As proved true for the mono-functionalized receptors **4.52a** and **4.52b**, receptor **4.53a** displayed only a weak affinity for the halide anions, Cl^-, Br^-, and I^-, but was found to complex carboxylate anions well. In DMSO-d_6, **4.53a** displayed a preference for acetate anion (K_a = 2200 M^{-1}) over other carboxylate anions studied, including benzoate (K_a = 290 M^{-1}), butyrate (K_a = 133 M^{-1}), and phthalate (K_a = 200–300 M^{-1}), binding this simple carboxylate species far more effectively than its mono-functionalized analogue **4.52a**.

Recently, receptor **4.53c**, bearing urea recognition units connected directly to the calixarene-backbone, was reported by Lhoták and Stibor.[85] The association constants were measured via NMR spectroscopic titration using a solvent mixture consisting of 4:1 $CDCl_3$:CD_3CN. Under these conditions, receptor **4.53c** was found to bind benzoate anion strongly, displaying a K_a value for this species of 161,000 M^{-1}, while displaying a reduced affinity for both chloride (K_a = 4640 M^{-1}) and acetate (K_a = 3940 M^{-1}). The high preference for benzoate anion was explained in terms of favourable π–π interactions between the benzoate phenyl and calixarene phenyl units.

After discovering that tetra (thio) urea-resorc[4]arenes can bind halide anions strongly in $CDCl_3$ solution, Ungaro and co-workers[86] prepared the tetrafunctionalized calix[4]arenes **4.54**. While receptor **4.54a** possesses a degree of conformational flexibility, receptors **4.54b** and **4.54c** are more rigid and constrained to the corresponding cone conformation due to the presence of a bis-crown ether unit in the lower rim. On the basis of anion-binding studies carried out in DMSO-d_6 using NMR titrations, it was concluded that the more rigid receptor **4.54b** displays relatively high affinities towards spherical anions, such as chloride (K_a = 120 M^{-1}) and bromide (K_a = 30 M^{-1}). By contrast, the less rigid receptor **4.54a** exhibits a preference for dihydrogen phosphate anion (K_a = 300 M^{-1}). Interestingly, receptor **4.54c** was found to bind only chloride anion and then only weakly among the halide anions (K_a = 35 M^{-1} in DMSO-d_6). For examples of calixarene receptor systems functionalized on the upper rim with a single urea/thiourea subunit, see references.[87]

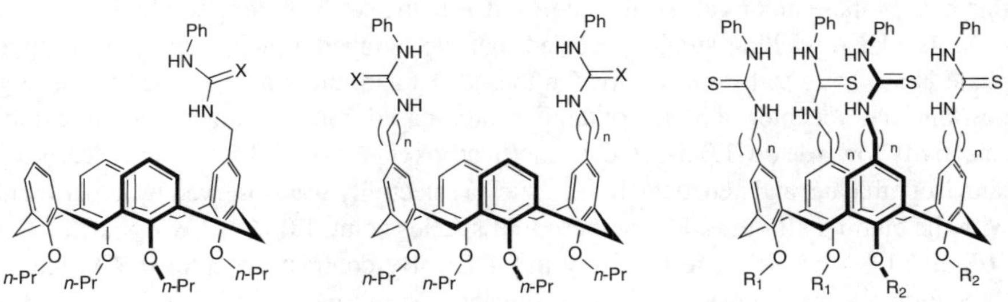

4.52a X = O
4.52b X = S

4.53a X = O, n = 1
4.53b X = S, n = 1
4.53c X = O, n = 0

4.54a $R_1 = R_2$ = *n*-Pr, n = 1
4.54b $R_1 = R_2$ = $-C_2H_4OC_2H_4-$, n = 1
4.54c $R_1 = R_2$ = $-C_2H_4OC_2H_4-$, n = 0

In another effort to improve the anion selectivity of urea-calix[4]arene receptors, Reinhoudt and co-workers[88] took *p-tert*-butylcalix[4]arene, a system expected to be more preorganized, and modified it with two and four (thio)urea units. In the case of the resulting bidentate receptor, **4.55**, a downfield shift in the position of the NH proton signals was observed in the ^{1}H NMR spectrum upon the addition of anions, such as chloride, bromide, iodide, cyanide, and thiocyanide in $CDCl_3$ (studied in the form of their tetrabutylammonium salts). Quantitative ^{1}H NMR spectral titrations were then performed, from which it was deduced that the receptor **4.55** binds anions preferentially in the order $Cl^- > Br^- > CN^- > I^- > SCN^-$.

The tetrafunctionalized-calixarenes **4.56a–4.56c** contain eight potential hydrogen-bond donor sites and show anion selectivities that are similar to those of **4.55**.[88] Unfortunately, they display general anion-binding affinities that are lower than this latter difunctionalized system receptor **4.55**. Presumably, this reflects the fact that system **4.55**, being less pre-organized, has fewer internal (*i.e.*, self-associative) hydrogen bond interactions to overcome prior to substrate binding as compared to the tetrasubstituted system **4.56a–4.56c**.

In an attempt to increase the acidity of the NH protons and thus enhance the anion recognition ability of systems **4.56**, thiourea units were attached to the calix[4]arene core instead of urea units. Disappointingly, the tetrakisthiourea receptor **4.56d** was found to display only a weak affinity for anions but did give rise to a different selectivity order, namely $CN^- > Br^- > Cl^- > I^- = SCN^-$. Consistent with what was proposed to rationalize the differences between **4.55** and **4.56a**–**4.56c**, the unexpectedly weak binding seen in the case of **4.56d** might reflect an enhanced level of internal (*i.e.*, intramolecular) hydrogen bonding present in the absence of any added anion. The need to overcome these interactions would then lead to reduced anion affinities, as would the need to compete with any solvent–receptor interactions that again might be increased as the result of using thiourea, rather than urea, subunits.

In 1995, Reinhoudt demonstrated that thiourea-containing calix[6]arenes, of general structure **4.57**, could serve as useful anion and cation receptors.[89] Both receptors, **4.57a** and **4.57b**, were found to bind halide and carboxylate anions in a 1:1 fashion and to show high selectivity for tricarboxylate anions. Quantitative analyses of the anion binding behaviour of **4.57a** and **4.57b** were made by recording the changes in the NH chemical shift seen in the ^{1}H NMR spectrum upon the addition of tetrabutlyammonium-anion salts in $CDCl_3$. However, no fluoride and dihydrogen phosphate binding was seen for the bis- and tetra-(thio)ureafunctionalized calixarenes, in spite of the fact that both of these anions are well recognized as being strong hydrogen-bond acceptors.

Molecular modelling studies revealed that the tetrahedral anion dihydrogen phosphate is too large to fit into cavity. On the other hand, these analyses led to the suggestion that receptor **4.57a**, containing urea-based binding units, would exhibit selectivity towards *cis*-1,3,5-cyclohexanetricarboxylate over 1,3,5-benzenetricarboxylate, isophthalate, and benzoate. In fact, such a selectivity was observed by experiment, with the binding affinities, K_a, for these four species being 1.0×10^5, 8.7×10^4, 6.9×10^4, and 1.6×10^4 M^{-1}, respectively in $CDCl_3$. By contrast, receptor **4.57b**, having three thiourea units in place of the urea linkages, was found to exhibit a slightly different selectivity order, specifically 1,3,5-benzenetricarboxylate ($K_a = 2.9 \times 10^5$ M^{-1}) > *cis*-1,3,5-cyclohexanetricarboxylate ($K_a = 2.9 \times 10^4$ M^{-1}) > isophthalate ($K_a = 6.4 \times 10^3$ M^{-1}) > benzoate ($K_a = 1.4 \times 10^3$ M^{-1}). In spite of these differences, it is important to appreciate that both **4.57a** and **4.57b** proved selective for tricarboxylate anions, presumably reflecting the three-fold symmetry of the molecular design.

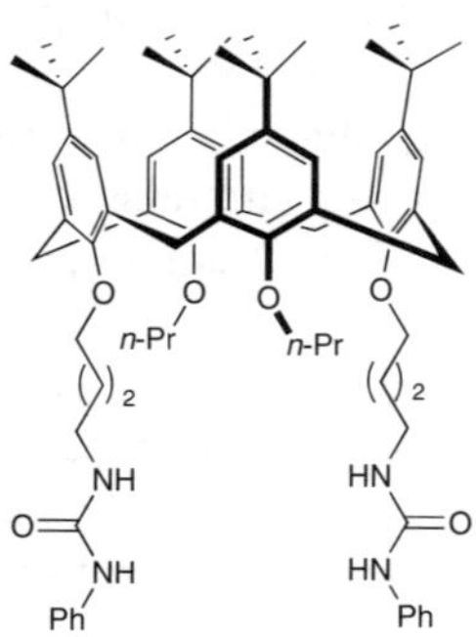

4.55

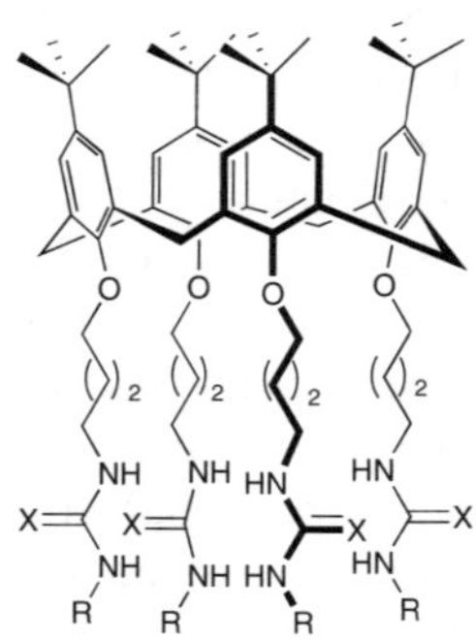

4.56a X = O, R = Ph
4.56b X = O, R = *n*-Pr
4.56c X = O, R = *t*-Bu
4.56d X = S, R = Ph

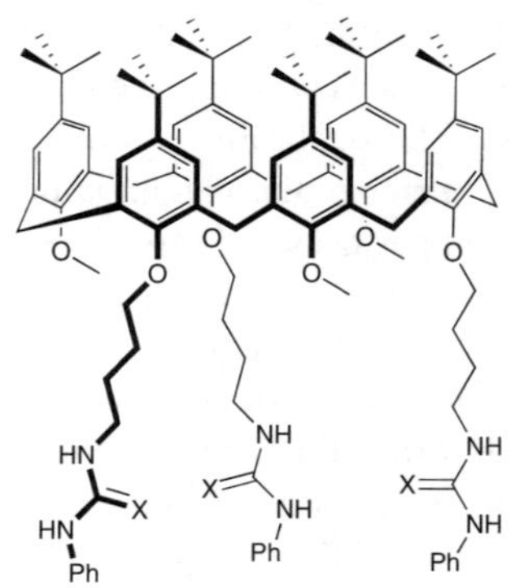

4.57a X = O
4.57b X = S

A novel series of (thio)urea cholapods, **4.58**, was reported in 2001.[90] These systems are based on a steroid core and incorporate both electron-withdrawing substitutents to (presumably) enhance the anion affinities and a C_{20} side chain to increase the solubility and lipophilicity. As expected given the choice of rigid backbone, no evidence of intrastrand hydrogen bonding among the four (thio)urea NH groups was found. Association constants for anion binding in water-saturated chloroform, as deduced using Cram's extraction method (water-$CHCl_3$), revealed that receptor **4.58b** and **4.58c**, containing the built-in electron-withdrawing groups, display affinities for chloride ($K_a = 2.83 \times 10^8$ and 4.77×10^8 M^{-1} for **4.58b** and **4.58c**, respectively) and bromide anion ($K_a = 1.84 \times 10^8$ and 2.26×10^8 M^{-1} for **4.58b** and **4.58c**, respectively) that are almost one order of magnitude higher than those displayed by receptor **4.58a** ($K_a = 1.62 \times 10^7$ and 9.79×10^6 M^{-1} for Cl^- and Br^-, respectively). The use of the thiourea receptor **4.58d** then leads to a further enhancement in affinities ($K_a = 1.05 \times 10^9$ and 3.24×10^8 M^{-1} for Cl^- and Br^-, respectively).

The use of a different set of steroid-based receptors, **4.59a–4.59e**, to effect the transport of chloride anions across vesicle and cell membranes has been demonstrated by A.P. Davis and B.D. Smith.[91] Initial studies in this context were made by measuring chloride anion efflux from unilamellar vesicles using a chloride anion-selective electrode. The rate of efflux was found to follow the order: **4.59e** > **4.59d** > **4.59c** > **4.59b** > **4.59a**, an order that was found to correspond with the association constants measured for this set of receptors. Taken in concert, these results provide support for the notion that it is the hydrogen-bond interactions between the neutral steroid-based receptor and the bound chloride anion that primarily controls the anion transport process. Further experiments with cholapod **4.59d** showed that it also transports nitrate anion effectively.

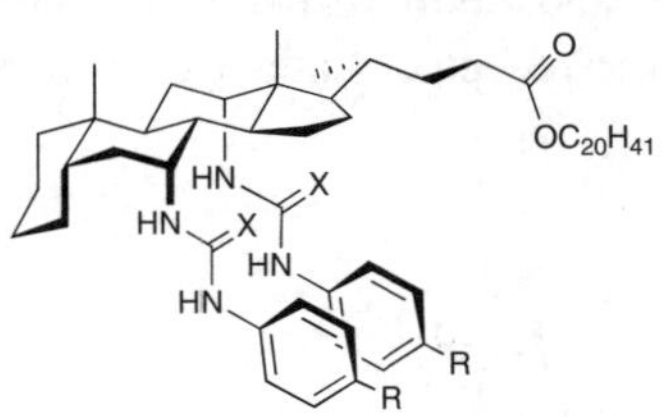

4.58a R = H, X = O
4.58b R = CF_3, X = O
4.58c R = NO_2, X = O
4.58d R = NO_2, X = S

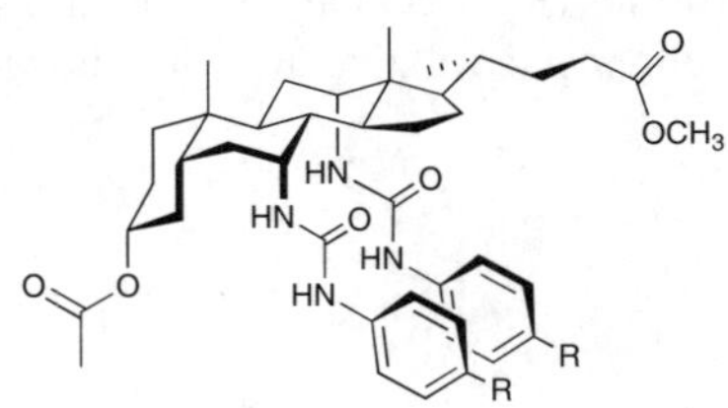

4.59a R = CH_3O
4.59c R = CH_3
4.59d R = H
4.59e R = NO_2

Anion receptors containing urea groups have also been employed in anion sensing receptors and devices (discussed in Chapter 8), in ion-pair receptors (presented in Chapter 6), and in receptors containing transition metals as organisational elements (*cf.* Chapter 7).

4.4 Alcohol-Based Anion Receptors

Hydroxyl groups are among the best known of all hydrogen-bonding donor groups. Interactions involving hydroxyl subunits play a critical role in a wide range of natural recognition processes, including those involving anions. For instance, carbohydrate-protein recognition often relies on specific hydrogen bonding interactions involving the hydroxyl groups present on a carbohydrate moiety and an anionic protein, the importance of which is supported by recent X-ray structural analyses carried out by Quiocho,[92] Honzatko,[93] Einspahr,[94] and Lemieux.[95] In separate structural work of considerable elegance, Remington reported the X-ray crystal structure of the iodide complex of the yellow fluorescent protein (*cf.* YFP-H148Q), solved to 2.1 Å resolution.[96] This structure reveals a bound iodide anion that is held in place via electrostatic interactions involving the guanidinium moiety of Arg 96 at a distance of 4.1 Å, as well as hydrogen bonding interactions with both phenolic hydroxyl groups of Tyr 203 and the amide subunits of Gln 69 at a distance of 3.3 and 3.2 Å, respectively (Figure 4.14). Comparing this X-ray structure to one of YFP-H148Q without the bound iodide reveals that the phenolic hydroxyl groups of Tyr 203 shift towards the iodide anion on binding. This result supports the hypothesis that the hydrogen bond between Tyr 203 and iodide plays a critical role in the anion binding process. Such a notion is consistent with the results of mutational analysis that independently indicate that the presence of the Tyr 203 residue is indispensable for tight binding.

Inspired, perhaps, by natural phenolic receptor systems, D.K. Smith and coworkers[97] have studied in great detail the anion binding properties of the synthetic phenolic receptors, **4.60–4.62**, using high-throughput NMR screening methods. As a result of these analyses, it was found that the dihydroxy systems, **4.61** and **4.62**, exhibit chloride anion-binding affinities (K_a = 145 and 1015 M^{-1} for **4.61** and **4.62** in CD_3CN, respectively) that are surprisingly high given the structural simplicity of the "receptors" in question. In the context of this work, it was also found that increasing the acidity of the phenolic OH protons enhanced the anion-binding ability. For example, systems **4.60c** and **4.60b**, containing electron withdrawing groups (fluorine and nitro, respectively) on their respective *para* positions, display chloride anion affinities (K_a = 555 and 95 M^{-1} for **4.60c** and **4.60b**, respectively, in CD_3CN) that are enhanced relative to what is seen for receptor **4.60a** (K_a = 48 M^{-1} in CD_3CN).

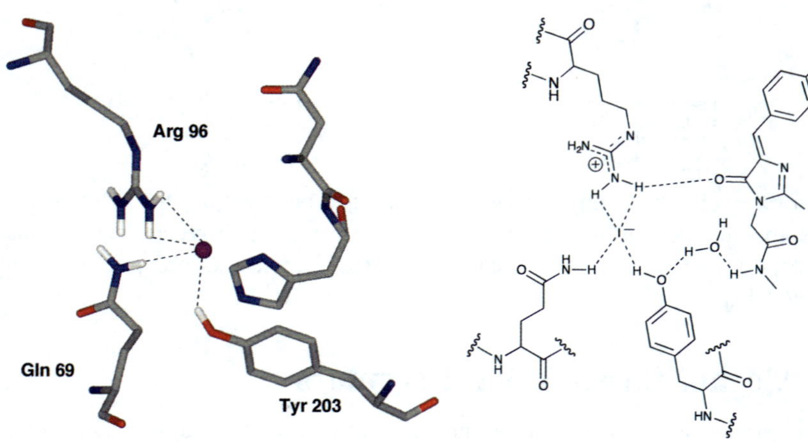

Figure 4.14 *Structure of the yellow fluorescent protein (YFP-H148Q) showing the interaction of the bound iodide anion with various amino acid residues. The chromophore has been removed for clarity*

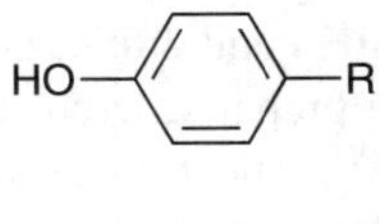

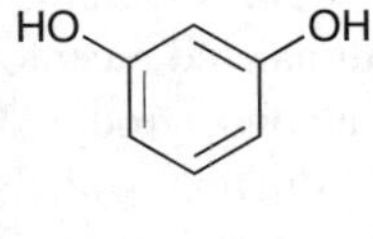

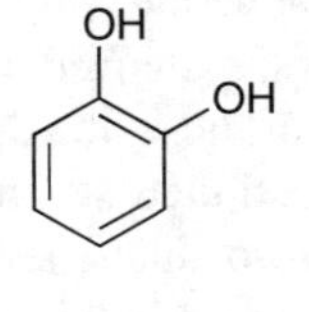

4.60a R = H
4.60b R = F
4.60c R = NO_2

4.61

4.62

In 1998, A.P. Davis and co-workers[98] reported the synthesis and anion-binding properties of the neutral steroid anion receptors of general structure **4.63**. The presence of hydroxyl hydrogen-bonding donor units in these preorganized receptors was considered to be an important design feature that was expected to allow for the effective binding of tridentate anions, such as sulfonates. In fact, the addition of *p*-toluenesulfonate anion to a solution of receptor **4.63a** in C_6D_6 induced shifts in the C*H* signals in the ^{1}H NMR spectrum that were consistent with anion binding. According to a Job plot, anion binding can take place in either a 1:1 and 2:1 (host-to-guest) manner. The calculated association constants (K_a) were 200 M^{-1} for the 1:1 process and 50 M^{-1} for association occurring in a 1:1 + 2:1 fashion, respectively.

Further anion-binding studies were performed with receptors **4.63b–4.63e** by observing their ability to solubilize TBA-hydrogenphenylphosphonate in a mixture of benzene-hexane.[98] In the case of **4.63b**, a system containing three hydroxyl groups, TBA-hydrogenphenylphosphonate was fully solubilized in this solution. By contrast, little or no solubilization was observed for system **4.63e** containing only one hydroxyl group. These results were taken as an indication that both hydroxyl groups present in **4.63a** and **4.63b** participate in anion binding and that increasing the number of hydroxyl units relative to, *e.g.*, **4.63d** improved the anion-binding affinity, at least in the case of these receptors.

Quite recently, Row and Maitra demonstrated selective fluoride anion-binding affinity of bile acid-derived cyclic dimer **4.64**, an interaction sustained by OH···F^- interactions. Quantitative measures of the association constant for what was determined to be a 1:2 stoichiometric process (**4.64** to fluoride anion) were made using ^{1}H NMR titration; this revealed K_{a1} = 1800 M^{-1} and K_{a2} = 250 M^{-1} in $CDCl_3$.[99] In the case of the larger anion, chloride, a relatively weak interaction ($K_a \approx$ 100 M^{-1}) was observed, along with the formation of a 1:1 complex.

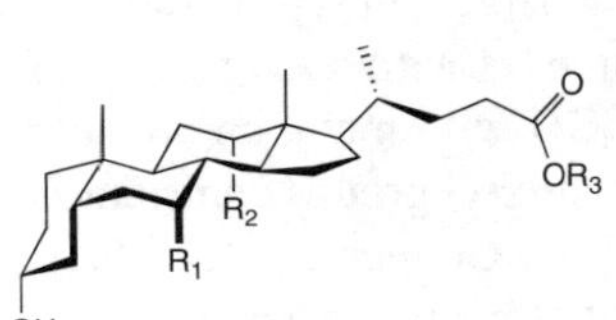

4.63a $R_1 = R_2$ = OH, R_3 = Me
4.63b $R_1 = R_2$ = OH, R_3 = *n*-Octyl
4.63c R_1 = H, R_2 = OH, R_3 = Me
4.63d R_1 = OH, R_2 = H, R_3 = Me
4.63e $R_1 = R_2$ = H, R_3 = Me

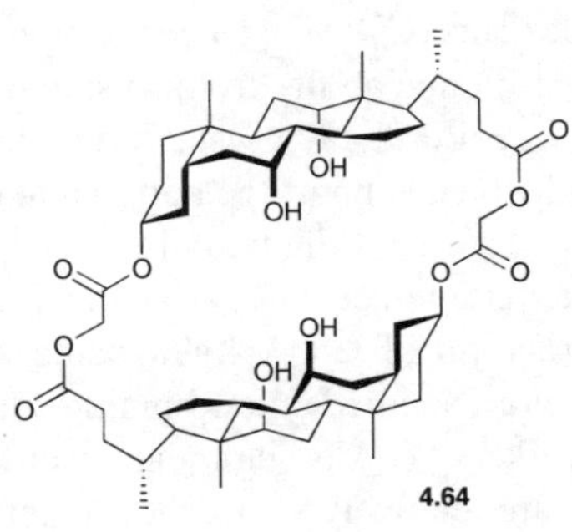

4.64

Slightly later, the Ungaro group[100] reported the synthesis of several modified calix[4]arenes, systems **4.65** and **4.66**, containing hydroxyl groups on the upper rims. The first set of these, the difunctionalized receptors **4.65a** and **4.65b**, bearing trifluoromethyl substituted secondary alcohol groups, were prepared from the corresponding hexafluoro diketones via reduction with $NaBH_4$. The two diastereoisomeric products produced in this way were easily separated by column chromatography. By contrast, receptors **4.66a** and **4.66b**, which contain 3 hydroxyl centres, were synthesized by the reaction of the tetraacyl chloride calixarene with R-I/Li in dry ethyl ether.

Quantitative measurements of the anion association constants revealed that the difunctionalized receptors **4.65** are selective for Y-shaped carboxylate anions over spherical anions (K_a = 435 and 200 M^{-1} for the acetate complex of **4.65a** and **4.65b**, respectively, in $CDCl_3$).[100] Moreover, the racemic receptor **4.65a** was found to display a higher affinity for the chiral guest, *N*-lauroyl-L-phenylalanine, than the corresponding meso receptor, **4.65b** (K_a = 165 *vs.* 40 M^{-1}, respectively, in $CDCl_3$). In the case of the receptor **4.66a**, which contains four hydroxyl groups but lacks the perfluoromethyl substituents, a preference for spherical anions, such as bromide anion (K_a = 480 M^{-1}), was observed relative to acetate ion (K_a = 480 *vs.* 90 M^{-1}; all studies in $CDCl_3$). Interestingly, however, receptor **4.66b** was found not to be an effective receptor for either bromide or acetate anions under analogous solution-phase conditions. These results provide support for the conclusion that the perfluorinated alcohol subunits play a key role in enhancing the anion binding seen in the case of receptors **4.65a** and **4.65b**.

The effect of lower rim modifications were demonstrated by Nam and co-workers[101] in 2003 via the synthesis of receptors **4.67a** and **4.67b**. Both these systems contain two hydroxyl group at the lower rim and display a strong selectivity for acetate over the spherical halide anions (*e.g.*, F^-, Cl^-, Br^-, and I^-) (K_a = 1200 and 5800 M^{-1} for **4.67a** and **4.67**, respectively, in $CDCl_3$).

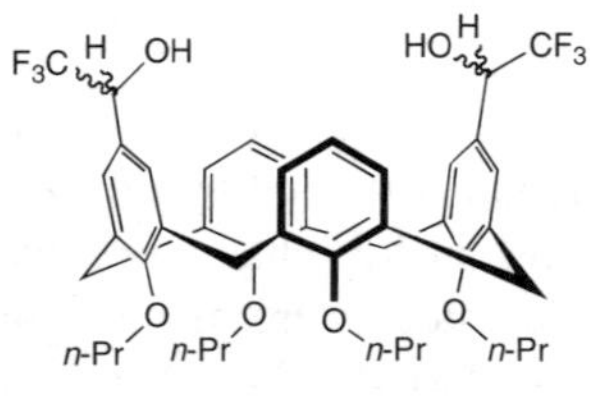

4.65a racemic mixture (*RR, SS*)
4.65b *meso* (*RS, SR*)

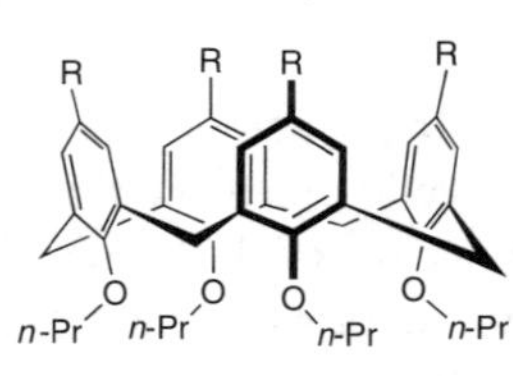

4.66a R = $C(C_6F_5)_2OH$
4.66b R = $C(n\text{-}Bu)_2OH$

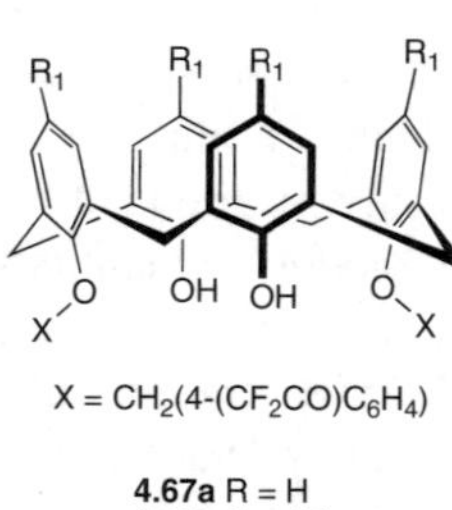

4.67a R = H
4.67b R = *t*-Bu

Cyclodextrins (CDs) have been exploited extensively in supramolecular chemistry due to their unique ability to form stable complexes with a range of guests in aqueous media. A few X-ray crystal structures of CD with metal-anion salts have been solved and which support the notion that the hydroxyl groups contribute to the binding process (*cf.*, *e.g.*, Figure 4.15).[102,103] In 1967, Cramer *et al.*,[104] found that perchlorate anion interacts with aromatic dyes, with the affinity constant corresponding to the formation of ClO_4^--**4.68a** being measured as 29 M^{-1} in aqueous solution. Later on, more detailed anion-binding studies of this CD were performed by Wojcik and co-workers.[105] Although the resulting association constants, summarized in Table 4.3, are relatively weak, they nonetheless serve to establish that these kinds of

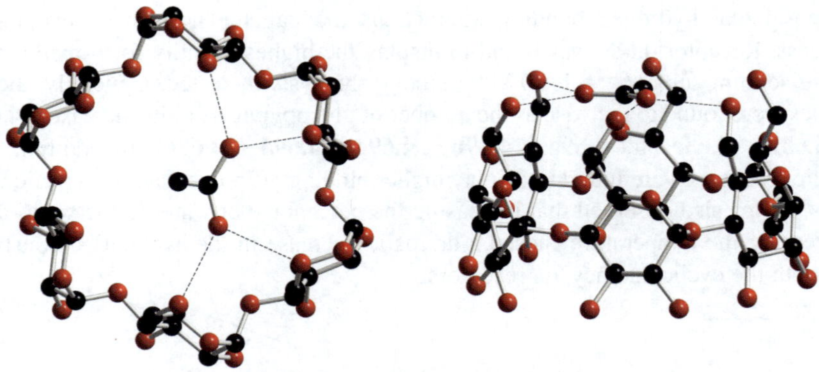

Figure 4.15 *Single crystal X-ray structure of the potassium acetate complex of α-cyclodextrin (α-CD)* ***4.68a*** *(top and side views). In order to improve the clarity of the presentation, the bound potassium cation has been removed from these representations*

Table 4.3 *Anion-binding affinities (K_a, M^{-1}) for receptors* ***4.68a*** *and* ***b*** *and selected anionic substrates as determined in aqueous solution*

	SCN⁻	*ClO_4^-*	*NO_3^-*	*Br⁻*	*I⁻*	*Cl⁻*
4.68a[a]	19	29	1	4	12	–
4.68b[b]	10	27	6	7	18	3

[a]Association constants were measured with potassium anion salts using conductance measurements in aqueous solution (ionic strength = 0.01–0.03 M)
[b]Association constants were measured with sodium anion salts using UV–Vis spectroscopic titrations in aqueous solution (ionic strength = 0.1 M)

hosts can be exploited as anion receptors. Indeed, the anion-binding properties of CD are currently being explored in several different groups.[106]

In 1990, Schwartz and co-workers[107] reported that receptor **4.68a** was able to bind dicarboxylate ions. This ability to form both mono- and dicarboxylate anion complexes was probed in aqueous solution using NMR spectroscopic,[108] calorimetric,[109] potentiometric,[107] and volumetric techniques.[110] However, in spite of considerable effort, questions remain as to whether carboxylate-anion recognition takes place primarily through an inclusion or non-inclusion binding mode. Here, the non-inclusion mode refers to interactions between the hydroxyl groups of the CDs in question and the targeted carboxylate ion that, in the limit, take place largely outside of the central binding cavity. By contrast, in-cavity binding is meant to imply an inclusion-like association where, presumably, hydrophobic interactions serve as the main stabilisation force.

In 2001, Kano *et al.*[111] demonstrated that CDs **4.68** and the acyclic dextrins **4.69** bind *p*-substituted benzoate and alkylcarboxylate anions in DMSO-d_6. In this case, anion binding was suggested to result from hydrogen bonding interactions involving the hydroxyl units of the CD and the targeted carboxylate anion. In the NMR spectra of these systems, only two of the CD hydroxyl protons are seen to undergo a shift in the presence of a bound carboxylate anion. This was considered as support for the proposal that hydrogen bonding interactions mediate the host-guest recognition process. Receptor **4.68c** was found to display the highest affinity for *p*-methylbenzoate sodium salts (K_a = 1300 M^{-1}) among the systems tested. Generally, the K_a values were found to increase as the number of glucopyranose units increased (**4.68a** < **4.68b** < **4.68c** and **4.69a** < **4.69b** < **4.69c** < **4.69d** < **4.69e**). In addition, the cyclic receptors were found to display higher binding affinities than the acyclic dextrins. It was also observed that increasing the percentage of water in the solution or increasing the temperature caused a dramatic decrease in the association constants for both the cyclic and acyclic receptors.

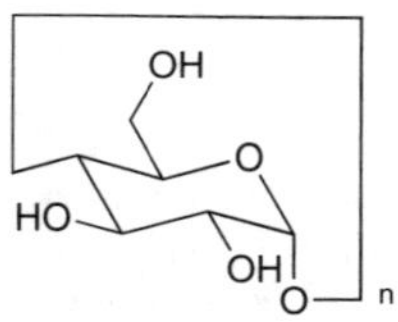

4.68a n = 6
4.68b n = 7
4.68c n = 8

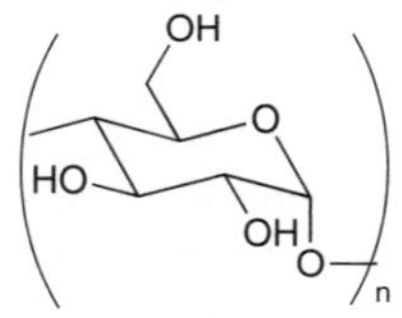

4.69a n = 3
4.69b n = 4
4.69c n = 5
4.69d n = 6
4.69e n = 7

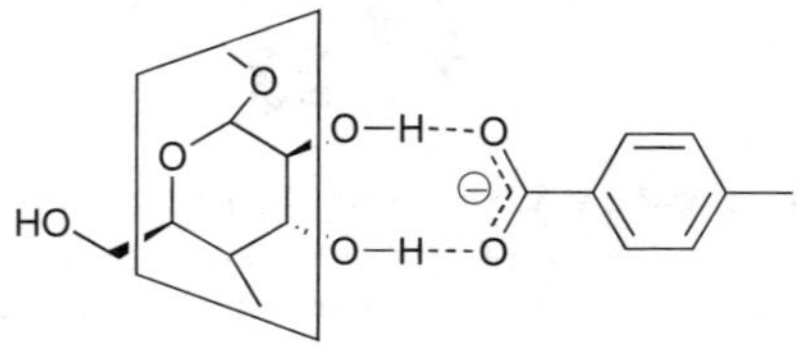

4.68b-*p*-methylbenzoate

4.5 Hybrid Receptors

In an effort to enhance the anion-binding efficiency of neutral receptors, a variety of hybrid systems containing at least two different anion recognition motifs (*e.g.*, urea, amide, hydroxyl group) have been prepared and studied. The advantage of this approach is that it allows, at least potentially, the inherent limitations of each constituent motif to be overcome such that systems with higher affinity or improved selectivity can be produced.

4.5.1 Amide-Urea Systems

In 1994, receptors **4.70a** and **4.71a**, containing simple amide-urea elements, as well as the sulfuryl analogue, **4.70b**, were prepared and studied by Morán and co-workers.[112] The ability of these chromenone-derived receptors to bind benzoate anion in DMSO-d_6 was studied by monitoring the concentration-dependent shift in the benzoate *ortho* proton signal by ^{1}H NMR spectroscopy. The results of these titrations revealed that the association constant of the sulfuryl oxygen receptor **4.70b**, 330 M^{-1}, is higher than that of receptor **4.70a** (20 M^{-1}), presumably because of the better geometry and higher acidity of the sulfuryl NH protons. However, receptor **4.71a**, which combines two amide units and one urea function within the same receptor framework, exhibits an affinity constant for benzoate-anion binding in DMSO (K_a = 15,000 M^{-1}) that is substantially enhanced relative to either **4.70a** or **4.70b**. This finding supports the notion that urea and amide recognition elements, when properly combined, can produce highly effective anion receptors.

Further modification of the bischromenone receptor concept embodied in **4.71a** led to the synthesis of the macrocyclic receptor **4.71b**, a system that was found to be capable of recognizing nonracemic hydroxycarboxylates, such as lactic and mandelic acids, in an enantioselective fashion.[113] The chiral recognition properties of system **4.71b** were initially assessed by carrying out a competition study in DMSO-d_6. In particular, a splitting in the signals ascribed to the racemic receptor was seen in the presence of enantiomerically pure acids such as naproxen, leucine-CBZ, and mandelic acid salts, as judged by ^{1}H NMR spectroscopy. Using this approach, the relative binding constant ratio (for the two enantiomers) was calculated to be 14 in the case of the hydroxyacid salts.

Another key finding supporting the proposed chiral recognition process, involved carrying out a TLC analysis of a mixture of racemic receptor **4.71b** and TMA-(*R*)-mandelic acid salt on silica gel plates. What was seen on upon eluting with $CHCl_3$/EtOAc (8/2) was the presence of two yellow diastereomeric spots with R_f of 0.07 and 0.16, respectively. (A similar experiment carried out with lactic acid salts gave concordant results.) Interestingly, a control experiment, carried out with receptor **4.71b** alone, revealed a larger $R_f = 0.49$ value, indicating that formation of the putative anion complex delays the elution of receptor **4.71b**.

Taking advantage of the difference in R_f values for the diastereomeric salts, the two enantiomers of **4.71b** were separated by column chromatography. Once this was done, ^{1}H NMR spectroscopic titrations carried out in DMSO-d_6 confirmed that the receptor recovered from the more polar fraction binds TMA-(*R*)-mandelic acid salt more strongly ($K_a = 2.8 \times 10^4$ M^{-1}) than that recovered from the less polar fraction ($K_a = 1.7 \times 10^3$ M^{-1}). The chiral differentiation seen for receptor **4.71b** can be attributed to steric effects. Probably, the most sterically favourable complex will be the one where the large phenyl group of the hydroxyacid anion points outwards from receptor cavity; the α-hydrogen is thus expected to be located close to the upper aromatic portion of the spirobifluorene, while the hydroxyl unit must lie over the lower aromatic portion of the spirobifluorene subunit.

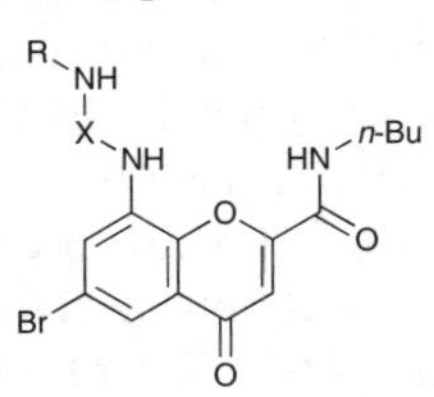

4.70a X = CO, X = $C(CH_3)_2CH_2C(CH_3)_3$
4.70b X = SO_2, X = cyclohexyl

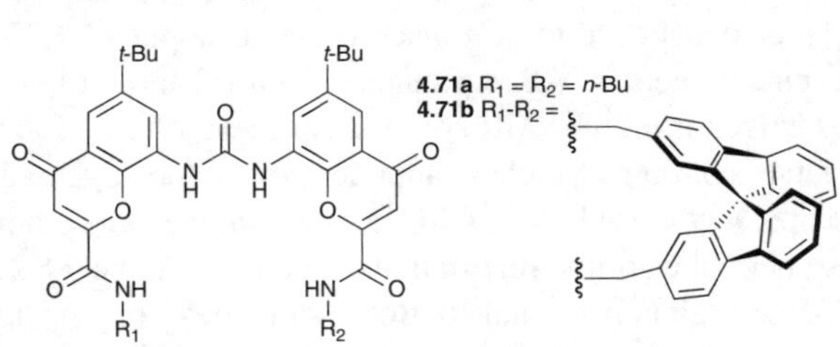

Kilburn and Mortishire-Smith[114] combined a thiourea group and an amide functionality to produce a bowl-shape macrocycle, **4.72a**, that was designed to be a receptor for various *N*-acetyl amino acid carboxylate anions (studied in the form of their tetrabutylammonium salts). Proton NMR spectroscopy-based anion-binding studies revealed that receptor **4.72a** has a high affinity for various amino acid salts in $CDCl_3$ (K_a = 68,600, 16,900, 14,600, 9600 and 6800 M^{-1} for *N*-Ac-Gly-CO_2^-, *N*-Ac-L-Ala-CO_2^-, *N*-Ac-D-Ala-CO_2^-, *N*-Ac-L-Asp-CO_2^-, and *N*-Ac-D-Asp-CO_2^-, respectively) but is not able to effect a differentiation based on the amino acid configurations (*R* or *S*). However, it was found that D-amino acid derivatives are bound on the outside of **4.72a** as the result of a strong interaction with the thiourea unit. This stands in the contrast to what is true for the corresponding L-amino acid derivatives, which are bound inside the receptor cavity in spite of having to adopt an unfavourable *cis* amide geometry. Modelling studies and comprehensive 2D NMR spectroscopic analyses (*viz*. DQF-COSY, TOCSY, ROESY, and NOESY) led the authors to propose that the strong, favourable interaction between the host amide and the *cis* amide functionality of the guest serves to overcome this energetic penalty. Recently, receptor systems, wherein a pyridine (**4.72b** and **4.72c**) unit serves to replace the benzene or, alternatively, the thiourea spacer of receptor **4.72a** have also been reported.[115]

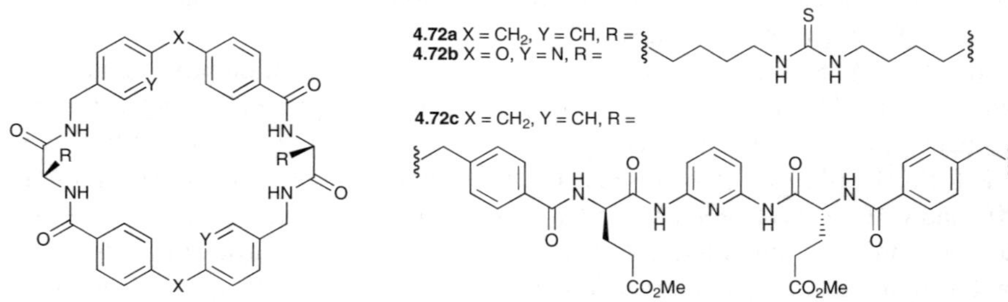

In an alternative approach to preparing a series of urea-amide hydride systems for enantioselective binding, a chiral phenylpropanamide "appendage" was used by Kilburn to produce the acyclic thiourea-bridged bispyridyl receptor **4.73**.[116] Binding studies with various TBA-amino acid salts were performed using standard ^{1}H NMR titrations carried out in $CDCl_3$. The pyridyl receptor **4.73** was found to display a strong selectivity for *N*-Ac-L-Trp-CO_2^-, which has electron-rich aromatic side chains (K_a = 12,400 M^{-1}), over *N*-Ac-L-Ser-CO_2^- (K_a = 380 M^{-1}). It also shows moderate enantioselectivity with a general preference for L-amino acids salts. For example, the association constants for the D- and L- isomers of *N*-Ac-Gln-CO_2^- were found to be 4520 and 9000 M^{-1}, respectively, in $CDCl_3$.

More recently, a more rigid enantioselective receptor, **4.74**, designed to be selective for *N*-Boc-glutamate, was prepared by Kilburn and co-workers.[117] This receptor, which can be considered to be a macrocyclic analogue of **4.73**, was found to bind the D- and L- enantiomers of *N*-Boc-glutamate with 1:2 and 1:1 binding stoichiometries in CD_3CN, as judged from ^{1}H NMR spectroscopic studies (*i.e.*, Job plots). Unfortunately, the association constants for both antipodes proved too large to determine by ^{1}H NMR spectroscopic methods ($K_a > 10^4$ M^{-1}); however, the association constants (K_a) obtained from isothermal calorimetric titrations were found to be 2.83×10^4 and 4.92×10^4 M^{-2} for *N*-Boc-L-glutamate and *N*-Boc-D-glutamate (as bis-tetrabutylammonium salts), in acetonitrile, respectively.

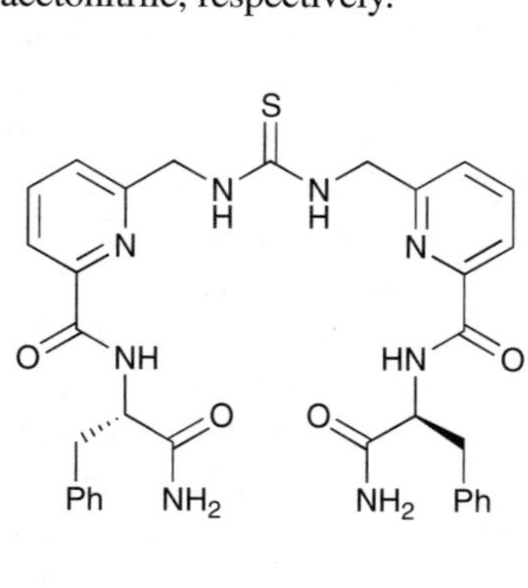

4.73

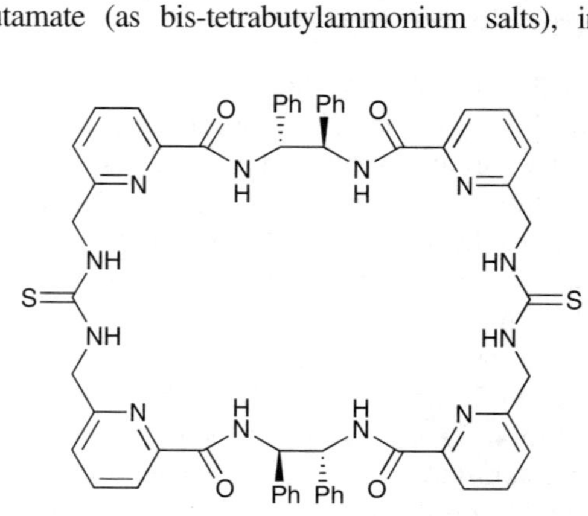

4.74

In 2001, A.P. Davis and co-workers[118] reported a new set of hybrid cholapod receptors **4.75a–4.75c**, that were expected to display a high affinity towards halide anions such as chloride and bromide. Modelling studies indicated the presence of internal interactions between the sulfonamide oxygen atoms and the NHAr moieties of the urea units. However, these interactions were considered likely to be weak due to structural distortion and thus not a major impediment to anion recognition. Actual halide-anion affinities (effective association constants) were measured by the classic extraction method of Cram using chloroform as the organic solvent. Based on such analyses, receptor **4.75c** was seen to be a highly effective receptor for halide anions (studied as TEA salts) ($K_a = 1.03 \times 10^{11}$ and 2.59×10^{10} M^{-1} for chloride and bromide, respectively), proving much more effective than either **4.75a** or **4.75b** (K_a = 4.58×10^9, 2.63×10^9, 6.60×10^{10}, and 1.68×10^{10} M^{-1} for **4.75a**·Cl^-, **4.75a**·Br^-, **4.75b**·Cl^-, **4.75b**·Br^-, respectively).

In 2003, Kilburn and co-workers[103] reported the synthesis and anion-binding properties of a calix[4]arene-based anion receptor **4.76** that is bridged by a strap containing both amide and thiourea-linking elements. The association constants (M^{-1}) for acetate, phenyl phosphinate, and diphenyl phosphate (studied as their respective tetrabutylammonium salts) were found to be 11,000, 24,000, and 1800 M^{-1} in CD_3CN, respectively, as judged from quantitative ^1H NMR titrations. While representing only a limited data set, these results provide support for the conclusion that receptor **4.76** binds the tetrahedral phenyl phosphinate anion more strongly than it does the Y-shaped acetate anion. The reader is referred to reference[119] for further examples of this generalized approach to receptor design.

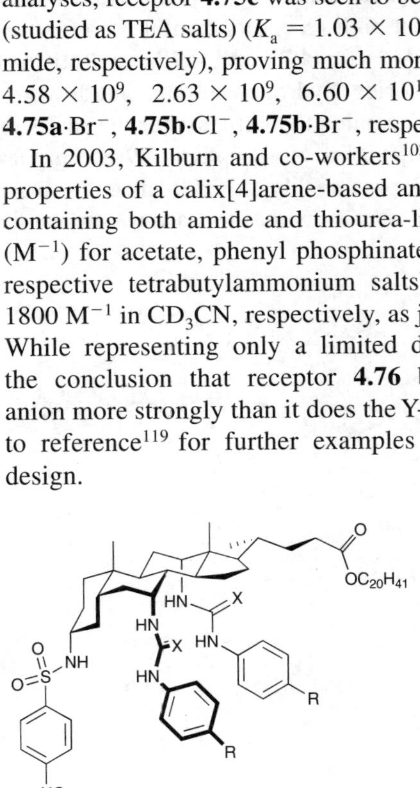

4.75a R = CF_3, X = O
4.75b R = NO_2, X = O
4.75c R = NO_2, X = S

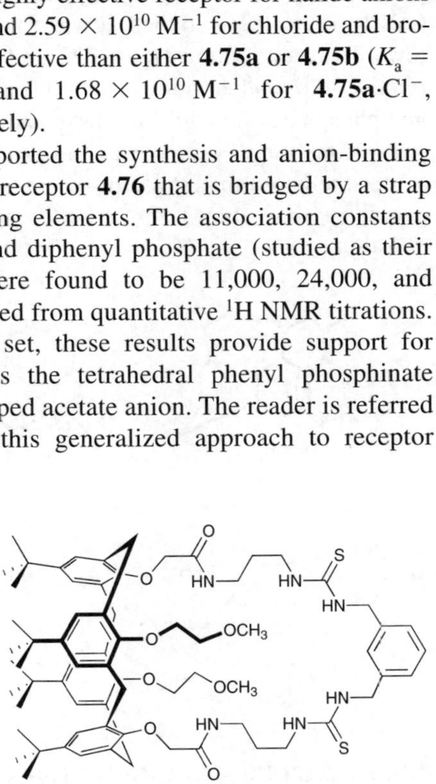

4.76

4.5.2 Urea-Alcohol Systems

In 2002, the calixarene-based urea-hydroxyl system **4.77** was reported by Nam and Jeon.[120] It was prepared from the simple reaction of a diamine with phenylthiocyanate. Receptor **4.77** was found to bind hydrogen sulfate anion with high selectivity over various anions, such as halides, carboxylate, and dihydrogen phosphate, as judged from ^1H NMR spectroscopic studies carried out in $CDCl_3$. In particular, the addition of an anion was found to induce a downfield shift in the urea NH and OH proton signals. This latter peak shift provides support for the supposition that the hydroxyl group, as well as the urea functionality, interacts with the bound anions. From the shift in this signal and that of the urea NH proton, association constants (K_a) for the interaction of receptor **4.77** with several anions were deduced; they were found to decrease in the following order: HSO_4^- (2990 M^{-1}) > $CH_3CO_2^-$ (414 M^{-1}) ≈ $H_2PO_4^-$ (410 M^{-1}) > Cl^- (80 M^{-1}).

Recently, bridged analogues of **4.77**, namely receptors **4.78** and **4.79**, were reported by Nam and co-workers.[121] These systems were designed to exist in their respective preorganized cone conformations. As with **4.77**, their anion binding properties were investigated by ^{1}H NMR spectroscopic titrations carried out in $CDCl_3$. From these studies, it was found that both receptors exhibit a preference towards acetate anion (K_a = 722 and 185 M^{-1} for **4.79** and **4.78,** respectively) relative to the other anions studied (*e.g.,* chloride, hydrogen sulfate, and dihydrogen phosphate). In addition, these analyses revealed that, at least for the series of test anions employed, the anion-binding affinities of receptor **4.79** are higher across the board than those of receptor **4.78**; presumably, this reflects an intrinsic geometry that allows for a better spatial match with the various anions in question.

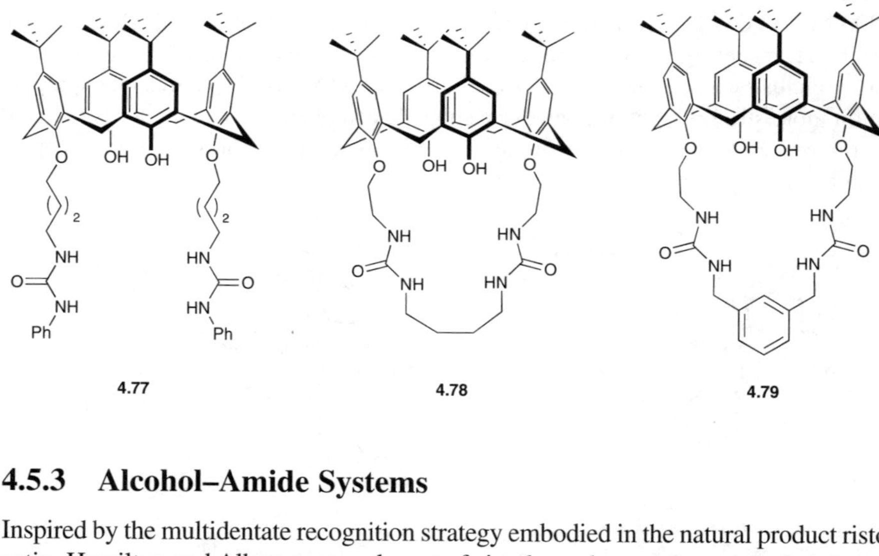

4.77 **4.78** **4.79**

4.5.3 Alcohol–Amide Systems

Inspired by the multidentate recognition strategy embodied in the natural product ristocetin, Hamilton and Albert prepared a set of simple analogues that might function as synthetic anion receptors.[122] The resulting products, receptors **4.80a** and **4.80b**, were found to bind acetate anion as the result of the built-in amide and hydroxyl groups acting as hydrogen-bond donors. It was noted that the receptors containing hydroxyl groups displayed higher binding constant values compared to the analogous systems that did not contain such functionality, underscoring the importance of the hydroxyl motif in terms of modulating the binding process. In the case of receptor **4.80a**, a system that contains two additional amide NH groups as compared to receptor **4.80b**, a 27-fold higher acetate-binding constant was observed (*i.e.,* $K_a = 2.7 \times 10^5$ M^{-1} in CD_3CN). This result supports the notion that the number of hydrogen-bond donors is strongly correlated to the overall anion-binding affinities, at least in systems of this basic structural type.

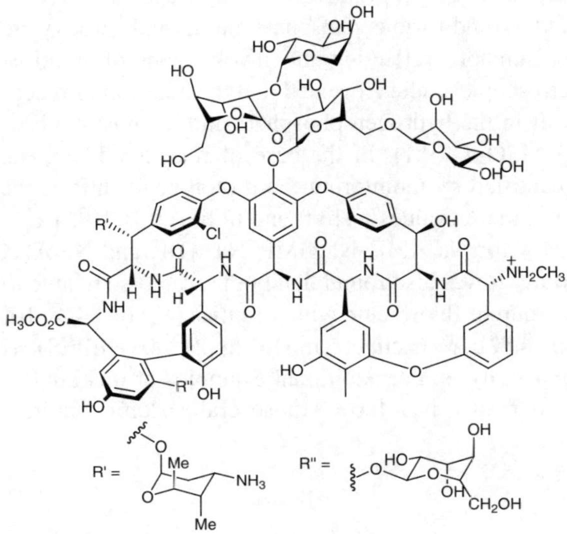

Ristocetin

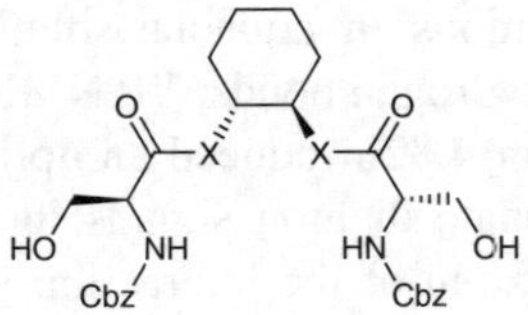

4.80a X = NH
4.80b X = O

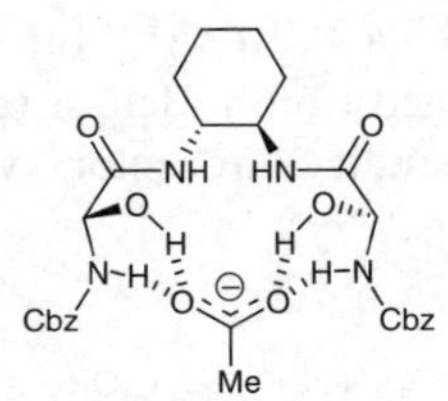

4.80a-acetate

In 1996, A.P. Davis and co-workers[123] designed a steroid-based cryptand **4.81** capable of recognizing halide anions. This macrodilactam, with four hydroxyl and two amide groups, was derived from cholic acid. It possesses a small-size cavity and a rigid framework that serves to maintain the putative anion-binding cavity while precluding intramolecular hydrogen-bonding interactions. Computer modelling (*e.g.*, Monte Carlo) studies reveal that the cavity of **4.81**, *ca.* 3.3 × 2.2 Å, is appropriately sized to accommodate a fluoride anion and, possibly, to bind the larger halides, chloride, and bromide. The results of anion-binding studies, performed using standard ^{1}H NMR spectroscopic titration studies carried out in $CDCl_3$, reveal that receptor **4.81**, as expected, is selective for fluoride anion over other halides (K_a = 3220, 990, and 250 M^{-1} for F^-, Cl^-, and Br^-, respectively).

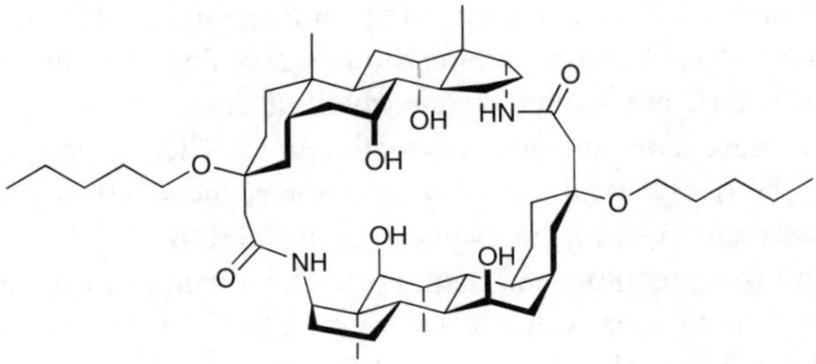

4.81

Oligosaccharides with their multiple hydroxyl groups are well known for the important role they play in cellular recognition processes. Recently, Aoyama and co-workers[124] attached oligosaccharide units to a calix[4]resorcarene bearing four long alkyl chains via an amide linkage. The resulting macrocylic oligosaccharide receptors, **4.82a** and **4.82b**, were found to bind various phosphate anions and carboxylate anions in aqueous solution, presumably reflecting the involvement of multiple hydrogen bonds. ^{31}P NMR spectroscopic studies revealed that the addition of receptor **4.82b** induced an upfield shift in the hydrogen phosphate and D-ribose-5-phosphate dianion signals in D_2O-DMSO-d_6 (9:1). In the case of receptor **4.82a**, the extent of the interaction was quantified by monitoring the proton peak shift of the ribose subunit. From this analysis, the K_a values were found to be 4.0×10^3, 1.6×10^4, 2.9×10^4, and 3.7×10^4 M^{-1} for Na_2AMP, Na_2GMP, Na_2ADP, and Na_2ATP, respectively, in this solvent mixture. It was also found that both receptors are able to bind the carboxylate functional group of fluorescein with affinities (K_a) of 1.1×10^4 and 5.4×10^4 M^{-1} for **4.82a** and **4.82b**, respectively, in H_2O at pH 7.0 (HEPES), as judged from a Benesi-Hildebrand analysis. For additional examples of this kind of generalized receptor system, the reader is referred to several specific publications.[125,126]

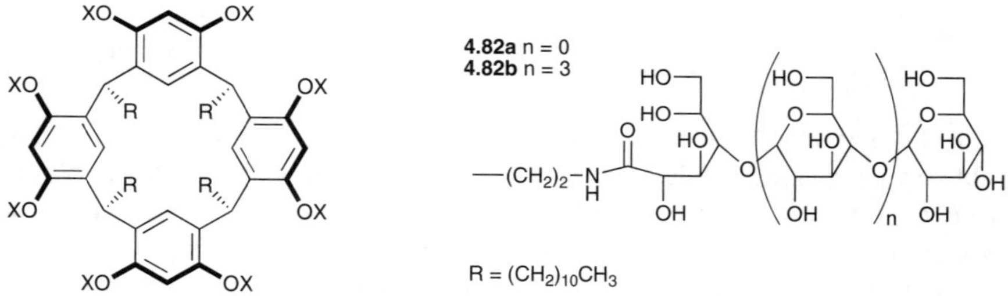

4.5.4 Amide-Hydroxy-Urea Systems

In an effort to develop specific receptors displaying greater affinity and selectivity towards the anions of α-heterocyclic and α-keto acids, the non-symmetrical receptor **4.83** was designed and synthesized.[126] This receptor contains three different hydrogen-bond donor groups, namely urea, amide, and hydroxyl. Standard ^1H NMR spectroscopy-based anion-binding studies revealed that this receptor binds the TEA salts of fusaric and 4-imidazole carboxylic acids ($K_a = 4.3 \times 10^5$ and 4.5×10^5 M^{-1} for these two species, respectively, in $CDCl_3$ containing 1% v/v CD_3OD) very effectively, while displaying a far reduced affinity for the corresponding benzoate anion salt in the same solvent mixture ($K_a = 1.1 \times 10^3$ M^{-1}). It was also found that receptor **4.83** displays a preference towards the α-keto carboxylate anion, 4-methyl oxovalerate ($K_a = 6.1 \times 10^4$ M^{-1}), over simple propionate anion ($K_a = 3.2 \times 10^3$ M^{-1}) under the same conditions. Taken together, these results underscore the important stabilizing role played by the "extra" hydrogen bond interaction arising from the hydroxamic functionality present in the host and the heterocyclic atom or keto subunit present in the guest as illustrated schematically in the complex of **4.83** with 2-carboxy-5-n-butylpyridine.

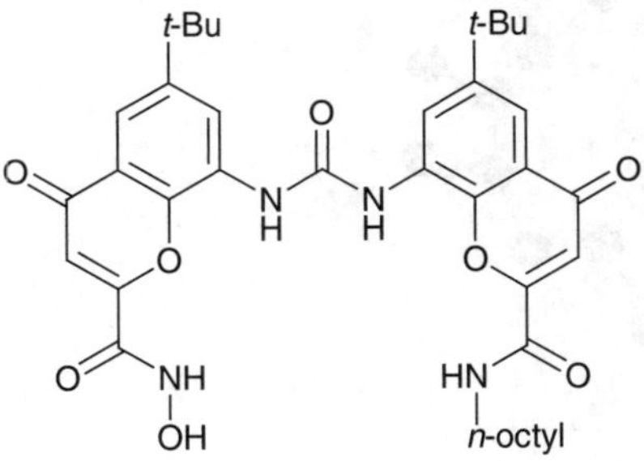

4.83

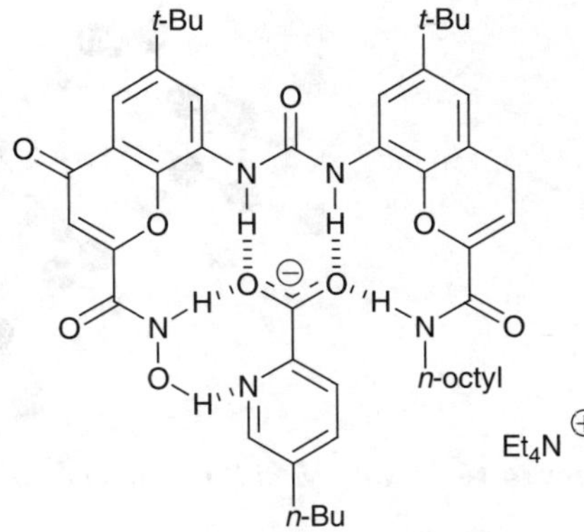

4.83-2-carboxy-5-*n*-butylpyridine

4.6 Other Systems

In 1990, Farnham and Dixon[127] reported the crystal structure of the fluoride-anion complex of the highly fluorinated macrocycle **4.84**. This 18-membered macrocyclic ether stabilises the formation of a fluoride anion complex wherein the anion is located in the middle of cavity. Based on a single crystal X-ray diffraction analysis (Figure 4.16), it was inferred that this complex is stabilized by the interaction between four CH protons and the fluoride ion (CH···F = 1.94–2.10 Å). This rare anion binding scenario was rationalized in terms of the polarization effect of the neighbouring CF_2 subunits, which are strongly withdrawing electron and endows the CH moieties with more positive character. It was also observed that the bound fluoride peak appears as broad singlet at −76.6 ppm in the low-temperature ^{19}F NMR spectrum of the anion complex.

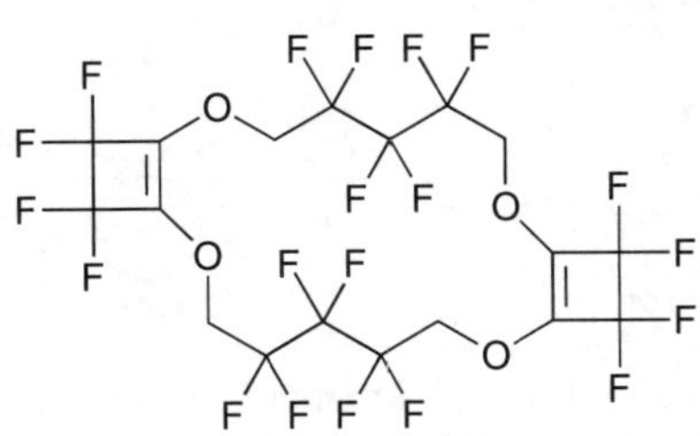

4.84

Gellman and co-workers[128] prepared macrocycle **4.85**, which contain phosphine oxide and disulfoxide motifs, in an effort to obtain a receptor system that would rely on mostly anion–dipole interactions. After reporting the cation and carbohydrate-binding properties of this system, evidence in support of anion complexation was obtained from ^{1}H NMR spectroscopic studies carried out in $CDCl_3$ containing 2%

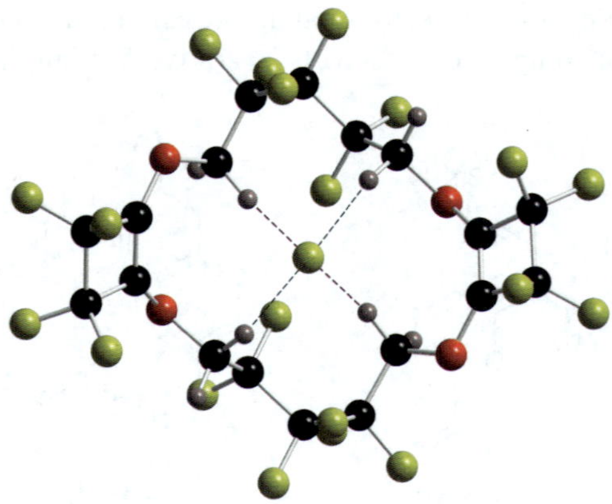

Figure 4.16 *Single crystal X-ray structure of the fluoride–anion complex of receptor* ***4.84***

CD_3OD. For instance, the benzylic proton was observed to shift in the presence of various anions.

Analysis of the X-ray crystal structure of **4.85** reveals that three oxygen atoms are oriented in the same direction. Presumably, the anion-binding ability of **4.85** is attributed to the preorganized structure, which serves to arrange three strong dipoles from P = O and S = O in such a way that they can interact effectively with a single anion (c.f. structure **4.85b**). The association constants for chloride and bromide were obtained by monitoring the changes in the chemical shift of the two benzylic proton peaks as a function of added anion. From this ^{1}H NMR spectroscopic study, essentially identical values of between 60 and 70 M^{-1} were deduced for these two anions, whereas the K_a value for iodide was found to be *ca.* 40 M^{-1}.

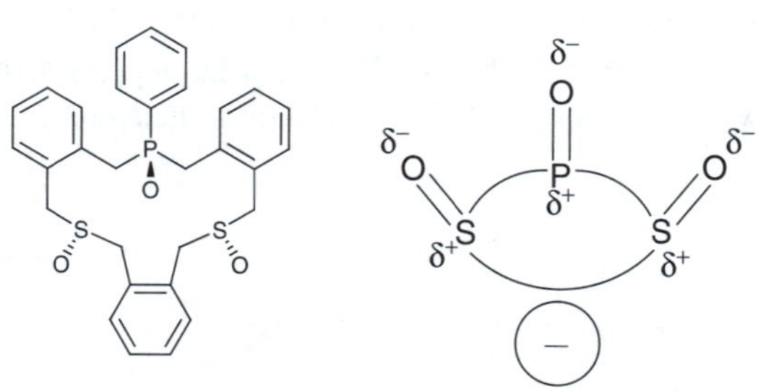

4.85 **4.85b**

In 1974, during a study of metal ion–nucleic acid interactions, Marzilli and Chang[129] found that guanosine **4.86** is capable of binding chloride-anions. For instance, the addition of alkaline earth metal chloride and bromide salts to a solution of neutral guanosine **4.86** in DMSO-d_6 was found to induce downfield shifts in the NH proton signals of 0.7 and 0.8 ppm, respectively. The same kind of downfield shift was observed in the presence of tetraethylammonium chloride. On the other hand, the addition of perchorate salts (Na^+, Li^+) and nitrate salts (Ca^{2+}, Mg^{2+}, Sr^{2+}, Ba^{2+}, Na^+) did not induce an appreciable shift. In order to quantify the putative chloride

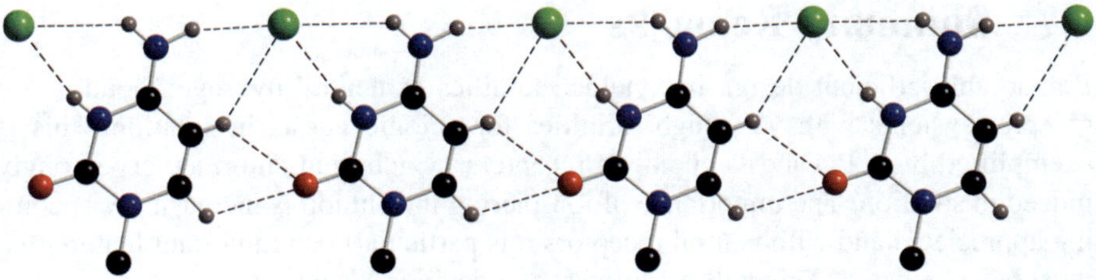

Figure 4.17 *Single crystal X-ray structure of the chloride-anion complex of* ***4.87a***

anion-binding interaction, NMR titrations were performed with $CaCl_2$ in DMSO-d_6. The resulting affinity constant was quite low ($K_a = 2.8\ M^{-1}$).

Supporting this proposed binding, is the 1972 crystal structure of the monoprotonated 1-methylcytosine (**4.87a**)-chloride anion salt reported by Trus and Marsh.[130] In the solid state, this oligomeric structure is stabilized by electrostatic interactions, strong hydrogen bonding interactions involving N-H···Cl^- (*cf.* 2.31–2.79 Å), and relatively weak hydrogen interactions of the C-H···O (*cf.* 2.60–2.64 Å) and C-H···Cl^- (2.95 Å) types (Figure 4.17).

Five years later, Mandel[131] revealed the result of the crystal structure of cytosine (**4.87b**) hydrochloride, which was also found to exist as an oligomer. The interactions between chloride and NH and the additional hydrogen-bond donor NH presumably contribute to the stabilization of the dimeric structure of cytosine. The cationic cytosine interacts with three chloride anions. Among these interactions, the one between the protonated NH and the bound chloride anion was found to be the shortest (namely 2.06 Å; *cf.* Figure 4.18). For further information the reader is directed to the bibliography.[132]

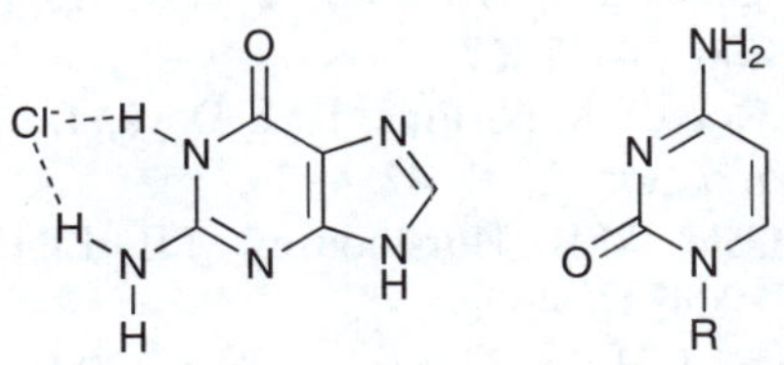

4.86-Cl^-

4.87a R = CH_3
4.87b R = H

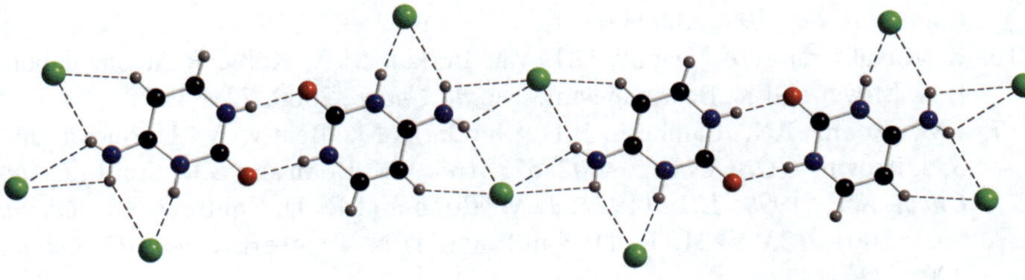

Figure 4.18 *Single crystal X-ray structure of the chloride–anion complex of* ***4.87b***

4.7 Summary Remarks

Remarkably efficient design has yielded families of neutral hydrogen-bond donor receptor systems with very high affinities for anionic guests in solution. This is exemplified by A.P. Davis's cholapod receptors, which bind chloride very strongly indeed in solution. The importance of ion-pairing in solution is increasingly becoming appreciated and with neutral receptors it is particularly an important feature that must be considered. This will be discussed in detail in Chapter 6.

References

1. J.W. Pflugrath and F.A. Quiocho, *Nature*, 1985, **314**, 257; J.W. Pflugrath and F.A. Quiocho, *J. Mol. Biol.*, 1988, **200**, 163.
2. K. Kavallieratos, S.R. de Gala, D.J. Austin and R.H. Crabtree, *J. Am. Chem. Soc.*, 1997, **119**, 2325.
3. K. Kavallieratos, C.M. Bertao and R.H. Crabtree, *J. Org. Chem.*, 1999, **64**, 1675.
4. M.P. Hughes and B.D. Smith, *J. Org. Chem.*, 1997, **62**, 4492.
5. S.J. Coles, J.G. Frey, P.A. Gale, M.B. Hursthouse, M.E. Light, K. Navakhun and G.L. Thomas, *Chem. Commun.*, 2003, 568.
6. S.J. Brooks, L.S. Evans, P.A. Gale, M.B. Hursthouse and M.E. Light, *Chem. Commun.*, 2005, 734.
7. S.-I. Kondo, Y. Hiraoka, N. Kurumatani and Y. Yano, *Chem. Commun.*, 2005, 1720.
8. R. Prohens, S. Tomàs, J. Morey, P.M. Deyà, P. Ballester and A. Costa, *Tetrahedron Lett.*, 1998, **39**, 1063.
9. C.P. Causey and W.E. Allen, *J. Org. Chem.*, 2002, **67**, 5963.
10. S. Camiolo, P.A. Gale, M.B. Hursthouse, M.E. Light and C.N. Warriner, *Tetrahedron Lett.*, 2003, **44**, 1367.
11. R. Prohens, M.C. Rotger, M.N. Piña, P.M. Deyà, J. Morey, P. Ballester and A. Costa, *Tetrahedron Lett.*, 2001, **42**, 4933.
12. S.J. Coles, P.A. Gale, M.B. Hursthouse, M.E. Light and C.N. Warriner, *Supramol. Chem.*, 2004, **16**, 469.
13. J. Kang, H.-K. Ju and J.-H. Jo, *Supramol. Chem.*, 2004, **16**, 175; J. Kang, J.-H. Jo, and S. In, *Tetrahedron Lett.*, 2004, **45**, 5225.
14. J. Kang and J. Kim, *Tetrahedron Lett.*, 2005, **46**, 1759.
15. S. Valiyaveettil, J.F.J. Engbersen, W. Verboom and D.N. Reinhoudt, *Angew. Chem. Int. Ed.*, 1993, **32**, 900.
16. K. Kavallieratos, A. Danby, G.J. Van Berkel, M.A. Kelly, R.A. Sachleben, B.A. Moyer and K. Bowman-James, *Anal. Chem.*, 2000, **72**, 5258.
17. J.M. Boon, T.N. Lambert, B.D. Smith, A.M. Beatty, V. Ugrinova and S.N. Brown, *J. Org. Chem.*, 2002, **67**, 2168; J.M. Boon and B.D. Smith, *J. Am. Chem. Soc.*, 1999, **121**, 11924; J. M. Boon and B. D. Smith, *J. Am. Chem. Soc.*, 2001, **123**, 6221; B. D. Smith and T. N. Lambert, *Chem. Commun.*, 2003, 2261.
18. J.M. Boon, R. Shukla, B.D. Smith, G. Licini and P. Scrimin, *Chem. Commun.*, 2002, 260.

19. C. Raposo, N. Pérez, M. Almaraz, M.L. Mussons, M.C. Caballero and J.R. Morán, *Tetrahedron Lett.*, 1995, **36**, 3255; C. Raposo, M. Almaraz, M. Martín, V. Weinrich, M.L. Mussóns, V. Alcázar, M.C. Caballero and J.C. Moran, *Chem. Lett.*, 1995, **24**, 759.
20. J.-H. Liao, C.-T. Chen and J.-M. Fang, *Org. Lett.*, 2002, **4**, 561.
21. R.J. Pieters, *Tetrahedron Lett.*, 2000, **41**, 7541.
22. N. Pant and A.D. Hamilton, *J. Am. Chem. Soc.*, 1988, **110**, 2002.
23. K. Choi and A.D. Hamilton, *J. Am. Chem. Soc.*, 2001, **123**, 2456; K. Choi and A.D. Hamilton, *J. Am. Chem. Soc.*, 2003, **125**, 10241.
24. K. Choi and A.D. Hamilton, *Angew. Chem., Int. Ed.*, 2001, **40**, 3912.
25. A. Szumma and J. Jurczak, *Eur. J. Org. Chem.*, 2001, 4031.
26. M. Chmielewski and J. Jurczak, *Tetrahedron Lett.*, 2004, **45**, 6007; M.J. Chmielewski, A. Szumma and J. Jurczak, *Tetrahedron Lett.*, 2004, **45**, 8699; M.J. Chmielewski and J. Jurczak, *Tetrahedron Lett.*, 2005, **46**, 3085; M. Chmielewski and J. Jurczak, *Chem. Eur. J.*, 2005, **11**, 6080.
27. M.A. Hossain, J.M. Llinares, D. Powell and K. Bowman-James, *Inorg. Chem.*, 2001, **40**, 2936; M.A. Hossain, S.O. Kang, D. Powell and K. Bowman-James, *Inorg. Chem.*, 2003, **42**, 1397.
28. A.P. Bisson, V.M. Lynch, M.-K.C. Monahan and E.V. Anslyn, *Angew. Chem. Int. Ed.*, 1997, **36**, 2340.
29. K. Niikura, A.P. Bisson and E.V. Anslyn, *J. Chem. Soc. Perkin Trans. 2*, 1999, 1111.
30. M.A. Hossain, S.O. Kang, J.M. Llinares, D. Powell and K. Bowman-James, *Inorg. Chem.*, 2003, **42**, 5043.
31. S.O. Kang, J.M. Llinares, D. Powell, D. VanderVelde and K. Bowman-James, *J. Am. Chem. Soc.*, 2003, **125**, 10152.
32. S.O. Kang, D. VanderVelde, D. Powell and K. Bowman-James, *J. Am. Chem. Soc.*, 2004, **126**, 12272.
33. S.O. Kang, M.A. Hossain, D. Powell and K. Bowman-James, *Chem. Commun.*, 2005, 328; S.O. Kang, D. Powell and K. Bowman-James, *J. Am. Chem. Soc.*, 2005, **127**, 13478.
34. Y. Morzherin, D.M. Rudkevich, W. Verboom and D.N. Reinhoudt, *J. Org. Chem.*, 1993, **58**, 7602.
35. B.R. Cameron and S.J. Loeb, *Chem. Commun.*, 1997, 573.
36. F. Sansone, L. Baldini, A. Casnati, M. Lazzarotto, F. Ugozzoli and R. Ungaro, *Proc. Natl. Acad. Sci. USA*, 2002, **99**, 4842.
37. R. Miao, Q.-Y. Zheng, C.-F. Chen and Z.-T. Huang, *Tetrahedron Lett.*, 2005, **46**, 2155.
38. V. Sidorov, F.W. Kotch, G. Abdrakhmanova, R. Mizani, J.C. Fettinger and J.T. Davis, *J. Am. Chem. Soc.*, 2002, **124**, 2267; V. Sidorov, F.W. Kotch, J. Kuebler, Y.-F. Lam and J.T. Davis, *J. Am. Chem. Soc.*, 2003, **125**, 2840.
39. A.P. Davis, J.J. Perry and R.P. Williams, *J. Am. Chem. Soc.*, 1997, **119**, 1793.
40. A.J. Ayling, S. Broderick, J.P. Clare, A.P. Davis, M.N. Pérez-Payán, M. Lahtinen, M.J. Nissinen and K. Rissanen, *Chem. Eur. J.*, 2002, **8**, 2197.
41. H.R. Perkins, *Biochem. J.*, 1969, **111**, 195.

42. D.H. Williams and M.S. Westwell, *Chem. Soc. Rev.*, 1998, **27**, 57; D.H. Williams and B. Bardsley, *Angew. Chem. Int. Ed.*, 1999, **38**, 1172.
43. M.H. McCormick, W.M. Stark, G.E. Pittenger, R.C. Pittenger and J. M. McGuire, *Antibiot. Annu.*, 1955–56, 606.
44. J.R. Kalman and D.H. Williams, *J. Am. Chem. Soc.*, 1980, **102**, 906; D.H. Williams, M.P. Williamson, D.W. Butcher and S.J. Hammond, *J. Am. Chem. Soc.*, 1983, **105**, 1332; J.C.J. Barna, D.H. Williams and M.P. Williamson, *J. Chem. Soc., Chem. Commun.*, 1985, 254.
45. M. Nieto and H.R. Perkins, *Biochem. J.*, 1971, **123**, 789; M. Nieto and H.R. Perkins, *Biochem. J.*, 1971, **123**, 773; M. Nieto and H.R. Perkins, *Biochem. J.*, 1971, **124**, 845.
46. J.P. Brown, J. Feeney and A.S.V. Burgen, *Mol. Pharmacol.*, 1975, **11**, 119; J.P. Brown, L. Terenius, J. Feeney and A.S.V. Burgen, *Mol. Pharmacol.*, 1975, **11**, 126; M.P. Williams and D.H. Williams, *Eur. J. Biochem.*, 1984, **138**, 345; M.P. Williams, D.H. Williamsand S.J. Hammond, *Tetrahedron*, 1984, **40**, 569; C.M. Pearce, U. Gerhard and D.H. Williams, *J. Chem. Soc. Perkin Trans. 2*, 1995, 159; D.H. Williams, J.P.L. Cox, A.J. Doig, M. Gardner, U. Gerhard, P.T. Kaye, A.R. Lal, I.A. Nicholls, C.J. Salter and R.C. Mitchell, *J. Am. Chem. Soc.*, 1991, **113**, 7020.
47. A. Rodríguez-Tebar, D. Vázquez, J.L.P. Velázquez, J.Laynez and I. Wadsö, *J. Antibiot.*, 1986, **39**, 1578.
48. J.P. Waltho and D.H. Williams, *J. Am. Chem. Soc.*, 1989, **111**, 2475.
49. P.J. Loll, A.E. Bevivino, B.D. Korty and P.H. Axelsen, *J. Am. Chem. Soc.*, 1997, **119**, 1516.
50. H. Shiozawa, B.C.S. Chia, N.L. Davies, R. Zerella and D.H. Williams, *J. Am. Chem. Soc.*, 2002, **124**, 3914.
51. H. Ishida, M. Suga, K. Donowaki and K. Ohkubo, *J. Org. Chem.* 1995, **60**, 5374.
52. D. Ranganathan, V. Haridas and I.L. Karle, *J. Am. Chem. Soc.*, 1998, **120**, 2695.
53. S. Kubik and R. Goddard, *Proc. Natl. Acad. Sci. USA*, 2002, **99**, 5127.
54. S. Kubik, R. Kirchner, D. Nolting and J. Seidel, *J. Am. Chem. Soc.*, 2002, **124**, 12752.
55. S. Kubik, R. Goddard, R. Kirchner, D. Nolting and J. Seidel, *Angew. Chem. Int. Ed.*, 2001, **40**, 2648.
56. S. Otto and S. Kubik, *J. Am. Chem. Soc.*, 2003, **125**, 7804.
57. G. Heinrichs, S. Kubik, J. Lacour and L. Vial, *J. Org. Chem.*, 2005, **70**, 4498; S. Kubik, R. Goddard, S. Otto, S. Pohl, C. Reyheller and S. Stüwe, *Biosens. Bioelectr.*, 2005, **20**, 2364.
58. D. Yang, J. Qu, W. Li, Y.-H. Zhang, Y. Ren, D.-P. Wang and Y.-D. Wu, *J. Am. Chem. Soc.*, 2002, **124**, 12410.
59. D. Yang, X. Li, Y. Sha and Y.-D. Wu, *Chem. Eur. J.*, 2005, **11**, 3005.
60. H. Huang, L. Mu, J. He and J.-P. Cheng, *Tetrahedron Lett.*, 2002, **43**, 2255.
61. W. Guo, J. Wang, J. He, Z. Li and J.-P. Cheng, *Supramol. Chem.*, 2004, **16**, 171.
62. M.C. Etter, Z. Urbañczyk-Lipkowska, M. Zia-Ebrahimi and T.W. Panunto, *J. Am. Chem. Soc.*, 1990, **112**, 8415.

63. M. Boiocchi, L.D. Boca, D. Esteban-Gómez, L. Fabbrizzi, M. Licchelli and E. Monzani, *J. Am. Chem. Soc.*, 2004, **126**, 16507.
64. D. Esteban-Gómez, L. Fabbrizzi and M. Licchelli, *J. Org. Chem.*, 2005, **70**, 5717; E.J. Cho, B.J. Ryu, Y.J. Lee and K.C. Nam, *Org. Lett.*, 2005, **7**, 2607; D. Esteban-Gómez, L. Fabbrizzi, M. Licchelli and E. Monzani, *Org. Biomol. Chem.*, 2005, **3**, 1495.
65. S. Camiolo, P.A. Gale, M.B. Hursthouse, M.E. Light and A.J. Shi, *Chem. Commun.*, 2002, 758; P.A. Gale, K. Navakhun, S. Camiolo, M.E. Light and M.B. Hursthouse, *J. Am. Chem. Soc.*, 2002, **124**, 11228; S. Camiolo, P.A. Gale, M.B. Hursthouse and M.E. Light, *Org. Biomol. Chem.*, 2003, **1**, 741.
66. T. Gunnlaugsson, P.E. Kruger, P. Jensen, F.M. Pfeffer and G.M. Hussey, *Tetrahedron Lett.*, 2003, **44**, 8909; T. Gunnlaugsson, P.E. Kruger, T.C. Lee, R. Parkesh, F.M. Pfeffer and G.M. Hussey, *Tetrahedron Lett.*, 2003, **44**, 6575.
67. M.P. Hughes, M. Shang and B.D. Smith, *J. Org. Chem.*, 1996, **61**, 4510.
68. E. Fan, S.A. Van Arman, S. Kincaid and A.D. Hamilton, *J. Am. Chem. Soc.*, 1993, **115**, 369.
69. B.R. Linton, M.S. Goodman, E. Fan, S.A. Van Arman and A.D. Hamilton, *J. Org. Chem.*, 2001, **66**, 7313.
70. K.-S. Jeong, J.W. Park and Y.L. Cho, *Tetrahedron Lett.*, 1996, **37**, 2795; A.J. Hall, L. Achilli, P. Manesiotis, M. Quaglia, E. de Lorenzi and B. Sellergren, *J. Org. Chem.*, 2003, **68**, 9132.
71. T.R. Kelly and M.H. Kim, *J. Am. Chem. Soc.*, 1994, **116**, 7072.
72. B.C. Hamann, N.R. Branda and J. Rebek Jr., *Tetrahedron Lett.*, 1993, **34**, 6837.
73. K.P. Xiao, P. Bühlmann and Y. Umezawa, *Anal. Chem.*, 1999, **71**, 1183 ; K.P. Xiao, P. Bühlmann, S. Nishizawa, S. Amemiya and Y. Umezawa, *Anal. Chem.*, 1997, **69**, 1038.
74. P. Bühlmann, S. Nishizawa, K.P. Xiao and Y. Umezawa, *Tetrahedron*, 1997, **53**, 1647; S. Nishizawa, P. Bühlmann, M. Iwao and Y. Umezawa, *Tetrahedron Lett.*, 1995, **36**, 6483.
75. S.J. Brooks, P.A. Gale and M.E. Light, *Chem. Commun.* 2005, 4696; S.J. Brooks, P.A. Gale and M.E. Light, *CrystEngComm*, 2005, **7**, 586; S.J. Brooks, P.R. Edwards, P.A. Gale and M.E. Light, *New J. Chem.*, 2006, **30**, 65.
76. F. Werner and H.-J. Schneider, *Helv. Chim. Acta*, 2000, **83**, 465.
77. S.-i. Sasaki, D. Citterio, S. Ozawa and K. Suzuki, *J. Chem. Soc. Perkin Trans. 2*, 2001, 2309.
78. H.-J. Choi, Y.S. Park, S.H. Yun, H.-S. Kim, C.S. Cho, K.Ko and K.H. Ahn, *Org. Lett.*, 2002, **4**, 795.
79. J.L. Jiménez Blanco, J.M. Benito, C.O. Mellet and J.M. García Fernández, *Org. Lett.*, 1999, **1**, 1217; J.M. Benito, M. Gómez-García, J.L. Jiménez Blanco, C.O. Mellet and J.M. García Fernández, *J. Org. Chem.*, 2001, **66**, 1366.
80. S.-i. Sasaki, M. Mizuno, K. Naemura and Y. Tobe, *J. Org. Chem.*, 2000, **65**, 275.
81. K.H. Lee and J.-I. Hong, *Tetrahedron Lett.*, 2000, **41**, 6083.
82. B.H.M. Snellink-Ruël, M.M.G. Antonisse, J.F.J. Engberson, P. Timmerman and D.N. Reinhoudt, *Eur. J. Org. Chem.*, 2000, 165.
83. D. Ranganathan and C. Lakshmi, *Chem. Commun.*, 2001, 1250.

84. A. Casnati, M. Fochi, P. Minari, A. Pochini, M. Reggiani and R. Ungaro, *Gazz. Chim. Ital.*, 1996, **126**, 99.
85. J. Budka, P. Lhoták, V. Michlová and I. Stibor, *Tetrahedron Lett.*, 2001, **42**, 1583.
86. A. Casnati, L. Pirondini, N. Pelizzi and R. Ungaro, *Supramol. Chem.*, 2000, **12**, 53.
87. A.M.A. van Wageningen, E. Snip, W. Verboom, D.N. Reinhoudt and H. Boerrigter, *Liebigs Ann.*, 1997, 2235; F. Sansone, E. Chierici, A. Casnati and R. Ungaro, *Org. Biomol. Chem.*, 2003, **1**, 1802; H. Boerrigter, L. Grave, J.W.M. Nissink, L.A.J. Chrisstoffels, J.H. van der Maas, W. Verboom, F. de Jong and D.N. Reinhoudt, *J. Org. Chem.*, 1998, **63**, 4174; J.W.M. Nissink, H. Boerrigter, W. Verboom, D.N. Reinhoudt and J.H. van der Maas, *J. Chem. Soc. Perkin Trans. 2*, 1998, 1671; J.W.M. Nissink, H. Boerrigter, W. Verboom, D.N. Reinhoudt and J.H. van der Maas, *J. Chem. Soc. Perkin Trans. 2*, 1998, 2541.
88. J. Scheerder, M. Fochi, J.F.J. Engberson and D.N. Reinhoudt, *J. Org. Chem.*, 1994, **59**, 7815.
89. J. Scheerder, J.F.J. Engberson, A. Casnati, R. Ungaro and D.N. Reinhoudt, *J. Org. Chem.*, 1995, **60**, 6448.
90. A.J. Ayling, M.N. Pérez-Payán and A.P. Davis, *J. Am. Chem. Soc.*, 2001, **123**, 12716.
91. T.N. Lambert, J.M. Boon, B.D. Smith, M.N. Pérez-Payán and A.P. Davis, *J. Am. Chem. Soc.*, 2002, **124**, 5276; A.V. Koulov, T.N. Lambert, R. Shukla, M. Jain, J.M. Boon, B.D. Smith, H. Li, D.N. Sheppard, J.-B. Joos, J.P. Clare and A.P. Davis, *Angew. Chem., Int. Ed.*, 2003, **42**, 4931; B.A. McNally, A.V. Koulov, B.D. Smith, J.-B. Joos and A.P. Davis, *Chem. Commun.*, 2005, 1087; J.P. Clare, A.J. Ayling, J.-B. Joos, A.L. Sisson, G. Magro, M.N. Pérez-Payán, T.N. Lambert, R. Shukla, B.D. Smith and A.P. Davis, *J. Am. Chem. Soc.*, 2005, **127**, 10739.
92. P.S. Vermersch, D.D. Lemon, J.J.G. Tesmer and F.A. Quiocho, *Biochemistry*, 1991, **30**, 6861; N.K. Vyas, M.N. Vyas and F.A. Quiocho, *Science*, 1988, **242**, 1290; J.C. Spurlino, G.-Y. Lu and F.A. Quiocho, *J. Biol. Chem.*, 1991, **266**, 5202.
93. E.M.S. Harris, A.E. Aleshin, L.M. Firsov and R.B. Honzatko, *Biochemistry*, 1993, **32**, 1618.
94. J.M. Rini, K.D. Hardman, H. Einspahr, F.L. Suddath and J.P. Carver, *J. Biol. Chem.*, 1993, **268**, 10126.
95. P.V. Nikrad, H. Beierbeck and R.U. Lemieux, *Can. J. Chem.*, 1992, **70**, 241.
96. R.M. Wachter, D. Yarbrough, K. Kallio and S.J. Remington, *J. Mol. Biol.*, 2000, **301**, 157.
97. D.K. Smith, *Org. Biomol. Chem.*, 2003, **1**, 3874.
98. A.P. Davis, J.J. Perry and R.S. Wareham, *Tetrahedron Lett.*, 1998, **39**, 4569.
99. S. Ghosh, A.R. Choudhury, T.N.G. Row and U. Maitra, *Org. Lett.*, 2005, **7**, 1441.
100. N. Pelizzi, A. Casnati and R. Ungaro, *Chem. Commun.*, 1998, 2607.
101. S.S. Whang, S.W. Ko, S.M. Oh, S. Cho and K.C. Nam, *Bull. Korean Chem. Soc.*, 2003, **24**, 165.

102. A. Hybl, R.E. Rundle and D.E. Williams, *J. Am. Chem. Soc.*, 1965, **87**, 2779; M. Noltemeyer and W. Saenger, *J. Am. Chem. Soc.*, 1980, **102**, 2710.
103. G. Tumcharern, T. Tuntulani, S.J. Coles, M.B. Hursthouse and J.D. Kilburn, *Org. Lett.*, 2003, **5**, 4971.
104. F. Cramer, W. Saenger and H.-C. Spatz, *J. Am. Chem. Soc.*, 1967, **89**, 14.
105. J.F. Wojcik and R.P. Rohrbach, *J. Phys. Chem.*, 1975, **79**, 2251; R.P. Rohrbach, L.J. Rodriguez, E.M. Eyring and J.F. Wojcik, *J. Phys. Chem.*, 1977, **81**, 944.
106. Á. Buvári and L. Barcza, *Inorg. Chim. Acta*, 1979, **33**, L179; H. Høiland, L.H. Hald and O.J. Kvammen, *J. Solution Chem.*, 1981, **10**, 775; R.I. Gelb, L.M. Schwartz, M. Radeos and D.A. Laufer, *J. Phys. Chem.*, 1983, **87**, 3349; S.E. Brown, J.H. Coates, P.A. Duckworth and S.F. Lincoin, *J. Chem. Soc. Faraday Trans.*, 1993, **89**, 1035.
107. A. Aversa, W. Etter, R.I. Gelb and L.M. Schwartz, *J. Inclusion Phenom.*, 1990, **9**, 277.
108. L.D. Wilson and R.E. Verrall, *Langmuir*, 1998, **14**, 4710; L.D. Wilson and R.E. Verrall, *Can. J. Chem.*, 1998, **76**, 25; S. Simova and H.-J. Schneider, *J. Chem. Soc., Perkin Trans. 2*, 2000, 1717.
109. I. Gómez-Orellana, D. Hallén and M. Stödeman, *J. Chem. Soc. Faraday Trans.*, 1994, **90**, 3397; Y. Inove, T. Hakushi, Y. Liu, L.-H. Tong, B.-J. Shen, and D.S. Jin, *J. Am. Chem. Soc.*, 1993, **115**, 475; G. Castronuovo, V. Elia, F. Velleca and G. Viscard, *Thermochim. Acta*, 1997, **292**, 31; M. Rekharsky and Y. Inove, *J. Am. Chem. Soc.*, 2002, **124**, 813.
110. L.D. Wilson and R.E. Verrall, *J. Phys. Chem. B*, 2000, **104**, 1880; L.D. Wilson and R.E. Verrall, *J. Phys. Chem. B*, 1998, **102**, 480; L.D. Wilson and F. Velleca, *J. Phys. Chem. B*, 1997, **101**, 9270.
111. K. Kano, N. Tanaka and S. Negi, *Eur. J. Org. Chem.*, 2001, 3689.
112. C. Raposo, M. Crego, M.L. Mussons, M.C. Caballero and J.R. Morán, *Tetrahedron Lett.*, 1994, **35**, 3409.
113. A. Tejeda, A.I. Oliva, L. Simón, M. Grande, M.C. Caballero and J.R. Morán, *Tetrahedron Lett.*, 2000, **41**, 4563.
114. G. Pernía, J.D. Kilburn, J.W. Essex, R.J. Mortishire-Smith and M. Rowley, *J. Am. Chem. Soc.*, 1996, **118**, 10220.
115. V. Jullian, E. Shepherd, T. Gelbrich, M.B. Hursthouse and J.D. Kilburn, *Tetrahedron Lett.*, 2000, **41**, 3963; P.D. Henley and J.D. Kilburn, *Chem. Commun.*, 1999, 1335.
116. G.M. Kyne, M.E. Light, M.B. Hursthouse, J. de Mendoza and J.D. Kilburn, *J. Chem. Soc. Perkin Trans. 1*, 2001, 1258.
117. S. Rossi, G.M. Kyne, D.L. Turner, N.J. Wells and J.D. Kilburn, *Angew. Chem. Int. Ed.*, 2002, **41**, 4233.
118. A.J. Ayling, M.N. Pérez-Payán and A.P. Davis, *J. Am. Chem. Soc.*, 2001, **123**, 12716.
119. M. Albert, J. Zauner, R. Burgert, H. Röttele and R. Fröhlich, *Mater. Sci. Eng. C*, 2001, **18**, 185; P. Zlatusková, I. Stibor, M. Tkadlecová and P. Lhoták, *Tetrahedron*, 2004, **60**, 11383; L.-H. Wei, Y.-B. He, J.-L. Wu, X.-J. Wu, L.-Z. Meng and X. Yang, *Supramol. Chem.*, 2004, **16**, 561; J.V. Hernández,

A.I. Oliva, L. Simón, F.M. Muñiz, M. Grande and J.R. Morán, *Tetrahedron Lett.*, 2004, **45**, 4831.
120. S.O. Kang, J.M. Oh, Y.S. Yang, J.C. Chun, S. Jeon and K.C. Nam, *Bull. Korean Chem. Soc.*, 2002, **23**, 145.
121. Y.S. Yang, S.W. Ko, I.H. Song, B.J. Ryu and K.C. Nam, *Bull. Korean Chem. Soc.*, 2003, **24**, 681.
122. J.S. Albert and A.D. Hamilton, *Tetrahedron Lett.*, 1993, **34**, 7363.
123. A.P. Davis, J.F. Gilmer and J.J. Perry, *Angew. Chem. Int. Ed.*, 1996, **35**, 1312.
124. O. Hayashida, M. Kato, K. Akagi and Y. Aoyama, *J. Am. Chem. Soc.*, 1999, **121**, 11597.
125. S.-I. Kondo, T. Suzuki and Y. Yano, *Tetrahedron Lett.*, 2002, **43**, 7059; N. Kameta and K. Hiratani, *Chem. Commun.*, 2005, 725; P.D. Beer, P.A. Gale and D. Hesek, *Tetrahedron Lett.*, 1995, **36**, 767; X. Hu, A.S.C. Chan, X. Han, J. He and J.-P. Cheng, *Tetrahedron Lett.*, 1999, **40**, 7115; H. Jeong, E.M. Choi, S.O. Kang, K.C. Nam and S. Jeon, *Bull. Korean Chem. Soc.*, 1999, **20**, 1232; H. Yoshida, K. Saigo and K. Hiratani, *Chem. Lett.*, 2000, 116.
126. M.F. de la Torre, E.G. Campos, S. González, J.R. Morán and M.C. Caballero, *Tetrahedron*, 2001, **57**, 3945.
127. W.B. Farnham, D.C. Roe, D.A. Dixon, J.C. Calabrese and R.L. Harlow, *J. Am. Chem. Soc.*, 1990, **112**, 7707.
128. P.B. Savage, S.K. Holmgren and S.H. Gellman, *J. Am. Chem. Soc.*, 1994, **116**, 4069.
129. C.-H. Chang and L.G. Marzilli, *J. Am. Chem. Soc.*, 1974, **96**, 3656.
130. B.L. Trus and R.E. Marsh, *Acta Cryst.*, 1972, **B28**, 1834.
131. N.S. Mandel, *Acta Cryst.*, 1977, **B33**, 1079.
132. J.S. Sherfinski and R.E. Marsh, *Acta Cryst.*, 1973, **B29**, 192; J.J. Guy, L.R. Nassimbeni, G.M. Sheldrick and R. Taylor, *Acta Cryst.*, 1976, **B32**, 2909; L.G. Marzill, C.-H. Chang, J.P. Caradonna and T.J. Kistenmacher, *Adv. Mol. Relaxation Interact. Processes*, 1979, **15**, 85; M. Rossi, J.P. Caradonna, L.G. Marzill and T.J. Kistenmacher, *Adv. Mol. Relaxation Interact. Processes*, 1979, **15**, 103.

CHAPTER 5

Neutral Pyrrole Systems

5.1 Introduction

As a general rule, neutral, organic-based anion receptors rely primarily on hydrogen-bonding interactions to stabilize their complexes with bound anions. This stands in contrast to cationic receptors, which benefit from both electrostatic and hydrogen-bonding interactions. Potential advantages of neutral receptors include the possibility of more selective binding. This is because, in contrast to binding effects due to positive charges, hydrogen-bonding interactions are generally anisotropic. Thus, the use of such interactions allows for the construction, at least in principle, of receptors that are selective for a particular anion. Another advantage associated with the use of a neutral system is that there is no inherent competition with a counter anion, an effect that often tends to complicate analysis of the systems in question. In the case of pyrrolic systems, the use of neutral receptors obviates the need for protonation, either prior to, or concurrent with, anion recognition. Not surprisingly, considerable effort has been devoted of late to the construction of neutral pyrrole-based anion receptors. However, most interestingly, the history of such systems dates back only to 1996 when the use of calix[4]pyrrole as an anion receptor was first reported.

5.2 Cyclic Receptors

As implied above, the calix[4]pyrroles (*e.g.*, **5.1a**) were the first neutral pyrrole-based anion-binding systems to be reported as such. They were considered attractive as receptors because they are not only easy to prepare, but also potentially "tunable" in terms of their inherent anion-binding characteristics. This promise which is still being realized in a number of laboratories world-wide, has made the generalized class of calix[*n*]pyrroles (where n may be ≥ 4) one of the more attractive approaches to anion-receptor generation currently being pursued.

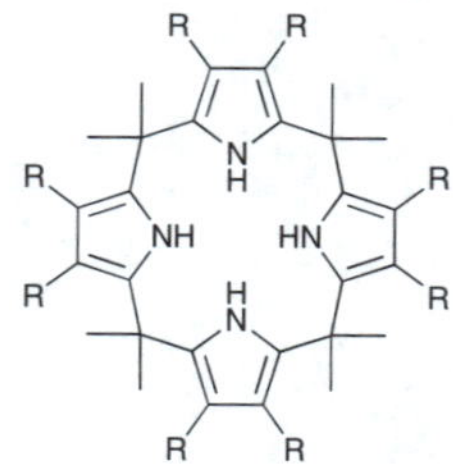

5.1a R = H
5.1b R = Br

The calix[4]pyrroles themselves, in contrast to their higher homologues (*vide infra*), are a venerable class of materials. They were first prepared by Baeyer[1] in 1886 but only put forward as possible anion-binding agents in 1996.[2] In the case of *meso*-octamethylcalix[4]pyrrole **5.1a**, it was found that both fluoride and chloride anions were bound in the solid state (Figure 5.1).[2,3] While these two structures are similar, in the case of the fluoride-anion complex, the average of N···F distance is 2.767 Å, while for the corresponding chloride complex the N···Cl distance is 3.303 Å. Thus, fluoride anion appears to be more tightly bound in the solid state. In both cases, it is important to appreciate that the cone conformation, seen in the presence of anions, is very different from the 1,3-alternate form seen in their absence.

In CD_2Cl_2 solution, calix[4]pyrrole **5.1a** was found to bind fluoride and chloride anions (studied in the form of their tetrabutylammonium salts) with affinities, K_a, of 1.7×10^4 and 3.5×10^2 M^{-1}, respectively, as judged from NMR experiments.[2,3] As part of this generalized study, several derivatives, including the β-octabromo-*meso*-octamethylcalix[4]pyrrole **5.1b**, were also prepared. In the case of **5.1b**, the synthesis was straightforward and simply involved reaction of calix[4]pyrrole **5.1a** with *N*-bromosuccinimide (NBS). The anion-binding ability of this compound was also measured by NMR under conditions analogues to those used for **5.1a**. On this basis, it was found that **5.1b** was an improved fluoride and chloride anion receptor (K_a = 2.7×10^4 and 4.3×10^3 M^{-1} for fluoride and chloride, respectively).[4] These higher anion-binding affinities were rationalized in terms of the eight bromine substituents; these electron-withdrawing atoms serve to increase the acidity of the pyrrolic NH, thereby enhancing the anion-binding affinities.

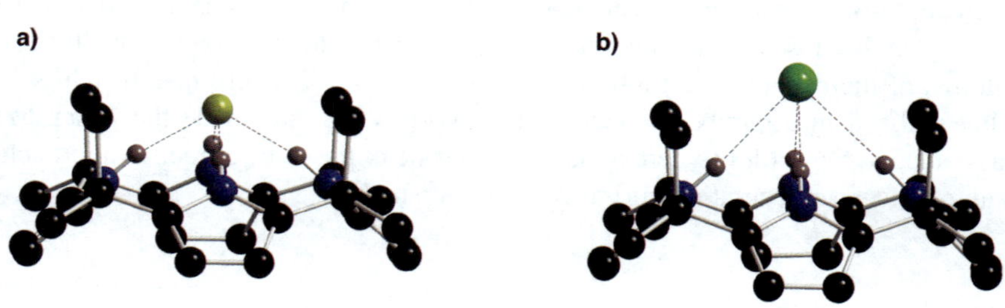

Figure 5.1 *(a) Single-crystal X-ray structure of the fluoride-anion complex of calix[4]pyrrole **5.1a** (b) single-crystal X-ray structure of the chloride-anion complex of calix[4]pyrrole **5.1a***

Recently, Schmidtchen[5] applied isothermal titration calorimetry (ITC) to the study of calix[4]pyrrole-anion recognition while using his report to highlight further the advantages of this method for studying problems in molecular recognition. These latter include an increased dynamic range, higher detection limits, and greater reproducibility as compared to NMR-spectroscopic methods. Table 5.1 summarizes the anion-affinity constants of calix[4]pyrrole **5.1a** measured by ITC that have proved reproducible in the authors' hands. An inspection of this table reveals that, in the case of systems with high affinities, the recorded K_a values are generally greater than those derived from NMR-spectroscopic methods. Nonetheless, at least among the set of anions studied, the same general selectivity trends are revealed by ITC as by NMR titrations.*

In a related work, the nucleoside-functionalized calix[4]pyrroles **5.2a** and **5.2b** were synthesized. As neutral receptors, the calix[4]pyrroles display much lower phosphate anion-binding affinities than do the protonated sapphyrins. On the other hand, the sapphyrins are flat, while the calixpyrroles adopt non-planar conformations. This means that a nucleobase substituent, if appropriately connected to the calix[4]pyrrole framework, could produce a system capable of effecting the cooperative recognition of a complementary mononucleotide substrate. In an effort to test this hypothesis, calix[4]pyrroles **5.2a** and **5.2b** were studied as potential nucleotide carriers using a U-tube model membrane in analogy to what was done with the sapphyrin systems (*cf.* Chapter 3). They were also studied as the key components in membrane-based ion-selective electrodes (ISEs) as detailed in Chapter 8.[6]

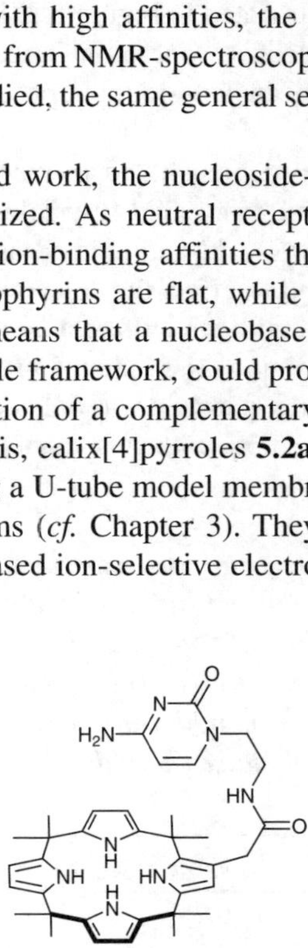

5.2a

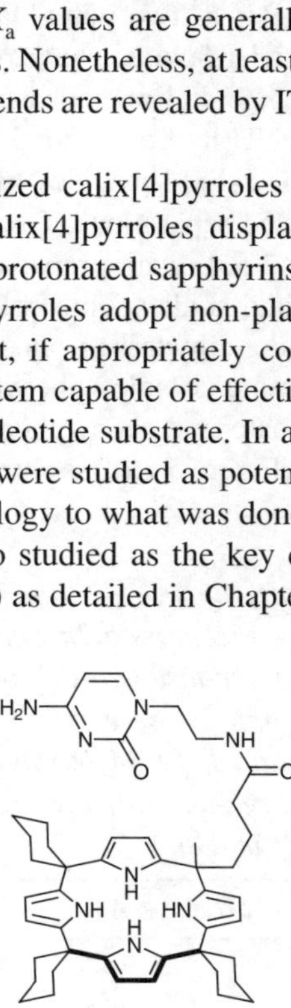

5.2b

Table 5.1 *Association constants, K_a, for compound **5.1a** (M^{-1}) for various anionic substrates as measured by ITC at 30 °C using the corresponding tetrabutylammonium salts*

Anion	Cl^-	Br^-	CN^-	NO_2^-	$CH_3CO_2^-$	$C_6H_5CO_2^-$	$H_2PO_4^-$
Solvent	CH_3CN	CH_3CN	CH_3CN	CH_3CN	CH_3CN	CH_3CN	DMSO
K_a	140,000	3400	9400	1700	290,000	120,000	5100

*In spite of considerable effort, the authors have been unable to obtain reliable K_a values for fluoride anion binding by ITC. As such, the claim made by Schmidtchen[5] that fluoride and chloride are both bound with approximately the same affinity in CH_3CN has yet to be independently substantiated, at least to the best of the authors' knowledge. On the other hand, this failure to obtain readily interpretable ITC data in the case of fluoride is an indication that the binding events as originally described, fluoride anion + calix[4]pyrrole → fluoride-anion complex, may be an oversimplification and that additional processes, such as ion-pairing or follow-up equilibria need to be considered.

Using a U-tube model membrane containing tetrabutylammonium chloride in the organic phase (added to provide a hydrophobic, charge-neutralizing countercation), good selectivity for 5′-GMP was seen for compound **5.2b** with the absolute rate of transport being 9.3×10^{-10} mol $cm^{-2}h^{-1}$. The use of carrier **5.2a** served to lower further the transport rate and led to selectivity for 5′-CMP rather than for 5′-GMP (Table 5.2).[6] These results were considered consistent with carrier **5.2b** acting as a *bona fide* ditopic receptor, capable of binding both the phosphate "head" and purine "tail" of 5′-GMP in a "two-point" fashion as illustrated in Figure 5.2. By contrast, it appears as if carrier **5.2a** acts more as a non-selective "one-point" receptor, binding both the phosphate and nucleobase portions of nucleotide monophosphates at the calix[4]pyrrole NH-hydrogen bond donor site.

The utility of calix[4]pyrroles has also been demonstrated in the area of chromatography-based separations. Towards this end, two calix[4]pyrrole-modified silica gels, **5.3** and **5.4**, were prepared by Sessler *et al.*[7] and studied as HPLC solid supports. In initial studies, both gels were found to effect the selective retention of fluoride, chloride, dihydrogen phosphate, and hydrogen sulfate anions (Table 5.3). In the case of more complex phenyl-based oxyanions (*i.e.*, phenyl arsenate, phenyl sulfonate, benzoate, and phenylphosphate), good separation was only achieved in the case of gel **5.3** (Figure 5.3a). As shown in Figure 5.3b, both gels, namely **5.3** and **5.4**,

Table 5.2 *Results of model through membrane transport experiments conducted using receptors **5.2a** and **5.2b**. The concentration of the carriers in the organic phase was 0.1 mM, and the pH of the initial and receiving aqueous phases were 6.0 and 12.5, respectively. The organic phase also contained 0.1 M tetrabutylammonium perchlorate, a species added as a charge-neutralizing agent. The tabulated k_T values are in units of 10^{-11} mol $cm^{-2}h^{-1}$*

Carrier	*Aq I (pH)*	*Aq II*	*k_T 5′-GMP*	*$k_{5'\text{-GMP}}/k_{5'\text{-CMP}}$*	*$k_{5'\text{-GMP}}/k_{5'\text{-AMP}}$*
5.2a	6.0	12.5	9.8	0.2	1.8
5.2b	6.0	12.5	93	7.8	1.9

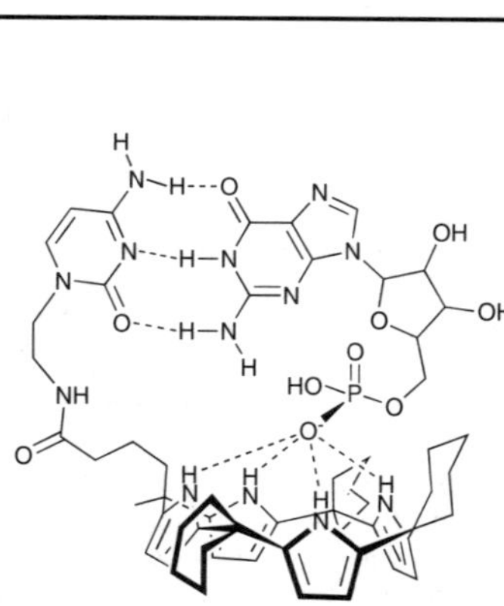

Figure 5.2 *Structure of the proposed supramolecular complex formed between cytosine appended calix[4]pyrrole **5.2b** and 5′-GMP*

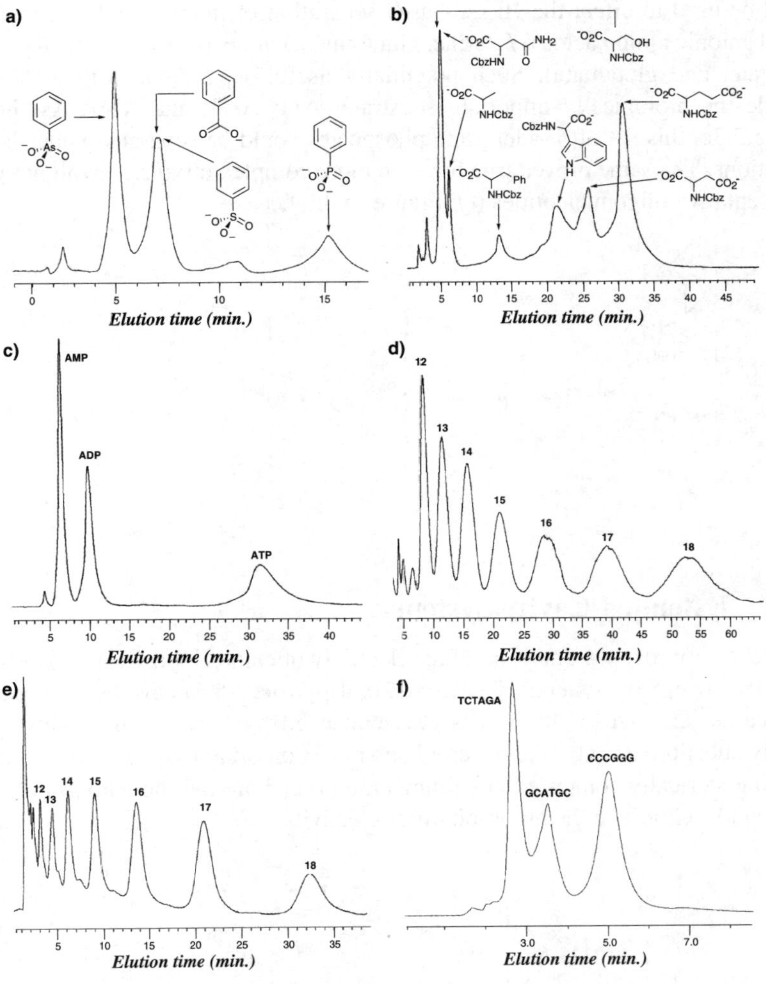

Figure 5.3 *HPLC-based separation of (a) various phenyl-substituted anions with **5.3** (flow rate, 0.3 ml min^{-1}; mobile phase, aqueous phosphate buffer (50 mM) at pH 7.0; UV detection at 254 nm), (b) N-carbobenzyloxy (Cbz) protected amino acids with **5.4** (flow rate, 0.3 ml min^{-1}; mobile phase, 3:1 (v/v) aqueous acetate buffer (30 mM) at pH 7.0 acetonitrile (isochratic); UV detection at 254 nm), (c) AMP, ADP, and ATP with **5.4** (flow rate, 0.3 ml min^{-1}; mobile phase, aqueous sodium phosphate buffer (105 mM) at pH 7.0 (isochratic); UV detection at 262 nm), (d) oligodeoxythymidylate fragments containing between 12 and 18 nucleotide subunits (dT_{12-18}) with **5.3** (flow rate, 0.4 ml min^{-1}; mobile phase, 1:1 (v/v) CH_3CN/aqueous sodium chloride (250 mM)-sodium phosphate (50 mM) at pH 7.0 (isochratic); UV detection at 265 nm), (e) dT_{12-18} with **5.4** (flow rate, 0.25 ml min^{-1}; mobile phase, 3:2 (v/v) CH_3CN/aqueous sodium chloride (50 mM)-sodium phosphate (40 mM) at pH 7.0 (isochratic); UV detection at 265 nm), and (f) oligonucleotide hexamers, TCTAGA, GCATGC, and CCCGGG, on a modified silica gel column derived from **5.3** (flow rate, 0.4 ml min^{-1}; mobile phase, 1:1 CH_3CN/50 mM sodium phosphate buffer at pH 7.0 (isochratic); UV detection at 265 nm).*

(Figure (d) reproduced with permission from P.A. Gale, J.L. Sessler and V. Král, *Chem. Commun.*, 1998, 1. Copyright 1998 Royal Society of Chemistry)

could be used to effect the HPLC-based separation of more complicated Cbz-protected anionic amino acids (*cf.* serine, glutamine, alanine, phenylalanine, tryptophan, aspartate, and glutamate). Such potentially useful separations were extended to include the biologically important substrates ATP, ADT, and AMP. As shown in Figure 5.3c, this set of 5′-adenosine phosphates could be separated under isocratic conditions. The same proved true for even more complex mixtures involving various representative oligonucleotides (*cf.* Figures 5.3d–f).

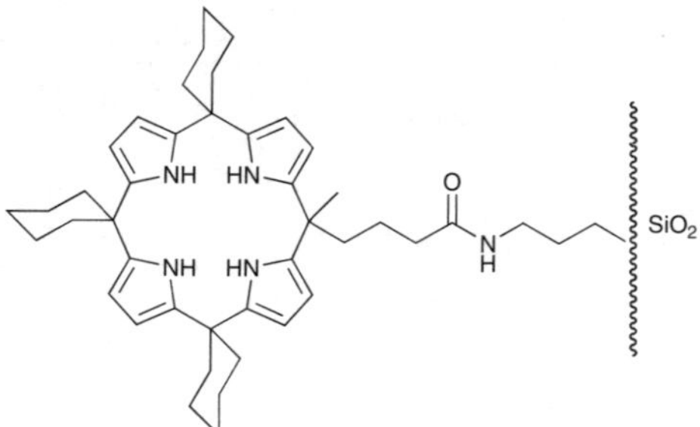

5.3

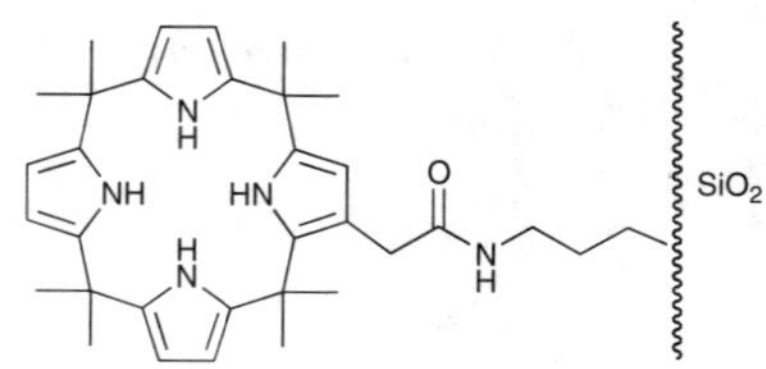

5.4

5.2.1 Extended Cavity Systems

In order to improve the anion-binding selectivity of calix[4]pyrrole, two kinds of the so-called "deep" or "extended" cavity calix[4]pyrroles (**5.5**) have been prepared. In both cases, the size of the cavity was enhanced by "appending" bulky groups, namely substituted aryl[8–11] and steroid subunits[12] onto the four *meso*-positions. The resulting sterically congested environment, enforced around the anion-binding site, was found to improve the anion-binding selectivity.

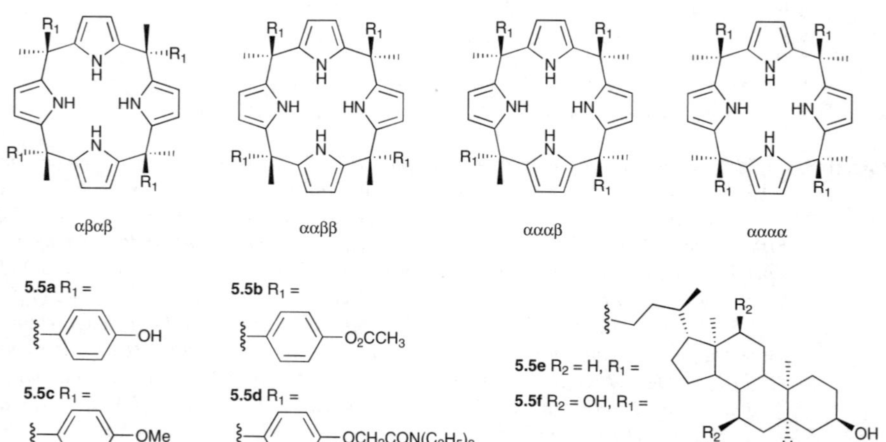

Both the *meso*-tetraaryl and tetrasteroidyl deep-cavity systems were prepared via acid-catalysed condensations involving pyrrole and the requisite ketone. Such reactions provide a mixture of four configurational isomers which are termed *αβαβ*, *ααββ*, *αααβ*, and *αααα*, respectively, in analogy to the porphyrin literature.

Table 5.3 *Retention times (min) for anions (studied as the corresponding tetrabutylammonium salts) on HPLC columns derived from modified silica gels* ***5.3*** *and* ***5.4***[a]

	F^-	Cl^-	$H_2PO_4^-$	HSO_4^-
5.3	16.4	15.2	20.1	16.2
5.4	16.9	17.9	22.0	16.2

[a]Anions were eluted as 1 mM CH_3CN solutions of their tetrabutylammonium salts under the following conditions (flow rate, 0.4 ml min^{-1}; mobile phase, CH_3CN; detection, conductivity).

X-ray structural analysis of the two aryl systems **5.5a** and **5.5b** reveal receptor-substrate complexes having high walls and well-defined binding cavities (Figure 5.4), at least in the case of the $\alpha\alpha\alpha\alpha$ isomers.[10,11]

Not surprisingly, the presence of a well-defined binding cavity resulted in a greatly enhanced selectivity for fluoride anion, relative to chloride anion and dihydrogen phosphate anion. Presumably, this enhancement reflects the fact that small guests such as fluoride can fit into the size-limited binding cavities. Consistent with this assumption, a significant difference in the anion-binding affinities for compounds **5.1a**, **5.5a**, and **5.5c** was observed in acetonitrile solution, as determined by standard ^{1}H NMR titration methods (*cf.* Table 5.4). For example, the $\alpha\alpha\beta\beta$ isomer of **5.5a**, the isomer having the most open binding cavity, was found to bind chloride anion roughly four times as well as the other isomers. Data from this same solution-phase binding study also revealed obvious electronic effects, in spite of the fact that the distance between the pyrrolic NH-donor atoms and the aryl substituents is rather considerable. For instance, each isomer of **5.5a** displays a notably higher anion affinity than do the corresponding isomers in the methoxy substituted series, **5.5c**.

In DMSO-d_6, a more polar solvent system than acetonitrile, the $\alpha\alpha\alpha\alpha$-isomers of the two extended cavity receptors **5.5b** and **5.5c** were found to bind only fluoride

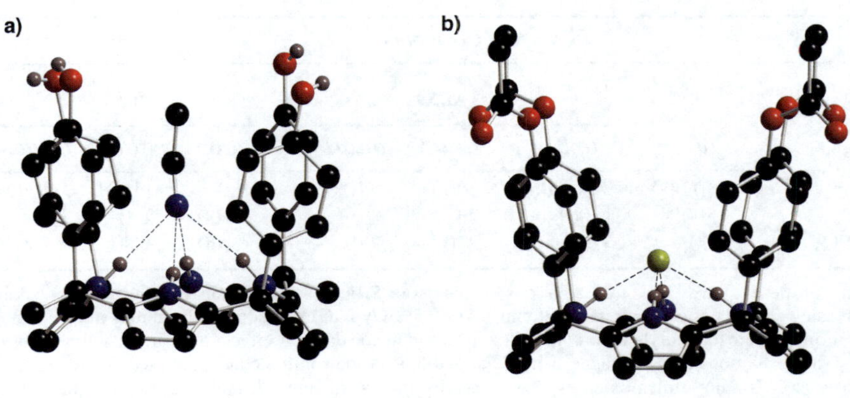

Figure 5.4 *(a) Single-crystal X-ray structure of the acetonitrile complex of calix[4]pyrrole* ***5.5a*** *and (b) Single-crystal X-ray structure of the chloride-anion complex of calix[4]pyrrole* ***5.5b***

anion within the limits of detection.[8,10] However, even for this anion, the binding affinities were low. For instance, in DMSO-d_6, the fluoride-anion-binding affinity of **5.5b** is *ca.* 74 M^{-1} as determined from ^1H NMR spectroscopic titrations.

The steroid-based calix[4]pyrrole systems **5.5e** and **5.5f** were introduced by Král and Sessler in 2002. In this case, a FAB-MS screening process was used to determine the ability of these systems to effect the enantioselective recognition of tartaric acid and mandelic acid (Figure 5.5).[12] Evidence for enantioselective binding was found in the case of the polyhydroxylated $\alpha\alpha\alpha\beta$ configurational isomer of **5.5f**. This finding is rationalized in terms of multiple substrate-receptor hydrogen-bonding interactions that involve not only specific anion-pyrrolic NH contacts, but also less well-defined steroid-substrate interactions. The importance of these latter ancillary contacts is underscored by the fact that **5.5e** proved far less effective than **5.5f**, as judged by both mass spectrometric (MS) screening and more direct extraction methods.

5.2.2 Higher Order Systems

Although not isolable from the normal acid-catalysed reaction of pyrrole and simple ketones, ever since Sessler's 1996 report, there has been interest in preparing the so-called higher order calix[*n*]pyrrole ($n > 4$) systems. Such systems, it could be expected, might display altered anion-binding affinities or display selectivities for larger anionic substrates. In recent years, several groups have, in fact, succeeded in preparing higher order calix[*n*]pyrroles (*i.e.*, **5.7–5.10**), bipyrrole-derived analogues (*e.g.*, **5.11** and **5.12**), bispyrroylcarbazole-derived analogues (**5.13**), and bispyrroylbenzene-derived analogues (**5.14**).

Table 5.4 *Association constants (K_a, M^{-1}) for compounds **5.1a**, **5.5a**, and **5.5c** with anionic substrates*[a] *as determined from ^1H NMR spectroscopic titrations carried out in acetonitrile-d_3 (0.5% v/v D_2O) at 22 °C*

	Compound						
			5.5a			**5.5c**	
Isomer	***5.1a***	$\alpha\alpha\beta\beta$	$\alpha\alpha\alpha\beta$	$\alpha\alpha\alpha\alpha$	$\alpha\alpha\beta\beta$	$\alpha\alpha\alpha\beta$	$\alpha\alpha\alpha\alpha$
F^-	>10,000	>10,000	5000[b]	>10,000[c]	460	1100[b]	>10,000
Cl^-	>5000	1400[d]	260	320	<100	220	300
$H_2PO_4^-$	1300	520[d]	230	500	<100	<80	<100

[a]Acetonitrile-d_3 (0.5% v/v D_2O) solutions of receptors **5.1a, 5.5a,** and **5.5c** were titrated by adding increasing aliquots of concentrated acetonitrile-d_3 (0.5% v/v D_2O) solutions of the anions in question (in the form of their tetrabutylammonium salts). To account for dilution effects and simplify the analyses, these latter solutions also contained **5.1a, 5.5a,** and **5.5c** at their initial concentrations. Estimated errors were <15%. Binding stoichiometries, determined by Job plots, were 1:1 unless otherwise indicated.
[b]Fit by following the change of two different β-pyrrolic CH resonances.
[c]At high [F^-]/[calixpyrrole] ratios, a second binding process, involving presumably interactions between the fluoride and the phenolic OH residues, is observed.
[d]Fit by following the change in both the *meso*-aryl CH and β-pyrrole CH resonances.

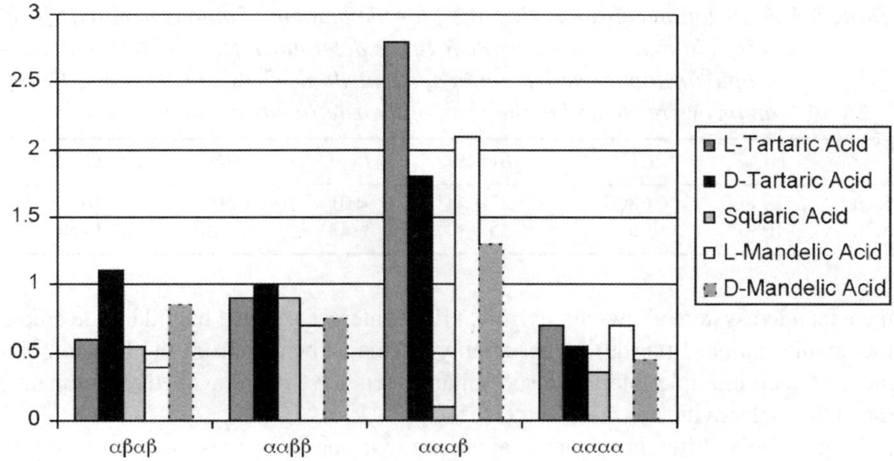

Figure 5.5 *Results of FAB-MS screening experiments designed to probe the interactions of **5.5f** with selected anions. For these studies, a large excess (>100-fold) of the carboxylic acid in question was added to a MeOH solution of the receptor and subjected to FAB-MS analysis. The ratios depicted are the ratios in percentages with relation to the parent ion peak. It should be noted that these results give only a semi-quantitative approximation of the relative concentrations of the species in question under more normal solution phase conditions.*
(Reproduced with permission from "Artificial Pyrrole-based Anion Receptors," in Functional Synthetic Receptors, J.L. Sessler, W.-S. Cho, T. Schrader and A.D. Hamilton (eds), Wiley–VCH, Weinheim, 2005)

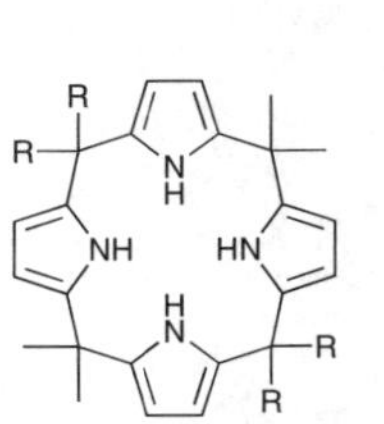

5.6a R = phenyl
5.6b R = 2-pyridyl
5.6c R-C-R = 9,9-fluorenediyl

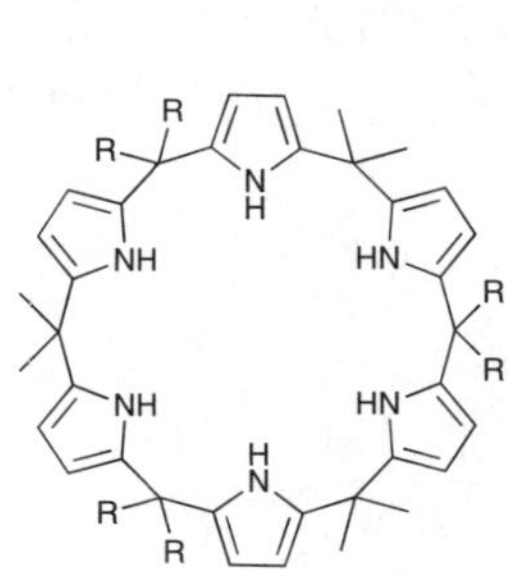

5.7a R = phenyl
5.7b R = 2-pyridyl

The first synthesis of free-standing higher order calix[6]pyrroles was achieved by Eichen and co-workers.[13] The resulting receptors, compounds **5.7a** and **5.7b**, have two more pyrrole units and possess an expanded core relative to calix[4]pyrrole. Table 5.5 lists the published association constants of calix[4]pyrrole (**5.1a**) and calix[6]pyrrole (**5.7a**), as derived from ^1H NMR titration studies carried out in acetone-d_6/$CDCl_3$ (1:9). In the case of the *meso*-octamethylcalix[4]pyrrole **5.1a**, a clear preference for fluoride is observed. In contrast, the extended calix[6]pyrrole system **5.7a** shows a moderate preference for iodide anion (the binding order is $I^- > F^- >> Cl^- > Br^-$ rather than $F^- > Cl^- > Br^- > I^-$ as seen for **5.1a**). The fact that higher affinities are found for larger anions in these systems underscores the importance of having a proper geometrical fit between the host and guest. While the core size of

Table 5.5 *Association constants (K_a, M^{-1}) for the binding of anions to calix[6]pyrrole **5.7a** and calix[4]pyrrole **5.1a**, as determined from ^1H NMR spectroscopic titrations carried out in acetonitrile-d_3-$CDCl_3$ (1:9) at 298 K. The anions were studied in the form of their tetrabutylammonium salts*

	F^-	Cl^-	Br^-	I^-	BF_4^-	$CF_3CO_2^-$
5.1a	23,800	6800	270	<10	<10	70
5.7a	1080	650	150	6600	2350	1150

the expanded system allows six pyrrolic NH to interact with the bound iodide anion, the smaller anion, fluoride, is presumably unable to benefit from the full complement of such multiple interactions. Similar effects have been invoked to rationalize the cation selectivities of crown ethers.[14]

Using a very different synthetic approach, but one that is also characterized by extreme elegance, Kohnke and his co-workers reported the synthesis of the *meso*-decamethylcalix[5]pyrrole and *meso*-dodecamethylcalix[6]pyrrole systems **5.8** and **5.9a**. These compounds were prepared from calix[*n*]furan (n = 5, 6) by using first *m*-CPBA to open the furan heterocycles, and then ammonium acetate to form the requisite pyrrole rings.[15] Single crystals of both the chloride and bromide complexes of **5.9a** were obtained and elucidation of the corresponding X-ray diffraction structures revealed that the anions are coordinated at the centre of the macrocycle via six hydrogen bonds (Figure 5.6). While these two structures are quite similar, in the case of the chloride-anion complex the N···Cl distances range from 3.265 to 3.305 Å (*vs.* a range of 3.264–3.331 Å in the corresponding calix[4]pyrrole-chloride structure[2]), while the N···Br distances range from 3.344 to 3.404 Å.[16]

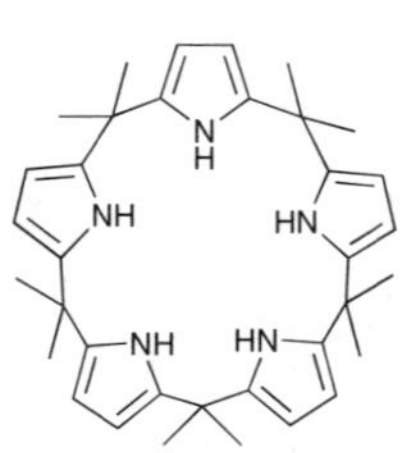

5.8

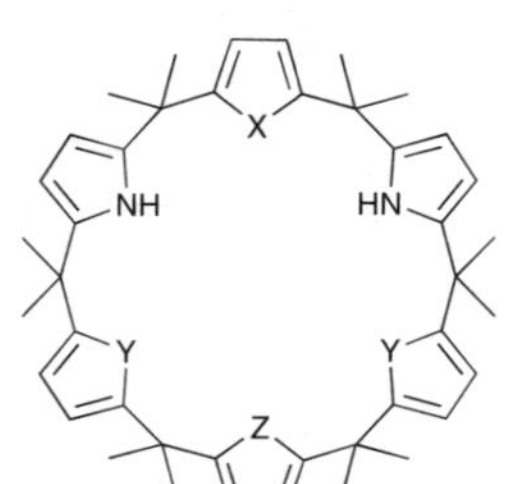

5.9a X = Y = Z =NH
5.9b X = Y = O, Z = NH
5.9c X = Z = O, Y = NH
5.9d X = O, Y = Z = NH

The result of anion-binding studies carried out with **5.8** and **5.9** are shown in Table 5.6.[15,17] It is difficult to make comparisons within this data set because not all of the K_a values were obtained under identical conditions or using the same method. Nonetheless, to the extent such comparisons may be made, it can be seen from Table 5.6 that for simple halide anions, the association constants increase with the number of pyrrolic NH units present in the macrocycles. Further, the inherent selectivity is seen to change. In the case of calix[4]pyrrole **5.1a**, for instance, the fluoride to chloride K_a ratio is roughly 59 in D_2O saturated dichloromethane. By contrast, the

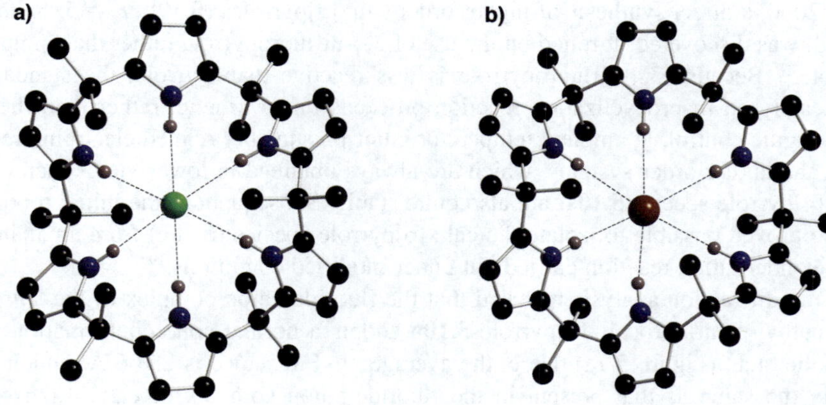

Figure 5.6 *(a) Single-crystal X-ray structure of the chloride anion complex of calix[6]pyrrole* ***5.9a*** *and (b) the single-crystal X-ray structure of the corresponding bromide anion complex*

Table 5.6 *Association constants (K_a, M^{-1}) for various 1:1 anion complexes. The countercation was n-Bu_4N^+ in all cases with the exact experimental conditions being indicated in the footnotes*

Anions	***5.1a***	***5.8***	***5.9a***	***5.9c***	***5.9d***
F^-	2700[a]	14,000[b]	*ca.* 320,000[g]	57,000[g]	f
Cl^-	46[a]	35[b]	12,000[e]	65[a]	5500[e]
Br^-	10[d]	–	710[c]	<10[a]	69[c]
I^-	<10[d]	–	i	i	i
$H_2PO_4^-$	97[d]	–	h	i	i
HSO_4^-	<10[d]	–	*ca.* 10[c]	i	i
NO_3^-	<10[d]	–	16[c]	i	<10[c]
CN^-	<10[a]	–	*ca.* 100[a]	<100[a]	h

[a]Measured by NMR spectroscopic titration in D_2O saturated CD_2Cl_2 at 20 °C.
[b]Measured by NMR spectroscopic titration in D_2O saturated CD_2Cl_2 at 22 °C.
[c]Measured by NMR spectroscopic titration in "dry" CD_2Cl_2 at 20 °C.
[d]Measured by NMR spectroscopic titration in "dry" CD_2Cl_2 at 25 °C.
[e]Determined using the Cram extraction method at 16 °C using D_2O/CD_2Cl_2.
[f]K_a proved too large to be determined by NMR spectroscopic titration methods. Moreover, the compound also proved ineffective as a phase transfer agent between D_2O and CD_2Cl_2, meaning the Cram extraction method could likewise not be used.
[g]Measured by competition experiment using calix[4]pyrrole **5.1a** as the competing species.
[h]An interaction is observed, but the NH resonances disappeared upon addition of the salt; thus, a K_a value could not be determined.
[i]No detectable binding observed in either dry or wet CD_2Cl_2.

corresponding ratio for calix[6]pyrrole **5.9a** is *ca.* 27, indicating a relatively lower fluoride selectivity than in the parent calix[4]pyrrole. On the other hand, receptors **5.8** and **5.9c** exhibit enhanced selectivity for fluoride over chloride, with the relevant K_a ratios being *ca.* 400 and 877, respectively.

In 2000, a direct synthesis of higher order calix[*n*]pyrroles (**5.10b**, $n = 5$; **5.10d**, $n = 8$) was discovered. It relied on the use of 3,4-difluoropyrrole rather than simple pyrrole.[18] Because 3,4-difluoropyrrole is less reactive than pyrrole, the standard acid-catalysed macrocyclization reaction proceeds under kinetic, rather than thermodynamic control, at ambient temperature. For presumably related electronic reasons, the higher order systems, which are always obtained in lower yields than the calix[4]pyrrole species **5.10a**, are also quite stable. Subsequent to the initial report, it also proved possible to isolate the calix[6]pyrrole species (**5.10c**) from an analogous condensation reaction carried out under modified conditions.[19]

X-ray diffraction analysis revealed that the fluoride-anion complex of the *meso*-octamethyl-octafluorocalix[4]pyrrole **5.10a** adopts a normal cone conformation in the solid state (Figure 5.7a). Here, the average N···F distance is 2.766 Å, which is nearly the same as that present in the fluoride-anion complex of calix[4]pyrrole **5.10a**. By contrast, considerable distortion is seen in the crystal structure of *meso*-decamethyl-decafluorocalix[5]pyrrole **5.10b**, although it is to be noted that this structure reflects the anion-free form (Figure 5.7b). A high level of distortion, coupled with the presence of interannular CF···H hydrogen bonds also characterizes the structure of **5.10d** (Figure 5.7c).

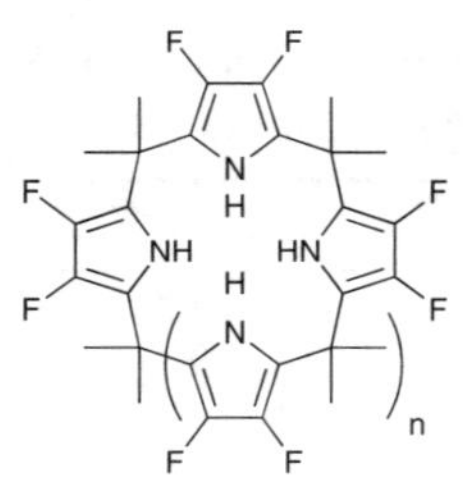

5.10a n = 1
5.10b n = 2
5.10c n = 3
5.10d n = 5

Quite recently, a comparable set of anion-binding affinities for the fluorinated calix[*n*]pyrroles were determined using ITC titration in solution under well-defined conditions. They were also studied via extraction experiments, which involved partitioning between an organic and aqueous layer. It was found that octafluorocalix[4]pyrrole **5.10a** exhibits little discrimination among chloride, bromide, and iodide anions. However, these halide anions were extracted more effectively than nitrate and fluoride anions. On the other hand, the decafluorocalix[5]pyrrole **5.10b** was found to exhibit a preference for nitrate and fluoride anions.[20]

The ITC titration studies, carried out under very different conditions (*e.g.*, dry acetonitrile and DMSO) revealed slightly different trends.[21] Here, it was found that the relative association constants ($K_{rel} = K_{a(Cl)}/K_{a(Br)}$) decrease with increasing macrocycle size (Table 5.7). This represents a reversal of what was expected based on anion electronegativity. Specifically, chloride anion, being more electronegative and possessing a higher charge density, was expected to be bound more strongly than bromide anion. While this is true in an absolute sense, in the case of **5.10c**, the K_{rel} value is rather small, presumably reflecting the fact that the relatively large cavity present in calix[6]pyrrole is better able to accommodate bromide anion than chloride anion. Interestingly, in the case of $H_2PO_2^-$ anion, an inherently less spherical anion, the

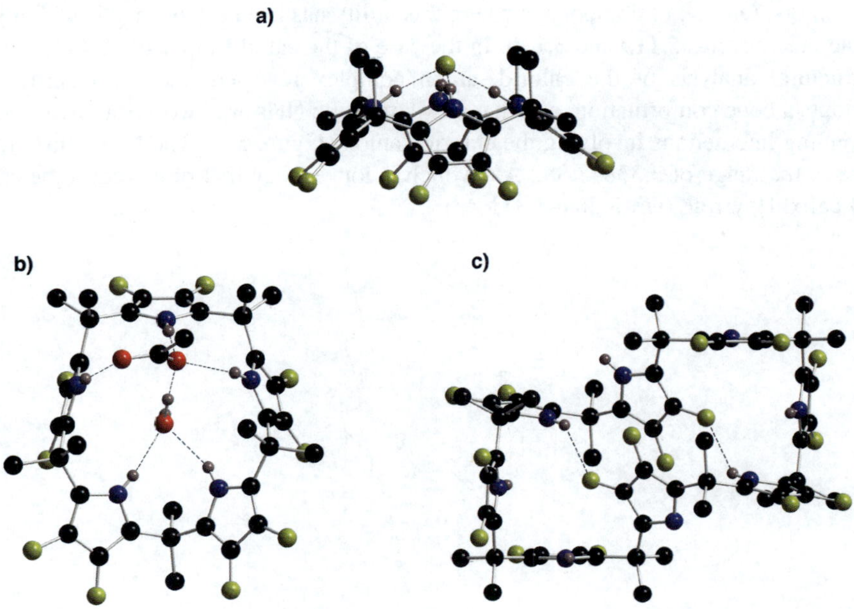

Figure 5.7 *X-ray structure of (a) the fluoride anion complex of octafluorocalix[4]pyrrole **5.10a**, (b) acetate-anion complex of decafluorocalix[5]pyrrole **5.10b**, and (c) neutral hexadecafluorocalix[8]pyrrole **5.10d***

Table 5.7 *Association constants (K_a, M^{-1}) determined in CH_3CN or DMSO solution by ITC at 30 °C using the n-Bu_4N^+ salts of the indicated anions*[a]

Anions	*Solvent*	***5.10a***	***5.10b***	***5.10c***
Cl^-	CH_3CN	530,000	41,000	280,000
Br^-	CH_3CN	8500	4500	110,000
I^-	CH_3CN	[b]	[b]	610
$H_2PO_4^-$	DMSO	17,000	9600	15,000
$H_2PO_2^-$	DMSO	3300	13,000	35,000
$C_6H_5CO_2^-$ [c]	CH_3CN	1,390,000	83,000	580,000
$CH_3CO_2^-$ [c]	CH_3CN	2,360,000	520,000	1,020,000
$K_{rel} = K_{a(Cl)}/K_{a(Br)}$		62.4	9.1	2.5

[a]The host (macrocycle) solution was titrated with the guest (anion) solution unless otherwise indicated.
[b]No evidence of anion binding was observed.
[c]The guest solution was titrated with the host solution (reverse titration).

association constants were seen to increase as the number of β-fluorinated pyrrolic subunits increased.

In order to make more efficient receptors for larger anions, several groups have begun working recently on the generation of new extended calixpyrrole-type systems that are based on various combinations of bipyrrole, furan, thiophene, and pyrrole. Not all of these systems have been demonstrated as being useful anion receptors. One set of compounds where this utility has been established is the bipyrrole macrocycles **5.11a** and **5.11b**. In the case of the calix[3]bipyrrole **5.11a**, X-ray structural analysis of the chloride-anion complex revealed that the macrocycle adopts a cone conformation, and that all six pyrrolic NHs are involved in hydrogen-bonding interactions involving the chloride anion (Figure 5.8). The N···Cl distances are in the range of 3.338–3.382 Å,[22] which is longer than that observed in the case of calix[4]pyrrole (*viz.* 3.264–3.331 Å[2]).

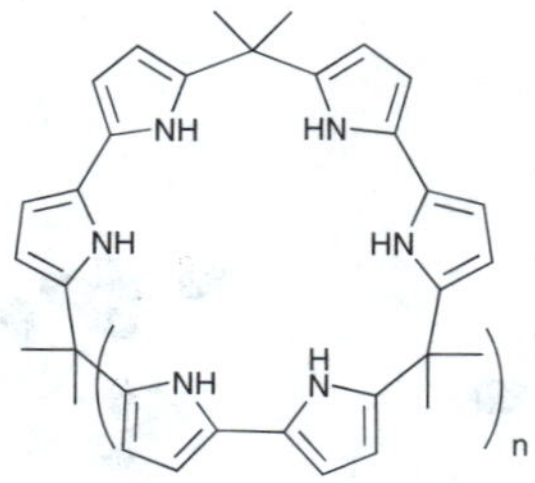

5.11a n = 1
5.11b n = 2

Recently, the X-ray structure of the chloride-anion complex of calix[4]bipyrrole **5.11b** was also solved. It reveals that a simple chloride anion is bound to the V-shaped cleft of this receptor and is held in place in the resulting "pocket" by interactions involving all eight pyrrolic NHs (Figure 5.9).[23] The N···Cl distances range from 3.422 to 3.572 Å.

The observation of bound anions in the solid state structures of the various calix[n]bipyrrole macrocyclic systems (n = 3, 4) led to considerations that these systems would be capable of acting as anion receptors in organic solution. It was also anticipated that these receptors, to the extent they bound anions, would favour larger species (*e.g.*, bromide anion over chloride anion). Table 5.8 provides support for this

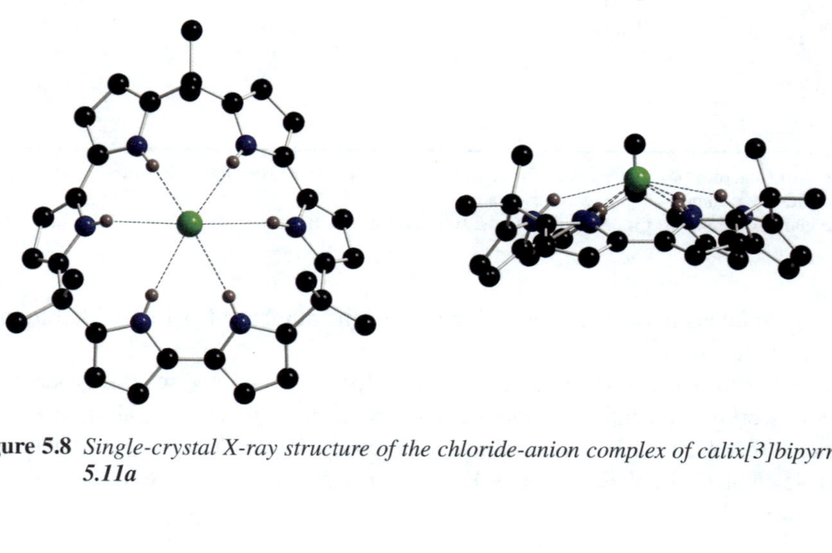

Figure 5.8 *Single-crystal X-ray structure of the chloride-anion complex of calix[3]bipyrrole 5.11a*

latter notion. Specifically, it contains data which highlights the fact that receptor **5.11a** displays bromide- and iodide-anion affinities that are greatly enhanced relative to those of calix[4]pyrrole. In the case of the larger calix[4]bipyrrole homologue **5.11b**, a receptor containing eight potential pyrrolic NH-donor groups, the association constant for chloride was observed to be much higher than that of bromide. The absolute anion affinities of this system were also found, for the most part, to be higher than those of calix[4]pyrrole. Presumably, this set of combined observations reflects the fact that both size- and geometry-based matching between the macrocyclic receptor and anionic substrates, as well as the total number of hydrogen-bond donors available for complexation (pyrrolic-NH in the present instance), are important in terms of regulating anion-binding affinities.

Recently, a new set of anion receptors, **5.12a–5.12c**, based on bipyrrole "combined" with either furan or thiophene was prepared.[24] These systems display good affinities for Y-shaped anions, such as benzoate and acetate, while binding such classic spherical anions as chloride and bromide less well. Two complementary single-crystal X-ray structure analyses (*cf.* Figure 5.10) served to show that **5.12b** adopts a cone conformation in the presence of acetate, but adopts a 1,3-alternate conformation with chloride anion in solid state. In the case of the acetate complex, both bipyrrole units bind to each oxygen atom of the bound acetate anion via NH···O hydrogen-bond interactions (*cf.* N···O = 2.85–2.86 Å). However, only one bipyrrole

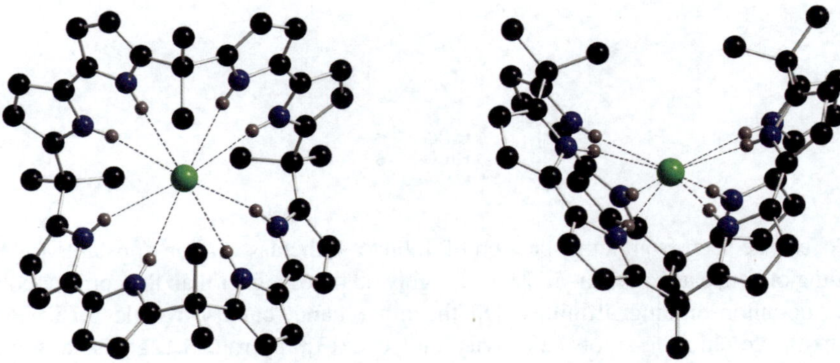

Figure 5.9 *Single-crystal X-ray structure of the chloride-anion complex of calix[4]bipyrrole 5.11b*

Table 5.8 *Association constants (K_a, M^{-1}) for the interaction of **5.11** with various anions in acetonitrile as determined from ITC titration carried out at 303 K*

Receptors	Cl^-	Br^-	I^{-a}	NO_3^{-a}
5.1	140,000	3400	17	52
5.11a	110,000	100,000	9300	11,000
5.11b	2,900,000	110,000	56	450

[a]Value obtained from ^{1}H NMR titration at 22 °C in CD_3CN.

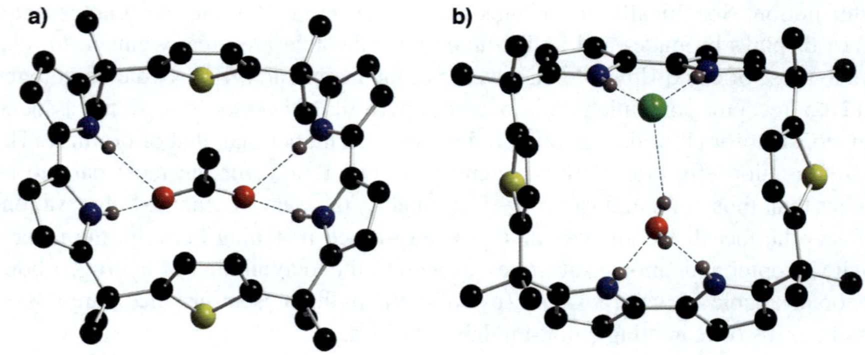

Figure 5.10 *Single-crystal X-ray structures of (a) the nitrate-anion and (b) the chloride-anion complex of compound **5.12b***

unit has a direct hydrogen-bond interaction with the chloride anion, while the other interacts through a bridging water molecule.

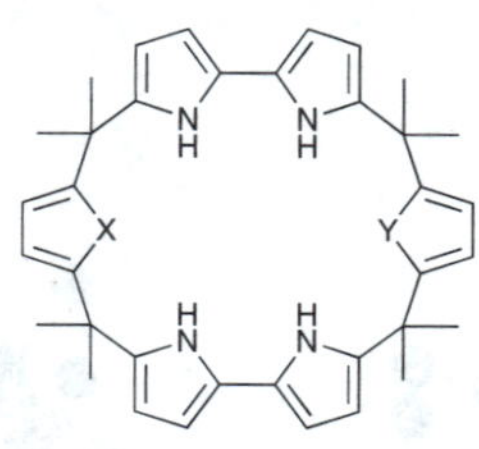

5.12a X = Y = O
5.12b X = Y = S
5.12c X = O, Y = S

As can be seen from an inspection of Table 5.9, the association constants for the binding of benzoate to **5.12a–5.12c** are roughly 60 times higher than the corresponding chloride-anion-binding affinities. On the other hand, calix[4]pyrrole **5.1a** shows no benzoate-chloride-anion selectivity and calix[3]bipyrrole **5.12a** displays only intermediate selectivity behaviour (approximately nine-fold). Taken in concert, these observations are consistent with the notion that controlling the internal cavity size, as well as the overall shape of pyrrole-derived receptors, can lead to systems whose selectivities are optimized for certain classes of anions.

In 2004, another type of expanded calixpyrrole, namely one containing carbazole subunits (**5.13**), was prepared by the Sessler group.[25] In the solid state, the benzoate anion complex of receptor **5.13** was found to adopt a wing-like conformation as shown in Figure 5.11. The complex structure was stabilized by four hydrogen-bond interactions involving the receptor **5.13** and the bound benzoate anion. The anion-binding properties in solution were then studied using a fluorescence-quenching-based titration method. The K_a values obtained in this way are summarized in Table 5.10; taken in concert, they reveal a clear preference for oxyanions.

Table 5.9 *Association constants (K_a, M^{-1}) for the interaction of* **5.12** *with various anions in acetonitrile as deduced from ITC titrations carried out at 303 K*

Receptors	Cl^-	Br^-	HSO_4^-	$H_2PO_4^-$	$C_6H_5CO_2^-$	$CH_3CO_2^-$
5.1a	140,000	3400	[b]	15,100	120,000	350,000
5.12a	960[a]	37[a]	130[a]	240[a]	63,000	78,000
5.12b	1540[a]	100[a]	28[a]	>10,000[a]	100,000	140,000
5.12c	6700[a]	150[a]	36[a]	[b]	670,000	710,000

[a]Value obtained from ^{1}H NMR titration at 25 °C in CD_3CN.
[b]No evidence of anion binding is observed.

Table 5.10 *Association constant (K_a, M^{-1}) for the interaction of* **5.13** *with different anions in dichloromethane as measured by fluorescence quenching*

	Acetate	*Benzoate*	*Oxalate*	*Succinate*	$H_2PO_4^-$	$HP_2O_7^{3-}$	Cl^-
5.13	230,000	77,000	31,000	9500	72,000	64,000	35,000

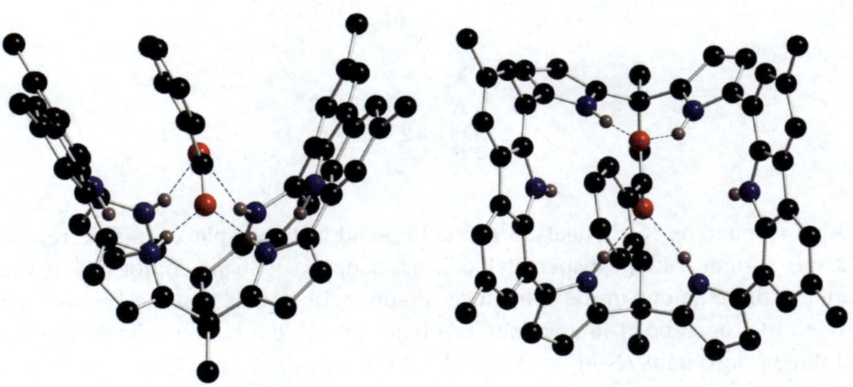

Figure 5.11 *Two views of the benzoate complex of calix[4]pyrrole[2]carbazole* **5.13** *as determined from a single-crystal X-ray diffraction analysis*

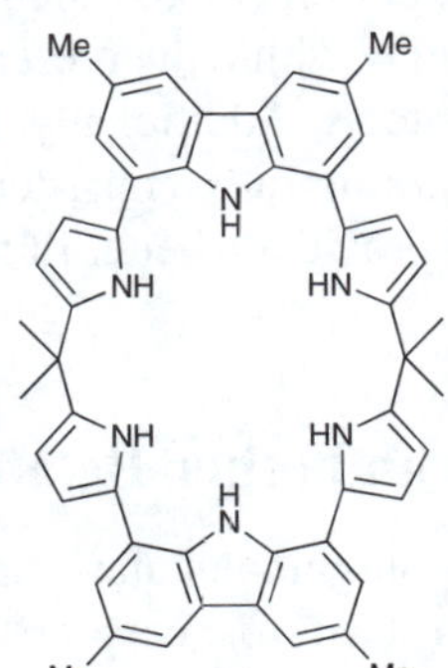

5.13

Quite recently, the 1,3-bispyrrolylbenzene was prepared as an alternative "building block" for use in the synthesis of a new type of expanded calixpyrrole, namely the calix[*n*]bispyrrolylbenzenes **5.14–5.16**.[26] In the case of calix[2]bipyrrolylbenzene, **5.14**, two diffraction-grade crystals were analysed by X-ray structural analysis and the resulting structures are shown in Figure 5.12. From these it can be seen that **5.14** adopts a cone conformation and binds anions (*i.e.*, chloride and nitrate) via four hydrogen bond interactions (N···Cl = 3.31–3.40 Å for the chloride anion complex and N···O = 2.96–3.41 Å for the nitrate-anion complex). While the four pyrrolic NH protons are involved in binding with a single chloride anion, only two of the oxygen atoms from the nitrate anion are involved in hydrogen bonds with the four pyrrolic NH protons.

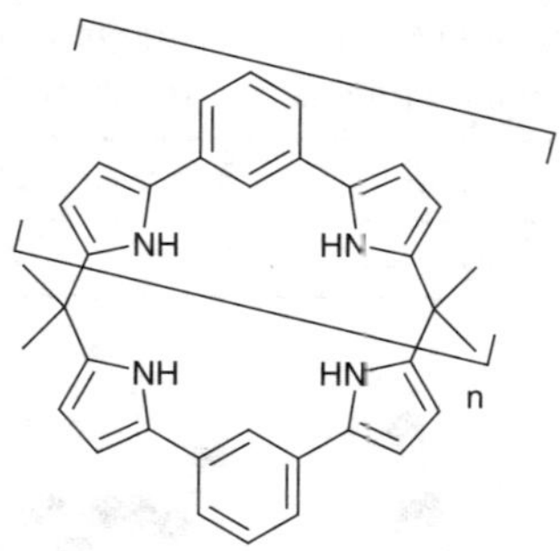

5.14 n = 1
5.15 n = 2
5.16 n = 3

A corresponding X-ray analysis of the PF_6^- and NO_3^- complexes of **5.15** revealed that this system, calix[3]bispyrrolylbenzene, adopts a V-shape conformation when bound to these anions in the solid state. Figure 5.13 reveals that the PF_6^- is bound through hydrogen-bond interactions involving five of the six pyrrolic NH protons and three fluoro units (N···F = 3.26–3.52 Å). Interestingly, all three oxygen atoms of the nitrate anion interact with the receptor via hydrogen bonds involving four of the six pyrrolic NH protons (N···O = 3.11–3.30 Å).

Quantitative association constants were measured using standard ^1H NMR spectroscopic methods in CD_2Cl_2 as well as via ITC titrations using 1,2-dichloroethane as the solvent. The results from both methods, summarized in Table 5.11, indicate that **5.14** binds anions most strongly within the present series of bispyrrolylbenzene-based expanded calixpyrrole systems. Additionally, the selectivity of **5.14** towards bromide anion over chloride anion is fully consistent with the hypothesis that the size of the cavity present in **5.14** provides a better fit for the bromide anion than does its congeners.

5.2.3 Strapped Systems and other Related Receptors

One of the guiding themes in supramolecular chemistry is that increasing the degree of preorganization can lead to receptors with improved selectivities,

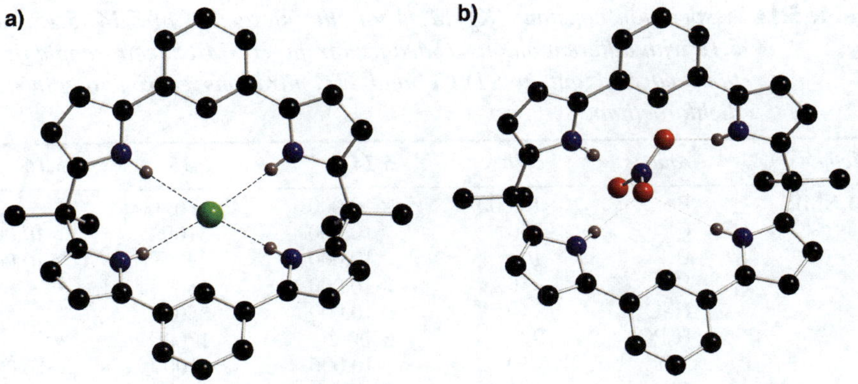

Figure 5.12 *Single-crystal X-ray structures of (a) the chloride-anion and (b) the nitrate-anion complexes of calix[2]bispyrrolylbenzene* ***5.14***

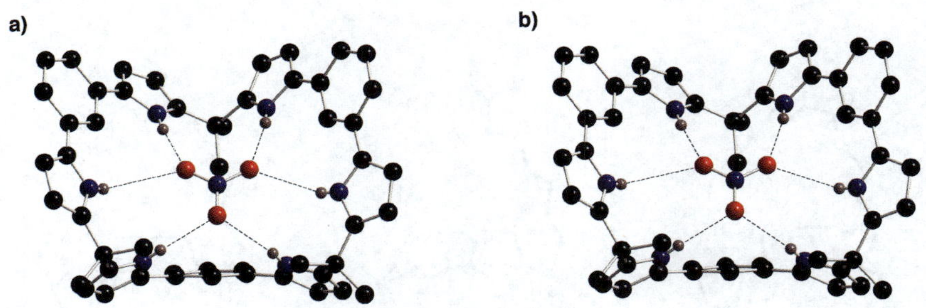

Figure 5.13 *Single-crystal X-ray structures of (a) the* PF_6^- *and (b) the* NO_3^- *complexes of calix[3]bispyrrolylbenzene* ***5.15***

enhanced affinities, or both. In the case of the calixpyrroles, this possibility has been explored by creating deep cavities as detailed in Section 5.2.1 or by expanding the core size as discussed in Section 5.2.2. It can also be achieved by increasing the dimensionality of the receptors. This latter approach, which actually defines one of the current frontiers in pyrrole-based anion-receptor design, is leading to the construction of strapped, capped, and cryptand-like systems. The first example of a "strapped" or "capped" calix[4]pyrrole was reported by Lee and coworkers[27] in 2002. This system, **5.17**, represents the vanguard of what is a growing class of calix[4]pyrrole receptors bearing *trans*-substituted straps on one face of the molecule. Qualitative assessments of anion-binding affinity were made using ^{1}H NMR spectroscopy, following the chemical shift changes or integrated intensity changes of various receptor-derived signals as a function of anion concentration.

Table 5.11 *Association constant (K_a, M^{-1}), for the interaction of **5.14**, **5.15**, and **5.16** with different anions as deduced from ^{1}H NMR spectroscopic titrations carried out in CD_2Cl_2 and ITC titrations carried out in 1,2-dichloroethane*

Method	*Anions*	***5.1a***	***5.14***	***5.15***	***5.16***
^{1}H NMR	F^-	17,000	>10,000	>10,000	[a]
	Cl^-	350	>10,000	3100	>10,000
	Br^-	10	>10,000	390	>10,000
	I^-	<10	>10,000	150	[a]
	HSO_4^-	<10	>10,000	850	[a]
	$H_2PO_4^-$	97	6300	1700	[a]
	NO_3^-	<10	>10,000	5100	>10,000
	ClO_4^-	[a]	7900	30	110
ITC	Cl^-	18,000	5,600,000	82,000	240,000
	Br^-	[a]	21,000,000	5600	44,000
	HSO_4^-	[a]	1,200,000	11,000	76,000
	NO_3^-	[a]	2,500,000	3800	220,000

[a]Value not determined.

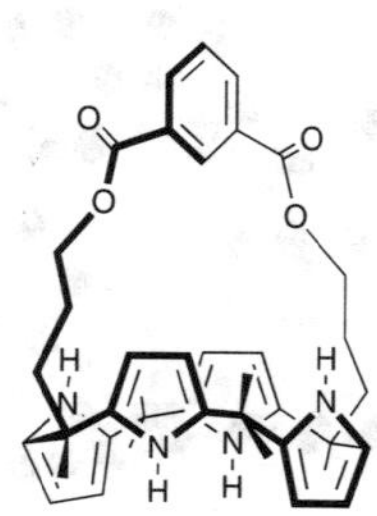

5.17

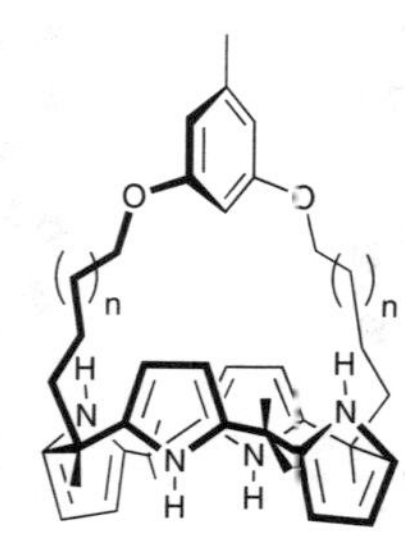

5.18a n = 1
5.18b n = 2
5.18c n = 3

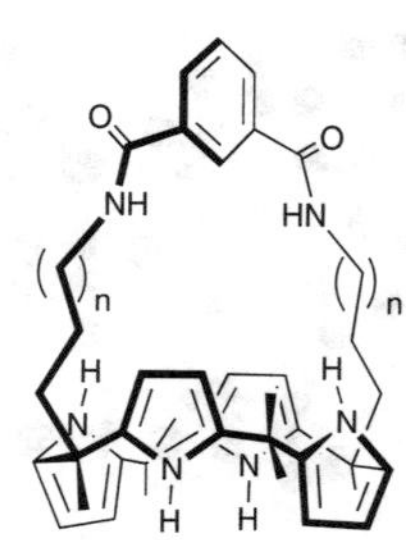

5.19a n = 1
5.19b n = 2
5.19c n = 3

Quantitative assessments of anion-binding affinities were made for receptor **5.17** using ITC methods. From these studies, it was determined that the chloride-anion-binding affinity is *ca.* 1.0×10^5 M^{-1} in DMSO. The corresponding fluoride-binding affinity, obtained from an ^{1}H NMR spectroscopic competition experiment, was found to be approximately 3.9×10^6 M^{-1}.[27] On the other hand, no appreciable binding interactions were observed in the presence of bromide, iodide, sulfate, or dihydrogen phosphate anions.

One year after Lee's original report, the X-ray crystal structure of the chloride-anion complex of **5.17** was obtained. The resulting structure, shown in Figure 5.14, reveals a cone-like conformation for the calix[4]pyrrole core and the presence of a chloride anion encapsulated within the three-dimensional binding cavity.

On a different level, it was recognized that by varying the nature of the strap, fundamental insights into the relationship between the structures of the receptor, the size and shape of the anion, and the binding affinities might be obtained. In 2002, Sessler and Lee[28] reported the synthesis of a new set of strapped calix[4]pyrroles, **5.18**, that

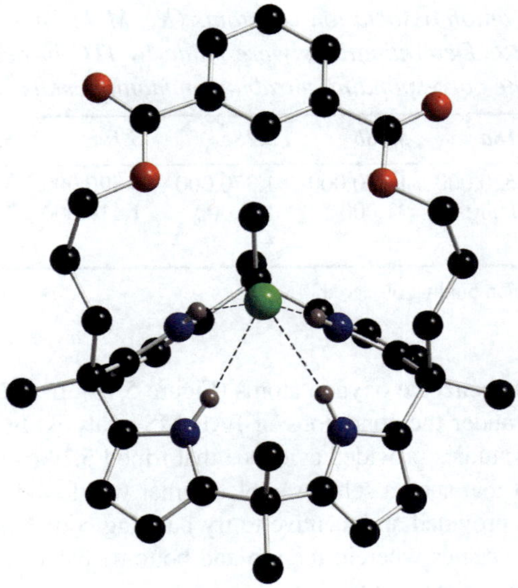

Figure 5.14 *Single-crystal X-ray structure of the chloride-anion complex of strapped-calix[4]pyrrole **5.17***

are bridged by ether-containing straps of varying lengths. They have recently extended this approach to include the amide-strapped analogues **5.19**.[29] Table 5.12 summarizes the affinity constants corresponding to the binding of chloride, bromide, and iodide anions to various strapped calix[4]pyrroles as determined by ITC in dry acetonitrile. Inspection of this table underscores the advantages that accrue as the result of "strapping" one face of a calix[4]pyrrole. For instance, the use of a tight C-4 ether strap (system **5.18a**) gives rise to an extremely high chloride anion binding affinity. By the contrast, the use of a C-6 ether strap (**5.18c**) produces a system with an enhanced bromide anion binding ability, at least relative to the parent calix[4]pyrrole system **5.1a**. In the case of the C-3, C-4, and C-5 amide strapped calix[4]pyrroles **5.19a–5.19c**, the chloride anion binding affinities were found to be almost identical but still generally higher than those of the other strapped systems. Rather, in contrast to what was found in the case of the analogous ether-linked strapped systems, all three new amide-strapped receptors, **5.19a–5.19c**, give rise to more or less the same binding behaviour and thus do not show any kind of anion-to-receptor size-matched selectivity. On the other hand, the dramatic increase in affinity observed relative to **5.1a** indicates that in the case of **5.19a–5.19c**, anion binding likely benefits from interactions involving both the amide NH and pyrrolic NH protons.

In 1996, Sessler and *et al.*[30] reported the first structurally characterized three-dimensional oligopyrrole system, namely the "tripod" **5.20**. It was prepared from the acid-catalysed condensation of mono-formylbipyrrole and mono-α-free bipyrrole. X-ray crystal structure analysis revealed that tripod **5.20a** exists as diastereogenic dimer in the solid state, wherein each monomer possesses a right- or left-handed twist. This dimerized structure is apparently stabilized by 12 hydrogen-bond interactions involving the

Table 5.12 *Halide-anion association constants (K_a, M^{-1}) for receptors* ***5.17****,* ***5.18****, and* ***5.19****. Determinations were made by ITC in acetonitrile at 30 °C using the corresponding tetrabutylammonium salts*

	5.17	*5.18a*	*5.18b*	*5.18c*	*5.19a*	*5.19b*	*5.19c*
Cl^-	1,380,000	3,630,000	1,370,000	1,370,000	3,890,000	3,350,000	3,240,000
Br^-	~ 0	30,000	31,000	120,000	1,410,000	1,250,000	700,000
I^-	–	–	–	–	–	2300	3000

N.D.: No evidence of anion binding observed.

pyrrolic NHs and the benzoate oxygen atoms (Figure 5.15). In the gas phase, a dimer peak was observed under the conditions of FAB-MS analysis. In protic solvents, ^1H NMR spectroscopic studies provided evidence that tripod **5.20a** also exists as a dimer in solution. The fact that such a self-assembled dimer was found to exist over such a range of conditions, provided an incentive to try building corresponding three-dimensional cryptand-like species wherein the top and bottom "halves", analogous to **5.20**, would be linked via covalent bonds.

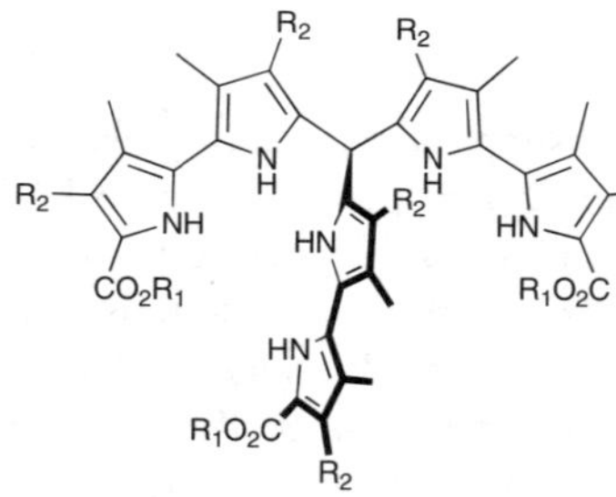

5.20a R_1 = Bn, R_2 = CH_3
5.20b R_1 = R_2 = C_2H_5

The first synthesis of an oligopyrrole-based cryptand was realized by Beer and co-workers[31] in 2001. Here, an α-formylated tripyrrolylmethane was condensed with either ethylenediamine or butylenediamine to produce the imine-linked products **5.21**. Single-crystal X-ray diffraction analysis revealed that the neutral diamine, used as a reactant, was trapped inside the three-dimensional cavity of cryptand **5.21a**, being held there via two different sets of hydrogen-bonding interactions involving the pyrrolic NHs, the Schiff base nitrogen atoms, and the diamine hydrogens (Figure 5.16). The inclusion of the diamine led to the suggestion that this species not only acts as a reactant but also as a template for the formation of the cryptand. While this hypothesis is yet to be tested, quantitative determinations of the stability constants for the binding of the diamine were made using ^1H NMR spectroscopic titration methods. Specifically, it was found that the neutral ethane-1,2-diamine and ethane-1,2-diol were bound with affinities of *ca.* 1500 and 1060 M^{-1}, respectively, in $CDCl_3$.

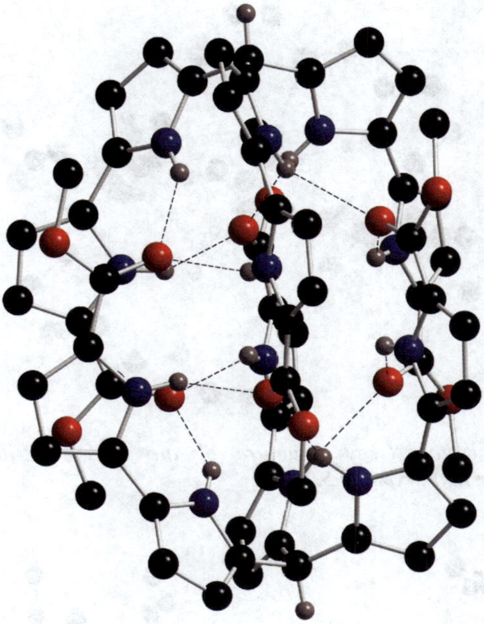

Figure 5.15 *Single-crystal X-ray structure of tripod **5.20a** showing the dimer formed between two molecules of **5.20a** in the solid state. In this view, the benzyl and methyl groups have been removed for clarity*

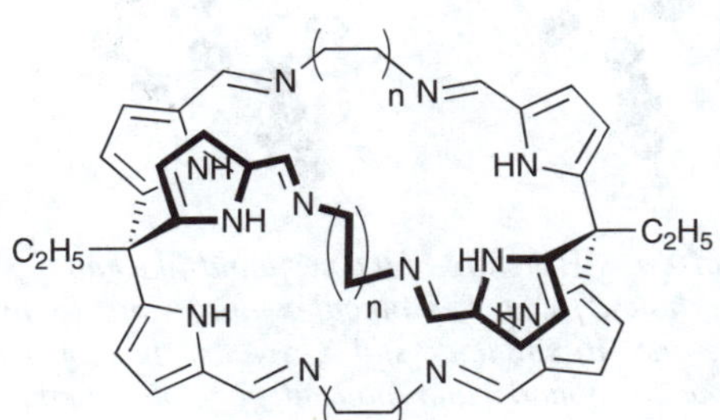

5.21a n = 1
5.21b n = 2

Nearly contemporaneous with the Beer report, Sessler and co-workers[32] published the first carbon—carbon-linked polypyrrole cryptand, namely the "3-D calixpyrrole" **5.22a**. Receptor **5.22a** is too small to allow for the internal binding of substrates (it contains essentially no cavity-like void). However, it presents three identical binding surfaces and was found to bind solvents in the solid state as shown in Figure 5.17. This led to the suggestion that it could stabilize anion-receptor complexes of 1:1, 1:2, or 1:3 binding stoichiometry in solution. In point of fact, ^{1}H NMR spectroscopic experiments, carried out in CD_2Cl_2 and THF-d_8, revealed host–guest stoichiometries that were even more complicated and depended directly

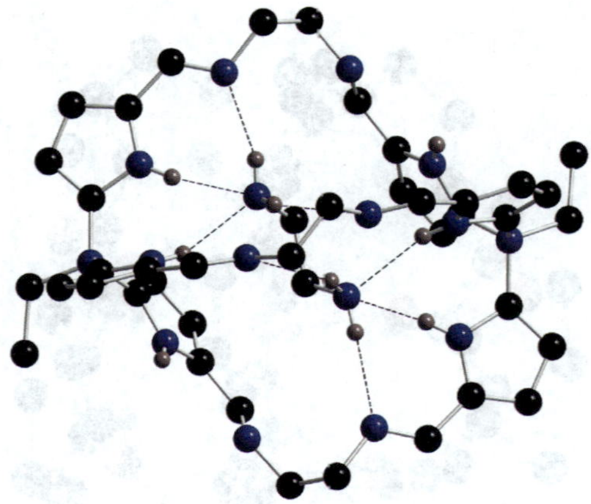

Figure 5.16 *Single-crystal X-ray structure of the ethane-1,2-diamine complex of bis(tripyrroyl)cryptand* ***5.21a***

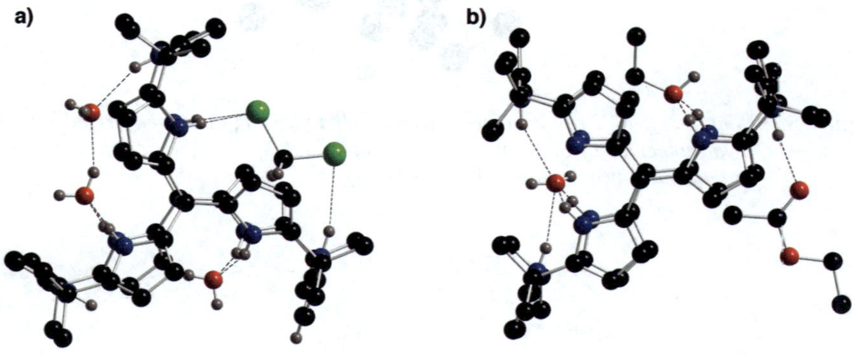

Figure 5.17 *(a) Single-crystal X-ray structure of the cryptand-like calixpyrrole* ***5.22a*** *showing the three bound water molecules and one bound molecule of dichloromethane and (b) single-crystal X-ray structure of the 3-D calixphyrin* ***5.22b*** *showing the one bound water molecule, one bound ethanol molecule, and the one bound ethyl acetate molecule*

on the choice of anion. For instance, fluoride, when added to **5.22a** as an anionic guest, was found to bind to six of the nine pyrrolic subunits, in an apparent 1:1 fashion. By contrast, analysis of the binding data revealed that a single chloride anion interacts with two molecules of receptor **5.22a** in solution (1:2 anion:receptor stoichiometry), being bound with a stability constant of 3.1×10^6 M^{-2} in CD_2Cl_2. Finally, a 2:1 anion:receptor stoichiometry was observed upon addition of tetrabutylammonium nitrate. In this case, the stability constants, K_{a1} and K_{a2}, were estimated to be *ca.* 1700 and 420 M^{-1}, respectively, in CD_2Cl_2 solution, as judged from ^1H NMR spectroscopic analyses.

Quite recently, the oxidized form of this receptor, **5.22b**, was also reported by Sessler and co-workers.[33] This interesting structure contains sp^2 bridging carbon atoms at the bridge-head positions. It may thus be viewed as being a "3-D calix-phyrin". As yet, however, the ability of this new system to act as a possible anion receptor has not been probed.

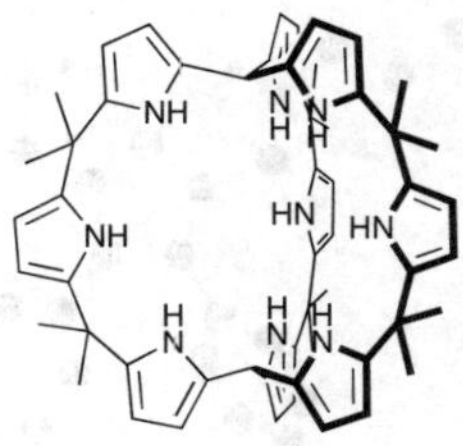

5.22a

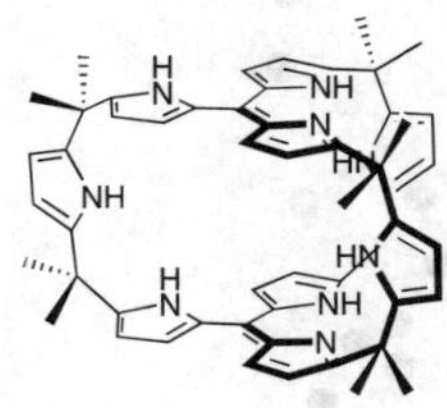

5.22b

5.3 Linear Receptors

While the anion-binding affinities of cyclic oligopyrrole receptors have been extensively studied over the last decade, until recently little attention was paid to linear (acyclic) pyrrole-based anion receptors due to a perception that they would show low anion-binding affinities. Belying this misconception, Gale and his group, building on precedent from Schmuck and Crabtree,[34] reported a series of highly effective acyclic mono- and dipyrrole-based anion receptors.[35]

This series of receptors, represented by structures **5.23**–**5.25**, were synthesized by the reaction of pyrrolic diacid chlorides and various amine derivatives. The crystal structure of the benzoate complex of the linear receptor **5.23a** was published in 2002. It revealed that all three hydrogen-bond donors are indeed involved in hydrogen bonding to this particular anionic guest (Figure 5.18a).[36] More recently, the single-crystal X-ray diffraction structure of the chloride-anion complex of the deprotonated receptor **5.23d** was reported. In this case, it was found that the deprotonated receptor binds the bound chloride anion via two amide-NH hydrogen bonds and two phenyl CH···Cl interactions (Figure 5.18b).[37]

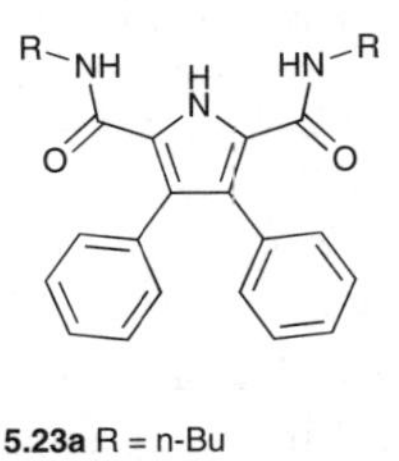

5.23a R = n-Bu
5.23b R = Ph
5.23c R = 4-nitrophenyl
5.23d R = 3,5-dinitrophenyl

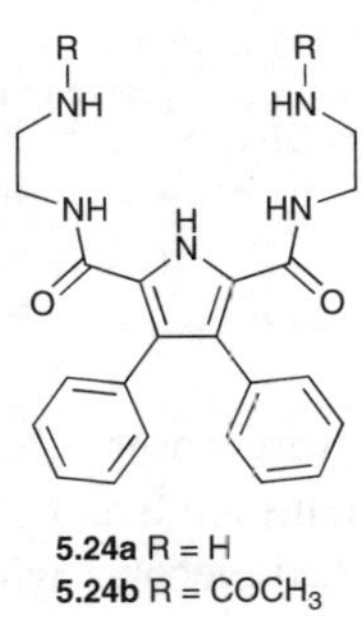

5.24a R = H
5.24b R = $COCH_3$

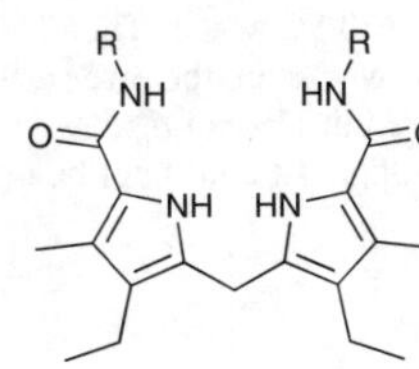

5.25a R = n-Bu
5.25b R = Ph

Table 5.13 summarizes the various anion-binding affinities of receptors **5.23** and **5.24**, as inferred from ^1H NMR spectroscopic analyses.[38] These data reveal that **5.23** and **5.24** act as selective receptors for oxo-anions in polar organic solvents (*e.g.*, CD_3CN or DMSO-d_6). However, within each receptor series some differences in the

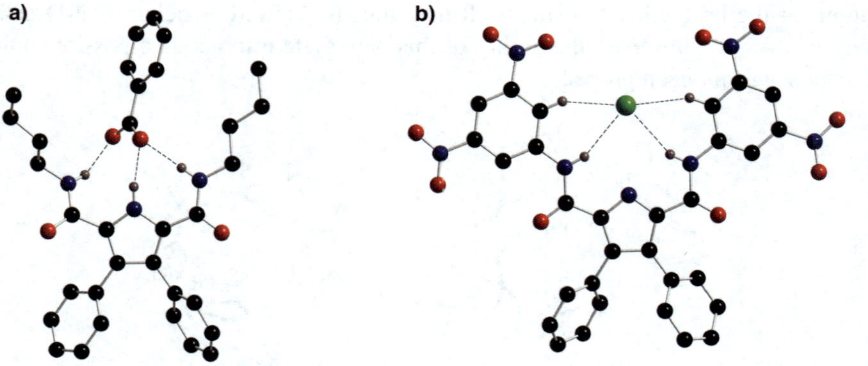

Figure 5.18 *(a) Single-crystal X-ray structure of the benzoate-anion complex of 2,5-diamidepyrrole **5.23a** and (b) single-crystal X-ray structure of the chloride-anion complex of the deprotonated form of the linear pyrrole-based anion receptor **5.23d**.*

Table 5.13 *Anion affinities (K_a, M^{-1}) of linear receptors as determined from ^1H NMR spectroscopic titrations carried out at 25 °C*[a]

	5.23a	*5.23b*	*5.24a*	*5.24b*	*5.25a*	*5.25b*
Anion	*CD_3CN*	*DMSO-d_6*	*DMSO-d_6*	*DMSO-d_6*	*DMSO-d_6*[d]	*DMSO-d_6*[d]
F^-	85	74	–	–	7560	8990
Cl^-	138	11	<20	<20	23	43
Br^-	<10	<10	[b]	[b]	13	10
$H_2PO_4^-$	357	1450	2050[c]	525	[e]	[e]
$C_6H_5CO_2^-$	2500	560	47.6	152	354	424
HSO_4^-	–	–	>10^4	[b]	44	128

[a]Anions added in the form of the corresponding tetrabutylammonium salts. Acetonitrile water content is 0.03% and DMSO water content 0.5%.
[b]No binding observed.
[c]High errors were obtained when fitting this data to a 1:1 binding model for titrations carried out in DMOS-d_6-0.5% water. Therefore, this titration was repeated in DMSO-d_6-5% water solution. Although a better fit was seen, the error (29%) remained large.
[d]Measured in DMSO-d_6-5% water.
[e]An adequate fit could not be obtained.

individual binding characteristics were observed. For instance, **5.23a** displayed a higher benzoate-anion association constant than its congeners.[38] In the case of **5.23c** and **5.23d**, receptors bearing attached-electron withdrawing groups on the amide-phenyl subunit, anion-binding affinities were recorded that are generally higher than those of **5.23b** (*viz.* K_a = 1250, 40, and 4150 M^{-1} for fluoride, chloride, and benzoate, respectively, in the case of **5.23c** and K_a = 53 and 4200 M^{-1} for chloride and benzoate in the case of **5.23d**; DMSO-d_6-0.5% water).[37]

Receptors **5.25a** and **5.25b** contain two pyrrole units and, not surprisingly, display anion-binding behaviour that differs from that of **5.23** and **5.24**. For instance, the halide anion association constants are higher and the oxyanion association constants are lower for both **5.25a** and **5.25b** than for receptors **5.23** and **5.24**. Further, these receptors were found to disply high selectivities for fluoride anion in partially mixed organic-aqueous solutions.[39]

Further modifications of the basic diamide pyrrole motif were reported by Gale and Brooker. In this study, receptors **5.26a–5.26d**, possessing a more flexible hydrogen-bonding array, were introduced. These systems were found to display anion-binding properties that made them more attractive as receptors than the parent system **5.23a**. As shown in Table 5.13, receptors **5.26a–5.26c** exhibit anion selectivity in the sequence benzoate > dihydrogen phosphate > chloride.[40] However, receptor **5.26a**, with extra saturated amine units, shows significantly enhanced affinities towards anions containing a proton source (*i.e.*, dihydrogen phosphate and hydrogen sulfate anions). These results can be rationalized in terms of proton transfer from these latter anions to the amine moieties in receptor **5.26a**, leading to enhanced electrostatic interactions between the then-protonated receptor and the resulting doubly negative anions (see Table 5.14).

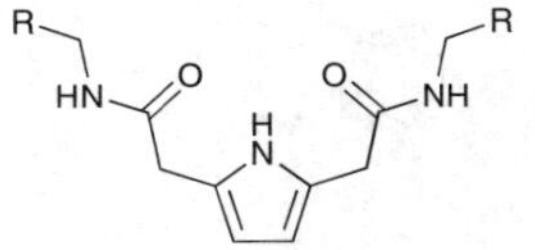

5.26a R = $CH_2N(CH_3)_2$
5.26b R = CH_2(2-pyridyl)
5.26c R = 2-pyridyl
5.26d R = Ph

Most recently, P.A. Gale and B.D. Smith have designed and synthesized an imidazole-functionalized amidopyrrole designed to cotransport HCl across lipid bilayer membranes. Receptor **5.27** contains two NH-hydrogen-bond donors and a pendant basic methylimidazole moiety and was designed as a synthetic analogue of prodigiosin. The structure of **5.27**·HCl revealed the formation of a "2+2" dimer in the solid state with each chloride bound by three hydrogen bonds; two from the pyrrole (N···Cl 3.24(2) Å) and amide (N···Cl 3.29(2) Å) groups of one receptor and one from the imidazolium group of another (N···Cl 3.10(2) Å) (Figure 5.19). With regard to the potential for membrane transport, a notable feature of the crystal structure is that all the polar and ionic functionality is inside the dimer, whereas the exterior projects primarily lipophilic groups. Studies in palmitoyl-oleoyl-phosphatidylcholine (POPC)/cholesterol vesicles under a range of different pH regimes have demonstrated that both chloride and protons are released from the vesicles upon addition of **5.27**. The fastest release of chloride was observed to occur from vesicles with a low

Table 5.14 *Anion affinity constants of linear receptors (K_a, M^{-1}) as determined from ^1H NMR spectroscopic titrations carried out in acetonitrile-d_3 at 25 °C*[a]

	F^-	Cl^-	$C_6H_5CO_2^-$	$H_2PO_4^-$	HSO_4^-
5.26a	430	135	1450	$>10^4$	$>10^4$
5.26b	425	150	3370	1080	420
5.26c	6060	190	5270	1590	250
5.26d	1500	190	2500	460	80

[a]Anions added as their tetrabutylammonium salts. Errors estimated to be less than 15%.

Figure 5.19 *The X-ray crystal structure of the dimer (5.27-HCl)$_2$*

pH (high HCl concentration) inside the vesicle and a near neutral pH outside the vesicle.[41]

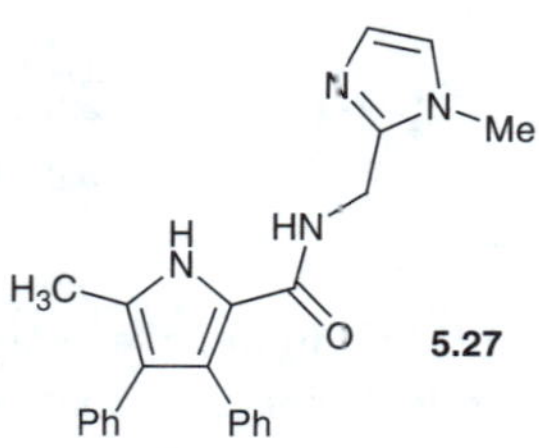

5.27

Since Gale's report, the anion-binding studies of simple linear pyrrolic systems **5.28–5.32** have begun to be studied in ernest.[24,26,42,43] For instance, two recent crystal structures of **5.30a**, in the form of its complex with fluoride and chloride anions (*cf.* Figure 5.20), have provided support for the notion that even simple linear pyrrolic systems can bind anions via pyrrolic NH-anion hydrogen-bond interactions.

Receptors **5.29** and **5.30a** have also been studied in the solution phase. As can be seen from the findings summarized in Table 5.15, these systems display selectivity for benzoate anion over spherical halide anions or tetrahedral-shaped anions. Interestingly, it was observed that receptor **5.31**, a system containing a rigid 1,3-benzene spacer between the two constituent pyrrolic subunits, binds chloride anion more strongly than bromide anion, in spite of possessing a size and shape that would perhaps favour binding of the latter. This result, which will require further study to analyse properly in the context of this class of receptors, nonetheless serves to underscore the potentially rich nature of this approach to anion recognition. In fact, the ease of synthesis associated with these and other open-chain pyrrole-based systems leads to the prediction that this is a direction of study that will be receiving considerable emphasis in the years to come.

Quite recently, Eichen and co-workers[43] described 1,3-di(2-pyrrolyl)azulene **5.32** as an efficient luminescent probe for fluoride anion. The addition of increasing

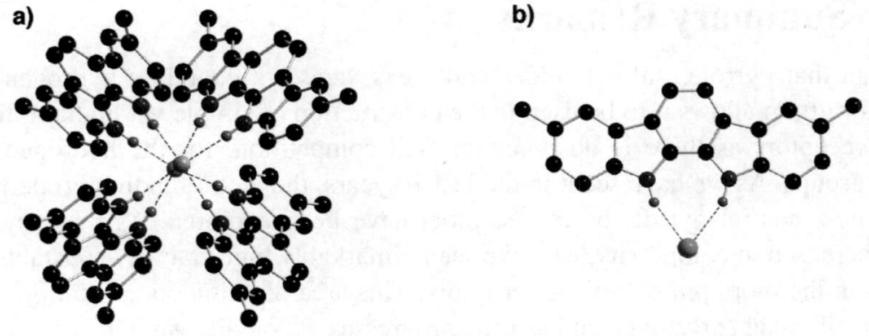

Figure 5.20 *Single-crystal X-ray structures of (a) the 1:2 complex formed between fluoride anion and receptor **5.30a** and (b) the chloride-anion complex of **5.30a***

Table 5.15 *Anion affinities of linear receptors **5.28–5.32** (K_a, M^{-1}) as determined using the corresponding tetrabutylammonium salts*

	F^-	Cl^-	Br^-	$C_6H_5CO_2^-$	$H_2PO_4^-$	HSO_4^-
5.28[a]	2100	110	19	–	310	<10
5.29	–	246[b]	53[c]	4090[b]	–	35[c]
5.30a[d]	50,000	32,000	–	200,000	79,000	–
5.31[a]	2300	4300	1100	–	1300	290
5.32[e]	11,000	110	100	–	–	–

[a]In CD_2Cl_2 as deduced from ^1H NMR spectroscopic titrations carried out at 298 K.
[b]In acetonitrile as determined by ITC titrations at 303 K.
[c]In CD_3CN as deduced from ^1H NMR spectroscopic titrations carried out at 298 K.
[d]In acetone as determined from UV–Vis spectroscopic titrations carried out at 298 K.
[e]In CH_2Cl_2 as determined from UV–Vis spectroscopic titrations carried out at 298 K.

amounts of fluoride anion to a solution of **5.32** in CH_2Cl_2 leads to the formation of complex inducing the change of absorption spectrum. Quantitative binding studies, the results of which are summarized in Table 5.15, illustrate the strong fluoride-anion selectivity of receptor **5.32**.

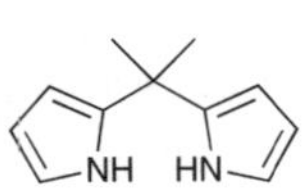

5.28

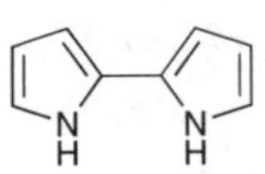

5.29

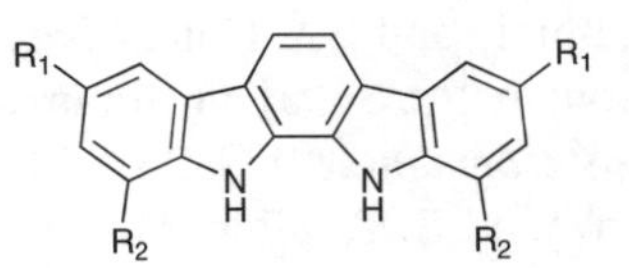

5.30a $R_1 = R_2 = H$
5.30b $R_1 = CH_3$, $R_2 = H$
5.30c $R_1 = H$, $R_2 = Br$
5.30d $R_1 = Br$, $R_2 = H$

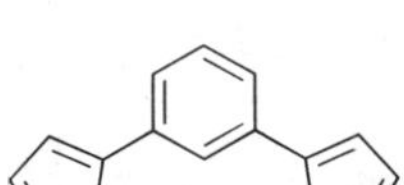

5.31

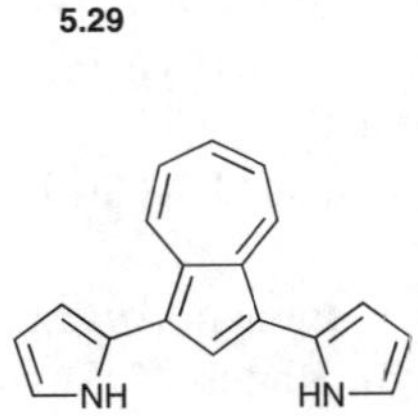

5.32

5.4 Summary Remarks

The fact that pyrrole, unlike amides and ureas, does not contain a hydrogen-bond acceptor group allows it to be used in the construction of simple yet highly effective anion receptors as there is no potential 'self-competition' for the hydrogen-bond donor groups. As we have seen, in the last 10 years, the anion-binding properties of many new neutral pyrrole-based receptors have been explored. As these systems have increased in complexity, we have seen remarkably high stability constants with anions in the more pre-organized receptors. This area of anion complexation chemistry is still at an early stage and as time progresses we predict we will see new generations of three-dimensional neutral pyrrole systems with high selectivity and affinity for anionic guests. This is, therefore, a very exciting area in which to work with new opportunities for growth and application expected to emerge in the coming years.

References

1. A. Baeyer, *Dtsch. Chem. Ges.*, 1886, **19**, 2184.
2. P.A. Gale, J.L. Sessler, V. Král and V. Lynch, *J. Am. Chem. Soc.*, 1996, **118**, 5140.
3. S. Camiolo, S.J. Coles, P.A. Gale, M.B. Hursthouse and J.L. Sessler, *Acta Cryst.*, 2001, **E57**, o816.
4. P.A. Gale, J.L. Sessler, W.E. Allen, N.A. Tvermoes and V. Lynch, *Chem. Commun.*, 1997, 665.
5. F.P. Schmidtchen, *Org. Lett.*, 2002, **4**, 431.
6. J.L. Sessler, V. Král, T.V. Shishkanova and P.A. Gale, *Proc. Natl. Acad. Sci. USA*, 2002, **99**, 4848.
7. J.L. Sessler, P.A. Gale and J.W. Genge, *Chem. Eur. J.*, 1998, **4**, 1095; J.L. Sessler, J.W. Genge, P.A. Gale and V. Král "Calix[4]pyrrole-Functionalized Silica Gels. Novel Supports for the HPLC-Based Separation of Anions," in *Calixarene-based Separtions* G.J. Lummetta, R.D. Rogers and A.S. Gopalan, Eds. ACS Symposium Series 757, ACS, Washington, D.C. 2000 Chapt. 18, pp. 238–254.
8. S. Camiolo and P.A. Gale, *Chem. Commun.*, 2000, 1129.
9. L. Bonomo, E. Solari, G. Toraman, R. Scopelliti, M. Latronico and C. Floriani, *Chem. Commun.*, 1999, 2413.
10. C.J. Woods, S. Camiolo, M.E. Light, S.J. Coles, M.B. Hursthouse, M.A. King, P.A. Gale and J.W. Essex, *J. Am. Chem. Soc.*, 2002, **124**, 8644.
11. P. Anzenbacher Jr., K. Jursíková, V.M. Lynch, P.A. Gale and J.L. Sessler, *J. Am. Chem. Soc.*, 1999, **121**, 11020.
12. M. Dukh, P. Drasar, I. Cerny, V. Pouzar, J.A. Shriver, V. Král and J.L. Sessler, *Supramol. Chem.*, 2002, **14**, 237.
13. B. Turner, M. Botoshansky and Y. Eichen, *Angew. Chem. Int. Ed.*, 1998, **37**, 2475; B. Turner, A. Shterenberg, M. Kapon, K. Suwinska and Y. Eichen, *Chem. Commun.*, 2001, 12.

14. J.L. Atwood, S.G. Bott, S. Harvey and P.C. Junk, *Organometallics*, 1994, **13**, 4151; A. Alvanipour, J.L. Atwood, S.G. Bott, P.C. Junk, U.H. Kynast and H. Prinz, *J. Chem. Soc. Dalton Trans.*, 1998, 1223.
15. G. Cafeo, F.H. Kohnke, G.L. La Torre, A.J.P. White and D.J. Williams, *Angew. Chem. Int. Ed.*, 2000, **39**, 1496; G. Cafeo, F.H. Kohnke, M.F. Parisi, R.P. Nascone, G.L. La Torre and D.J. Williams, *Org. Lett.*, 2002, **4**, 2695.
16. G. Cafeo, F.H. Kohnke, G.L. La Torre, A.J.P. White and D.J. Williams, *Chem. Commun.*, 2000, 1207.
17. G. Cafeo, F.H. Kohnke, G.L. La Torre, M.F. Parisi, R.P. Nascone, A.J.P. White and D.J. Williams, *Chem. Eur. J.*, 2002, **8**, 3148.
18. J.L. Sessler, P. Anzenbacher Jr., J.A. Shriver, K. Jursíková, V.M. Lynch and M. Marquez, *J. Am. Chem. Soc.*, 2000, **122**, 12061; P. Anzenbacher Jr., A.C. Try, H. Miyaji, K. Jursíková, V.M. Lynch, M. Marquez and J.L. Sessler, *J. Am. Chem. Soc.*, 2000, **122**, 10268.
19. J.A. Shriver, Ph.D. Dissertation, The University of Texas at Austin, 2002.
20. T.G. Levitskaia, M. Marquez, J.L. Sessler, J. A. Shriver, T. Vercouter and B.A. Moyer, *Chem. Commun.*, 2003, 2248.
21. J.L. Sessler, W.-S. Cho, D.E. Gross, J.A. Shriver, V.M. Lynch and M. Marquez, *J. Org. Chem.*, 2005, **70**, 5982.
22. J.L. Sessler, D. An, W.-S. Cho and V. Lynch, *Angew. Chem. Int. Ed.*, 2003, **42**, 2278.
23. J.L. Sessler, D. An, W.-S. Cho, V. Lynch and M. Marquez, *Chem. Commun.*, 2005, 540.
24. J.L. Sessler, D. An, W.-S. Cho and V.M. Lynch, *J. Am. Chem. Soc.*, 2003, **125**, 13646; J.L. Sessler, D. An, W.-S. Cho, V. Lynch, D.-W. Yoon, S.-J. Hong and C.-H. Lee, *J. Org. Chem.*, 2005, **70**, 1511.
25. P. Piatek, V.M. Lynch and J.L. Sessler, *J. Am. Chem. Soc.*, 2004, **126**, 16073.
26. J.L. Sessler, D. An, W.-S. Cho, V. Lynch and M. Marquez, *Chem. Eur. J.*, 2005, **11**, 2001.
27. D.-W. Yoon, H. Hwang and C.-H. Lee, *Angew. Chem. Int. Ed.*, 2002, **41**, 1757.
28. C.-H. Lee, H.-K. Na, D.-W. Yoon, D.-H. Won, W.-S. Cho, V.M. Lynch, S.V. Shevchuk and J.L. Sessler, *J. Am. Chem. Soc.*, 2003, **125**, 7301.
29. C.-H. Lee, J.-S. Lee, H.-K. Na, D.-W. Yoon, H. Miyaji, W.-S. Cho and J.L. Sessler, *J. Org. Chem.*, 2005, **70**, 2067.
30. J.L. Sessler, M.C. Hoehner, D.W. Johnson, A. Gebauer and V.M. Lynch, *Chem. Commun.*, 1996, 2311.
31. O.D. Fox, T.D. Rolls, M.G.B. Drew and P.D. Beer, *Chem. Commun.*, 2001, 1632.
32. C. Bucher, R.S. Zimmerman, V. Lynch and J.L. Sessler, *J. Am. Chem. Soc.*, 2001, **123**, 9716.
33. C. Bucher, R.S. Zimmerman, V. Lynch and J.L. Sessler, *Chem. Commun.*, 2003, 1646.
34. C. Schmuck and L. Geiger, *Curr. Org. Chem.*, 2003, **7**, 1485; C. Schmuck and M. Heil, *Org. Lett.*, 2001, **3**, 1253; C. Schmuck, *Chem. Eur. J.*, 2000, **6**, 709; C. Schmuck, *Tetrahedron*, 2001, **57**, 3063; C. Schmuck and J. Lex, *Eur. J. Org.*

Chem., 2001, 1519; C. Schmuck and J. Lex, *Org. Lett.*, 1999, **1**, 1779; C. Schmuck, *Eur. J. Org. Chem.*, 1999, 2397; C. Schmuck, *Chem. Commun.*, 1999, 843; C. Schmuck and W. Wienand, *J. Am. Chem. Soc.*, 2003, **125**, 452; C. Schmuck and V. Bickert, *Org. Lett.*, 2003, **5**, 4579; K. Kavallieratos, S.R. de Gala, D.J. Austin and R.H. Crabtree, *J. Am. Chem. Soc.*, 1997, **119**, 2325; K. Kavallieratos, C.M. Bertao and R.H. Crabtree, *J. Org. Chem.*, 1999, **64**, 1675; C. Schmuck and M. Heil, *Org. Biomol. Chem.*, 2003, **1**, 633; C. Schmuck and L. Geiger, *Chem. Commun.*, 2004, 1698.

35. P.A. Gale, *Chem. Commun.*, 2005, 3761.
36. S. Camiolo, P.A. Gale, M.B. Hursthouse and M.E. Light, *Tetrahedron Lett.*, 2002, **43**, 6995.
37. S. Camiolo, P.A. Gale, M.B. Hursthouse and M.E. Light, *Org. Biomol. Chem.*, 2003, **1**, 741.
38. P.A. Gale, S. Camiolo, C.P. Chapman, M.E. Light and M.B. Hursthouse, *Tetrahedron Lett.*, 2001, **42**, 5095; P.A. Gale, S. Camiolo, G.J. Tizzard, C.P. Chapman, M.E. Light, S.J. Coles and M.B. Hursthouse, *J. Org. Chem.*, 2001, **66**, 7849; K. Navakhun, P.A. Gale, S. Camiolo, M.E. Light and M.B. Hursthouse, *Chem. Commun.*, 2002, 2084.
39. I.E.D. Vega, S. Camiolo, P.A. Gale, M.B. Hursthouse and M.E. Light, *Chem. Commun.*, 2003, 1686.
40. R. Li, L.S. Evans, D.S. Larsen, P.A. Gale and S. Brooker, *New J. Chem.*, 2004, **28**, 1340.
41. P.A. Gale, M.E. Light, B. McNally, K. Navakhun, K.E. Sliwinski and B.D. Smith, *Chem. Commun.*, 2005, 3773.
42. D. Curiel, A. Cowley and P.D. Beer, *Chem. Commun.*, 2005, 236.
43. H. Salman, Y. Abraham, S. Tal, S. Meltzman, M. Kapon, N. Tessler, S. Speiser and Y. Eichen, *Eur. J. Org. Chem.*, 2005, 2207.

CHAPTER 6

Receptors for Ion-Pairs

6.1 Introduction

Anions are always accompanied by their counter cations, and one approach to the design of new selective anion receptors is to synthesize a receptor that encapsulates both the cationic and anionic components of a salt. As mentioned in the introduction, competition from ion-pairing must be considered when designing these types of receptor. This is because such effects can be quite substantial. One example of the importance of ion-pairing comes from the work of B.D. Smith and co-workers,[1] which involved the study of the urea containing anion receptor **6.1**. They found that the complexation of dihydrogen phosphate by **6.1** in CD_3CN differs in the presence and absence of potassium tetraphenylborate ($KBPh_4$). As illustrated by the NMR spectroscopic titration shown in Figure 6.1, in the presence of $KBPh_4$, the potassium ions in solution sequester the initial aliquots of dihydrogen phosphate and no shift is seen in the aryl NH resonance. However, when the free cations have been exhausted, the added anions interact with the urea and the NH resonance shifts downfield.

Several receptors containing appended crown ethers were also synthesized by Smith and co-workers[1] (**6.2**–**6.4**). In this case, it might be expected that the presence of a cation bound to the crown ether portion of these receptors would serve to enhance the anion-binding affinity due to the presence of a positive charge. However, Smith *et al.* found that the binding ability of the receptor was either *suppressed* or enhanced depending upon the ion-pairing ability of the cationic guest and the nature of the particular receptor. The ion-sequestering ability of the Group 1 metal cations was in the order $Cs^+ < K^+ < Na^+$, matching their ion-pairing ability. Thus, very careful designing is required to produce successfully a receptor capable of binding an ion-pair. Indeed, receptors that seem to contain the necessary components to bind an ion-pair will often fail to function as desired due to this ion-pairing competition.

This generally undesirable competition is illustrated by receptors **6.5**–**6.6**,[2] which although containing both cation-binding crown ether groups and pyrrolic anion complexation sites, fail to function successfully as ion-pair complexation agents.

In this chapter we will look at receptors designed to bind cation–anion pairs (ditopic receptors), generate cascade complexes, and complex zwitterionic guests. While considerable overlap often exists in practice, the basic design concepts underlying these receptor systems generally differ, as illustrated in Figure 6.2.

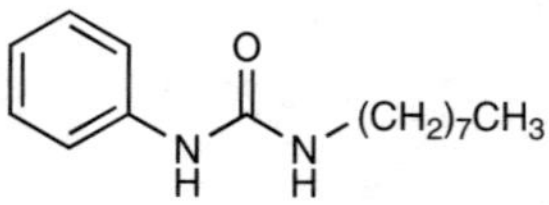

6.1

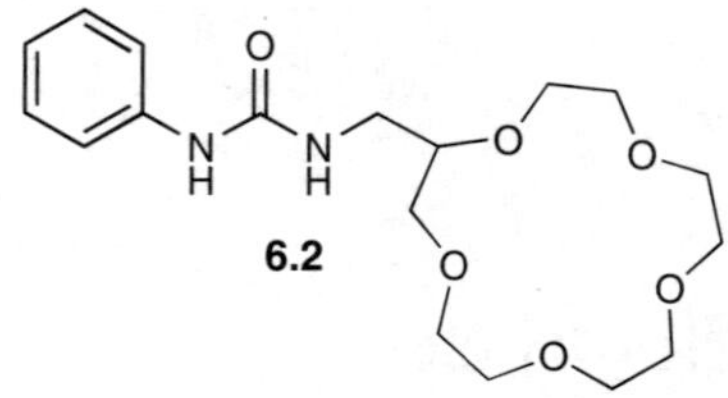

6.2

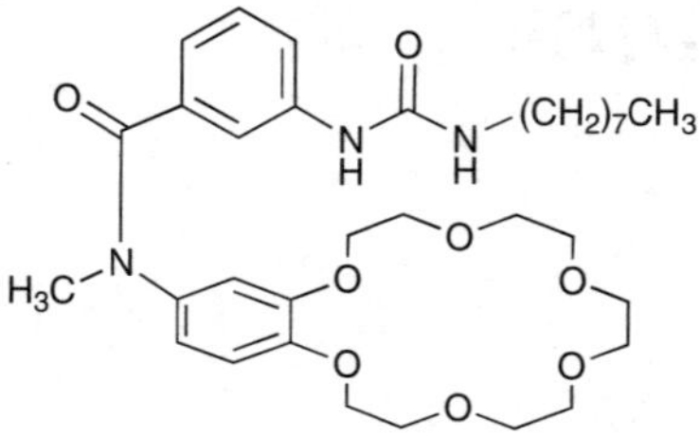

6.3

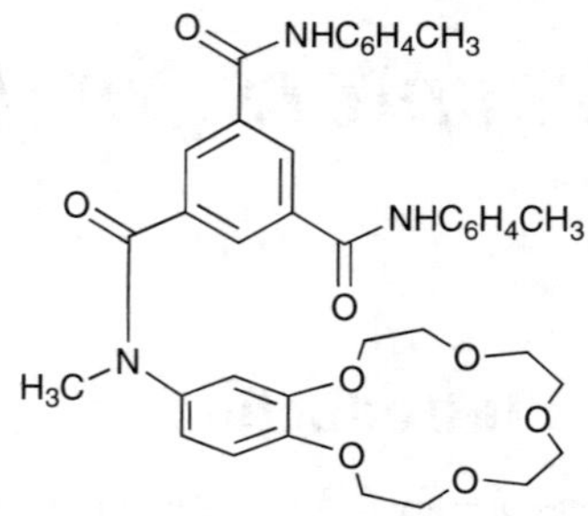

6.4

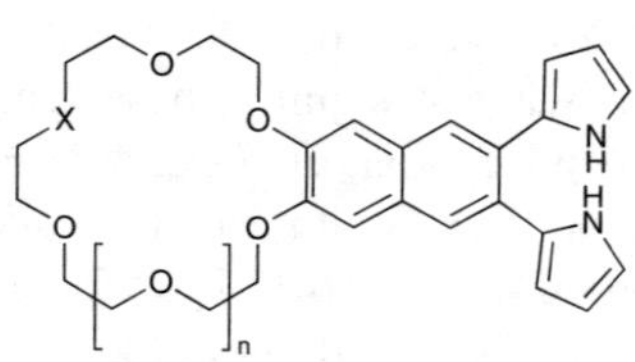

6.5a X=O, n=1
6.5b X=O, n=0
6.5c X=NH, n=1
6.5d X=NH, n=0

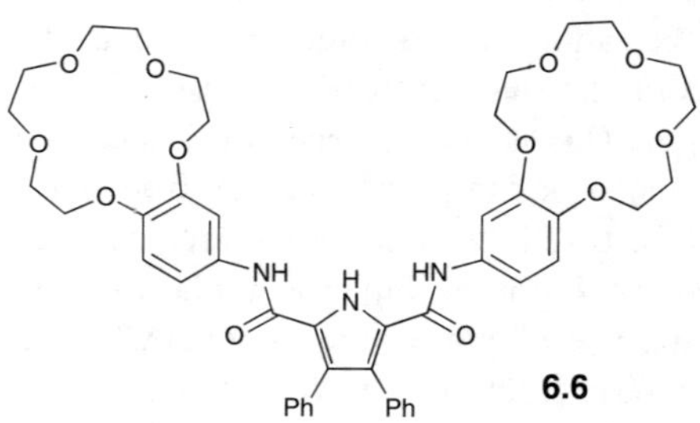

6.6

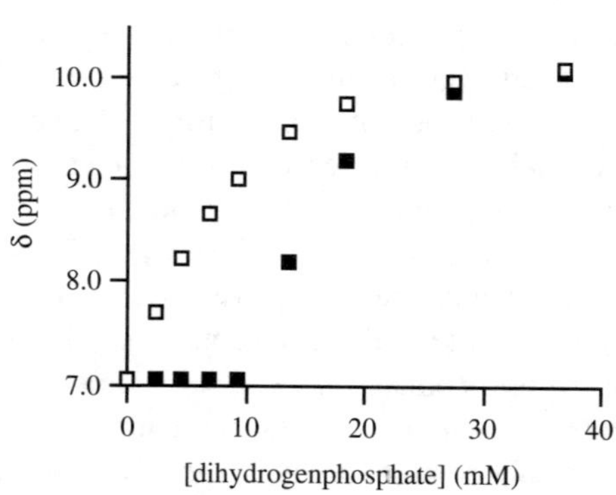

Figure 6.1 *Chemical shift for aryl-NH proton in **6.1** (initially 10 mM) as observed by ^1H NMR spectroscopy in CD_3CN and 295 K upon addition of tetrabutylammonium dihydrogen phosphate: presence (solid square), and absence (outline square) of potassium tetraphenylborate (initially 10 mM). The signal for the alkyl-NH in **6.1** shows the same behaviour. (Reproduced with permission from ref. [1]. Copyright 2000, American Chemical Society.)*

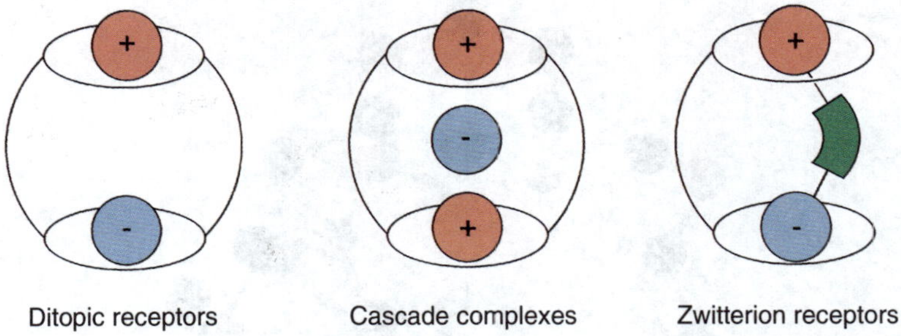

Figure 6.2 *Three types of ion-pair receptors: ditopic receptors, cascade complexes, and receptors for zwitterions*

6.2 Ditopic Receptors

Reetz and co-workers[3] reported one of the first examples of an ion-pair receptor. These workers synthesized a crown ether with a Lewis-acidic boron centre (**6.7a)** that was found to complex potassium and fluoride ions simultaneously. The crystal structure (Figure 6.3) reveals the potassium ion bound within the macrocyclic cavity, while the Lewis-acidic boron atom is bound to the fluoride anion. Evidence for the existence of the complex in solution was provided by ^{13}C- and ^{11}B-NMR spectroscopy. They also showed that the 18-membered phenolic crown ether analogue of **6.7a** can be metalated with trimethylaluminium to give **6.7b**, which was shown to form a ditopic complex with LiCl, both in solution and the solid state.[4]

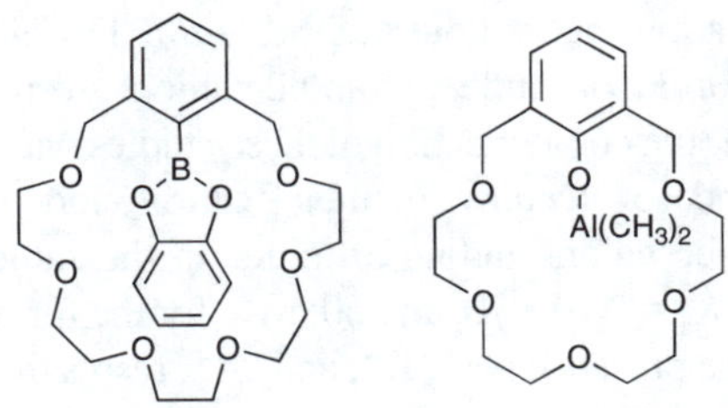

6.7a **6.7b**

In 1993, Kilburn and co-workers[5] reported the synthesis and binding properties of the pyridine-strapped bis-amide **6.8**. It was found that this cryptand-like system does not bind the ammonium salts of amino acids, but rather that it binds the mono-potassium salts of dicarboxylic acids well, presumably as the result of acting as a ditopic (cation + anion) receptor. Qualitative binding studies were carried out by observing the shifts in the amide NH peak in the ^{1}H NMR spectrum engendered by the addition of mono-potassium dicarboxylic acid salts; these studies were supported by extraction experiments and by FAB mass spectrometric analyses.

As part of a focused effort to prepare amide-based ditopic receptors, the new macrobicyclic systems **6.9a** and **6.9b** were synthesized and evaluated for their ability to bind concurrently both alkali halide salts and neutral molecules by B.D. Smith and co-workers.[6] X-ray diffraction analysis revealed that in the case of the sodium chloride

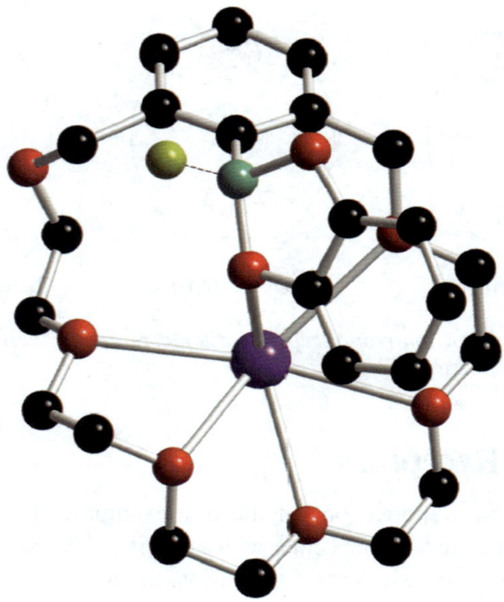

Figure 6.3 *Single-crystal X-ray structure of the potassium fluoride complex of receptor **6.7a***

complex of **6.9a** the sodium cation is held within the dibenzo-18-crown-6 core, while the chloride anion is hydrogen bonded to the two amide NH residues (Figure 6.4). A chloroform solvent molecule was found between the cation and the anion separating the ion-pair in the solid state. The association constants of **6.9a**, determined by ^{1}H NMR spectroscopic titrations carried out in DMSO-d_6/CD_3CN (3:1), proved to be 50, 9, and <1 M^{-1} for chloride, bromide, and iodide anions, respectively, in the absence of an added metal anion (the counter cation in these studies was tetrabutylammonium). However, in the presence of 1 equiv. of metal cation, added as the corresponding tetraphenylborate salt, the anion-binding affinities in the same solvent system were increased dramatically ($K_a = 410$, 470, and 60 M^{-1} for the formation of the **6.9a**-chloride anion complex in the presence Na^+, K^+, and Cs^+, respectively). These results provide support for the notion that cooperative anion *plus* cation binding takes place in solution and that these systems are *bona fide* ditopic receptors.

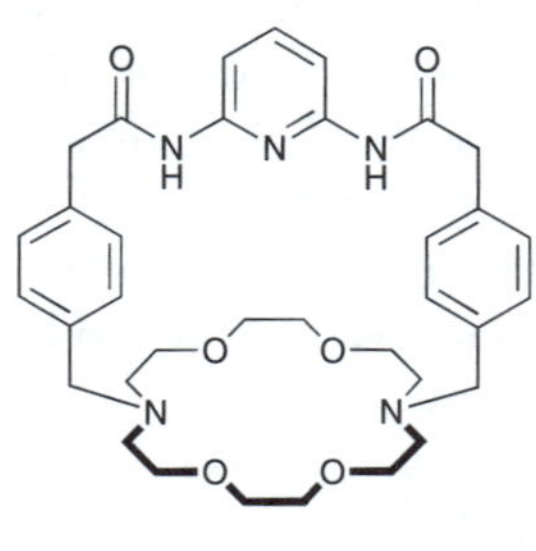

6.8

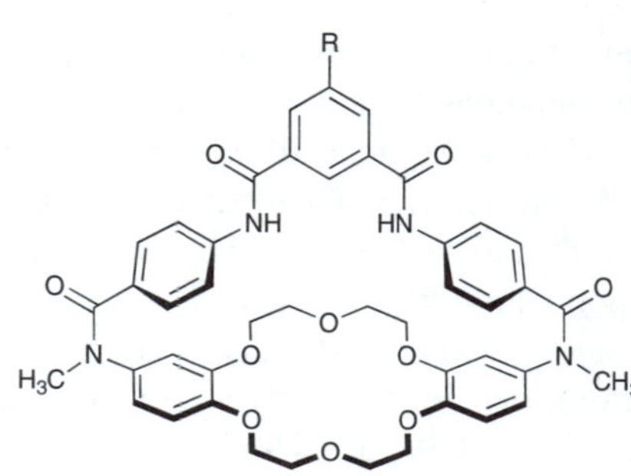

6.9a R = H
6.9b R = *t*-Bu

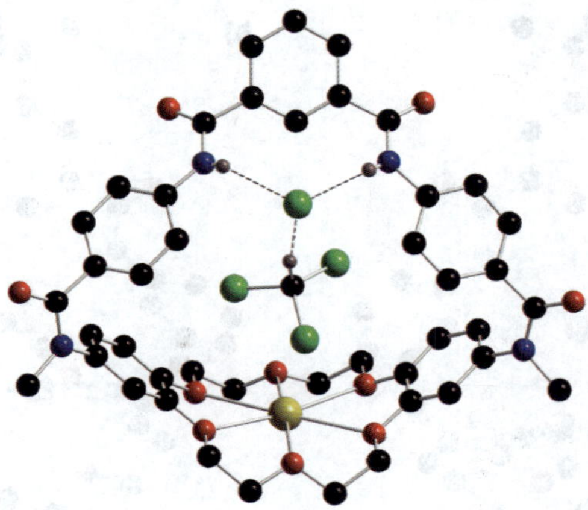

Figure 6.4 *Single-crystal X-ray structure of the solvent separated sodium chloride complex of receptor* ***6.9a****. A chloride anion is bound to the isophthalamide moiety, while a sodium cation is bound to the crown ether. A chloroform molecule is bound between the cation and anion, held in place by a single hydrogen bond to the chloride*

B.D. Smith and co-workers[7] have also reported the synthesis of ion-pair receptors capable of binding *contact* ion-pairs. Receptor **6.10** contains a slightly smaller cavity than receptors **6.9a** and **6.9b**. The association constant corresponding to the binding of chloride by receptor **6.10**, (*i.e.*, 35 M^{-1}) in DMSO-d_6 at 295 K, is enhanced to 460 M^{-1} in the presence 1 equiv. of potassium tetraphenylborate. Addition of sodium to the receptor only serves to increase the chloride anion affinity to a K_a of 50 M^{-1} under the same conditions. X-ray crystal structures of the NaCl and KCl complexes reveal that the ion-pairs are in contact in the solid state (Figure 6.5) and that the sodium cation is bound more closely to the crown ether than is the potassium cation. Presumably, this serves to increase ion-dipole repulsions between the chloride anions and the crown ether oxygen atoms, thus lowering the affinity for NaCl.

The interactions of **6.10** with the potassium and sodium salts of trigonal oxyanions (*e.g.*, nitrate and acetate) were also characterized using NMR spectroscopy in the solution phase and single-crystal X-ray diffraction analysis in the solid state.[8] Transport experiments, involving the use of a liquid organic membrane, were also carried out, as were solid–liquid extraction studies.[9] Taken in concert, this work provided further evidence for ion-pair binding in the case of this receptor system.

In 2003, P.A. Gale, B.D. Smith and co-workers[10] reported an analogue of **6.10** that contains a 2,5-diamidopyrrole unit instead of the isophthalamide moiety present in the original receptor. The pyrrole containing receptor **6.11** binds chloride anions with a stability constant of 109 M^{-1} in DMSO-d_6 at 298 K. In the presence of 1 equiv. of potassium tetraphenylborate the chloride binding is enhanced to 540 M^{-1}, while 1 equiv. of sodium tetraphenylborate enhances the chloride affinity by only a small amount (K_{Cl} = 128 M^{-1}). The enhancement in affinity of **6.11** as compared to **6.10** is presumably due to the extra hydrogen bond donor present in the pyrrole-based receptor (as shown in Figure 6.6a).

An alternative modification was reported by Tuntulani and co-workers. In this instance, the pyrrole and/or benzene units were replaced by the electrochemically active

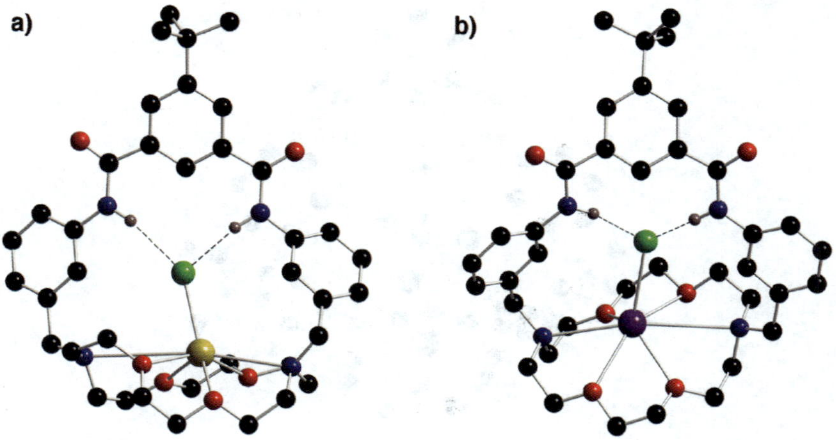

Figure 6.5 *Single-crystal X-ray structures of (a) the sodium chloride and (b) the potassium chloride complex of **6.10***

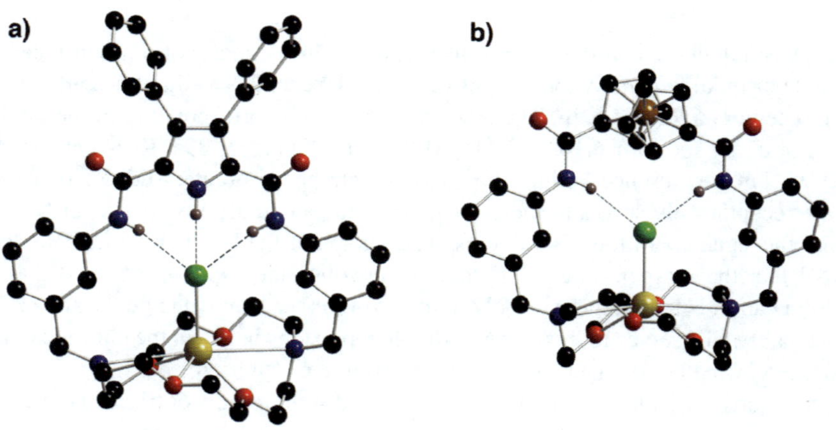

Figure 6.6 *Single-crystal X-ray structures of (a) the sodium chloride complex of **6.11** and (b) the sodium chloride complex of **6.12***

ferrocene "reporter group". It was found that the ditopic receptor **6.12**, a species that contains a crown ether as well as amidoferrocene groups, shows high selectivity towards bromide anion in the presence of a bound sodium cation, but only weak binding in its absence.[11] The association constant for the cooperative binding of bromide in the presence of sodium cation was found to be $K_a = 16{,}100\ M^{-1}$, as judged from ^1H NMR spectroscopic titrations carried out in $CDCl_3$ containing 5% CD_3CN. Unfortunately, the corresponding study with chloride anion could not be carried out due to ion-pair formation between the bound sodium cation and the chloride anion. A solid-state analysis of the putative NaCl complex revealed encapsulation of the sodium cation within the crown ether unit, while the chloride counter anion is seen to be bound, as expected, to two amide NH protons via hydrogen bond interactions (Figure 6.6b).

Solution phase cyclic voltammometric studies (CH_3CN/CH_2Cl_2 (40:60) containing 0.1 M $TBAPF_6$) were also carried out. These served to confirm that, as expected, the presence of metal cations results in a cathodic shift in the redox potential upon binding anions (ΔE (mV) = 107(28), 122(46), and 153(not determined)) for **6.12**, **6.12•**Na^+, and **6.12•**K^+, respectively, in the presence of Cl^- and Br^- (values in parentheses).

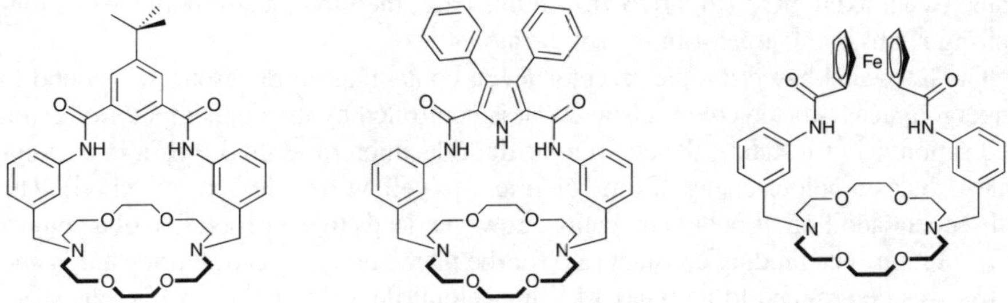

6.10 **6.11** **6.12**

In 1999 Beer and co-workers[12] reported the synthesis of a compound **6.13**, a tripodal tris(amido benzo-15-crown-5). In the presence of 1 equiv. of sodium picrate, **6.13** acts as a ditopic receptor, showing enhanced anion binding for Cl^-, I^-, and ReO_4^- ions. Receptor **6.13** was also found to be capable of extracting sodium pertechnetate ($NaTcO_4$) efficiently under conditions designed to simulate aqueous waste streams containing this radioactive species. Other related tripodal systems have also been reported. The interested reader is referred to reference 13.

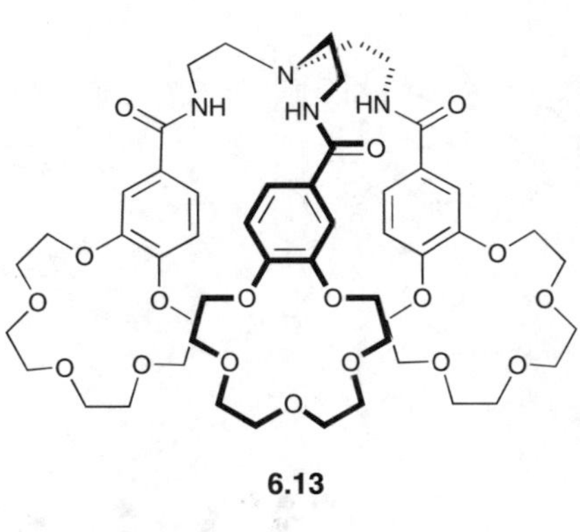

6.13

Teramae and co-workers[14] have demonstrated cooperative binding for a series of anions in the presence of sodium cations utilizing a thiourea-functionalized benzo-15-crown-5 system (**6.14a**). In the presence of 2 equiv. of $Na(BPh_4)$ (conditions under which Na^+ is over 95% bound) receptor **6.14a** in acetonitrile-d_3 exhibits an approximate 10-fold increase in anion-binding affinity for NO_3^- (K_{NO_3} = 6.0 M^{-1} *vs.* K_{NO3}(Na) = 66 M^{-1}) and Br^- (K_{Br} = 25 M^{-1} *vs.* K_{Br}(Na) = 260 M^{-1}), and a five-fold increase for I^- (K_I = 4.3 M^{-1} *vs.* K_I(Na) = 20 M^{-1}).

Slightly later, the anion- and cation-binding properties of ditopic receptor **6.14b**, analogues of **6.14a**, were studied by Barboiu and co-workers[15] using single-crystal

X-ray diffraction analysis. As shown in Figure 6.7a, the crystal structure of (**6.14b**)$_2$•NaCl reveals that the sodium cation is sandwiched by two crown ether units and that the chloride anion is bound to four urea NH groups. The long distance between the Na^+ and Cl^- ions is consistent with little or no interaction between the two species. In the case of $NaNO_3$, the structure of the 2:2 complex confirmed that the sodium cation is not only complexed within the crown ether but also is coordinated by an axial NO_3^- (Figure 6.7b). In this case, therefore, a strong degree of ion-pairing is observed, at least in the solid state.

The ferrocene-based ditopic receptor **6.15a** containing a urea group was found to undergo an easy-to-see colour change that is controlled by the dual "inputs" of anion and cation.[16] For instance, the addition of fluoride anion to receptor **6.15a** in acetonitrile induces a colour change from colourless to yellow (*i.e.*, it is switched on). The subsequent addition of potassium cation, however, leads to the generation of a colourless solution. The binding constant (K_a) for the interaction between fluoride anion and **6.15a** was determined to be 9340 M^{-1} in acetonitrile as judged from UV–Vis spectroscopic titrations. The binding constant (K_a) for the interaction between **6.15a** and K^+ in the presence of 10 equiv. F^- was found to be 1460 M^{-1} in acetonitrile (again, as determined from UV–Vis spectroscopic titrations). By contrast, analogue **6.15b** was observed to bind K^+ with an affinity of only 230 M^{-1} under identical conditions. The higher value of binding constant for potassium seen in the case of **6.15a** is thought to reflect a direct interaction between the potassium cation and the crown ether unit, which serves to enhance the intrinsic ion-pairing effects that contribute to binding in the case of **6.15b**. A similar strategy was used to design several other ostensibly analogous ditopic receptors, as detailed in the recent literature.[17]

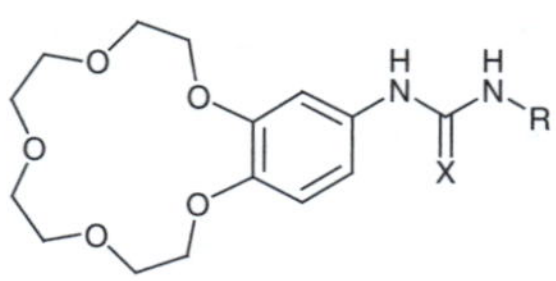

6.14a X = S, R = CH3
6.14b X = O, R = Ph

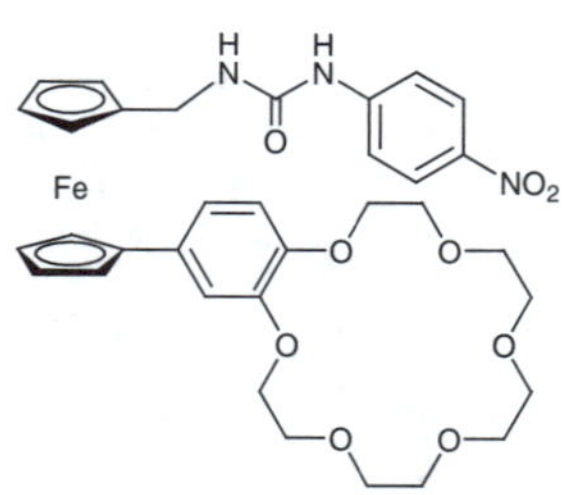

6.15a

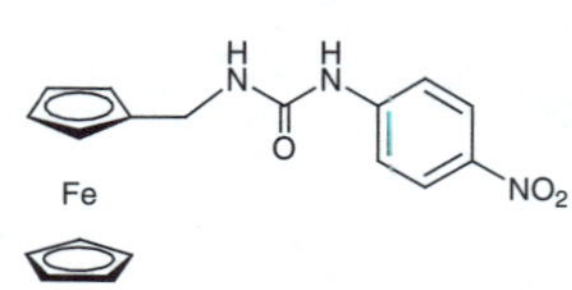

6.15b

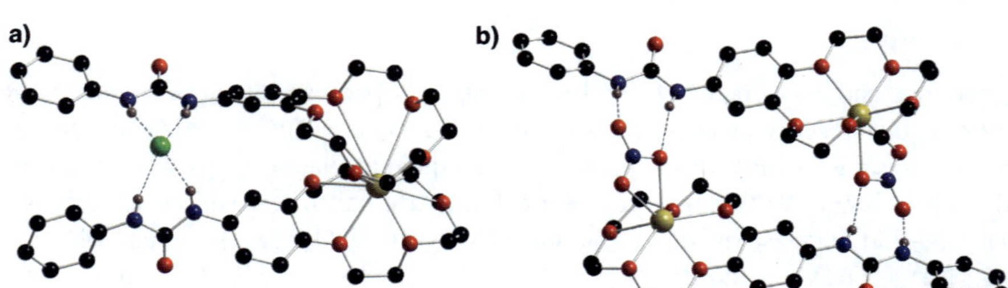

Figure 6.7 *Single-crystal X-ray structures of (a) the 2:1 complex of* ***6.14b*** *with NaCl and (b) the 2:2 complex of* ***6.14b*** *with* $NaNO_3$

Crown ethers have been widely studied as receptors for cations. In 2000, Kubo and co-workers[18] used a crown ether as a backbone for attaching two thiourea groups. In this way, he generated the ditopic receptor **6.16**, a species that was designed to bind anions through interactions with the thiourea subunits, while concurrently complexing a potassium cation within the dibenzo-diaza-30-crown-10 core (Figure 6.8). It was found that the addition of K^+ induced an overall upfield shift in the signals ascribed to the crown ether moiety when the addition process was studied by ^1H NMR spectroscopy in CD_3CN solution. By carrying out these studies in a quantitative manner and employing a standard non-linear curve-fitting procedure, the association constant for K^+ complexation was calculated to be 5600 M^{-1}. Receptor **6.16** was also found to exhibit a high selectivity for certain oxyanions, binding such species in a 1:2 host-guest ratio. Further, as would be expected in light of the design suggestions, this anion-binding ability was found to be tunable via the addition of potassium cation. For instance, the association constant for diphenyl phosphate binding in the presence of potassium cation was found to be higher than in its absence by a factor of 19. Presumably, the addition of K^+ and its expected complexation by the crown ether moiety, not only induces a conformational change that produces a structure more favourable for anion binding, but also increases the acidity of the thiourea protons as the result of electrostatic effects. Schröder and Steed[19] have also reported similar ditopic receptors.

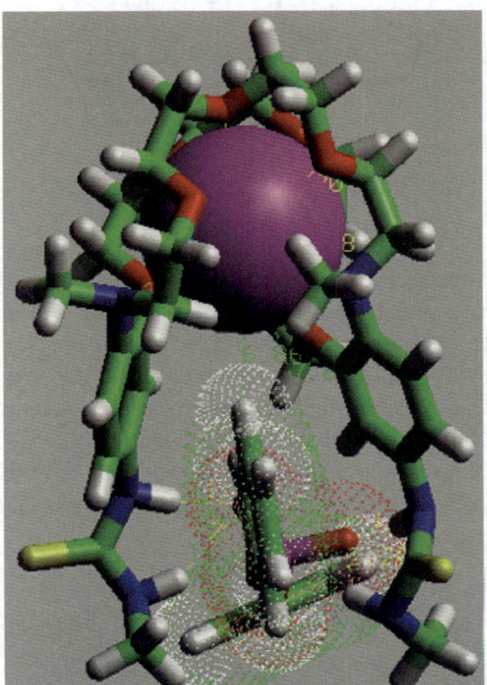

Figure 6.8 *Energy minimized structure of the complex formed between* ***6.16*** *and* $K(PhO)_2P(O)O$.
(Reproduced with the permission from reference [17]. Copyright 2000 Elsevier.)

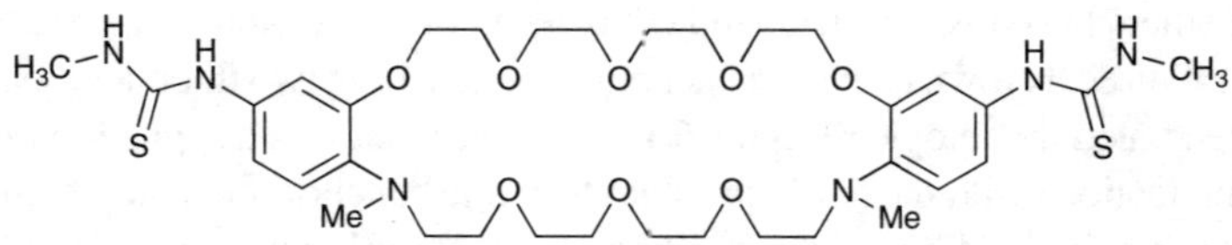

6.16

An early example of an ion-pair receptor for ammonium fluoride was reported by the Sessler group in 1995. The HCl salt of the sapphyrin-crown ether conjugate **6.17** was shown to complex simultaneously ammonium and fluoride ions, in the crown and sapphyrin units, respectively.[20] Unfortunately, the proposed concurrent cation- and anion-binding behaviour could not be substantiated by X-ray structural analyses. Thus, this approach to ion-pair receptor generation has not received much in the way of follow-up attention.

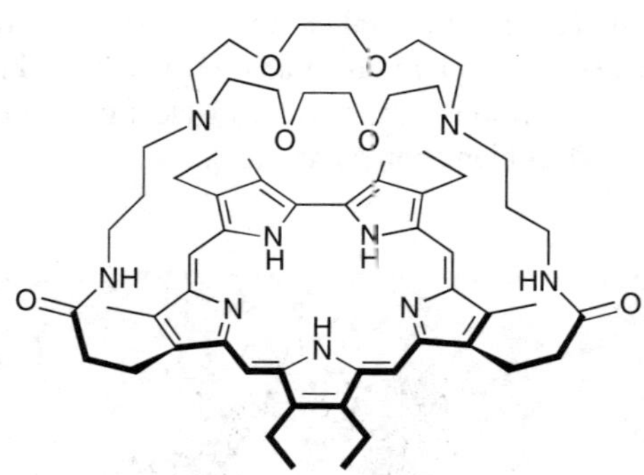

6.17

Reinhoudt and co-workers[21] have described the use of compounds **6.18a** and **6.18b** as ditopic receptors that simultaneously recognize potassium and dihydrogen phosphate ions. The first of these receptors consists of a Lewis-acidic uranyl (UO_2^{2+}) centre bound to two benzo[15]crown-5 units, while the second contains the crown ether moiety in the form of a modified calix[4]arene. In both cases, the uranyl group functions as a binding site for the anion, while the potassium cation is bound by the crown ether moieties. As discussed later (*cf.* Chapter 8), Reinhoudt and co-workers[22] have shown that uranyl-containing clefts are capable of binding dihydrogen phosphate and chloride anions.

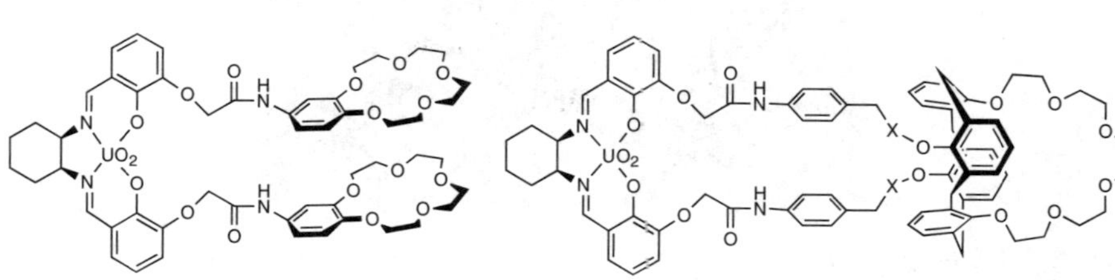

6.18a **6.18b** X = $(CH_2)_4$

Compound **6.18b**, which contains a rigid Cs^+-selective calix[4]arene crown-6 platform, was also synthesized.[22] This receptor was assessed for its ability to transport caesium chloride and caesium nitrate through a supported liquid membrane (SLM). Although chloride is much more hydrophilic than nitrate, a higher rate of flux was observed through the hydrophobic membrane for the chloride salt (1.20×10^{-7} mol m^{-2} s^{-1}) than for caesium nitrate (0.89×10^{-7} mol m^{-2} s^{-1}) in the presence of **6.18b**. However, not only was the substrate selectivity demonstrated more significantly but also the caesium chloride transport rate was found to exceed greatly that observed when mono-functional analogues of compound **6.18b** were used. Such findings were considered consistent with both binding sites being necessary to achieve efficient complexation and transport. A more detailed investigation into the facilitated transport of salts through SLMs revealed that the bis(thiouredio)-calix[4]-crown-6 **6.19a** serves to effect the symport of caesium and chloride more efficiently than its mono-functional calix[4]-crown-6 analogue **6.20a**.[23] However, with higher concentrations of caesium chloride in the source phase ([CsCl] $>$ 0.3 M), a mixture of cation-binding **6.20a** and anion-binding bis(thiouredio)calix[4]arene **6.21** receptors was found to give higher CsCl transport rates than the ditopic carrier **6.19a**. Similar results were obtained for KCl transport with the calix[4]-crown-5, **6.19b**; again, this system was found to be less effective than a mixture of its mono-functional analogues **6.20b** and **6.21**. These findings were ascribed to a lower diffusion coefficient for the ditopic carriers than the monotopic systems. Similar conclusions were also drawn recently by Tuntulani and Pappalardo and Parisi.[24]

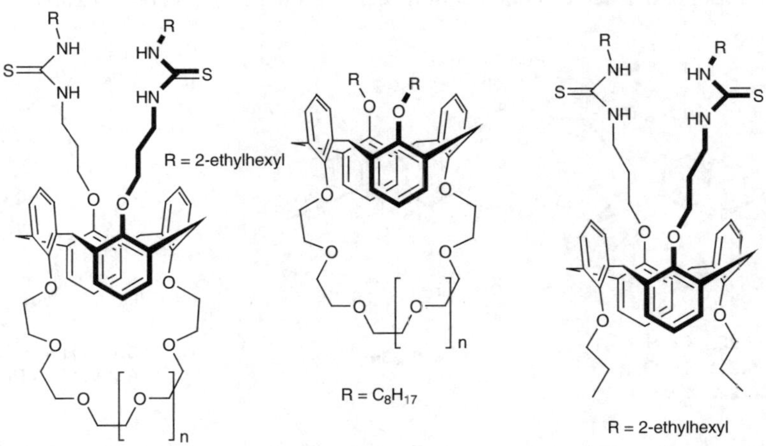

6.19a n = 1
6.19b n = 0

6.20a n = 1
6.20b n = 0

6.21

Ungaro and co-workers[25] prepared the (thio)urea functionalized calixarene-based ditopic receptors **6.22**. These systems were designed to complex cations at their lower rim through interactions with the four appended amide groups, while recognizing anions at the upper rim *via* the attached (thio)urea groups. The results of ^1H NMR spectroscopic titrations, carried out in DMSO-d_6, revealed that the complexation of sodium ion at the lower rim produced different anion-binding behaviour in the case of receptors **6.22a** and **6.22c**. In particular, **6.22c** displayed an increased affinity for various anions in the presence of Na^+, whereas receptor **6.22a** was found to bind anions slightly less well in the presence of this cation, with the reference point in each case being the corresponding free receptor. These results were rationalized in terms of the methylene unit between the calixarene phenyl moieties and the thiourea subunits present in **6.22a** serving to "insulate" the electron-withdrawing effect of the bound metal cation. The spacer thus serves to mitigate the anion-binding enhancing effects that cation complexation might otherwise be expected to provide.

Slightly before the report of **6.22** by Ungaro and co-workers, the synthesis of the bis- and tetraurea functionalized calixarenes **6.23** was reported by Reinhoudt and co-workers.[26] It was found that in the absence of a coordinating cation, system **6.23a** did not bind anions with appreciable affinity at room temperature in $CDCl_3$. Presumably, this reflects the fact that in the absence of added guests this system exists in the form of a "pinched" conformation that is stabilized via strong intramolecular hydrogen bond interactions between the urea groups. However, a conformation change from this pinched to a more normal cone conformation was inferred in the presence of sodium ion. Specifically, in the presence of this cation, ^1H NMR spectral shifts were observed upon the addition of anions to solutions of **6.23a** in $CDCl_3$. In the case of **6.23b**, quantitative measurements of several anion affinities were carried out in the presence and absence of Na^+ using standard ^1H NMR spectral methods. From these, it was determined that **6.23b** binds chloride and bromide with association constants (K_a) of $>1.1 \times 10^4$ and 1.3×10^3 M^{-1} in $CDCl_3$, respectively, in the presence of Na^+. On the other hand, no evidence of anion binding was observed in the absence of added cation (*i.e.*, with the free receptor **6.23b**).

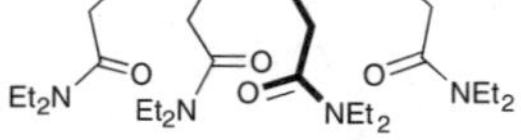

6.22a X = S, n =1
6.22b X = O, n =1
6.22c X = S, n =0

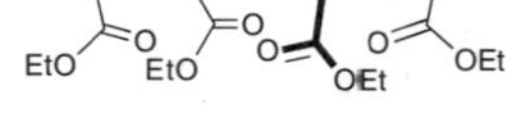

6.23a

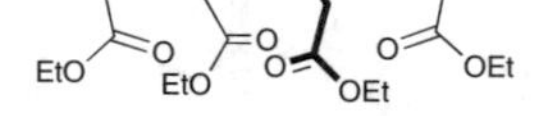

6.23b R = *n*-Oct
6.23c R = *t*-Bu

In 2001, a new rigid calixarene-based receptor, **6.24a**, containing four hydroxyl groups, was introduced by Pochini and co-workers.[27] This heterotopic receptor contains a cavity suitable for the complexation of cationic quaternary ammonium cations, as well as four hydroxyl groups attached to the upper rim designed to support anion binding. This receptor was thus expected to be capable of binding tetramethylammonium (TMA) salts of, *e.g.*, tosylate, chloride, acetate, trifluoroacetate, and picrate readily, without breaking fully the ion-pair interactions seen in the salts themselves. To the extent this proved true, cation binding *per se* would benefit from a positive allosteric effect arising from the pair-wise binding of a counter anion.

Support for the proposed inside-the-cage cation binding came from a ROESY spectral analysis; here, a close correlation between the protons of the calixarene upper benzene subunits and the methyl protons of the TMA cation was seen. Evidence consistent with the binding of TMA salts was then obtained from standard NMR spectroscopic titrations carried out in $CDCl_3$. These experiments revealed that receptor **6.24a** has a strong preference for TMA-acetate ion ($K_a > 10{,}000\ M^{-1}$) relative to other species. In this solvent, the anion-binding order was found to be acetate > tosylate > chloride > trifluoroacetate > picrate. The addition of polar deuterated solvents, such as CD_3OD, CD_3CN, or DMSO-d_6 to the host–guest complex in $CDCl_3$ solution, was found (as expected) to result in the release of the TMA cation from within the receptor cavity. In the case of **6.24b** having four methoxy groups instead of OH group, little evidence of cation binding was seen, a result that was taken in support of the proposed positive anion-induced allosteric effect seen in its analogue, **6.24a**.

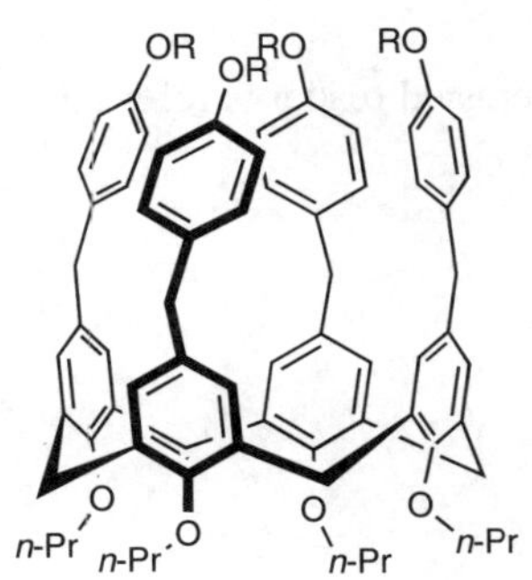

6.24a R = H
6.24b R = Me

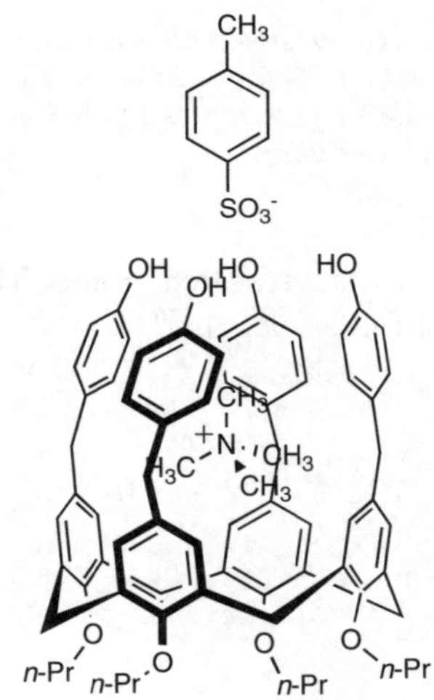

6.24a -TMATs

Atwood and Szumna[28] have reported the synthesis of a molecular capsule for the recognition of TMA halide salts that function in competitive solvents such as methanol. A first-generation system, compound **6.25a**, was originally reported in 2002. In this case, the TMA cation was found to sit within the aromatic cavity of the receptor, being held in place as the result of multiple CH–π interactions. Meanwhile, the anion was seen to be bound to the appended amide groups. Atwood and Szumna[29] then synthesized the second-generation receptors **6.25b** and **6.25c** in which the phenyl groups attached to the amide serve to isolate the anion from the solvent. This structural modification allows the ion-pair (TMACl or TMABr) complex formed with receptor **6.25b** to remain stable in solutions containing high concentrations of methanol – conditions under which the halide bound to **6.25a**–TMA^+ would dissociate from the complex. The crystal structure of the TMACl complex of **6.25c** reveals encapsulation of the ion-pair in a manner that largely shields it from the environment outside the capsule (Figure 6.9). Subsequent to this initial work, related studies were

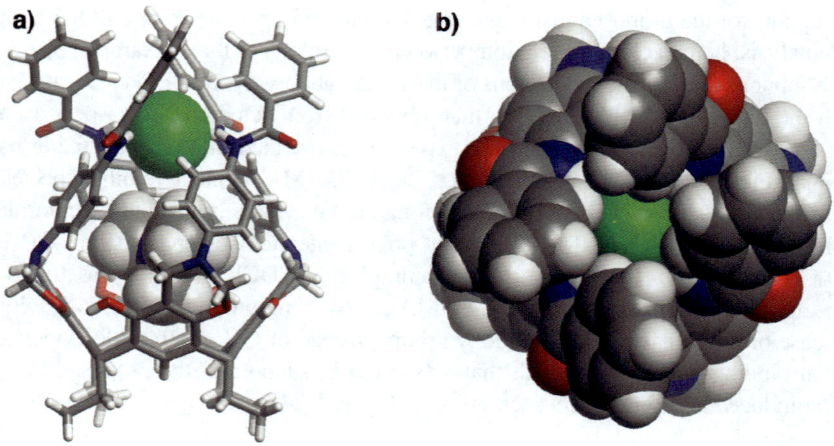

Figure 6.9 *Single-crystal X-ray structure of* **6.25c***-TMACl (obtained from a mixture of* $CHCl_3$ *and nitrobenzene as* **6.25c***-TMACl·4.5*$PhNO_2$*). (a) Side view with receptor molecule in stick representation, guest as van der Waals sphere and (b) top view – space filling*

carried out by other research groups. The interested reader is referred to the original literature for further details.[30]

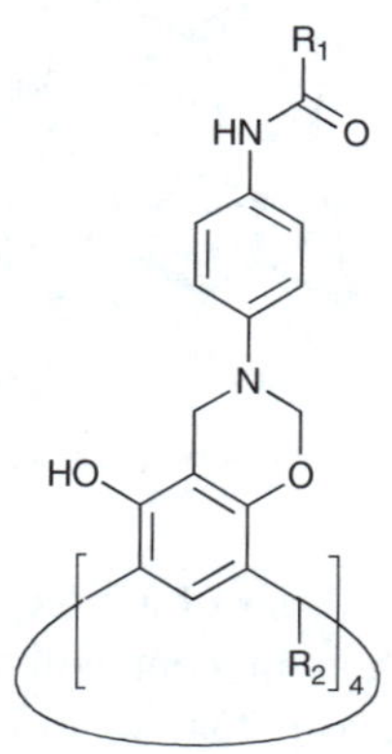

6.25a $R_1 = CH_3$, $R_2 = i$-Bu
6.25b $R_1 = Ph$, $R_2 = i$-Bu
6.25c $R_1 = Ph$, $R_2 = Et$

J.T. Davis and co-workers[31] have recently reported the synthesis and complexation properties of a self-assembled ion-pair receptor. This receptor, the functionalized calix[4]arene **6.26**, was found to be only slightly soluble in dry $CDCl_3$. It thus gave rise to a poorly resolved ^1H NMR spectrum, consistent with non-specific aggregation. However, in wet $CDCl_3$ a set of well-resolved resonances was observed corresponding to the dimeric species shown in Scheme 6.1. This water-stabilized dimer contains a G-quartet capable of alkali metal complexation and amide groups capable of anion recognition. Indeed, the dimer is capable of extracting alkali metal halide salts such as NaCl from water into organic solution. Moreover, the bound salt appears to increase the thermal stability of the assembly.

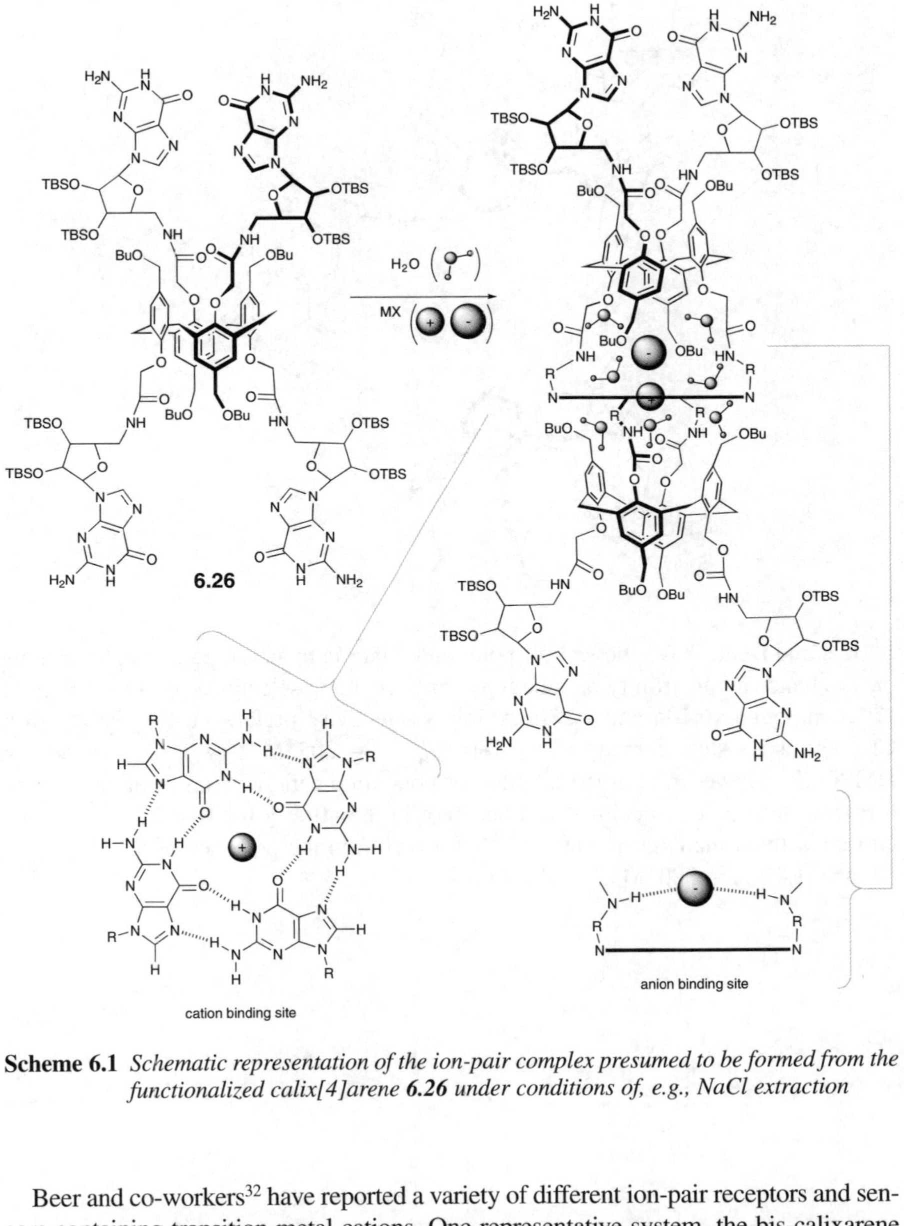

Scheme 6.1 *Schematic representation of the ion-pair complex presumed to be formed from the functionalized calix[4]arene* **6.26** *under conditions of, e.g., NaCl extraction*

Beer and co-workers[32] have reported a variety of different ion-pair receptors and sensors containing transition metal cations. One representative system, the bis-calixarene **6.27**, was found to bind alkali metal cations at the lower rims of the two calix[4]arene units and to complex an anion in the amide cleft of the ferrocene or bipyridyl subunits. The binding of the metal cations serves to enhance the anion affinity of the receptor. For example, the affinity of **6.27** for iodide increases from 40 to 305 M^{-1} upon the addition of lithium cations in acetonitrile-d_3. This enhanced affinity was attributed to metal complexation inducing rigidity into the calixarene scaffold, thereby preorganizing the central cavity for iodide binding.

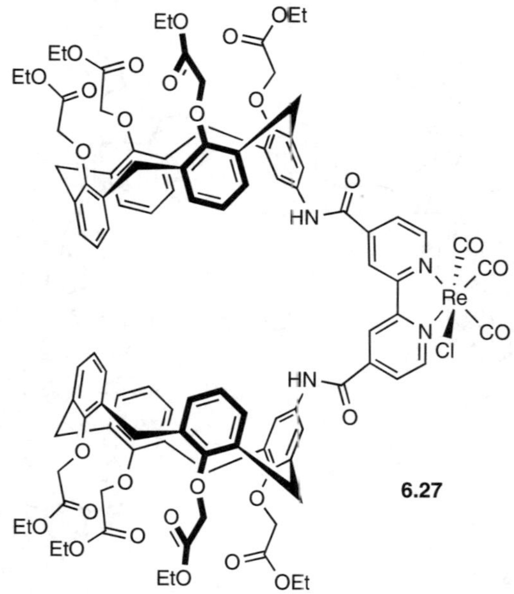

6.27

Beer and Dent[33] have shown how potassium binding to an ion-pair receptor can not only enhance anion affinity but effect a change in anion selectivity. In the absence of K^+, compounds **6.28a** and **6.28b** exhibit a selectivity preference for $H_2PO_4^-$ over Cl^- with a 1:1 stoichiometry (for **6.28b**, $K_{H_2PO_4} = 205\ M^{-1}$ *vs.* $K_{Cl} = 55\ M^{-1}$ in DMSO-d_6). However, after the addition of potassium cations, both receptors exhibit a reverse selectivity. Specifically, a decrease in the affinity for $H_2PO_4^-$ is seen concurrent with an increase in that for Cl^- (for **6.28b**, in the presence of K^+, $K_{H_2PO_4} = 35\ M^{-1}$ *vs.* $K_{Cl} = 300\ M^{-1}$ in DMSO-d_6).

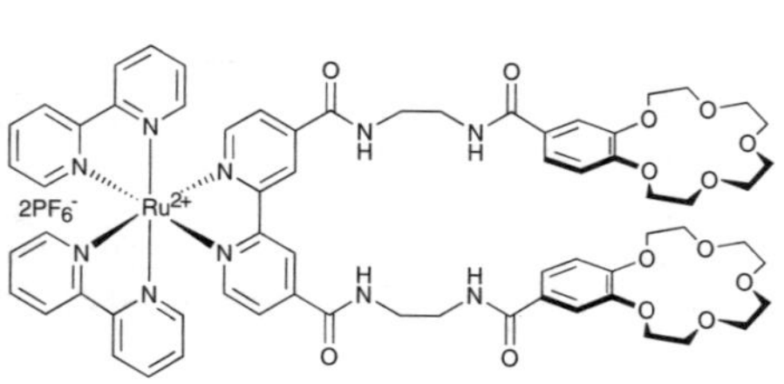

6.28a

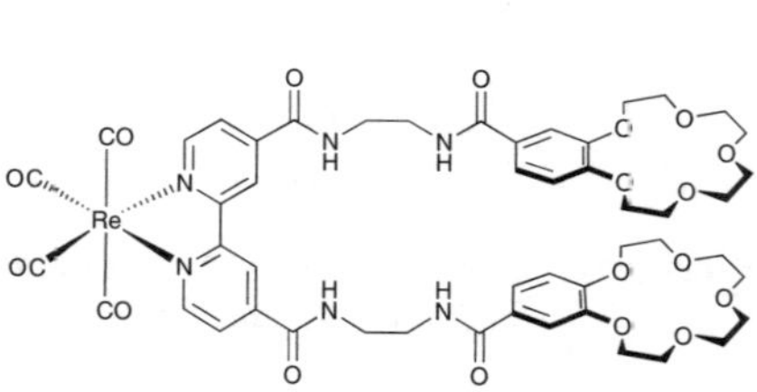

6.28b

Tasker and co-workers[34] have produced a family of ditopic receptors, such as **6.29** and **6.30**, that contain a dianionic-binding site for transition metal cations and a dicationic-binding site for anions. Receptor **6.29** (Scheme 6.2) contains a salen-based binding site capable of coordinating to transition metal cations such as Cu(II) or Ni(II). Upon complexation, the phenolic protons transfer to the morpholine nitrogen atoms and the receptor is forced into a conformation such that the protonated morpholine groups come into close proximity, thus defining an anion-binding site. As a consequence of this presumed motion, metal binding enhances dramatically the anion-binding affinity of

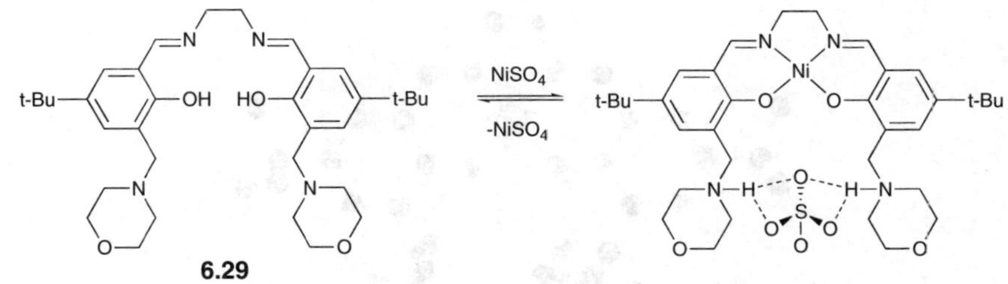

Scheme 6.2 *Formation of an ion-pair complex from system* ***6.29***

the receptor. Extraction experiments at pH 3.8 demonstrated that system **6.29** is capable of extracting $CuSO_4$ into chloroform with a near 100% receptor loading efficiency. The crystal structure of the nickel sulfate complex of **6.29** is shown in Figure 6.10.

Very recently, Tasker[35] has reported the synthesis of macrocyclic analogues of the above systems. The resulting compounds receptors **6.30a** and **6.30b**, both contain an hexyl linker between the two amine groups. In anion transport experiments, metal complexes of these receptors were found to successfully extract sulfate anion from aqueous solution to organic solvents. Under different conditions, organic solutions of these systems were found to effect the stripping of Cu(II) from test materials in the presence of sulfate. In the case of the copper complex of receptor **6.30b**, it was found that more than 90% exists in the form of [Cu(**6.30b**)(SO_4)] over a pH range of 3–5. However, less than 10% of the corresponding free ligand exists as the sulfate complex in this pH range. These results were attributed to the cooperative effects of metal(II) and sulfate anion binding, which, in turn, provides an explanation for the efficient sulfate anion transport observed in the extraction experiments.

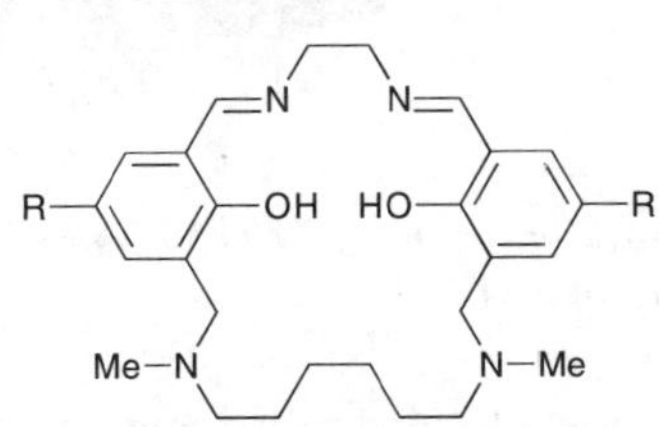

6.30a R = t-Bu
6.30b R = nonyl

We have seen in Chapter 5 that calix[4]pyrroles such as **6.31** bind anions *via* four NH···anion hydrogen bonds. The anion complexes adopt the so-called cone conformation wherein the pyrrole NH groups are oriented towards the anionic guests preorganizing the pyrrole rings to form an electron-rich cup-shaped cavity. Recently, Moyer, Sessler, Gale and co-workers[36] have shown that large charge diffuse cations such as caesium and imidazolium may be included in the anion-induced bowl in both solution and the solid state. For example, the caesium carbonate complex of **6.31** is shown in Figure 6.11a. Here two calixpyrroles encapsulate the carbonate anion via eight hydrogen bonds, while the caesium counter cations are bound in the electron-rich calixpyrrole bowl. Crystal structures of a variety of caesium halides have also been elucidated which demonstrate ion-pair complexation by **6.31**. Preliminary evidence from titration NMR experiments supports the suggestion that caesium inclusion also occurs in solution.

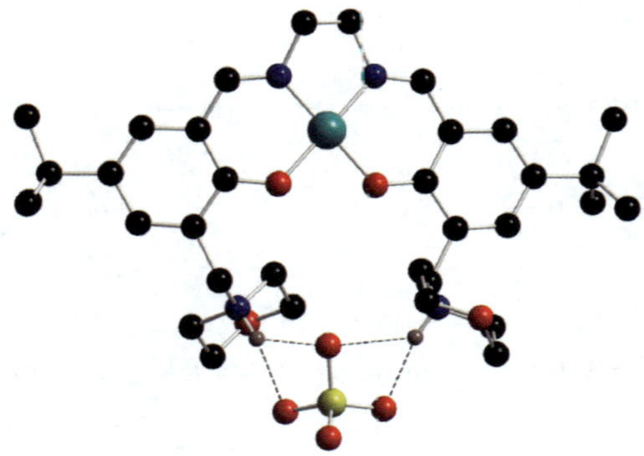

Figure 6.10 *Single-crystal X-ray structure of the complex formed from receptor **6.29** as determined from crystals grown in the presence of $NiSO_4$*

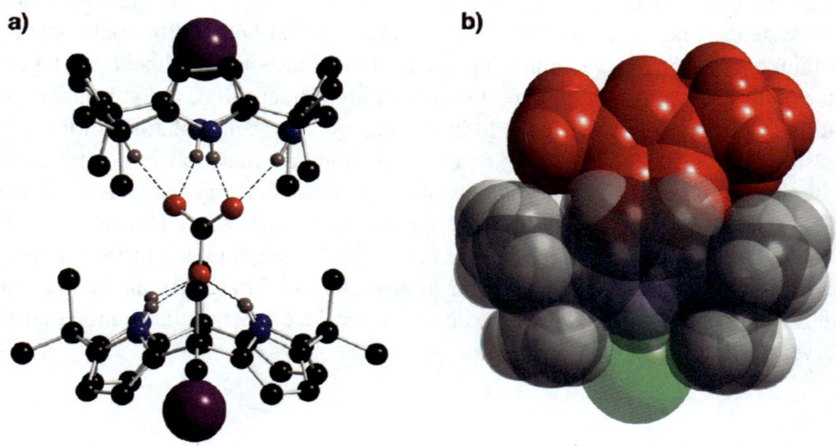

Figure 6.11 *Single-crystal X-ray structures of (a) the $Cs_2(CO_3)$ complex and (b) the imidazolium chloride salt complex of **6.31***

Similarly, organic cations have been shown to bind in the anion-induced calixpyrrole cup. A variety of imidazolium halide and oxo-anion complexes have been prepared and crystallographically characterized, demonstrating that **6.31** can also encapsulate organic salts. For example, the crystal structure of the 1-butyl-3-methylimidazolium chloride complex of **6.31** is shown in Figure 6.11b and shows the 4- and 5-positions of the imidazolium cation included in the cup. Solution studies with a variety of imidazolium salts shows that in the presence of a preorganizing anion such as chloride, imidazolium binding occurs in the calixpyrrole cup, while in the presence of "innocent" anions such as BF_4^-, which do not bind to **6.31** and hence do not preorganize the calixpyrrole cavity, imidazolium cation inclusion is not observed.

Recent studies by Gale have shown that alkylammonium salts are also bound by **6.31**. The crystal structures of the tetramethyl-, tetraethyl- and tetrabutylammonium chloride salts are shown in Figure 6.12. The TMA cation is oriented in the cavity such that one methyl group points down towards the bound chloride anion. The remaining three

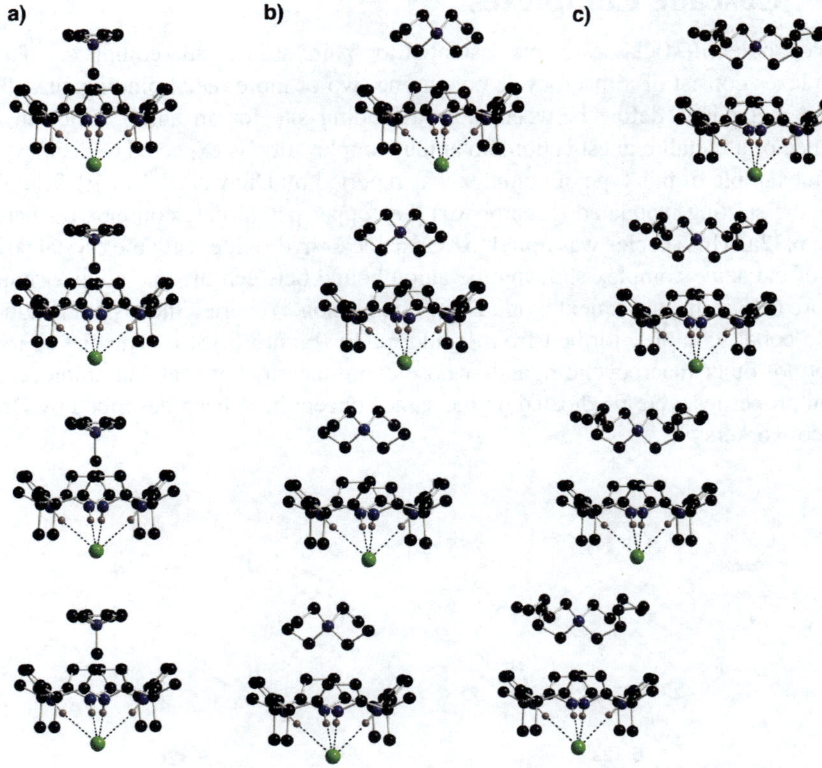

Figure 6.12 *Single-crystal X-ray structures of (a) $(Me)_4N^+Cl^-$ complex, (b) $(Et)_4N^+Cl^-$ complex, and (c) $(nBu)_4N^+Cl^-$ complex of* ***6.31***

methyl groups are disordered. The tetraethylammonium complex shows the cation oriented differently such that the S_4-axis present in the cation lies along the C_4-axis through the calix[4]pyrrole complex. One measure of the degree of cation inclusion in the cup is to define the centroid between the four nitrogen atoms in the macrocycle and measure the distance between the centroid and the nitrogen atom in the ammonium cation. These distances for the TMA, tetraethylammonium, and tetrabutylammonium chloride complexes are 3.906, 4.361, and 4.445 Å, respectively illustrating that as the size of the alkyl groups increases the positively charged ammonium centre is included in the calixpyrrole cup to a progressively lower degree.[37]

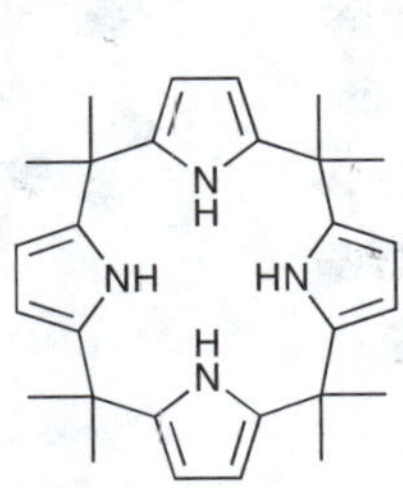

6.31

6.3 Cascade Complexes

One of the earliest classes of encapsulated ion-pairs are cascade complexes. These complexes consist of a macrocycle containing two or more metal binding sites. The bound metal ions define between them a binding site for an anion. Thus, in the absence of a metallic guest cation, no anion complexation is expected to occur.

An example of this type of complex was reported by Drew *et al.*[38] in 1979. In this case, the starting metalated receptor was the copper perchlorate complex of macrocycle **6.32a**. This species was found to bind azide or hydroxide, with the crystal structure of the azide complex showing the anion bound between the two copper centres (Figure 6.13a). In subsequent work, Drew and Nelson[39] reported the bis-azide adduct of the cobalt complex formed from ligand **6.32b** (Figure 6.13b). Over the years, a variety of other macrocyclic ligands whose coordination chemical and anion recognition properties were predicated on the same concept have been described by Drew and co-workers.[40]

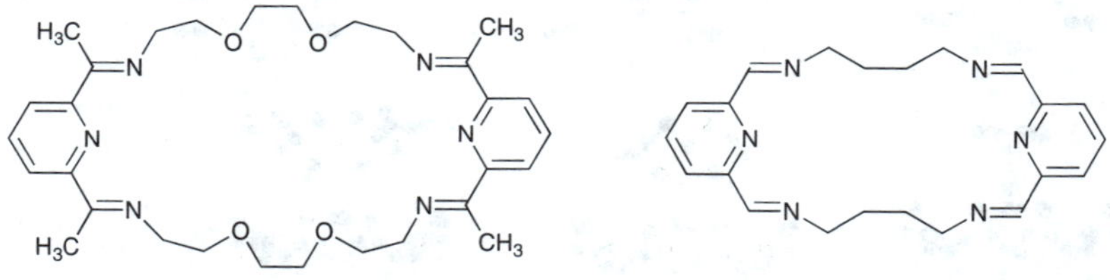

6.32a **6.32b**

In 1979 R. Weiss and co-workers[41] also reported the synthesis and binding studies of macrocycle **6.33a**. While this work was primarily directed towards the production of active-site mimics of copper containing proteins, it has historical implications in terms of anion-binding chemistry. Weiss showed that when a solution of sodium azide in water was added to one of **6.33a** and $Cu(NO_3)_2$ in methanol, dark-green crystals of

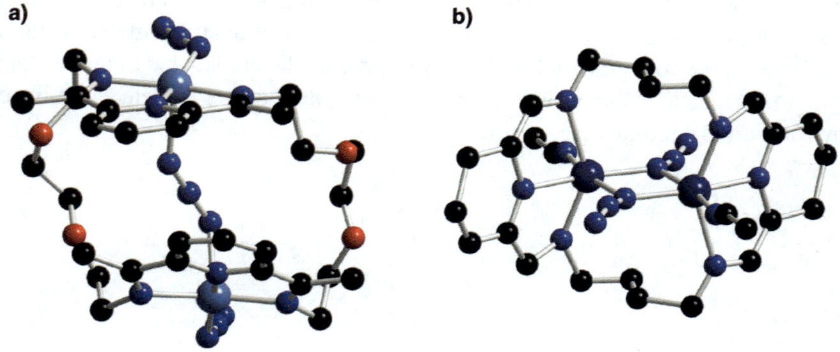

Figure 6.13 *Single-crystal X-ray structures of (a) the mixed copper azide complex of* **6.32a** *and (b) the mixed cobalt azide – acetonitrile complex of* **6.32b**

the bis-copper tetrakis-azide cascade complex were obtained. X-ray analysis of the crystals revealed that two of the azide anions are bridged between the copper ions, while the other two azides are bound, one to each copper, in a monodentate fashion (Figure 6.14a). In contrast to what is observed in this latter structure, Lehn *et al.*[42] found that macrocycles **6.33b** and **6.33c** bind copper azide with two bridging azide anions end-on, *i.e.*, coordinated to the two coppers through the same atoms as seen in the structure of the cobalt-derived cascade complex **6.32b** discussed above. Crystal structures of the copper azide cascade complexes of **6.33b** and **6.33c** are shown in Figure 6.14b and c.

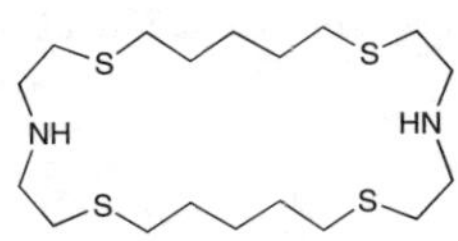

6.33a

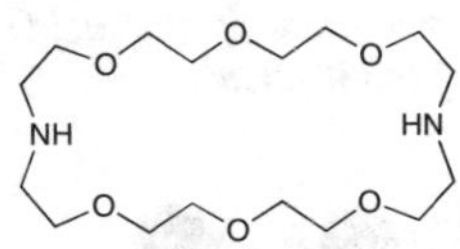

6.33b

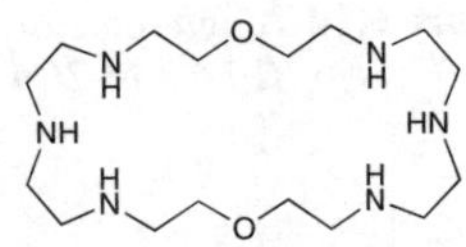

6.33c

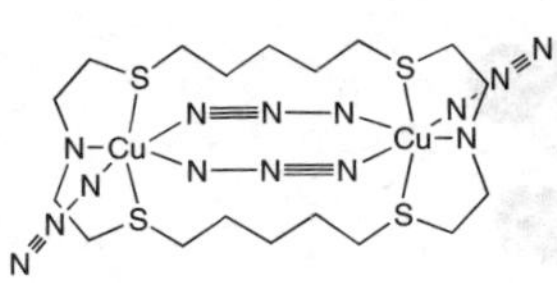

6.33a-($Cu_2(N_3)_4$)

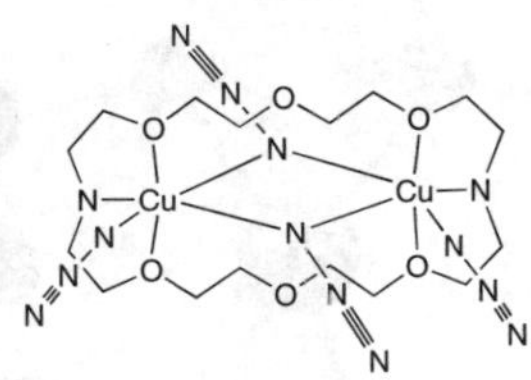

6.33b-($Cu_2(N_3)_4$)

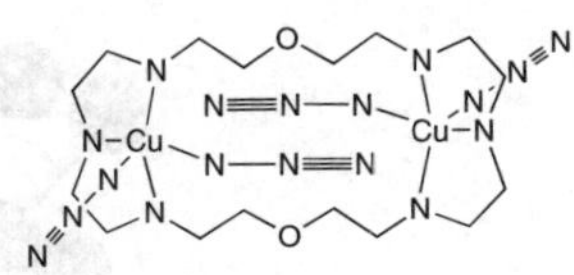

6.33c-($Cu_2(N_3)_4$)

Another interesting cascade complex from Lehn's group was published in 1985. In this complex, derived from macrocycle **6.34**, three copper ions serve to define a binding site for two hydroxide anions (Figure 6.15).[43] Later, Martell and co-workers[44] described similar macrocyclic ligands and their corresponding complexation properties.

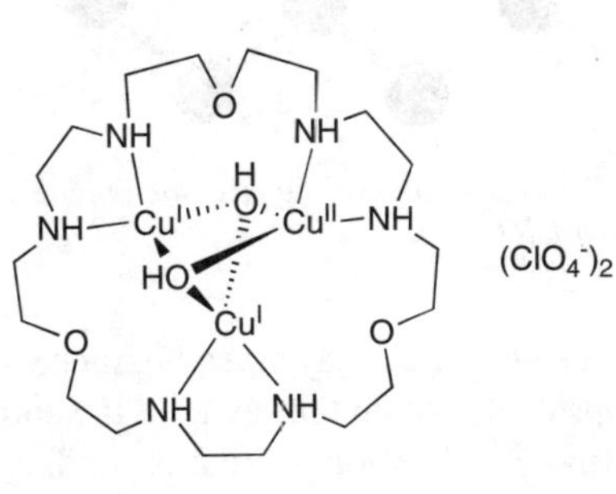

6.34-($Cu_3(OH)_2$)(ClO_4)$_2$

In 1995, Fabbrizzi and co-workers[45] reported the anion-binding properties of the dicopper(II) complex of the bistren cage **6.35a** in aqueous solution buffered at pH 8 (0.1 M CF_3SO_3H/morpholine buffer). According to the results of spectrophotometric titrations, the dimetallic cryptate [Cu_2**6.35a**]$^{4+}$ provided a binding site that was well suited for the subsequent binding of the linear anions N_3^- and NCO^- (log K_a = 4.78 and 4.60,

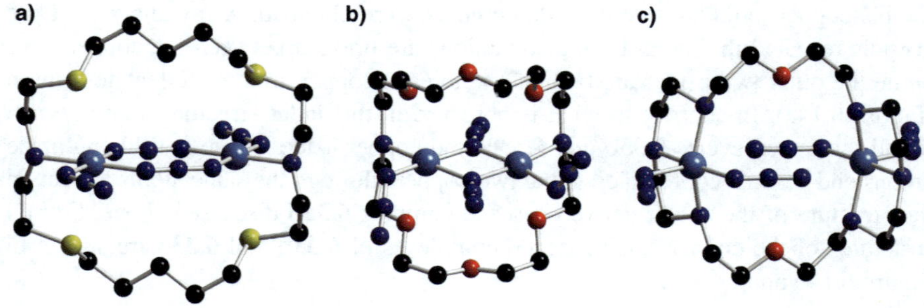

Figure 6.14 *Single-crystal X-ray structures of the azide complexes of (a)* ***6.33a***-$2Cu^{2+}$ *(b)* ***6.33b***-$2Cu^{2+}$ *and (c)* ***6.33c***-$2Cu^{2+}$

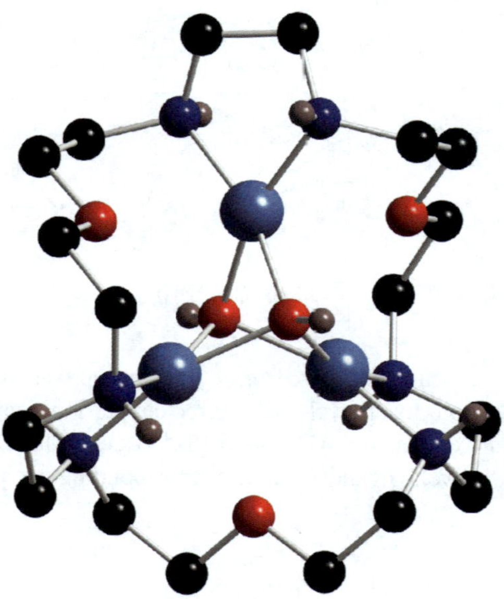

Figure 6.15 *Single-crystal X-ray structure of the cascade complex formed from* ***6.34****, namely* ***6.34***-$(Cu_3(OH)_2)(ClO_4)_2$

respectively). A year later, a solid-state X-ray crystallographic analysis revealed that the azide anion is indeed encapsulated within the cavity of the corresponding bis-nickel(II) crypate **6.35a** (Figure 6.16a).[46,47] The log K_a, corresponding to azide (added as the sodium salt) binding to this bis-Ni complex, was determined to be 3.1 at pH 9.6 (0.1 M $HClO_4$/*N*-methyl piperazine buffer) via spectrophotometric titrations.[46]

In subsequent work, the crystal structure of a bis-copper(II) complex of the furan-containing cryptand **6.35b** was obtained (Figure 6.16b).[48] The crystal structure reveals a bromide anion bound between the two copper ions. Association constants were determined by UV/Vis titrations carried out in buffered (pH 5.2) aqueous solution. On this basis, the bis-copper complex was found to be selective for chloride anion binding this particular anionic substrate (added as the sodium salt) with a log K_a = 3.98.

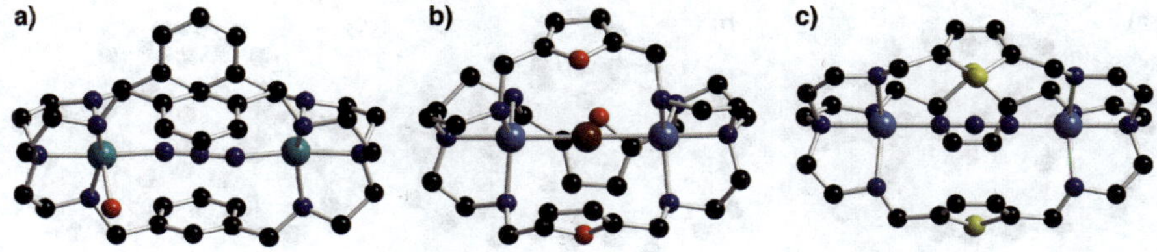

Figure 6.16 *Single-crystal X-ray structures of various cascade complexes: (a) [**6.35a**-Ni(II)$_2$N$_3$OH]$^{3+}$, (b) [**6.35b**-Cu(II)$_2$Br]$^{3+}$, and (c) [**6.35c**-Cu(II)$_2$N$_3$]$^{3+}$ (perchlorate counter anions are omitted for clarity)*

In the context of this general program, the cryptand **6.35c**, containing a thiophene spacer, was prepared and studied. The crystal structure, shown in Figure 6.16c, reveals the presence of an azide anion that bridges the two copper cations in a linear fashion in the solid state.[49] The anion-binding properties in solution were determined by means of spectrophotometric titrations carried out at pH 6.9. These latter revealed that the bis-copper complex shows a preference for linear anions (log K_a = 6.75, 4.79, and 2.72 for N_3^-, NCO^-, and NCS^-, respectively).

In a set of complementary studies, involving an analogous ligand set, the groups of Nelson and McKee[50] investigated the interaction of various metal complexes with several oxyanions, including carbonate for which several structures were obtained (*cf.* Figure 6.17) and the cyanide anion.

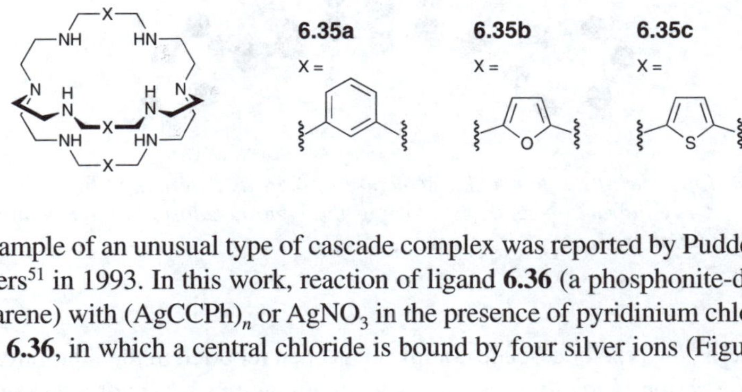

An example of an unusual type of cascade complex was reported by Puddephatt and co-workers[51] in 1993. In this work, reaction of ligand **6.36** (a phosphonite-derivatized resorcinarene) with $(AgCCPh)_n$ or $AgNO_3$ in the presence of pyridinium chloride gave complex **6.36**, in which a central chloride is bound by four silver ions (Figure 6.18).

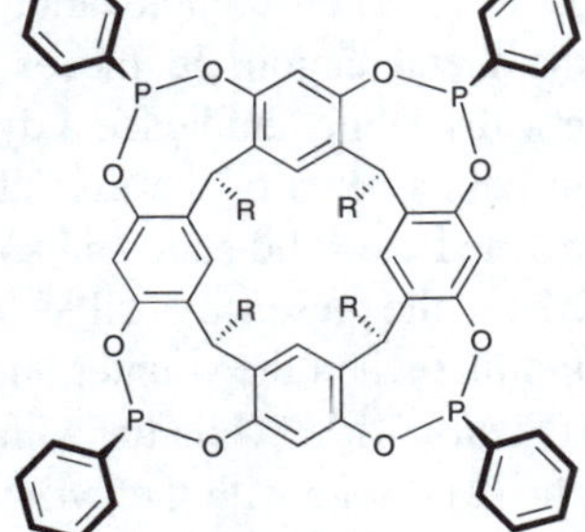

6.36

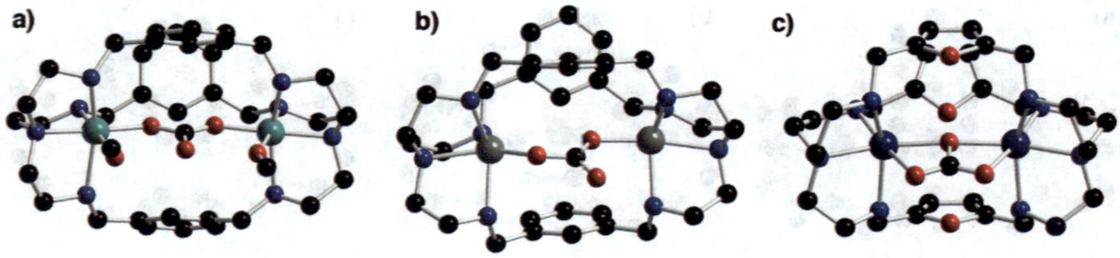

Figure 6.17 *Single-crystal X-ray structures of various cascade complexes: (a) [**6.35a**-$Ni_2CO_3(CH_3OH)_2$]$^{2+}$, (b) [**6.35a**-Zn_2CO_3]$^{2+}$, and (c) [**6.35b**-Co_2CO_3]$^{2+}$ (counter anions are omitted for clarity)*

Figure 6.18 *Top view of the molecular structure of **6.36** as determined from single-crystal X-ray diffraction analysis showing the central chloride anion bound between four silver ions*

Recently, a new type of cascade complex has been reported in the literature. In these systems, the cation binds to an organic ligand containing NH hydrogen bond donor groups that are, in turn, capable of anion complexation. Anion complexation only occurs in the presence of the metal cation. In the examples described here, the metal–ligand complex is kinetically labile and hence a dynamic equilibrium exists in solution between the different components of the assembly and the assembly itself. Steed and co-workers have reported the solid state and solution-binding properties of the pyridylureas **6.37a** and **6.37b** in the presence of silver(I) nitrate, sulfate, and triflate salts.[52] These researchers found that when the counter anion was nitrate, a 1:1 receptor:salt complex was formed (Figure 6.19), while the sulfate and triflate salts serve to stable extended structures in the solid state with the pyridyl urea ligands. On the basis of ^{1}H NMR spectroscopic studies, the authors concluded that the simple silver nitrate

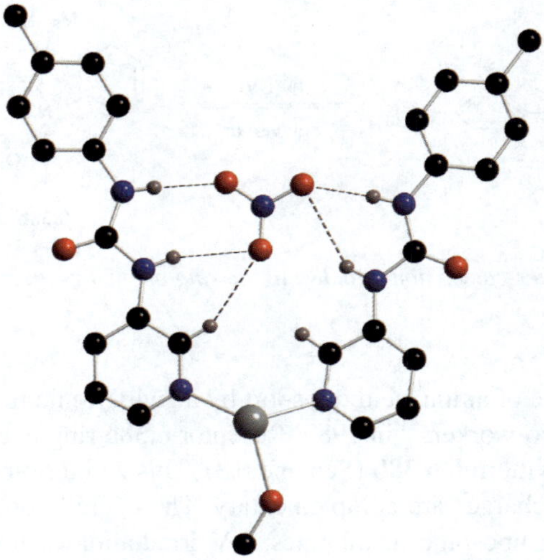

Figure 6.19 *Single-crystal X-ray diffraction structure of the discrete [Ag(**6.37a**)$_2$]NO$_3$·MeOH complex formed between **6.37a** and silver nitrate*

complex also forms in solution. These same researchers subsequently put forward other examples of this paradigm.[53]

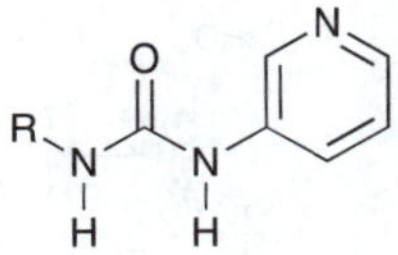

6.37a R = p-tolyl
6.37b R = n-octyl

6.4 Receptors for Zwitterions

Perhaps the most important zwitterions are the amino acids, which exist in predominantly zwitterionic forms under physiological conditions. Not surprisingly, therefore, considerable work has been directed towards the recognition of these species. Many of the receptors developed for this purpose are themselves zwitterionic with the resulting charge complementarity being used to effect binding of the guest. Alternatively, a number of amino acid receptors are positively charged; these systems rely for the most part on a neutral cation-binding group, such as crown ether, to bind the positively charged terminus of the guest. As most amino acids are chiral, enantioselective recognition has been a recurring theme in this area of supramolecular chemistry. Transport experiments, where a receptor is used to extract an amino acid or short peptide from an aqueous solution through an organic phase, to an aqueous receiving phase are often used to assess the complexation ability and selectivity of this type of receptor. Alternatively, liposomes or vesicles may be prepared and the release of an encapsulated species from within the vesicle monitored as a zwitterionic receptor is introduced into the membrane.

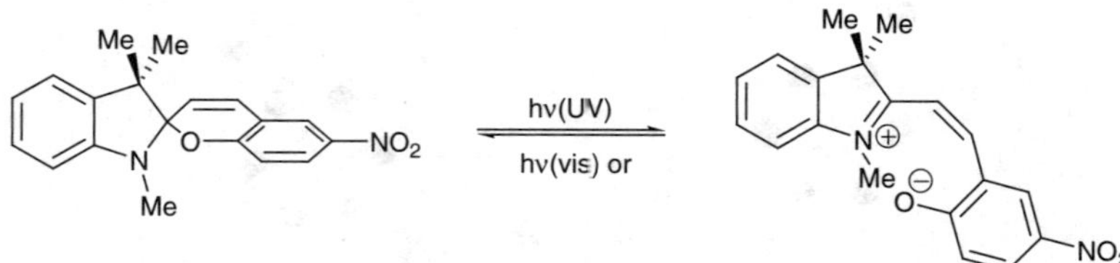

Scheme 6.3 *Inter-conversion of* ***6.38a*** *to the ring opened zwitterionic form* ***6.38b***

An early example of amino acid transport by a zwitterionic receptor was reported by Sunamoto and co-workers[54] in 1982. Receptor **6.38a** ring opens upon UV irradiation to give the zwitterion **6.38b** (Scheme 6.3). This zwitterion can then bind to an amino acid as the charges are complementary. The overall complex is neutral and hence is soluble in liposome membranes. UV irradiation of liposomes containing **6.38** and phenylalanine was then used to effect release of the amino acid to the external solution (receiving phase). This transport and release process did not occur in the absence of **6.38** or in the dark.

A charge complementarity strategy was also used by Rebek and co-workers[55] to bind and extract a variety of amino acids from water into chloroform solution using receptor **6.39**.

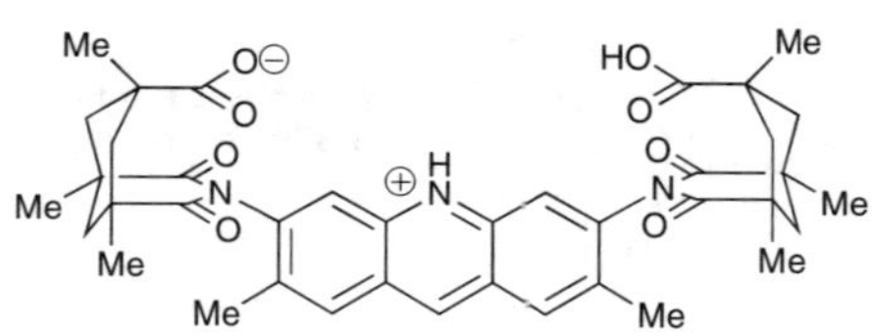

6.39

Quite early on, Tabushi and co-workers[56] employed the hydrophobic cavities of cyclodextrins to create zwitterionic receptors for amino acids. These researchers demonstrated that D-tryptophan was bound by a modified β-cyclodextrin **6.40** (shown schematically in Figure 6.20) with a stability constant of $54\pm8\ \mathrm{M}^{-1}$ at 25°C in aqueous solution. In the resulting complex, the indole ring is thought to be encapsulated within the cavity of the cyclodextrin, while the carboxylate and ammonium groups of this amino acid guest are bound by the protonated ammonium and carboxylate groups appended to the cyclodextrin, respectively.

Schmidtchen[57] has combined his tetrahedral quaternary ammonium cages that were discussed in Chapters 1 and 2 with a triaza-18-crown-6 to give the ditopic receptor **6.41**. This receptor binds zwitterionic γ-amino butyric acid (GABA) with the carboxy terminus of the guest held within the tetrahedral positively charged cage and the ammonium group bound by the triaza-18-crown-6 group.

Barboiu and co-workers[58] used an interesting approach to complex amino acids in the solid state. Specifically, these researchers prepared receptor **6.42** (Scheme 6.4),

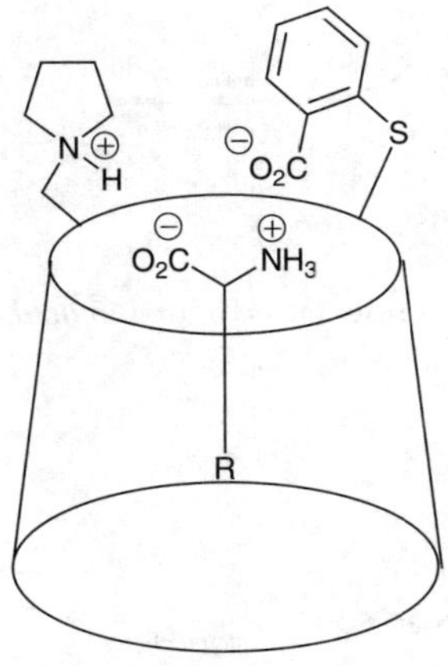

6.40

Figure 6.20 *Schematic representation showing the presumed binding interactions between* D*-trypophan and the cyclodextrin receptor* ***6.40***

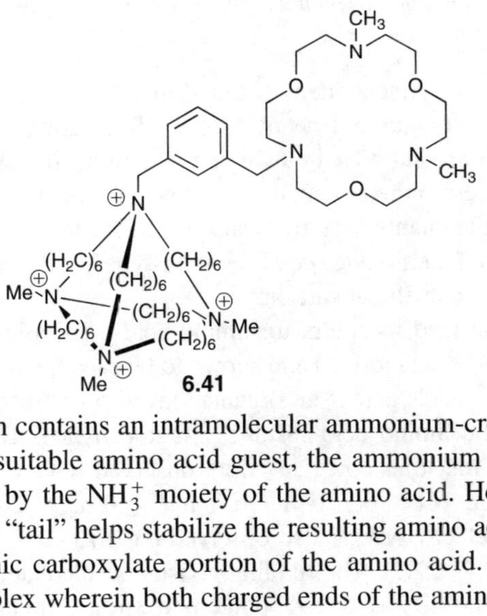

6.41

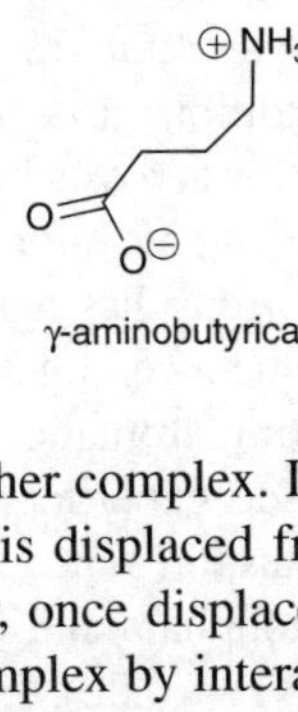

γ-aminobutyricacid(GABA)

which contains an intramolecular ammonium-crown ether complex. In the presence of a suitable amino acid guest, the ammonium group is displaced from the crown ether by the NH_3^+ moiety of the amino acid. However, once displaced this ammonium "tail" helps stabilize the resulting amino acid complex by interacting with the anionic carboxylate portion of the amino acid. The net result is a supramolecular complex wherein both charged ends of the amino acid are bound by the receptor, as illustrated in Scheme 6.4.

As we have seen in Chapter 2, bicyclic guanidinium groups can be used for enantioselective recognition of carboxylates. Aware of this, de Mendoza and co-workers[59] appended an anthracene and a mono-aza-18-crown-6 subunit to a chiral bicyclic guanidinium group to produce receptor **6.43**, the (*S,S*)-isomer of which is shown in Scheme 6.5. Receptor **6.43** was designed to bind L-tryptophan or L-phenylalanine via a three-point interaction, involving (1) π stacking with the aromatic amino acid side chain, (2) binding of the carboxylate group to the bicyclic guanidinium moiety, and (3) interactions between the ammonium group of the amino acid and the crown ether. This receptor was thus expected to be capable of effecting enantioselective amino

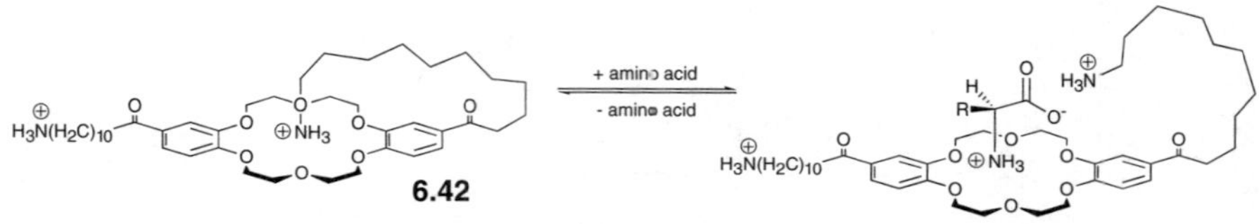

Scheme 6.4 *Schematic representation of amino acid binding by receptor* **6.42**

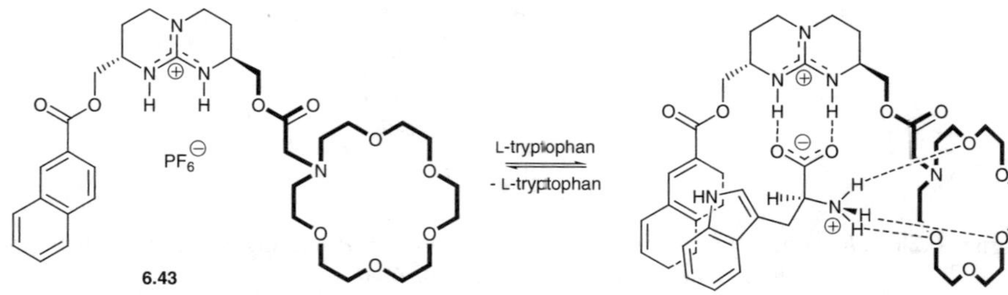

Scheme 6.5 *Schematic representation of* L-*tryptophan binding to receptor* **6.43**

acid recognition. In fact, extraction experiments demonstrated that this receptor can extract these two L-amino acids (tryptophan and phenylalanine) from aqueous into dichloromethane solution (removing around 40% in a single extraction). In contrast, only negligible amounts of the corresponding D-enantiomers were extracted under these same conditions. In addition, the enantioselective transport of zwitterionic aromatic amino acids across a bulk model membrane (*i.e.*, Pressman-type U-tube set-up) was achieved using receptor **6.43** as an artificial carrier.[60]

The Sessler group has also synthesized receptors for amino acids. Specifically, a sapphyrin–lasalocid conjugate **6.44** was synthesized and shown to be capable of selectively transporting aromatic α-amino acids across an organic "layer" in U-tube type model membrane experiments.[61] The amino acid residue was shown to influence strongly the transport rate, *i.e.*, different transport rates were observed for L-phenylalanine, L-tryptohphan, and L-tyrosine (k_t(=10^{-5} mol cm^{-2} h^{-1}): L-Phe = 20.0, L-Trp = 5.0, L-Tyr = 0.02). Enantioselectivity was also observed *e.g.*, (k_t(=10^{-5} mol cm^{-2} h^{-1}): L-Phe = 20.0 *vs.* D-Phe = 12.7). The binding constants calculated provided support for the conclusion that L-Phe was indeed bound more strongly thanL-Trp or L-Tyr. On the other hand, equilibrium analyses revealed that L-Phe and D-Phe were bound with roughly the same affinity. It was thus suggested that off rates play a critical role in modulating the transport dynamics and that carrier-based transport selectivity cannot necessarily be correlated with the corresponding binding constants.

Recently, a ferrocene-based receptor, **6.45**, containing built-in multipoint recognition sites was synthesized by Roy and co-workers. This system was found to effect the recognition of the zwitterionic forms of Gly and Gln at pH 8.0. Under these conditions, the ester unit in receptor **6.45** not only binds to the NH_3^+ end of the amino acid, the two hydroxyl groups also act as hydrogen bond donors, stabilizing interactions with the amino acid carboxylate moieties. Cyclic voltammetric (CV) titration experiments, carried out in a mixture of MeCN–H_2O (55:45, v/v), allowed the association constants for Gln and Gly to be determined; the resulting K_a values were found to be 1,540,000 and 890,000 M^{-1} for these two amino acids, respectively.[62]

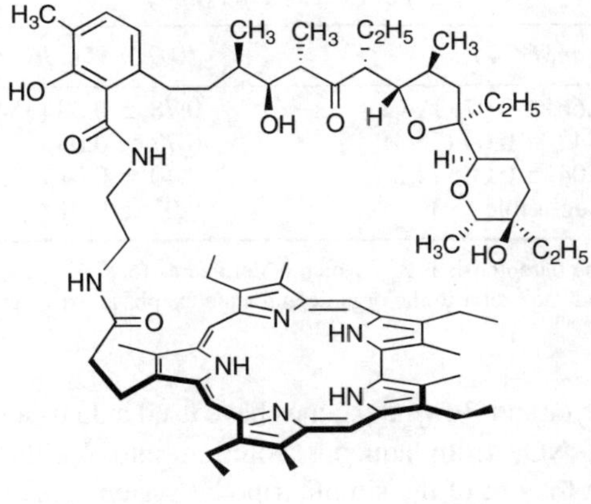

6.44

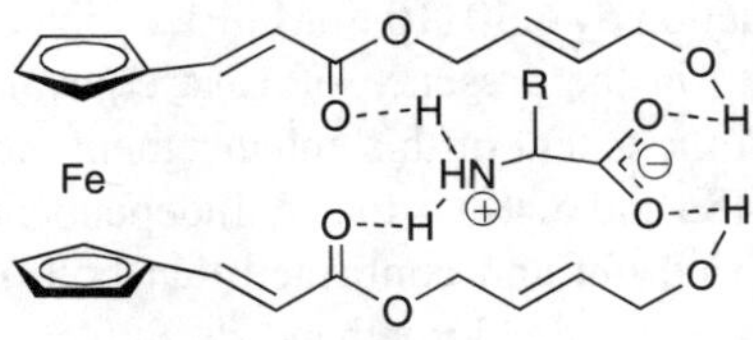

6.45-α-aminoacids

6.5 Dual-Host Extraction of Salts

The so-called dual-host approach to salt extraction is one of the more attractive means of effecting the ion-pair-based extraction and transport of inorganic salts. This approach relies on the judicious combination of individual anion and cation receptors that are presumably more accessible synthetically than more elaborate ditopic systems. In one important study, published in 1999, Hayashita and Teramae[63] demonstrated an efficient synergistic effect when the thiourea receptor (**6.46**) was used in conjunction with dicyclohexyl-18-crown-6 (**6.47**) to effect the ion-pair extraction of potassium salts. Typical results for potassium salt extraction from aqueous solution to nitrobenzene observed in the presence and absence of **6.46** are summarized in Table 6.1. As can be seen from inspection of this table, a greatly increased apparent extraction constant (K_{ex}), indicating an enhanced level of potassium extraction, was observed upon the addition of **6.46**.

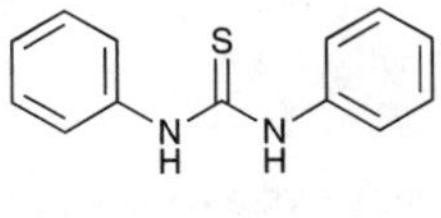

6.46

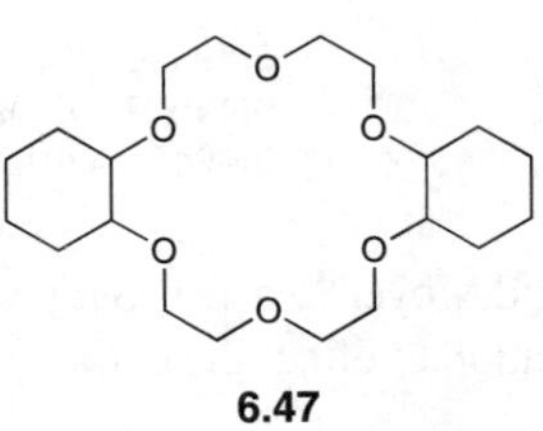

6.47

Table 6.1 *Ion-pair extraction of potassium salts (0.01 M) from an aqueous phase into nitrobenzene containing **6.47** (1.0 mM)*

	[K$^+$] back-extracted (mM)		
Anion species	*0 mM **6.46***	*10.0 mM **6.46***	*K_{ex} (**6.46**)/K_{ex}*
NO_3^-	0.66 ± 0.07 (194.1)	0.78 ± 0.08 (354.5)	1.8
Br^-	0.43 ± 0.04 (75.4)	0.72 ± 0.05	3.4
Cl^-	0.04 ± 0.006 (4.2)	0.40 ± 0.01 (66.7)	15.9
OAc^-	Negligible (−)	0.03 ± 0.01 (3.1)	–

Note: The value inside the parenthesis is K_{ex}, which is defined as [**6.47**·K^+·X^-]$_{org}$/[**6.47**]$_{org}$[K^+]$_{aq}$[X^-]$_{aq}$, where the terms "org" and "aq" refer to the organic and aqueous phases, respectively.

The groups of Moyer and Bowman-James have used a dual-host approach to effect the extraction of $CsNO_3$ from aqueous solution into 1,2-dichloroethane. These researchers relied on the use of the simple tripodal systems, **6.48** and **6.49**, as receptors for nitrate and tetrobenzo-24-crown-8 (**6.50**) as a complexant for Cs^+.[64] By monitoring the distribution of $^{137}Cs^+$, it was found that the efficiency of $CsNO_3$ (10 mM in aqueous solution) extraction by **6.50** (10 mM in 1,2-dichloroethane) was improved by a factor of 1.9 and 1.3 in the presence of 1 M equivalent of **6.48a** and **6.48b**, respectively. As expected, the extent of this enhancement increases as the concentration of anion receptors **6.48a** and **6.48b** is raised. Independent ^{1}H NMR spectroscopic titration studies were carried out and confirmed that both receptors are capable of forming a nitrate complex in 1,2-dichloroethane-d_4. In fact, the association constants for the binding of nitrate (tetrabutylammonium salt) by **6.48a** and **6.48b** were determined to be 250 and 23 M^{-1}, respectively. Further studies with more flexible tripodal systems **6.49a** and **6.49c** were performed using the same extraction method. In this case, the extraction of $CsNO_3$ (10 mM in aqueous solution) was found to be enhanced by a factor of 2.4, 1.7, and 4.4, respectively, when 50 mM solutions of **6.49a**, **6.49b**, and **6.49c** in 1,2-dichloroethane were used.

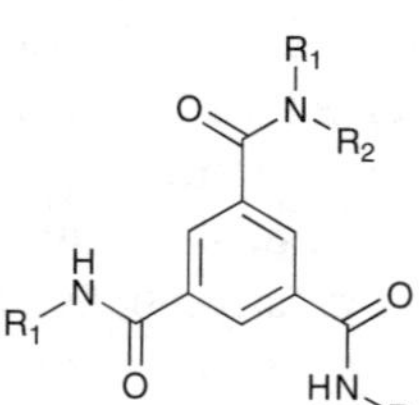

R$_1$ = 2,5-di(tBu)phenyl

6.48a R$_2$ = H
6.48b R$_2$ = CH$_3$

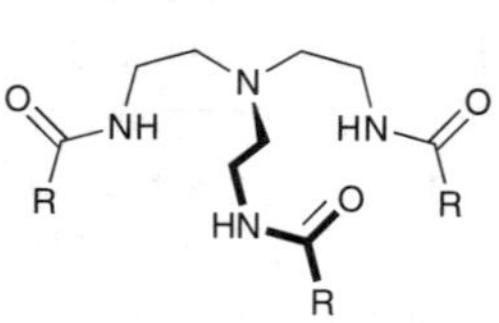

6.49a R = C_5H_{11}
6.49b R = C_7H_{15}
6.49c R = 4-tBu phenyl

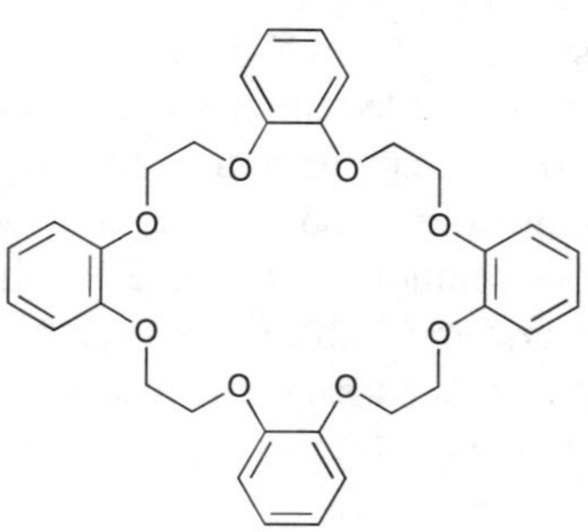

6.50

Kavallieratos and Moyer[65] have also reported the use of a calix-crown – disulfoamide dual-host combination to effect the extraction of caesium salts into 1,2-dichloroethane.

As shown in Table 6.2, the receptors in question, namely **6.51–6.53**, all displayed a propensity to bind more charge-dense anions, such as chloride and acetate, over less charge-dense anions (*e.g.*, nitrate and perchlorate). Strong anion dependent extraction results were thus expected. The actual extraction studies, carried out by monitoring the partition of a radioactive Cs^+ tracer between an aqueous phase consisting of 0.10 M NaX (not extractable) and 5×10^6 M^{-1} CsX (X = Cl, Br, I, OAc, NO_3, and ClO_4) and an organic phase containing 0.01 M of **6.54** and 0.035 M of **6.51–6.53**, served to confirm this expectation. The disulfonamide anion receptor **6.51** was found to be particularly effective, acting to synergize strongly the extraction of CsCl and CsOAc when used in conjunction with cation receptor **6.54** (*cf.* Table 6.3). Similar dual-host systems have also been described by Parisi[66] and Reinhoudt.[23,67]

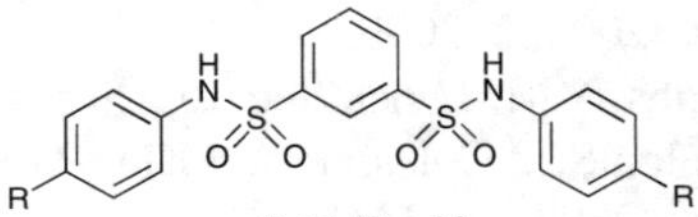

6.51 R = H
6.52 R = *t*-Bu

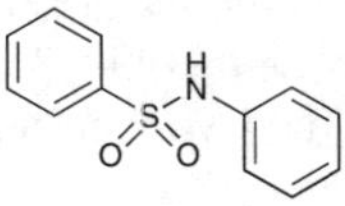

6.53

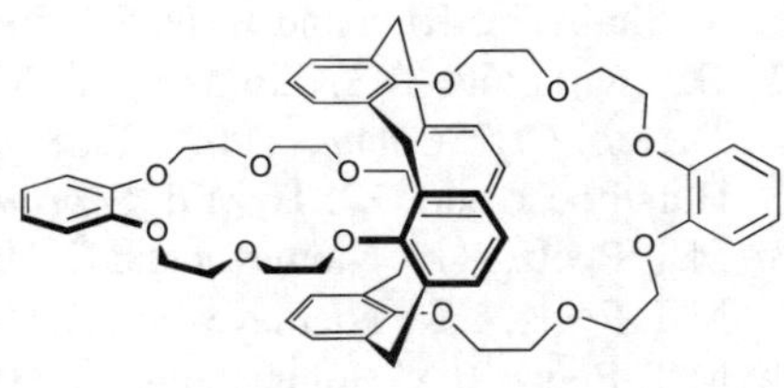

6.54

Table 6.2 *Association constant (K_a, M^{-1}) for the formation of 1:1 and 1:2 complexes **6.51**, **6.52**, and **6.53** with anions (tetrabutylammonium salts) in 1,2-dichloroethane-d_4, as determined from ^1H NMR spectroscopic titrations*

	6.51		*6.52*		*6.53*
	K_{a1}	K_{a2}	K_{a1}	K_{a2}	K_{a1}
OAc^-	19,500	73	13,500	70	750
Cl^-	50,000	5.8	32,500	3.6	410
Br^-	10,600	1.7	8900	1.2	150
NO_3^-	4300	2	1800	1.2	55
I^-	1400	–	690	–	19
ClO_4^-	81	–	48	–	<3

Table 6.3 *Caesium distribution ratios for calixarene **6.54** and for calixarene **6.54** plus sulfonamides **6.51–6.53***

	6.54	*6.54+6.51*	*6.54+6.52*	*6.54+6.53*	*6.54/(6.54+6.51)*
OAc^-	0.471	291	62.3	1.45	618
Cl^-	0.357	42.2	11.8	0.88	118
Br^-	2.64	125	48.5	6.32	47.3
NO_3^-	8.44	180	73.0	17.0	21.3
I^-	77.6	644	284	113	8.3
ClO_4^-	850	1770	1070	1020	2.1

6.6 Summary Remarks

In this chapter, an effort has been made to detail a number of approaches that have proved effective for the binding of zwitterionic substrates or anion–cation salt systems within the confines of a single receptor. Although impressive results are seen for a number of the systems, especially for certain, appropriately selected guests, it is nonetheless clear that there is a tremendous amount that is left to be done. Thus, this is viewed as an area of anion recognition that is likely to see exceptional attention and growth in the years to come.

References

1. R. Shukla, T. Kida and B. D. Smith, *Org. Lett.*, 2000, **2**, 3099.
2. G.J. Kirkovits, R.S. Zimmerman, M.T. Huggins, V.M. Lynch and J.L. Sessler, *Eur. J. Org. Chem.*, 2002, 3768; S. Camiolo, S.J. Coles, P.A. Gale, M.B. Hursthouse and G.J. Tizzard, *Supramol. Chem.*, 2003, **15**, 231.
3. M.T. Reetz, C.M. Niemeyer and K. Harms, *Angew. Chem. Int. Ed.*, 1991, **30**, 1472; M.T. Reetz, C.M. Niemeyer and K. Harms, *Angew. Chem. Int. Ed.*, 1991, **30**, 1474.
4. M.T. Reetz, B.M. Johnson and K. Harms, *Tetrahedron Lett.*, 1994, **35**, 2525.
5. S.S. Flack, J.-L. Chaumette, J.D. Kilburn, G.J. Langley and M. Webster, *J. Chem. Soc. Chem. Commun.*, 1993, 399.
6. M.J. Deetz, M. Shang and B.D. Smith, *J. Am. Chem. Soc.*, 2000, **122**, 6201.
7. J.M. Mahoney, A.M. Beatty and B.D. Smith, *J. Am. Chem. Soc.*, 2001, **123**, 5847; J.M. Mahoney, J.P. Davis, A.M. Beatty and B.D. Smith, *J. Org. Chem.*, 2003, **68**, 9819.
8. J.M. Mahoney, K.A. Stucker, H. Jiang, I. Carmichael, N.R. Brinkmann, A.M. Beatty, B.C. Noll and B.D. Smith, *J. Am. Chem. Soc.*, 2005, **127**, 2922.
9. J.M. Mahoney, G. Nawaratna, A.M. Beatty, P.J. Duggan and B.D. Smith, *Inorg. Chem.*, 2004, **43**, 5902; J.M. Mahoney, A.M. Beatty and B.D. Smith, *Inorg. Chem.*, 2004, **43**, 7617.
10. J.M. Mahoney, R.A. Marshall, A.M. Beatty, B.D. Smith., S. Camiolo and P.A. Gale, *J. Supramol. Chem.*, 2001, **1**, 289.
11. C. Suksai, P. Leeladee, D. Jainuknan, T. Tuntulani, N. Muangsin, O. Chailapakul, P. Kongsaeree and C. Pakavatchai, *Tetrahedron Lett.*, 2005, **46**, 2765.
12. P.D. Beer, P.K. Hopkins and J.D. McKinney, *Chem. Commun.*, 1999, 1253.
13. K.-S. Jeong, K.H. Shin and S.-H. Kim, *Chem. Lett.*, 2002, **31**, 1166; J. Morey, M. Orell, M.À. Barceló, P.M. Deyà, A. Costa and P. Ballester, *Tetrahedron Lett.*, 2004, **45**, 1261.
14. S. Nishizawa, K. Shigemori and N. Teramae, *Chem. Lett.*, 1999, **28**, 1185; K. Shigemori, S. Nishizawa, T. Yokobori, T. Shioya and N. Teramae, *New J. Chem.*, 2002, **26**, 1102.
15. M. Barboiu, G. Vaughan and A. van der Lee, *Org. Lett.*, 2003, **5**, 3073.
16. H. Miyaji, S.R. Collinson, I. Prokes and J.H.R. Tucker, *Chem. Commun.*, 2003, 64.
17. A.P. de Silva, G.D. McClean and S. Pagliari, *Chem. Commun.*, 2003, 2010; P. Gunning, A.C. Benniston and R.D. Peacock, *Chem. Commun.*, 2004, 2226;

S.J.M. Koskela, T.M. Fyles and T.D. James, *Chem. Commun.*, 2005, 945; A.C. Benniston, P. Gunning and R.D. Peacock, *J. Org. Chem.*, 2005, **70**, 115.

18. T. Tozawa, Y. Misawa, S. Tokita and Y. Kubo, *Tetrahedron Lett.*, 2000, **41**, 5219.
19. J.B. Love, J.M. Vere, M.W. Glenny, A.J. Blake and M. Schröder, *Chem. Commun.*, 2001, 2678; A. Channa and J.W. Steed, *J. Chem. Soc. Dalton Trans.*, 2005, 2455.
20. J.L. Sessler and E.A. Brucker, *Tetrahedron Lett.*, 1995, **36**, 1175.
21. D.M. Rudkevich, Z. Brzozka, M. Palys, H.C. Visser, W. Verboom and D.N. Reinhoudt, *Angew. Chem. Int. Ed.*, 1994, **33**, 467.
22. D.M. Rudkevich, J.D. Mercer-Chalmers, W. Verboom, R. Ungaro, F. de Jong and D.N. Reinhoudt, *J. Am. Chem. Soc.*, 1995, **117**, 6124.
23. L.A.J. Christoffels, F. de Jong, D.N. Reinhoudt, S. Sivelli, L. Gazzola, A. Casnati and R. Ungaro, *J. Am. Chem. Soc.*, 1999, **121**, 10142.
24. P. Tongraung, N. Chantarasiri and T. Tuntulani, *Tetrahedron Lett.*, 2003, **44**, 29; B. Tomapatanaget, T. Tuntulani and O. Chailapakul, *Org. Lett.*, 2003, **5**, 1539; D. Garozzo, G. Gattuso, A. Notti, A. Pappalardo, S. Pappalardo, M.F. Parisi, M. Perez and I. Pisagatti, *Angew. Chem. Int. Ed.*, 2005, **44**, 4892.
25. N. Pelizzi, A. Casnati, A. Friggeri and R. Ungaro, *J. Chem. Soc. Perkin Trans. 2*, 1998, 1307.
26. J. Scheerder, J.P.M. van Duynhoven, J.F.J. Engbersen and D.N. Reinhoudt, *Angew. Chem. Int. Ed.*, 1996, **35**, 1090.
27. A. Arduini, G. Giorgi, A. Pochini, A. Secchi and F. Ugozzoli, *J. Org. Chem.*, 2001, **66**, 8302.
28. J.L. Atwood and A. Szumna, *J. Am. Chem. Soc.*, 2002, **124**, 10646.
29. J.L. Atwood and A. Szumna, *Chem. Commun.*, 2003, 940.
30. T. Tuntulani, S. Poompradub, P. Thavornyutikarn, N. Jaiboon, V. Ruangpornvisuti, N. Chaichit, Z. Asfari and J. Vicens, *Tetrahedron Lett.*, 2001, **42**, 5541; T. Nabeshima, T. Saiki, J. Iwabuchi and S. Akine, *J. Am. Chem. Soc.*, 2005, **127**, 5507.
31. F.W. Kotch, V. Sidorov, Y.-F. Lam, K.J. Kayser, H. Li, M.S. Kaucher and J.T. Davis, *J. Am. Chem. Soc.*, 2003, **125**, 15140.
32. J.B. Cooper, M.G.B. Drew and P.D. Beer, *J. Chem. Soc. Dalton Trans.*, 2000, 2721; J.B. Cooper, M.G.B. Drew and P.D. Beer, *J. Chem. Soc. Dalton Trans.*, 2001, 392.
33. P.D. Beer and S.W. Dent, *Chem. Commun.*, 1998, 825.
34. D.J. White, N. Laing, H. Miller, S. Parsons, S. Coles and P.A. Tasker, *Chem. Commun.*, 1999, 2077; H. Miller, N. Laing, S. Parsons, A. Parkin, P.A. Tasker and D.J. White, *J. Chem. Soc. Dalton Trans.*, 2000, 3773.
35. P.G. Plieger, P.A. Tasker and S.G. Galbraith, *J. Chem. Soc. Dalton Trans.*, 2004, 313.
36. R. Custelcean, L.H. Delmau, B.A. Moyer, J.L. Sessler, W.-S. Cho, D. Gross, G.W. Bates, S.J. Brooks, M.E. Light and P.A. Gale, *Angew. Chem. Int. Ed.*, 2005, **44**, 2537.
37. J.L. Sessler, F.P. Schmidtchen, P.A. Gale, M.E. Light, G.W. Bates and W.-S. Cho, unpublished results.
38. M.G.B. Drew, M. McCann and S.M. Nelson, *J. Chem. Soc. Chem. Commun.*, 1979, 481.
39. M.G.B. Drew, C.J. Harding and J. Nelson, *Inorg. Chim. Acta*, 1996, **246**, 73.

40. M.G.B. Drew, C. Cairns, A. Lavery and S.M. Nelson, *J. Chem. Soc. Chem. Commun.*, 1980, 1122; M.G.B. Drew, J. Nelson, F. Esho, V. McKee and S.M. Nelson, *J. Chem. Soc. Dalton Trans.*, 1982, 1837; B.P. Murphy, J. Nelson, S.M. Nelson, M.G.B. Drew and P.C. Yates, *J. Chem. Soc. Dalton Trans.*, 1987, 123; M.G.B. Drew, B.P. Murphy, J. Nelson and S.M. Nelson, *J. Chem. Soc. Dalton Trans.*, 1987, 873; A. Lavery, S.M. Nelson and M.G.B. Drew, *J. Chem. Soc. Dalton Trans.*, 1987, 2975; R. Menif, A.E. Martell, P.J. Squattrito and A. Clearfield, *Inorg. Chem.*, 1990, **29**, 4723; D.A. Nation, A.E. Martell, R.I. Carroll and A. Clearfield, *Inorg. Chem.*, 1996, **35**, 7246; Q. Lu, J.H. Reibenspies, R.I. Carroll, A.E. Martell and A. Clearfield, *Inorg. Chim. Acta*, 1998, **270**, 207; B. Kersting, *Angew. Chem. Int. Ed.*, 2001, **40**, 3988; B. Verdejo, J. Aguilar, A. Doménech, C. Miranda, P. Navarro, H.R. Jiménez, C. Soriano and E. García-España, *Chem. Commun.*, 2005, 3086.
41. Y. Agnus, R. Louis and R. Weiss, *J. Am. Chem. Soc.*, 1979, **101**, 3381.
42. J. Comarmond, P. Plumeré, J.-M. Lehn, Y. Agnus, R. Louis, R. Weiss, O. Kahn and I. Morgenstern-Badarau, *J. Am. Chem. Soc.*, 1982, **104**, 6330.
43. J. Comarmond, B. Dietrich, J.-M. Lehn and R. Louis, *J. Chem. Soc. Chem. Commun.*, 1985, 74.
44. R.J. Motekaitis and A.E. Martell, *Inorg. Chem.*, 1991, **30**, 694; N.D. Rosso, B. Szpoganicz, R.J. Motekaitis and A.E. Martell, *Inorg. Chim. Acta*, 1994, **227**, 49.
45. L. Fabbrizzi, P. Pallavicini, L. Parodi and A. Taglietti, *Inorg. Chim. Acta*, 1995, **238**, 5.
46. L. Fabbrizzi, P. Pallavicini, L. Parodi, A. Perotti, N. Sardone and A. Taglietti, *Inorg. Chim. Acta*, 1996, **244**, 7.
47. A. Escuer, C.J. Harding, Y. Dussart, J. Nelson, V. McKee and R. Vicente, *J. Chem. Soc. Dalton Trans.*, 1999, 223.
48. V. Amendola, E. Bastianello, L. Fabbrizzi, C. Mangano, P. Pallavicini, A. Perotti, A.M. Lanfredi and F. Ugozzoli, *Angew. Chem. Int. Ed.*, 2000, **39**, 2917.
49. V. Amendola, L. Fabbrizzi, C. Mangano, P. Pallavicini and M. Zema, *Inorg. Chim. Acta*, 2002, **337**, 70.
50. C. Harding, V. McKee, J. Nelson and Q. Lu, *J. Chem. Soc. Chem. Commun.*, 1993, 1768; Q. Lu, J.-M. Latour, C.J. Harding, N. Martin, D.J. Marrs, V. McKee and J. Nelson, *J. Chem. Soc. Dalton Trans.*, 1994, 1471; C.J. Harding, F.E. Mabbs, E.J.L. MacInnes, V. McKee and J. Nelson, *J. Chem. Soc. Dalton Trans.*, 1996, 3227; J. Nelson, G.G. Morgan and V. McKee, *Prog. Inorg. Chem.*, 1998, **47**, 167; Y. Dussart, C. Harding, P. Dalgaard, C. McKenzie, R. Kadirvelraj, V. McKee and J. Nelson, *J. Chem. Soc. Dalton Trans.*, 2002, 1704; S. Derossi, A.D. Bond, C.J. Mckenzie and J. Nelson, *Acta. Cryst.*, 2005, **E61**, m1379; A.D. Bond, S. Derossi, C.J. Harding, E.J.L. McInnes, V. McKee, C.J. Mckenzie, J. Nelson and J. Wolowska, *J. Chem. Soc. Dalton Trans.*, 2005, 2403; A. Escuer, V. McKee, J. Nelson, E. Ruiz, N. Sanz and R. Vicente, *Chem. Eur. J.*, 2005, **11**, 398.
51. W. Xu, J.J. Vittal and R.J. Puddephatt, *J. Am. Chem. Soc.*, 1993, **115**, 6456.
52. D.R. Turner, E.C. Spencer, J.A.K. Howard, D.A. Tocher and J.W. Steed, *Chem. Commun.*, 2004, 1352.
53. D.R. Turner, M.B. Hursthouse, M.E. Light and J.W. Steed, *Chem. Commun.*, 2004, 1354; L. Applegarth, A.E. Goeta and J.W. Steed, *Chem. Commun.*, 2005,

2405; D.R. Turner, B. Smith, E.C. Spencer, A.E. Goeta, I.R. Evans, D.A. Tocher, J.A.K. Howard and J.W. Steed, *New J. Chem.*, 2005, **29**, 90; N.L.S. Yue, D.J. Eisler, M.C. Jennings and R.J. Puddephatt, *Inorg. Chem.*, 2004, **43**, 7671.

54. J. Sunamoto, K. Iwamoto, Y. Morhi and T. Kominato, *J. Am. Chem. Soc.*, 1982, **104**, 5502.
55. J. Rebek Jr, B. Askew, D. Nemeth and K. Parris, *J. Am. Chem. Soc.*, 1987, **109**, 2432.
56. I. Tabushi, Y. Kuroda and T. Mizutani, *J. Am. Chem. Soc.*, 1986, **108**, 4514.
57. F.P. Schmidtchen, *J. Org. Chem.*, 1986, **51**, 5161.
58. M. Barboiu, C.T. Supuran, A. Scozzafava, F. Briganti, C. Luca, G. Popescu, L. Cat and N. Hovanian, *Liebigs Ann. Recueil*, 1997, 1853.
59. A. Galán, D. Andreu, A.M. Echavarren, P. Prados and J. de Mendoza, *J. Am. Chem. Soc.*, 1992, **114**, 1511.
60. P. Breccia, M. van Gool, R. Pérez-Fernández, S. Martín-Santamaría, F. Gago, P. Prados and J. de Mendoza, *J. Am. Chem. Soc.*, 2003, **125**, 8270.
61. J.L. Sessler and A. Andrievsky, *Chem. Commun.*, 1996, 1119.
62. P. Debroy, M. Banerjee, M. Prasad, S.P. Moulik and S. Roy, *Org. Lett.*, 2005, **7**, 403.
63. M.M. Murad, T. Hayashita, K. Shigemori, S. Nishizawa and N. Teramae, *Anal. Sci.*, 1999, **15**, 1185.
64. K. Kavallieratos, R.A. Sachleben, G.J. van Berkel and B.A. Moyer, *Chem. Commun.*, 2000, 187; K. Kavallieratos, A. Danby, G.J. van Berkel, M.A. Kelly, R.A. Sachleben, B.A. Moyer and K. Bowman-James, *Anal. Chem.*, 2000, **72**, 5258.
65. K. Kavallieratos and B.A. Moyer, *Chem. Commun.*, 2001, 1620.
66. G. Cafeo, C. Gargiulli, G. Gattuso, F.H. Kohnke, A. Notti, S. Occhipinti, S. Pappalardo and M.F. Parisi, *Tetrahdron Lett.*, 2002, **43**, 8103; G. Cafeo, G. Gattuso, F.H. Kohnke, A. Notti, S. Occhipinti, S. Pappalardo and M.F. Parisi, *Angew. Chem. Int. Ed.*, 2002, **41**, 2122.
67. H.C. Visser, D.M. Rudkevich, W. Verboom, F. de Jong and D.N. Reinhoudt, *J. Am. Chem. Soc.*, 1994, **116**, 11554; L.A.J. Chrisstoffels, F. de Jong and D.N. Reinhoudt, *Chem. Eur. J.*, 2000, **6**, 1376.

CHAPTER 7

Metal and Lewis Acid-Based Receptors

In addition to the cascade complexes discussed in the previous chapter, there is a wide range of anion-binding receptors that contain metal atoms. These include systems that contain Lewis acidic anion-binding sites as electron-withdrawing groups or as organizational elements in the structural framework. Metal centres have also been used extensively as redox-active groups or as luminescent reporter sites within complexes designed to sense the presence of an anion. These latter two cases will be discussed in Chapter 8, which will cover a variety of approaches to anion sensing. Here, we will look at receptors that employ metals as organizational elements, electron-withdrawing groups, and systems that use Lewis acidic metals (and non-metals) as anion coordination sites.

7.1 Lewis Acidic Receptors

As mentioned in the introduction, Shriver and Biallas (1967)[1] reported the anion complexation properties of simple boron chelates. Receptor **7.1** was found to have a higher affinity for methoxide anions than monodentate boron trifluoride. Piers and co-workers[2] have reported the synthesis of the highly electrophilic borane **7.2** that contains perfluorinated aryl groups. The distance between two boron atoms (*i.e.*, 3.138 Å) in this receptor makes it ideal as a chelate for small anions, such as F^- and OH^-. The complexation of these two anions was verified by the ^{11}B NMR spectra (signals at 14.0 and 2.8 ppm for **7.2**·F^- and **7.2**·OH^-, respectively).

Prior to Piers' report, Katz[3–5] reported a variety of chelating boron-based receptors, **7.3** and **7.4**, designed for anion complexation. For example, the anion-coordinating ability of receptor **7.4a** (with hydride, fluoride, and hydroxide) was compared with the monodentate analogue **7.3**. It was found that the "hydride sponge" ligand **7.4a** could extract hydride or fluoride that was bound to compound **7.3**.[4] The crystal structure of the hydride complex of **7.4a** revealed that the anion is bound between the boron atoms (Figure 7.1a). Receptor **7.4b** provided the first example of a chelated chloride by a di-boron receptor (Figure 7.1b).[3]

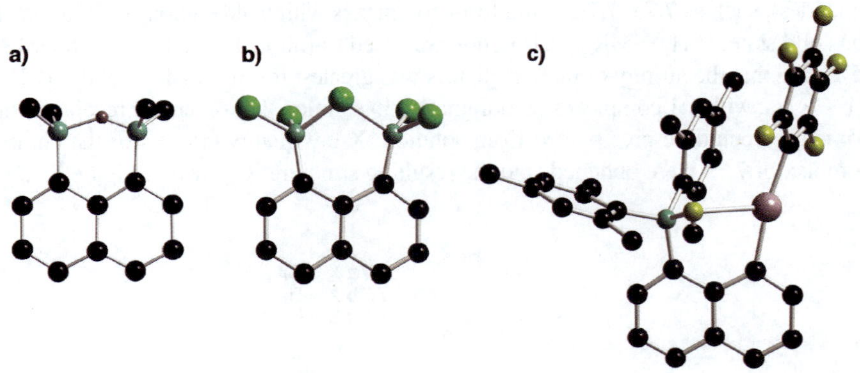

Figure 7.1 *Single crystal X-ray structures of (a) the hydride complex of receptor **7.4a**, (b) the chloride complex of receptor **7.4b**, and (c) the fluoride complex of receptor **7.6***

In order to obtain quantitative assessments of binding constants, the bright-yellow bidentate diborane **7.5** containing 9-thia-10-boraanthracene was prepared by Solé and Gabbaï.[6] The addition of fluoride anion led to loss of the yellow colour, while no colour change was observed in the presence of chloride, bromide, and iodide anions. As judged by UV–Vis titration studies, the strength of complexation with fluoride anion was found to be high ($K_a > 5 \times 10^9$ M^{-1} in THF).

Quite recently, a heteronuclear bidentate Lewis acid **7.6** was prepared as a phosphorescent fluoride sensor.[7] As deduced from an X-ray structural analysis, the fluoride anion is bound to both Lewis acid atoms (*i.e.*, B and Hg), as shown in Figure 7.1c. Receptor **7.6** displays high selectivity for fluoride ($K_a > 10^8$ M^{-1} in THF) over other halide, acetate, nitrate, sulfate, and phosphate anions in THF solution.

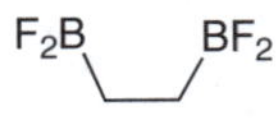

7.1

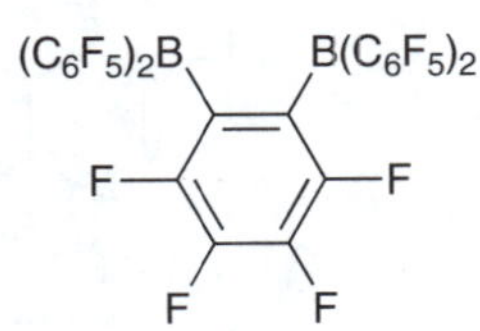

7.2

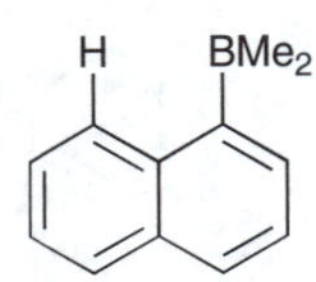

7.3

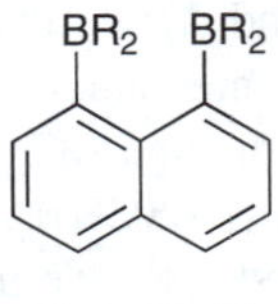

7.4a R = Me
7.4b R = Cl

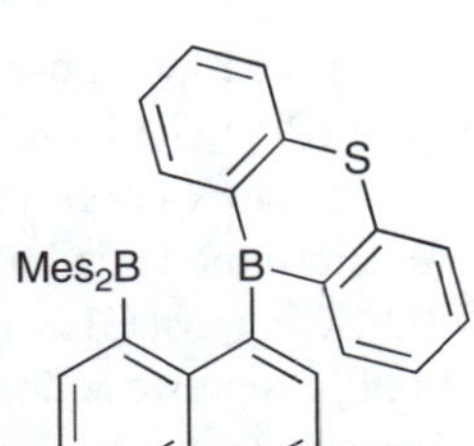

7.5

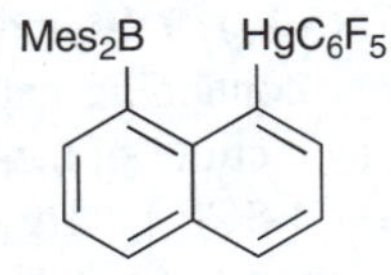

7.6

Mercury has been used as a Lewis acidic centre in a number of anion-receptor species. Wuest and Zacharie[8] demonstrated that simple mercury chelates (1,2-phenylene mercury dihalides), such as **7.7a–7.7c**, could form complexes with halide anions in solution and the solid state. ^{199}Hg NMR spectroscopy was used to follow halide binding in DMSO, revealing that the affinity of these receptors was greatest for the smaller halides ($Cl^- > Br^- > I^-$), with 1:1 complexes predominating in solution. However, 2:1 receptor/anion complexes could be precipitated from solution. X-ray quality crystals of the chloride complex of **7.7a** were obtained, and the resulting structure is shown in Figure 7.2.[8,9]

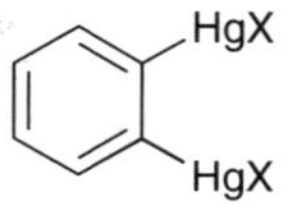

7.7a X = Cl
7.7b X = Br
7.7c X = I

Shur and co-workers[10] have studied the anion complexation properties of mercury containing macrocyclic species, such as **7.8** and **7.9**. Receptor **7.9** is notable in that it provided the first example of an anion sandwich complex. In particular, crystallization of the tetraphenylphosphonium bromide complex of **7.9** from dibromomethane allowed diffraction grade crystals to be obtained. The resulting structure, shown in Figure 7.3, revealed that each bromide anion is in contact with six mercury atoms of two neighbouring molecules of macrocycle **7.9**. The net result is an alternating stack of macrocycles and bromide anions, which provides a sandwich-like arrangement about the bromide centres. Shur and Tikhonova[11] have recently reviewed the application of macrocycles such as **7.9** in catalysis.

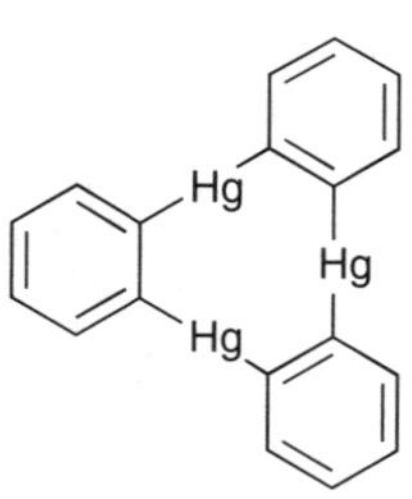

7.8

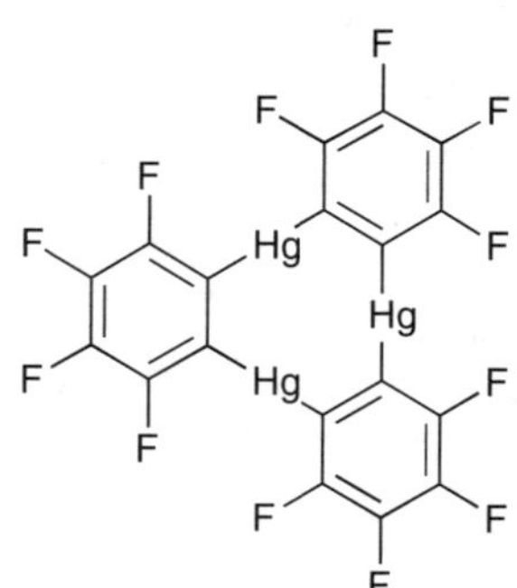

7.9

Peringer and co-workers[12] have prepared the sub-valent mercury clusters $Hg_3(\mu$-dppm$)_3(SiF_6)_2$ **7.10** and $Hg_3(\mu$-dppm$)_3(PF_6)_4$ **7.11** (Scheme 7.1) and shown that the mercury-containing cations in these clusters coordinate their counter anions. The crystal structures of these salts were determined in addition to the structures of $Hg_3(\mu$-dppm$)_3(O_3SCF_3)_4$, $Hg_3(\mu$-dppm$)_3(O_3SCF_3)_4{\cdot}MeOH$, $Hg_3(\mu$-dppm$)_3(O_3SCH_3)_4$, and $Hg_3(\mu$-dppm$)_3$ $(O_3SCH_3)_4{\cdot}4H_2O$. In all cases two anions were found to reside inside the cavity formed by the 12 phenyl groups (*e.g.*, see Figure 7.4). Solution phase NMR spectroscopic studies of the salts in dichloromethane-d_2 revealed a counter anion dependent shift in the ^{31}P resonance (Table 7.1).

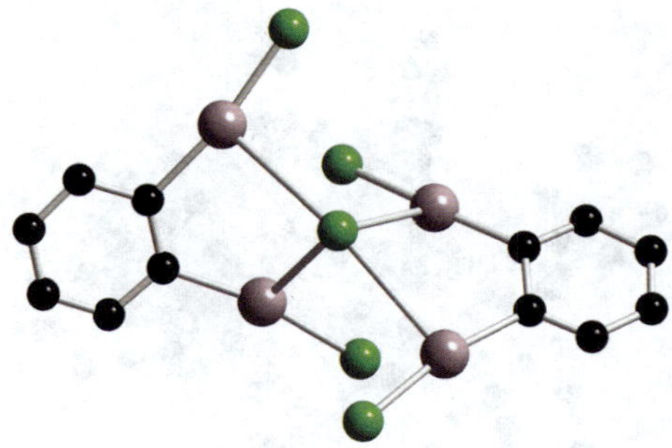

Figure 7.2 *Single crystal X-ray structure of the 2:1 complex formed from dichloro-1,2-phenylenedimercury (**7.7a**) and tetraphenylphosphonium chloride. The tetraphenylphosphonium cation and all hydrogen atoms have been omitted for clarit*

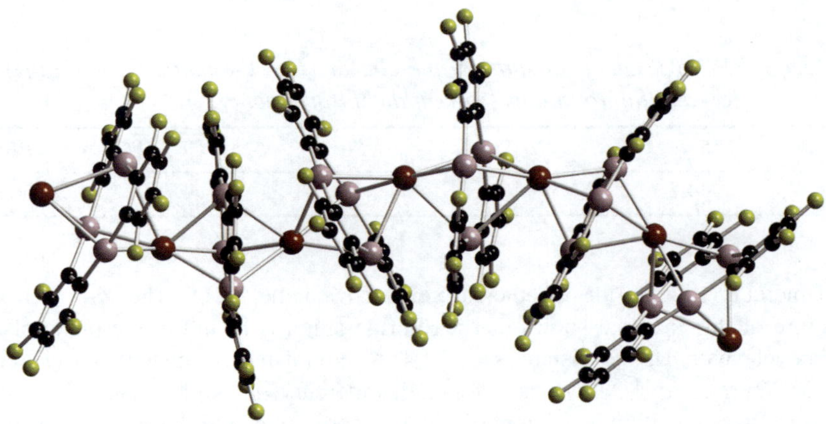

Figure 7.3 *Single crystal X-ray structure of the tetraphenylphosphonium bromide complex of receptor **7.9***

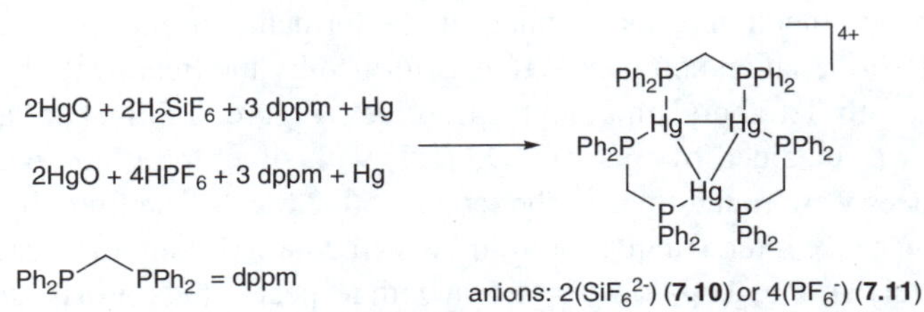

Scheme 7.1 *Synthesis of sub-valent mercury clusters, $Hg_3(\mu\text{-dppm})_3(SiF_6)_2$ (**7.10**), and $Hg_3(\mu\text{-dppm})_3(PF_6)_4$ (**7.11**)*

Some of the most elegant examples of mercury-based receptors are the mercuracarboranes produced by Wedge and Hawthorne.[13] Hawthorne's mercuracarborands consist of carborane cages linked via Lewis acidic mercury atoms. For example, macrocycle **7.15** was synthesized in 75% yield by reacting 1,2-dilithio-1,2-$C_2B_{10}H_{10}$

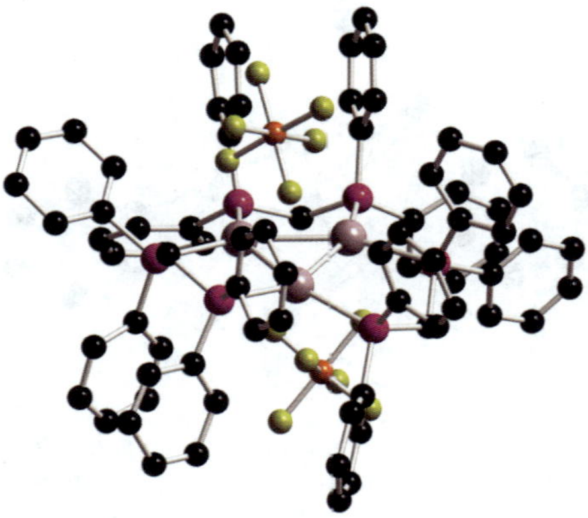

Figure 7.4 *Single crystal X-ray structure of the cluster* $Hg_3(\mu\text{-dppm})_3(SiF_6)_2$ ***(7.10)***

Table 7.1 *^{31}P NMR chemical shifts of the cluster $[Hg_3(\mu\text{-dppm})_3]^{4+}$ with different oxy- and fluoro-anions (solvent dichloromethane-d_2)*

Anion	SO_4^{2-}	$CH_3SO_3^-$	SiF_6^{2-}	$CF_3SO_3^-$	PF_6^-
$\delta\,^{31}P$	44.7	49.4	49.9	54.7	60.5

with mercury(II) chloride in ethereal solvent (Scheme 7.2).[14] The X-ray crystal structure of this material shows that a chloride anion is bound in the plane of the macrocycle with Hg–Cl distances of 2.94 Å, equidistant from the four mercury atoms (Figure 7.5a).[14,15] It was later found that mercury(II) iodide could also be used in this reaction resulting in a complex with two core-bound iodides perching on each face of the macrocycle (Figure 7.5b).[16] In the absence of halide anions (*e.g.*, if mercury(II) acetate is used as the source of mercury(II)), a trimeric macrocycle **7.13** forms in 65% yield,[17] with no formation of the tetramer. Thus, it is likely that the halide anions are functioning as templates in the formation of the tetrameric macrocycle. The fluoride-anion complex **7.16** was formed by the treatment of the macrocycle **7.15** with tetrabutylammonium fluoride (Figure 7.5c). The anion free macrocycle was obtained by treating **7.15**·$(LiI)_2$with silver tetrafluoroborate.[18] The Hg–F distances were found to be in the range 2.56–2.66 Å. The four Hg atoms and F are coplanar but, interestingly, the four cages adopt a 1,2-alternate arrangement (*i.e.*, alternately arranged up-down-up-down with respect to the four Hg atom plane).

In fact, Hawthorne and co-workers have recently shown that trimeric mercuracarborands can form sandwich complexes with anions. Specifically, a sandwich complex consisting of two trimeric hexamethyl-9-mercuracarborand-3 $[9,12\text{-}(CH_3)_2\text{-}C_2B_{10}H_8Hg]_3$ receptors (*i.e.*, **7.14**) and halide anions (*i.e.*, chloride, bromide, and iodide)[19] were formed by reaction of the receptor with bis(triphenyl phosphoranylidene) ammonium chloride (PPNCl) $MePPh_3Br$, or LiI. The crystal structure of the complex, shown in Figure 7.6, reveals that the iodide anion is bound to six mercury atoms with Hg···I distances ranging between *ca.* 3.25 and 3.27 Å.

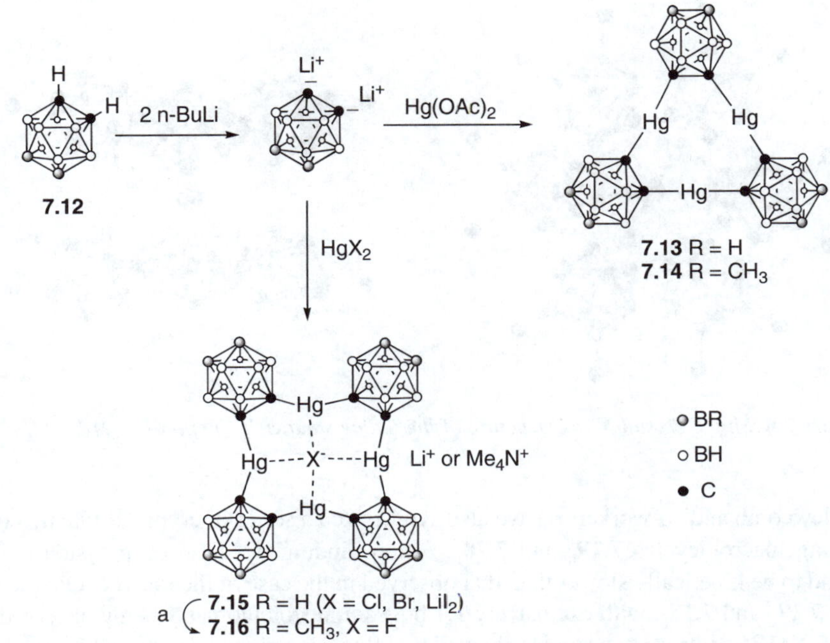

Scheme 7.2 *Synthesis of mercuracaborands (a: 1. $AgBF_4$, 2. Me_4NF)*

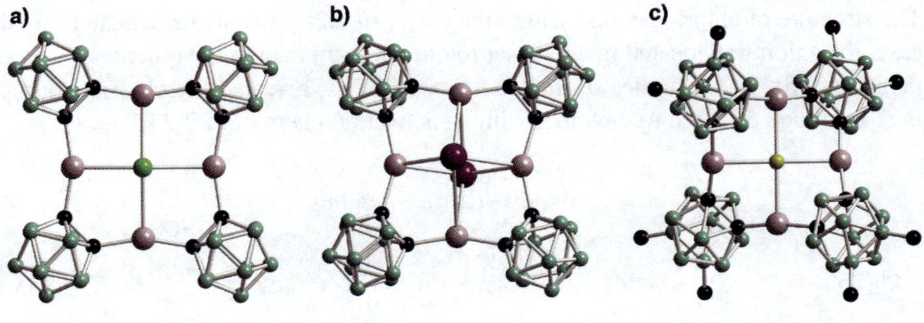

Figure 7.5 *Single crystal X-ray structures of (a) the chloride anion complex of* ***7.15****, (b) the iodide anion complex of* ***7.15****, and (c) the fluoride anion complex of* ***7.16***

A variety of Lewis acidic group 14 elements have also been used in macrocyclic and macrobicylic anion receptor systems. Newcomb and co-workers pioneered this area with an exploration of the synthetic anion receptor chemistry of tin containing macrocycles and macrobicycles in the 1980s. In early work, Azuma and Newcomb[20] reported the synthesis of a series of tin-based macrocycles **7.17_n** and **7.18_n**. Coordination studies were conducted on some of these systems (**7.17_8**, **7.17_{10}**, and **7.17_{12}**) using ^{119}Sn NMR; such analyses revealed that this family of macrocycles is capable of forming 1:1 and 1:2 host/anion complexes with chloride in acetonitrile solution.[21] For example, **7.17_8** was found to bind two equivalents of chloride with $K_{a1} = 814\ M^{-1}$ and $K_{a2} = 863\ M^{-1}$.

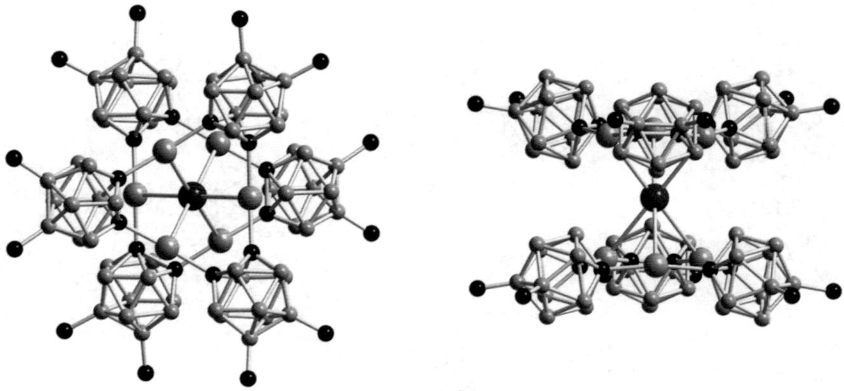

Figure 7.6 *Single crystal X-ray structure of the iodide sandwich complex $(7.14)_2I^-$*

Newcomb and co-workers[22] have also synthesized a series of cryptand-like tin containing macrobicycles **7.19$_n$** and **7.20$_n$**. Anion binding by these cage systems was found to be kinetically slower than that observed in the case of the macrocyclic receptors **7.17$_n$** and **7.18$_n$**, with exclusively 1:1 host/anion stoichiometry being observed.[23] ^{119}Sn NMR spectroscopic studies showed that fluoride anions were bound five orders of magnitude more strongly than chloride anions in chloroform solution.[24] Crystallography later revealed that the fluoride anion was bound within the cavity of **7.20$_6$** and coordinated to both the tin atoms (r_{Sn-F} = 2.12 and 2.28 Å) (Figure 7.7). The structure of chloride bound within the cavity of **7.20$_8$** was also elucidated. In this case, the anion was located much closer to one tin atom than to the other.[25] Low temperature ^{119}Sn NMR studies of this complex in $CFCl_3$ revealed the chloride "jumping" from one tin atom to the other with an activation energy of 22.2 kJ mol^{-1}.

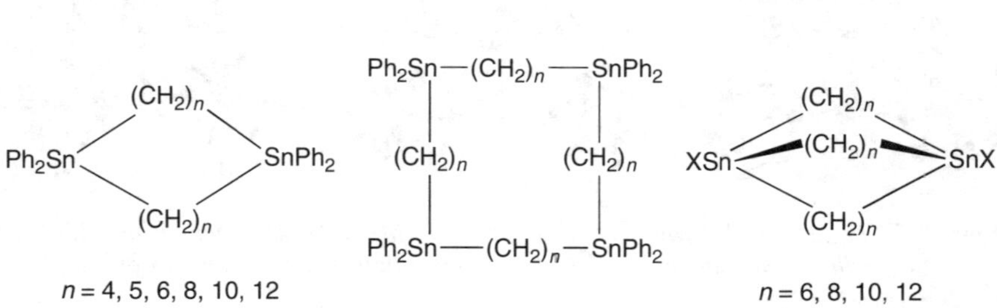

7.17$_n$ **7.18$_n$** **7.19$_n$** X = Ph **7.20$_n$** X = Cl

By hydrolysing $[CH_2[SnPh_2CH_2Si(O^iPr)Me_2]_2]$ under acid conditions and subsequently treating with mercuric chloride, Jurkschat and co-workers[26] succeeded in isolating a new bis-tin bis-silicon macrocycle **7.21** (Scheme 7.3). The crystal structure of this material is shown in Figure 7.8. The macrocycles form dimers in the solid state linked via a weak Cl–Sn interaction (3.684(1) Å). The chloride complexation properties of the receptor were followed by means of ^{119}Sn{^{1}H} NMR spectroscopic titration

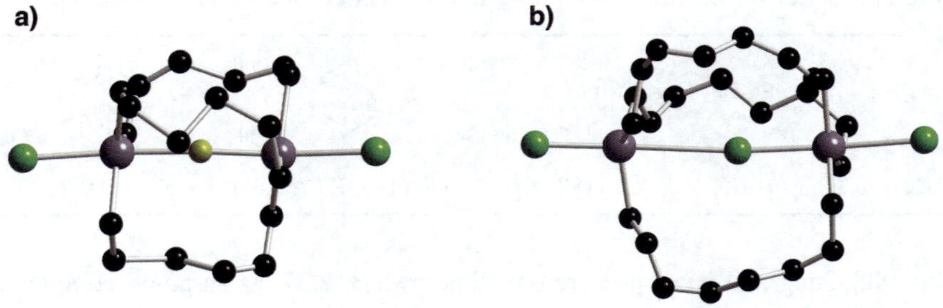

Figure 7.7 *Single crystal X-ray structures of (a) the fluoride complex of **7.19**$_6$ and (b) the chloride complex of **7.20**$_8$*

7.21

Scheme 7.3 *Synthesis of the eight-membered ring **7.21** (i) 0.5 M H_2SO_4, Et_2O (−2 i−PrOH); (ii) 4 $HgCl_2$, acetone (−4 PhHgCl)*

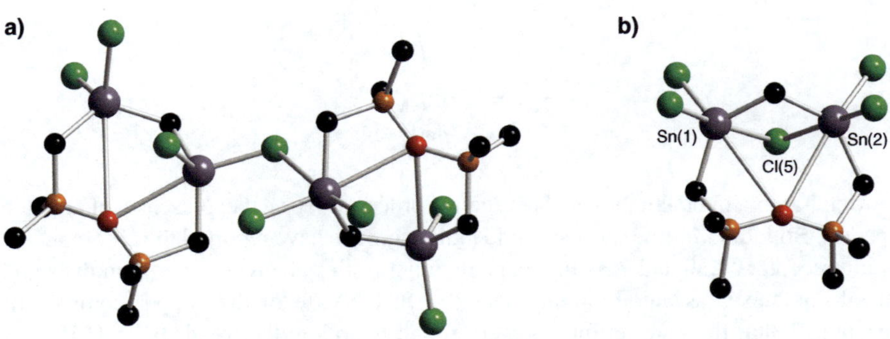

Figure 7.8 *Single crystal X-ray structures of (a) **7.21** showing dimerization via a Cl···Sn interaction and (b) the 1:1 chloride complex of **7.21***

in dichloromethane. On the basis of these studies, the formation of a 1:1 complex in solution was inferred. The crystal structure of the chloride complex, shown in Figure 7.8b, confirmed this stoichiometry in the solid state and provided the following bond distances: Sn(1)–Cl(5) 2.891(1) Å and Sn(2)–Cl(5) 2.781(1) Å.

A variety of silicon-containing chelates have been synthesized and their fluoride-binding abilities compared to those of the corresponding monodentate silanes. The results shown in Table 7.2 reveal that the most Lewis acidic chelating ligand, **7.26**, shows a remarkably high affinity for fluoride in acetone-d_6 (log $K_a > 9$).[27]

Table 7.2 *Order of fluoride ion binding affinity as determined in acetone-d_6*

$PhMeSiF_2$	<	Ph_2SiF_2	<	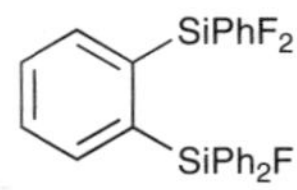	<	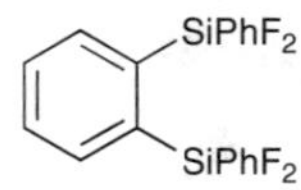	<	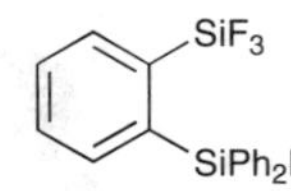
log K_a	3.3 (291 K)		4.2 (188 K)	5.8 (183 K)	>	7.8 (183 K)	>	9.0 (193 K)

In 1988, Jung and Xia[28] prepared 12-silacrown-3, **7.27**, as an anion complexing agent. In U-tube experiments using a saturated aqueous solution of the tetramethylammonium halide in one arm, pure water in the other, and a dichloromethane solution of the receptor as the central phase, this cyclic polysilane receptor (**7.27**) was seen to transport chloride and bromide efficiently. In contrast, no transport of fluoride and iodide anions was observed. These results support the postulate that chloride and bromide salts form strong complexes with **7.27**, as shown in Scheme 7.4, at least in dichloromethane solution.

Germanium-based macrocycles *e.g.*, **7.28** and **7.29** have also been prepared,[29] with receptor **7.28** having been demonstrated as transporting chloride anions more efficiently than bromide across organic phases.[30]

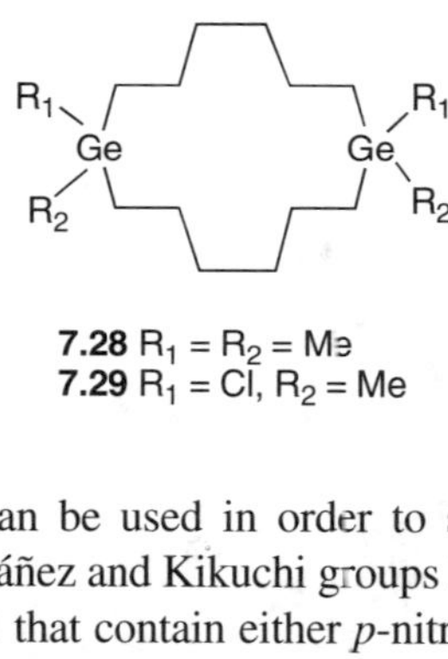

7.28 $R_1 = R_2 = Me$
7.29 $R_1 = Cl$, $R_2 = Me$

Metal-anion interactions can be used in order to sense the presence of anionic species. Both the Martínez-Máñez and Kikuchi groups have reported the chromogenic anion receptors **7.30** and **7.31** that contain either *p*-nitrophenyl or coumarin dyes and metal complexes as anion-binding sites.[31,32] In the case of the copper complex of receptor **7.30a**, the anion affinities were found to follow the trend: $F^- > H_2PO_4^- > NO_3^- > Cl^- = I^- > HSO_4^- = Br^-$ in acetonitrile solvent. By contrast, receptors **7.30b** and **7.30c** both show a more selective optically detectable response in the presence of nitrate, forming the nitrate complex presumed responsible for this signalling with association constants that are in the same order as those of the copper receptors **7.30a**. Specifically, for the binding of nitrate in acetonitrile, log K_a values of 6.2, 6.2, and 5.7 for **7.30a**, **7.30b**, and **7.30c**, respectively, were determined on the basis of UV–Vis spectroscopic titrations. Under these same experimental conditions, no appreciable response was observed when complexes **7.30b** or **7.30c** were exposed to other anions with the exception of **7.30c** + I^-, which produced a complex formed with a stability constant of log $K_a = 4.8$.[31]

The anion-binding properties of the macrocyclic polyamine-Cd^{2+} complex **7.31** were measured in aqueous solution containing 100 mM HEPES buffer. The interaction

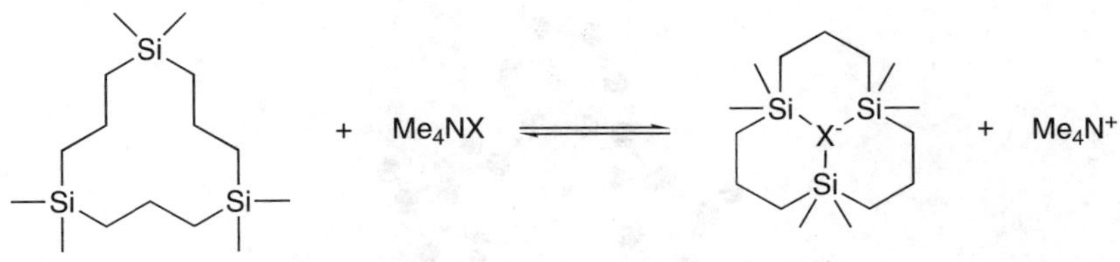

7.27 **7.27-X⁻**

Scheme 7.4 *Schematic representation of proposed anion-binding events*

between Cd^{2+} and anions was found to perturb the fluorescence of the metal complex. Quantitative association constants of complex **7.31** were measured using fluorescence spectroscopic titration methods; these analyses revealed that this receptor displays a preference towards citrate and phosphate anions ($K_a = 1.1 \times 10^4$, 1.3×10^4, 7.1×10^4, and 3.8×10^4 M^{-1} for citrate, pyrophosphate, ATP, and ADP, respectively, in the form of their potassium salts).[32]

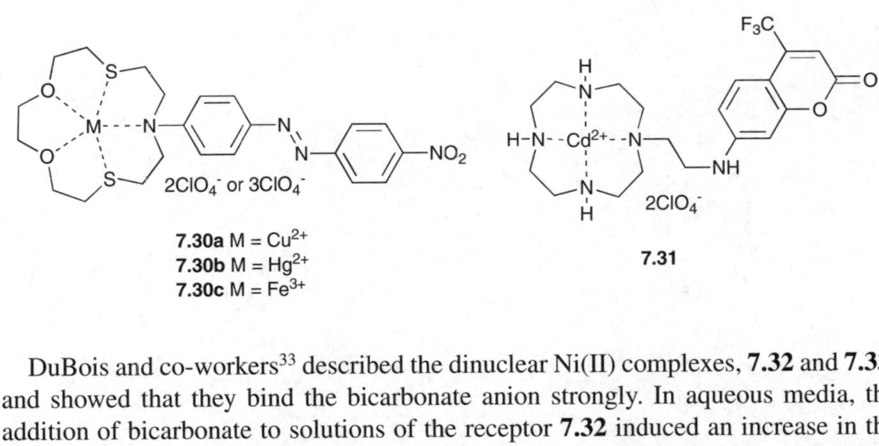

7.30a M = Cu^{2+}
7.30b M = Hg^{2+}
7.30c M = Fe^{3+}

7.31

DuBois and co-workers[33] described the dinuclear Ni(II) complexes, **7.32** and **7.33**, and showed that they bind the bicarbonate anion strongly. In aqueous media, the addition of bicarbonate to solutions of the receptor **7.32** induced an increase in the intensity of the absorption bands at 350 and 545 nm, a finding that was rationalized in terms of the formation of a $\mu-\eta^2,\eta^2$–carbonate complex. Quantitative analyses, using UV–Vis titrations, allowed the association constant for the complexation of **7.32** with bicarbonate to be obtained, yielding log $K_a = 4.39$ at pH 7.4. In contrast, no evidence of bicarbonate or carbonate complexation in aqueous solution was seen in the case of receptor **7.33**. However, the reaction of **7.33** with bicarbonate was found to lead rapidly to the $\mu-\eta^1,\eta^1$–carbonate complex, as evidenced by X-ray crystallographic analysis of the reaction product (Figure 7.9).

In the context of extending metal-containing receptors to the recognition of tricarboxylate anions, three copper-cycloamine subunits were attached synthetically to a 1,3,5-trimethylbenzene unit. The resulting sexicationic receptor, **7.34**, was found to form strong 1:1 complexes with tricarboxylate anions in an aqueous solution buffered to pH 7 (HEPES 0.01 M), including citrate (log $K_a = 5.59$), 1,3,5-benzene-tricarboxylate (log $K_a = 5.81$), and *cis,cis*-1,3,5-cyclohexanetricarboxylate

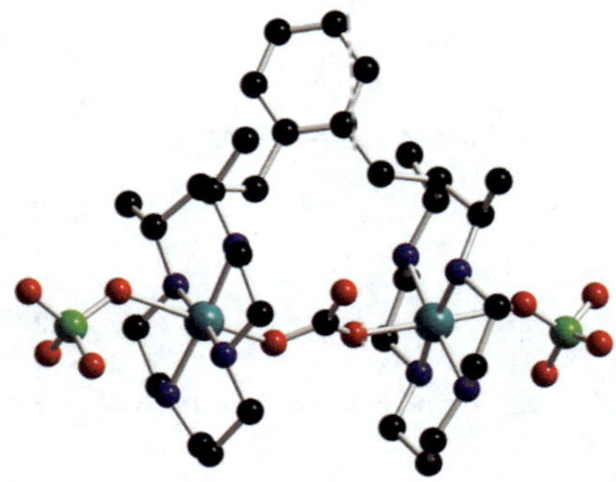

Figure 7.9 *Single crystal X-ray structure of receptor **7.33** complexed to a carbonate anion and two perchlorate anions*

(log K_a = 5.00), displaying a preference for these analytes over dicarboxylate anions (log K_a = 4.5, 4.1, and 3.8 for maleate, tartrate, and succinate, respectively).[34] Other examples of this type of receptor are listed in the bibliography.[35]

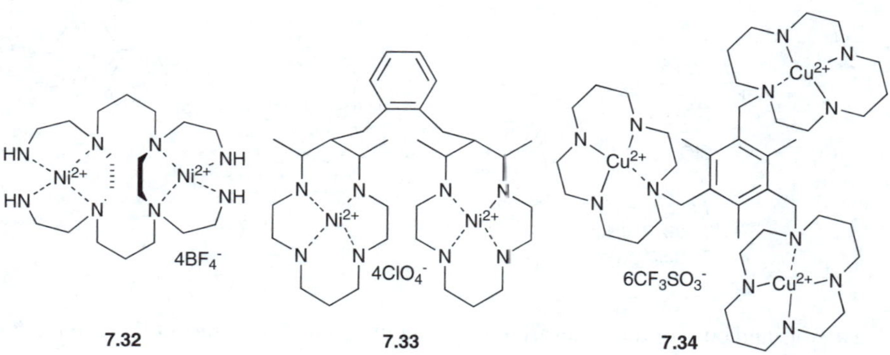

7.32 **7.33** **7.34**

Murase and co-workers[36] have reported the crystalline dinuclear complex of [**7.35**($(CH_3CO_2)_2$]·ClO_4 and [**7.35**(CH_3CO_2)]·$(ClO_4)_2$, while Chin and co-workers[37] have described phosphonate anion binding to the dinuclear complex **7.36**. In the solid state, the structure of [**7.36**($O_3P(Ph)(OH_2)(OH)$]·$(ClO_4)_2$, shown in Figure 7.10a, revealed the presence of two inequivalent cobalt units. At one of the cobalt centres, the tertiary alkylamine is coordinated *trans* to the coordinated phosphonate oxygen, while at the other cobalt centre the alkylamine is coordinated *trans* to the water or hydroxy group.

Inspired by Chin's work, M.S. Han and D.H. Kim prepared the dinuclear Zn(II) complex, **7.37**·$(ClO_4)_2$, as a phosphate-anion sensor for use in aqueous solution. Using the receptor in a displacement assay with pyrocatechol violet as a reporter dye, the addition of phosphate anions to an aqueous solution of the ensemble resulted in a colour change from blue to yellow. The thermodynamic parameters and quantitative association constant of **7.37**·$(ClO_4)_2$ and HPO_4^{2-} were determined in aqueous solution

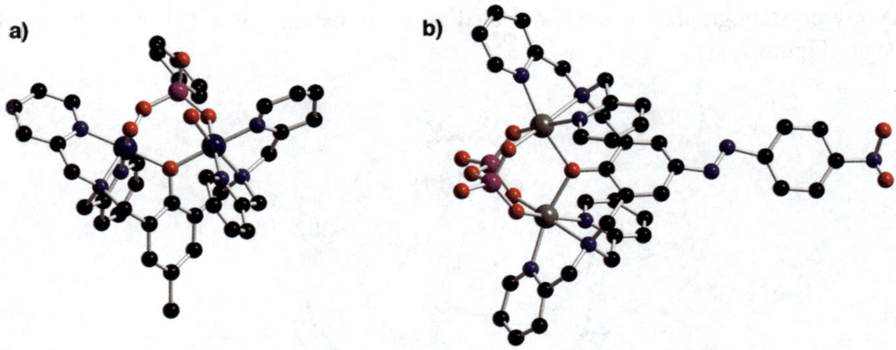

Figure 7.10 *Single crystal X-ray structures of (a) the complex of **7.36** with phenylphosphate and (b) the complex of **7.38** with pyrophosphate*

of pH 7.0 (HEPES) using ITC methods at 303K; these analyses revealed $\Delta H = -31.6$ kJ mol^{-1}, $T\Delta S = -7.48$ kJ mol^{-1}, and $K_a = 1.1 \times 10^5$ M^{-1}.[38]

In an effort to develop a Zn complex-based anion sensor, **7.38**·$(NO_3)_2$, that would function without the need for a displacement assay, a colour-inducing *p*-nitrophenylazo group was appended covalently onto the para position of the azophenol moiety.[39] The interaction between pyrophosphate and **7.38** was confirmed by X-ray crystallography. As shown in Figure 7.10b, the pyrophosphate anion is bound to the dinuclear Zn(II) complex (**7.38**) by bridging between the two metal cations. In 10 mM HEPES buffered aqueous solution (pH 7.4), changes in the absorption spectrum of receptor **7.38**·$(NO_3)_2$ were observed upon the addition of $P_2O_7^{4-}$, from which a $K_a = 6.6 \times 10^8$ M^{-1} could be derived. However, no appreciable colour change was detected with other anions (*i.e.*, $CH_3CO_2^-$, F^-, HCO_3^-, Cl^-, HPO_4^{2-}, and $H_2PO_4^-$). Inspired by the powerful anion-binding abilities of Zn^{2+}–pyridine complexes, several research groups have designed similar style receptors,[40] some of which employ different metal–pyridine complexes.[41]

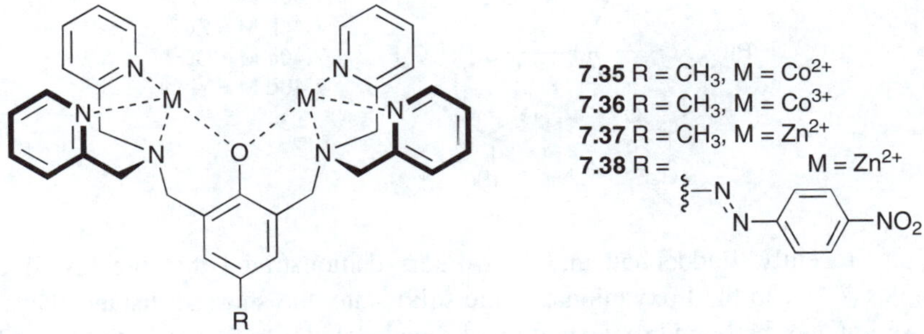

The Li^+–metallocrown complex **7.39** displays good selectivity for fluoride anions over other putative halide (chloride and bromide) and oxyanions (*cf.* nitrate, hydrogen sulfate, and perchlorate) guests. This selectivity was rationalized via the hypothesis that the cavity of receptor **7.39** only provides the small fluoride anion to access the Li^+ centre, the other anions being too large to coordinate to the metal centre.

X-ray crystallographic analysis revealed an Li–F distance of 1.755 Å in the solid state (Figure 7.11).[42]

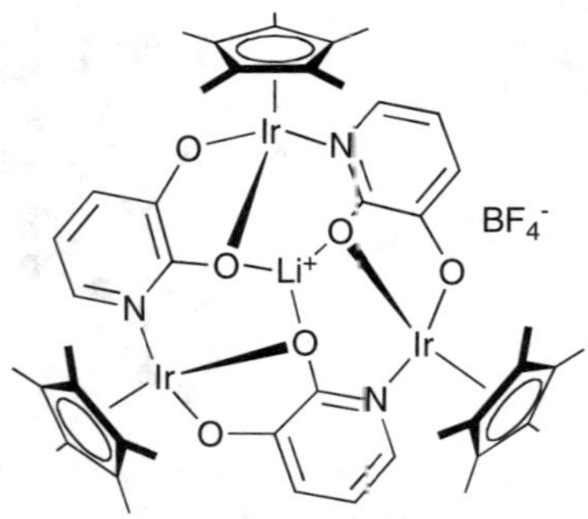

7.39

Interestingly, simple lanthanide tris(fluorinated *β*-diketonates) complexes, **7.40a–7.40d**, function as effective chloride-anion receptors and luminescent sensors via metal centred coordination.[43] In $CDCl_3$, ^1H NMR titration methods revealed that the estimated K_a values for chloride anion depended on the nature of lanthanide centres while exhibiting a trend, which deviates from those that are expected based on the considerations of ionic radii. For instance, association constants, K_a, = 450, 600, 610, and 94 were determined for **7.40a**, **7.40b**, **7.40c**, and **7.40d**, respectively. However, the ionic radii of the lanthanide centres in an octacoordinated environment is: Pr^{3+} (1.13 Å) > Eu^{3+} (1.07 Å) > Dy^{3+} (1.03 Å) > Yb^{3+} (0.99 Å). Presumably, the smaller lanthanide centres provided a shorter coordination bond but higher degree of steric repulsion between the *β*-diketonates and the chloride anion.

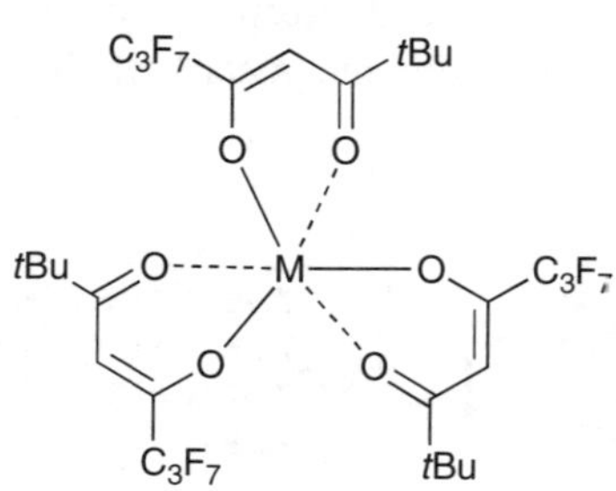

7.40a M = Pr^{3+}
7.40b M = Eu^{3+}
7.40c M = Dy^{3+}
7.40d M = Yb^{3+}

Quite recently, Puddephatt and co-workers demonstrated that the bis-silver(I) complex **7.41** can bind oxyanions. In the solid state, the shortest distance between the Ag ion and the bound oxygen atom of the chelated nitrate was found to be 2.53 Å (Figure 7.12a).[44] In the case of the trifluoroactate complex, the structure of which is shown in Figure 7.12b, the shortest distance was found to be 2.61 Å (in this case two unchelated anions are associated with the receptor). While this latter distance is slightly longer than that seen in the case of the nitrate complex, solution-phase anion-exchange experiments, complemented by ESI–MS analyses, revealed that the anion affinities followed the sequence $CF_3CO_2^-$ > NO_3^- > $CF_3SO_3^-$.

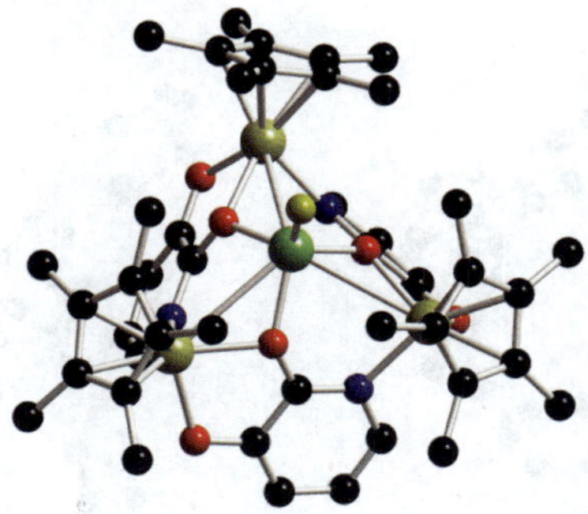

Figure 7.11 *Single crystal X-ray structure of the fluoride anion complex of* ***7.39****. A water molecule has been removed for clarity*

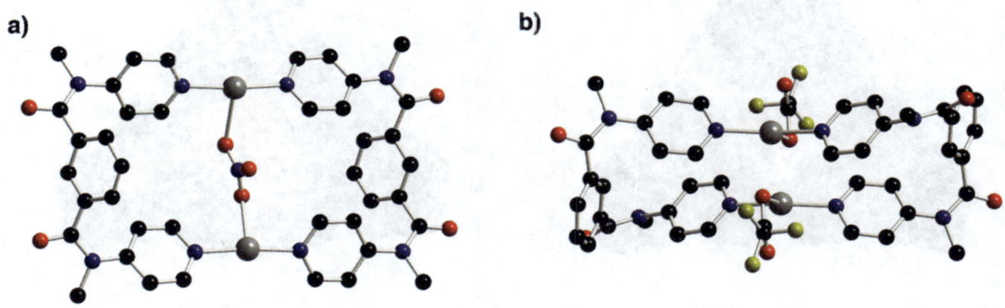

Figure 7.12 *Single crystal X-ray structures of (a) the nitrate complex of* ***7.41*** *(only the bound nitrate anion is shown) and (b) the trifluoroacetate complex of* ***7.41***

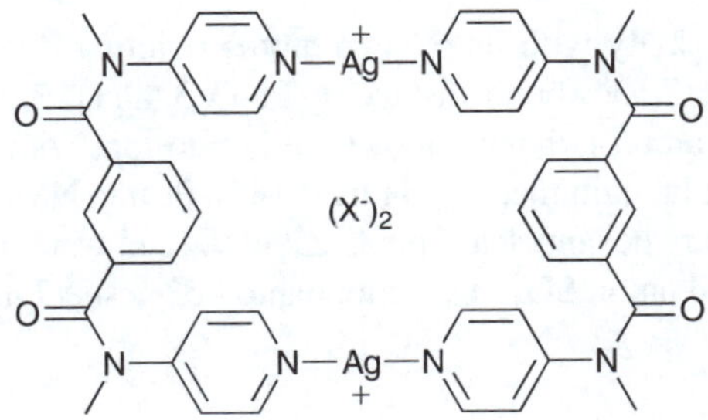

7.41

7.2 Metals as Organizers

Palladium and platinum metal ions have been used by Lippert and co-workers[45] to construct a highly charged (+12) anion receptor. The crystal structure of $\mathbf{7.42}^{12+}$ reveals that the platinum atoms form an equilateral triangle with side lengths of 7.88(1) Å. The top and side views of the cation are shown in Figures 7.13a and b respectively, and reveal that the Pd(en) groups define calixarene-like cavities on both the top and bottom faces of the receptor. The X-ray crystal structure also reveals a nitrate ion bound

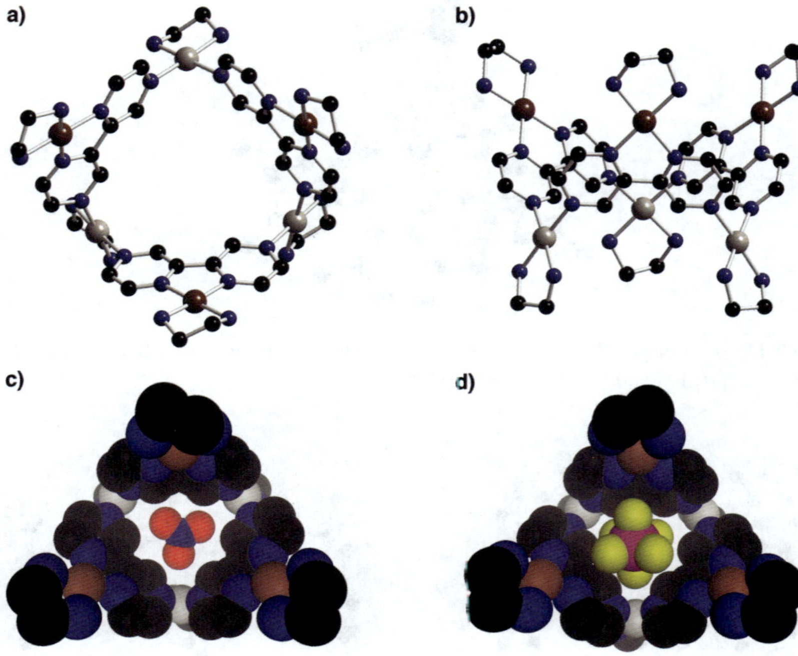

Figure 7.13 *Single crystal X-ray structures of (a) and (b) top and side views of the crystal structure of cation* **7.42**; *(c) space filling model of the cation* **7.42** *with an* NO_3^- *ion at the bottom of the calixarene-like cavity, and (d) spacefilling model of the encapsulation of a* PF_6^- *ion within the cavity of* **7.42** *(*NO_3^- *omitted for clarity)*

at the bottom of the cavity with its oxygen atoms pointing directly at the platinum metal centres (Figure 7.13c; bond distances: Pt–O 3.49(1), 3.26(1), 3.39(1) Å). A hexafluorophosphate anion is bound above the nitrate ion, with three of its fluorine atoms directed at the platinum metals (Figure 7.13d). Proton NMR spectroscopic titration analyses of the pure hexanuclear nitrate salt of **7.42** in water revealed a selectivity for the doubly charged anion SO_4^{2-} over monoanionic guests (Table 7.3).

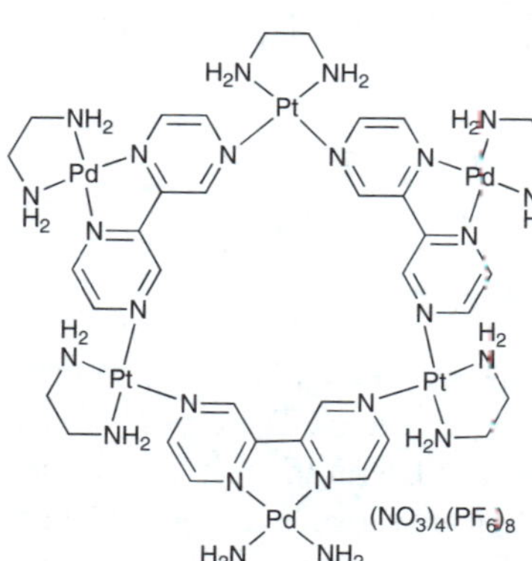

7.42

Table 7.3 *Association constants (K_a, M^{-1}) for receptor **7.42** (nitrate salt) with a variety of anions in water at 20 °C pD = 2.9. TSP = Sodium-3-(trimethylsilyl) propanesulfonate*

Anions	PF_6^-	ClO_4^-	BF_4^-	SO_4^{2-}	*TSP*
K_a (M^{-1}) ± 3σ	10.6 ± 3.9	9.6 ± 4.5	4.1 ± 1.1	255.8 ± 57.3	0.0

In a different approach to ditopic receptor construction, Hamilton and co-workers[46] used a metal centre as a template to combine two thiourea functionalized terpyridine units, thus creating a recognition site for dicarboxylate anions. In the case of receptor **7.43b**, the addition of dicarboxylate anions caused a displacement of the terpyridine ligands of the Fe(II) complex. Therefore, reliable association constants were not obtained. More accurate association constants were thus determined in 5% D_2O-DMSO-d_6. In this solvent, it was observed that receptor **7.43a** binds several alkyl dicarboxylate anions, such as glutarate, adipate, pimelate, isophthalate, and 1,3-phenylenediacetate strongly (*i.e.*, K_a values in the range of 2.9–8.3 × 10^3 M^{-1}). This strong binding is thought to reflect the stabilizing electrostatic effect of the doubly charged metal on a bound dianionic guest. Because of its high affinity and its perceived potential utility, the ruthenium(II)-derived receptor **7.43a** has also been applied to the design convergent combinatorial libraries.[47]

A different metal-based assembly approach was reported by J. Weiss and I. Prevot-Halter[48] who reported the synthesis of receptor **7.44**. In this case, the final anion receptor is prepared from two catechols bearing thioureas that are brought together via complexation with a transition metal such as molybdenum(VI). The ability of receptor **7.44** to bind dicarboxylate anions was probed via UV–Vis spectroscopic titrations. The resulting binding isotherms, as well as absorbance diagrams, were consistent with a 1:1 binding stoichiometry and, on the basis of these analyses, the association constants (log K_a) were determined to be in the range of 5.7–6.9 in acetonitrile for various alkyl biscarboxylates whose size was varied sequentially from succinate to pimelate.

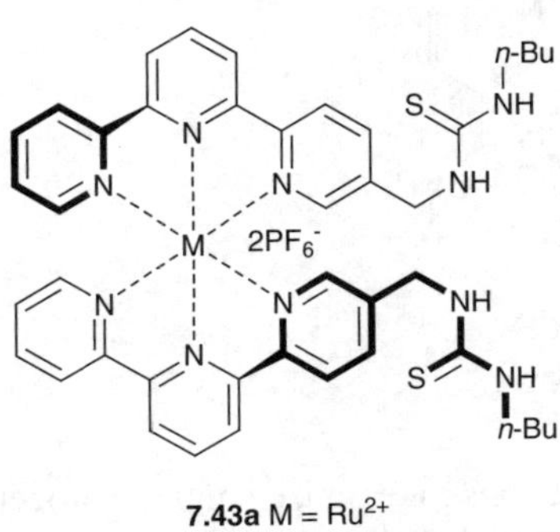

7.43a M = Ru^{2+}
7.43b M = Fe^{2+}

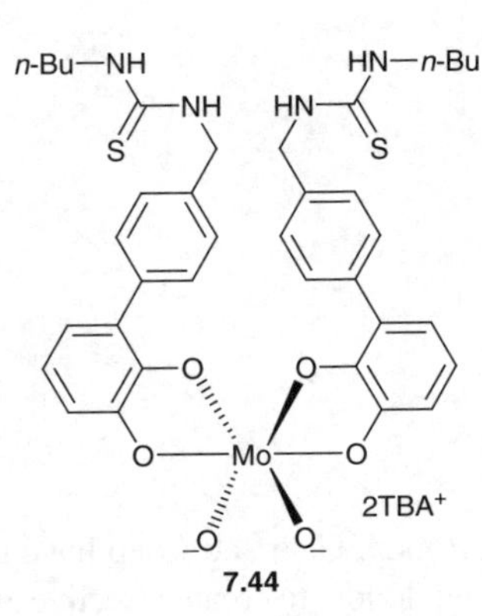

7.44

The synthesis and anion-binding studies of the ruthenium(II) tris(5,5′-diamide-2,2′-bipyridine) functionalized receptor **7.45** were reported by Beer and co-workers.[49] It was demonstrated that the anion-binding strength and selectivity depends critically on the solvent mixture ratio (CH_2Cl_2/CH_3OH) and the nature of the amide substituent. In the solid state, an X-ray crystal-structure analysis of **7.45a** served to reveal that two chloride anions are encapsulated within two triangular-faced cavities that, in turn, are defined by the three bidentate ligands, as illustrated in Figure 7.14. Quantitative measures of the association constants were made by carrying out UV–Vis titrations in dichloromethane/methanol at different solvent ratios. All association constants were found to decrease as the methanol fraction increased. However, the selectivites were found to change. For instance, in both 9:1 and 7:3 CH_2Cl_2/CH_3OH solvent mixtures, the selectivity of receptor **7.45** was found to follow the sequence $Cl^- > NO_3^- > AcO^-$. However, receptor **7.45c** was found to display a different preference, namely $NO_3^- > Cl^- > AcO^-$ when the CH_2Cl_2/CH_3OH ratio was 1:1. This difference was rationalized in terms of solvation effects. The hydration enthalpies of nitrate, chloride, and acetate anions are calculated to be 293, 335, and 402 kJ mol^{-1}, respectively. To the extent such hydration enthalpies reflect the relative ease of desolvation in the more polar mixture, an inherent selectivity of $NO_3^- > Cl^- > AcO^-$ would be expected absent other effects (*e.g.*, compensating solvation involving the receptor or complex). This limiting approximation should be more valid in the case of receptor **7.45c**, where the large lipophilic substituents serve to limit the access of solvent to the amide groups. By contrast, such differential solvation effects are likely to be less important for the more open receptor systems, especially when studied in less polar media; in such cases, the binding selectivities are expected to reflect more closely such "normal" considerations as receptor–anion matching in terms of, *e.g.*, size, shape, and number of expected hydrogen bonds.

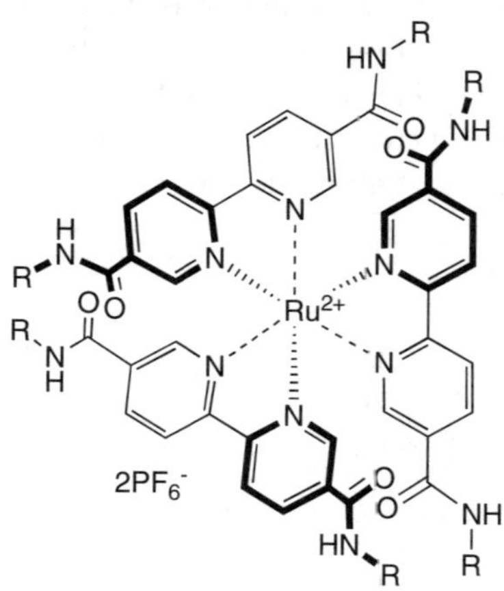

7.45a R = $CH_2CH_2OCH_3$
7.45b R = $(CH_2)_3CH_3$
7.45c R = $(CH_2)_{11}CH_3$

Bondy, Gale and Loeb have used platinum(II) cations to organize amide-hydrogen-bond donors to create effective anion-binding receptors.[50] For instance, in receptor **7.46**,

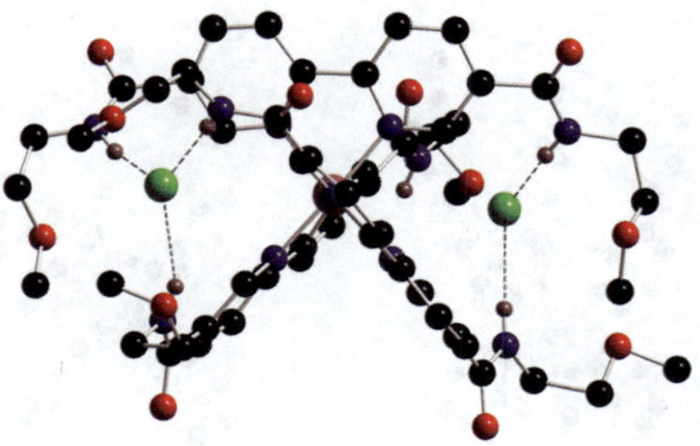

Figure 7.14 *Single crystal X-ray structure of the bis-chloride anion complex of receptor* ***7.45a***

four ligands containing amide residues are brought together via coordination to a square planar Pt(II) metal center. Complex **7.46**, in the form of its PF_6^- salt, was found to adopt a 1,2-alternate conformation in the solid state (Figure 7.15a). Interestingly, hydrogen-bonding interactions between PF_6^- and amide were not observed; however, electrostatic interactions with the metal were inferred. The X-ray structure of **7.46**·ReO_4^- was also solved (cf. Figure 7.15b) and it strongly supports the notion that receptor **7.46** should be able to bind anions in solution. In fact, association constants, determined by standard ^{1}H NMR spectroscopic titrations carried out in CD_3CN or CD_3CN/DMSO-d_6 show that receptor **7.46** is an effective receptor for oxyanions, such as $CH_3CO_2^-$ (K_{a1} = 230 M^{-1} and K_{a2} = 491 M^{-1} in CD_3CN/DMSO-d_6 = 1/9 v/v), HSO_4^- (K_a = 149 M^{-1} in CD_3CN/DMSO-d_6 = 3/1 v/v), ReO_4^- (K_a = 150 M^{-1} in CD_3CN), NO_3^- (K_{a1} = 562 M^{-1} and K_{a2} = 132 M^{-1} in CD_3CN), and $CF_3SO_3^-$ (K_a = 129 M^{-1} in CD_3CN).

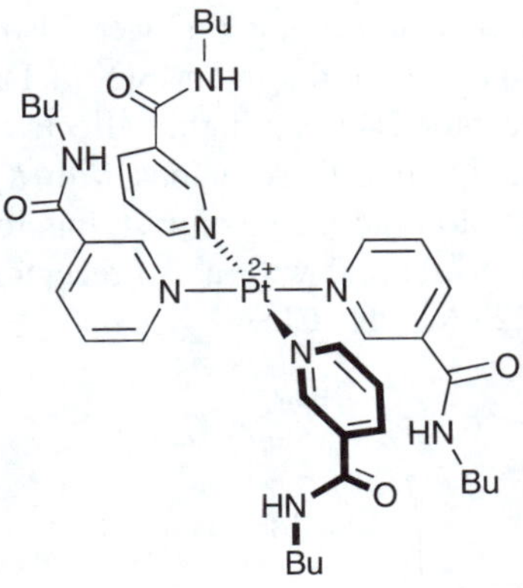

7.46

Quite recently, this group reported the synthesis of a new type of metal-organic anion receptor, **7.47**, that displays a high affinity for sulfate.[51] In this case, a single crystal X-ray structure analysis shows that the receptor **7.47** can adopt two different conformations in the solid state when bound to anions. In particular, the solid-state structure of the bis-chloride anion complex reveals that receptor **7.47** exists in a 1,2-alternate

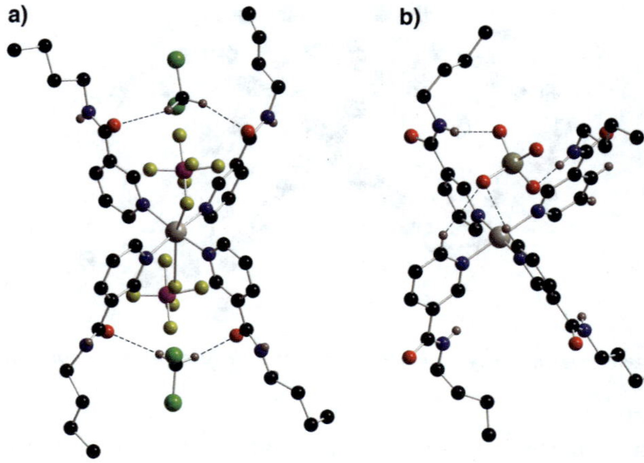

Figure 7.15 *(a) Single crystal X-ray structure of the bis-PF_6^- salt of receptor **7.46**. Two dichloromethane molecules are also included in the binding cavities defined by the receptor and are shown in this structure. (b) Single crystal X-ray structure of ReO_4^- anion complex of receptor **7.46***

conformation and stabilizes the bound chloride anions via two different types of hydrogen-bonding interactions, namely NH···Cl^- and CH···Cl^- (Figure 7.16a). However, in the corresponding sulfate-anion complex, receptor **7.47** exists in a cone conformation, which permits more complete encapsulation of the bound anion (Figure 7.16b). In this case, there are eight stabilizing hydrogen-bonding interactions between the eight-urea NH protons and three of the sulfate oxygen atoms.

The anion-binding properties of receptor **7.47** were also studied in solution via ^1H NMR spectroscopic titrations carried out in DMSO-d_6. These studies provided support for the notion that receptor **7.47** is a highly effective receptor for both sulfate and dihydrogen phosphate, binding these anions with K_a values in excess of 10^5 M^{-1} and in a 1:1 receptor/anion ratio. By contrast, this receptor was found to bind halide anions far less effectively and with a 1:2 receptor/anion ratio (*e.g.*, K_{a1} = 11,693 M^{-1} and K_{a2} = 2223 M^{-1} for Cl^-).

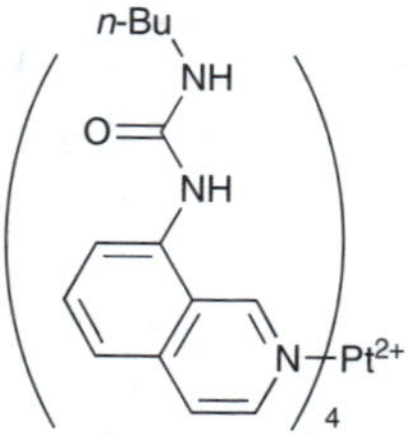

7.47

Halcrow and co-workers[52] have employed a similar strategy in the production of anion receptors from zinc 3{5}-*tert*-butylpyrazole (HpztBu) complexes. Complexation of ZnX_2 (X = Cl, Br, I) by HpztBu yields zinc complexes [ZnX(HpztBu)$_3$]X (X = Cl(**7.48**), Br(**7.49**), I(**7.50**)). Crystal structure elucidations of these complexes reveal a

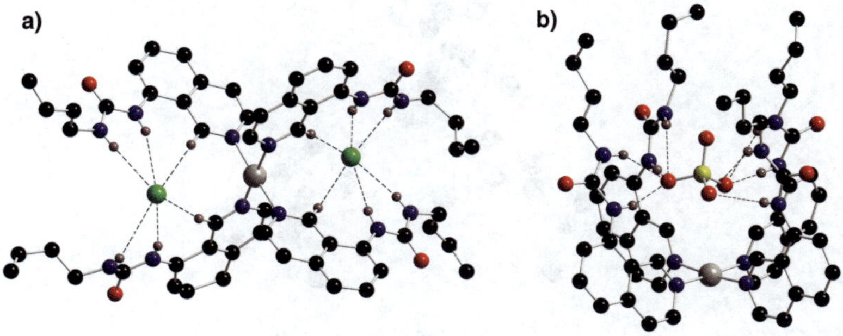

Figure 7.16 *(a) Single crystal X-ray structures of (a) the bis-chloride-anion complex and (b) sulfate-anion complex of receptor **7.47***

distorted Zn(II) complex cation with the *tert*-butyl groups of the attached ligands oriented such that the zinc complex offers a convergent array of three NH-hydrogen-bond donors to the non-metal coordinated halide anion (Figure 7.17). The hydrogen-bonded chloride in **7.48** may be exchanged for other anions by treatment with Ag(I) or Tl(I) salts affording the BF_4^-, ClO_4^-, NO_3^-, $CF_3SO_3^-$, and PF_6^- complexes, all of which were crystallographically characterized. Recently, Pérez and co-workers[53] also reported *fac*–tris(pyrazol) complexes as anion receptors.

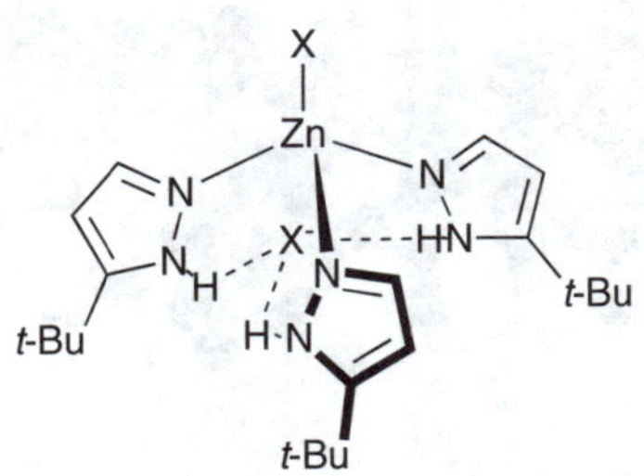

7.48 X = Cl
7.49 X = Br
7.50 X = I

Quite recently, Raptis and co-workers[54] reported the results of a single crystal X-ray diffraction analysis of the chloride-anion complex of the neutral supramolecular ensemble formed from the cyclic Cu(II) complexes **7.51**, a species designated as **7.51a·7.51d**-chloride. The metal-containing ensemble, which contains both **7.51a** and **7.51d** as subunits, is stabilized by the interaction between the six inward-pointing hydroxyl group of **7.51d** and all six Cu-centres of the inner six-membered ring **7.51a** (Cu···O = 2.399(6)–2.464(6) Å). The resulting dimer **7.51a·7.51d**, is neutral in the absence of a bound anion, and contains six hydroxyl (hydrophilic) and pyrazole (hydrophobic) groups. This complex entity, in turn, is able to encapsulate a bound chloride anion by forming a sandwich-type dimer that is stabilized by 12 hydrogen-bond interactions (Cl···O = 3.306(7)–3.698(7) Å), six on each side (Figure 7.18). These same authors also described two other complex, anion-containing structures, namely **7.51a·7.51d**-carbonate-**7.51c** and **7.51b·7.51e**-sulfate-**7.51c**. In conjunction with the description of metallacyclic isomers, the formation of these complexes clearly supports the notion that a μ-OH subunit constrained within neutral receptors is able to act as a good hydrogen-bond donor and anion recognition element.

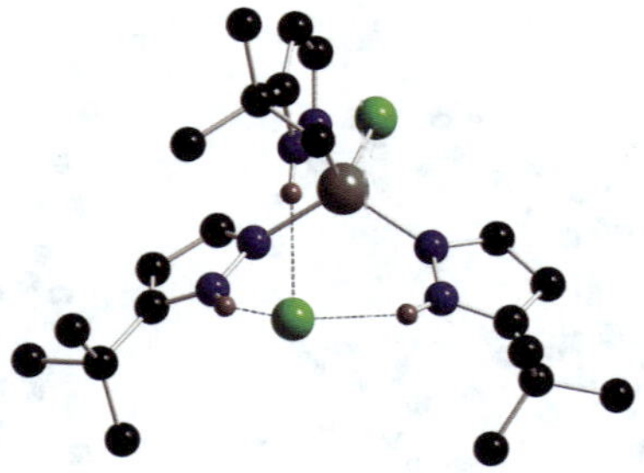

Figure 7.17 *Single crystal X-ray structure of complex* ***7.48****. Certain hydrogen atoms have been omitted for clarity*

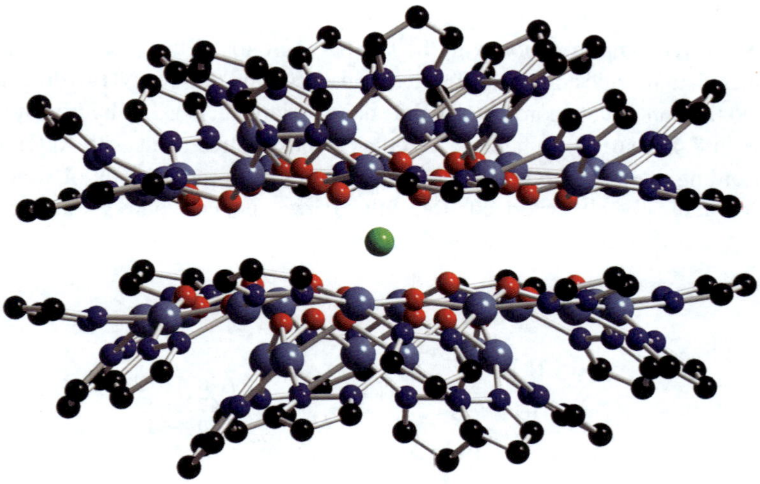

Figure 7.18 *Single crystal X-ray structure of the chloride-anion complex formed from the metal-ligated mixed receptor dimer* ***7.51a–7.51d****. Hydrogen atoms have been omitted for clarity*

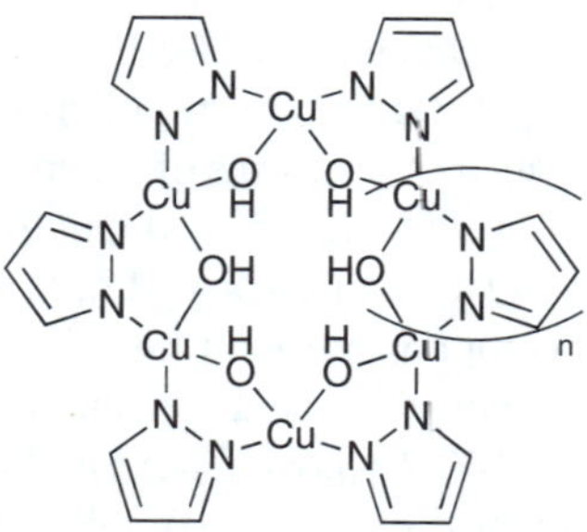

7.51a n = 6
7.51b n = 8
7.51c n = 9
7.51d n = 12
7.51e n = 14

7.3 Other Anion Receptors Containing Metals

Atwood and Steed[55] have synthesized a family of calixarene and cyclotriveratrylene-based anion receptors in which transition metals have been appended to the outside of the aromatic cavity in order to reduce the electron density present in the "cup" making it more amenable to anionic guest species (*e.g.*, **7.52–7.53**). "Extended cavity" systems, such as **7.54**, have also been synthesized.[56]

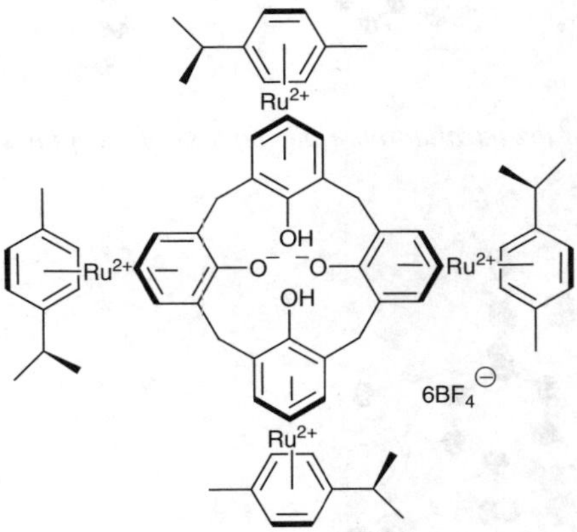

7.52

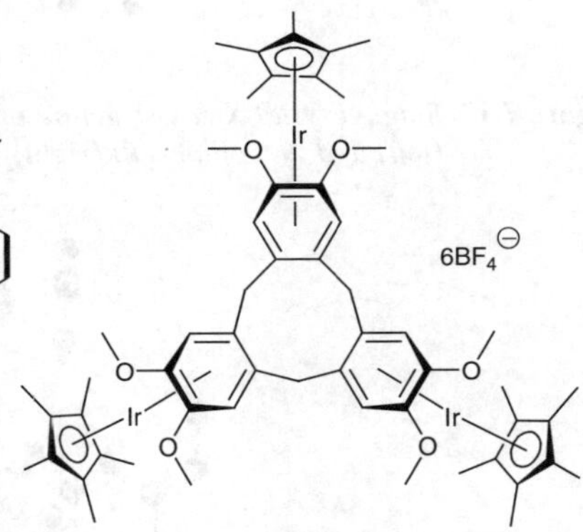

7.53

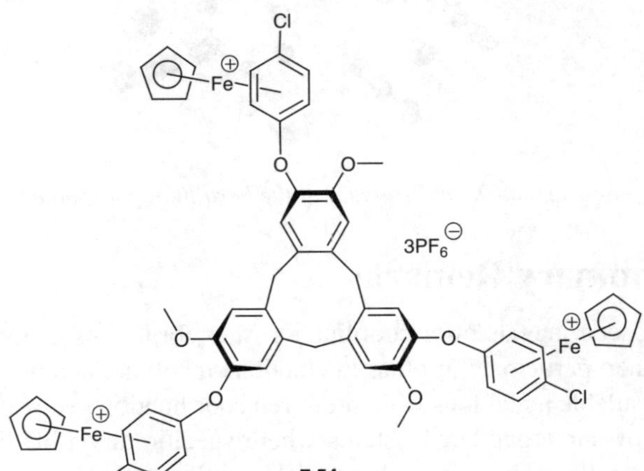

7.54

A variety of anion complexes have been isolated and crystallographically characterized. For example, the tetrafluoroborate and iodide salts of **7.52**, are shown in Figure 7.19. In both cases, one of the counter ions is encapsulated within the cup of the calixarene. The extended cavity system **7.54** also encapsulates anionic guests (the hexafluorophosphate salt of this receptor is shown in Figure 7.20).

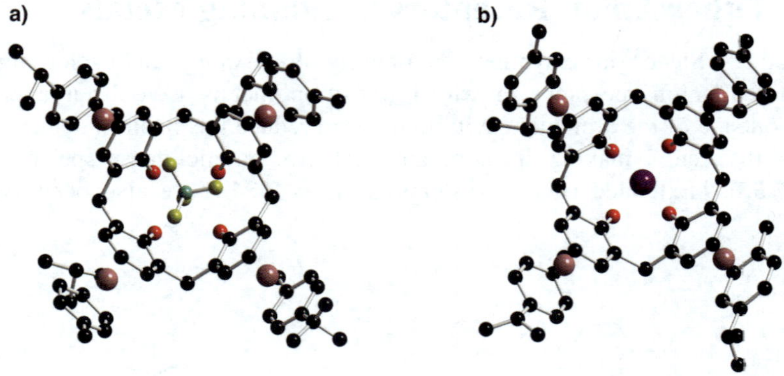

Figure 7.19 *Single crystal X-ray structure of the tetrafluoroborate complex of receptor* **7.52** *(left) and the iodide salt (right)*

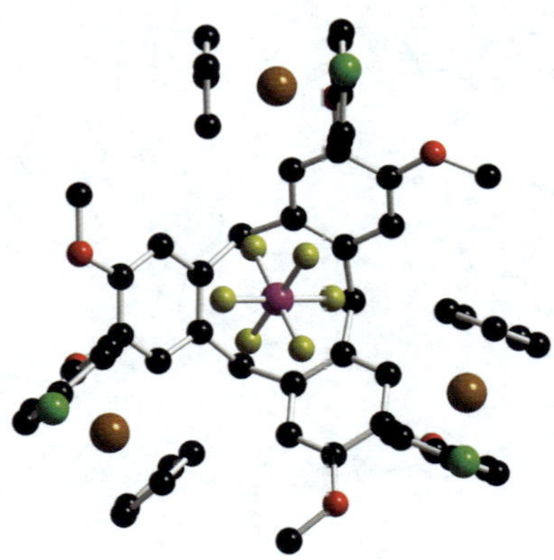

Figure 7.20 *Single crystal X-ray structure of the hexafluorophosphate salt of* **7.54**

7.4 Summary Remarks

As we have seen, metals offer coordination sites for anions or for organic ligands, which can then participate in other interactions involving anionic guests. The huge variety of available metal ions with preferred coordination geometries allows for the conception of numerous new systems where specific anion-binding environments are established through rational design. This ability to create systems with unique and interesting anion-binding properties should make this an area of special interest to both inorganic- and organic-oriented supramolecular chemists alike.

References

1. D.F. Shriver and M.J. Biallas, *J. Am. Chem. Soc.*, 1967, **89**, 1078.
2. V.C. Williams, W.E. Piers, W. Clegg, M.R.J. Elsegood, S. Collins and T.B. Marder, *J. Am. Chem. Soc.*, 1999, **121**, 3244.

3. H.E. Katz, *Organometallics*, 1987, **6**, 1134.
4. H.E. Katz, *J. Org. Chem.*, 1985, **50**, 5027.
5. H.E. Katz, *J. Am. Chem. Soc.*, 1985, **107**, 1420.
6. S. Solé and F.P. Gabbaï, *Chem. Commun.*, 2004, 1284.
7. M. Melaimi and F.P. Gabbaï, *J. Am. Chem. Soc.*, 2005, **127**, 9680.
8. J.D. Wuest and B. Zacharie, *Organometallics*, 1985, **4**, 410.
9. A.L. Beauchamp, M.J. Olivier, J.D. Wuest and B. Zacharie, *J. Am. Chem. Soc.*, 1986, **108**, 73.
10. V.B. Shur, I.A. Tikhonova, P.V. Petrovskii and M.E. Vol'pin, *Metalloorg. Khim.*, 1989, **2**, 1431; V.B. Shur, I.A. Tikhonova, A.I. Yanovsky, Y.T. Struchkov, P.V. Petrovskii, S.Y. Panov, G.G. Furin and M.E. Vol'pin, *J. Organomet. Chem.*, 1991, **418**, C29.
11. V.B. Shur and I.A. Tikhonova, *Russ. Chem. Bull. Int. Ed.*, 2003, **52**, 2539.
12. A. Knoepfler-Mühlecker, B. Scheffter, H. Kopacka, K. Wurst and P. Peringer, *J. Chem. Soc. Dalton Trans.*, 1999, 2525.
13. T.M. Wedge and M.F. Hawthorne, *Coord. Chem. Rev.*, 2003, **240**, 111.
14. X. Yang, C.B. Knobler and M.F. Hawthorne, *Angew. Chem. Int. Ed.*, 1991, **30**, 1507.
15. Z. Zheng, C.B. Knobler and M.F. Hawthorne, *J. Am. Chem. Soc.*, 1995, **117**, 5105.
16. X. Yang, C.B. Knobler and M.F. Hawthorne, *J. Am. Chem. Soc.*, 1992, **114**, 380; Z. Zheng, X. Yang, C.B. Knobler and M.F. Hawthorne, *J. Am. Chem. Soc.*, 1993, **115**, 5320; Z. Zheng, C.B. Knobler, M.D. Mortimer, G. Kong and M.F. Hawthorne, *Inorg. Chem.*, 1996, **35**, 1235.
17. X. Yang, Z. Zheng, C.B. Knobler and M.F. Hawthorne, *J. Am. Chem. Soc.*, 1993, **115**, 193.
18. M.J. Bayer, S.S. Jalisatgi, B. Smart, A. Herzog, C.B. Knobler and M.F. Hawthorne, *Angew. Chem. Int. Ed.*, 2004, **43**, 1854.
19. H. Lee, M. Diaz, C.B. Knobler and M.F. Hawthorne, *Angew. Chem. Int. Ed.*, 2000, **39**, 776; H. Lee, C.B. Knobler and M.F. Hawthorne, *J. Am. Chem. Soc.*, 2001, **123**, 8543.
20. Y. Azuma and M. Newcomb, *Organometallics*, 1984, **3**, 9.
21. M. Newcomb, A.M. Madonik, M.T. Blanda and J.K. Judice, *Organometallics*, 1987, **6**, 145.
22. M. Newcomb, M.T. Blanda, Y. Azuma and T.J. Delord, *J. Chem. Soc. Chem. Commun.*, 1984, 1159.
23. M. Newcomb, J.H. Horner and M.T. Blanda, *J. Am. Chem. Soc.*, 1987, **109**, 7878.
24. M. Newcomb and M.T. Blanda, *Tetrahedron Lett.*, 1988, **29**, 4261.
25. M. Newcomb, J.H. Horner, M.T. Blanda and P.J. Squattrito, *J. Am. Chem. Soc.*, 1989, **111**, 6294.
26. M. Schulte, M. Schürmann and K. Jurkschat, *Chem. Eur. J.*, 2001, **7**, 347.
27. K. Tamao, T. Hayashi and Y. Ito, *J. Organomet. Chem.*, 1996, **506**, 85.
28. M.E. Jung and H. Xia, *Tetrahedron Lett.*, 1988, **29**, 297.
29. K. Ogawa, S. Aoyaki and Y. Takeuchi, *J. Chem. Soc. Perkin Trans. 2*, 1993, 2389; S. Aoyagi, K. Tanaka, I. Zicmane and Y. Takeuchi, *J. Chem. Soc. Perkin Trans. 2*, 1992, 2217.
30. S. Aoyagi, K. Tanaka and Y. Takeuchi, *J. Chem. Soc. Perkin Trans. 2*, 1994, 1549.

31. F. Sancenón, R. Martínez-Máñez and J. Soto, *Angew. Chem. Int. Ed.*, 2002, **41**, 1416.
32. S. Mizukami, T. Nagano, Y. Urano, A. Odani and K. Kikuchi, *J. Am. Chem. Soc.*, 2002, **124**, 3920.
33. R. Newell, A. Appel, D.L. DuBois and M.R. DuBois, *Inorg. Chem.*, 2005, **44**, 365.
34. L. Fabbrizzi, F. Foti and A. Taglietti, *Org. Lett.*, 2005, **7**, 2603.
35. B. García-Acosta, X. Albiach-Martí, E. García, L. Gil, R. Martínez-Máñez, K. Rurack, F. Sancenón and J. Soto, *Chem. Commun.*, 2004, 774; L.-L. Zhou, H. Sun, H.-P. Li, H. Wang, X.-H. Zhang, S.-K. Wu and S.-T. Lee, *Org. Lett.*, 2004, **6**, 1071.
36. M. Suzuki, H. Kanatomi and I. Murase, *Chem. Lett.*, 1981, **10**, 1745.
37. J.S. Seo, N.-D. Sung, R.C. Hynes and J. Chin, *Inorg. Chem.*, 1996, **35**, 7472.
38. M.S. Han and D.H. Kim, *Angew. Chem. Int. Ed.*, 2002, **41**, 3809.
39. D.H. Lee, J.H. Im, S.U. Son, Y.K. Chung and J.-I. Hong, *J. Am. Chem. Soc.*, 2003, **125**, 7752.
40. B.P. Munch, F.C. Bradley, P.D. Boyle, V. Papaefthymiou and L. Que Jr., *J. Am. Chem. Soc.*, 1987, **109**, 7993; S. Torelli, C. Belle, I. Gautier-Luneau, J.L. Pierre, E. Saint-Aman, J.M. Latour, L. le Page and D. Luneau, *Inorg. Chem.*, 2000, **39**, 3526; P. Molenveld, J.F.J. Engbersen and D.N. Reinhoudt, *Chem. Soc. Rev.*, 2000, **29**, 75; L.A. Cabell, M.D. Best, J.J. Lavigne, S.E. Schneider, D.M. Perreault, M.-K. Monahan and E.V. Anslyn, *J. Chem. Soc. Perkin Trans. 2*, 2001, 315; H. Aït-Haddou, S.L. Wiskur, V.M. Lynch and E.V. Anslyn, *J. Am. Chem. Soc.*, 2001, **123**, 11296; H. Aït-Haddou, J. Sumaoka, S.L. Wiskur, J.F. Folmer-Andersen and E.V. Anslyn, *Angew. Chem. Int. Ed.*, 2002, **41**, 4014; A. Ojida, Y. Mito-oka, M.-a. Inoue and I. Hamachi, *J. Am. Chem. Soc.*, 2002, **124**, 6256; J.F. Folmer-Andersen, H. Aït-Haddou, V.M. Lynch and E.V. Anslyn, *Inorg. Chem.*, 2003, **42**, 8674; A. Ojida, M.-a. Inoue, Y. Mito-oka and I. Hamachi, *J. Am. Chem. Soc.*, 2003, **125**, 10184; A. Ojida, Y. Mito-oka, K. Sada and I. Hamachi, *J. Am. Chem. Soc.*, 2004, **126**, 2454; I. Yoshimura, Y. Miyahara, N. Kasagi, H. Yamane, A. Ojida and I. Hamachi, *J. Am. Chem. Soc.*, 2004, **126**, 12204; R.G. Hanshaw, S.M. Hilkert, H. Jiang and B.D. Smith, *Tetrahedron Lett.*, 2004, **45**, 8721; J.C. Mareque-Rivas, R.T.M. de Rosales and S. Parsons, *Chem. Commun.*, 2004, 610; M. Livieri, F. Mancin, U. Tonellato and J. Chin, *Chem. Commun.*, 2004, 2862; H.K. Cho, D.H. Lee and J.-I. Hong, *Chem. Commun.*, 2005, 1690; H. Jiang, E.J. O'Neil, K.M. DiVittorio and B.D. Smith, *Org. Lett.*, 2005, **7**, 3013; R. Cacciapaglia, A. Casnati, L. Mandolini, D.N. Reinhoudt, R. Salvio, A. Sartori and R. Ungaro, *J. Org. Chem.*, 2005, **70**, 624; R. Cacciapaglia, A. Casnati, L. Mandolini, D.N. Reinhoudt, R. Salvio, A. Sartori and R. Ungaro, *J. Org. Chem.*, 2005, **70**, 5398; V.W.-W. Yam, K.H.-Y. Chan, K.M.-C. Wong and N. Zhu, *Chem. Eur. J.*, 2005, **11**, 4535; C. Bazzicalupi, A. Bencini, A. Bianchi, A. Danesi, C. Giorgi, C. Lodeiro, F. Pina, S. Santarelli and B. Valtancoli, *Chem. Commun.*, 2005, 2630.
41. M. Montalti, L. Prodi, N. Zaccheroni, L. Charbonnière, L. Douce and R. Ziessel, *J. Am. Chem. Soc.*, 2001, **123**, 12694; L. Charbonnière, R. Ziessel, M. Montalti, L. Prodi, N. Zaccheroni, C. Boehme and G. Wipff, *J. Am. Chem. Soc.*, 2002, **124**, 7779; M.S. Han and D.H. Kim, *Tetrahedron*, 2004, **60**, 11251; Z. Zhong, B.J. Postnikova, R.E. Hanes, V.M. Lynch and E.V. Anslyn, *Chem.*

Eur. J., 2005, **11**, 2385; T. Yamada, S. Shinoda and H. Tsukube, *Chem. Commun.*, 2002, 1218.
42. M.-L. Lehaire, R. Scopelliti, H. Piotrowski and K. Severin, *Angew. Chem. Int. Ed.*, 2002, **41**, 1419.
43. R.K. Mahajan, I. Kaur, R. Kaur, S. Uchida, A. Onimaru, S. Shinoda and H. Tsukube, *Chem. Commun.*, 2003, 2238.
44. N.L.S. Yue, M.C. Jennings and R.J. Puddephatt, *Inorg. Chem.*, 2005, **44**, 1125.
45. R.-D. Schnebeck, E. Freisinger and B. Lippert, *Angew. Chem. Int. Ed.*, 1999, **38**, 168.
46. M.S. Goodman, V. Jubian and A.D. Hamilton, *Tetrahedron Lett.*, 1995, **36**, 2551.
47. M.S. Goodman, V. Jubian, B. Linton and A.D. Hamilton, *J. Am. Chem. Soc.*, 1995, **117**, 11610.
48. I. Prevot-Halter and J. Weiss, *New J. Chem.*, 1998, **22**, 869.
49. L.H. Uppadine, M.G.B. Drew and P.D. Beer, *Chem. Commun.*, 2001, 291.
50. C.R. Bondy, P.A. Gale and S.J. Loeb, *Chem. Commun.*, 2001, 729; C.R. Bondy, P.A. Gale and S.J. Loeb, *J. Supramol. Chem.*, 2002, **2**, 93.
51. C.R. Bondy, P.A. Gale and S.J. Loeb, *J. Am. Chem. Soc.*, 2004, **126**, 5030.
52. X. Liu, C.A. Kilner and M.A. Halcrow, *Chem. Commun.*, 2002, 704; S.L. Renard, C.A. Kilner, J. Fisher and M.A. Halcrow, *J. Chem. Soc. Dalton Trans.*, 2002, 4206.
53. S. Nieto, J. Pérez, V. Riera, D. Miguel and C. Alvarez, *Chem. Commun.*, 2005, 546.
54. G. Mezei, P. Baran and R.G. Raptis, *Angew. Chem., Int. Ed.*, 2004, **43**, 574.
55. K.T. Holman, M.M. Halihan, J.W. Steed, S.S. Jurisson and J.L. Atwood, *J. Am. Chem. Soc.*, 1995, **117**, 7848; K.T. Holman, M.H. Halihan, S.S. Jurison, J.L. Atwood, R.S. Burkhalter, A.R. Mitchell and J.W. Steed, *J. Am. Chem. Soc.*, 1996, **118**, 9567.
56. K.T. Holman, G.W. Orr, J.W. Steed and J.L. Atwood, *Chem. Commun.*, 1998, 2109; J.L. Atwood, K.T. Holman and J.W. Steed, *Chem. Commun.*, 1996, 1401.

CHAPTER 8

Sensors

8.1 Introduction

Sensors for anions must not only bind an anionic guest selectively but also report the anion-recognition event via an electrochemical or optical macroscopic physical response. Several different strategies have been employed for the electrochemical detection of anions; two of the most important involve (1) extraction of a charged guest into a membrane by a non-electroactive host and detection of the resulting membrane potential (ion-selective electrodes (ISEs) and chemically modified field effect transistors (CHEMFETs)) and (2) synthesis of a receptor containing a redox-active group and detection of a current/potential perturbation on complex formation (voltammetric/amperometric sensors). Similarly, a number of different strategies have also been employed in the production of optical sensors for anions. Included among these are optodes, which may be regarded as the optical equivalent of ISEs in that a change in the properties of a receptor-doped membrane film is used to signal the presence of a targeted guest species. Distinct from optodes are discrete molecular receptors containing fluorescent groups that serve to signal the presence of an anion via a change in the emission (colour or intensity) as the result of, *e.g.*, photo-electron-transfer quenching. Simple colorimetric sensors have also been reported that undergo a colour change in the presence of a particular anion – a change that is detectable by the use of a spectrometer or, more preferably by eye. Ensembles of molecules have also been used as "displacement assays" wherein, for example, a dye is bound to a receptor that is displaced by the anionic guest. The displaced dye has different optical properties from the bound dye and hence anion complexation may be detected. Aggregation effects have also been used to affect anion sensing. Chemical systems that signal the presence of an anion through discrete bond-making or bond-forming processes (*i.e.*, by undergoing a generally irreversible chemical reaction) are also known. In this chapter an effort is made to review these strategies in some detail.

8.2 Devices that Employ Anion-Selective Membranes

ISEs are electrodes that contain a guest-selective membrane between the analyte solution and the electrode, and give a selective response to the presence of a particular species. A two-electrode set-up for a generic ISE is illustrated in Figure 8.1.

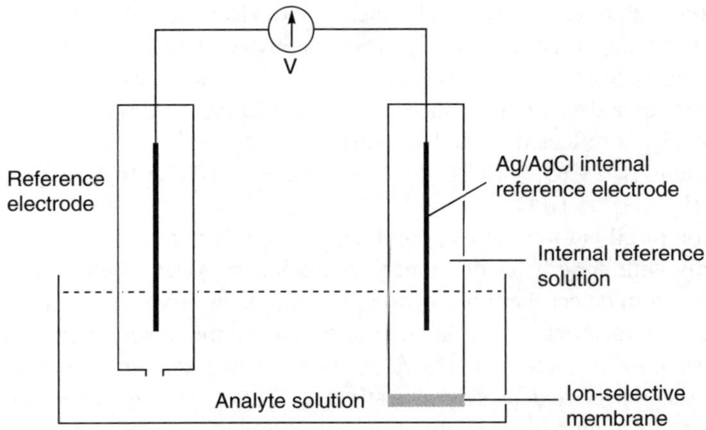

Figure 8.1 *A generic ion-selective electrode set-up with reference electrode*

The ISE measures the potential of the target ion in solution, and compares this potential to that of a reference electrode. The potential difference is related to the activity of the ion in solution, which in turn is related to its concentration. Hence this type of device may be used to measure ion concentrations. The selectivity of the electrode is dictated by the membrane, which is designed to extract with high specificity the target guest from solution. For example, a widely used fluoride ISE consists of a lanthanum fluoride (LaF_3) crystal membrane (which may be doped with a small quantity of europium fluoride (EuF_2) to reduce resistivity).[1] The shorthand notation for a typical fluoride electrode is Ag/AgCl/0.1 M KCl, 0.1 M NaF/LaF_3.

The crystal contains fluoride vacancies (the number of which is increased by the EuF_2 dopant); hence, fluoride anions can be conducted through the crystal to the electrode. Other anions are too large to pass through the crystal membrane hence there is no interference from other anionic species. The exception is hydroxide, a species that can interfere by reacting with the lanthanum fluoride, thereby releasing fluoride anions. This unwanted release may be countered by buffering the solutions that are the subject to analysis so as to limit the effective hydroxide-anion concentration.

Many commercially available chloride selective electrodes are based upon silver chloride/silver sulfide solid-state membranes. Unfortunately, these systems suffer from interference from silver and sulfide ions as well as from common anions that may be present in the environment (*e.g.*, iodide and bromide).

Other types of ISE contain polymer membranes that contain an ionophore. These membranes consist mainly of poly(vinyl chloride) (PVC) mixed with a plasticizer such as bis(2-ethylhexyl) sebacate or 2-nitrophenyl octyl ether.[2] While problems can arise in these systems as a result of either the plasticizer or ionophore leaching from the membrane into the test solution, inspiring current work on the development of new types of more durable membrane system,[3] these kinds of systems remain among the most useful and promising of known anion-sensing systems. Early polymer membrane anion-selective electrodes were based on anion-exchange principles. In these systems, as well as many currently under development, the ionophore has generally been an ammonium cation, with salts such as tridodecylmethylammonium chloride (TDDMACl) being used extensively. The selectivity of these electrodes is determined by the hydrophobicity of the anionic guests and hence by their relative dehydration energies and ease of exchange into the membrane. As discussed earlier, anion hydrophobicity is reflected in the Hofmeister series[4]:

organic anions > ClO_4^- > SCN^- > I^- > salicylate$^-$ > NO_3^- > Br^- > Cl^- > HCO_3^- > $H_2PO_4^-$ > F^-, SO_4^{2-} > HPO_4^{2-}

Detection of all but the most lipophilic anions could not be achieved completely successfully with these types of ammonium ionophore-based electrodes as attempts to use them in to detect the more hydrophilic anions in mixtures of anions (*e.g.*, in blood) were characterized by strong interference from the more lipophilic species in solution, such as bromide and salicylate.[5] Hence, ISEs containing synthetic-anion receptors have been extensively investigated. These devices often show "anti-Hofmeister" behaviour as the result of the selectivity of the receptor, which if properly chosen can serve to modulate or even overcome the intrinsic bias towards lipophilic anionic analytes. These types of membrane may also contain positively charged additives such as alkylammonium cations to aid extraction of the anion into the lipophilic membrane. This is essential if the ionophore is neutral. For a comprehensive overview of this area up to 1998 the reader is directed to Bühlmann and co-workers'[6] excellent review in *Chemical Reviews*. Antonisse and Reinhoudt's[7] review of potentiometric anion sensors in *Electroanalysis* in 1999 is also recommended.

To date, a wide variety of both charged and neutral-anion receptors have been incorporated into polymer matrix membrane ISEs. For example, Umezawa and co-workers have reported a number of ISEs that employ amide- and urea-containing anion receptors and which show selective responses with a variety of different anionic guests. For instance, receptor **8.1** was incorporated into PVC membranes containing 2-nitrophenyl octyl ether as a plasticizer and TDDMACl as an additive. The resulting ISEs were found to respond selectively to sulfate anion and display little interference from SCN^-, NO_3^-, Br^-, or Cl^-.[8] In contradistinction to these results, ISEs prepared with similar membranes containing **8.2** showed chloride-selective behaviour.[9] In these systems, the selectivity preference of the ionophore is the determining factor in the selectivity of the membrane. Bachas and co-workers[10] have also studied the anion-detection properties of ISEs based on urea-containing receptor (**8.3**),[11] which contain aminochromenone moieties linked to various scaffolds. A comparison of electrodes containing these receptors showed that membranes incorporating compound **8.3** displayed a remarkable selectivity for sulfate.

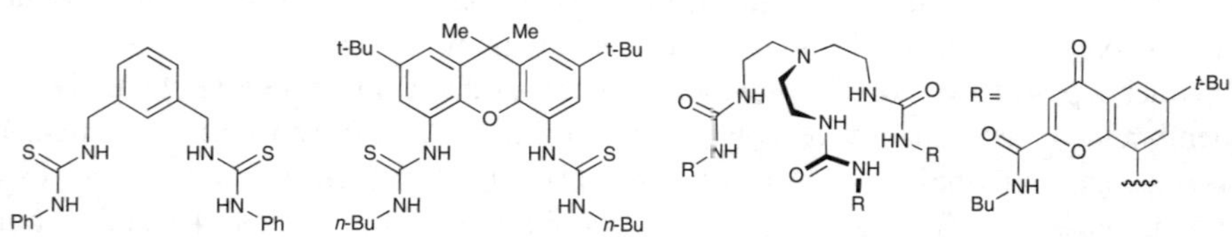

8.1 **8.2** **8.3**

Umezawa and co-workers also pioneered the use of expanded porphyrins as sensor elements in ISEs. For instance, in early work, these researchers showed that the organic soluble sapphyrin **8.4**, when incorporated in a dioctyl phthalate (DOP)–polyvinylchloride

(PVC) membrane (3:75:22 **8.4**/DOP/PVC wt% ratio) gave an anionic response for both 5′-GMP and 5′-AMP from a concentration of 1×10^{-3} M at pH 6.6, displaying a slight selectivity for the latter species.[12] More recently, the Umezawa group has shown that both sapphyrin **8.5** and rubyrin **8.5** can be used to produce ISEs that are selective for 3,5-dinitrobenzoate anion at or below neutral pH.[13] In all cases, the key to the observed response is that the expanded porphyrins are partially protonated under the conditions of the experiment and thus function as anion-selective lipophilic cations.

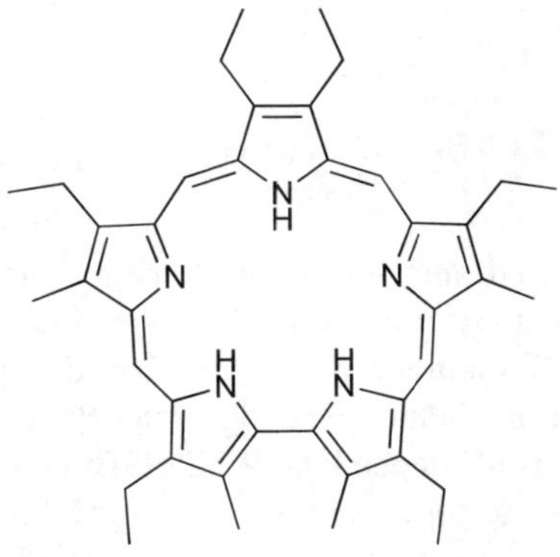

8.4

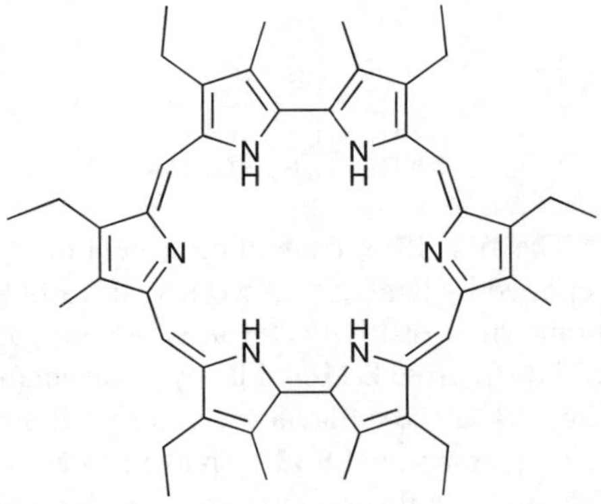

8.5

Ion-selective membranes have also been employed in CHEMFETs to give selective sensing devices. Reinhoudt and co-workers have incorporated a variety of uranyl salophane derivatives, such as **8.6a–8.6i**, into CHEMFET membranes and shown that these devices are capable of detecting a range of anions selectively (*e.g.*, fluoride may be sensed in the presence of a 150-fold excess of thiocyanate). These systems offer the further advantage that they are built on receptors with tunable binding properties, *i.e.*, whose selectivity is dependent on the nature of the lipophilic and hydrogen-bond donor/acceptor substituents present near the uranyl-binding site.[2,14]

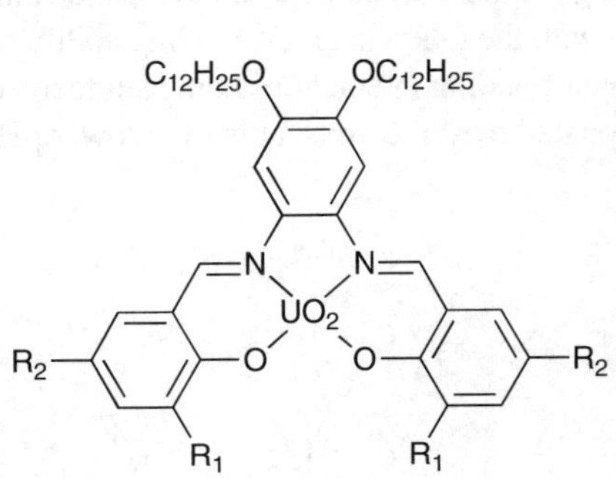

8.6a $R_1 = R_2 = H$
8.6b $R_1 = OMe, R_2 = H$
8.6c $R_1 = NHC(O)Me, R_2 = t\text{-}Bu$
8.6d $R_1 = F, R_2 = H$
8.6e $R_1 = C_6H_5, R_2 = H$
8.6f $R_1 = R_2 = t\text{-}Bu$
8.6g $R_1 = OCH_2C(O)NH(4\text{-}Me\text{-}C_6H_4), R_2 = H$
8.6h $R_1 = NHC(O)C_7H_{15}, R_2 = t\text{-}Bu$
8.6i $R_1 = NHC(O)NHC_3H_7, R_2 = t\text{-}Bu$

Radecki and co-workers[15] have employed aza-crowns **8.7–8.10** in PVC-membrane electrodes and used these systems to sense a variety of carboxylate anions. These compounds showed responses based on the shape of the carboxylate anion in question, with differing responses being observed for maleate and fumarate, as well as for isophthalate and terephthalate.

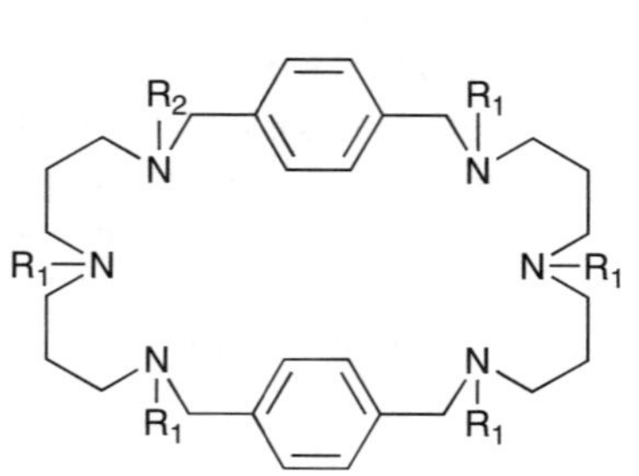

8.7 $R_1 = R_2 = C_{16}H_{33}$
8.8 $R_1 = H$, $R_2 = C_{10}H_{21}$

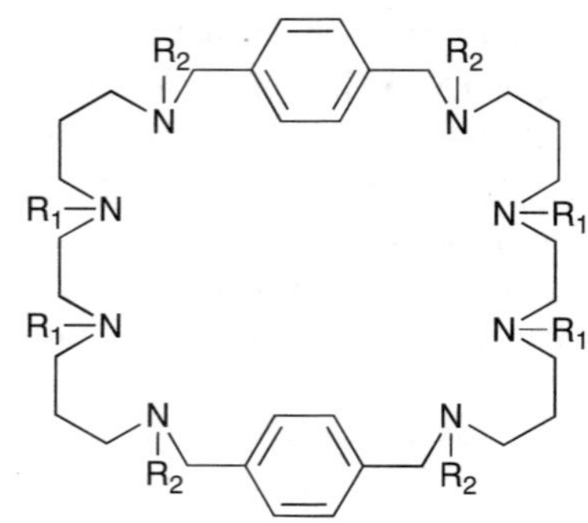

8.9 $R_1 = R_2 = C_{16}H_{33}$
8.10 $R_1 = H$, $R_2 = C_{10}H_{21}$

The first ISEs containing a neutral pyrrole-based anion-receptor element were reported by Král and co-workers in 1999.[16] In this study, it was found that the potentiometric sensitivity towards anions of PVC membrane-type ISEs based on calix[4]pyrrole **8.11** and its pyridine analogues[17] (*i.e.*, dichlorcalix[2]pyrrole[2]pyridine **8.12a** and tetrachlorocalix[4]pyridine **8.12b**) is pH dependent. While ISEs containing compound **8.11** displayed an anionic response for F^- and HPO_4^{2-} at pH 3.5, 5.5, and 9.0, this same system was found to produce a cationic response to chloride and bromide at higher pH. Presumably, such findings reflect the fact that **8.11** has a high selectivity for OH^- and a rather low affinity for Cl^- or Br^- at higher pH. Consistent with such a hypothesis, a selectivity order of $Br^- < Cl^- < OH^- \approx F^- < HPO_4^{2-}$, which represent a deviation from Hofmeister behaviour, was observed at pH 9.0. This result supports the notion that calix[4]pyrrole **8.11** acts as a pure anion receptor at lower pH, whereas it shows two different responses, reflecting both anion and counter-cation binding, at higher pH.

ISEs derived from **8.12a** and **8.12b** display selectivities that are fully reversed ($F^- < OH^- < Cl^- < Br^-$) compared to those based on **8.11** at higher pH. However, at lower pH, both systems (**8.12a** and **8.12b**) produce a strong response when exposed to fluoride and phosphate anions. Such findings are consistent with the inherent anion-binding ability of **8.12a** derived from neutral pyrrolic NH-hydrogen-bonding interactions being augmented by electrostatic interactions involving the protonated pyridine units, at least at lower pH.

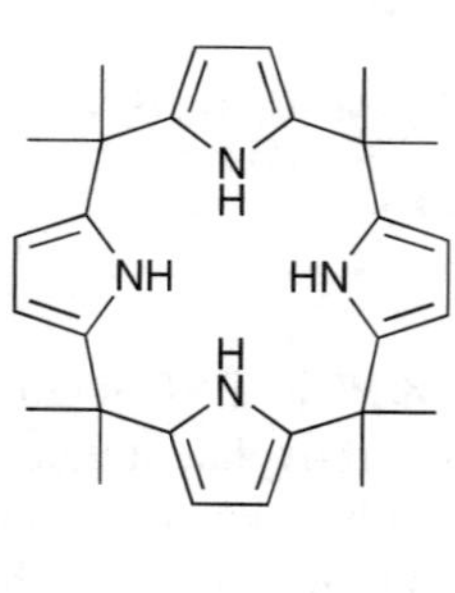

8.11

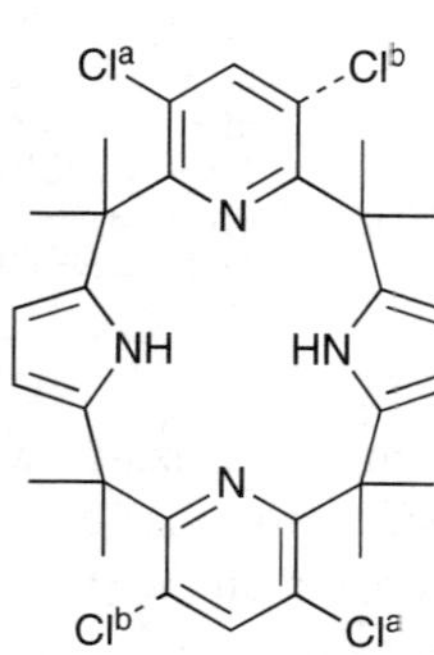

8.12a

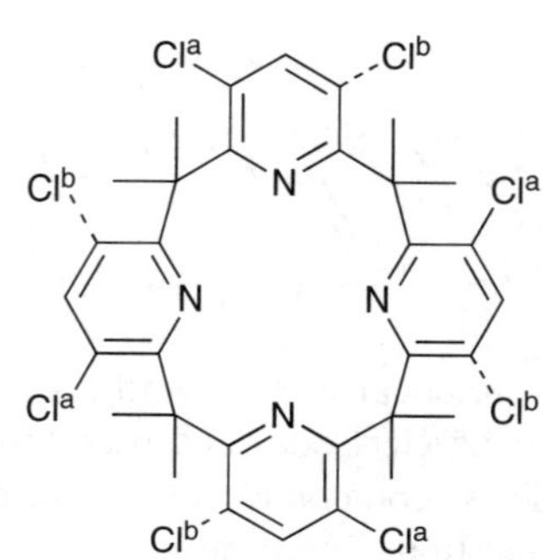

8.12b

In an effort to design nucleotide-specific ISEs, the nucleoside-functionalized calix[4]pyrroles **8.14** and **8.15** were tested as membrane elements.[18] However, prior to studying these systems, the potentiometric sensitivities and selectivities of ISEs containing the analogous unfunctionalized compounds (**8.11** and **8.13**) were tested without added tridodecylmethylammonium chloride (TDDMACl). At pH 6.6 (a pH at which the phosphate moiety exists mostly in its dianionic form) both systems (*i.e.*, those derived from **8.11** as well as those derived from **8.13**) were found to display an inherent selectivity for 5′-AMP < 5′-GMP ≈ 5′-CMP < 5′-UMP ≈ 5′-TMP, as judged from the observed anionic response. On the other hand, it was found that system **8.11** shows a slightly different selectivity pattern (5′-AMP < 5′-UMP < 5′-CMP < 5′-GMP) once the electrically charged species TDDMACl is added to the PVC membranes (Table 8.1). The fact that this additive has such an effect reflects the need to neutralize the negative charge present in the calix[4]pyrrole–nucleotide complex.

In the case of the more hydrophobic cyclohexyl-substituted calix[4]pyrrole **8.13**, a selectivity pattern of 5′-AMP (no carbonyl) < 5′-CMP (one carbonyl) < 5′-GMP < 5′-TMP (two carbonyl) ≈5′-UMP was observed using a membrane made up with TDDMACl. This result led to the suggestion that ISE-response selectivity is dominated by interactions between the nucleobase and the calix[4]pyrrole, even when the substrate is a nucleotide containing a negatively charged phosphate subunit (that could be a source for an independent anionic response).

Functionalized calix[4]pyrroles **8.14** and **8.15** were found display 5′-CMP and 5′-GMP selectivities in model membrane transport experiments involving an initial aqueous phase (Aq I) held at pH 6.0 and a receiving phase of pH 12.5 (Aq II). From these and related studies, it was surmised that the pH of the aqueous phases could play an important role in defining the transport selectivity. Therefore, for ISEs made up from these receptors, it was considered important to measure the potentiometric response at both low and high pH. Table 8.1 summarizes the potentiometric selectivity of PVC membranes that are derived from **8.14** and **8.15** and which also contain TDDMACl. In the case of system **8.14**, a selectivity order of 5′-AMP < 5′-GMP < 5′-CMP was observed at both pH 6.6 and 8.5. On the other hand, the selectivity sequence for system **8.15** was found to be a more sensitive function of pH (*viz.* 5′-CMP < 5′-AMP < 5′-GMP at pH 6.6; 5′-GMP < 5′-AMP < 5′-CMP at pH 8.5).[19] The high selectivity for 5′-GMP observed at lower pH supports the notion that ditopic interactions play a key role in mediating the receptor-nucleotide binding events at or below pH 6.6. On the other hand, at high pH, the nucleobase is expected to exist as a

Table 8.1 *Potentiometric selectivity of PVC membranes containing unfunctionalized calix[4]pyrrole **8.11*** (log $K^{sel}_{5'\text{-AMP}/5'\text{-XMP}}$)*, the cytosine-functionalized calix[4]pyrroles **8.14*** (log $K^{sel}_{5'\text{-UMP}/5'\text{-XMP}}$)*, and analogue **8.15*** (log $K^{sel}_{5'\text{-CMP}/5'\text{-XMP}}$)*. These membranes contained TDDMACl*

	At pH 6.6			*At pH 8.5*		
	5′-AMP	*5′-CMP*	*5′-GMP*	*5′-AMP*	*5′-CMP*	*5′-GMP*
8.11	0.00	+0.24	+0.76	–	–	–
8.14	−0.07	+0.97	+0.89	−0.20	+0.58	+0.10
8.15	+1.53	0.00	+2.23	−0.68	0.00	−1.23

dianion, making the selectivity advantage imparted by the Watson–Crick base-pairing interactions less significant. While not yet fully established, such rationales can help explain the low potentiometic selectivity for 5′-GMP that is observed at pH 8.5.

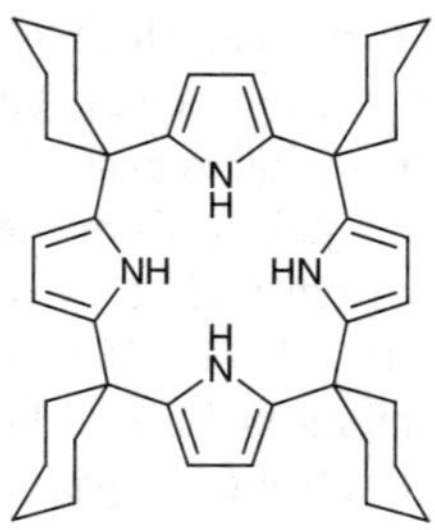

8.13

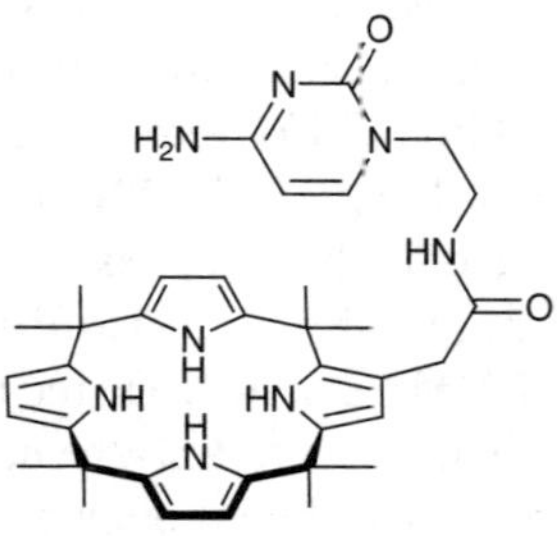

8.14

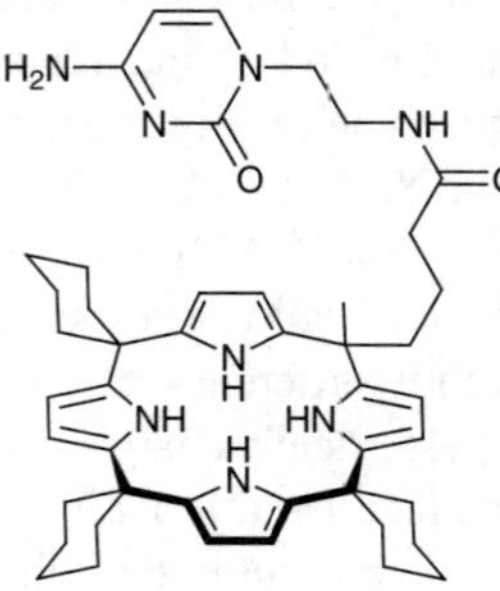

8.15

Optodes are similar to ISEs in that they employ a guest-selective membrane. However, unlike ISEs, the change that is detected is not in electrochemical potential but rather in the optical properties of the membrane. Metalloporphyrins or corrins have been used in a number of anion-selective electrodes and optodes.[20,21] For example, a number of nitrite selective optodes containing a vitamin B_{12} derivative (aquo-cyano-cobyrinic acid-heptakis-(2-phenyl ethyl ester))[22] (nitrite ionophore I) are known. Likewise, several chloride selective optodes have been made based on indium(III)tetraethylporphyrin chloride.[21,23] Kopelman and co-workers[24] have employed both these systems as nitrite- and chloride-selective nano-optodes and applied them to the *in vitro* analysis of bio-samples.

Bachas and co-workers[25] have made chloride selective optodes that contain Hawthorne's mercuracarborand chloride receptors. A PVC optode membrane was prepared by combining the chloride-selective [9]mercuracarborand-3 macrocycle **8.16** with a proton-sensitive molecule, proton chromoionophore VI (**8.17**), together with a plasticizer and TDDMACl. When chloride is extracted into the membrane and bound by **8.16**, protons are also extracted to maintain charge neutrality. This leads to the protonation of compound **8.17** and produces an optical response. This optode is particularly selective for chloride, and does not suffer from interference from salicylate, thiocyanate, bromide, or nitrate.

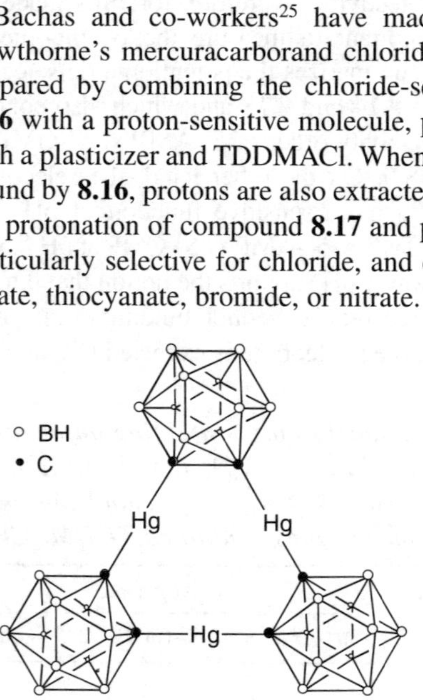

8.16

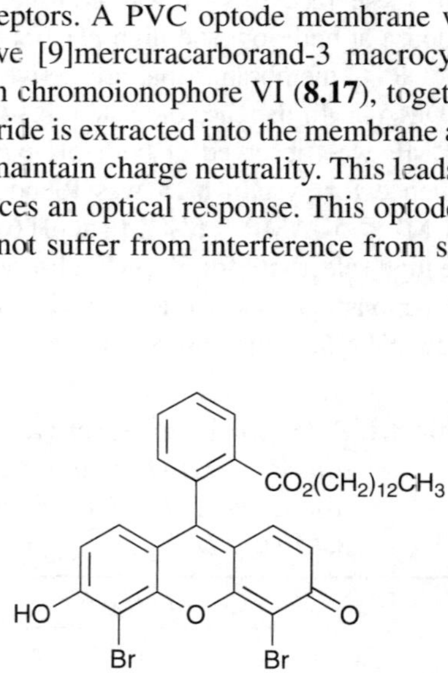

8.17

8.3 Discrete Molecular Electrochemical-Anion Sensors

Molecular electrochemical sensors consist of a redox-active subunit linked to a guest-binding site. When a guest binds to the receptor, the electrochemical properties of the redox-active entity (or "reporter" group) change and may be detected using cyclic voltammetric or other electrochemical techniques (Scheme 8.1). It is essential at the design stage of this type of sensor to ensure that the binding site and reporter group are arranged such that the binding event perturbs the redox properties of the receptor.

Such systems can be described by the scheme of one square as shown below (Scheme 8.2).

As noted in the caption to this scheme, H, G, and HG in normal or subscript positions represent the host, guest, and complex species, respectively; the subscripts "ox" and "red" indicate that the corresponding molecules or parameters are in their oxidized or reduced states; E° is the formal potential of the electron-transfer reaction and K is the stability constant. From this scheme it is possible to derive the following expression:

$$(E^\circ{}_{\mathrm{H}} - E^\circ{}_{\mathrm{HG}})nF/RT = \ln(K_{\mathrm{ox}}/K_{\mathrm{red}}) \qquad (8.1)$$

This equation (Equation (8.1)) links the stability constants K_{ox} and K_{red} of a complex in two different oxidation states with the experimentally measurable redox potentials

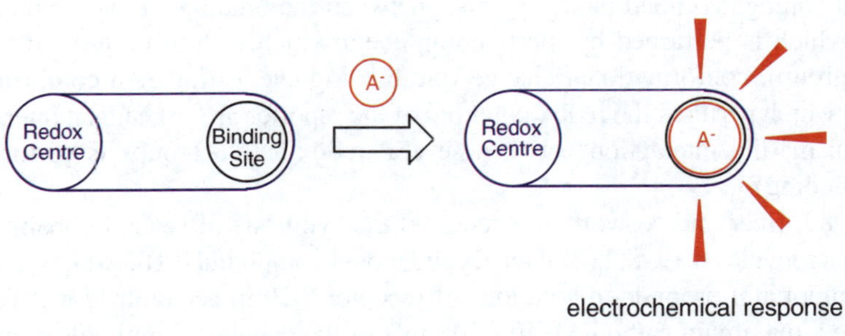

Scheme 8.1 *A guest-binding event triggers an electrochemical response in a redox-active receptor*

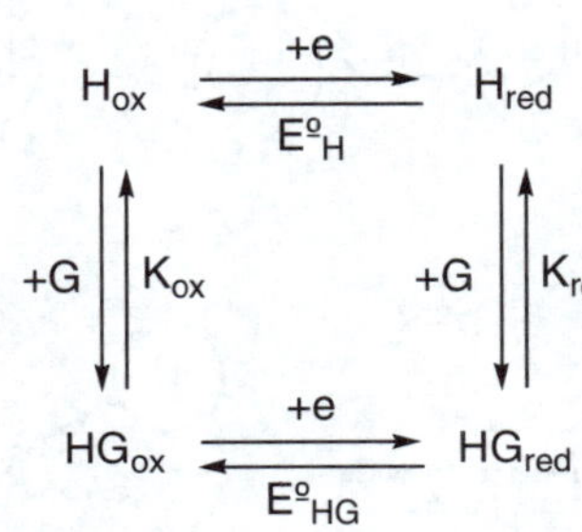

Scheme 8.2 *Electrochemical cycle which defines the intrinsic behaviour of redox-active-anion sensors. This cycle, the terms of which are defined in the text, is often referred to as the "scheme of one square". Here, H, host; G, guest; and HG, complex*

E_H and E_{HG}. Therefore, it provides an easy means of obtaining the ratio of K_{ox}/K_{red}, which is a useful parameter known as the binding enhancement factor (BEF) or reaction-coupling efficiency (RCE). This expression also allows the calculation of K_{ox} if K_{red} is known or *vice versa*.

Receptors designed to recognize guest molecules electrochemically must couple the complexation process to the redox reaction, *i.e.*, the two reactions must mutually influence each other. Addition of an electron (reduction) to, or its withdrawal (oxidation) from, a host molecule will change the stability constant of the complex formed, leading to a change in the ratio of K_{ox}/K_{red}. Equation (8.1) predicts that this change in stability constant will cause a change in the redox potential of the host. The magnitude and the direction of the potential change will depend primarily on the reaction-coupling mechanism and the properties of the complexed guest molecule. The expected variations in the redox properties of the system as a whole can be measured, for example, by voltammetric means.

Beer and co-workers have been pioneers in this area of supramolecular chemistry. They have used ferrocene, cobaltocenium, ruthenium bipyridyl, and other redox-active groups to produce anion receptors that display an electrochemical response. Further, these workers have identified five mechanisms by which the binding of a guest species may perturb a redox-active centre in a redox-active receptor (Figure 8.2). These include through space (the binding site is very close to the redox-active group), direct coordination (the guest binds directly to the redox centre), through bond (a conjugated bond pathway exists between the binding site and redox-active group which is perturbed by the bonding event which in turn perturbs the redox-active group), conformational change (the binding event triggers a conformational change which perturbs the redox potential of the reporter group), and an interference mechanism (the interaction between several redox-active groups is perturbed by guest binding).

In 1989, Beer and co-workers[26] reported the synthesis of several cobaltocenium ester macrocycles **8.19–8.22** and an acyclic model compound **8.18**. Addition of tetrabutylammonium bromide to solutions of receptor **8.20** in acetonitrile was found to produce a maximum cathodic shift of 45 mV of the cobaltocenium redox couple.

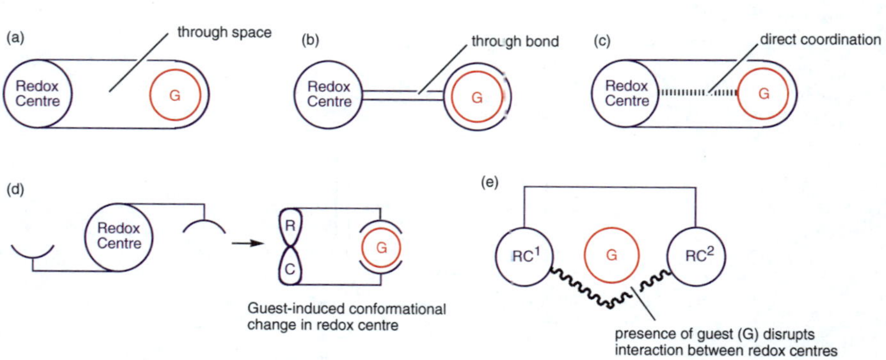

Figure 8.2 *Mechanisms identified by Beer and co-workers*

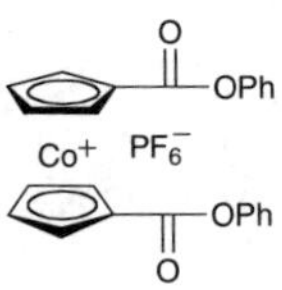

8.18

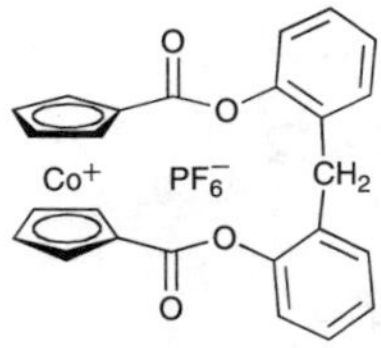

8.19

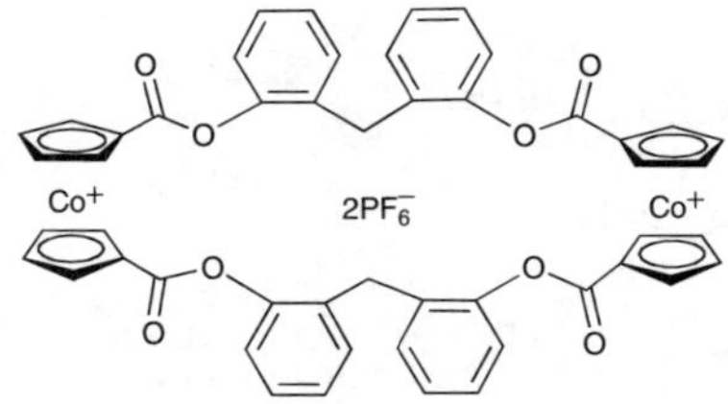

8.20

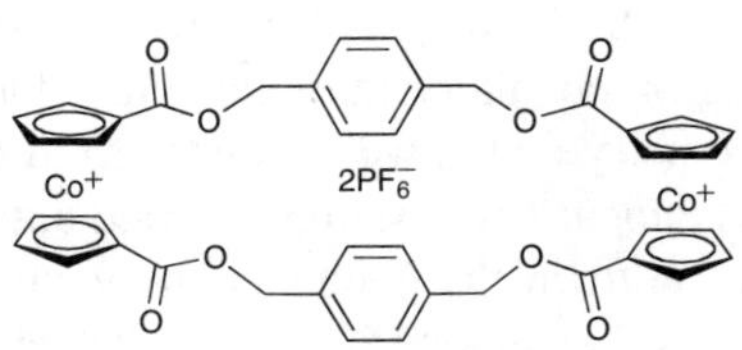

8.21

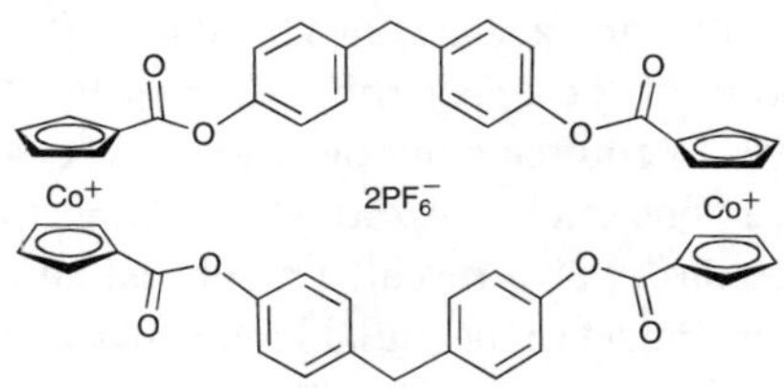

8.22

This discovery inspired the Beer[27] group to develop more stable redox-responsive anion receptors containing amide-linked cobaltocenium subunits. These second-generation systems combined electrostatic interactions with hydrogen bonding to produce a variety of receptors. In the case of receptor **8.23**, a downfield shift in the NH peak was observed upon the addition of halide anions (studied as their corresponding tetrabutylammonium salts) in DMSO-d_6. A significant cathodic shift in the reduction potential was also observed in the presence of anions, as judged by cyclic voltammetry.

Later on, the Beer group reported the synthesis of several bis-cobaltocenium diamide-bridged receptors (*i.e.*, **8.24**).[28] As part of this work, the anion-binding affinities were determined in CD_3CN using standard ^{1}H NMR spectroscopic titration methods. On this basis, it was concluded that receptor **8.24a**, linked by an ethylenediamine-derived spacer, exhibits a preference for chloride anion over bromide and iodide anions (K_a = 2500, 330, and 450 M^{-1} for chloride, bromide, and iodide, respectively). Increasing the length of the linker (*cf.* **8.24b** and **8.24c**) leads to a dramatic reduction in both the anion-binding affinity and selectivity. For instance, chloride and bromide are bound to **8.24b** with affinities of 1300 and 270 M^{-1}, respectively, *vs*. 280 and 260 M^{-1} for **8.24c**. As found for system **8.23**, substantial cathodic shifts in the cobaltocenium/cobaltocene redox couple were observed in the presence of chloride and bromide anions ($\Delta E(Cl^-)$ = 60, 45, and 40 mV for **8.24a**, **8.24b**, and **8.24c**, respectively, in CH_3CN).

The electrochemical properties of the triscobaltocenium receptor **8.25** were also studied in organic solvents.[27,29] In the case of fluoride anion, a cathodic shift (ΔE = 60 mV in DMSO) was observed upon the addition of the corresponding tetrabutylammonium salt. In accord with such findings, downfield shifts for the amide protons were seen in the ^{1}H NMR spectrum of **8.25** upon the addition of chloride, bromide, and nitrate anions.

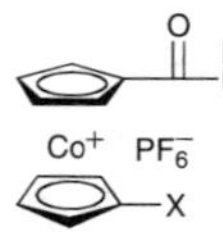

8.23a R = n-Bu, X = H
8.23b R = Ph, X = CONHPh

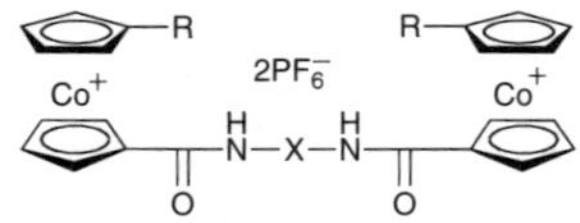

8.24a R = $CO_2C_2H_5$, X = $-CH_2CH_2-$
8.24b R = $CO_2C_2H_5$, X = $-CH_2CH_2CH_2-$
8.24c R = $CO_2C_2H_5$, X = $-CH_2CH_2CH_2CH_2-$

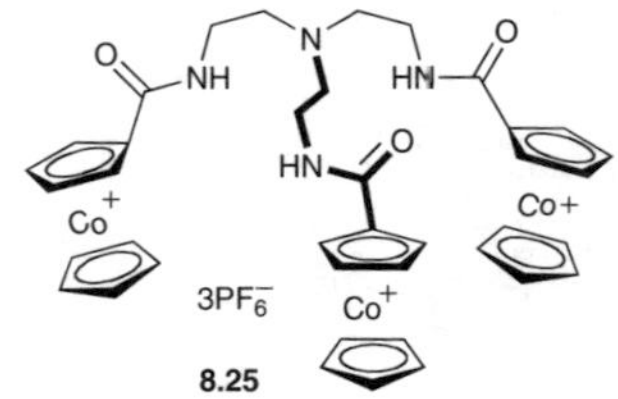

8.25

In 1997, these workers reported a new upper-rim cobaltocenium diamide-bridged calix[4]arene, **8.26**.[30] This receptor exhibits a high affinity for carboxylate anions, in particular for acetate, in DMSO-d_6. Presumably, this binding affinity reflects the rigidity of the upper rim, enforced by the bridging cobaltocenium subunit, which serves to enforce a ditopic "binding pocket" geometry that is well suited for bidentate anions, such as carboxylate anions. Electrochemical-anion studies, carried out in acetonitrile, also revealed a substantial cathodic shift in the cobaltocenium/cobaltocene reduction potential in the presence of carboxylate anions. These results led to the suggestion that receptor **8.26** is a prototype of a generalized class of amperometic sensing agents that are selective for carboxylate anions.

First reported in 1999, the bis-cobaltocenium systems **8.27a** and **8.27b** are elaborated analogues of **8.26**.[31] These two isomeric receptors display different anion selectivities and stability constants. As a general rule, receptor **8.27b** was found to give rise to lower binding constants than **8.27a**. Other differences were also observed. For instance, while receptor **8.27b** was found to exhibit a preference for dihydrogen phosphate in DMSO-d_6, receptor **8.25a** was found to display a high selectivity towards acetate anion. Molecular modelling calculations and analysis of CPK models indicated that the distance between the amide groups is closer in **8.27a** than in **8.27b**, a result that is ascribed to the presence of the two lower-rim bulky tosyl groups para to that of the upper-rim amide group. Independent of this rationalization, these results are important because they provide important experimental support for the notion that the inherent selectivities of receptors can be varied by making appropriate geometric changes. Reports of further examples of cobaltocenium receptors are listed in the reference section.[32]

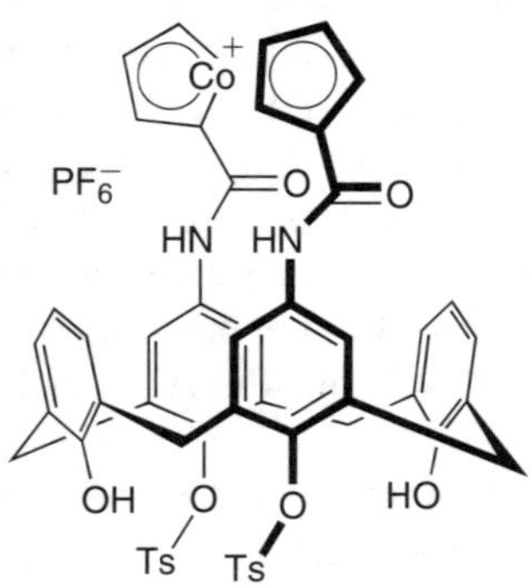

8.26

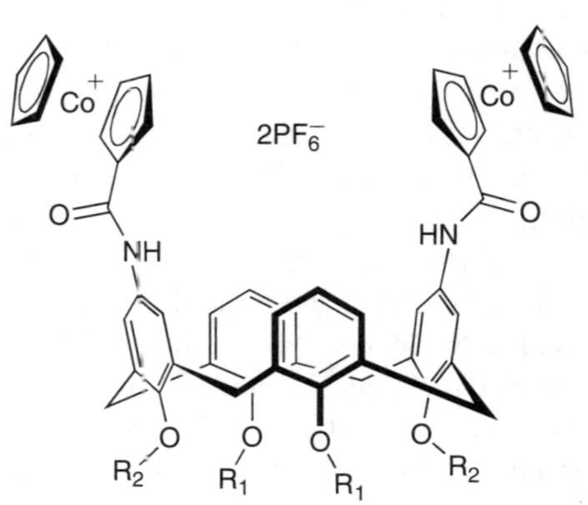

8.27a R_1 = H, R_2 = Ts
8.27b R_1 = Ts, R_2 = H

In addition to their work on cobaltocenium-containing anion receptors, the Beer group has played a pioneering role in developing ferrocene-based systems that can sense anions in both organic and aqueous media through electrochemical means.[33] As part of this effort, a family of compounds containing ferrocene units appended to secondary amides were prepared. In contrast to the cobaltocenium-containing receptors described above, these receptors have no inherent electrostatic attraction for anions. This makes their inherent affinity for anions lower than the analogous positively charged systems. However, oxidation of the ferrocene group to ferrocenium can "switch-on" electrostatic interactions and consequently these molecules function as redox sensors for anions. For example, compounds **8.28–8.32** were found capable of sensing $H_2PO_4^-$ anions in acetonitrile solution via the large cathodic shifts (up to 240 mV) produced in the presence of a 10-fold excess of either HSO_4^- or Cl^- ions. In contradistinction to these results, receptor **8.28** was found to bind HSO_4^- selectively in the presence of $H_2PO_4^-$. The basic amine functionality of **8.28** is protonated by the hydrogen sulfate anion and the resulting positively charged receptor strongly binds the SO_4^{2-} anion produced as the result of this proton transfer.

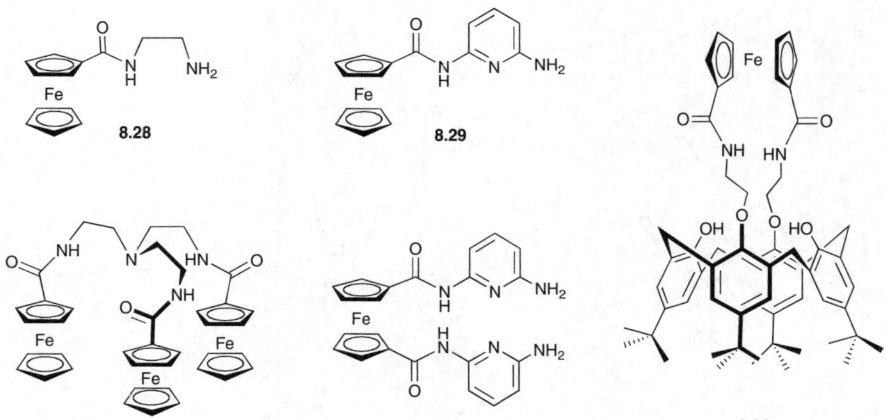

8.28 **8.29** **8.30** **8.31** **8.32**

Astruc and co-workers have produced dendrimers containing 3, 9, and 18 ferrocene groups (systems **8.33**, **8.34**, and **8.35**, respectively). Electrochemical studies provided evidence for a dendritic effect playing a role in anion recognition.[34] The largest induced cathodic shifts were observed upon the addition of $H_2PO_4^-$, with the perturbation caused by this and other anions increasing as the size of the dendrimer increased. For instance, in the presence of excess $H_2PO_4^-$, the ferrocene/ferrocenium redox couple was seen to shift by 110, 220, and 315 mV for compounds **8.33**, **8.34**, and **8.35**, respectively.

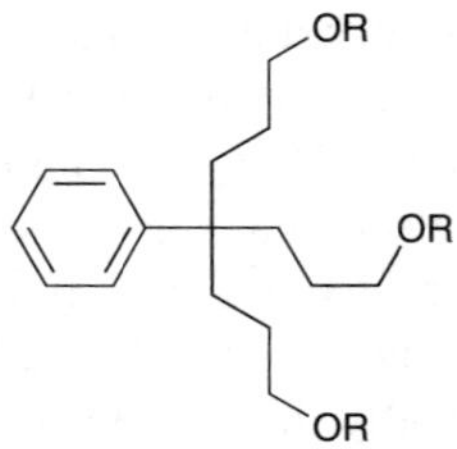

8.33

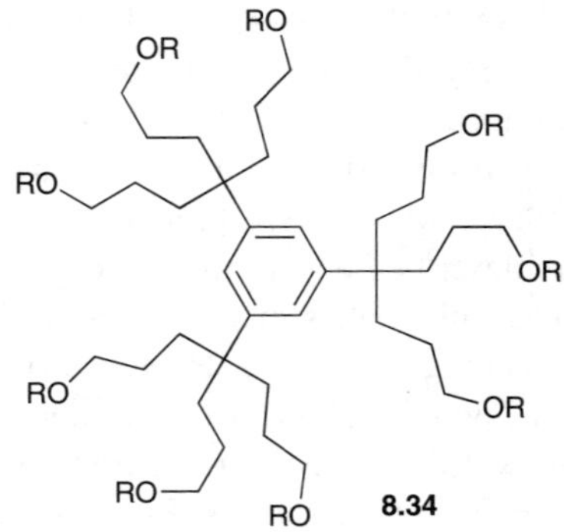

8.34

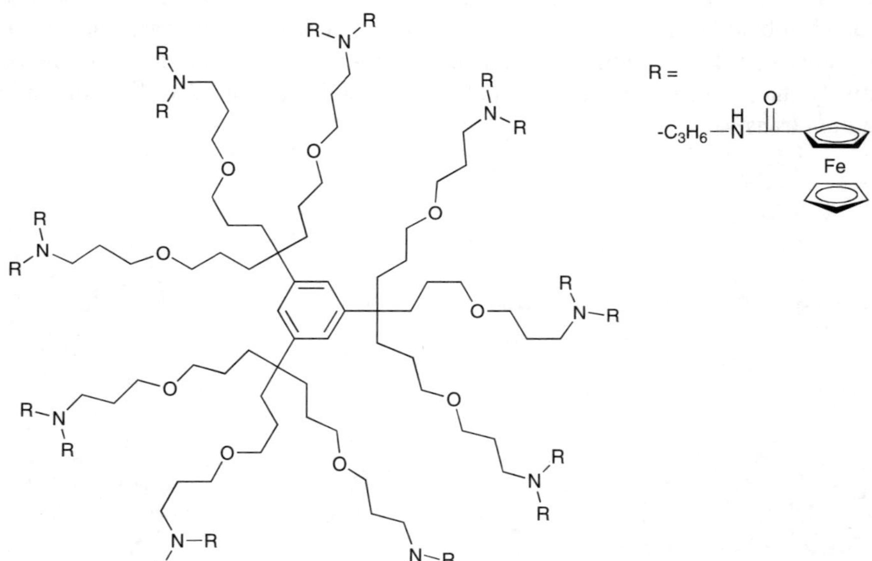

8.35

Beer and co-workers have also synthesized a series of acyclic and macrocyclic ferrocene amine ligands, *e.g.*, **8.36** and **8.37**, that can selectively bind and electrochemically detect phosphate and sulfate (in various states of protonation) and nucleotide anions in aqueous solution.[35] For example, compound **8.36** under neutral conditions senses phosphate via a cathodic shift of 50 mV whereas sulfate does not trigger an electrochemical response. However, for receptor **8.37** in aqueous THF solution at pH 4 the reverse is true; this latter system serves to discriminate sulfate over phosphate electrochemically via the production of a cathodic shift of 54 mV. Calibration curves of the change in the half-wave potential ΔE *vs.* $[A^-]/[L]$ ratio constructed at well-defined pH values for **8.36** and **8.37** allowed the concentrations of phosphate and sulfate to be determined quantitatively in the presence of competing anions.

Ferrocene-containing receptors bearing fully charged positive centres have also been prepared as electrochemical-anion sensors. For example, the ferrocene appended guanidinium receptor **8.38** selectively senses pyrophosphate in methanol/water mixtures, exhibiting limiting cathodic shifts of 70 mV in the presence of this anion.[36]

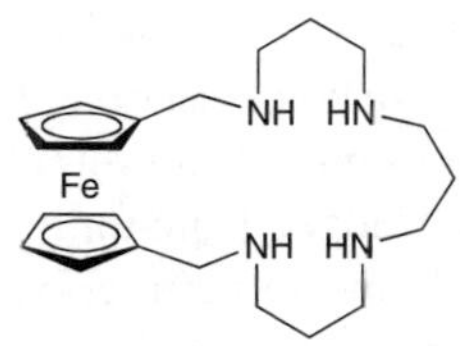

8.36

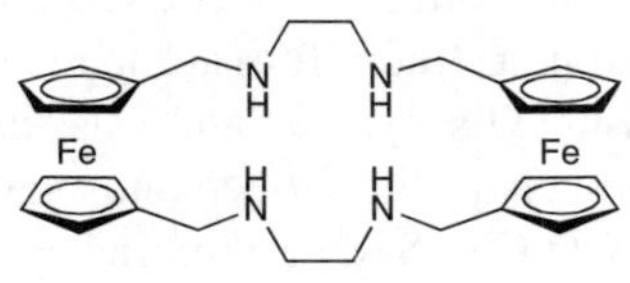

8.37

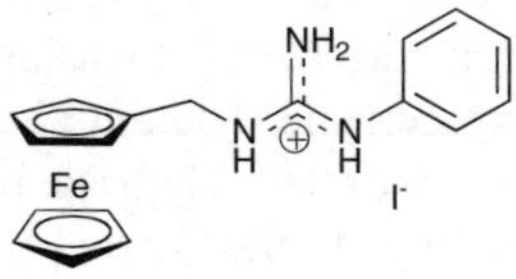

8.38

Shinkai and co-workers[37] have shown that very simple ferrocene species can function as selective sensors. Specifically, this group has shown that ferroceneboronic acid **8.39** is capable of acting as a redox sensor for fluoride. This receptor has excellent selectivity for fluoride in the presence of other halides and anions such as SCN^-, SO_4^{2-}, and $H_2PO_4^-$. A K_{ox} value of 1000 M^{-1} in MeOH–H_2O (1:9) for fluoride, compared to values of less than 2 M^{-1} for chloride and bromide, was found. The interaction between the boronic acid group and fluoride is attributed to the hardness of the boron atom, which strongly interacts with fluoride (a hard base). Upon oxidation, the ferrocene group becomes more electron withdrawing. This decreases the electron density of the boron atom and increases the strength of the fluoride-anion complex.

Jurkschat and co-workers[38] have incorporated ferrocene groups into silicon- and tin-containing macrocycles forming new electrochemical-anion sensors such as, *e.g.*, **8.40**. In CH_2Cl_2, receptor **8.40** displays anion-dependent cathodic shifts in the presence of certain anions (*e.g.*, Cl^- $\Delta E = 130$ mV, F^- $\Delta E = 210$ mV; $H_2PO_4^-$ $\Delta E = 480$ mV). Here, in analogy to other metal- and metalloid-based anion receptors (*cf.* Chapter 7), it is thought that the presence of the electron-deficient centres abets the anion-recognition process.

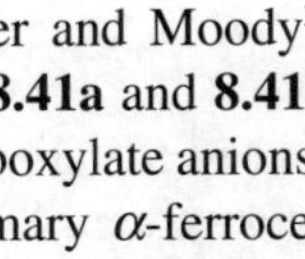

8.39

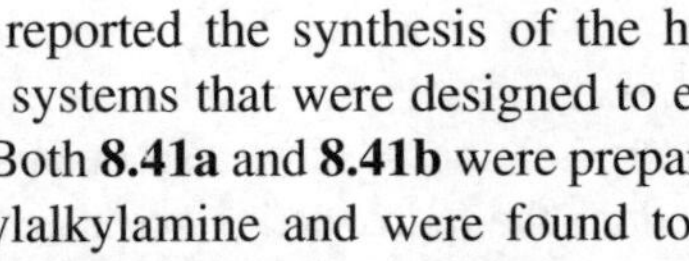

8.40

In 2002 Tucker and Moody[39] reported the synthesis of the homochiral redox-active receptors **8.41a** and **8.41b**, systems that were designed to effect the recognition of chiral carboxylate anions. Both **8.41a** and **8.41b** were prepared from the same non-racemic primary α-ferrocenylalkylamine and were found to display a strong affinity for the 2-phenylbutyrate anion. In the case of **8.41a**, the association constant for racemic 2-phenylbutyrate was calculated to be 2080 M^{-1} in CD_3CN, as judged from standard ^1H NMR spectroscopic titrations. On the other hand, receptor **8.41b** was found to bind this substrate too strongly in this solvent to allow an accurate binding constant to be determined by NMR methods. Therefore, quantitative assessments of the association constants were made via UV–Vis spectroscopic titrations carried out in DMSO. It was found that **8.41b** binds the (*S*) and (*R*) antipodes of 2-phenylbutyrate with K_a values of 2350 and 2910 M^{-1}, respectively, in this solvent at 293 K. While not a particularly large difference in absolute terms, this degree of differentiation was noteworthy in that it showed the potential power of this approach.

One year later, the bis-urea substituted ferrocene receptor **8.42** was reported by Pratt and Beer.[40] It was prepared in a straightforward manner by the condensation of ferrocenemethyldiamine and *n*-hexylisocyanate and was shown to possess a preference for relatively basic anions such as dihydrogen phosphate, as judged from initial binding studies carried out in CD_3CN. Such a preference was found to be in good agreement with the results of an electrochemical study that revealed a significant cathodic shift (ΔE = 150 mV; CH_3CN/0.1 M $TBABF_4$) in the presence of dihydrogen phosphate.

Most recently, a new aza-ferrocenophane receptor, **8.43**, was synthesized by the Molina group. It was designed to include both electrochemical (ferrocene unit) and fluorescent chemosensor (naphthalene unit) subunits and show selectivity towards fluoride anion.[41] The results from quantitative binding studies, carried out in DMSO-d_6 using standard ^{1}H NMR spectroscopic titration methods, revealed that fluoride ion formed a considerably stronger complex (K_a>10^4 for a 1:2 host/guest binding stoichiometry), than dihydrogen phosphate (K_a = 405 M^{-1} for a 1:1 complex). In addition, no appreciable interaction was observed with chloride, bromide, and hydrogen sulfate anions. These results match the results from cyclic voltammetric studies. In the presence of two equivalents of fluoride and dihydrogen-phosphate anions, a clear cathodic shift was observed for both anions ($\Delta E_{1/2}$ = 190 and 125 mV for F^- and $H_2PO_4^-$, respectively, in DMSO/0.1 M $TBAPF_6$), with the BEFs being calculated as 1628 (F^-) and 130 ($H_2PO_4^-$). Consistent with these results was the finding that the addition of fluoride and dihydrogen phosphate induced a strong fluorescence enhancement (up to 1186% and 172%, respectively) in DMF. Such a "switching on" of fluorescence represents an easy-to-detect response that is considered highly favourable in the context of substrate-specific sensor development.

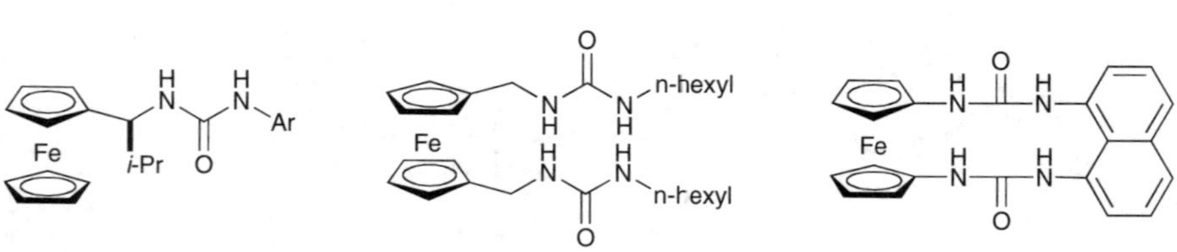

8.41a Ar = Ph
8.41b Ar = 4-NO_2-Ph

8.42

8.43

Ferrocene "read out" elements have also been attached to calixpyrroles by Gale and Sessler[42] to produce the redox-active anion-receptor systems **8.44** and **8.45**. The solution-phase anion-association constants of **8.44** and **8.45** were determined in CD_2Cl_2 using standard ^{1}H NMR titration methods and are summarized in Table 8.2. In general, these systems were found to display affinities and selectivities (*i.e.*, F^- > Cl^- > $H_2PO_4^-$) analogous to those seen for other alkylated calix[4]pyrroles in dichloromethane. The electrochemical properties of **8.44** and **8.45** were determined using cyclic voltammetry (CV). For instance, reversible ferrocene/ferrocenium

Table 8.2 *Affinity constants corresponding to the interaction of anionic substrates with **8.44** and **8.45** with anionic substrates and electrochemical parameters obtained from CV measurements*

		No anion	F^-	Cl^-	Br^-	$H_2PO_4^-$	HSO_4^-
8.44[a]	K_a (M^{-1})	N/A	1496	444	–	40	–
	$E_{1/2}$ (mV)[b]	503	566	481	–	534	–
	ΔE (mV)	N/A	63	−22	–	31	–
8.45a[c]	K_a (M^{-1})	N/A	3375	3190	50	304	–
	$E_{1/2}$ (mV)[d]	444	368	408	432	350	436
	ΔE (mV)	N/A	−76	−36	−12	−100	<10
8.45b[a]	K_a (M^{-1})	N/A	–	202	–	40	–
	$E_{1/2}$ (mV)[b]	511	525	718	–	502	–
	ΔE (mV)	N/A	14	207	–	−9	–

[a]Measured in dichloromethane (deuterated solvent for K_a determinations).
[b]Determined in dichloromethane containing 0.1 M n-Bu_4NPF_6 as the supporting electrolyte. The potentials were determined with reference to Ag/AgCl.
[c]Measured in CH_3CN/DMSO 9:1 (deuterated solvents for K_a determinations).
[d]Values obtained from square-wave voltammograms.

waves were observed at $E_{1/2}$ = 503 and 511 mV (*vs.* Ag/AgCl) for **8.44** and **8.45a**, respectively. Upon the addition of excess dihydrogen phosphate and fluoride ion to receptor **8.44**, cathodic shifts of 31 and 63 mV, respectively, were observed in the Fc/Fc^+ couple. However, addition of chloride anion was found to engender an anodic shift of –22 mV. Further complicating a straightforward analysis was the finding that cathodic shifts of 14 and 207 mV, respectively, were produced upon the addition of fluoride and chloride anions to receptor **8.45b**, while the addition of excess dihydrogen phosphate was found to lead to an anodic shift of −9 mV.[42] By contrast, the Fc/Fc^+ couple of receptor **8.45a** underwent an anodic shift after the addition of fluoride, chloride, bromide, and dihydrogen-phosphate anions.[43] Thus, while receptors **8.44** and **8.45** serve to demonstrate that metallocene elements may be combined with calixpyrroles to produce redox-active sensors, these first-generation systems fall short of being fully functional in this regard.

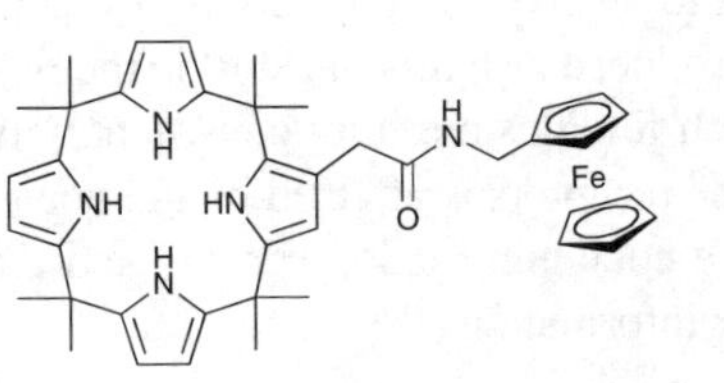

8.44

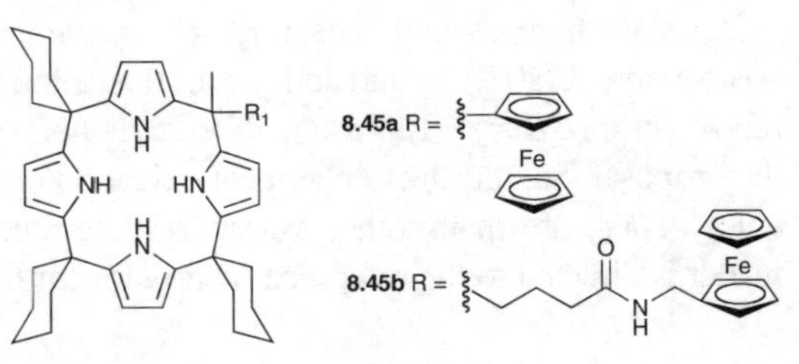

The *ansa*-ferrocenes **8.46**, containing two pyrrolic NH and two amidic NH groups, represent a different class of ferrocene–pyrrole conjugates produced by the Sessler group.[44] In this case, quantitative ^1H NMR spectroscopic titration studies revealed that receptor **8.46a** binds fluoride, chloride, and dihydrogen phosphate with high affinity in CD_3CN (*cf.* $K_a \geq 10^5$, 9030, and 11,300 M^{-1} for F^-, Cl^-, and $H_2PO_4^-$, respectively). These results were found to be consistent with electrochemistry studies, carried out concurrently, and led to the suggestion that this receptor acts as an effective redox-based sensor for F^- (ΔE = 80 mV) and $H_2PO_4^-$ (ΔE = 136 mV).[44] In working with this series of receptors, it was observed that the binding affinities for $H_2PO_4^-$ increased in a stepwise fashion as the number of oxygen atoms in the linking bridge was increased from 0 to 2 (*cf.* K_a = 4050, 13,200, and 81,400 M^{-1}, respectively, for **8.46b**, **8.46c**, and **8.46d** in CD_2Cl_2 containing 2% DMSO-d_6). Such an observation is consistent with the ether-type oxygen atoms being involved directly in the anion-binding process. On the other hand, the degree of the anion-induced cathodic shift in the Fc/Fc$^+$ potentials was not found to correlate fully with the number of oxygen atoms in the bridging tether (*cf.* ΔE = 128, 140, and 140 mV for the binding of dihydrogen phosphate to **8.46b**, **8.46c**, and **8.46d**, respectively).[45] As a result, the interplay between structure and electrochemical sensing function in these systems remains less than fully understood.

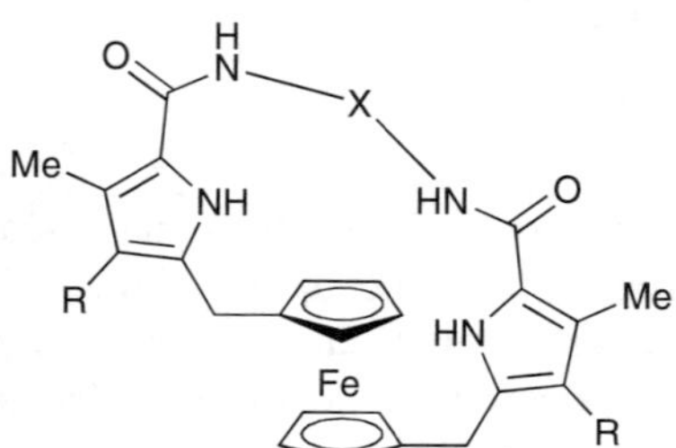

8.46a X = $CH_2(CH_2OCH_2CH_2OCH_2)CH_2$, R = C_2H_5
8.46b X = $CH_2(CH_2)_3CH_2$, R = CH_3
8.46c X = $CH_2(CH_2OCH_2)CH_2$, R = CH_3
8.46d X = $CH_2(CH_2OCH_2)_2CH_2$, R = CH_3

Quite recently, Gale and co-workers[46] reported two new ferrocene-appended amidopyrroles that show considerable promise as electrochemical-anion sensors. The electrochemical behaviour of receptors, **8.47**, in the presence and absence of a variety of anions were determined by CV methods in dichloromethane using a platinum microdisc electrode. Upon the addition of fluoride anion to **8.47b**, a large anodic shift in the Fc/Fc$^+$ couple was seen (*cf.* ΔE = −130 mV). In the case of **8.47a**, addition of fluoride anion was found to yield two waves while producing significant anodic shifts (ΔE = −125 and −255 mV). Interestingly, receptor **8.47a** was found to display a strong affinity for benzoate anion (1820 M^{-1}) and addition of this anion produced a significant shift in the first redox wave (ΔE = −120 mV). Taken together, such findings are fully consistent with the proposal that the electrochemical oxidation of the receptors is affected by the bound anion. There are many other examples of ferrocene-containing redox sensors and the reader is directed to the reference section for further information.[47,48]

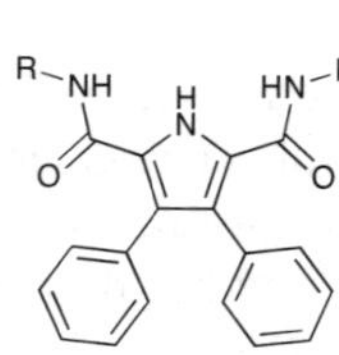

8.47a R =

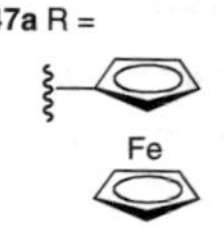

8.47b R =

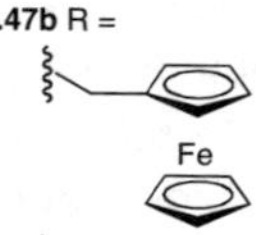

Another strategy that has attracted attention as part of an effort to prepare anion sensors has involved the synthesis of systems containing ruthenium bipyridyl moieties, such as **8.48a–8.48d**.[49,50] The ability of these receptors, prepared by Beer and Maestri groups, to sense chloride, acetate, and dihydrogen-phosphate anions was tested via electrochemical, optical, and spectroscopic means. The X-ray crystal structure of **8.48a**–2AcO$^-$ was also determined (*cf.* Figure 8.3).[50] It shows that the receptor contains a 26-membered macrocyclic core and that two acetate anions are associated with it. These latter are held in place via two sets of multiple hydrogen-bonding interactions, wherein each oxygen atom forms three hydrogen bonds, one to an amide NH group and two to an aromatic CH moiety. The results of ^{1}H NMR spectroscopic titrations, carried out in DMSO-d_6, led the authors to suggest that the anion selectivities of receptors of general structure **8.48** is determined by the cavity size and shape. For example, receptors **8.48a**, **8.48b**, and **8.48d**, possessing a relatively large core, were found to be selective for $H_2PO_4^-$ over Cl$^-$. By contrast, receptor **8.48c**, containing a smaller core, displays little if any affinity for dihydrogen phosphate anion, yet binds chloride anion with an association constant of 40,000 M^{-1} in this solvent (DMSO-d_6). It was also found that receptor **8.48a**, with its built-in phenyl spacers, binds two acetate anions in DMSO-d_6 solution (*i.e.*, forming a 2:1 anion:receptor complex), which is in good agreement with the results of the solid-state X-ray diffraction study. Electrochemical studies of **8.48c**, carried out in acetonitrile, in the presence and absence of added anions reveal a similar anion selectivity trend, with the observed $\Delta E_{1/2}$ being 110, 15, and < 5 mV for Cl$^-$, AcO$^-$, and $H_2PO_4^-$, respectively. Other examples of this type of receptor are listed in the reference section.[48,51]

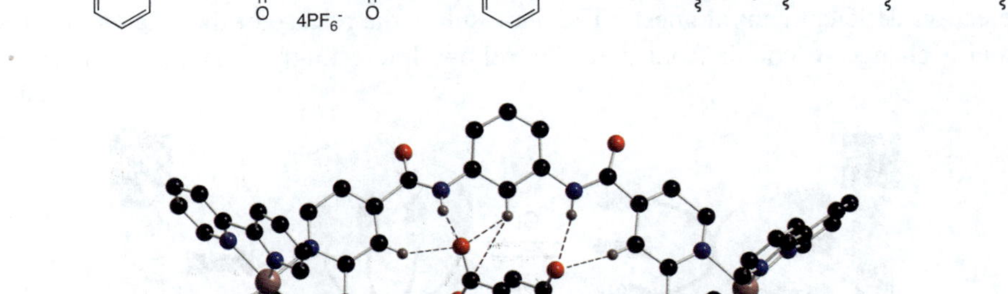

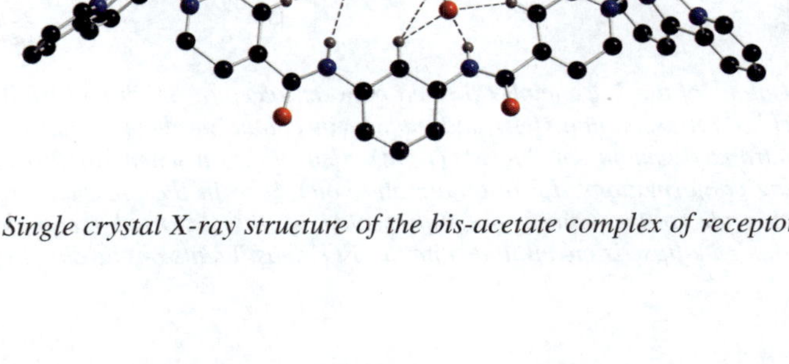

Figure 8.3 *Single crystal X-ray structure of the bis-acetate complex of receptor **8.48a***

A different electrochemical sensor element was introduced by Becher and co-workers[52] in 2003 when they reported the synthesis of the mono-tetrathiafulvalene (TTF) calix[4]pyrrole, **8.49a**. This system displayed an affinity constant for bromide anion of 7.6×10^3 M^{-1} in CD_3CN containing 0.5% D_2O, as determined from standard ^1H NMR titrations. In the case of chloride and fluoride anion, the corresponding K_a values were determined via ^1H NMR spectroscopic competition experiments and were found to be 1.2×10^5 and 2.1×10^6 M^{-1}, respectively. The electrochemical properties of receptor **8.49a**, containing a TTF unit, were monitored by CV methods in acetonitrile solution. In the presence of anions, this system displays cathodic shifts, with these values being −34 and −43 mV for chloride and bromide anions, respectively.

Recently, Sessler and Becher[53] demonstrated controlled complexation and decomplexation of electron-deficient neutral guests using the tetra-TTF calix[4]pyrrole **8.49b**. In the solid and solution states, this receptor stabilizes the formation of charge-transfer complexes involving trinitrobenzene (TNB), tetrafluoroquinone (TFQ), and tetrachloroquinone (TCQ). These species are bound in a 1:2 (host:guest) ratio within the two "pockets" defined by the calix[4]pyrrole in its 1,3-alternate form, as deduced from X-ray structural analyses and ^1H NMR and UV−Vis spectroscopy. In the presence of neutral guests, the colour of **8.49b** in dichloromethane underwent a change from yellow to green, presumably as the result of the proposed charge-transfer interaction. The associated complexation also induced peak shifts in the ^1H NMR spectrum in $CDCl_3$, which both helped confirm the structure of the complex in solution and permitted quantitative association constants to be determined ($K_{a1} < 20$ M^{-1} for TNB, TFQ and $K_{a2} = 900$ and 3500 M^{-1} for TNB, TFQ).

A particularly noteworthy feature of these systems is that the addition of chloride anion to chloroform solutions containing the host–neutral guest complex induced decomplexation, as the result of the strong interaction between this anionic substrate and the calixpyrrole ($K_a = 2.5 \times 10^6$ M^{-1} for TBACl in dichloroethane as determined by ITC), an interaction that served to induce a conformational switching from the 1,3-alternate form capable of binding the electron-deficient substrates to the classic cone form known to be stabilized in the presence of anions (Scheme 8.3). This latter form is unable to stabilize effectively charge-transfer interactions with the electron-deficient neutral guests. The net result is the release of these latter species and a change in colour from green to yellow. Interestingly, washing with water,

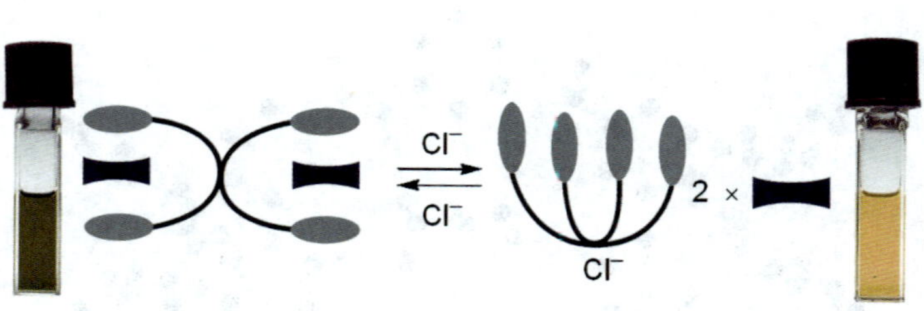

Scheme 8.3 *Colour of the 1:2 complex formed between receptor **8.49b** and trinitrobenzene in chloroform solution (left) and change in colour produced upon the addition of tetrabutylammonium chloride (right). Also shown in schematic form are the limiting conformations, 1,3-alternate and cone, seen in the presence and absence of chloride anion. Guests are represented by the distorted rectangle, where the shaded ellipse is meant to denote the TTF substituents on the calix[4]pyrrole*

which serves to effect decomplexation of chloride anion from calix[4]pyrrole, serves to restore the green colour of the 1:2 receptor–neutral substrate complex. The reader is also referred to Fourmigué and Batail's[54] work for another approach to TTF-based anion-receptor design.

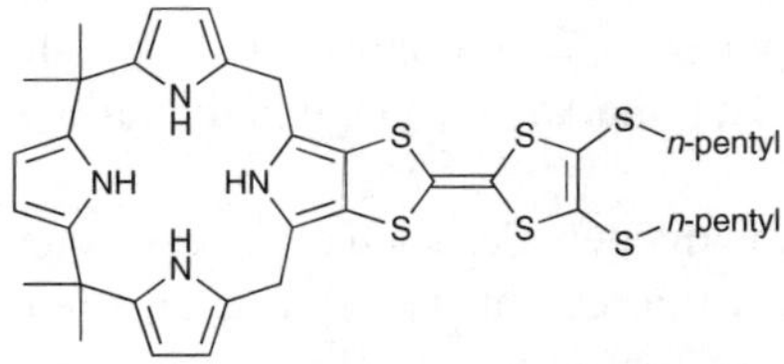

8.49a

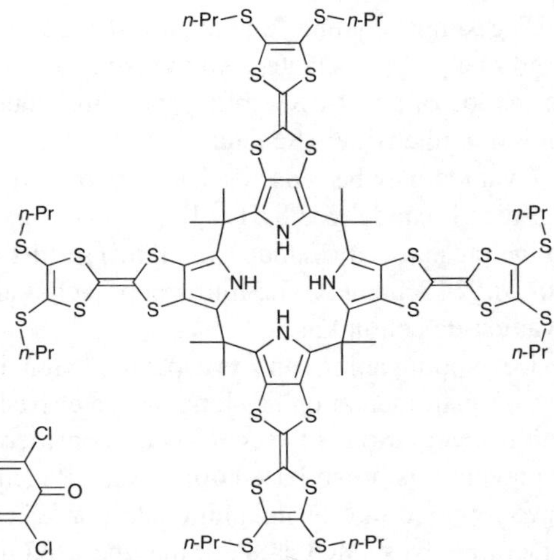

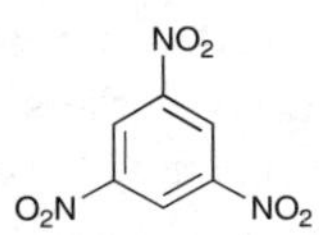

TNB

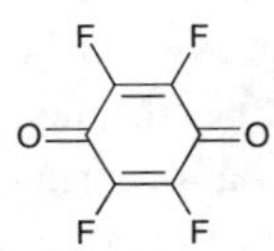

TFQ

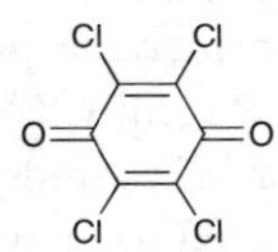

TCQ

8.49b

8.4 Discrete Molecular Optical Anion Sensors

There are two main approaches to the design of discrete molecular optical sensors. One is similar to the electrochemical sensors described above, in that a receptor containing a fluorescent or coloured group is synthesized that upon guest binding undergoes a change in fluorescence emission or visible colour as the result of anion-induced perturbations to the electronic (or other) features of this response element. The other is to synthesize a compound, which reacts with the targeted anion to form a new compound possessing different fluorescence properties or a different colour. This latter non-reversible "chemodosimeter" approach often permits very selective anion determinations.

Several optical anion-sensor systems are commercially available at present. For example, Molecular Probes, Inc. sells a number of indicators that may be used to monitor intracellular and extracellular chloride.[55] These sensors include 6-methoxy-*N*-(3-sulfopropyl)quinolinium, *N*-(ethoxycarbonylmethyl)-6-methoxyquinolinium bromide, 6-methoxy-*N*-ethylquinolinium iodide and lucigenin (shown below).

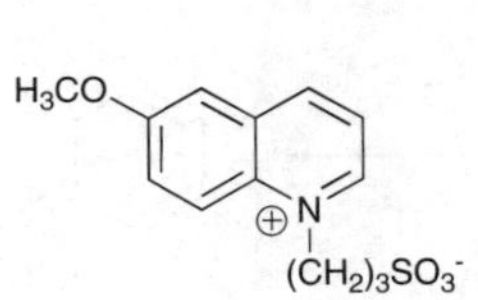

6-methoxy-*N*-(3-sulfopropyl)quinolinium

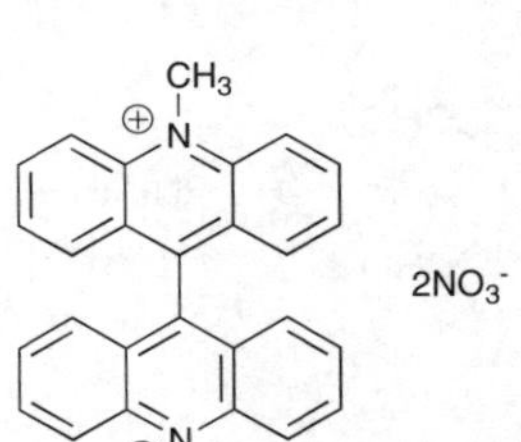

lucigenin

These compounds are thought to function as fluorescent sensors via a diffusion-limited collisional quenching mechanism.[56] In fact, other anions (such as bromide or iodide) also quench the fluorescence of these indicators. Fortunately, however, under physiological conditions the concentrations of these latter potential interfering species are low and they do not normally interfere with analyses of chloride anion.

The simple pyrene-functionalized mono-guanidinium receptor **8.50** can also be used as a pyrophosphate-anion sensor.[57] In the presence of pyrophosphate anion, the formation of a self-assembling structure induced a remarkable change in the ratio of emission intensities (Scheme 8.4).

Cyanide may be sensed selectively by aromatic dialdehydes such as naphthalene-2,3-dicarboxaldehyde **8.51**.[58] Reaction of this non-fluorescent species with an amine in the presence of cyanide (or a thiol) results in the formation of a fluorescent isoindole **8.52** (Scheme 8.5). Similar approaches are employed in commercially available cyanide-detection kits.[55]

Developing anion sensors capable of signalling the presence of anions without the use of instrumentation has long been a cherished goal in the area of supramolecular-anion recognition. In recent years, considerable progress has been made towards achieving this broad-based objective.[59] Recently, Martínez-Máñez and co-workers[60] have reported that nucleophilic addition of cyanide to appropriately functionalized squaraine dyes, such as **8.53**, may be used to determine cyanide anion selectively. Receptor **8.53** is blue coloured due to charge transfer from the anilinium groups to the electron-deficient central ring. The central ring in this system is susceptible to nucleophilic attack and these workers proposed that nucleophilic attack by cyanide on the central ring would allow for the selective detection of this notoriously toxic

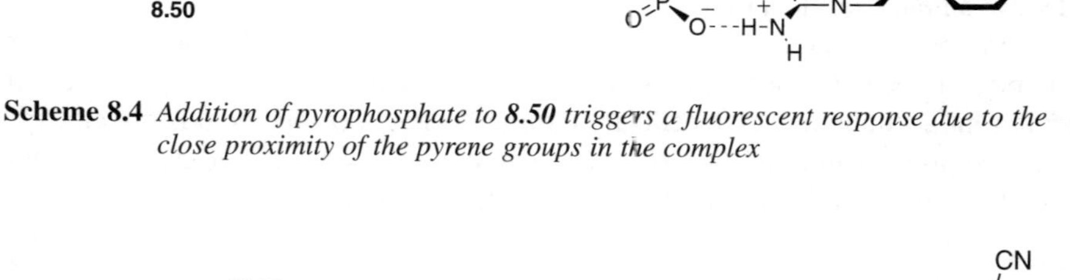

Scheme 8.4 *Addition of pyrophosphate to 8.50 triggers a fluorescent response due to the close proximity of the pyrene groups in the complex*

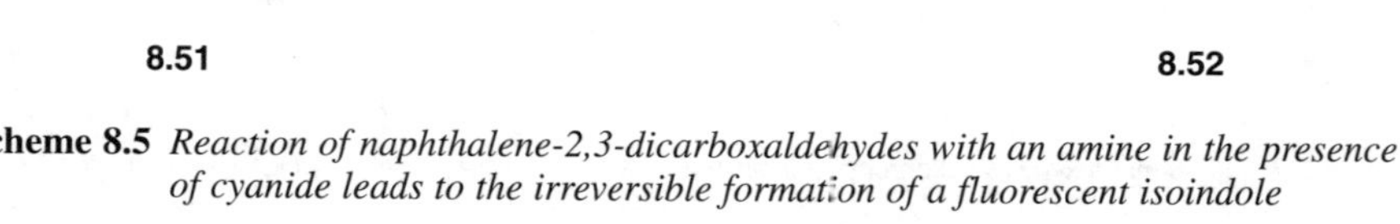

Scheme 8.5 *Reaction of naphthalene-2,3-dicarboxaldehydes with an amine in the presence of cyanide leads to the irreversible formation of a fluorescent isoindole*

species. This proved to be the case with the blue colour disappearing selectively in the presence of CN^- (Figure 8.4).

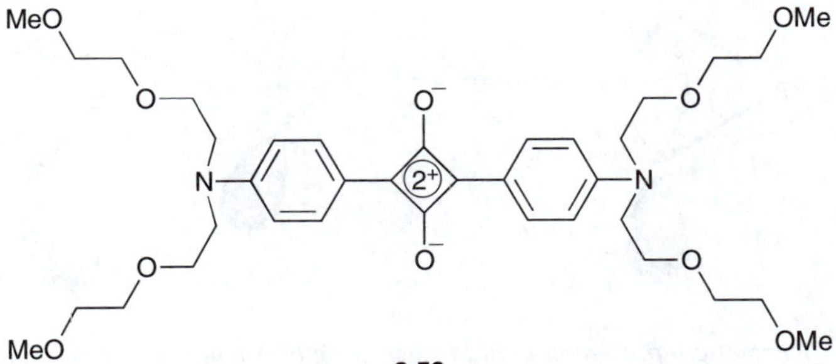

8.53

There are a number of different commercially available tests for nitrite NO_2^-. Typically, these work by exploiting an irreversible, nitrite-promoted reaction that leads to a demonstrable colour change. For instance, one popular nitrite-detection system employs 2,3-diaminonaphthalene which reacts with nitrite to form the highly fluorescent species, 1*H*-naphthotriazol (Scheme 8.6).[61]

By inserting Lewis acidic boron[62] or silicon atoms[63] in extended π-conjugated systems, Yamaguchi and co-workers produced several new colorimetric sensors for fluoride ion. In the presence of fluoride, triarylboranes, such as **8.54**, are converted to borates. This conversion disrupts the π-electron delocalization in the receptor (Scheme 8.7) causing a colour change (from orange to colourless for **8.54** in THF). In the case of receptor **8.55**, the UV-Vis spectra in THF revealed the formation of three fluoride complexes corresponding to fluoride anions binding to the three outer

Figure 8.4 *Receptor **8.53** in acetonitrile ([**8.53**] = 3.0 × 10^{-5} M) in the presence of 5 mol equiv. of various anions; from left to right, F^-, Cl^-, Br^-, I^-, NO_3^-, $H_2PO_4^-$, HSO_4^-, Ac^-, Bz^-, SCN^-, CN^-, and no anion. (Reproduced with permission from ref. 60 Copyright 2002 Royal Society of Chemistry)*

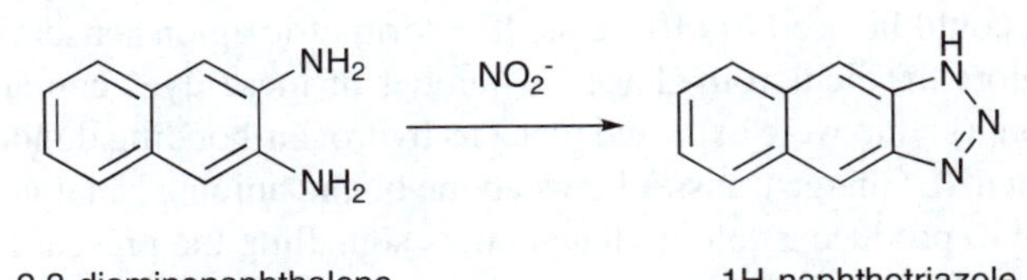

Scheme 8.6 *Nitrite detection by 2,3-diaminonaphthalene*

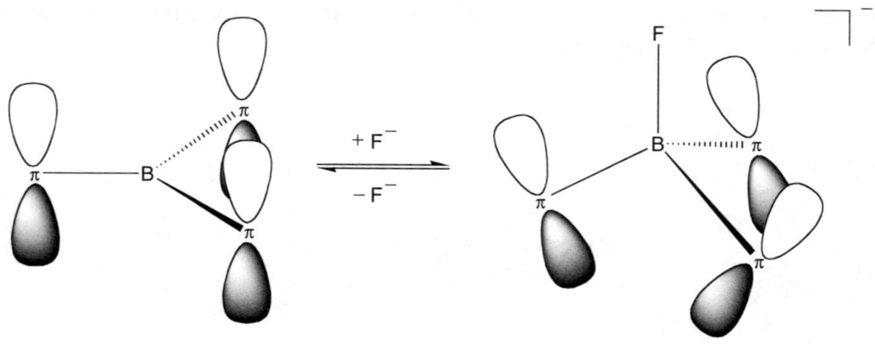

Scheme 8.7 *Schematic representation of the switching of π-conjugation in the LUMO of boron-based sensor systems such as* ***8.54*** *and* ***8.55****. In this figure, π represents the point of attachment of a π-conjugated moiety*

boron atoms. In the silicon-based receptors produced by the same group, the addition of fluoride to trianthrylfluorosilanes *e.g.*, **8.56** in THF (Scheme 8.8) causes a change in the through space interactions between the anthracene moieties. This results in a change in the fluorescence properties of the receptor, thereby allowing for fluoride-anion sensing. More examples of this type of sensor are listed in the reference section (see ref.64).

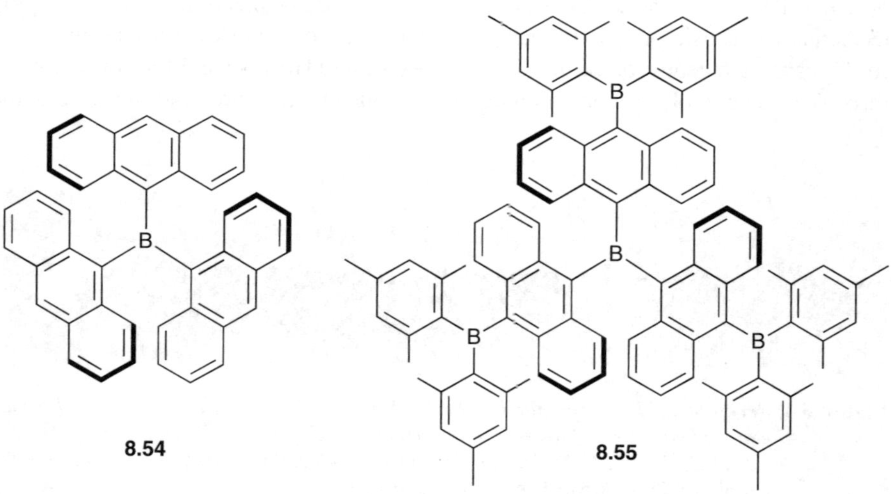

In 2001 Sessler and Miyaji[65] demonstrated that the commercially available dyes **8.57a–8.57k** could be used as off-the-shelf colorimetric anion sensors, or more precisely indicators, in dichloromethane. A number of these dyes contain one or two hydroxyl group(s) that were expected to act as hydrogen-bonding donors (*i.e.*, anion-binding elements). Charge transfer between the bound anionic "analyte" and the dye was expected to produce a colour change, thus signalling the presence of the bound anion. The results of anion-sensing tests involving dyes **8.57a–8.57k** are summarized in Table 8.3.

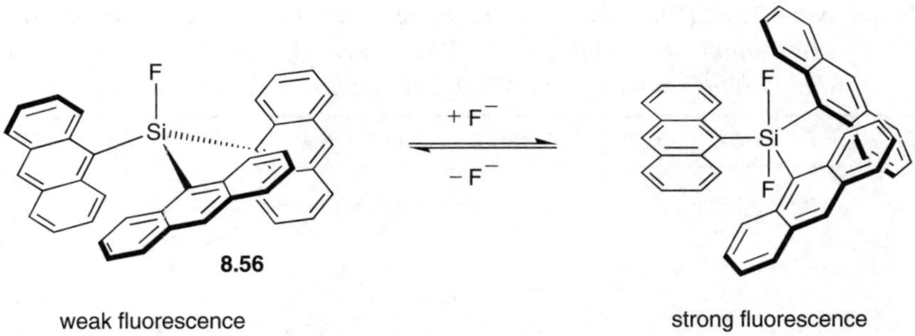

Scheme 8.8 *Fluoride addition to the silicon atom increases the fluorescence of **8.56***

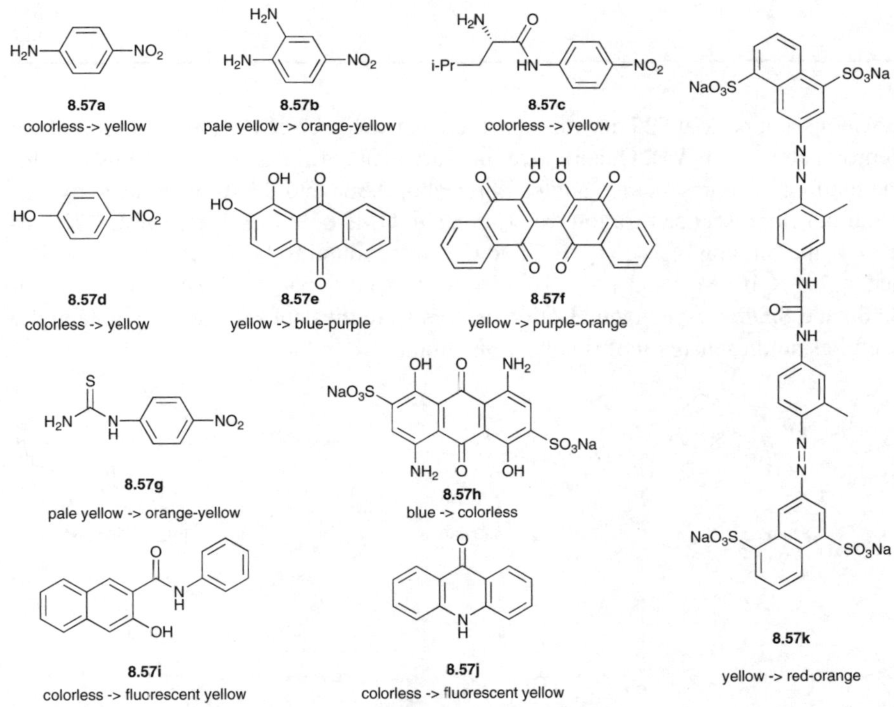

In early 2003, Wang and co-workers[66] reported the new hydroxyl-containing anion-sensing systems, **8.58a** and **8.58b**, along with the results of quantitative anion-binding studies. These latter revealed a high selectivity towards dihydrogen-phosphate and fluoride anions. For instance, in the case of the mono-hydroxyl receptor **8.58a** in DMF, the addition of $H_2PO_4^-$ (as the tetrabutylammonium salt) was found to give rise to a colour change from colourless to yellow. A strong quenching of the emission at 448 nm and an enhancement of the emission at 354 nm was also observed. These results were rationalized in terms of an equilibrium shift between the tautomeric forms **III** → **II**, a change that, in turn, was attributed to the hydrogen-bonding interactions between the $H_2PO_4^-$ analyte and the hydroxyl group of the receptor. In the presence of excess dihydrogen-phosphate or excess fluoride anion, a

Table 8.3 *Naked-eye detectable colour change for alcohol-based anion sensors seen upon the addition of 100 equiv. of the indicated anion in dichloromethane and/or DMSO. Shaded boxes indicate a colour change*

8.57	F^-	Cl^-	Br^-	$H_2PO_4^-$	HSO_4^-	AcO^-	BzO^-	CN^-	SCN^-	*Solvent*
a	■			■		■	■	■		CH_2Cl_2
b	■	■	■	■	■	■	■	■	■	CH_2Cl_2
c	■							■		CH_2Cl_2
d	■	■		■		■	■	■		CH_2Cl_2
e	■			■		■	■	■		CH_2Cl_2
f	■	■		■	■	■	■	■	■	CH_2Cl_2
g	■			■		■	■	■		DMSO
h	■									DMSO
i	■			■		■	■	■		CH_2Cl_2
j	■			■		■		■		DMSO
k	■			■				■		DMSO

new emission peak at 525 nm was observed that was ascribed to the formation of the deprotonated form (**VI**). Quantitative measurements of the association constants for the binding of representative anions to receptor **8.58a** and **8.58b** were made by carrying out fluorescence titration experiments in DMF solution. The resulting K_a values for the binding of $H_2PO_4^-$, F^-, and Cl^- were found to be 7.9×10^5, 8.6×10^4, and $< 1.0 \times 10^3$ M^{-1} and 1.8×10^6, 4.1×10^4, and $< 5.0 \times 10^2$ M^{-1} in the case of **8.58a** and **8.58b**, respectively. Both receptors were thus considered to be potentially useful as anion sensors in this solvent medium.

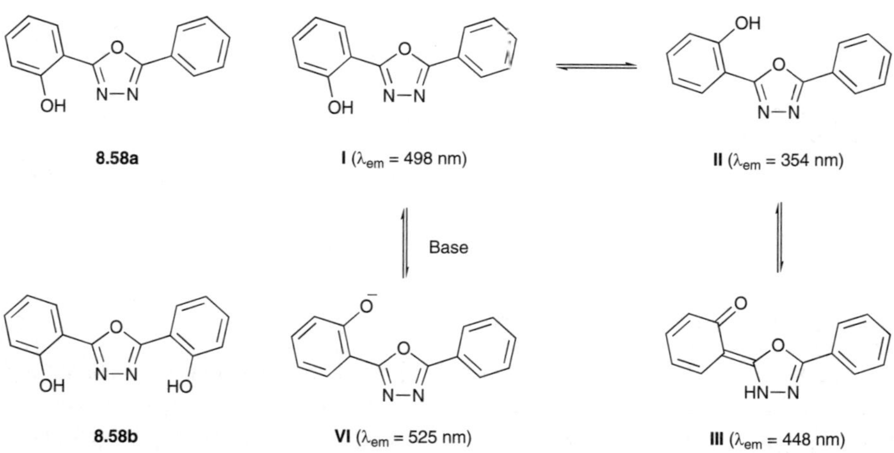

In 2001, Hong and co-workers[67] reported the synthesis of a series of thiourea-based anion sensors of general structure **8.59**. A special feature of these receptors is that they contain two separate chromophores, namely azophenol and nitrophenyl. In the case of **8.59b**, the presence of this dual sensor motif enabled discrimination between anions of similar basicity, namely $H_2PO_4^-$, F^-, and AcO^-, as evidenced by the different colorimetic responses produced for each of these species in chloroform. By contrast, in the case of sensors **8.59a** and **8.59c**, little in the way of observable colour difference was seen in the presence of these three test anions. Nor, was an appreciable change seen in the maximum absorbance values (either λ_{max} or ε). On the other hand, with receptor **8.59b** the absorption maximum was found to shift to the red upon the addition of selected anions, with the maximum shift being observed in the presence of dihydrogen-phosphate anion. Presumably, the colour change seen for this system in the presence of anions reflects the interaction between the azophenolic OH group present in **8.59b** (only) and the targeted anions.

A year later, the Hong[68] group reported that the indoaniline–thiouea system **8.60** acts as a highly effective anion sensor that allows for the selective colorimetric detection of tetrahedral oxoanions such as hydrogen sulfate and dihydrogen phosphate. This selective sensing ability was ascribed to the presence of complementary hydrogen-bonding interactions between this receptor and these specific anions. In the event, the colour of solution of the host (**8.60**) in chloroform was found to change from blue-green to deep blue upon the addition of $H_2PO_4^-$ or HSO_4^-. However, the addition of more basic acetate or fluoride anion gave rise to a less intense colour change. Moreover, detectable colour changes were not observed for Cl^-, Br^-, and I^-. Quantitative anion-binding studies, involving standard UV–Vis titrations, were carried out from which association constants of 1.1×10^4 and 2.5×10^4 M^{-1} were derived for $H_2PO_4^-$ and HSO_4^-, respectively, in $CHCl_3$.

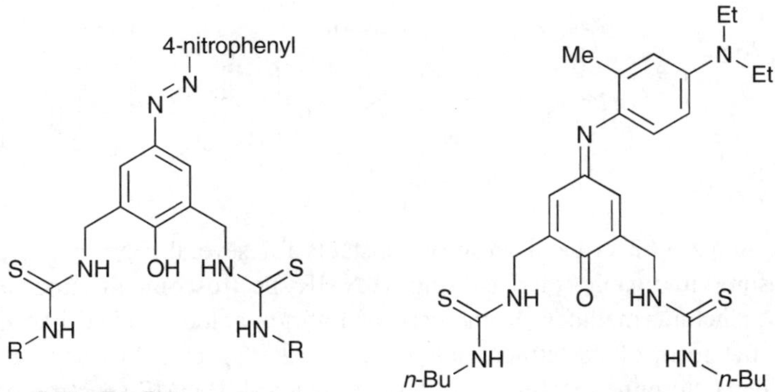

8.59a R = Ph
8.59b R = 4-nitrophenyl
8.59c R = *n*-Bu

8.60

The modified thiourea receptor **8.61** represents another colorimetic-based approach to anion sensing. This system, which was prepared by Fabbrizzi and Bermejo,[69] permits the selective detection of fluoride anion via a naked eye-detectable colour change, an effect that is not observed in the presence of other anions. Quantitative studies of these systems were carried out using UV–Vis titrations. In the presence of fluoride anion, it was observed that the intensity of the absorption band at 324 nm seen in acetonitrile decreased, while that of the absorption band at 412 nm increased. From these spectral changes, an association constant of log K_a = 4.14 was derived for the binding of fluoride anion in this solvent. As expected, lower values (*i.e.*, log $K_a < 2$) were found for various other anions, including in particular dihydrogen phosphate, chloride, bromide, and iodide. The strong interaction between fluoride anion and the NH protons of the receptor are thought to reflect the improved π-electron delocalization that is induced upon anion binding, a change that, in turn, is expected to engender the large red-shift seen in the UV–Vis absorbance spectrum upon fluoride-anion binding. Other examples of (thio)urea[70,71] and (sulfon)amide-based[72] anion sensors are listed in refs. 70–72.

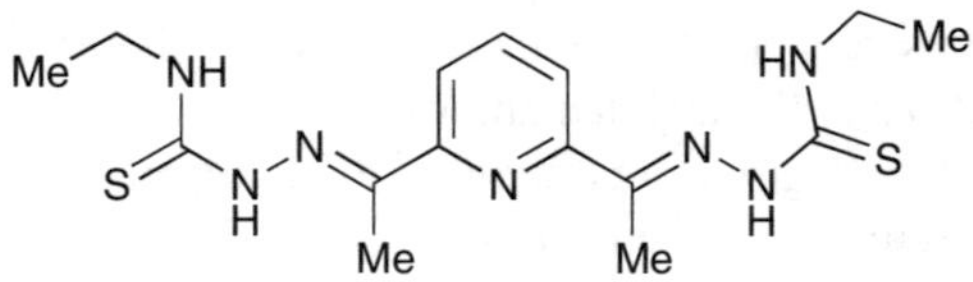

8.61

One of the methods used to prepare oligopyrrole-based optical sensors has involved the covalent attachment of a chromophore or fluorophore to a calixpyrrole skeleton. Perturbation of the electronic properties of these attached groups upon anion complexation (either directly or indirectly) then produces a response detectable by the naked eye or via changes in the fluorescence spectrum, or both. Here, the first systems reported in the literature were covalently linked, calixpyrroles bearing anthracene subunits (*i.e.*, calix[4]pyrroles **8.62**). These systems were obtained from the coupling of a calixpyrrole mono-acid and various aminoanthracenes.[73]

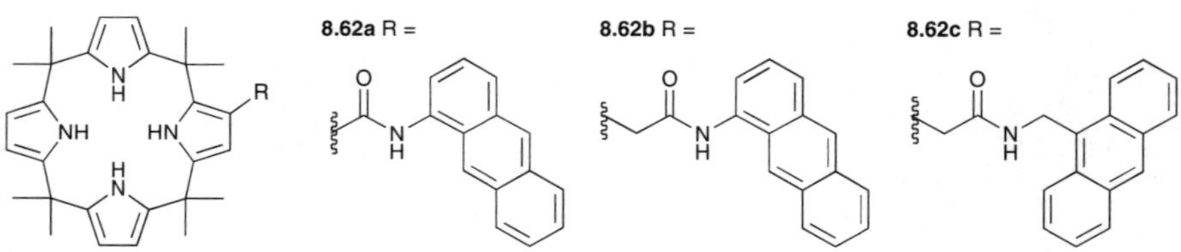

With systems **8.62** in hand, affinity constants for several representative anions were measured in organic solvents using ^{1}H NMR spectroscopic titrations and fluorescence-quenching methods. In the case of fluoride anion (studied like the other anions in the form of its tetrabutylammonium salt), spectral broadening made it impossible to determine a stability constant by standard ^{1}H NMR spectroscopic titration methods. On the other hand, the fluorescence intensities of these sensors (*i.e.*, **8.62a–8.62c**) were found to be quenched significantly in the presence of certain anionic guests, with the degree of quenching depending on the specific choice of anions. The greatest quenching was seen for fluoride and the associated changes in the fluorescence spectrum for sensor **8.62a** are reproduced in Figure 8.5. From these and related quenching experiments, the affinity constant for fluoride, chloride, bromide, and dihydrogen phosphate anions were calculated for the full family of receptors **8.62**. The resulting values are listed in Table 8.4.

In a related work, the second-generation calix[4]pyrrole-anion sensors **8.63** were produced. Here, dansyl, Lissamine™-rhodamine B, and fluorescein moieties were attached to the calix[4]pyrrole framework to serve as the fluorescent reporter groups, while an amide or thiourea residue was used as the covalent linker between these fluorophores and the calix[4]pyrrole backbone. This choice of linker also acts to introduce an additional anion-recognition motif that was expected to enhance the anion affinities, at least for anionic guests of appropriate size and shape.

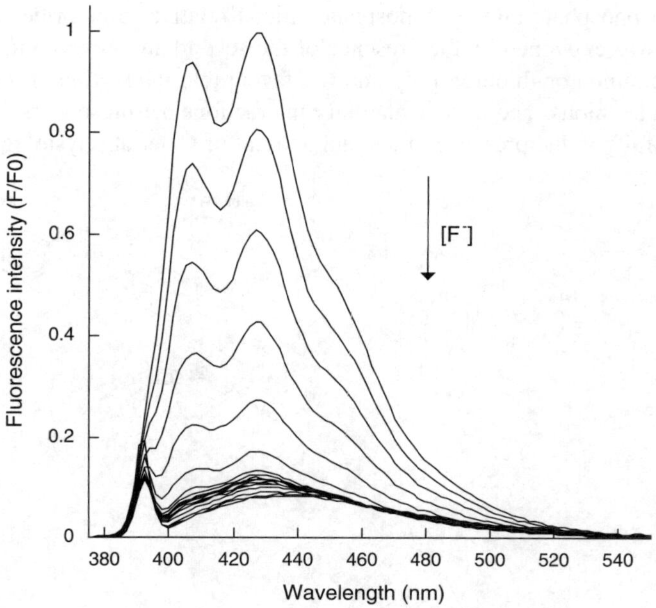

Figure 8.5 *Fluorescence spectra of sensor* ***8.62a*** *in CH_2Cl_2 (0.05 mM) excited at 378 nm showing the changes in the emission spectrum induced upon the addition of increasing quantities of tetrabutylammonium fluoride.*
(Reproduced in modified form with permission from reference [73]. Copyright 1999 Royal Society of Chemistry)

Table 8.4 *Stability constants for anion sensors* ***8.62*** *with various anions (studied as their tetrabutylammonium salts) in CH_3CN and CH_2Cl_2 as determined from fluorescence-quenching analyses at 25 °C. Sensor* ***8.62a*** *was excited at 378 nm and emission was monitored at the λ_{max} of 446 nm; for sensors* ***8.62b*** *and* ***8.62c****, the corresponding values were 393 and 429 nm and 387 and 418 nm, respectively*

	log K_a in CH_2Cl_2			*log K_a in CH_3CN*		
	8.62a	***8.62b***	***8.62c***	***8.62a***	***8.62b***	***8.62c***
F^-	4.94	4.52	4.49	5.17	4.69	4.69
Cl^-	3.69	2.96	2.79	4.87	3.81	3.71
Br^-	3.01	a	a	3.98	2.86	a
$H_2PO_4^-$	4.20	3.56	a	4.96	3.90	a

[a]The extent of quenching proved insufficient to allow calculation of an accurate value.

The changes observed upon anion binding, specifically the decrease in fluorescence emission intensity seen when sensor **8.63b** is titrated with an increasing concentration of tetrabutylammonium fluoride, are reproduced in Figure 8.6.[74] Quantitative assessments of the anion-induced spectral changes were used to determine the affinity constants for this and other representative anions and are summarized in Table 8.5. It was found that sensors **8.63** show considerable selectivity for dihydrogen phosphate and pyrophosphate anions relative to chloride anions. This selectivity was explained by the presence of the second anion-recognition element, namely the amide or thiourea link, and the favourable interactions it permits with non-spherical anions. These same ancillary interactions permit sensors **8.63** to operate successfully in the presence of a small amount of water at physiological pH.

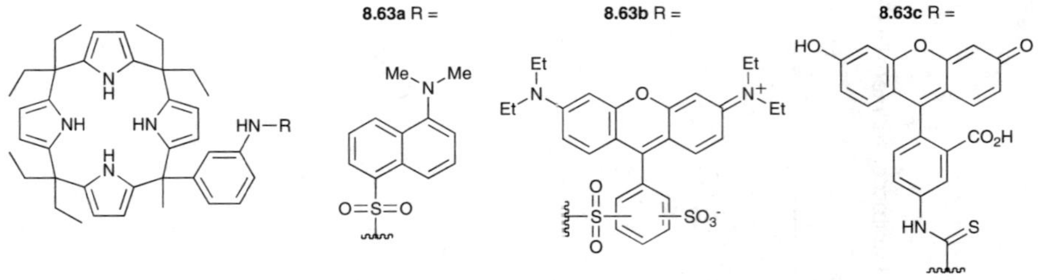

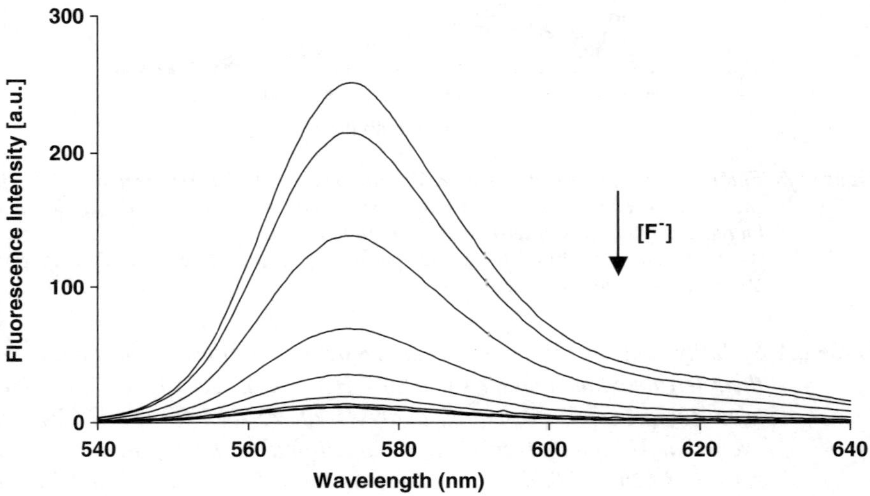

Figure 8.6 *Decrease in fluorescence emission intensity observed when sensor* ***8.63b*** *(0.1 μM in acetonitrile containing 0.01% v/v water) is titrated with increasing concentrations of tetrabutylammonium fluoride. From highest to lowest curve,* $[F^-]=0$, *0.30, 0.76, 1.53, 2.30, 3.06, 3.83, 4.60 μM.*
(Reproduced in modified form with permission from reference [74]. Copyright 2000 American Chemical Society)

Table 8.5 *Association constants* (M^{-1}) *for sensors* ***8.63*** *and representative anionic substrates (studied as their tetrabutylammonium salts) as determined by fluorescence emission quenching in acetonitrile (0.01% v/v water) for sensors* ***8.63a*** *and* ***8.63b*** *and in acetonitrile–water (94:4, pH 7.0± 0.1) for sensor* ***8.63c***

	F^-	Cl^-	$H_2PO_4^-$	$HP_2O_7^{3-}$
8.63a	222,500	10,500	168,300	131,000
8.63b	>1,000,000	18,200	446,000	170,000
8.63c	>2,000,000	<10,000	682,000	>2,000,000

In 2000, Sessler and co-workers[75] reported a set of powerful calixpyrrole-based naked-eye sensors for selected anions such as F^-, Cl^-, and $H_2PO_4^-$ (*cf.* structures **8.64**). As shown in Figure 8.7a, the addition of tetrabutylammonium fluoride to dichloromethane solutions of receptor **8.64c** leads to a noticeable red-shift (λ_{max} from 441 to 498 nm) and a broadening of the absorption maximum. Reflecting this anion-induced spectral difference, the colour of the solution changes from yellow to red. On

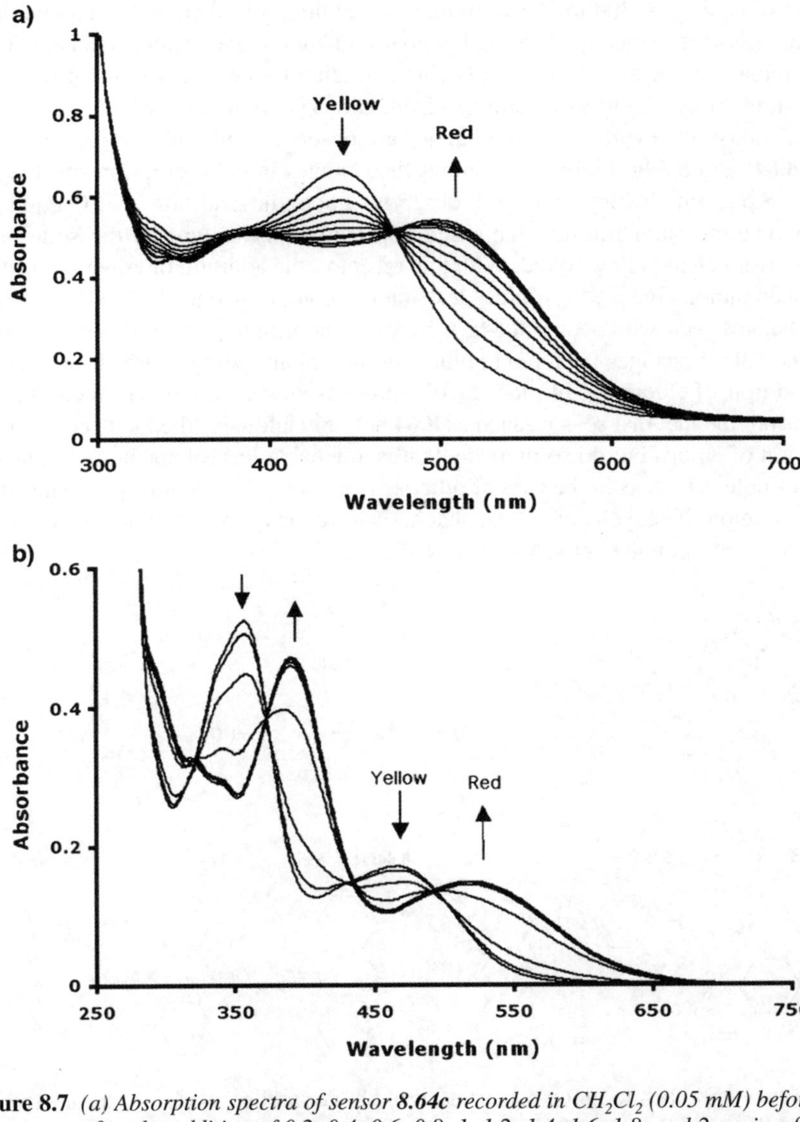

Figure 8.7 *(a) Absorption spectra of sensor* ***8.64c*** *recorded in* CH_2Cl_2 *(0.05 mM) before and after the addition of 0.2, 0.4, 0.6, 0.8, 1, 1.2, 1.4, 1.6, 1.8, and 2 equiv. of tetrabutylammonium fluoride. (b) Absorption spectra of* ***8.64f*** *recorded in* CH_2Cl_2 *(0.05 mM) before and after the addition of 0, 1, 2, 4, 6, 8, and 10 equiv. of tetrabutylammonium fluoride.*
(Reproduced with permission from reference [75]. Copyright 2000 Elsevier (Fig. a) and 2000 Wiley–VCH (Fig. b))

the other hand, the addition of either chloride or dihydrogen-phosphate anion (as the tetrabutylammonium salts) caused the colour of the solution to change from yellow to orange. In the case of the mono-nitrobenzene conjugated calixpyrrole **8.64b**, again in dichloromethane, the maximum absorption peak was seen to shift from 391 nm (pale yellow colour) to 433 nm (intense yellow colour) upon the addition of fluoride anion. Receptor **8.64a** is colourless and was used as a control. It was observed that the absorption maximum (λ_{max} = 308 nm) of this latter species underwent a slight shift (from a λ_{max} = 308 to 321 nm) upon the addition of 20 equiv. of tetrabutylammonium fluoride, indicating binding but no useful "naked eye" optical read out. Table 8.6 summarizes the affinity constants corresponding to the interactions of receptors **8.64** with various anions as determined from UV–Vis titration studies.

Particularly noteworthy colour changes are observed with anthraquinone derivatives **8.64f** and **8.64g**. Figure 8.7b shows the changes in the absorption spectrum of sensor **8.64f** (in dichloromethane) observed upon the addition of 10 equiv. of tetrabutylammonium fluoride. This addition caused the colour of the solution to change from clear yellow to red. With this receptor, the addition of either chloride or phosphate anions induced a slightly less intense colour change (*i.e.*, to orange-red). The changes seen with sensor **8.64g** were even more dramatic. In this case, anion-induced colour changes from red to blue, purple, or dark purple were observed upon the addition of fluoride, chloride, or dihydrogen-phosphate anions. Such changes underscore the fact that sensors such as **8.64** not only allow for the facile colorimetric detection of anions but do so in a way that is amenable to "colour tuning". The one issue of note is that, as in the case of other sensors, the effect is most pronounced for fluoride anion. Nishiyabu and Anzenbacher[76] have also reported colorimetric anion sensors based upon the calix[4]pyrrole skeleton.

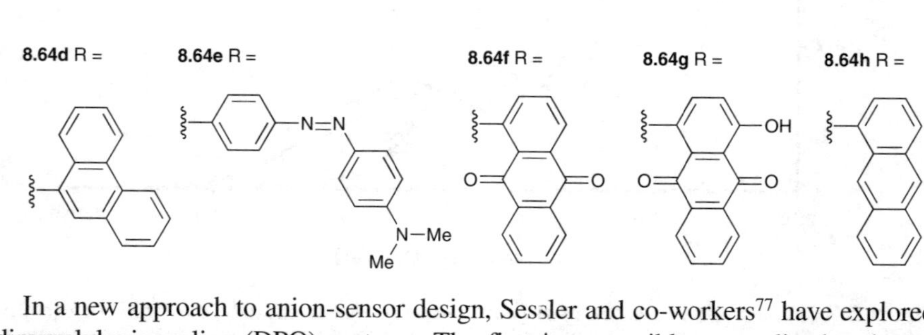

In a new approach to anion-sensor design, Sessler and co-workers[77] have explored dipyrrolylquinoxaline (DPQ) systems. The first in a possibly generalized series of anion sensors where two (or more) pyrrolic-recognition elements are bridged by a rigid, aromatic chromophore, these acyclic systems were prepared in only two steps; specifically, they were made from simple compounds, namely pyrrole, oxalyl chloride,

Table 8.6 *Anion-binding affinity constants (log K_a) for sensors **8.64** and selected anionic substrates as deduced from UV–Vis titration experiments carried out in CH_2Cl_2[a]*

	8.64a	8.64b	8.64c	8.64f	8.64g	8.64h
F^-	4.04	4.23	4.51	3.71	3.43	3.66
Cl^-	3.36	3.67	3.84	3.16	3.13	3.16
$H_2PO_4^-$	2.68	3.03	3.28	3.04	2.76	2.86

[a]All anions were studied in the form of their tetrabutylammonium salts.

and various phenylenediamines[78] using a modification of a procedure first reported by Oddo in 1911.[79] In the absence of anions, the colours of dichloromethane or DMSO solutions of DPQs **8.65a**, **8.65b**, and **8.65c** ranged between pale yellow and deep yellow. Addition of fluoride anion to these solutions in the form of tetrabutylammonium fluoride resulted in an immediate quenching of the fluorescence intensity and a naked eye-detectable response that was manifest in terms of a dramatic colour change from yellow to orange-red in the case of **8.65a** and **8.65c** and from yellow to dark purple in the case of **8.65b** (Figures 8.8a and b). By contrast, the addition of chloride or phosphate anion did not induce appreciable fluorescence quenching; nor, did it lead to a noticeable change in colour except in the case of sensor **8.65c**. In the case of **8.65c**, the addition of dihydrogen phosphate anion induced a colour change similar to that produced by fluoride anion, but only at a considerably higher anion concentration.

Quantitative measurements of the anion-binding affinities were made using standard fluorescence quenching or UV–Vis absorbance titration techniques. These analyses helped confirm that the observed colorimetric change did indeed reflect an anion-binding process. For instance, as would be expected based on its colorimetric behaviour, receptor **8.65c** showed a significantly enhanced affinity for $H_2PO_4^-$ anion relative to **8.65b** ($K_a = 1.73 \times 10^4\ M^{-1}$ *vs.* 80 M^{-1} for **8.65c** and **8.65b**, respectively; CH_2Cl_2, tetrabutylammonium salt in both cases).[77,80] Further, the more dramatic colour changes seen for **8.65b** and **8.65c** in the presence of fluoride anion (tetrabutylammonium salt) are also reflected in higher K_a values (1.2×10^5 and $6.2 \times 10^4\ M^{-1}$ for **8.65b** and **8.65c**, respectively, in CH_2Cl_2 *vs.* $1.8 \times 10^4\ M^{-1}$ for **8.65a**). These findings were rationalized in terms of an increased NH-hydrogen-bond donor ability in the case of **8.65b** and **8.65c** that result from the electron-withdrawing groups present on the DPQ skeleton.

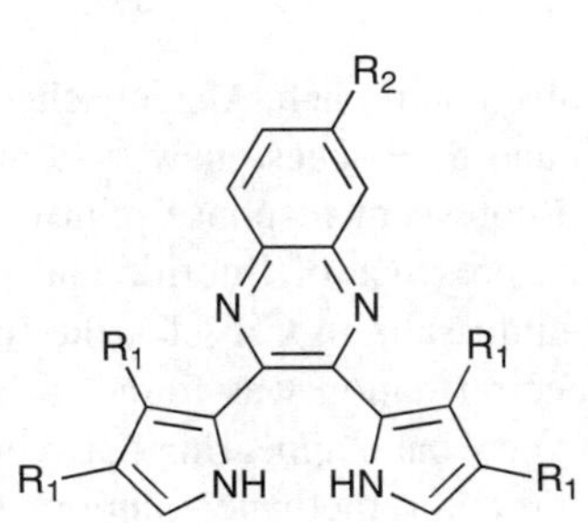

8.65a $R_1 = R_2 = H$
8.65b $R_1 = H$, $R_2 = NO_2$
8.65c $R_1 = F$, $R_2 = H$

The success of these initial systems led to the synthesis of the fused DPQ phenanthroline conjugates **8.66** and their metal complexes **8.67**. In the case of the latter systems, it was expected that appending cationic charges to a DPQ framework would withdraw electron density from the NH bonds leading to an increase in the anion-binding affinity, just as proved true for **8.65b** and **8.65c**. These expectations were in fact realized. In particular, while the electron-rich, free phenanthroline DPQ system **8.66** displayed a rather low fluoride-anion affinity (K_a = 440 M^{-1} in DMSO for **8.66a** and 68,000 M^{-1} in CH_2Cl_2 containing 2% CH_3CN for **8.66b**; tetrabutylammonium salts in both the cases), the two Ru(II) complexes **8.67a** and **8.67c** displayed higher fluoride anion affinities (K_a = 640,000 M^{-1} in CH_2Cl_2 containing 2% CH_3CN for **8.67a** and K_a = 12,000 M^{-1} in DMSO for **8.67c**) compared to DPQ **8.65a** (K_a < 100 M^{-1} in DMSO). In the case of DPQ **8.67a**, a red-shift in the emission spectrum from 594 to 610 nm was also observed as a function of an increased CN^- concentration. The binding affinity for cyanide anion could thus be calculated and was found to be 428,000 M^{-1} in CH_2Cl_2 containing 2% CH_3CN.[81] The Co(III) complex **8.67b**, with its incrementally greater charge (+3 *vs.* +2 for **8.67c**), displayed one of the highest affinities for fluoride anion yet recorded in DMSO (K_a = 54,000 M^{-1}).[81,82]

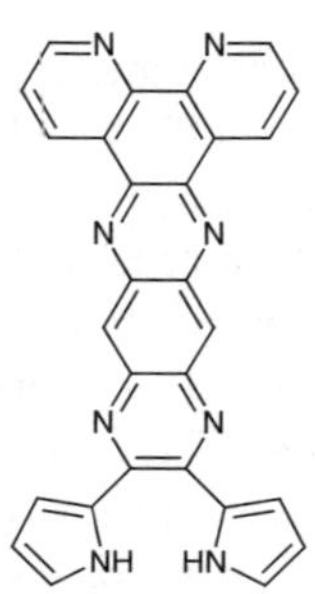

8.66a

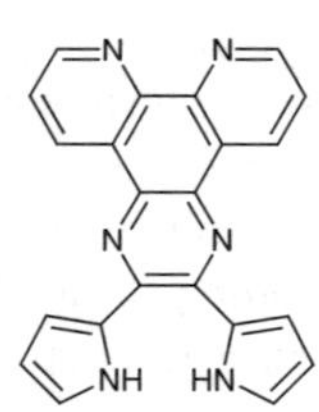

8.66b

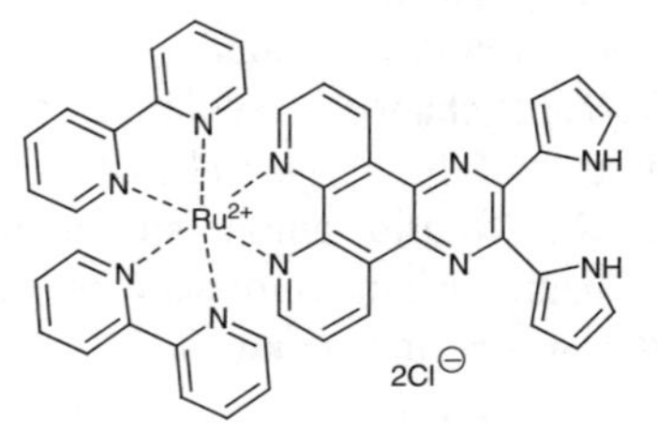

8.67a

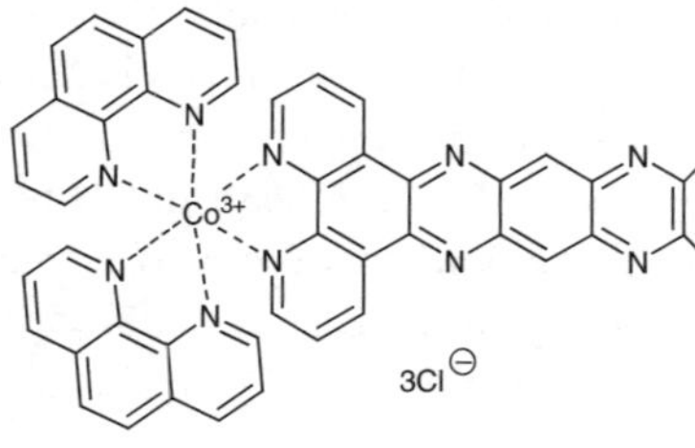

8.67b

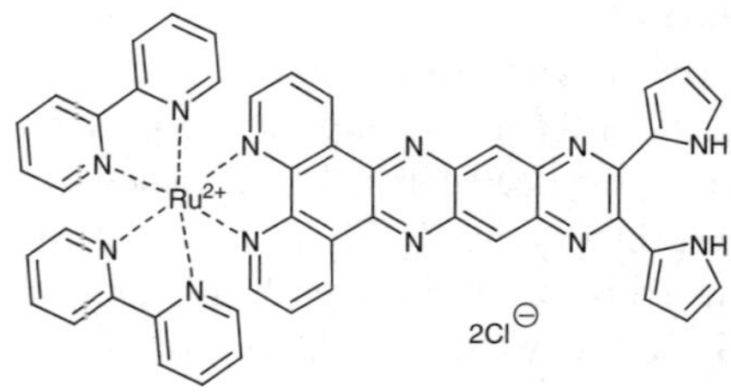

8.67c

In an elegant elaboration of this generalized approach, Anzenbacher and co-workers reported the new DPQ systems **8.68** and **8.69**. These new systems contain aryl subunits and demonstrate an improved fluorescent response relative to the parent DPQ system **8.65a**.[83] For instance, in the presence of fluoride and pyrophosphate anions (studied as their tetrabutylammonium salts in CH_2Cl_2), the intensity of the absorption bands in the 400–450 nm spectral region was found to be reduced for both systems **8.68** and **8.69**, with a strong new band appearing between 500 and 550 nm. The addition of these two anions to dichloromethane solutions of this pair of

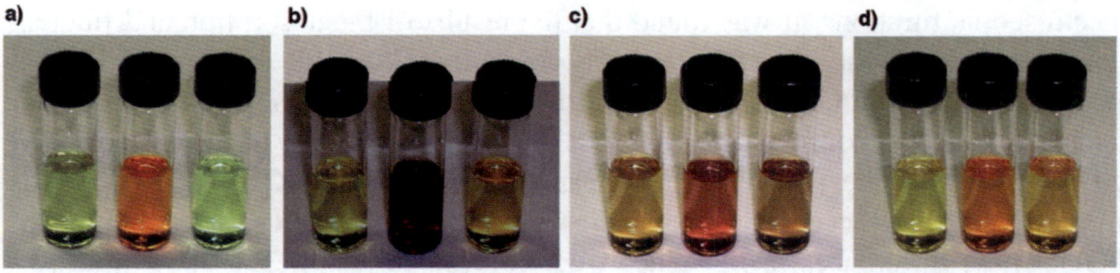

Figure 8.8 *Colour changes induced by the addition of anions (studied as the corresponding tetrabutylammonium salts) to the CH_2Cl_2 solution of receptor* ***8.65a, 8.65b, 8.74b,*** *and* ***8.75a****. (a) From left to right:* ***8.65a****;* ***8.65a****+F^-;* ***8.65a****+$H_2PO_4^-$. (b) From left to right* ***8.65b****;* ***8.65b****+F^-;* ***8.65b****+$H_2PO_4^-$. (c) From left to right* ***8.74b****;* ***8.74b****+malonate;* ***8.74b****+succinate. (d) From left to right* ***8.75a****;* ***8.75a****+F^-;* ***8.75a****+$H_2PO_4^-$.* (Reproduced with permission from Artificial pyrrole-based anion receptors, in *Functional Synthetic Receptors*, T.Schrader and A.D. Hamilton (eds), Wiley–VCH, Weinheim, 2005)

sensors also gave rise to significant fluorescence quenching. Accompanying quantitative studies confirmed that the affinities for fluoride and pyrophosphate anions in CH_2Cl_2 were indeed substantially enhanced relative to those of DPQ **8.65a**. Further anion-sensing studies involving polymer-based DPQ systems were also carried out by the Anzenbacher group.[84]

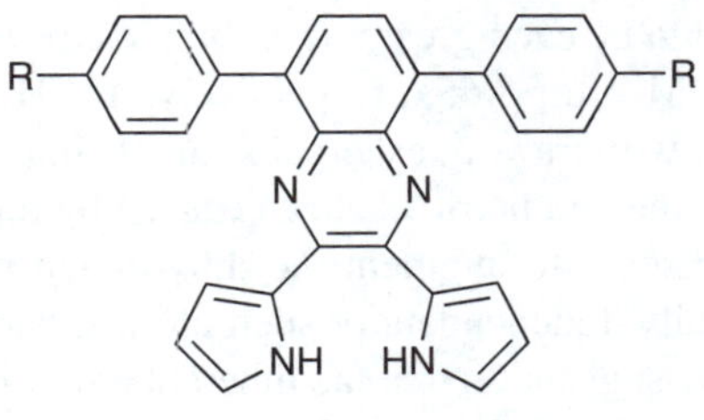

8.68a R = H
8.68b R = OCH_3
8.68c R = $N(CH_3)_2$

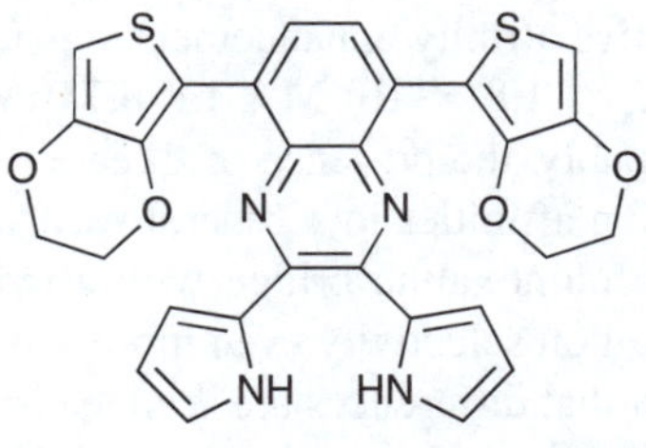

8.69

An alternative approach to the design new pyrrole-based DPQ sensors involves modifying the basic chromophore bridging used to connect the two pyrrole-recognition subunits. One recently reported system that falls within this paradigm is 2,7-bis(1*H*-pyrrol-2-yl)ethynyl-1,8-naphthyridine (BPN) **8.70**, a system that contains two hydrogen-bonding donors (pyrrolic NH) linked by a dialkynyl naphthyridine spacer.[85] Here, the specific arrangement of the constituent-recognition elements, involving an organized array of hydrogen-bonding donor–acceptor-acceptor–donor subunits, was found to favour saccharide binding. These binding events, in turn, produced colour changes that led to the suggestion that systems such as **8.70** could be used for saccharide sensing. Dichloromethane solutions of BPN were found, for instance, to change from cyanine to green upon the addition of octyl *β*-D-glucopyranoside (OGU).

Quantitative assessments of the binding affinities were made using three different methods, namely UV–Vis absorption, fluorescence quenching, and ^{1}H NMR spectroscopic titrations. It was found that the results of the absorption and fluorescence titrations measurements were nearly coincident ($K_a = 4.8 \times 10^3$ and 5.5×10^3 M^{-1}, respectively, in CH_2Cl_2). However, the ^1H NMR titrations gave rise to higher calculated stability constants ($K_a = 2.0 \times 10^4$ M^{-1} in $CDCl_3$). In the case of octyl β-D-galactpyranoside (OGA), the corresponding K_a value ($K_a = 1600$ M^{-1}, as inferred from fluorescence titration studies) is roughly three times lower than that of OGU. These differences in K_a values were considered significant since these two saccharides differ only in the orientation of the 4-hydroxyl group.

To date, system **8.70** does not appear to have been studied as an anion-binding sensor. However, the ability to bind and sense anions was demonstrated in the case of another DPQ derivative containing an "alternative" bridging subunit, namely the pyrrole-linked system **8.71**. This system was originally obtained as the side product during the iodination of pyrrole. It was then subsequently synthesized in excellent yield using a stepwise approach. Its unique structure and electronic configuration led to the consideration that it might prove useful as an anion receptor and sensor.[86] In fact, the addition of tetrabutylammonium fluoride to CH_2Cl_2 solutions of compound **8.71** was found to induce a colour change from pale brown to yellow. Moreover, the anion affinities of compound **8.71** were found to be substantially enhanced relative to those of DPQ **8.65a**. Whereas this latter system displays a K_a of *ca.* 1.8×10^4 M^{-1} for the binding of fluoride anion in CH_2Cl_2, the pyrrole-linked system **8.71** was found to display an affinity for this species that was enhanced by roughly 10-fold ($K_a = 1.8 \times 10^5$ M^{-1} in CH_2Cl_2). In the case of dihydrogen phosphate anion, the extent of relative affinity enhancement was found to be even greater (*i.e.*, by a factor of *ca.* 220; $K_a = 1.8 \times 10^4$ M^{-1} for **8.71** *vs.* 80 M^{-1} for **8.65a**, respectively, in CH_2Cl_2). Presumably, the presence of three electron-withdrawing ester substituents improves the anion affinities in a general way, while the geometric changes caused by replacing the quinoxaline bridge with a pyrrole serve to augment the dihydrogen phosphate-anion selectivity even more specifically. Independent of such rationalizations, the fact that differences are seen leads to the suggestion that the inherent anion-binding affinities and selectivities of DPQ systems may be modulated by changing the nature of the bridging subunit. To the extent this prediction holds true, new analogues of **8.70** and **8.71** can be expected that might display very interesting substrate-recognition properties.

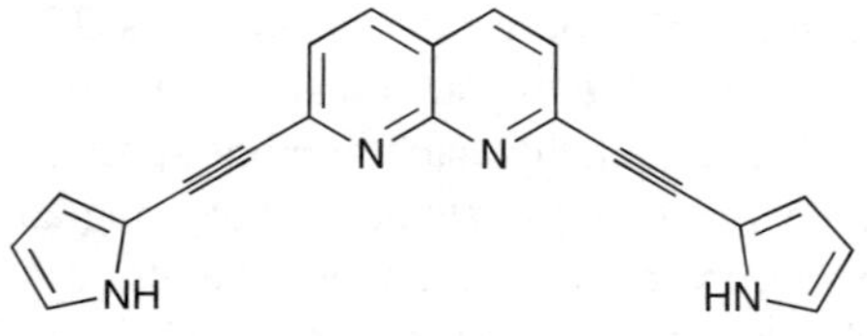

8.70

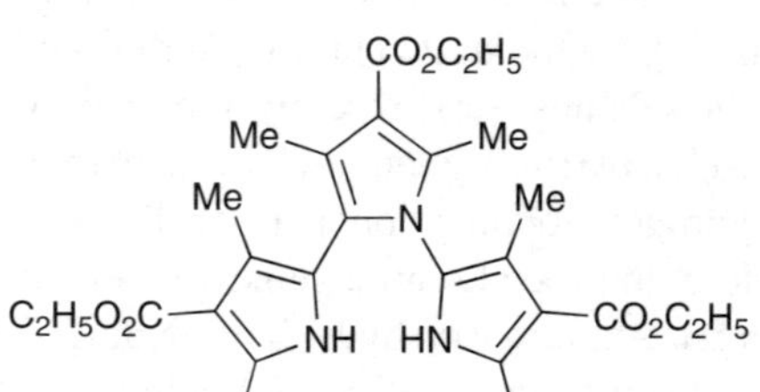

8.71

Another strategy being pursued in an effort to extend the DPQ-based approach to anion sensing and, in particular, to produce systems with enhanced anion affinities, has involved appending additional pyrroles to the anion-binding portion of the DPQ backbone. So far, application of this appealing strategy has led to the synthesis of systems **8.72**, **8.73**, and **8.74**, all of which contain built-in pyrrole-derived NH "claws".

The bowl-like conformations of DPQ **8.72** and **8.74** were expected to favour the binding of larger anions such phosphate, acetate, oxalate, malonate, and succinate. However, the fluoride-anion affinities were also expected to remain high. In the case of **8.72a** and **8.74a**, UV–Vis spectroscopic titrations, carried out in CH_2Cl_2 solution, revealed that the absolute fluoride- and phosphate-anion affinities were increased relative to **8.59a** ($K_a(F^-)$ = 32,000 and >1,000,000 M^{-1} for **8.72a** and **8.74a**, respectively, and K_a ($H_2PO_4^-$) = 4300 and 300,000 M^{-1} for **8.72a** and **8.74a**, respectively).[87]

More recently, colour changes from dark yellow to purple or brown upon the addition of malonate or succinate anions to CH_2Cl_2 solutions of DPQ **8.74** have been observed (Figure 8.8c).[88]

UV–Vis spectroscopic titrations were used to study the binding of mono- and dicarboxylate anions (tetrabutylammonium salts; CH_2Cl_2). These analyses confirmed that DPQ **8.74b** binds oxalate anion with good selectivity relative to malonate and succinate anion (the ratio of the respective K_a values is 450:40:1 in CH_2Cl_2). However, these same studies also revealed that DPQ **8.74b** binds the mono-carboxylate anion, acetate, even more effectively than these dianionic species. By contrast, in the case of DPQ **8.74a**, the binding of acetate anion was not found to be enhanced relative to oxalate, malonate, or succinate.[88]

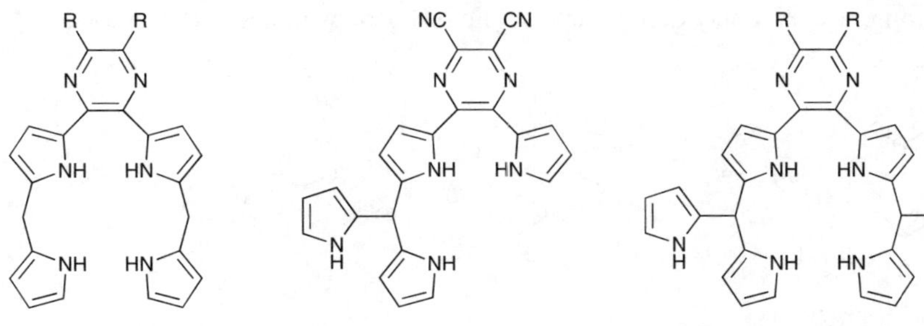

8.72a R = $-C_4H_4-$
8.72b R = CN

8.73

8.74a R = $-C_4H_4-$
8.74b R = CN

In order to create potential "molecular cages" for anionic substrates, a first macrocyclic DPQ systems **8.75** was prepared recently by Sessler and Furuta.[89] A prototype of what could emerge as a potentially large class of quinoxaline-bridged porphyrinoids, this target was synthesized from formyl functionalized DPQ units. The absorption spectrum of **8.75a**, recorded in dichloromethane, is characterized by two bands with λ_{max} of 367 and 427 nm, respectively. Upon the addition of an increasing concentration of tetrabutylammonium fluoride, new peaks at 329 and 480 nm were seen to grow in. It was also found that the colour of CH_2Cl_2 solutions of **8.75a** underwent a change from yellow to orange or dark yellow in the presence of fluoride and dihydrogen phosphate anions, respectively (Figure 8.8d). Based on a quantitative analysis of these spectral changes, it was deduced that receptor **8.75a** binds fluoride and phosphate anions in a cooperative 2:1 fashion and does so with rather high affinity ($K_a = 3 \times 10^5$ and 80 M^{-2} for fluoride and phosphate, respectively).[89]

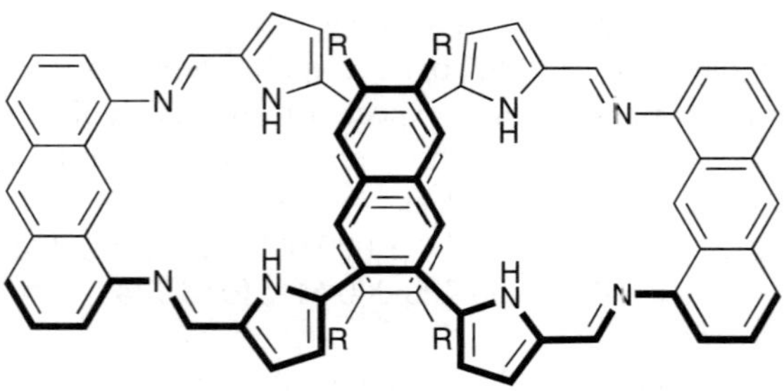

8.75a R = H
8.75b R = OCH_3
8.75c R = OCH_2CH_3

Recently, Gale and co-workers and Gunnlaugsson and co-workers have independently demonstrated that fluoride anions can deprotonate hydrogen-bond donor receptors, which can in some cases lead to a colour change. Gunnlaugsson and co-workers[71] have observed deprotonation of an amino group in 4-amino-1,8-naphthalimide-based chemosensors, *e.g.*, **8.76** and **8.77** upon addition of fluoride, while Gale and co-workers[90] have observed deprotonation in a variety of 2,5-diamidopyrrole systems functionalized with electron-withdrawing groups including **8.78**, a compound that undergoes a colourless to deep blue colour change upon addition of fluoride anion. Care must be taken in the attribution of the cause of colour changes in such systems on addition of fluoride as, in the absence of water, fluoride functions as a strong base and may deprotonate acidic NH groups in neutral receptors.

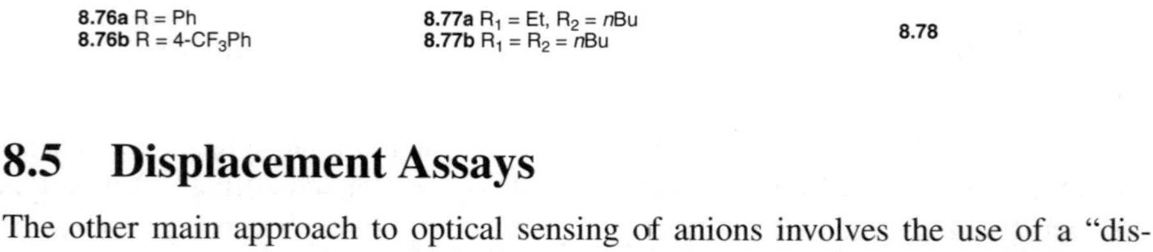

8.76a R = Ph
8.76b R = 4-CF_3Ph

8.77a R_1 = Et, R_2 = *n*Bu
8.77b R_1 = R_2 = *n*Bu

8.78

8.5 Displacement Assays

The other main approach to optical sensing of anions involves the use of a "displacement assay" (Scheme 8.9). This method has been pioneered by Anslyn and co-workers. An optical displacement assay consists of a receptor bound to a fluorescent or coloured molecule such as a dye. The microenvironment of the bound dye is different from that in bulk solution. Hence, if an anionic guest is added to the ensemble it will displace the dye triggering a fluorescence or colour change.

Anslyn and co-workers[91–93] have published a number of papers on the recognition of tri-carboxylate and tri-phosphate polyanions by tris-guanidinium receptor

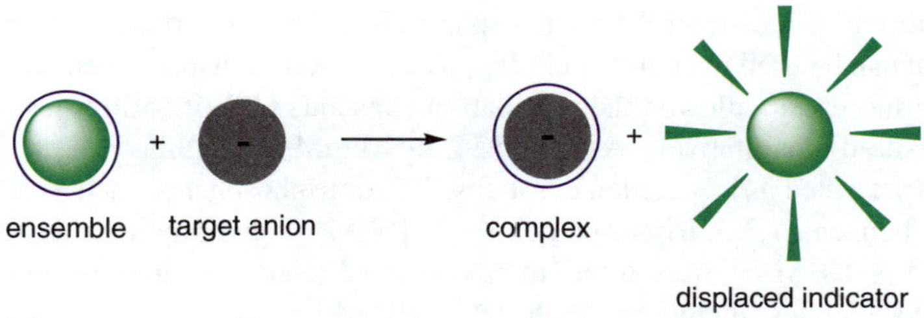

Scheme 8.9 *A schematic representation of a displacement assay*

species, *e.g.*, **8.79**. This receptor contains three guanidinium subunits and is therefore complementary to guests containing three carboxylate groups. Predicative analyses of stability constant data revealed that guests containing three anionic moieties, such as citrate, are bound more strongly than those with fewer anionic groups (*e.g.*, acetate). Consistent with this finding, it is proved possible to obtain diffraction grade crystals of the tricarballate complex of **8.79**; the resulting structure is shown in Figure 8.9.

This receptor was used by Metzger and Anslyn[92] to produce a chemosensor for citrate in beverages. This sensor is based on the displacement of 5-carboxyfluorescein from receptor **8.79** upon the addition of an appropriately chosen anionic analyte. 5-Carboxyfluorescein (**8.80**) is a commercially available fluorescent probe containing two carboxylate groups. Its fluorescence is particularly sensitive to changes in pH, with the intensity decreasing as a function of increasing protonation. The two carboxylate groups present in **8.80** coordinate to **8.79** forming a complex. Under these conditions, the pK_a of the phenol group of carboxyfluorescein is lowered due to the presence of a positively charged microenvironment within the binding cleft. More strongly bound citrate displaces the carboxyfluorescein from the complex, thus shifting the pK_a of the phenol anion such that **8.80** is more fully protoned once decomplexed (*i.e.*, released into the surrounding solvation environment) (Scheme 8.10). Since the fluorescence intensity, as well as inherent absorptivity, of carboxyfluorescein decreases with increasing protonation, the result of this citrate-based displacement process is an easy-to-monitor decrease in fluorescence. Further, the Anslyn group found that this anion-induced reduction in intensity may be calibrated against standard solutions of citrate to produce an analytically useful sensor capable of providing accurate quantitative assessments of citrate concentration in, for instance, a wide variety of popular soft drinks and other beverages.

Subsequent to this initial work, Anslyn and co-workers have worked to generalize their displacement-based approach to sensor production, reporting, for instance, systems **8.81–8.82** that were designed to be selective for other analytes. These new systems were made by modifying the initial 1,3,5-triethyl-2,4,6-tri substituted benzene scaffold so as to include, *e.g.*, guanidinium and boronic acid subunits. For instance, the cleft-like receptor **8.81** has six guanidinium groups. These provide a cavity suitable for polyphosphate guests. In fact, this system was specifically designed for the recognition of inositol-1,4,5-trisphosphate (IP_3),[93] an important secondary messenger that can be used to analyse cellular process.[94] A dye-displacement assay using 5-carboxyfluorescein allowed the association constants (K_a) of various test analytes to be measured quantitatively. As expected, in 10 mM HEPES buffer (pH 7.4), the affinities (presented as association constants, K_a) for triphosphates, such as IP_3 (4.7×10^5 M^{-1}), benzene-1,3,5,-triphosphate (5.0×10^5 M^{-1}), and phytic acid (hexaphosphate) (7.5×10^5 M^{-1}), were found to be enhanced relative to those of other highly oxygenated analytes, including citrate (8×10^3 M^{-1}), ATP (2.3×10^4 M^{-1}), and fructose-1,6-diphosphate (2.2×10^4 M^{-1}). In methanol, an association constant as high as 1.0×10^8 M^{-1} was found for IP_3 using this receptor–dye ensemble, meaning that this targeted anion could be detected at the 2 nM level in such solutions.

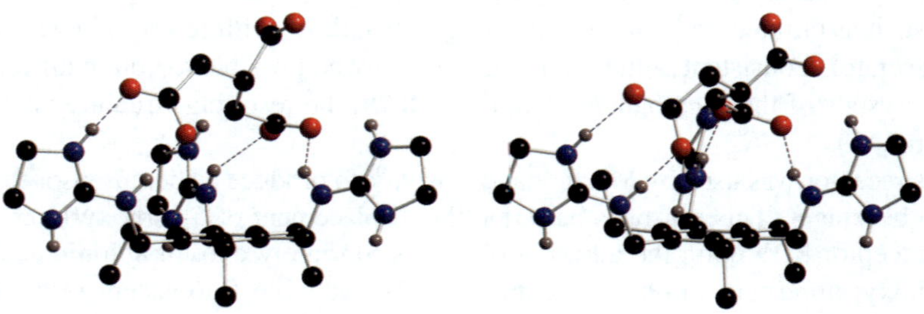

Figure 8.9 *Two different host/guest complexes are seen in the solid-state structure of the tricarballate complex of **8.79**, as determined by single crystal X-ray diffraction analysis*

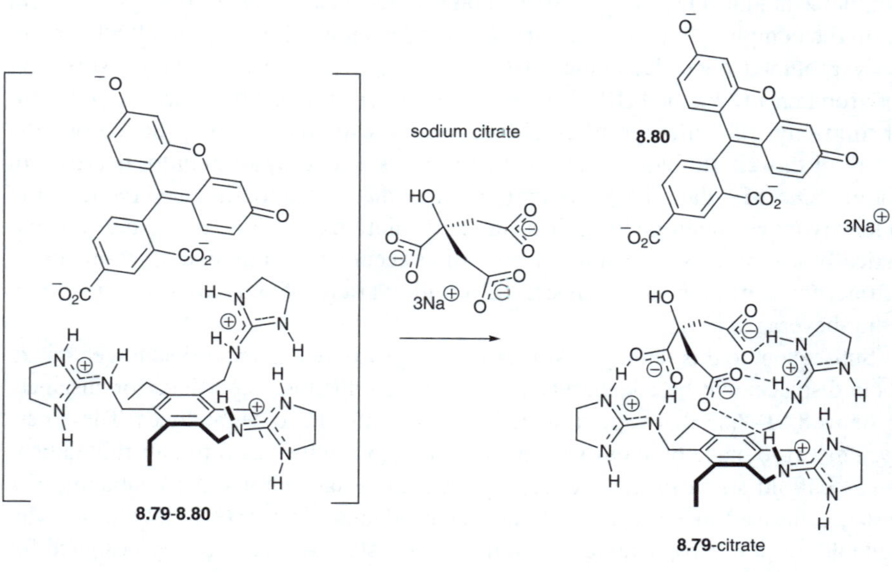

Scheme 8.10 *Ensemble **8.79–8.80** functions as a displacement assay for citrate*

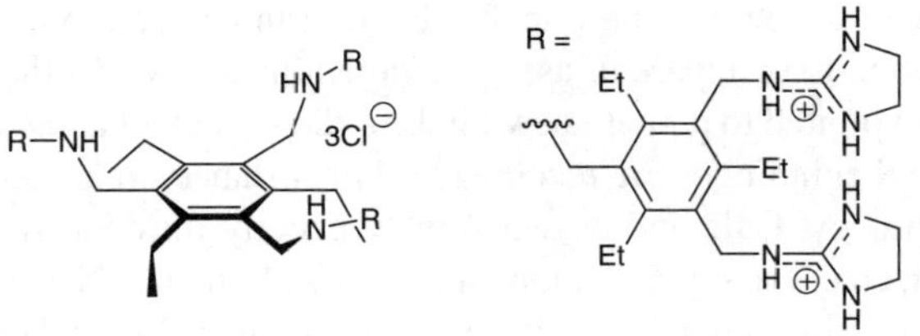

8.81

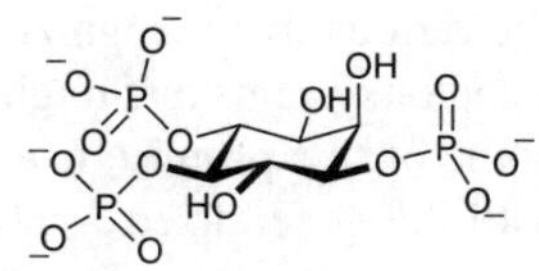

inositol-1,4,5-trisphosphate (IP_3)

Appreciating that boronic acids are capable of binding 1,2- and 1,3-diols,[95,96] the Anslyn group prepared the host systems **8.82a** and **8.82b**. These new systems contain different combinations of guanidinium and boronic acid subunits and were expected to allow for the recognition of the carboxylate-/diol-containing guests, such as tartarte, malate, and gallate.[96–100] By using alizarin as the displayed chromophore and monitoring the changes in the UV–Vis spectrum under standard dye-displacement conditions, it was found that **8.82a** binds tartrate and malate with affinities of 5.5×10^4 and $4.8 \times 10^4\,M^{-1}$, respectively, in a methanol/water mixture (25% water; buffered with 10 mM HEPES).[97] In associated work, the percentage of tartaric and malic acids in commercial wines and other grape-derived beverages were also determined. It was also found that receptor **8.82b**, containing a single guanidinium and two boronic acid motifs, shows a high selectivity towards gallic acid over other ostensibly similar analytes, such as caffeic and 4-hydroxycinnamic acids.[101] This selectivity permitted the age of various commercial Scotch whiskeys to be analysed, with the results being calibrated by, *e.g.*, HPLC methods.

The association constant (K_a) of **8.82c** towards tartrate were also measured in 25% water/methanol (10 mM HEPES buffer at pH 7.4) using an indicator-displacement assay involving alizarin and was found to be $4.0 \times 10^4\,M^{-1}$.[98] A thermodynamic analysis of the interaction between **8.80**, **8.82–8.82c** with various anionic analytes containing carboxylates and hydroxyl motifs were also carried out using ITC technique; these studies revealed that the binding of citrate and tartrate by all three receptors in aqueous solution is exothermic, with favourable entropy.[102]

In an other effort to exploit further the utility of these receptors as anion sensors, Anslyn and McDevitt demonstrated that a pattern-recognition protocol involving the use of a multilayer perception-based artificial neural network allowed structurally similar guests, such as tartrate and malate, to be detected with great reliability in the case of **8.82a** and **8.82b**.[99] Slightly later, chip-based array using a combinatorial library of sensors was developed to differentiate structurally similar analytes such as ATP, AMP, and GTP.[103]

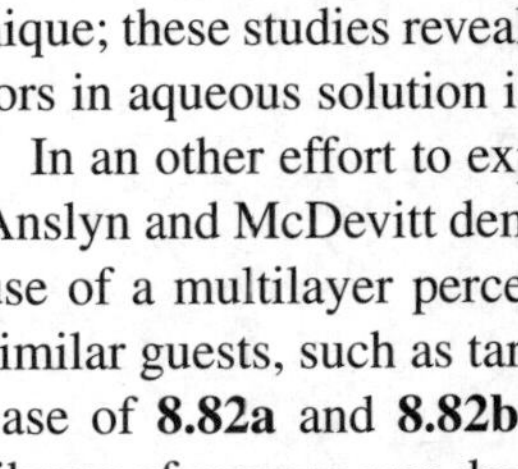

8.82a

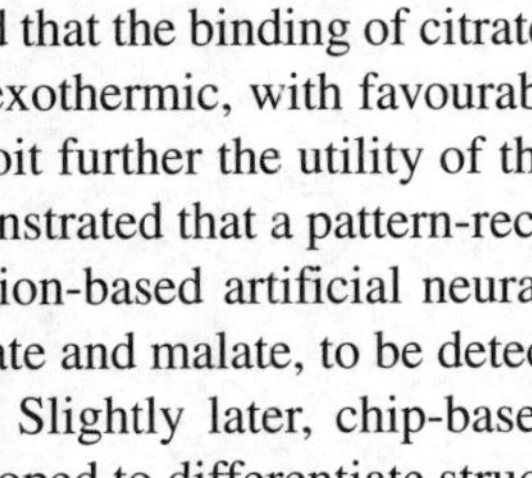

8.82b

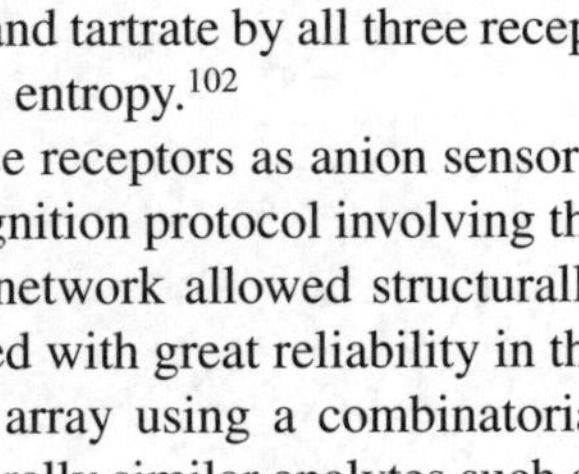

8.82c

Given this success, the value of developing new scaffolds that can be used to orient various dye-recognizing subunits is apparent, as is the potential utility of other receptor systems that might be exploited to prepare new kinds of displacement assays. Some of this promise is already beginning to be recognized. For instance, in accord with the generalized Anslyn strategy, Gale and co-workers[104] recently introduced a very simple colorimetric displacement assay for anions that is based on calix[4]pyrrole and 4-nitrophenolate anion. This anion loses its intense yellow colour when bound to *meso*-octamethylcalix[4]pyrrole **8.11**. This allows the calix[4]pyrrole–4-nitrophenolate complex to be used as a colorimetric displacement assay in a manner that bears analogy to what is seen in the case of Anslyn's systems (Scheme 8.11). The primary difference is that in the case of calix[4]pyrrole, it is small anions, such as fluoride, that serve to displace the 4-nitrophenolate anion from the complex thus enhancing the absorbance of the 4-nitrophenolate anion. Anion recognition was thus observed (*i.e.*, "sensed") as a colourless-to-yellow colour change.

A very different kind of displacement assay is embodied in the luminescent systems developed by Parker. These workers reported that lanthanide(III) complexes of ligand **8.83**, specifically (Eu–**8.83**)$^{3+}$ and (Tb–**8.83**)$^{3+}$, can be used as sensors for anions in aqueous environments.[105] In water, both complexes contain two coordinated water molecules. These molecules quench the luminescence properties of the lanthanide cations. However, the addition of certain anions can cause displacement of these coordinated waters leading to increased luminescence. For instance, fluoride, acetate, and sulfate were found to displace one water molecule from the metal coordination sphere, thus increasing the luminescence of the metal centres. Likewise, hydrogen carbonate acted to displace both the initial water molecules by binding in a bidentate fashion; again, this produced a large change in the luminescent lifetime of the complex. By contrast, chloride, bromide, iodide, and nitrate did not produce a change in the optical signature, presumably because they did not act to displace the bound water molecules.

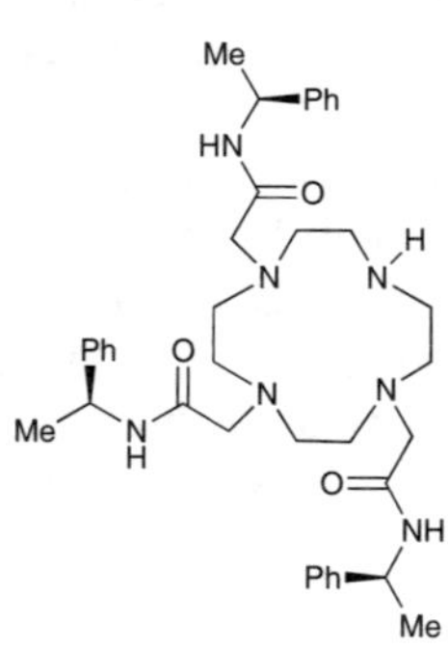

8.83a

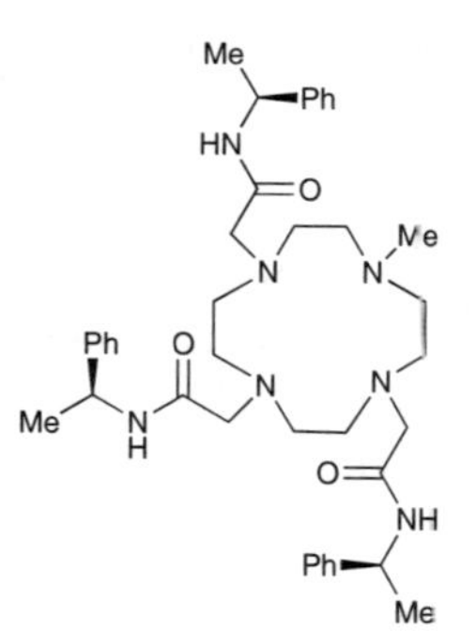

8.83b

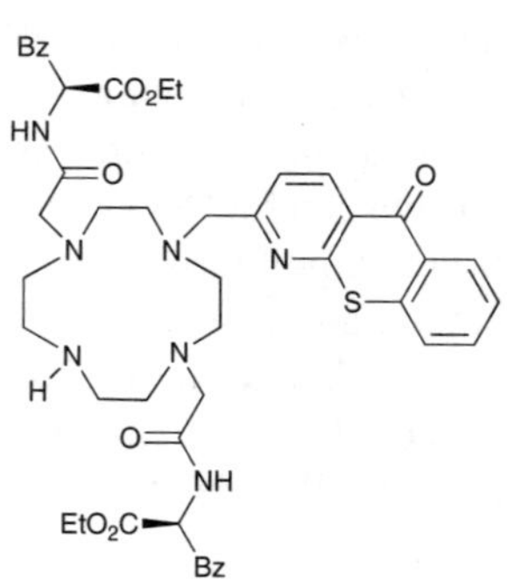

8.83c

8.6 Assays Based on Deaggregation Phenomena

One final approach to the optical sensing of anions involves the use of a chromophore- or fluorophore-containing systems, whose aggregation properties change

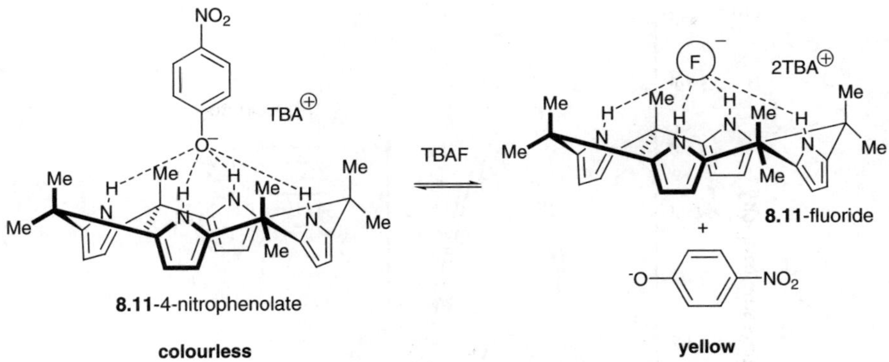

Scheme 8.11 *4-Nitrophenolate-calix[4]pyrrole is colourless but upon addition of fluoride the nitrophenolate anion is displaced and the yellow colour of the nitrophenolate anion restored*

as a function of added anion concentration. The potential utility of this approach, which has yet to be widely exploited in the anion-sensing area, was recently demonstrated by Sessler and co-workers.[106] These researchers employed a water-soluble sapphyrin, such as **8.84**, that was found to be highly aggregated (and monoprotonated) in aqueous media at neutral pH. However, addition of inorganic phosphate was found to effect deaggregation. Since the resulting deaggregated forms, particularly the monomer produced at high phosphate-to-sapphyrin ratios, was found to be highly fluorescent in marked contrast to the initial aggregate, the increase in fluorescence could be used to track the change in phosphate-anion concentration, as illustrated in Figure 8.10. Similar approaches, relying on changes in the UV–Vis absorption spectrum, were used to monitor the interaction of sapphyrin with pertechnetate, a radioactive species of considerable environmental concern.[107]

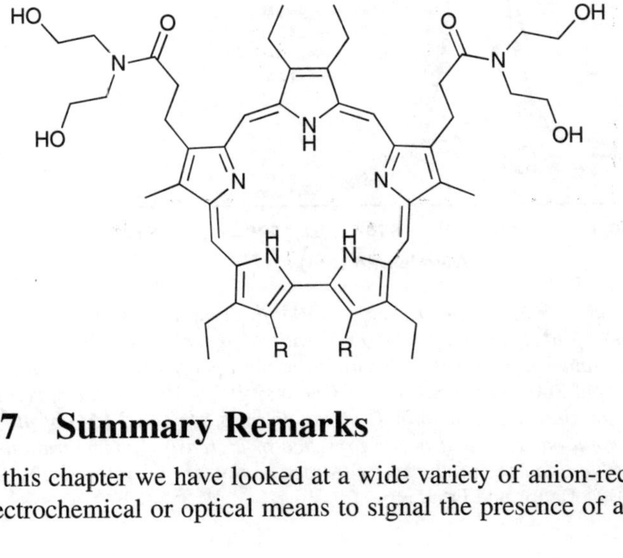

8.84a R = Me
8.84b R = Et

8.7 Summary Remarks

In this chapter we have looked at a wide variety of anion-receptor systems that use electrochemical or optical means to signal the presence of an anionic guest. These

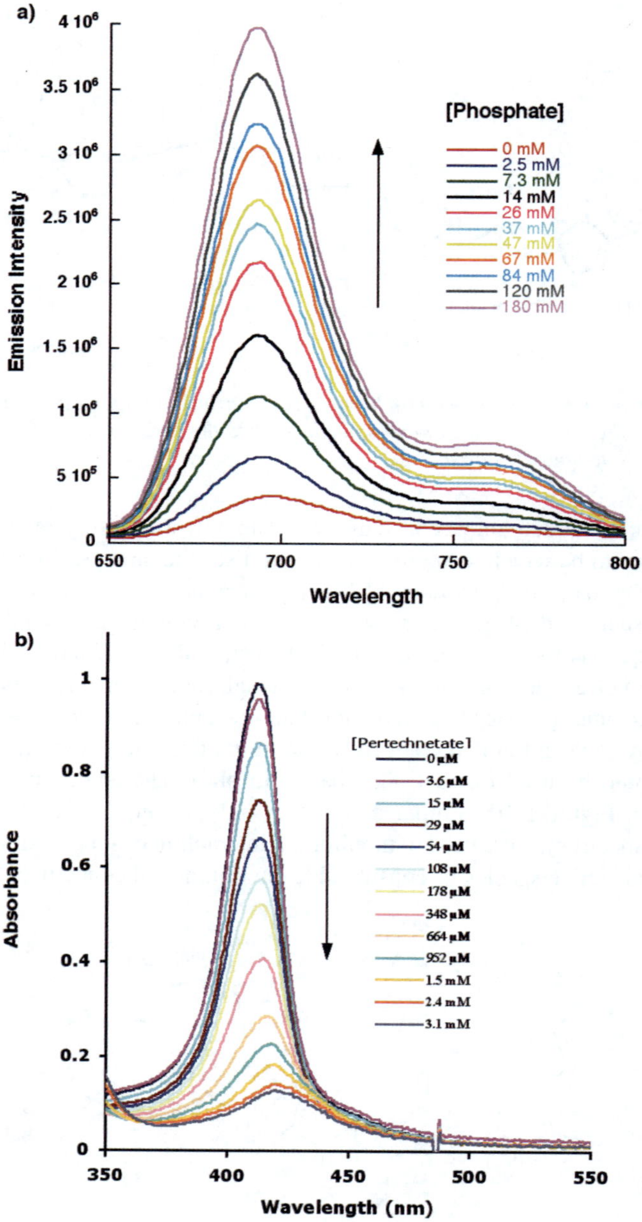

Figure 8.10 *(a) Fluorescence emission spectra (λ_{ex} = 450 nm) of aqueous solutions of sapphyrin* ***8.84a*** *(2.7 μM) containing 0–50 mM sodium phosphate at pH 7.0, 25 mM PIPES buffer, and 150 mM NaCl.* (Reproduced with permission from reference [106]. Copyright 2003 Royal Society of Chemistry) *(b) Absorbance spectra of an aqueous solution (2.5% methanol) of sapphyrin* ***8.84b*** *(46.5 μM) at pH 7.0 (unbuffered solution) recorded in the presence of increasing concentrations of pertechnetate anion.* (Reproduced with permission from reference [107]. Copyright 2004 Taylor and Francis)

sensors have been incorporated into membranes to form ISEs or optodes, or may be used in solution as discrete redox-active, fluorescent, or colorimetric molecular sensors. Anslyn's development of the displacement assay approach has been a leap forward in sensor design as it allows the construction and design of an anion-selective binding site to be unencumbered by a covalently linked reporter group. Since this latter group is now non-covalently associated with the receptor, this strategy allows for optimal design of the receptor and simple formation of the sensing ensemble. We anticipate that variations on this theme as well as more developments in anion-sensor design will emerge as work in this all-important area continues to progress.

References

1. M.S. Frant and J.W. Ross, *Science*, 1966, **154**, 1553.
2. M.M.G. Antonisse, B.H.M. Snellink-Ruël, A.C. Ion, J.F.J. Engbersen and D.N. Reinhoudt, *J. Chem. Soc. Perkin Trans. 2*, 1999, 1211.
3. M.M.G. Antonisse, R.J.W. Lugtenberg, R.J.M. Egberink, J.F.J. Engbersen and D.N. Reinhoudt, *Anal. Chim. Acta*, 1996, **332**, 123.
4. F. Hofmeister, *Arch. Exp. Patol. Phamakol.*, 1888, **24**, 247.
5. U. Oesch, D.A. Ammann and W. Simon, *Clin. Chem.*, 1986, **32**, 1448.
6. P. Bühlmann, E. Pretsch and E. Bakker, *Chem. Rev.*, 1998, **98**, 1593.
7. M.M.G. Antonisse and D.N. Reinhoudt, *Electroanalysis*, 1999, **11**, 1035.
8. S. Nishizawa, P. Bühlmann, K.P. Xiao and Y. Umezawa, *Anal. Chim. Acta*, 1998, **358**, 35.
9. K.P. Xiao, P. Bühlmann, S. Nishizawa, S. Amemiya and Y. Umezawa, *Anal. Chem.*, 1997, **69**, 1038.
10. M.J. Berrocal, A. Cruz, I.H.A. Badr and L.G. Bachas, *Anal. Chem.*, 2000, **72**, 5295.
11. C. Raposo, M. Almaraz, M. Martín, V. Weinrich, M.L. Mussons, V. Alcázar, M.C. Caballero and J.R. Morán, *Chem. Lett.*, 1995, **24**, 759; C. Raposo, M. Crego, M.L. Mussons, M.C. Caballero and J.R. Morán, *Tetrahedron Lett.*, 1994, **35**, 3409; C. Raposo, N. Pérez, M. Almaraz, M.L. Mussons, M.C. Caballero and J.R. Morán, *Tetrahedron Lett.*, 1995, **36**, 3255.
12. K. Tohda, R. Naganawa, X.M. Lin, M. Tange, K. Umezawa, K. Odashima, Y. Umezawa, H. Furuta and J.L. Sessler, *Sensor. Actuat., B*, 1993, **14**, 669.
13. K. Umezawa, K. Tohda, X.M. Lin, J.L. Sessler and Y. Umezawa, *Anal. Chim. Acta*, 2001, **426**, 19.
14. M.M.G. Antonisse, B.H.M. Snellink-Ruël, J.F.J. Engbersen and D.N. Reinhoudt, *Sensor. Actuat., B*, 1998, **47**, 9; M.M.G. Antonisse, B.H.M. Snellink-Ruël, J.F.J. Engbersen and D.N. Reinhoudt, *J. Chem. Soc. Perkin Trans. 2*, 1998, 773; M.M.G. Antonisse, B.H.M. Snellink-Ruël, I. Yigit, J.F.J. Engbersen and D.N. Reinhoudt, *J. Org. Chem.*, 1997, **62**, 9034.
15. I. Szymanska, H. Radecka, J. Radecki, M. Pietraszkiewicz and O. Pietraszkiewicz, *Electroanalysis*, 2003, **15**, 294.
16. V. Král, J.L. Sessler, T.V. Shishkanova, P.A. Gale and R. Volf, *J. Am. Chem. Soc.*, 1999, **121**, 8771.

17. V. Král, P.A. Gale, P. Anzenbacher Jr., K. Jursíková, V. Lynch and J.L. Sessler, *Chem. Commun.*, 1998, 9.
18. J.L. Sessler, V. Král, T.V. Shishkanova and P.A. Gale, *Proc. Natl. Acad. Sci. USA*, 2002, **99**, 4848.
19. V. Král, M. Valík, T.V. Shishkanova and J.L. Sessler, in *Dekker Encyclopedia of Nanoscience and Nanotechnology*, J.A. Schwarz, C.I. Contescu and P. Karol (eds), Marcel Dekker, New York, 2004.
20. E. Bakker, E. Malinowska, R.D. Schiller and M.E. Meyerhoff, *Talanta*, 1994, **41**, 881; E. Malinowska and M.E. Meyerhoff, *Anal. Chim. Acta*, 1995, **300**, 33; E. Wang and M.E. Meyerhoff, *Anal. Lett.*, 1993, **26**, 1519.
21. S.B. Park, W. Matuszewski, M.E. Meyerhoff, Y.H. Liu and K.M. Kadish, *Electroanalysis*, 1991, **3**, 909.
22. P. Schulthess, D. Ammann, W. Simon, C. Caderas, R. Stepánek and B. Kräutler, *Helv. Chim. Acta*, 1984, 67, 1026; R. Stepánek, B. Kräutler, P. Schulthess, B. Lindemann, D. Ammann and W. Simon, *Anal. Chim. Acta*, 1986, **182**, 83.
23. W. Zhang, E. Roznieckia, P.G. Parzuchowski, E. Malinowska and M.E. Meyerhoff, *Anal. Chem.*, 2002, **74**, 4548.
24. S.L.R. Barker, B.A. Thorsrud and R. Kopelman, *Anal. Chem.*, 1998, **70**, 100.
25. I.H.A. Badr, R.D. Johnson, M. Diaz, M.F. Hawthorne and L.G. Bachas, *Anal. Chem.*, 2000, **72**, 4249.
26. P.D. Beer and A.D. Keefe, *J. Organomet. Chem.*, 1989, **375**, C40.
27. P.D. Beer, D. Hesek, J. Hodacova and S.E. Stokes, *J. Chem. Soc. Chem. Commun.*, 1992, 270.
28. P.D. Beer, D. Hesek, J.E. Kingston, D.K. Smith, S.E. Stokes and M.G.B. Drew, *Organometallics*, 1995, **14**, 3288.
29. P.D. Beer, C. Hazlewood, D. Hesek, J. Hodacova and S.E. Stokes, *J. Chem. Soc. Dalton Trans.*, 1993, 1327.
30. P.D. Beer, M.G.B. Drew, D. Hesek and K.C. Nam, *Chem. Commun.*, 1997, 107.
31. P.D. Beer, D. Hesek, K.C. Nam and M.G.B. Drew, *Organometallics*, 1999, **18**, 3933.
32. P.D. Beer and S.E. Stokes, *Polyhedron*, 1995, **14**, 873; P.D. Beer, M.G.B. Drew, J. Hodacova and S.E. Stokes, *J. Chem. Soc. Dalton Trans.*, 1995, 3447; P.D. Beer, M.G.B. Drew, A.R. Graydon, D.K. Smith and S.E. Stokes, *J. Chem. Soc. Dalton Trans.*, 1995, 403; P.D. Beer, M.G.B. Drew and A.R. Graydon, *J. Chem. Soc. Dalton Trans.*, 1996, 4129.
33. P.D. Beer, P.A. Gale and G.Z. Chen, *Adv. Phys. Org. Chem.*, 1998, **31**, 1; P.D. Beer, P.A. Gale and G.Z. Chen, *J. Chem. Soc. Dalton Trans.*, 1999, 1897; P.D. Beer, P.A. Gale and G.Z. Chen, *Coord. Chem. Rev.*, 1999, **185–186**, 3; P.D. Beer, Z. Chen, A.J. Goulden, A.R. Graydon, S.E. Stokes and T. Wear, *J. Chem. Soc. Chem. Commun.*, 1993, 1834; P.D. Beer, A.R. Graydon, A.O.M. Johnson and D.K. Smith, *Inorg. Chem.*, 1997, **36**, 2112.
34. C. Valério, J.-L. Fillaut, J. Ruiz, J. Guittard, J.-C. Blais and D. Astruc, *J. Am. Chem. Soc.*, 1997, **119**, 2588.

35. P.D. Beer, Z. Chen, M.G.B. Drew, J. Kingston, M.I. Ogden and P. Spencer, *J. Chem. Soc. Chem. Commun.*, 1993, 1046; P.D. Beer, Z. Chen, M.G.B. Drew, A.O.M. Johnson, D.K. Smith and P. Spencer, *Inorg. Chim. Acta*, 1996, **246**, 143; P.D. Beer, J. Cadman, J.M. Lloris, R. Martinéz-Máñez, M.E. Padilla, T. Pardo, D.K. Smith and J. Soto, *J. Chem. Soc. Dalton Trans.*, 1999, 127.
36. P.D. Beer, M.G.B. Drew and D.K. Smith, *J. Organomet. Chem.*, 1997, **543**, 259.
37. C. Dusemund, K.R.A.S. Sandanayake and S. Shinkai, *J. Chem. Soc. Chem. Commun.*, 1995, 333; H. Yamamoto, A. Ori, K. Ueda, C. Dusemund and S. Shinkai, *Chem. Commun.*, 1996, 407.
38. R. Altmann, O. Gausset, D. Horn, K. Jurkschat and M. Schürmann, *Organometallics*, 2000, **19**, 430.
39. P. Laurent, H. Miyaji, S.R. Collinson, I. Prokes, C.J. Moody, J.H.R. Tucker and A.M.Z. Slawin, *Org. Lett.*, 2002, **4**, 4037.
40. M.D. Pratt and P.D. Beer, *Polyhedron*, 2003, **22**, 649.
41. F. Otón, A. Tárraga, M.D. Velasco, A. Espinosa and P. Molina, *Chem. Commun.*, 2004, 1658.
42. J.L. Sessler, A. Gebauer and A. Gale Philip, *Gazz. Chim. Ital.*, 1997, **127**, 723.
43. P.A. Gale, M.B. Hursthouse, M.E. Light, J.L. Sessler, C.N. Warriner and R.S. Zimmerman, *Tetrahedron Lett.*, 2001, **42**, 6759.
44. M. Scherer, J.L. Sessler, A. Gebauer and V. Lynch, *Chem. Commun.*, 1998, 85.
45. J.L. Sessler, R.S. Zimmerman, G.J. Kirkovits, A. Gebauer and M. Scherer, *J. Organomet. Chem.*, 2001, **637–639**, 343.
46. G. Denuault, P.A. Gale, M.B. Hursthouse, M.E. Light and C.N. Warriner, *New J. Chem.*, 2002, **26**, 811.
47. Z. Chen, A.R. Graydon and P.D. Beer, *J. Chem. Soc. Faraday Trans.*, 1996, **92**, 97; P.D. Beer and M. Shade, *Chem. Commun.*, 1997, 2377; P.A. Gale, Z. Chen, M.G.B. Drew, J.A. Heath and P.D. Beer, *Polyhedron*, 1998, **17**, 405; J.E. Kingston, L. Ashford, P.D. Beer and M.G.B. Drew, *J. Chem. Soc. Dalton Trans.*, 1999, 251; M. Buda, A. Ion, J.-C. Moutet, E. Saint-Aman and R. Ziessel, *J. Electroanal. Chem.*, 1999, **469**, 132; J.F. Gallagher, P.T.M. Kenny and M.J. Sheehy, *Inorg. Chem. Commun.*, 1999, 2, 200; K. Kavallieratos, S. Hwang and R.H. Crabtree, *Inorg. Chem.*, 1999, **38**, 5184; O. Reynes, F. Maillard, J.-C. Moutet, G. Royal, E. Saint-Aman, G. Stanciu, J.-P. Dutasta, I. Gosse and J.-C. Mulatier, *J. Organomet. Chem.*, 2001, **637–639**, 356; G. Cooke, F.M.A. Duclairoir, A. Kraft, G. Rosair and V.M. Rotello, *Tetrahedron Lett.*, 2004, **45**, 557.
48. I. Dumazet and P.D. Beer, *Tetrahedron Lett.*, 1999, **40**, 785.
49. P.D. Beer and F. Szemes, *J. Chem. Soc. Chem. Commun.*, 1995, 2245.
50. P.D. Beer, F. Szemes, V. Balzani, C.M. Salà, M.G.B. Drew, S.W. Dent and M. Maestri, *J. Am. Chem. Soc.*, 1997, **119**, 11864.
51. P.D. Beer, C.A.P. Dickson, N.C. Fletcher, A.J. Goulden, A. Grieve, J. Hodacova and T. Wear, *J. Chem. Soc. Chem. Commun.*, 1993, 828; P.D. Beer, Z. Chen, A.J. Goulden, A. Grieve, D. Hesek, F. Szemes and T. Wear, *J. Chem. Soc. Chem. Commun.*, 1994, 1269; F. Szemes, D. Hesek, Z. Chen, S.W. Dent,

M.G.B. Drew, A.J. Goulden, A.R. Graydon, A. Grieve, R.J. Mortimer, T. Wear, J.S. Weightman and P.D. Beer, *Inorg. Chem.*, 1996, **35**, 5868; P.D. Beer and N.C. Fletcher, *Polyhedron*, 1996, **15**, 1339; P.D. Beer, S.W. Dent and T. Wear, *J. Chem. Soc. Dalton Trans.*, 1996, 2341; P.D. Beer, N.C. Fletcher, M.G.B. Drew and T. Wear, *Polyhedron*, 1997, **16**, 815; P.D. Beer, S.W. Dent, G.S. Hobbs and T. Wear, *Chem. Commun.*, 1997, 99; P.D. Beer and M. Shade, *Gazz. Chim. Ital.*, 1997, **127**, 651; P.D. Beer, V. Timoshenko, M. Maestri, P. Passaniti and V. Balzani, *Chem. Commun.*, 1999, 1755; S.-S. Sun and A.J. Lees, *Chem. Commun.*, 2000, 1687; S. Camiolo, S.J. Coles, P.A. Gale, M.B. Hursthouse, T.A. Mayer and M.A. Paver, *Chem. Commun.*, 2000, 275; P.D. Beer, N. Berry, M.G.B. Drew, O.D. Fox, M.E. Padilla-Tosta and S. Patell, *Chem. Commun.*, 2001, 199; L.H. Uppadine, F.R. Keene and P.D. Beer, *J. Chem. Soc. Dalton Trans.*, 2001, 2188.

52. K.A. Nielsen, J.O. Jeppesen, E. Levillain and J. Becher, *Angew. Chem. Int. Ed.*, 2003, **42**, 187.
53. K.A. Nielsen, W.-S. Cho, J.O. Jeppesen, V.M. Lynch, J. Becher and J.L. Sessler, *J. Am. Chem. Soc.*, 2004, **126**, 16296.
54. K. Heuzé, C. Mézière, M. Fourmigué, P. Batail, C. Coulon, E. Canadell, P. Auban-Senzier and D. Jérome, *Chem. Mater.*, 2000, **12**, 1898.
55. R.P. Haugland, in *Handbook of Fluorescent Probes and Research Products*, J. Gregory (ed), Molecular Probes Inc, Eugene Oregon, 2002.
56. S. Jayaraman and A.S. Verkman, *Biophys. Chem.*, 2000, **85**, 49.
57. S. Nishizawa, Y. Kato and N. Teramae, *J. Am. Chem. Soc.*, 1999, **121**, 9463.
58. P. de Montigny, J.F. Stobaugh, R.S. Givens, R.G. Carlson, K. Srinivasachar, L.A. Sternson and T. Higuchi, *Anal. Chem.*, 1987, **59**, 1096.
59. R. Martínez-Máñez and F. Sancenon, *Chem. Rev.*, 2003, **103**, 4419.
60. J.V. Ros-Lis, R. Martínez-Máñez and J. Soto, *Chem. Commun.*, 2002, 2248.
61. T.P. Misko, R.J. Schilling, D. Salvemini, W.M. Moore and M.G. Currie, *Anal. Biochem.*, 1993, **214**, 11.
62. S. Yamaguchi, S. Akiyama and K. Tamao, *J. Am. Chem. Soc.*, 2001, **123**, 11372.
63. S. Yamaguchi, S. Akiyama and K. Tamao, *J. Am. Chem. Soc.*, 2000, **122**, 6793.
64. C.R. Cooper, N. Spencer and T.D. James, *Chem. Commun.*, 1998, 1365; M. Nicolas, B. Fabre and J. Simonet, *Chem. Commun.*, 1999, 1881; C.J. Ward, P. Patel and T.D. James, *Chem. Lett.*, 2001, **30**, 406; C.J. Ward, P. Patel and T.D. James, *Org. Lett.*, 2002, **4**, 477; S. Yamaguchi, T. Shirasaka, S. Akiyama and K. Tamao, *J. Am. Chem. Soc.*, 2002, **124**, 8816; Y. Kubo, M. Yamamoto, M. Ikeda, M. Takeuchi, S. Shinkai, S. Yamaguchi and K. Tamao, *Angew. Chem. Int. Ed.*, 2003, **42**, 2036; A. Coskun and E.U. Akkaya, *Tetrahedron Lett.*, 2004, **45**, 4947.
65. H. Miyaji and J.L. Sessler, *Angew. Chem. Int. Ed.*, 2001, **40**, 154.
66. H. Tong, G. Zhou, L. Wang, X. Jing, F. Wang and J. Zhang, *Tetrahedron Lett.*, 2003, **44**, 131.
67. D.H. Lee, K.H. Lee and J.-I. Hong, *Org. Lett.*, 2001, **3**, 5; D.H. Lee, H.Y. Lee, K.H. Lee and J.-I. Hong, *Chem. Commun.*, 2001, 1188.
68. D.H. Lee, H.Y. Lee and J.-I. Hong, *Tetrahedron Lett.*, 2002, **43**, 7273.

69. M. Vázquez, L. Fabbrizzi, A. Taglietti, R.M. Pedrido, A.M. González-Noya and M.R. Bermejo, *Angew. Chem. Int. Ed.*, 2004, **43**, 1962.
70. R.C. Jagessar, M. Shang, W.R. Scheidt and D.H. Burns, *J. Am. Chem. Soc.*, 1998, **120**, 11684; T. Hayashita, T. Onodera, R. Kato, S. Nishizawa and N. Teramae, *Chem. Commun.*, 2000, 755; C. Lee, D.H. Lee and J.-I. Hong, *Tetrahedron Lett.*, 2001, **42**, 8665; T. Gunnlaugsson, A.P. Davis and M. Glynn, *Chem. Commun.*, 2001, 2556; R. Kato, S. Nishizawa, T. Hayashita and N. Teramae, *Tetrahedron Lett.*, 2001, **42**, 5053; M. Mei and S. Wu, *New J. Chem.*, 2001, **25**, 471; G. Hennrich, H. Sonnenschein and U. Resch-Genger, *Tetrahedron Lett.*, 2001, **42**, 2805; S.K. Kim and J. Yoon, *Chem. Commun.*, 2002, 770; D.H. Lee, J.H. Im, J.-H. Lee and J.-I. Hong, *Tetrahedron Lett.*, 2002, **43**, 9637; D. Jiménez, R. Martínez-Máñez, F. Sancenón and J. Soto, *Tetrahedron Lett.*, 2002, **43**, 2823; T. Gunnlaugsson, A.P. Davis, J.E. O'Brien and M. Glynn, *Org. Lett.*, 2002, **4**, 2449; F.-Y. Wu, Z. Li, Z.-C. Wen, N. Zhou, Y.-F. Zhao and Y.-B. Jiang, *Org. Lett.*, 2002, **4**, 3203; M. Dudic, P. Lhoták, I. Stibor, K. Lang and P. Prosková, *Org. Lett.*, 2003, **5**, 149; S.-I. Kondo, M. Nagamine and Y. Yano, *Tetrahedron Lett.*, 2003, **44**, 8801; E.J. Cho, J.W. Moon, S.W. Ko, J.Y. Lee, S.K. Kim, J. Yoon and K.C. Nam, *J. Am. Chem. Soc.*, 2003, **125**, 12376; J.Y. Lee, E.J. Cho, S. Mukamel and K.C. Nam, *J. Org. Chem.*, 2004, **69**, 943; D.A. Jose, D.K. Kumar, B. Ganguly and A. Das, *Org. Lett.*, 2004, **6**, 3445; P. Manesiotis, A.J. Hall, M. Emgenbroich, M. Quaglia, E. de Lorenzi and B. Sellergren, *Chem. Commun.*, 2004, 2278; Z.-C. Wen and Y.-B. Jiang, *Tetrahedron*, 2004, 60, 11109; Z.-Y. Zeng, Y.-B. He, J.-L. Wu, L.-H. Wei, X. Liu, L.-Z. Meng and X. Yang, *Eur. J. Org. Chem.*, 2004, 2888; Z.-Y. Zeng, Y.-B. He, J.-L. Wu, L.-H. Wei, S.-Y. Liu, Y.-Y. Huang, L.-Z. Meng and L. Hu, *Supramol. Chem.*, 2004, **16**, 233; J.-L. Wu, Y.-B. He, L.-H. Wei, S.-Y. Liu, L.-Z. Meng and L. Hu, *Supramol. Chem.*, 2004, **16**, 353; M. Boiocchi, L.D. Boca, D. Esteban-Gómez, L. Fabbrizzi, M. Licchelli and E. Monzani, *Chem. Eur. J.*, 2005, **11**, 3097; D.H. Burns, K. Calderon-Kawasaki and S. Kularatne, *J. Org. Chem.*, 2005, **70**, 2803; V. Thiagarajan, P. Ramamurthy, D. Thirumalai and V.T. Ramakrishnan, *Org. Lett.*, 2005, **7**, 657; A.B. Descalzo, K. Rurack, H. Weisshoff, R. Martinéz-Máñez, M.D. Marcos, P. Amorós, K. Hoffmann and J. Soto, *J. Am. Chem. Soc.*, 2005, **127**, 184; D.A. Jose, D.K. Kumar, B. Ganguly and A. Das, *Tetrahedron Lett.*, 2005, **46**, 5343.
71. T. Gunnlaugsson, P.E. Kruger, P. Jensen, F.M. Pfeffer and G.M. Hussey, *Tetrahedron Lett.*, 2003, **44**, 8909; T. Gunnlaugsson, P.E. Kruger, T.C. Lee, R. Parkesh, F.M. Pfeffer and G.M. Hussey, *Tetrahedron Lett.*, 2003, **44**, 6575.
72. P. Piatek and J. Jurczak, *Chem. Commun.*, 2002, 2450; S. Watanabe, M. Sonobe, M. Arai, Y. Tazume, T. Matsuo, T. Nakamura and K. Yoshida, *Chem. Commun.*, 2002, 2866; C.-F. Chen and Q.-Y. Chen, *Tetrahedron Lett.*, 2004, 45, 3957; R. Miao, Q.-Y. Zheng, C.-F. Chen and Z.-T. Huang, *Tetrahedron Lett.*, 2004, **45**, 4959; Q.-Y. Chen and C.-F. Chen, *Eur. J. Org. Chem.*, 2005, 2468; R. Miao, Q.-Y. Zheng, C.-F. Chen and Z.-T. Huang, *Tetrahedron Lett.*, 2005, **46**, 2155; J.-L. Fillaut, J. Andriès, L. Toupet and J.-P. Desvergne, *Chem. Commun.*, 2005, 2924.

73. H. Miyaji, P. Anzenbacher Jr., J.L. Sessler, E.R. Bleasdale and P.A. Gale, *Chem. Commun.*, 1999, 1723.
74. P. Anzenbacher Jr., K. Jursíková and J.L. Sessler, *J. Am. Chem. Soc.*, 2000, **122**, 9350.
75. H. Miyaji, W. Sato, J.L. Sessler and V.M. Lynch, *Tetrahedron Lett.*, 2000, **41**, 1369; H. Miyaji, W. Sato and J.L. Sessler, *Angew. Chem. Int.* Ed., 2000, **39**, 1777; H. Miyaji, W. Sato, D. An and J.L. Sessler, *Collect. Czech. Chem. Commun.*, 2004, **69**, 1027.
76. R. Nishiyabu and P. Anzenbacher Jr., *J. Am. Chem. Soc.*, 2005, **127**, 8270.
77. C.B. Black, B. Andrioletti, A.C. Try, C. Ruiperez and J.L. Sessler, *J. Am. Chem. Soc.*, 1999, **121**, 10438.
78. B. Oddo, *Gazz. Chim. Ital.*, 1911, **41**, 248.
79. B. Lindstom, *Acta Chem. Scand.*, 1973, **27**, 2411.
80. P. Anzenbacher Jr., A.C. Try, H. Miyaji, K. Jursíková, V.M. Lynch, M. Marquez and J.L. Sessler, *J. Am. Chem. Soc.*, 2000, **122**, 10268.
81. P. Anzenbacher Jr., D.S. Tyson, K. Jursíková and F.N. Castellano, *J. Am. Chem. Soc.*, 2002, **124**, 6232.
82. T. Mizuno, W.-H. Wei, L.R. Eller and J.L. Sessler, *J. Am. Chem. Soc.*, 2002, **124**, 1134.
83. D. Aldakov and P. Anzenbacher Jr., *Chem. Commun.*, 2003, 1394.
84. D. Aldakov and P. Anzenbacher Jr., *J. Am. Chem. Soc.*, 2004, **126**, 4752; P. Anzenbacher Jr., K. Jursikova, D. Aldakov, M. Marquez and R. Pohl, *Tetrahedron*, 2004, **60**, 11163.
85. J.-H. Liao, C.-T. Chen, H.-T. Chou, C.-C. Cheng, P.-T. Chou, J.-M. Fang, Z. Slanina and T.J. Chow, *Org. Lett.*, 2002, **4**, 3107.
86. S.V. Shevchuk, V.M. Lynch and J.L. Sessler, *Tetrahedron*, 2004, **60**, 11283.
87. J.L. Sessler, H. Maeda, T. Mizuno, V.M. Lynch and H. Furuta, *Chem. Commun.*, 2002, 862.
88. J.L. Sessler, G.D. Pantos, E. Katayev and V.M. Lynch, *Org. Lett.*, 2003, **5**, 4141.
89. J.L. Sessler, H. Maeda, T. Mizuno, V.M. Lynch and H. Furuta, *J. Am. Chem. Soc.*, 2002, **124**, 13474.
90. S. Camiolo, P.A. Gale, M.B. Hursthouse and M.E. Light, *Org. Biomol. Chem.*, 2003, **1**, 741.
91. A. Metzger, V.M. Lynch and E.V. Anslyn, *Angew. Chem. Int. Ed.*, 1997, **36**, 862; M. Rekharsky, Y. Inoue, S.L. Tobey, A. Metzger and E.V. Anslyn, *J. Am. Chem. Soc.*, 2002, **124**, 14959; S.C. McCleskey, A. Metzger, C.S. Simmons and E.V. Anslyn, *Tetrahedron*, 2002, **58**, 621; S.C. McCleskey, P.N. Floriano, S.L. Wiskur, E.V. Anslyn and J.T. McDevitt, *Tetrahedron*, 2003, **59**, 10089.
92. A. Metzger and E.V. Anslyn, *Angew. Chem. Int. Ed.*, 1998, **37**, 649.
93. B.V.L. Potter and D. Lampe, *Angew. Chem. Int. Ed.*, 1995, **34**, 1933; M.J. Berridge, *Nature*, 1993, **361**, 315.
94. K. Niikura, A. Metzger and E.V. Anslyn, *J. Am. Chem. Soc.*, 1998, **120**, 8533.
95. J.P. Lorand and J.O. Edwards, *J. Org. Chem.*, 1959, **24**, 769.
96. Z. Zhong and E.V. Anslyn, *J. Am. Chem. Soc.*, 2002, **124**, 9014.
97. J.J. Lavigne and E.V. Anslyn, *Angew. Chem. Int. Ed.*, 1999, **38**, 3666.

98. A.M. Piatek, Y.J. Bomble, S.L. Wiskur and E.V. Anslyn, *J. Am. Chem. Soc.*, 2004, **126**, 6072.
99. S.L. Wiskur, P.N. Floriano, E.V. Anslyn and J.T. McDevitt, *Angew. Chem. Int. Ed.*, 2003, **42**, 2070.
100. B.T. Nguyen, S.L. Wiskur and E.V. Anslyn, *Org. Lett.*, 2004, **6**, 2499.
101. S.L. Wiskur and E.V. Anslyn, *J. Am. Chem. Soc.*, 2001, **123**, 10109.
102. S.L. Wiskur, J.J. Lavigne, A. Metzger, S.L. Tobey, V. Lynch and E.V. Anslyn, *Chem. Eur. J.*, 2004, **10**, 3792.
103. S.C. McCleskey, M.J. Griffin, S.E. Schneider, J.T. McDevitt and E.V. Anslyn, *J. Am. Chem. Soc.*, 2003, **125**, 1114.
104. P.A. Gale, L.J. Twyman, C.I. Handlin and J.L. Sessler, *Chem. Commun.*, 1999, 1851.
105. R.S. Dickins, T. Gunnlaugsson, D. Parker and R.D. Peacock, *Chem. Commun.*, 1998, 1643; J. Bruce, R.S. Dickins, L.J. Govenlock, T. Gunnlaugsson, S. Lopinski, M.P. Lowe, D. Parker, R.D. Peacock, J.J.B. Perry, S. Aime and M. Botta, *J. Am. Chem. Soc.*, 2000, **122**, 9674; D. Parker and J. Yu, *Chem. Commun.*, 2005, 3141; P. Atkinson, Y. Bretonnière, D. Parker and G. Muller, *Helv. Chim. Acta*, 2005, **88**, 391.
106. J.L. Sessler, J.M. Davis, V. Král, T. Kimbrough and V.M. Lynch, *Org. Biomol. Chem.*, 2003, **1**, 4113.
107. A.E.V. Gorden, J.M. Davis, J.L. Sessler, V. Král, D.W. Keogh and N.L. Schroeder, *Supramol. Chem.*, 2004, **16**, 91.

CHAPTER 9

Anion-Controlled Assembly and Template-Based Synthesis

9.1 Introduction

The use of anions as templates for the formation of new molecules or supermolecules is a relatively new and relatively unexplored area of supramolecular chemistry. Over the last few decades, many self-assembling systems have been reported that rely on π-stacking, hydrogen bonding, or transition metal ions to direct the formation of the assembled structure.[1] The directional nature of hydrogen bonds make this interaction ideal for assembling supermolecules (in both biotic and abiotic systems). Similarly, the directional nature of the coordination sphere of certain transition metals also makes these species useful for directing self-assembly processes (producing self-assembled complexes with varying degrees of lability). As we have seen, anions have many shapes, sizes, and charges and hence, as a group, possess a variety of features that make them potentially attractive for the assembly of new molecular and supramolecular species. Anions may also be used to interfere with other self-assembling systems by binding to their component parts and, for instance, disrupting the assembly process. In the previous chapters in this book, we classified anion receptors according to the nature of the interaction used to coordinate the anion (hydrogen bonding, electrostatics, electron pair donation, *etc.*). However, in this chapter we will look at the assembly processes from the point of view of the anion, starting with halide-controlled assemblies, moving through oxyanion-controlled assemblies and ending up with examples of the formation of molecular structures controlled by the normally weakly coordinating fluorinated anions, BF_4^- and PF_6^-.

9.2 Halide-Controlled Assemblies

Halide anions are spherical, with no directional preferences in their coordination spheres. As we saw in Chapter 1, chloride anions have been used by Hawthorne to template the assembly of mercuracarborand-anion receptors, while Sessler has used both nitrate and chloride to template the synthesis of expanded porphyrins (Chapter 3). The examples of halide-controlled assemblies in this section concern the assembly of positively charged structures around a negative halide core.

As was discussed briefly in the introduction, Lehn and co-workers[2] have discovered a striking example of anion-directed assembly (Scheme 9.1). The tris-bipyridine ligand **9.1** self-assembles in the presence of iron(II) to form a variety of circular double-helicate complexes with sizes that depend upon the nature of the iron(II) salt used (Table 9.1). In the presence of an equimolar quantity of $FeCl_2$, the ligands and iron(II) centres form a pentameric complex **9.2** that encapsulates a chloride anion at its core (see Chapter 1). However, in the presence of $FeSO_4$, an hexameric circular double helicate **9.3** is formed that does not encapsulate an anion. The chloride anion directs the assembly of the pentamer and is consequently trapped within the structure, with no exchange observed when exposed to other anions such as hexafluorophosphate or triflate. In the presence of $FeBr_2$, a mixture of pentamer and hexamer is formed. There are, therefore, two distinct assembly processes occurring

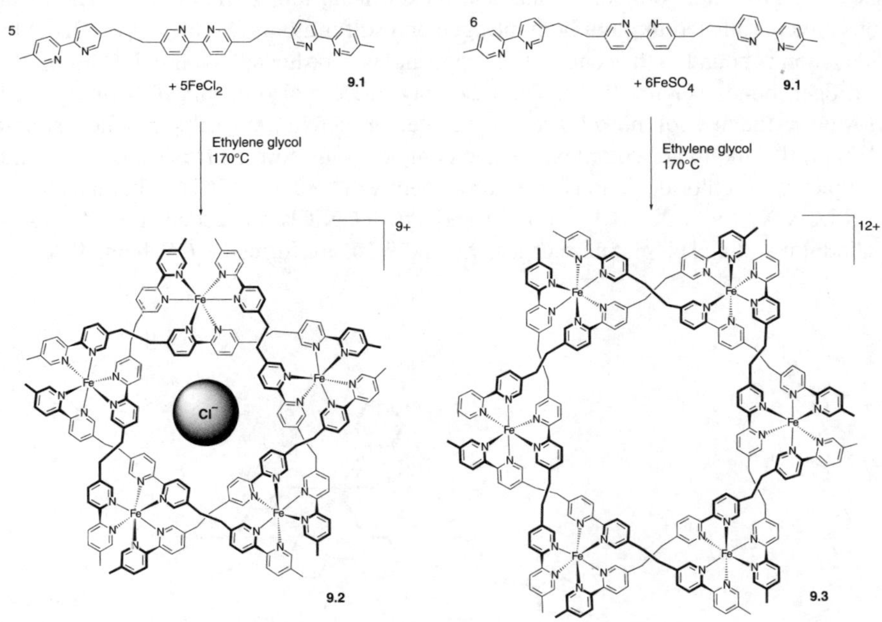

Scheme 9.1 *Formation of the chloride-templated pentameric circular helicate* ***9.2*** *and the untemplated hexameric circular helicate* ***9.3***

Table 9.1 *Compounds formed by complexation of ligand* ***9.1*** *with a variety of iron(II) salts*

Iron (II) salt	*Complex*
FeF_2	Insoluble
$FeCl_2$	Pentamer **9.2**
$FeBr_2$	Pentamer **9.2** + Hexamer **9.3**
FeI_2	Insoluble
$Fe(BF_4)_2$	Hexamer **9.3**
$FeSO_4$	Hexamer **9.3**
$FeSiF_6$	Hexamer **9.3**

during the formation of the pentamer complex. First, the iron centres form precursor helicate structures with the ligands. Next, a torus is formed from these precursor complexes with, in the case of the pentamer, its size being defined by the templating chloride anion.

In contradistinction to ligand **9.1**, the longer ligand **9.4** forms a tetrameric double helicate structure **9.5** with a variety of iron(II) salts irrespective of the identity of the counter anion (Scheme 9.2). In this case, the formation of the circular double helicate is not anion controlled. Nevertheless, electrospray mass spectrometric studies provide evidence that this complex may act as a receptor for chloride anion.

Halide anions also control the assembly of cage complexes formed from nickel(II) salts and amidinothiourea.[3] Mingos and co-workers have discovered that the reaction of $NiCl_2$ with amidinothiourea in methanol produces the cage complex **9.8** that consists of eight amidinothiourea units and nickel metal ions (Scheme 9.3). The metal ions are coordinated through both nitrogen and sulfur atoms (Figure 9.1a). The chloride anion is bound in the centre of the cage and is coordinated to eight NH groups by hydrogen bonds (Figure 9.1b). The assembly process also occurs if $NiBr_2$ is used. However, the use of nickel acetate, nitrate, or perchlorate salts produce simple $[Ni(atu)_2]^{2+}$ monomer complexes. Cage complexes are formed from these monomer complexes if chloride anions are subsequently added (as KCl). Alternatively, if $Pd(PhCN)X_2$ (with X = Cl^- or Br^-) and $Pt(MeCN)Cl_2$ are added, mixed nickel–palladium and nickel–platinum cages, **9.9** and **9.10**, are formed (*cf.* Scheme 9.3).[4]

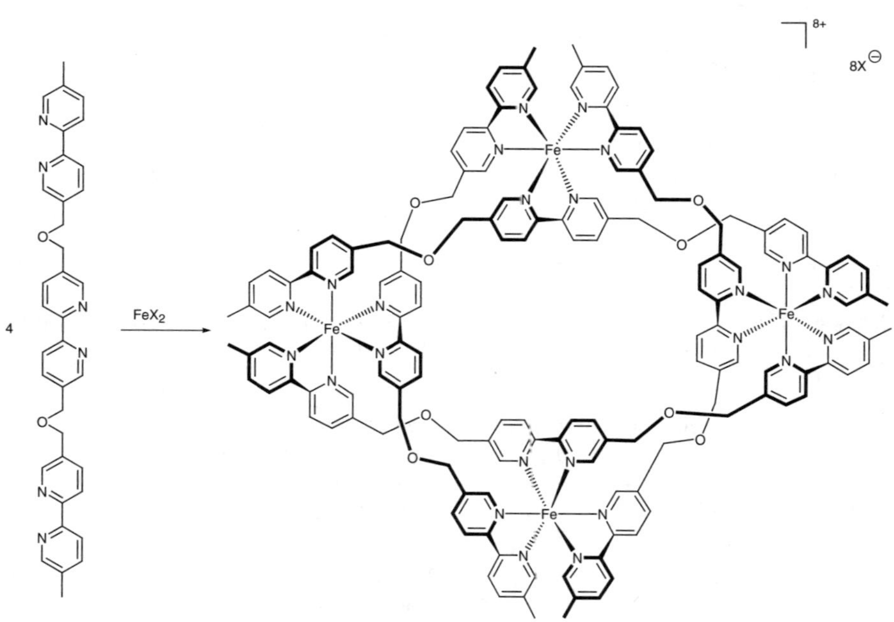

Scheme 9.2 *Formation of the tetrameric helicate* **9.5**. *Counter ions are defined by the salts listed in Table 9.1.*

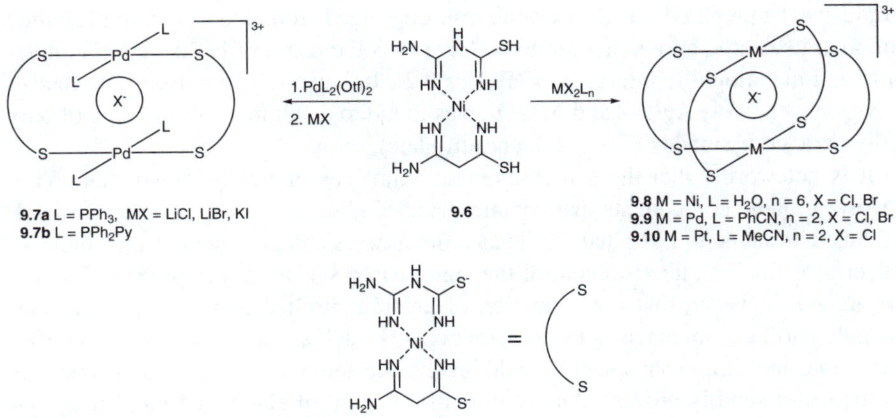

Scheme 9.3 *Synthesis of a mixed nickel/palladium anion-templated cage and metallamacrocycles*

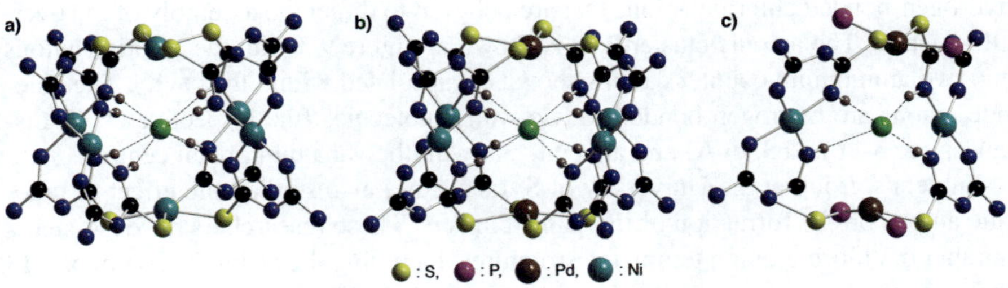

Figure 9.1 *Single crystal X-ray diffraction structures of the chloride-templated (a) nickel cage* ***9.8****, (b) nickel–palladium complex* ***9.9****, and (c) nickel–palladium macrocycle* ***9.7a****. For the sake of clarity, the phenyl groups have been omitted from these representations*

Slightly later, Vilar and co-workers described the anion-templated synthesis of corresponding macrocycles **9.7a** and **9.7b**. Again, in the presence of large anions such as triflate and nitrate, only monometallic species were observed. These species could be converted to the macrocycles via the addition of halide anions (Scheme 9.3). In the solid state, the chloride anion was found encapsulated within the cavity, being held in place via a combination of hydrogen bonds and Lewis acid–base interactions (Figure 9.1c).[5]

The work of Vilar and co-workers provides further examples of the use of anions to template the formation of inorganic structures. These researchers have reported the synthesis of $[Ag_{14}(C{\equiv}C^tBu)_{12}Cl]OH$, from a solution of $AgBF_4$ appropriately mixed with t-BuC≡CH and NEt_3. Initially, the product of the reaction is an insoluble polymeric species, $[Ag(C{\equiv}Ct\text{-}Bu)]_n$, which, upon the addition of chloride, is transformed into the metallacage **9.11**.[6] More recently, a more direct route to product **9.11**, from $AgBF_4$, t-BuC≡CH, NEt_3, and NMe_4Cl, has been developed.[7] Via this latter route, cages of general structure $[Ag_{14}(C{\equiv}C^tBu)_{12}X][BF_4]$ (X = F, Cl, Br) could also be prepared. In all cases, the resulting cage is found to have a single halide anion at its centre, and is thought to be formed as the result of halide-anion templation and metallophilic interactions (Figure 9.2). Interestingly, the use of alternative silver salts, namely AgOTs and $AgNO_3$, was found to result in the production of only polymeric compounds (Ts = *p*-toluenesulfonate).

It is noteworthy that the self-assembled complexes of Lehn, Mingos, and Vilar involve organic components that contain metal-binding sites in specific positions. It is thus reasonable to infer that the ligands themselves contain much of the "molecular information" required to control the anion-based self-assembly process. It could be argued, however, that the formation of anion-controlled inorganic cluster compounds starting from much "simpler" components, such as metal salts, water, hydroxide anion, and simple phosphates would involve the anion in a more fundamental way in the self-assembly process. An example of this kind of elaborated molecular construction was reported by Müller and his group in the early 1990s.[8] In this case, the reaction of ammonium metavanadate with phenylphosphonic acid and hydrazine hydrate in the presence of dimethylammonium chloride in water/DMF was found to form a cage complex **9.12**. This system consists of an anionic cluster containing two hydrogen-bonded chloride anions that are believed to direct the assembly of the overall structure. The anionic cluster **9.12** is shown in Figure 9.3 with the chloride anions and two ammonium counter cations shown encapsulated within the cavity. The chloride anions are hydrogen bonded to two water molecules (the respective O···Cl distances are 3.11 and 3.16 Å) and also interact with the vanadium metal centres.

Zubieta's group at the University of Syracuse has also studied the effect of chloride anions on the formation of inorganic clusters. These researchers have isolated a number of chloride-anion templated structures, including the beautiful complex **9.13** that encapsulates a single chloride anion (Figure 9.4). This cage was formed by the solvothermal reaction of *t*-$BuPO_3H_2$ with $[Ph_4P][VO_2Cl_2]$ in acetonitrile and was isolated as lustrous green crystals.[9] This group also succeeded in isolating a variety of other chloride-anion templated cages as the result of making very minor changes to the starting materials or reaction conditions.

Another very elegant example of halide-controlled assembly has been reported by Zheng and his co-workers. These researchers have relied on chloride anions to direct

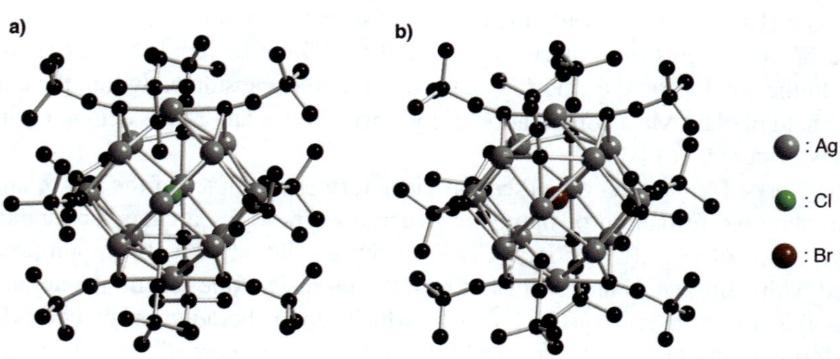

Figure 9.2 *Structures of the silver-cage complexes **9.11** formed with chloride and bromide anions as determined from single crystal X-ray diffraction analyses*

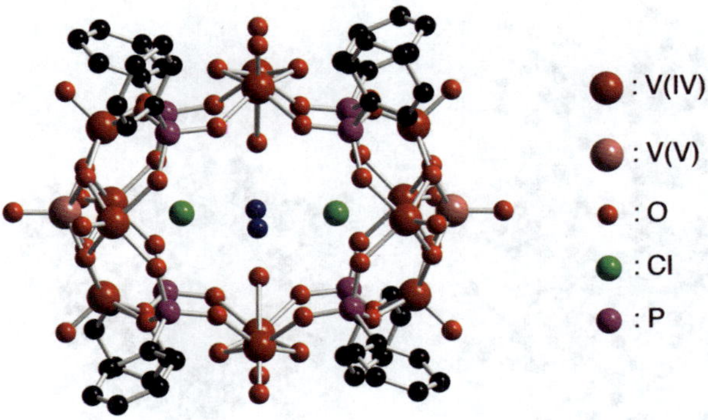

Figure 9.3 *Structure of **9.12** as determined from a single crystal X-ray diffraction analysis*

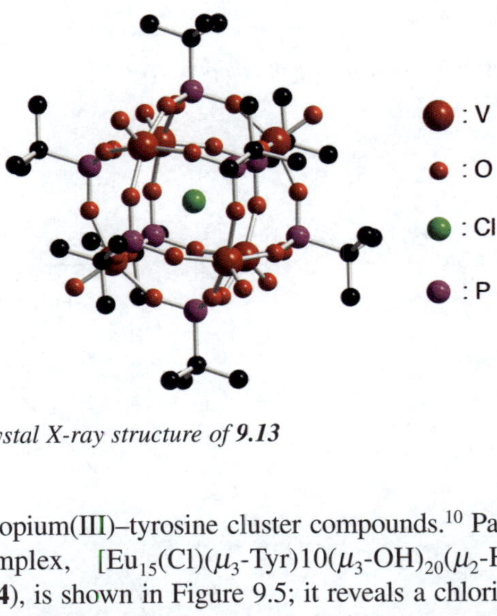

Figure 9.4 *Single crystal X-ray structure of **9.13***

the assembly of europium(III)–tyrosine cluster compounds.[10] Part of the structure of the resulting complex, $[Eu_{15}(Cl)(\mu_3\text{-Tyr})10(\mu_3\text{-OH})_{20}(\mu_2\text{-}H_2O)_5(OH)_{12}(H_2O)_8][ClO_4]_2\cdot 56H_2O$ (**9.14**), is shown in Figure 9.5; it reveals a chloride anion surrounded by a wheel of Eu–O cubane sub-structures such that an average Eu–Cl distance of 3.31 Å is established. The europium ions are assembled into three layers, each of which contains five metal centres. Slightly later, a series of pentadecanuclear lanthanide-hydroxo complexes (*i.e.*, Nd, Gd, Pr, Dy, and Er) were reported. These species were also thought to be formed via a halide-templated self-assembled pathway.[11]

Wright and co-workers have investigated the anion-templated synthesis of a series of macrocycles containing main group elements. They found that the reaction of $[ClP(\mu\text{-N}^t\text{Bu})]_2$ (**9.15**) with $[NH_2P(\mu\text{-N}^t\text{Bu})]_2$ (**9.16**) in the presence of a base produces a $\{[P(\mu\text{-N}^t\text{Bu})_2](\mu\text{-NH})\}_n$ framework.[12] In the presence of NEt_3, the tetrameric species **9.17** is formed (*i.e.*, $n = 4$).[13] However, when the reaction is performed in the presence of excess LiX (X = Cl, Br, and I), the major product obtained is the pentameric product $\{[P(\mu\text{-N}^t\text{Bu})_2](\mu\text{-NH})\}_5(HX)$ **9.18** (X = Cl, Br, and I), as shown in Scheme 9.4. The halide ions were found at the centre of the macrocycle, bound by five hydrogen bonds from the NH groups (Figure 9.6). ^{31}P NMR spectroscopic studies of the

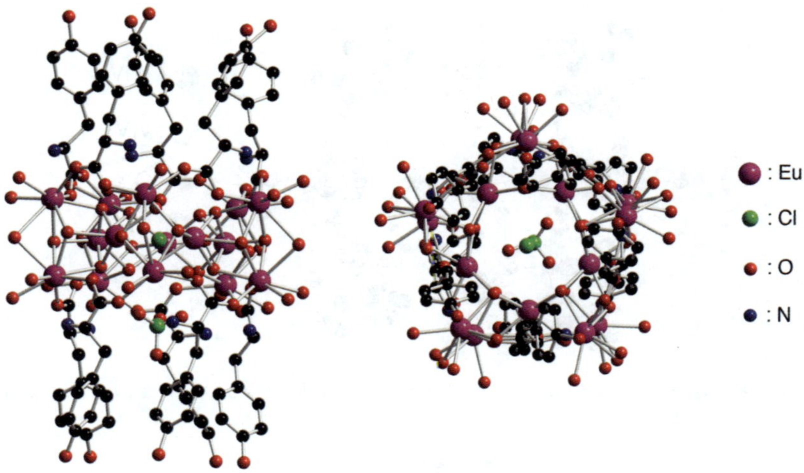

Figure 9.5 *Single crystal X-ray structure of complex* ***9.14***

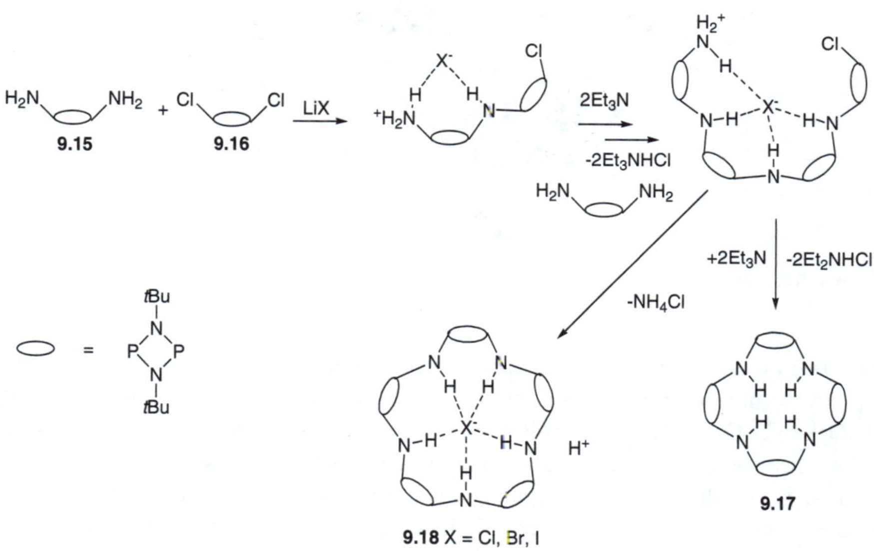

Scheme 9.4 *Preparation of both the tetra- (**9.17**) and pentameric (**9.18**) forms of* $\{[P(\mu\text{-}N^tBu)_2](\mu\text{-}NH)\}_n$

reaction products also showed that the formation of **9.18** can be templated by excess halide ions, with iodide being better than bromide, which is in turn better than chloride, in terms of effecting this process. Presumably, this reflects the fact that iodide provides a better size match than bromide or chloride for the cavity present in **9.18** or the incipient species leading to its formation.

Steel and Sumby[14] have employed fluoride as a template for the formation of a silver-containing cage. In particular, these workers found that complexation of

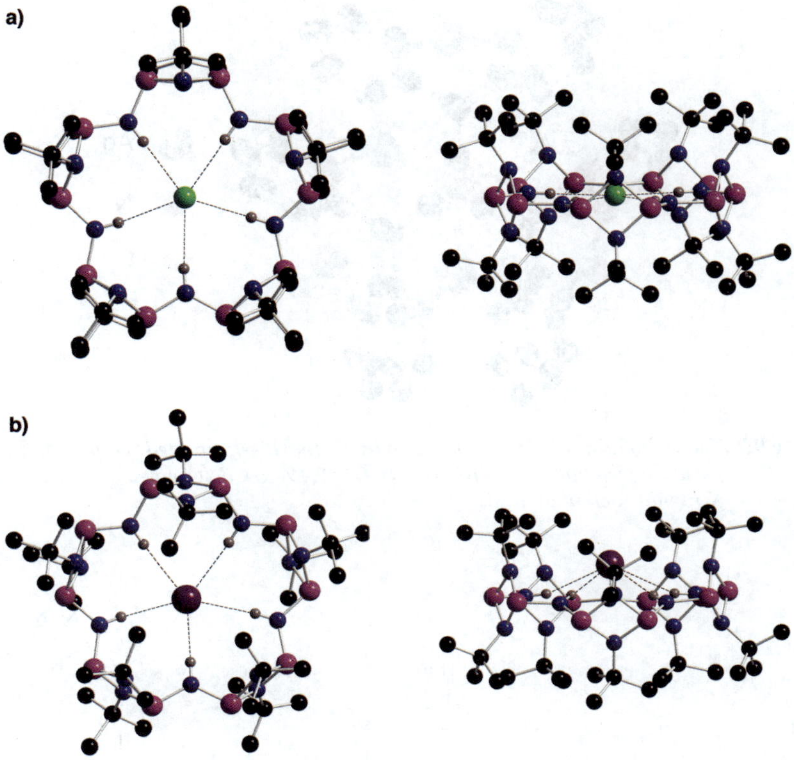

Figure 9.6 *Two views of the single crystal X-ray structure of complex **9.18** with (a) chloride anion and (b) iodide anion*

hexa(2-pyridyl)[3]radialene[15] with silver tetrafluoroborate produces [Ag_6(hexa(2-pyridyl)[3]radialene)$_2$)F$(BF_4)_5$·11H_2O] (**9.19**). This product represents a hexametallic cage that contains a single fluoride ion bound to three silver atoms (*cf.* Figure 9.7). In contradistinction to these results, reaction of this ligand with silver nitrate yields a coordination polymer, wherein four coordinated silver atoms help stabilize a twisted helical arrangement.

The use of hydrogen-bonded anion complexation to control self-assembly processes represents a fascinating new approach to the construction of elaborate molecular architectures. An interesting example of this paradigm is illustrated by the synthesis of dicationic [1_4]imidazoliophane, a process that Alcalde[16] found to be enhanced in the presence of halide anions. Performing the requisite "3 + 1" convergent macrocyclization reaction under standard conditions in the absence of a templating anion gave yields for **9.20**·2Cl, **9.21**·2Cl, and **9.22**·2Cl of 42, 42, and 50%, respectively (Scheme 9.5). Adding extra chloride anions during the last stage of the synthesis enhanced all yields (the most dramatic being that of **9.21**·2Cl, for which the yield increased from 42 to 83%). Interestingly, addition of TBABr was found to enhance the formation of **9.21**·2X, but suppress that of **9.20**·2X. It is believed that the anions assist the synthesis by forming a hydrogen-bonded intermediate, which is

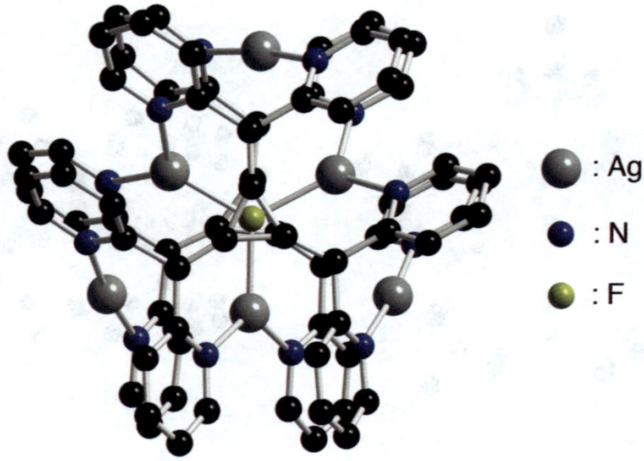

Figure 9.7 *The structure of Steel and Sumby's fluoride-templated cage* $[Ag_6(hexa(2\text{-}pyridyl)[3]radialene)_2)F(BF_4)_5{\cdot}11H_2O]$, ***9.19***, *as determined by single crystal X-ray diffraction analysis*

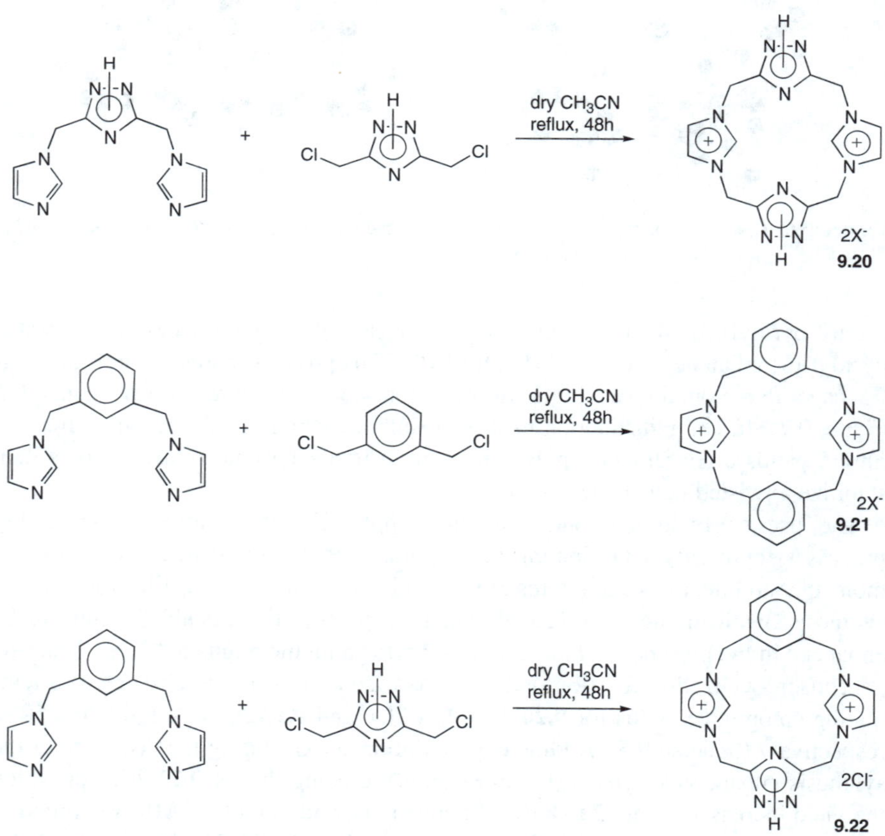

Scheme 9.5 *Synthesis of imidazolium macrocycles* ***9.20–9.22***

stabilized by C–H···Cl^- interactions. The conformations of the intermediate chloride complexes presumably favour cyclization, thus enhancing the yields of the macrocyclic products (Table 9.2).

The formation of an hexameric cyclic aromatic amide structure has been reported by Y.H. Kim and co-workers[17] (Scheme 9.6). Reaction of isophthalic acid chloride and *m*-phenylenediamine in the presence of $CaCl_2$ and DMAc (dimethyl acetamide) under conditions of high dilution produced the cyclic product **9.23**. Interestingly, a pseudo-octahedral anion, $[CaCl_3(DMAc)_3]^-$, was found to be present in the centre of the cavity, with each chloride interacting with two hydrogen atoms from the hexamer. In the absence of the anionic template, the yield of this macrocycle is considerably reduced, while polymerization and the production of macrocycles of different sizes is observed.

Rotaxanes (from Latin: *rota* (wheel) and *axis* (axle)) are composed of a macrocycle through which a rod is threaded.[18] In a true rotaxane, the ends of the axle are stoppered with bulky groups that prevent the macrocycle slipping-off. In the case of pseudorotaxanes, the ends of the axle are generally left unstoppered. Strategies for rotaxane synthesis consist of two main approaches: *threading* and *clipping*. Threading involves mixing chemical entities that can act as self-assembling axles and macrocycles to form a pseudorotaxane that is then stoppered to prevent the macrocycle from slipping off the rod. The clipping approach involves using a "pre-stoppered" linear component that is mixed with a self-assembling component that then undergoes macrocyclic ring closure around it.

Table 9.2 *Enhanced yields are obtained when the syntheses of **9.20**, **9.21**, and **9.22** are carried out in the presence of chloride anions*

Anion	***9.20***×*2X(%)*	***9.21***×*2X(%)*	***9.22***×*2Cl(%)*
None	42	42	50
TBA-Cl	70	83	67
TBA-Br	34	75	

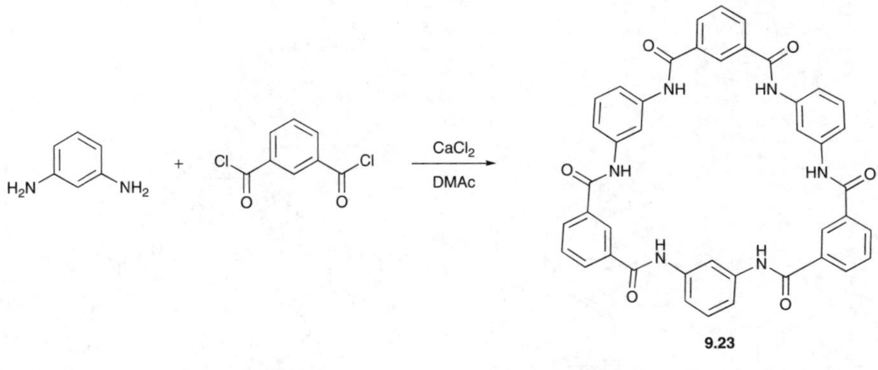

9.23

Scheme 9.6 *An early example of the use of an isophthalamide in an anion-templated macrocycle synthesis*

In chemistry that is of a very different nature from that already described in this chapter, Montalti and Prodi[19] have shown that chloride anions are able to control the threading and unthreading of a pseudorotaxane formed between (9-anthrylmethyl) methylammonium hexafluorophosphate **9.24a** and dibenzo-[24]crown-8 **9.24b** (Scheme 9.7). Here, the basic strategy relies on the fact that chloride anions are capable of forming strong ion-pairs with the ammonium group of (9-anthrylmethyl)methylammonium hexafluorophosphate. This prevents the threading process required to form a pseudorotaxane. The presence of chloride anion, therefore, *effectively* breaks up the pseudorotaxane ensembles present in solution (since the assembly–disassembly of the pseudorotaxane is a dynamic process). Subsequent addition of tributylammonium cations (that can compete for the chloride anions bound in the ion-pairs) serves to drive the equilibrium back in favour of the pseudorotaxane. Therefore, in this case the chloride anion controls the assembly process in a negative sense by interfering with it in a direct, competitive manner. Recently, the use of an analogous concept was demonstrated by Al-Sayah and Branda.[20]

In work that provides an intellectual complement to the above, Beer and co-workers[21] have demonstrated how chloride anions may be used to template the

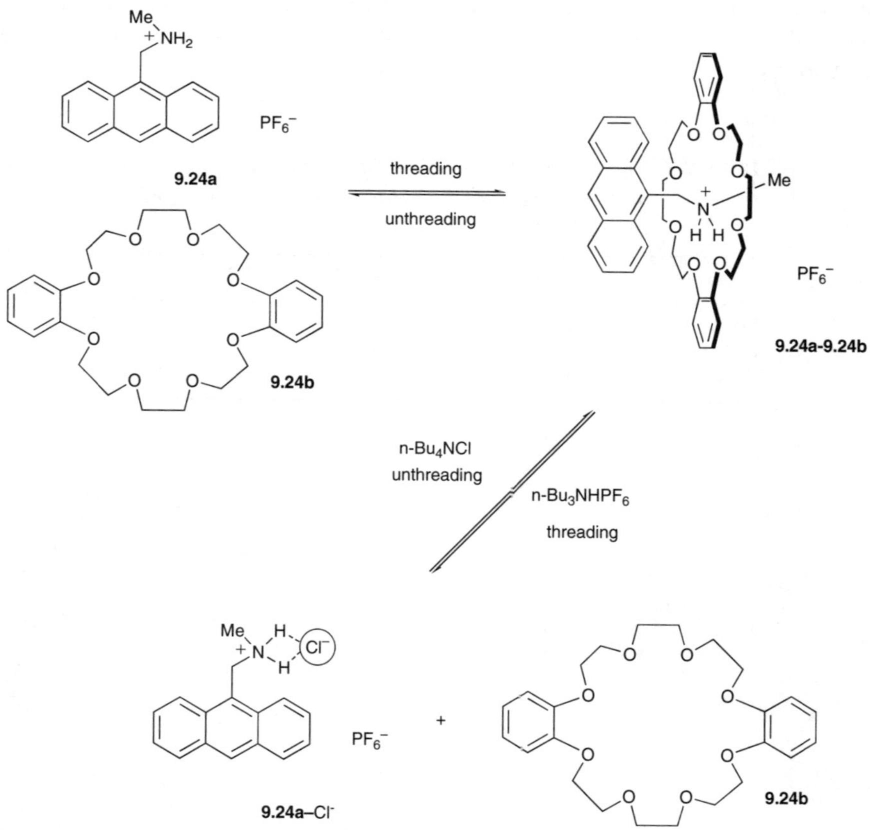

Scheme 9.7 *Anion control of the threading–dethreading of a pseudorotaxane*

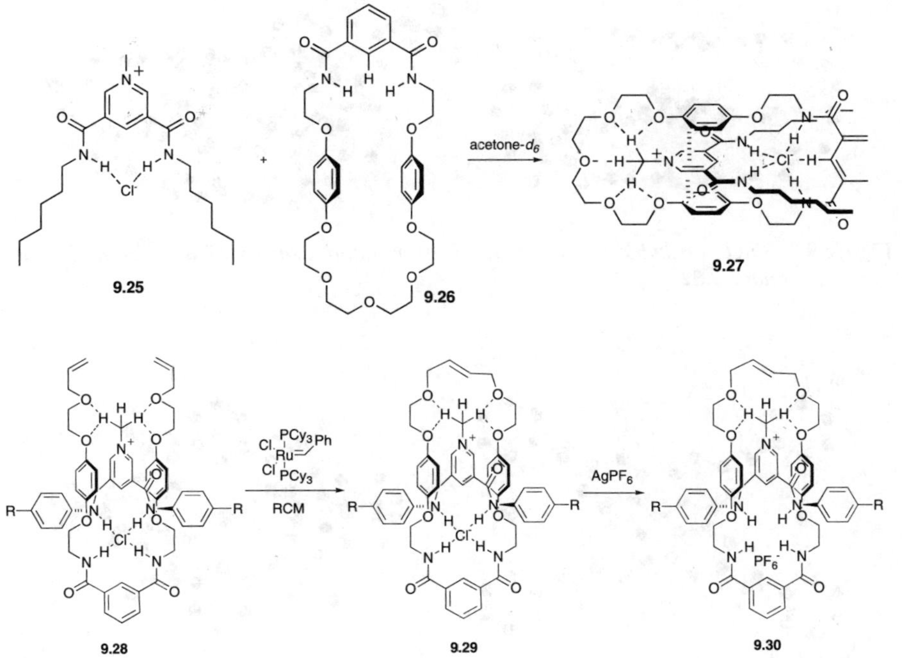

Scheme 9.8 *Beer's anion-templated synthesis of (a) pseudorotaxane* ***9.27*** *and (b) rotaxane* ***9.29***

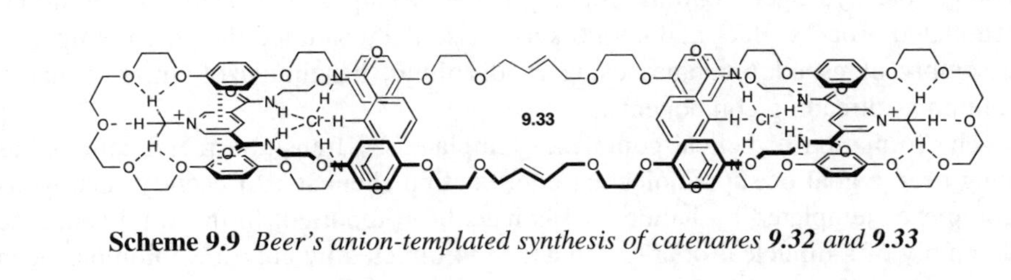

Scheme 9.9 *Beer's anion-templated synthesis of catenanes* ***9.32*** *and* ***9.33***

formation of catenanes, rotaxanes, and pseudorotaxanes. Beer's methodology relies on the use of an isophthalamide-like methylated 3,5-diamidopyridinium chloride salt (*e.g.*, **9.25**), which forms a very tight ion-pair. The chloride anion is not coordinatively saturated and so may hydrogen bond to an isophthalamide unit (*e.g.*, **9.25**), thereby bringing the two components of the ensemble together (*cf.* **9.27**, **9.28**, and **9.31** in Schemes 9.8 and 9.9; see also Figure 9.8). A metathesis reaction, involving complexes **9.28** and **9.31**, then leads to the formation of the

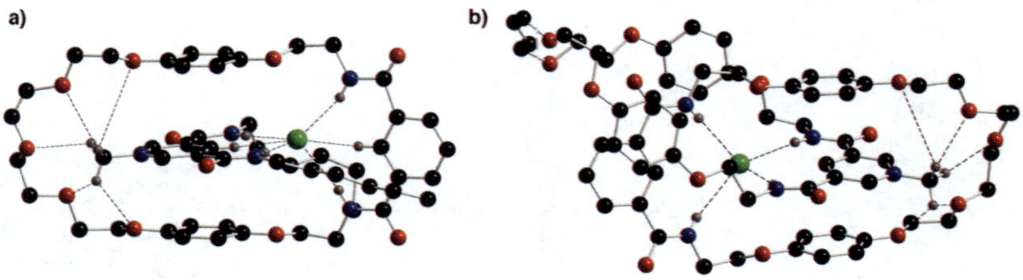

Figure 9.8 *Single crystal X-ray structure of (a) pseudorotaxane* ***9.27*** *and (b) the [2]catenane* ***9.32***

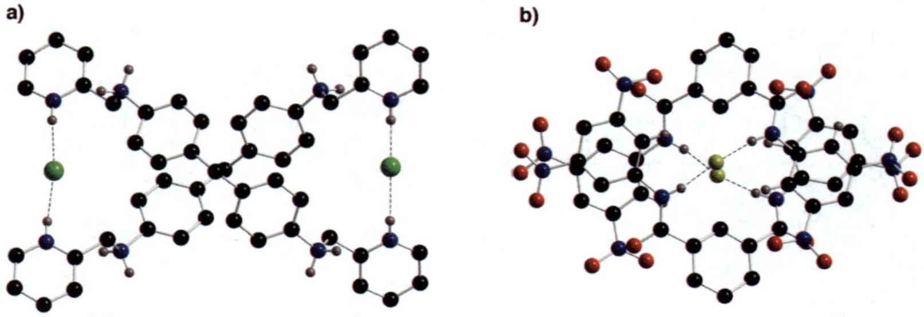

Figure 9.9 *(a) The dinuclear double helicate* ***9.34***$\cdot 2Cl^-$ *and (b) the fluoride complex of a 3,5-dinitrophenyl-substituted isophthalamide (****9.35****)*

rotaxane and catenane products, as shown in Schemes 9.8b and 9.9, respectively. In the case of the rotaxane product **9.29**, the chloride anion could be removed by treatment with silver hexafluorophosphate. The resulting chloride-free rotaxane (**9.30**) was found to have an enhanced affinity for chloride anion as compared to the starting components, presumably due to the presence of the chloride-compatible hydrogen-bonding pocket in **9.30** set up as the result of the anion-templated process. Beer and co-workers[22] also demonstrated the anion-templated assembly of pseudorotaxanes using imidazolium, benzimidazolium, and guanidinium as threading components.

The formation of helical complexes templated by transition metal cations has long been a goal of supramolecular coordination chemists.[23] Recently, analogous complexes templated by halide anions have been reported. In the solid state, the assembly of a dinuclear double helicate **9.34** directed by chloride anion has been reported by Kruger, Martin, and co-workers.[24] In this case, assembly of two ligands of a diammonium-bispyridinium salt in the presence of dilute hydrochloric acid was shown to afford the helical structure shown in Figure 9.9a. Gale and co-workers[25] have shown that simple neutral isophthalamide molecules can also assemble around anions. For example, a 3,5-dinitrophenyl derivative has been shown to form a double helix around two fluoride anions in the solid state (producing complex **9.35**; *cf.* Figure 9.9b). The interested reader is referred to related work by Yam.[26]

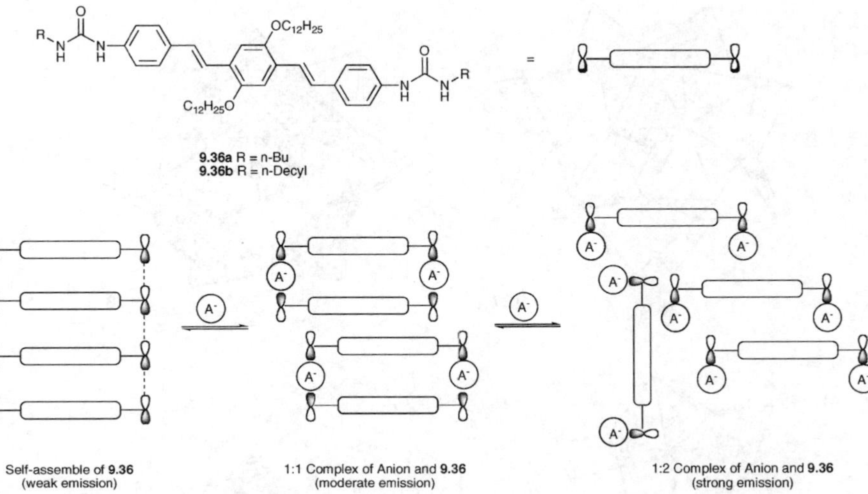

Scheme 9.10 *Interaction of halide with self-assembled ensemble **9.36** leads to break up of the aggregated material and produces an enhancement in the fluorescence-emission intensity*

While the above examples serve to illustrate how anions may be used to generate self-assembled structures, Ajayaghosh[27] has demonstrated an almost orthogonal approach to controlling supramolecular architectures through the use of halide anions. In this work it was noted that the bis-urea end-capped oligo(phenylenevinylene)s **9.36a** and **9.36b** exist as the strong self-assembled ensembles that are characterized by essentially no single-state emission. Presumably, as the result of cooperative interactions involving both hydrogen bonding between the urea groups and π-staking of the aromatic moieties these one-dimensional arrays remain stable in solution at concentrations = 10^{-7} M in chloroform. However, the addition of halide anions (*i.e.*, F^-, Cl^-, Br^-, and I^-) induces destruction of these ensembles and leads, as expected for less aggregated entities, to a remarkable enhancement in the emission intensity (Scheme 9.10). Applications of this principle to sensor design are discussed in Chapter 8.

9.3 Oxyanion-Directed Assemblies

As has been underscored by the discussions throughout this book, oxyanions are characterized by a variety of sizes and, more importantly, shapes. They, therefore, could allow for a more sophisticated level of assembly, *i.e.*, directional control, than the simple spherical halides. In this section, we will examine the role oxyanions play in the assembly of both inorganic cluster compounds and organic supramolecular architectures.

The role of shape in directing the formation of new molecular architectures is perhaps most clearly illustrated in the work of Müller and co-workers[28] This group has shown that anionic (including oxyanionic) templates can control the linking of $V^{n+}O_x$ polyhedra by forming either shell-like clusters, where the directing anion is

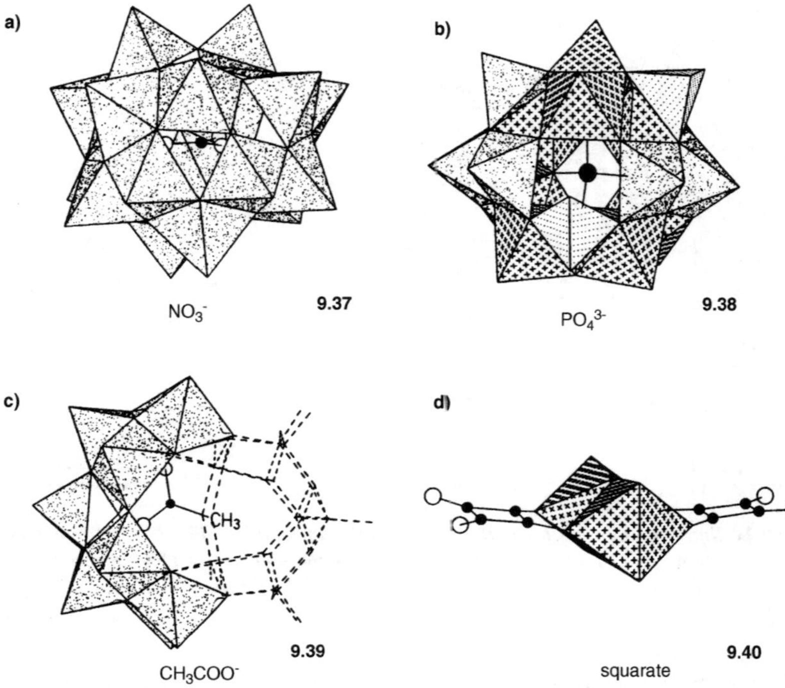

Figure 9.10 *Schematic representations of various oxyanion-templated V-O tetrahedra.* (This figure was reproduced with permission from reference [28]. Copyright 1993, Wiley-VCH)

encapsulated within an inorganic molecular container, or V–O aggregates in which the anionic directing agent acts on the exterior of the cluster. Examples of encapsulated anions include nitrate in **9.37** ($K_{10}[HV_{18}O_{44}(NO_3 \cdot 14.5H_2O)]$), phosphate in **9.38** ($K_7[H_4V_{15}O_{40}(PO_4)] \cdot 10H_2O$), and acetate in **9.39** ($(NEt_4)_5(NH_4)_2[H_2V_{22}O_{54}(CH_3COO)]7H_2O$), while one of the more fascinating of the external templates is squarate in **9.40** ($Na[V_2O_2(OH_2)_2(\mu\text{-}OH)(\mu\text{-}OH_2)(C_4O_4)_2] \cdot 6H_2O$). The structures of the anions of these complexes are shown in Figure 9.10; taken together they illustrate the great structural diversity that may be accessed by the use of oxyanionic templating agents. Particularly interesting is cluster **9.37**, a species that possesses an inorganic "shell" that is almost perfectly complementary to the encapsulated nitrate guest.

Chen and his group[29] have used the trigonal planar carbonate anion as a bridging ligand around which quadruply bonded Mo_2 units may be arranged to form hexanuclear molybdenum complexes. One such complex, shown schematically as **9.41** (Figure 9.11), was prepared via the reaction of $[trans\text{-}Mo_2(O_2CCF_3)_2(MeCN)_6][BF_4]_2$ and potassium carbonate with *N,N′*-bis(diphenyl phosphino)amine (dppa) in dichloromethane. Here, the chloride anions in the product presumably originate from the dichloromethane solvent. Indeed, when the analogous reaction was carried out in

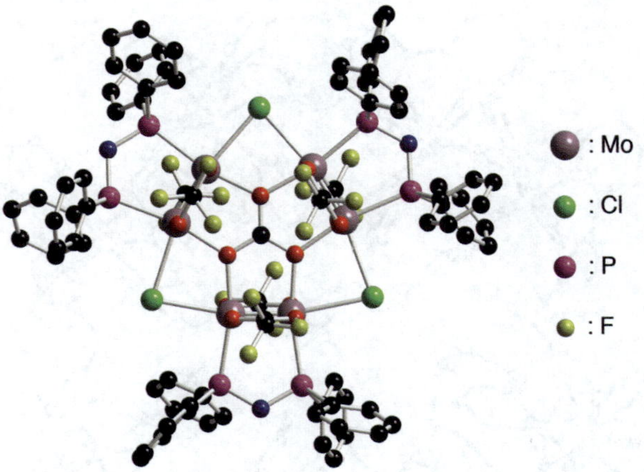

Figure 9.11 *Single crystal X-ray structure of the carbonate-templated molybdenum cluster* ***9.41***

acetonitrile, the addition of ZnX_2 (X = Br or I) was found to afford the isostructural bromide or iodide complexes (shown schematically as **9.42** or **9.43**), respectively.

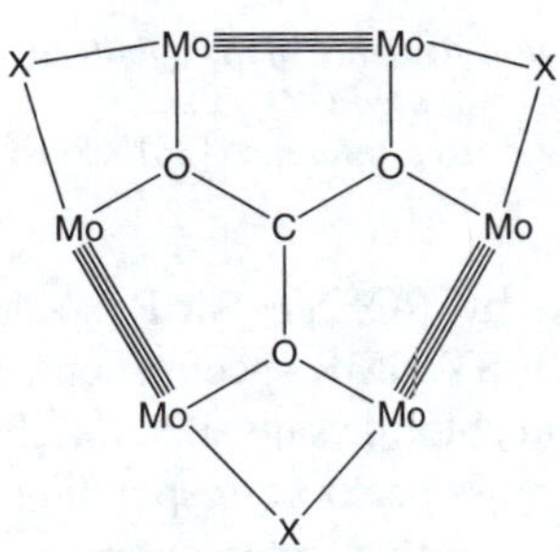

9.41 X = Cl
9.42 X = Br
9.43 X = I

McCleverty and co-workers[30] have used cobalt in combination with a bis{3-(2-pyridyl)pyrazol-1-yl}dihydroborate ligand to form a cyclic structure encapsulating a perchlorate anion. Specifically, it was found that reaction of [bis{3-(2-pyridyl)pyrazol-1-yl}dihydroborate] with cobalt(II) produced the cyclic product **9.44** shown in Figure 9.12. In this structure, each ligand bridges between two adjacent metal ions, with an alternating pattern of one and then two ligands. The perchlorate anion sits in the middle of the resulting metal-containing annulus.

Moving away from the oxyanion-directed assembly of inorganic clusters to a more organic-based approach, de Mendoza and co-workers[31] have synthesized the tetraguanidinium strand (**9.45**) and studied how it self-assembles in the presence of sulfate anions. The strand consists of four chiral bicyclic guanidinium units that are linked by short CH_2SCH_2 spacer units. Upon addition of SO_4^{2-}, the short spacer units prevent two (or more) guanidinium units from wrapping around a single sulfate anion in an orthogonal manner. Instead these spacers "force" two of the guanidinium ligands to adopt a chiral double helical structure with the "handedness" of the helix being determined by the absolute configuration of the starting ligand. Evidence for

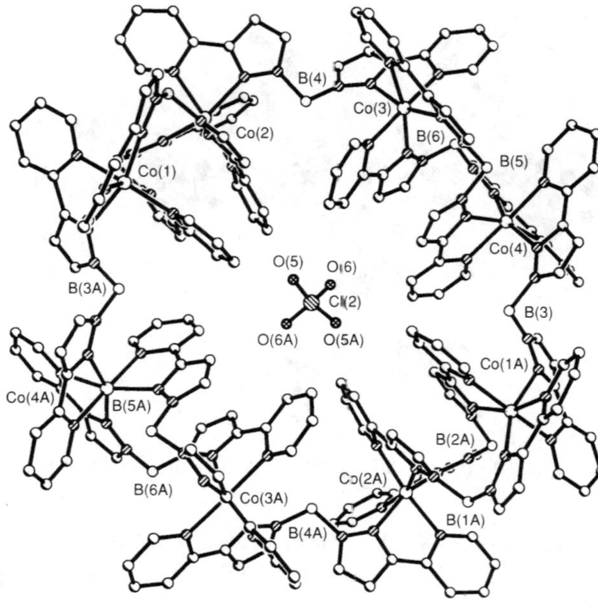

Figure 9.12 *Single crystal X-ray diffraction structure of the cobalt cage **9.44** of McCleverty and co-workers showing the encapsulated ClO_4^- ion.*
(Reproduced with permission from reference [30]. Copyright 1997, Royal Society of Chemistry.)

the helical structure was provided by ROESY NMR spectroscopy and circular dichroism studies. For example, ROESY NMR spectroscopic analyses were used to study the structures of the sulfate and chloride salts of **9.45**. The ROESY NMR spectra of the sulfate complex revealed cross peaks corresponding to through space couplings that were too far separated to reflect interactions between the monomers. Accordingly, they were considered as originating from couplings involving different strands of the receptor. A CPK model of the double helix assembled from two (*R,R*) subunits is shown in Figure 9.13.

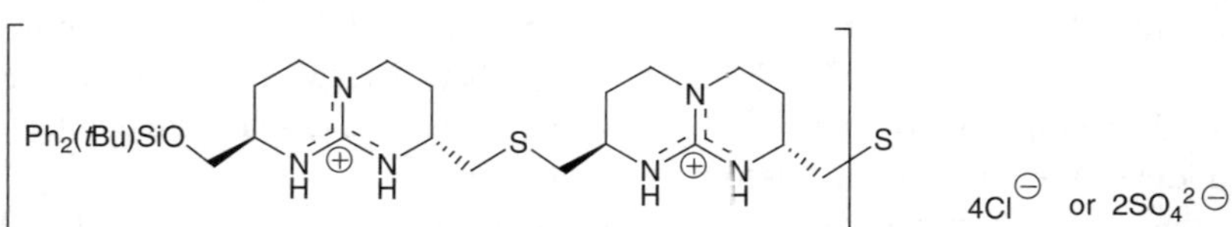

9.45

Tubular structures such as carbon nanotubes are attracting a great deal of research interest at the moment due to their potential to form new, superstrong materials, molecular reaction vessels, and traps for immobilising biomolecules, among other things.[32] In this context, it is noteworthy that Fujita and co-workers[33] have used linear aromatic guest species (including carboxylate-substituted derivatives) to control the formation of *coordination*-based nanotubes such as **9.47**, **9.48**, and **9.49** (Scheme 9.11). For example, pentakis(3,5-pyridine) **9.46** quantitatively forms a

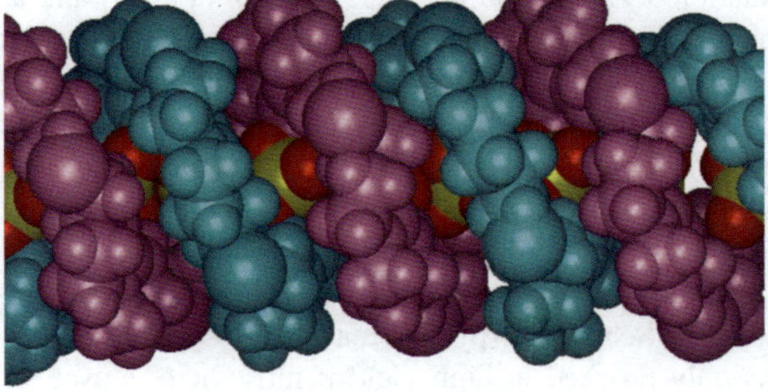

Figure 9.13 *CPK model of a polyguanidinium right-handed double helix derived from two strands of **9.45** of individual (R) configuration.*
(We thank Prof. de Mendoza for providing with colour copy of this figure, which originally appeared in reference [31]. Copyright 1996, American Chemical Society.)

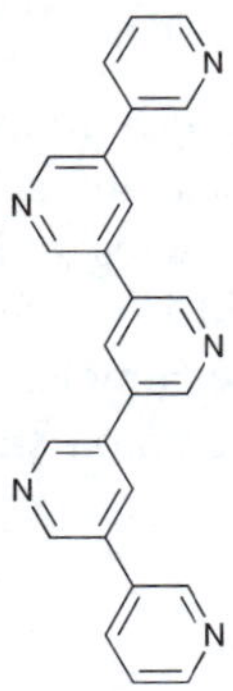

9.46

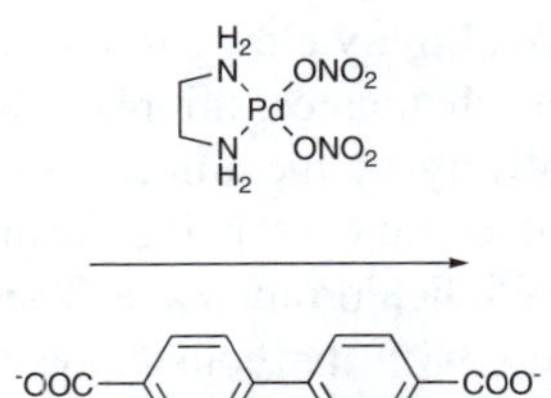

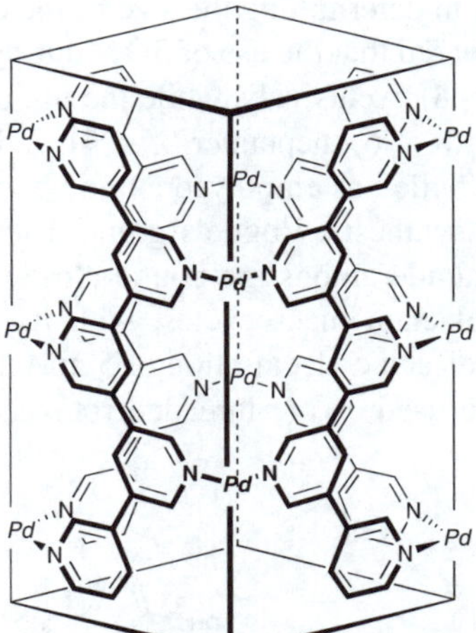

9.47

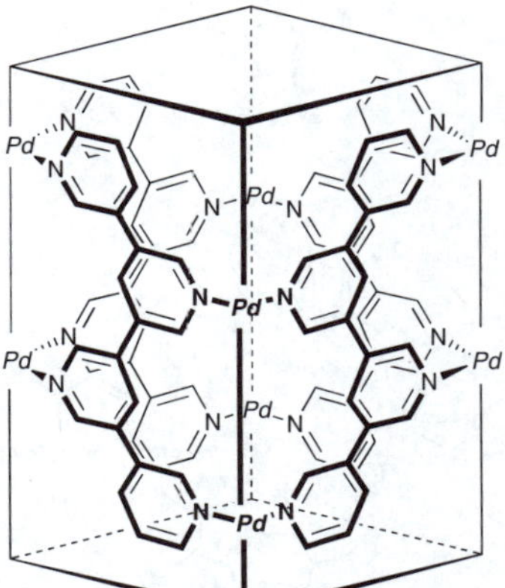

9.48

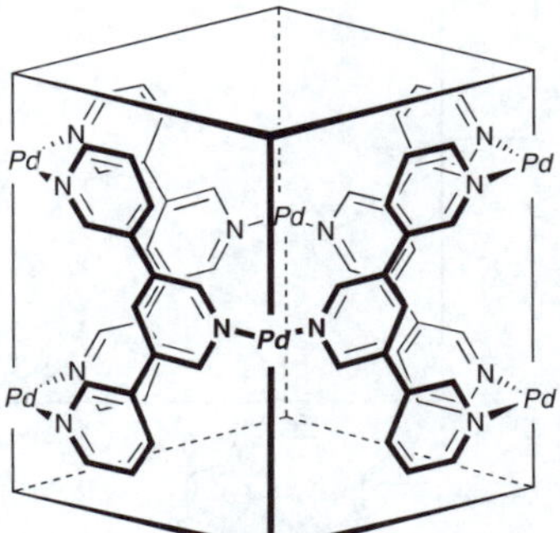

9.49

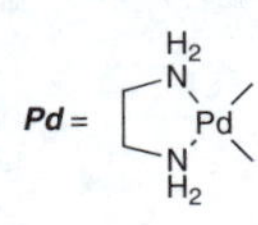

Scheme 9.11 *Synthesis of the coordination-chemistry-based nanotubes **9.47–9.49***

coordination nanotube **9.47** in the presence of (en)Pd$(NO_3)_2$ and a linear template such as 4,4′-biphenylenedicarboxylate. Neutral linear templates such as biphenyl and *p*-terphenyl were also effective in producing such structures. However in the absence of any template, the nanotubes did not form. The crystal structure of **9.49**–(4,4′-biphenylenedicarboxylate)$_2$$(NO_3^-)_{10}$ has been elucidated and reveals that the anionic template is bound within the tube by a combination of π–π and CH–π interactions.

Sessler and co-workers[34,35] have shown that oxyanions may play a templating role in the synthesis of the Schiff base-pyrrole macrocycles **9.50** and **9.51** (Scheme 9.12). These reactions only proceed in high yield if nitric acid is used as the catalyst. If other acids, such as HCl are used, lower yields are obtained, a finding that led Sessler to suggest that nitrate anion may be acting as a template in this reaction.

As mentioned earlier, the first reports of cyclo[n]pyrrole synthesis have appeared recently (*cf.* Chapter 3).[36] These macrocycles consist solely of pyrrole rings and do not contain any bridging *meso*-carbon atoms. It appears that anions play a decisive role in determining the size of the cyclo[n]pyrrole formed. In particular, it has been observed that the use of SO_4^{2-} during the synthesis appears to give rise to the octamer ($n = 8$) exclusively, while the use of chloride (as HCl) results in a mixture of hexamer ($n = 6$), heptamer ($n = 7$), and octamer.[37]

Vögtle has employed anion coordination in a high-yielding *threading type* rotaxane synthesis. Vögtle's group discovered that phenolates, thiophenolates, and sulfonamide anions are bound close to quantitatively by the amide groups present in tetralactam macrocycles, with the associated complexes being formed in >95% yields at a concentration of 5 mM in deuteriated dichloromethane. These complexes were used as pre-threaded rotaxane precursors with the bound organic anion still

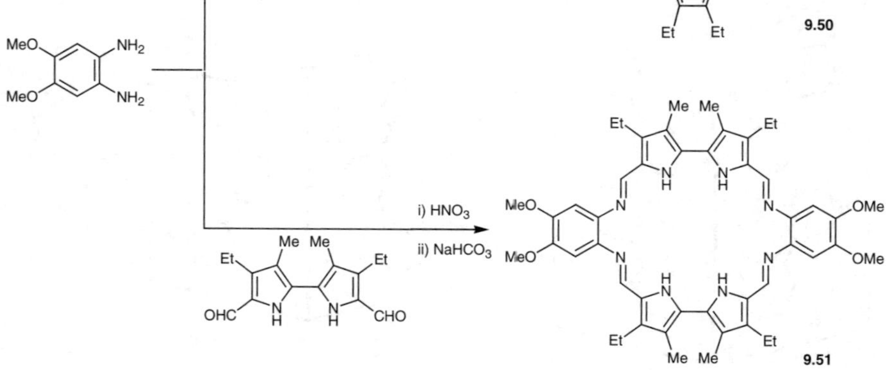

Scheme 9.12 *Synthesis of anion-templated pyrrolic macrocycles **9.50** and **9.51***

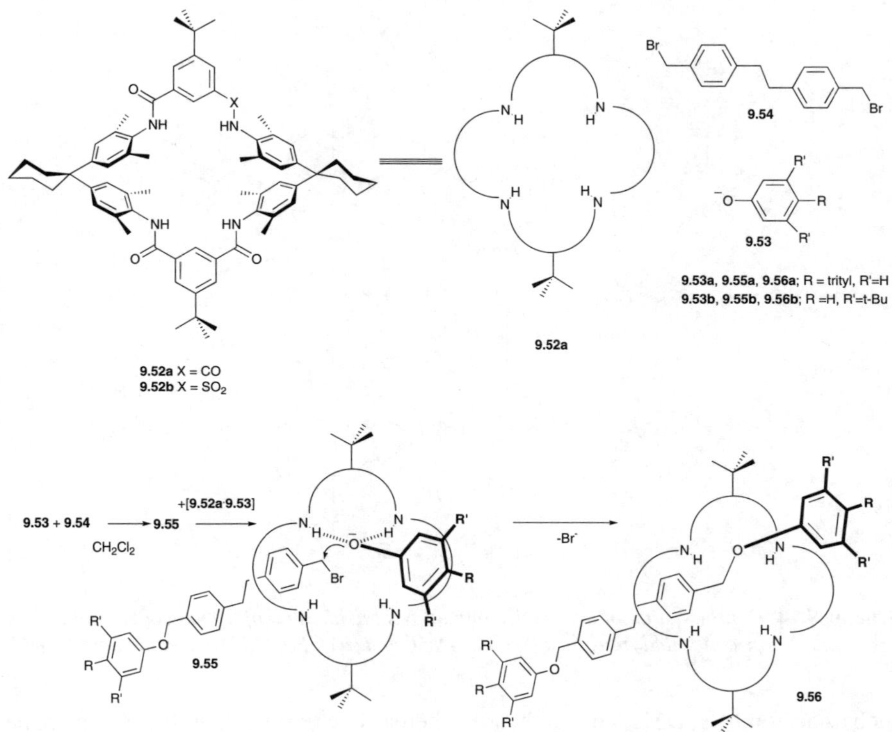

Scheme 9.13 *Efficient formation of a rotaxane effected via pre-coordination of a nucleophilic anion*

capable of reacting as a nucleophile. For example, the complex **9.52–9.53** (Scheme 9.13) reacts with molecule **9.54** to form rotaxane **9.56** in 95% yield (probably one of the highest yields ever reported for this kind of synthesis).[38] The generality of this approach has been shown by the synthesis of a series of rotaxanes with differing axle groups and stoppers using this same basic methodology.[39]

Keaveney and Leigh[40] have recently extended their studies of amide-based rotaxanes and used anion complexation to control the position of a benzylic amide macrocycle on a thread containing a phenol group (Scheme 9.14). The thread in rotaxane **9.57** contains two potential "stations" for the macrocycle, a succinamide group (shown in green in the scheme) and a phenolate group (shown in magenta). The macrocycle in this system contains two isophthalamide groups, one of which forms a strong complex with the phenolate anion. Upon protonation of the thread, however, the phenol group (shown in red) does not bind to the macrocycle. Thus, the macrocycle moves to the alternate succinamide station (shown in green in Scheme 9.14). Interestingly, this shuttling process works best in polar competitive solvents, such as acetonitrile-d_3, DMF-d_7, or methanol-d_4. In less polar solvents, such as chloroform-*d* or dichloromethane-d_2, intramolecular folding occurs with the result that the macrocycle remains at the succinimide station. This presumably reflects the fact that the second isophthalamide group is adequately solvated in more competitive solvents,

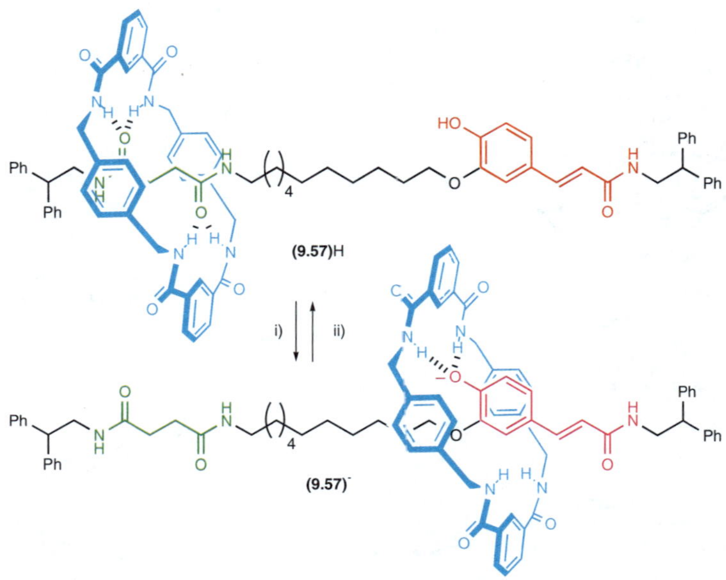

Scheme 9.14 *Deprotonation of a phenol group allows the location of a "molecular shuttle" to be controlled; reagents: (i) base, DMF and (ii) CF_3CO_2H (one equivalent), DMF*

such as acetonitrile, DMF, and methanol, whereas in chloroform or dichloromethane the most stable arrangement is for both isophthalamide groups to remain hydrogen bonded to the succinimide station. Interestingly, in DMF-d_7, the addition of competitive anions such as F^-, Cl^-, Br^-, I^-, OH^-, NO_3^-, or AcO^- does not disrupt or otherwise hamper the shuttling process.

Other examples of rotaxane systems whose chemistry is anion related are known; the interested reader is referred to the references.[41]

Král and co-workers[42] recently reported the assembly properties of cationic porphyrins **9.58** containing various numbers (1, 2, or 4) of bicyclic guanidinium groups. These cations form highly ordered chiral supramolecular structures in aqueous solution and in the presence of anionic counterparts, such as organic dicarboxylates and disulfonates (Figure 9.14). The chirality (measured by circular dichroism) was found to be controlled by the nature of the anion present.

The effect of anion templation on the photodimerization of a thymine functionalized isothiouronium receptor has been reported by Teramae *et al.*[43] In the absence of an anion and upon being irradiated with UV light, the thymine moiety forms a *cis-anti* photodimer (**9.60**), a species that is not predisposed for binding pyrophosphate. In the presence of pyrophosphate, the photodimer was again produced. This time, however, the *cis-syn* and *trans-syn* dimers **9.62** and **9.63** are produced in preference; both of these products contain two thiouronium groups oriented in the same direction and predisposed for pyrophosphate complexation (Scheme 9.15).

Quite a number of reports in recent years have involved anion-controlled "crystal engineering". Crystal engineering has attracted a tremendous amount of interest over the last decade and it would be impossible in this book to give a complete review of

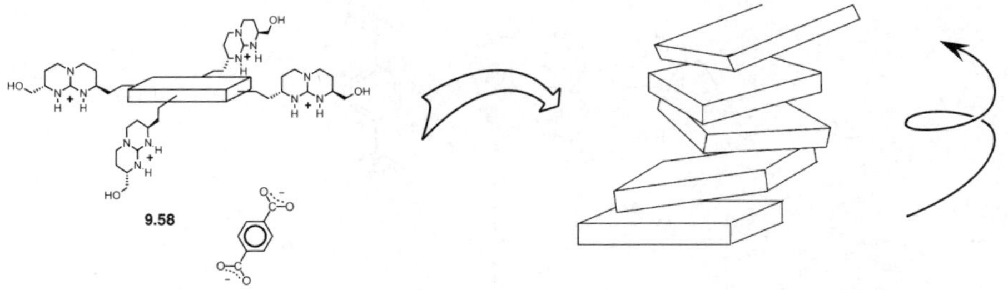

Figure 9.14 *Schematic representation showing how the addition of anions can serve to template the formation of chiral assemblies of porphyrins in aqueous solution.* (Reproduced with permission from reference [42]. Copyright 2002, American Chemical Society.)

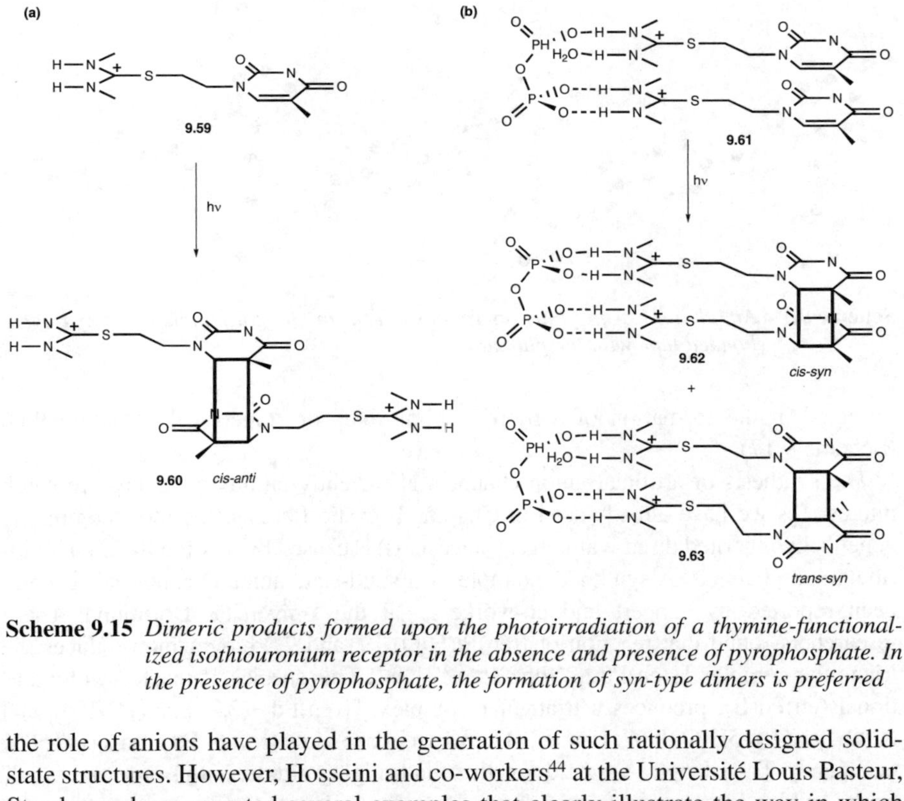

Scheme 9.15 *Dimeric products formed upon the photoirradiation of a thymine-functionalized isothiouronium receptor in the absence and presence of pyrophosphate. In the presence of pyrophosphate, the formation of syn-type dimers is preferred*

the role of anions have played in the generation of such rationally designed solid-state structures. However, Hosseini and co-workers[44] at the Université Louis Pasteur, Strasbourg have reported several examples that clearly illustrate the way in which the shape of the anion may be used to control the formation of a crystalline assembly. For instance, this group prepared a variety of bis-amidinium receptors (*e.g.*, **9.64**; Scheme 9.16) and obtained crystals of their isophthalate and terephthalate salts. In the case of the terephthalate salt **9.65**, a single crystal X-ray diffraction analysis revealed the formation of a hydrogen-bonded, rod-like one-dimensional coordination polymer. The isophthalate salt also formed a coordination polymer in the solid state. However, the non-linear geometry of the isophthalate anion is

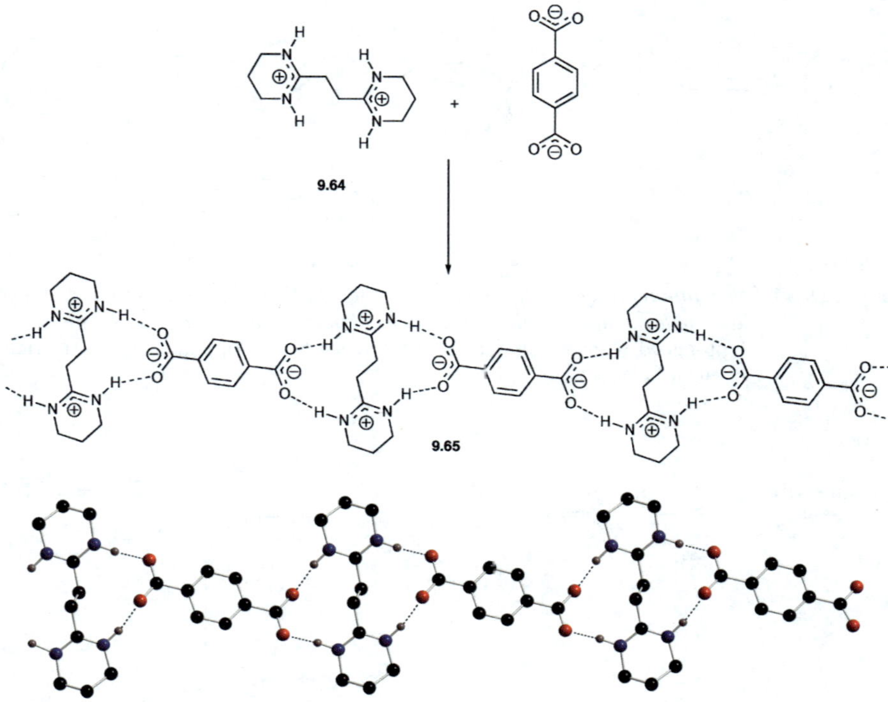

Scheme 9.16 *Amidinium-carboxylate interactions lead to the formation of a hydrogen-bonded tape with terephthalate*

reflected in the formation of a more "ribbon-like" or "tape-type" structure **9.66** (Scheme 9.17).

The synthesis of artificial anion channels is currently an area of intense research interest. As we have already seen in Chapter 1, cystic fibrosis (the most commonly genetically inherited disease amongst Caucasians) is caused by mis-regulation of chloride-anion channels. A synthetic example of a solid-state-anion channel has recently been reported by Lippert and co-workers[45] at the Universität Dortmund. These researchers found that reaction of $[(en)Pt(H_2O)_2]^{2+}$ and 2,2-bipyrazine produces the triangular complex $[\{(en)Pt(2,2'\text{-bpz-}N^4,N^{4'})\}_3]^{6+}$. This species, if reacted with additional (en)Pd(II), produces a triangular complex, $[\{(en)Pd_{2.5}(2,2'\text{-bpz})_3\{(NH_3)_2Pt\}_3]$ $(ClO_4)_6(NO_3)_5{\cdot}5H_2O$ (**9.67**; Figure 9.15), wherein one of the corner Pd atoms is at 50% occupancy. The triangles are held together via bridging nitrate anions and water molecules, which leads to the formation of channels (this is shown in Figure 9.15c with the perchlorate anions omitted for clarity). A perchlorate anion is bound at the centre of the triangle in the channel. A side view is shown in Figure 9.15d.

9.4 Polyfluoro-Anion Directed Assemblies

Fluorinated anions such as tetrafluoroborate BF_4^- and hexafluorophosphate PF_6^- are usually assumed to be innocent in that they are weakly coordinating. Indeed, as we

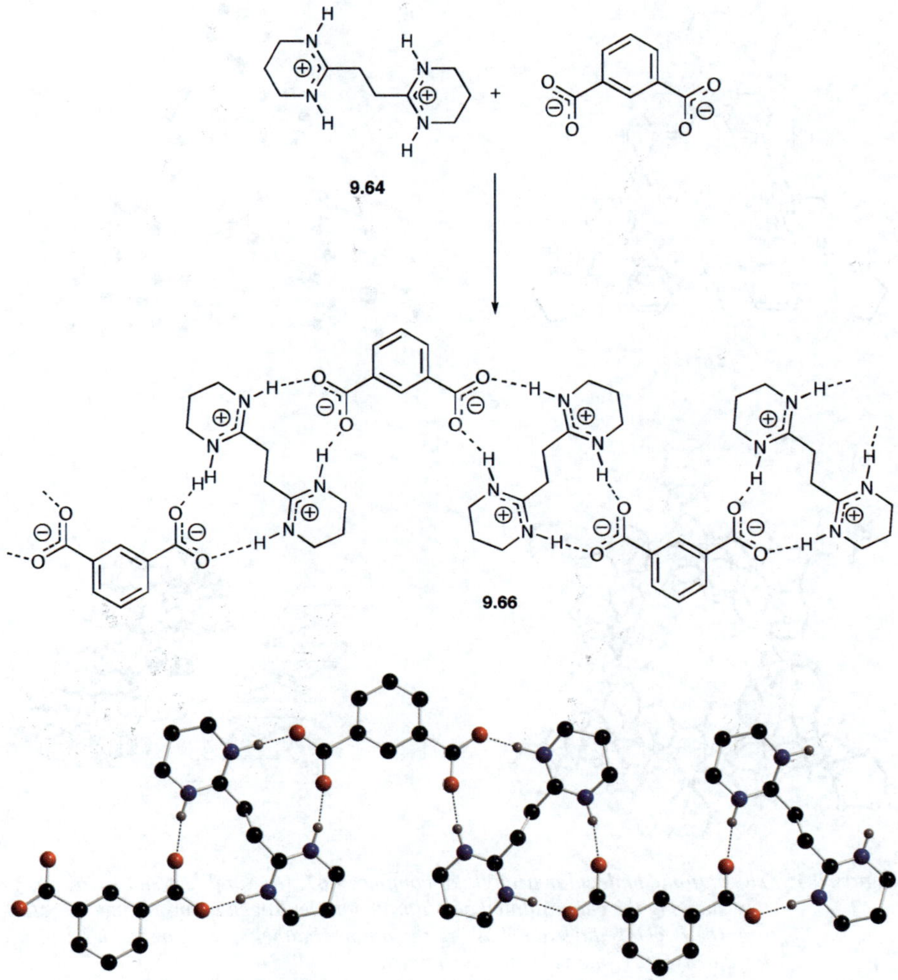

Scheme 9.17 *Amidinium carboxylate interactions lead to the formation of a zig-zag hydrogen-bonded tape with isophthalate*

have seen in previous chapters, they are frequently used as non-interfering counter anions in the case of positively charged synthetic anion receptors. However, there have recently been several examples of such anions playing roles in the assembly of non-covalently linked supramolecular ensembles and metal-coordinated cage complexes. For example, Stoddart and co-workers[46] have employed hexafluorophosphate anions to effect the self-assembly of a pseudorotaxane. The pseudorotaxane in question, **9.68**, is formed from four dibenzylammonium ions and tetrakis-*p*-phenylene[68]crown. The crystal structure of this ensemble revealed an ordered hexafluorophosphate anion bound within the centre of the assembly (Figure 9.16). Hexafluorophosphate anions are usually disordered in crystal structures, *i.e.*, the octahedrally disposed fluorine atoms are not constrained to point in particular

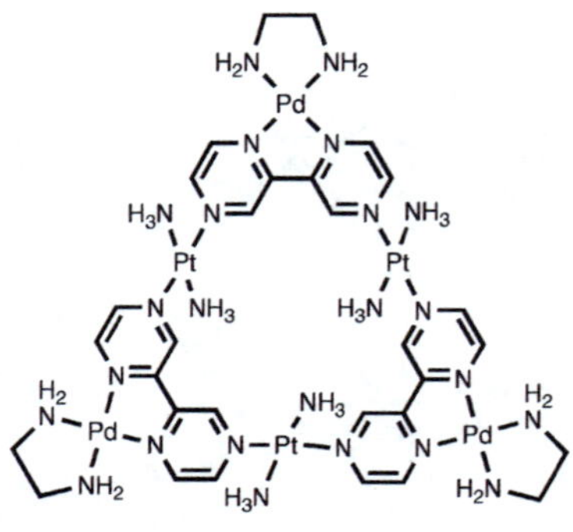

9.67

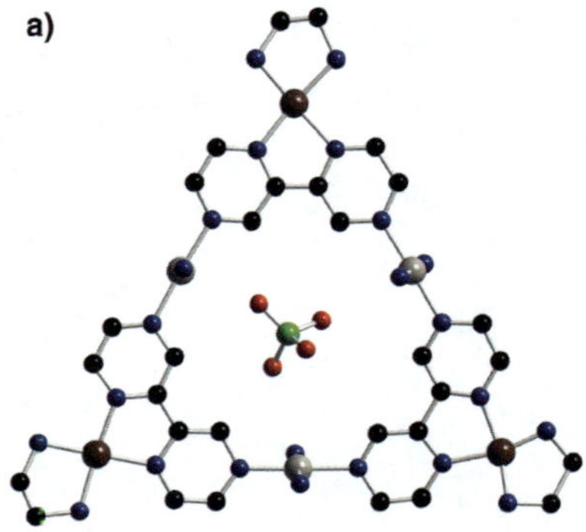

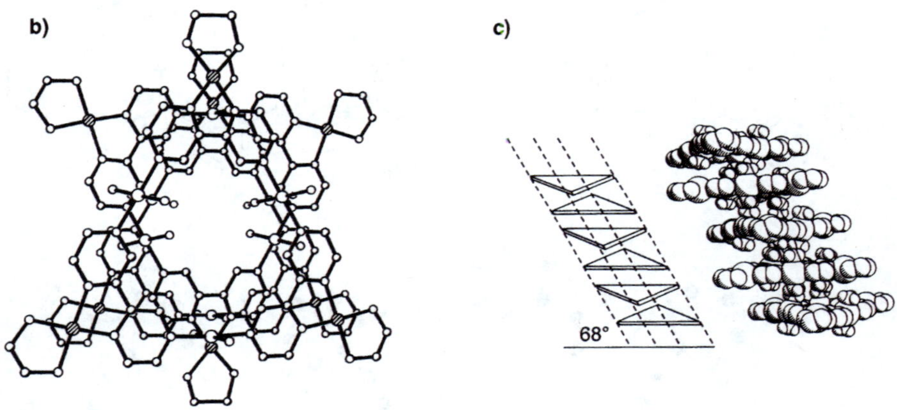

Figure 9.15 *The cationic molecular triangle of complex* ***9.67****. (a) Single crystal X-ray structure showing the encapsulated perchlorate ion (hydrogen atoms omitted for clarity). (b) Packing diagram showing the anion channel (perchlorate ions omitted for clarity). (c) Side view of the channel.*
(Reproduced with permission from reference [45]. Copyright 1999, The Royal Society of Chemistry.)

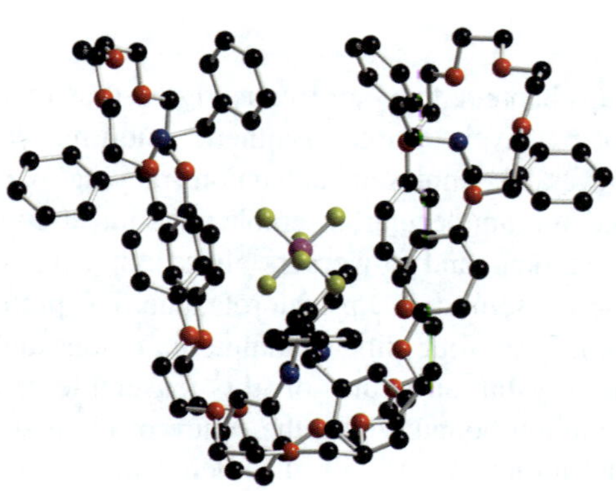

Figure 9.16 *Single crystal X-ray structure of* ***9.68***

directions. However, in Stoddart's pseudo-rotaxane the ordering of the anion suggests the formation of C–H···F hydrogen bonds with the rest of the pseudorotaxane, locking the anion into a single orientation. The presence of the anion is, therefore, likely to reduce any electrostatic repulsion between the dibenzylammonium cations in the pseudorotaxane, thereby assisting the assembly process.

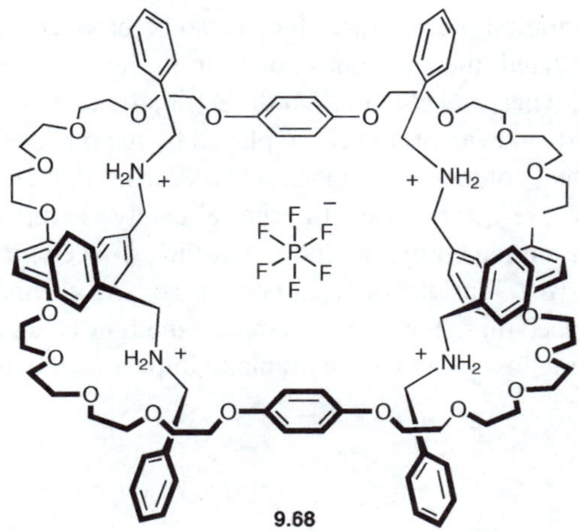

9.68

Cobalt-based tetrahedral supramolecular arrays have been reported by Ward and co-workers.[47] Cage complexes have been synthesized from a pyrazoyl-pyridine ligand by reaction of the metal-free species with an appropriate metal acetate hydrate in methanol, followed by addition of aqueous sodium tetrafluoroborate. This results in the complex precipitating from solution and facilitates its isolation. As shown in Figure 9.17, a BF_4^- anion is located at the centre of the tetrahedral cage (**9.70**) formed by the four cobalt(II) metal ions at the vertices and six bridging ligands **9.69**. Each of

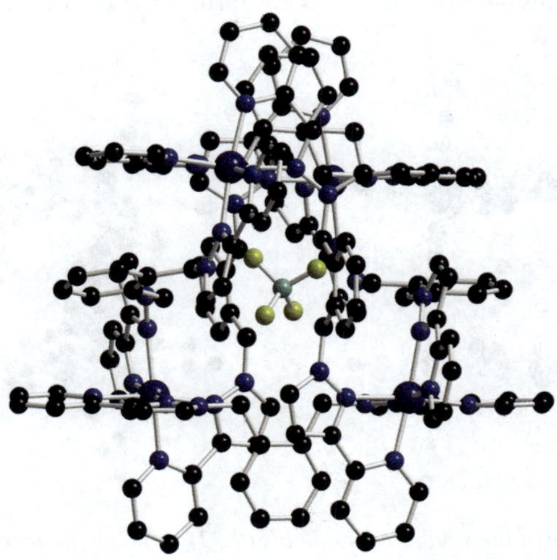

Figure 9.17 *Single crystal X-ray structure of the BF_4^- anion-templated cage complex **9.70** (hydrogen atoms have been omitted for clarity)*

the fluorine atoms of the encapsulated anion is directed towards the triangular faces of the cluster rather than the metal ions. This leads to the inference that electrostatic interactions are the driving force for this templation process, with the anion specifically serving to balance in part the 8^+ charge of the ring. Presumably, the anion adopts an orientation dictated by the geometry of the internal cavity. Multiple π-stacking interactions between the ligands also assist the assembly of the structure.

This basic configuration present in **9.70** can also be produced via templation with ClO_4^-. On the other hand, the cage does not form in the absence of any templating anion. Interestingly, when nickel is used, a different structure is formed (a dinuclear structure species with no evident role being played by the putative templating anion). If an analogous ligand, containing a biphenyl spacer is used, then a larger tetrahedral complex forms. However, in this case the central cavity is empty, meaning that the system is likely not templated by an anion.[48] In fact, in acetonitrile solution, it has been shown that the BF_4^- ions diffuse in and out of the cavity. This example illustrates how delicately balanced these types of system are, and how changing only one aspect of the system in a small way can have a dramatic impact on the assembled structure.

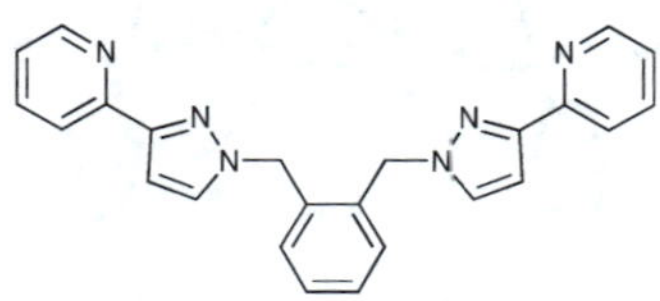

9.69

The anion-templated assembly of a 3,6-bis(2-pyridyl)-1,2,4,5-tetrazine (bptz **9.71**) nickel complex **9.72** has been reported by Dunbar and co-workers.[49] These researchers reacted $[Ni(CH_3CN)_6][BF_4]_2$ with bptz in a 1:1 molar ratio to form a dark green solid that was recrystallized by diffusion of toluene into an acetonitrile solution of the complex. The complex formed, **9.72**, was a partially solvated square consisting of four nickel(II) ions and four bptz units ($[Ni_4(bptz)_4(CH_3CN)_8[BF_4]_8 \cdot 4CH_3CN$ (Figure 9.18a). Analogous complexes do not form if the bis-hexafluorophosphate salt

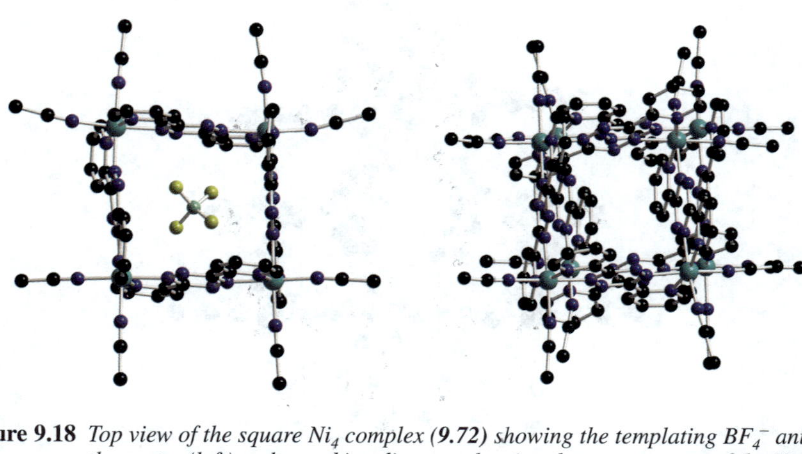

Figure 9.18 *Top view of the square Ni_4 complex (**9.72**) showing the templating BF_4^- anion in the centre (left) and a packing diagram showing the arrangement of the Ni_4 complexes in the solid state*

of nickel(II) hexaacetonitrile **9.72** (shown in Figure 9.18b) is used. This led these workers to suggest that assembly of the complex is templated by the tetrafluoroborate anions. The complexes form channel-like stacks in the solid state that do indeed contain BF_4^- anions (Figure 9.18b).

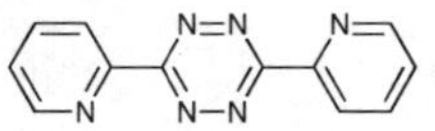

9.71

The first example of an inorganic "tennis-ball" motif encapsulating an anion has been reported by K.M. Kim and co-workers.[50] Specifically, these researchers found that dissolution of the complex {(dach)Pt(BETMP)}$_2$Cu(BF_4)$_2$ (dach = *trans*(±) –1,2-diaminocyclohexane, BETMP = bis(ethylthio)methylenepropanedioate) in D_2O and CD_3OD resulted in the formation of complex **9.73**, as shown in Scheme 9.18. The starting material contains hydrogen-bond acceptors (carboxylate) and donor (amine) groups. This allows the complex to dimerize around a central BF_4 anion, giving rise to the tennis-ball motif.

McMorran and Steel have investigated the assembly properties of the ligand 1,4-bis(3-pyridyloxy)benzene with metal and anions. Complexation of the ligand with [$PdCl_2(PPh_3)_2$] and [$PdI(py)_2$] in the presence of silver triflate results in the formation of dimeric complexes.[51] However, crystallization of the latter dimeric species in the presence of ammonium hexafluorophosphate resulted in a reorganization of the complex and the formation of a M_2L_4 helical cage (**9.74**) that encapsulates a hexafluorophosphate anion between the two palladium metal centres (Scheme 9.19 and Figure 9.19).

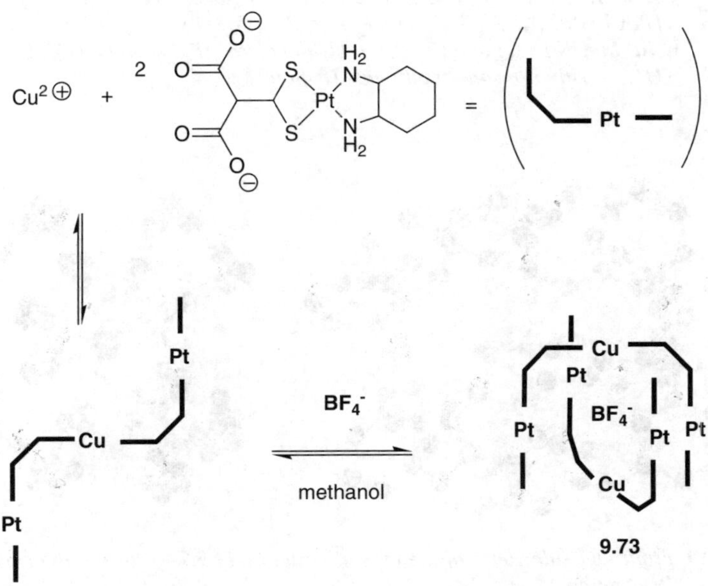

9.73

Scheme 9.18 *Schematic representation of the arrangement of the platinum, copper, and BF_4^- components found in the inorganic "tennis-ball" of Kim and co-workers*

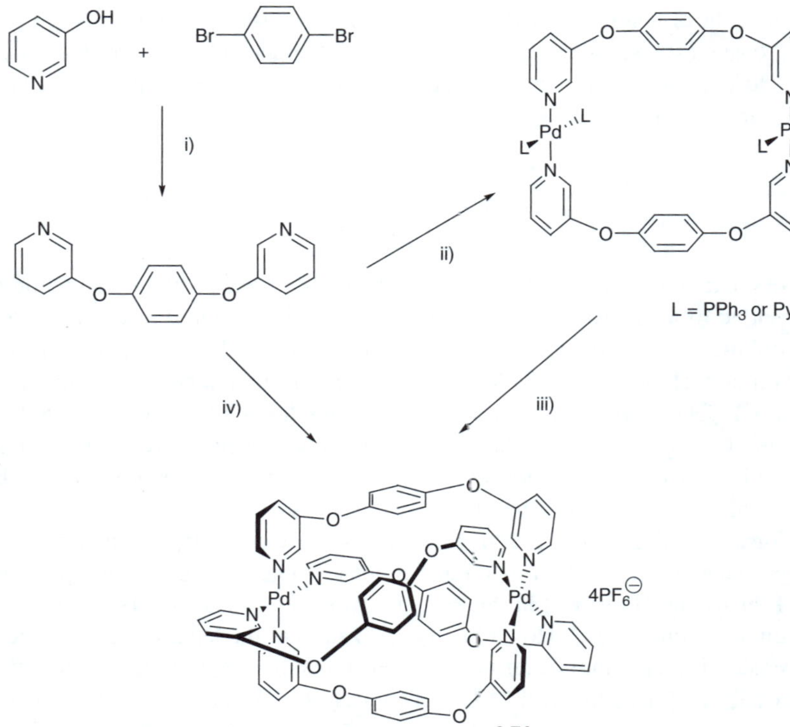

9.74

Scheme 9.19 *Template formation of the hexafluorophosphate-containing quadruple helix* ***9.74****. Reagents: (i) K_2CO_3, Cu(0), dimethylacetamide, heat; (ii) L = PPh_3: [$PdCl_2(PPh_3)_2$], AgOTf, CH_2Cl_2, L = Py: [$PdI_2(Py)_2$, AgOTf, acetone, NH_4PF_6; (iii) MeCN/Et_2O; and (iv) 0.5 equivalent [$PdI_2(Py)_2$], AgOTf, MeCN, NH_4PF_6, OTf = trifluoromethanesulfonate (triflate)*

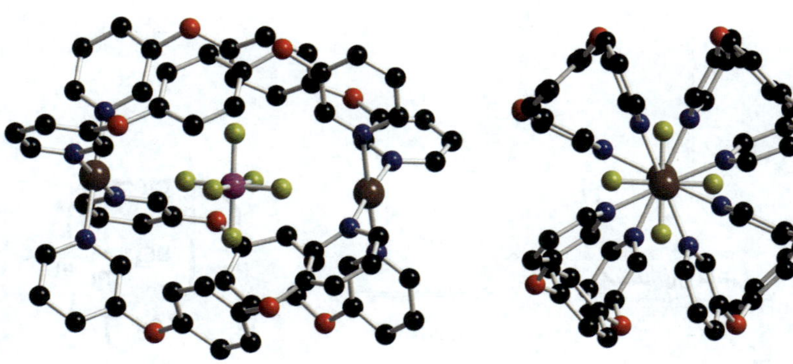

Figure 9.19 *Front and side views of the crystal structure of* ***9.74*** *showing the encapsulated PF_6^- anion*

9.5 Summary Remarks

We have seen in this chapter that anions are capable of controlling the assembly of metal complexes of organic ligands, inorganic cluster compounds, as well as either stabilizing or destabilizing the formation of non-covalently linked supramolecular arrays. It is perhaps remarkable that anion-controlled assembly is still a relatively unexplored facet of anion-receptor chemistry and these examples are small in number when compared to the examples of transition metal-directed assemblies that have been reported.

Biological systems use anionic peptides to control biomineralization processes and the prospect of generating simple synthetic analogues of these complex entities to control molecular and nanoscale assembly is an exciting one. The results discussed in this chapter thus likely represent the vanguard of even more exciting developments yet to come.

References

1. D. Philp and J.F. Stoddart, *Angew. Chem. Int. Ed.*, 1996, **35**, 1154.
2. B. Hasenknopf, J.-M. Lehn, B.O. Kneisel, G. Baum and D. Fenske, *Angew. Chem. Int. Ed.*, 1996, **35**, 1838; B. Hasenknopf, J.-M. Lehn, N. Boumediene, A. Dupont-Gervais, A. van Dorsselaer, B. Kneisel and D. Fenske, *J. Am. Chem. Soc.*, 1997, **119**, 10956.
3. R. Vilar, D.M.P. Mingos, A.J.P. White and D.J. Williams, *Angew. Chem. Int. Ed.*, 1998, **37**, 1258.
4. R. Vilar, D.M.P. Mingos, A.J.P. White and D.J. Williams, *Chem. Commun.*, 1999, 229; P. Diaz, D.M.P. Mingos, R. Vilar, A.J.P. White and D.J. Williams, *Inorg. Chem.*, 2004, **43**, 7597.
5. S.-T. Cheng, E. Doxiadi, R. Vilar, A.J.P. White and D.J. Williams, *J. Chem. Soc. Dalton Trans.*, 2001, 2239; E. Doxiadi, R. Vilar, A.J.P. White and D.J. Williams, *Polyhedron*, 2003, **22**, 2991.
6. D. Rais, J. Yau, D.M.P. Mingos, R. Vilar, A.J.P. White and D.J. Williams, *Angew. Chem. Int. Ed.*, 2001, **40**, 3464.
7. D. Rais, D.M.P. Mingos, R. Vilar, A.J.P. White and D.J. Williams, *J. Organomet. Chem.*, 2002, **652**, 87.
8. A. Müller, K. Hovemeier and R. Rohlfing, *Angew. Chem. Int. Ed.*, 1992, **31**, 1192.
9. J. Salta, Q. Chen, Y.-D. Chang and J. Zubieta, *Angew. Chem. Int. Ed.*, 1994, **33**, 757.
10. R. Wang, Z. Zheng, T. Jin and R.J. Staples, *Angew. Chem. Int. Ed.*, 1999, **38**, 1813.
11. R. Wang, H.D. Selby, H. Liu, M.D. Carducci, T. Jin, Z. Zheng, J.W. Anthis and R.J. Staples, *Inorg. Chem.*, 2002, **41**, 278.
12. A. Bashall, A.D. Bond, E.L. Doyle, F. Garcia, S. Kidd, G.T. Lawson, M.C. Parry, M. McPartlin, A.D. Woods and D.S. Wright, *Chem. Eur. J.*, 2002, **8**, 3377; F. García, J.M. Goodman, R. Kowenicki, I. Kuzu, M. McPartlin, M.A. Silva, L. Riera, A.D. Woods and D.S. Wright, *Chem. Eur. J.*, 2004, **10**, 6066.
13. A. Bashall, E.L. Doyle, C. Tubb, S.J. Kidd, M. McPartlin, A.D. Woods and D.S. Wright, *Chem. Commun.*, 2001, 2542.

14. P.J. Steel and C.J. Sumby, *Inorg. Chem Commun.*, 2002, **5**, 323; P.J. Steel and C.J. Sumby, *Chem. Commun.*, 2002, 322.
15. T. Enomoto, N. Nishigaki, H. Kurata, T. Kawase and M. Oda, *Bull. Chem. Soc. Jpn.*, 2000, **73**, 2109; T. Enomoto, T. Kawaso, H. Kurata and M. Oda, *Tetrahedron Lett.*, 1997, **38**, 2693.
16. E. Alcalde, C. Alvarez-Rúa, S. García-Granda, E. García-Rodriguez, N. Mesquida and L. Pérez-García, *Chem. Commun.*, 1999, 295; E. Alcalde, S. Ramos and L. Pérez-García, *Org. Lett.*, 1999, **1**, 1035; E. Alcalde, M. Gisbert, C. Alvarez-Rúa and S. García-Granda, *Tetrahedron*, 1996, **52**, 15189; E. Alcalde, M. Alemany and M. Gisbert, *Tetrahedron*, 1996, **52**, 15171.
17. Y.H. Kim, J. Calabrese and C. McEwen, *J. Am. Chem. Soc.*, 1996, **118**, 1545.
18. J.-P. Sauvage and C. Dietrich-Buchecker, *Molecular Catenanes, Rotaxanes and Knots*, Wiley-VCH, Weinheim, 1999; V. Balzani, M. Venturi and A. Credi, *Molecular Devices and Machines*, Wiley-VCH, Weinheim, 2003.
19. M. Montalti and L. Prodi, *Chem. Commun.*, 1998, 1461.
20. M.H. Al-Sayah and N.R. Branda, *Org. Lett.*, 2002, **4**, 881.
21. J.A. Wisner, P.D. Beer and M.G.B. Drew, *Angew. Chem. Int. Ed.*, 2001, **40**, 3606; J.A. Wisner, P.D. Beer, M.G.B. Drew and M.R. Sambrook., *J. Am. Chem. Soc.*, 2002, **124**, 12469; M.R. Sambrook., P.D. Beer, J.A. Wisner, R.L. Paul and A.R. Cowley, *J. Am. Chem. Soc.*, 2004, **126**, 15364; G. Cafeo, F.H. Kohnke, G.L.L. Torre, A.J.P. White and D.J. Williams, *Angew. Chem. Int. Ed.*, 2000, **39**, 1496.
22. J.A. Wisner, P.D. Beer, N.G. Berry and B. Tomapatanaget, *Proc. Natl. Acad. Sci. USA*, 2002, **99**, 4983; B. Tomapatanaget, T. Tuntulani, J.A. Wisner and P.D. Beer, *Tetrahedron Lett.*, 2004, **45**, 663; M.R. Sambrook., P.D. Beer, J.A. Wisner, R.L. Paul, A.R. Cowley, F. Szemes and M.G.B. Drew, *J. Am. Chem. Soc.*, 2005, **127**, 2292.
23. M.J. Hannon and L.J. Childs, *Supramol. Chem.*, 2004, **16**, 7.
24. J. Keegan, P.E. Kruger, M. Nieuwenhuyzen, J. O'Brien and N. Martin, *Chem. Commun.*, 2001, 2192.
25. S.J. Coles, J.G. Frey, P.A. Gale, M.B. Hursthouse, M.E. Light, K. Navakhun and G.L. Thomas, *Chem. Commun.*, 2003, 568.
26. X.-X. Lu, H.-S. Tang, C.-C. Ko, J.K.-Y. Wong, N. Zhu and V.W.-W. Yam, *Chem. Commun.*, 2005, 1572.
27. R. Varghese, S.J. George and A. Ajayaghosh, *Chem. Commun.*, 2005, 593.
28. A. Müller, R. Rohlfing, E. Krickemeyer and H. Bögge, *Angew. Chem. Int. Ed.*, 1993, **32**, 909.
29. M.-C. Suen, G.-W. Tseng, J.D. Chen, T.-C. Keng and J.-C. Wang, *Chem. Commun.*, 1999, 1185.
30. P.L. Jones, K.J. Byrom, J.C. Jeffery, J.A. McCleverty and M.D. Ward., *Chem. Commun.*, 1997, 1361.
31. J. Sánchez-Quesada, C. Seel, P. Prados, J.de Mendoza, I. Dalcol and E. Giralt, *J. Am. Chem. Soc.*, 1996, **118**, 277.
32. J. Cook, J. Sloan and M.L.H. Green, *Chem. Ind.*, 1996, 600; J.J. Davis, M.L.H. Green, H.A.O. Hill, Y.C. Leung, P.J. Sadler, J. Sloan, A.V. Xavier and S.C. Tsang, *Inorg. Chim. Acta.*, 1998, **272**, 261.
33. M. Aoyagi, K. Biradha and M. Fujita, *J. Am. Chem. Soc.*, 1999, **121**, 7457.

34. J.L. Sessler, T.D. Mody and V. Lynch, *Inorg. Chem.*, 1992, **31**, 529.
35. J.L. Sessler, T.D. Mody and V. Lynch, *J. Am. Chem. Soc.*, 1993, **115**, 3346.
36. D. Seidel, V.M. Lynch and J.L. Sessler, *Angew. Chem. Int. Ed.*, 2002, **41**, 1422.
37. T. Köhler, D. Seidel, V. Lynch, F.O. Arp, Z. Ou, K.M. Kadish and J.L. Sessler, *J. Am. Chem. Soc.*, 2003, **125**, 6872.
38. G.M. Hübner, J. Gläser, C. Seel and F. Vögtle, *Angew. Chem. Int. Ed.*, 1999, **38**, 383.
39. C. Reuter, W. Wienard, G.M. Hübner, C. Seel and F. Vögtle, *Chem. Eur. J.*, 1999, **5**, 2692; G.M. Hübner, C. Reuter, C. Seel and F. Vögtle, *Synthesis*, 2000, 103; C. Seel and F. Vögtle, *Chem. Eur. J.*, 2000, **6**, 21.
40. C.M. Keaveney and D.A. Leigh, *Angew. Chem. Int. Ed.*, 2004, **43**, 1222.
41. J.W. Jones, L.N. Zakharov, A.L. Rheingold and H.W. Gibson, *J. Am. Chem. Soc.*, 2002, **124**, 13378; E. Arunkumar, C.C. Forbes, B.C. Noll and B.D. Smith, *J. Am. Chem. Soc.*, 2005, **127**, 3288.
42. V. Král, F.P. Schmidtchen, K. Lang and M. Berger, *Org. Lett.*, 2002, **4**, 51.
43. Y. Kato, S. Nishizawa and N. Teramae, *Org. Lett.*, 2002, **4**, 4407.
44. M.W. Hossieni, R. Ruppert, P. Schaeffer, A. de Cian, N. Kyritsaka and J. Fischer, *J. Chem. Soc. Chem. Commun.*, 1994, 2135; G. Brand, M.W. Hossieni, R. Ruppert, A. de Cian, J. Fischer and N. Kyritsaka, *New J. Chem.*, 1995, **19**, 9; M.W. Hosseini, G. Brand, P. Schaeffer, R. Ruppert, A. de Cian and J. Fischer, *Tetrahedron lett.*, 1996, **37**, 1405.
45. R.-D. Schenbeck, E. Freisinger and B. Lippert, *Chem. Commun.*, 1999, 675.
46. M.C.T. Fyfe, P.T. Glink, S. Menzer, J.F. Stoddart, A.J.P. White and D.J. Williams, *Angew. Chem. Int. Ed.*, 1997, **36**, 2068.
47. J.S. Fleming, K.L.V. Mann, C.-A. Carraz, E. Psillakis, J.C. Jeffery, J.A. McCleverty and M.D. Ward, *Angew. Chem. Int. Ed. Engl.*, 1998, **37**, 1279.
48. R.L. Paul, S.M. Couchman, J.C. Jeffery, J.A. McCleverty, Z.R. Reeves and M.D. Ward, *J. Chem. Soc. Dalton Trans.*, 2000, 845.
49. C.S. Campos-Fernández, R. Clérac and K.R. Dunbar, *Angew. Chem. Int. Ed.*, 1999, **38**, 3477.
50. K.M. Kim, J.S. Park, Y.-S. Kim, Y.J. Jun, T.Y. Kang, Y.S. Sohn and M.-J. Jun, *Angew. Chem. Int. Ed.*, 2001, **40**, 2458.
51. D.A. McMorran and P.J. Steel, *Angew. Chem. Int. Ed.*, 1998, **37**, 3295.

CHAPTER 10

Afterword

This book was "conceived" at the Calix[99] meeting in Perth, Western Australia in 1999. The fact that it was not completed until 2005 (co-incidentally at the Calix 2005 meeting in Prague) reflects the fact that there has been an explosion of interest in anion receptor systems over the last six years, with many publications appearing in the literature. What we have attempted to do in the book was not to be comprehensive but rather illustrate particular concepts in design with one or two examples. Even so it has been difficult for us to keep up with the many new systems that have appeared. So, while it has been an excellent time for anion receptor chemistry, it has been a challenging time to write a book!

Cation coordination chemistry and cation sensing has become a mature field and it is possible to go to the literature for a selective receptors for most metal cations. By contrast, the range of highly selective anion receptors is more limited reflecting that there is much more work to be done in production of anion systems. The added dimension that ion-pair receptor systems have to offer in improving selectivity is only just starting to be appreciated by the anion coordination community and we predict that this area, in particular, will see much effort in the coming years. There are also surprisingly few receptors for zwitterionic guests and this is one area where biological applications may drive research forward. The use of anion receptors in cell membranes as anion or ion-pair transport agents is only just starting to be explored and again we believe that this area which will see considerable effort and growth over the coming years, as these compounds will offer new ways to treat a variety of diseases. Other applications, including those associated with separation, extraction, and sensing, are beginning to see the results of sustained activity in recent years. However, much more work in the area is needed to transition from accomplishments of academic interest into working devices and systems that will see use in the "real world" of everyday utility. Another area that has blossomed recently involves the use of anions as templates for the construction of new supramolecular architectures. Perhaps the next stage of this research will be to take inspiration from biomineralization processes in nature and design larger anionic molecules that can control nucleation of a variety of crystalline forms to produce new materials.

The very factors that have made bringing this book to completion difficult serve to highlight in a most dramatic way just what an exciting time it now is to be working in the area of anion receptor chemistry. It is our hope that this book will give workers new to the field a reasonably concise overview of this rapidly growing research area. We also hope it will inspire our colleagues who have contributed so much to the growth of the anion recognition field to take their research to the "next level" such that much of the promise now envisioned becomes expressed as concrete reality.

Subject Index